Lecture Notes in Comput 4

Edited by G. Goos, J. Hartmanis and

Jieh Hsiang (Ed.)

Rewriting Techniques and Applications

6th International Conference, RTA-95
Kaiserslautern, Germany, April 5-7, 1995
Proceedings

Springer

Series Editors

Gerhard Goos
Universität Karlsruhe
Vincenz-Priessnitz-Straße 3, D-76128 Karlsruhe, Germany

Juris Hartmanis
Department of Computer Science, Cornell University
4130 Upson Hall, Ithaca, NY 14853, USA

Jan van Leeuwen
Department of Computer Science, Utrecht University
Padualaan 14, 3584 CH Utrecht, The Netherlands

Volume Editor

Jieh Hsiang
Department of Computer Science, National Taiwan University
Taipei, Taiwan

CR Subject Classification (1991): D.3, F.3.2, F.4, I.1, I.2.2-3

ISBN 3-540-59200-8 Springer-Verlag Berlin Heidelberg New York

CIP data applied for

Printed in Germany

Typesetting: Camera-ready by author
SPIN: 10485757 06/3142-543210 - Printed on acid-free paper

Preface

This volume contains the proceedings of the *Sixth International Conference on Rewriting Techniques and Applications* (RTA-95), held April 5-7, 1995, at Kaiserslautern, Germany. The previous five RTA proceedings were also published as part of the Lecture Notes in Computer Science Series of Springer-Verlag. They were

volume 202 (held at Dijon, France, 1985),
volume 256 (held at Bordeaux, France, May 1987),
volume 355 (held at Chapel Hill, U.S.A., May 1989),
volume 488 (held at Como, Italy, April 1991),
volume 690 (held at Montreal, Canada, June 1993)

There were 87 submissions to RTA-95, with authors from Algeria, Austria, Belgium, Brasil, China, France, Germany, Italy, Japan, the Netherlands, Norway, Paraguay, Spain, United Kingdom, and the United States of America. On December 8, 1994, the Program Committee met at Kaiserslautern and chose 27 papers for presentation and 8 papers as system descriptions. The average number of reviews for each paper was four, with at least three for each paper.

Following the tradition of the two previous RTAs, this volume contains two problem sets, one contributed by Nachum Dershowitz, Jean-Pierre Jouannaud and Jan Willem Klop, and another one by Mark Stickel and Hantao Zhang.

In addition to paper presentations and system demonstrations, there were also three invited talks. They were given by Professor Yuri Matiyasevich, on early rewriting work done in the former Soviet Union, Professor Tobias Nipkow on higher order rewriting, and Dr. Mark Stickel on recent developments on automated deduction related to rewriting.

I would like to take this opportunity to thank Professors Jürgen Avenhaus and Klaus Madlener for the wonderful local arrangements, the program committee and the reviewers for the excellent and conscientious job of refereeing all the papers. Claude Kirchner generously provided the shell scripts from the previous RTA which made preparing the proceedings much easier. Special thanks also go to the students and assistants, Martin Anlauf, Juin Jieh Chen, Thomas Deiß, Birgit Reinert, Hsieh-Chang Tu, and I-Fei You. Without their help with the preparation of the proceedings and the conference program, the meeting would not have been possible.

RTA-95 was sponsored by

DFG Deutsche Forschungsgemeinschaft
Land Rheinland-Pfalz
Universität Kaiserslautern
Fachbereich Informatik, Universität Kaiserslautern

Taipei, Taiwan
April 1995

Jieh Hsiang
Chairman of RTA-95

Conference Organization

RTA-95 Program Chair: Jieh Hsiang

RTA-95 Local Arrangement Chairs: Jürgen Avenhaus, Klaus Madlener

RTA-95 Program Committee:

Leo Bachmair (Stony Brook)
Hubert Comon (Orsay)
Nachum Dershowitz (Urbana)
Hoon Hong (Linz)
Jieh Hsiang (Taipei)
Joxan Jaffar (Yorktown Heights)
Deepak Kapur (Albany)
Klaus Madlener (Kaiserlautern)
José Meseguer (Menlo Park)
Theo Mora (Genova)
Tobias Nipkow (München)
Michael Rusinowitch (Nancy)
Manfred Schmidt-Schauss (Frankfurt)
Val Tannen (Philadelphia)
Yoshihito Toyama (Ishikawa)
Hantao Zhang (Iowa City)

RTA-95 Organizing Committee:

Ronald Book (Santa Barbara)
Nachum Dershowitz (Urbana)
Jieh Hsiang (Taipei)
Jean-Pierre Jouannaud (Orsay)
Deepak Kapur (Albany)
Claude Kirchner (Nancy)
Klaus Madlener (Kaiserlautern)
David Plaisted (Chapel Hill)

List of Referees

The program committee gratefully acknowledges the help of the following referees:

Y. Akama
Rajendra Akerkar
Siva Anantharaman
S.Antoy
Egidio Astesiano
J.Avenhaus
Franz Baader
Peter Baumgartner
F. Bellegarde
Piergiorgio Bertoli
Susanne Biundo
Alexander Bockmayr
Maria Paola Bonacina
Harald Boley
Adel Bouhoula
Alexandre Boudet
S.Brookes
F. Bronsard
Olga Caprotti
Barbara Catania
Ta Chen
Wei-Ngan Chin
Pierre Cregut
Vincent Danos
Max Dauchet
Thomas Deiß
Jörg Denzinger
Roberto Di Cosmo
Maribel Fernandez
John Field
Harald Ganzinger
Stephen Garland
Jürgen Giesl
Georges Gonthier
E. Contejean
Erwan David
Fer-Jan de Vries
R. Echahed
Amy Felty
Roland Fettig
Charles Hoot
Stefan Gerberding
Bernhard Gramlich
Miki Hermann
Maria Huber
F. Jacquemard
Dalibor Jakuš
C. Barry Jay
Bharat Jayaraman
Yuichi Kaji
Delia Kesner
Helene Kirchner
J.W. Klop
Yuri Kobayashi
Michael Kohlhase
Masahito Kurihara
Arne Kutzner
T. K. Lakshman
John Lamping
Julia Lawall
Pierre Lescanne
Patrick Lincoln
Steven Lindell
Denis Lugiez

Chris Lynch
Harry Mairson
C. Marche
Aart Middeldorp
Umberto Modigliani
Sanjai Narain
Andreas Neubacher
E. Nocker
Eugenio G. Omodeo
Frank Pfenning
C.R. Ramakrishnan
Uday Reddy
Christophe Ringeissen
Albert Rubio
Rosario F. Salamone
K.U. Schulz
Wayne Snyder
Joachim Steinbach
Wolfgang Stoecher
C Talcott
Tomas E. Uribe
Vincent van Oostrom
Daniela Vasaru
L. Vigneron
Claus-Peter Wirth
Roland Yap
Louxin Zhang
Michael Maher
Ken Mano
W. McCune
Dale Miller
Dimitri Naidich
Paliath Narendran
Robert Nieuwenhuis
Mizuhito Ogawa
F. Otto
Zhenyu Qian
Stefan Ratschan
Michael M.Richter
Laurent Rosaz
M.Sakai
Patrick Salle
G. Sivakumar
Rolf Socher-Ambrosius
Mark E. Stickel
Ramesh Subrahmanyam
Val Tannen
Jaco Van de Pol
Stefano Varricchio
Rakesh Verma
Christoph Weidenbach
Junnosuke Yamada
Kathy Yelick
H. Zantema

Table of Contents

Regular Papers

Invited Talk: On Some Mathematical Logic Contributions to Rewriting Techniques: Lost Heritage
Yuri Matiyasevich 1

Modularity of Completeness Revisited
Massimo Marchiori 2

Automatic Termination Proofs with Transformation Orderings
Joachim Steinbach 11

A Termination Ordering for Higher Order Rewrite Systems
Olav Lysne and Javier Piris 26

A Complete Characterization of Termination of $0^p1^q \to 1^r0^s$
Hans Zantema and Alfons Geser 41

On Narrowing, Refutation Proofs and Constraints
Robert Nieuwenhuis 56

Completion for Multiple Reduction Orderings
Masahito Kurihara, Hisashi Kondo and Azuma Ohuchi 71

Towards an Efficient Construction of Test Sets for Deciding Ground Reducibility
Klaus Schmid and Roland Fettig 86

Invited Talk: Term Rewriting in Contemporary Resolution Theorem Proving
Mark E. Stickel 101

$\delta o!\epsilon = 1$: Optimizing Optimal λ-Calculus Implementations
Andrea Asperti 102

Substitution Tree Indexing
Peter Graf 117

Concurrent Garbage Collection for Concurrent Rewriting
Ilies Alouini 132

Lazy Rewriting and Eager Machinery
J.F.Th. Kamperman and H.R. Walters 147

A Rewrite Mechanism for Logic Programs with Negation
Siva Anantharaman and Gilles Richard 163

Level-Confluence of Conditional Rewrite Systems with
Extra Variables in Right-Hand Sides
Taro Suzuki, Aart Middeldorp and Tetsuo Ida 179

A Polynomial Algorithm Testing Partial Confluence of
Basic Semi-Thue Systems
Géraud Sénizergues 194

Problems in Rewriting Applied to Categorical Concepts
by the Example of a Computational Comonad
Wolfgang Gehrke 210

Relating Two Categorical Models of Term Rewriting
A. Corradini, F. Gadducci and U. Montanari 225

Towards a Domain Theory for Termination Proofs
Stefan Kahrs 241

Invited Talk : Higher-Order Rewrite Systems
Tobias Nipkow 256

Infinitary Lambda Calculi and Böhm Models
Richard Kennaway, Jan Willem Klop, Ronan Sleep
and Fer-Jan de Vries 257

Proving the Genericity Lemma by Leftmost Reduction is Simple
Jan Kuper 271

(Head-)Normalization of Typeable Rewrite Systems
Steffen van Bakel and Maribel Fernández 279

Explicit Substitutions with de Bruijn's Levels
Pierre Lescanne and Jocelyne Rouyer-Degli 294

A Restricted Form of Higher-Order Rewriting Applied to an HDL Semantics
Richard J. Boulton 309

Rewrite Systems for Integer Arithmetic
H.R. Walters and H. Zantema 324

General Solution of Systems of Linear Diophantine Equations and Inequations
H. Abdulrab and M. Maksimenko 339

Combination of Constraint Solving Techniques: An Algebraic Point of View
Franz Baader and Klaus U. Schulz 352

Some Independence Results for Equational Unification (Extended Abstract)
Friedrich Otto, Paliath Narendran and Daniel J. Dougherty 367

Regular Substitution Sets: A Means of Controlling E-Unification
Jochen Burghardt 382

System Description

DISCOUNT: A System for Distributed Equational Deduction
Jürgen Avenhaus, Jörg Denzinger, Matthias Fuchs 397

ASTRE: Towards a Fully Automated Program Transformation System
Françoise Bellegarde 403

Parallel ReDux → PaReDuX
Reinhard Bündgen, Manfred Göbel, Wolfgang Küchlin 408

STORM: A Many-to-One Associative-Commutative Matcher
Ta Chen, Siva Anantharaman 414

LEMMA: a System for Automated Synthesis of Recursive Programs in Equational Theories
Jacques Chazarain and Serge Muller 420

Generating Polynomial Orderings for Termination Proofs
Jürgen Giesl 426

Disguising Recursively Chained Rewrite Rules as Equational Theorems, as Implemented in the Prover EFTTP Mark 2
M. Randall Holmes 432

Prototyping Completion with Constraints Using Computational Systems
Hélène Kirchner and Pierre-Etienne Moreau 438

Guiding Term Reduction Through A Neural Network: Some Preliminary Results for the Group Theory
Alberto Paccanaro 444

Problem Sets

Studying Quasigroup Identities by Rewriting Techniques: Problems and First Results
Mark E. Stickel and Hantao Zhang 450

Problems in Rewriting III
Nachum Dershowitz, Jean-Pierre Jouannaud and Jan Willem Klop 457

Author Index 473

On some mathematical logic contributions to rewriting techniques: lost heritage

Yuri Matiyasevich
Steklov Institute of Mathematics
Saint-Petersburg Branch
27, Fontanka
POMI
St.Petersburg, 191011, Russia

Abstract. A considerable amount of research closely related to rewriting techniques had been developed in the frameworks of mathematical logic. However, it seems that not all achievements have been inhereted by computer science. Some results were later rediscovered independently, and some others still remain little known to computer scientists. The aim of the lecture will be to survey certain of such results with the emphathis on investigations performed in the USSR; papers published in Russian are often little known even among the world logical community.

Modularity of Completeness Revisited

Massimo Marchiori

Department of Pure and Applied Mathematics
University of Padova
Via Belzoni 7, 35131 Padova, Italy
max@hilbert.math.unipd.it

Abstract. One of the key results in the field of modularity for Term Rewriting Systems is the modularity of completeness for left-linear TRSs established by Toyama, Klop and Barendregt in [TKB89]. The proof, however, is quite long and involved. In this paper, a new proof of this basic result is given which is both short and easy, employing the powerful technique of 'pile and delete' already used with success in proving the modularity of $UN^{\rightarrow}$. Moreover, the same proof is shown to *extend* the result in [TKB89] proving modularity of termination for left-linear and consistent with respect to reduction TRSs.

1 Introduction

The property of completeness for a Term Rewriting System, that is being Church-Rosser and terminating, is of fundamental importance in every application of rewriting. One of the cornerstones as far as completeness is concerned is the smart result obtained by Toyama, Klop and Barendregt in [TKB89] (see also [TKB94]) asserting the modularity of completeness for left-linear TRSs (a property is modular provided it is valid for two TRSs if and only if it holds for their disjoint sum). However, as stated by the same authors in their paper, the proof used to obtain such result is 'rather intricate and not easily digested'.

In this paper we show that the same result can be obtained making use of the technique of 'pile and delete' developed in [Mar93] to prove the modularity of $UN^{\rightarrow}$ (uniqueness of normal form with respect to reduction) for left-linear TRSs. The proof obtained is extremely short and rather intuitive, thus providing new insights in the study of completeness.

The established link between proof methods of completeness and $UN^{\rightarrow}$ should not come as a total surprise, however, since it is easily seen that:

$$\text{Completeness} = \text{Church-Rosser} + \text{Termination} = UN^{\rightarrow} + \text{Termination}.$$

In fact, we will see that the given proof even *extends* the result in [TKB89], proving modularity of termination for left-linear and *consistent with respect to reduction* TRSs (a TRS is consistent with respect to reduction if a term cannot be rewritten to two different variables).

The paper is organized in the following way: Section 2 gives the necessary preliminaries, Section 3 states the main theorem about modularity of completeness for left-linear TRSs, and finally Section 4 shows how exactly the same proof yields the stronger result of modularity of termination for left-linear and consistent with respect to reduction TRSs.

2 Preliminaries

We assume the reader to be familiar with the basic notions regarding Term Rewriting Systems: the notation used is essentially the one in [Klo92] and [Mid90].

The properties of being confluent (Church-Rosser) and terminating (Strongly Normalizing) will be indicated respectively with CR and SN. Contexts will be denoted as usual with $\Box$ and with square brackets ($C[\cdots]$). Throughout the paper we will indicate with $\mathcal{A}$ and $\mathcal{B}$ the two TRSs to operate on, their corresponding sets of function symbols with $\mathcal{F}_{\mathcal{A}}, \mathcal{F}_{\mathcal{B}}$, the variables set as $\mathcal{V}$, and the set of terms built from some set of function symbols $\mathcal{F}$ and $\mathcal{V}$ as $\mathcal{T}(\mathcal{F})$.

When not otherwise specified, all symbols and notions not having a TRS label are to be intended operating on the disjoint sum $\mathcal{A} \oplus \mathcal{B}$. For better readability, we will talk of function symbols belonging to $\mathcal{A}$ and $\mathcal{B}$ like *white* and *black* functions, indicating the first ones with upper case functions, and the second ones with lower case. Variables, instead, have no colour.

Definition 2.1 The *root* symbol of a term $t \in \mathcal{T}(\mathcal{F}_{\mathcal{A}} \cup \mathcal{F}_{\mathcal{B}})$ is f provided $t = f(t_1, \ldots, t_n)$, and t itself otherwise. $\Box$

Let $t = C[t_1, \ldots, t_n] \in \mathcal{T}(\mathcal{F}_1 \cup \mathcal{F}_2)$ and $C \not\equiv \Box$; we write $t = C[\![t_1, \ldots, t_n]\!]$ if $C[\ldots]$ is an $\mathcal{F}_{\mathcal{A}}$-context and each of the t_i has $root(t_i) \in \mathcal{F}_{\mathcal{B}}$, or vice versa (exchanging $\mathcal{A}$ and $\mathcal{B}$). The *topmost homogeneous part* (briefly *top*) of a term $C[\![t_1, \ldots, t_n]\!]$ is the context $C[\![\cdots]\!]$.

Definition 2.2 The *rank* of a term $t \in \mathcal{T}(\mathcal{F}_{\mathcal{A}} \cup \mathcal{F}_{\mathcal{B}})$ is 1 if $t \in \mathcal{T}(\mathcal{F}_{\mathcal{A}})$ or $t \in \mathcal{T}(\mathcal{F}_{\mathcal{B}})$, and $\max_{i=1}^{n}\{rank(t_i)\} + 1$ if $t = C[\![t_1, \ldots, t_n]\!]$ $(n > 0)$. $\Box$

The following well known lemma will be implicitly used in the sequel:

Lemma 2.3 ([Toy87]) $s \twoheadrightarrow t \Rightarrow rank(s) \geq rank(t)$

Proof. Clear. $\Box$

Definition 2.4 The multiset $S(t)$ of the *special subterms* of a term t is

1. $S(t) = \begin{cases} \{t\} \text{ if } t \in (\mathcal{T}(\mathcal{F}_{\mathcal{A}}) \cup \mathcal{T}(\mathcal{F}_{\mathcal{B}})) \backslash \mathcal{V} \\ \emptyset \quad \text{if } t \in \mathcal{V} \end{cases}$
2. $S(t) = \cup_{i=1}^{n} S(t_i) \cup \{t\}$ if $t = C[\![t_1, \ldots, t_n]\!]$ $(n > 0)$ $\Box$

Note that this definition is slightly different from the usual ones in the literature (for example in [Mid90]), since here variables are not considered special subterms.

If $t = C[\![t_1, \ldots, t_n]\!]$, the t_i are called the *principal* special subterms of t. Furthermore, a reduction step of a term t is called *outer* if the rewrite rule isn't applied in the principal special subterms of t.

A (strict) *partial order* on the special subterms of a term can be naturally given defining $t_1 \succ t_2$ iff t_2 is a proper special subterm of t_1 (that is $t_1 = C[\![\ldots t_2 \ldots]\!]$).

The following proposition will reveal useful:

Proposition 2.5 *If $\mathcal{A}$ and $\mathcal{B}$ are left-linear, then rewrite rules that have the possibility to act outer on a special subterm t are exactly those that have the possibility to act on its top.*

Proof. Let $t = C[\![t_1, \ldots, t_n]\!]$: since $t_1, \ldots, t_n$ have a root belonging to the other TRS (with respect to C), they are matched by variables from any rewrite rule applicable to C, and for the left-linearity assumption these variables are independent each other. □

2.1 Marking and Collapsing

To be able to describe the special subterms of a given term throughout a reduction, it is natural to develop a concept of (modular) marking. A first, naïve approach of modular marking for a term is to take an assignment from the multiset of its special subterms to a (fixed) set of markers. So, for instance, given the term $F(f(G,a),H)$, we could mark $F(\Box, H)$ to m_1, $f(\Box, a)$ to m_2, G to m_3. Then reductions steps, as usual, should preserve the markers. However, this simple definition presents a problem, since for one case there is ambiguity: when a collapsing rule makes an *inner* top vanish. In this case, we have the following situation:

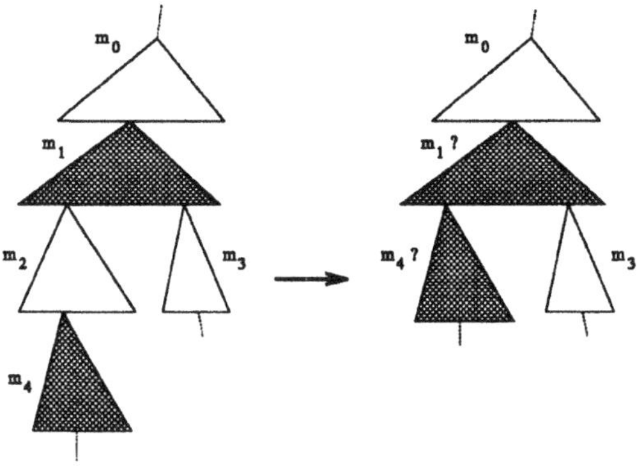

and we have a conflict between m_1 and m_4.

This situation is dealt with by defining a *modular marking* for a term to be an assignment from the multiset of its special subterms to *sets* of markers, and taking in the ambiguous case just described the union of the marker sets of the two special subterms involved.

Thus, the previous example would give (singletons like $\{m_3\}$ are written simply m_3):

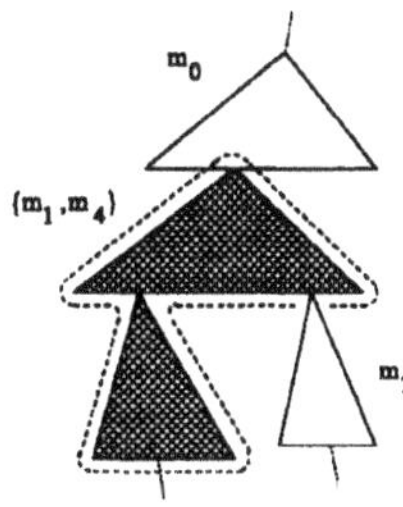

When this situation occurs, we say that the special subterm m_4 has been *absorbed* by m_1, and the special subterm m_2 has had a *modular collapsing* (briefly *m-collapsing*).

When dealing with reductions $t \twoheadrightarrow t'$ we will always assume, in order to distinguish all the special subterms, that the initial modular marking of t is injective and maps special subterms to singletons.

Inside a reduction a notion of descendant for every special subterm can be defined: in a reduction a special subterm is a *descendant* (resp. *pure descendant*) of another if the set of markers of the former contains (resp. is equal to) the set of markers of the latter. Note, en passant, that due to the presence of duplicating rules, there may be more than one descendant, or even none (due to erasing rules). Observe also that, since in a reduction without m-collapsings all the descendants are pure, the first special subterm to m-collapse in a generic reduction is a pure descendant. Hence it readily holds the following:

Fact: *A reduction has m-collapsings iff a pure descendant m-collapses.*

One of the reasons cited in [TKB89] for the difficulty in treating the problem of the completeness modularity is the *nondeterministic* collapsing behaviour given by ambiguities among rewrite rules. The following result shows a case in which things behave nicely:

Proposition 2.6 *Let $\mathcal{A}$ be left-linear and $UN^{\rightarrow}$, and $t = C[\![t_1, \ldots, t_n]\!]$ a top white special subterm: if t m-collapses into t_i $(1 \leq i \leq n)$ via a white reduction (i.e. using only rules from $\mathcal{A}$), then the index i is unique.*

Proof. Since $\mathcal{A}$ is left-linear, by Proposition 2.5 the white reduction depends only on the top of t. Hence, if we take instead of $t = C[\![t_1, \ldots, t_n]\!]$ a term $t' = C[\![X_1, \ldots, X_n]\!]$ (with $X_1, \ldots, X_n$ new fresh variables), then every previous white reduction that m-collapsed t to t_i can be repeated on t' to reduce it to X_i, and if the index i were not unique t' could be reduced to different normal forms, contradicting the fact $\mathcal{A}$ is $UN^{\rightarrow}$. □

3 The Main Theorem

We have arrived to the main theorem: the essence of the new proof is to use the pile and delete technique developed in [Mar93] to transform every infinite reduction into an infinite reduction without m-collapsings, and then prove this kind of reductions cannot exist.

Theorem 3.1 *Completeness is a modular property for left-linear TRSs.*

Proof. Since CR is a modular property ([Toy87]), it only remains to prove that if $\mathcal{A}$ and $\mathcal{B}$ are left-linear and complete, then $\mathcal{A}\oplus\mathcal{B}$ is SN (the reverse implication is trivial). So, suppose ab absurdo there exists an infinite reduction. Then take an infinite reduction $u \to u_1 \to u_2 \to \ldots$ that has the starting term u with *minimum rank* among the terms with an infinite reduction.

First, note how the number of reduction rules acting on the top of u are infinite: if it were not so, then we would have a finite number of proper special subterms, and since all of these are SN by the rank minimality of u no infinite reduction would be possible.

We will show that from this reduction a new infinite one can be obtained, which is without m-collapsings (employing the same 'pile and delete' technique as in [Mar93]).

If the reduction is already without m-collapsings, the assertion is trivially satisfied.

Thus, suppose in the reduction some m-collapsing is performed. Select a special subterm of u that has rank minimal amongst the ones with a pure descendant that m-collapses in the reduction itself: say $t = \tau[\![t_1, \ldots, t_n]\!]$. Because of its rank minimality, t must m-collapse by Proposition 2.6 into a fixed principal subterm, namely t_i. The top of t, say τ, might be absorbed from other tops of $(\prec)$-greater special subterms in the reduction. All of these special subterms are descendants of some special subterms of the start $r_1, \ldots, r_k$.

We now modify u via the 'pile and delete' process.

First, we 'pile' $\tau[\![t_1, \ldots, t_{i-1}, \Box, t_{i+1}, \ldots, t_n]\!]$ just below the tops of the $r_1, \ldots, r_k$, that is to say if $r_i = r_i[\![s_1, \ldots, s_v]\!]$ and t is in s_j (viz. $t \prec s_j$), then r_i is replaced with

$$r_i[\![s_1, \ldots, s_{j-1}, \tau[t_1, \ldots, t_{i-1}, s_j, t_{i+1}, \ldots, t_n], s_{j+1} \ldots, s_v]\!]$$

The situation is shown in Figure 1.

The intuition is that we inserted copies of t where needed later in the reduction for absorption. So now we can 'delete' it replacing t by t_i (see Figure 2).

How can we get an infinite reduction from this modified starting term? We simply use the previous infinite reduction, with the following modifications:

- We drop the rules from the original reduction acting on pure descendants of t but not on pure descendants of t_i.
- When a descendant of t was absorbed by, say, $\bar{r}_q$, we piled to its ancestor r_p (and so to its descendant $\bar{r}_q$) in that place $\tau[\![t_1, \ldots, t_{i-1}, \Box, t_{i+1}, \ldots, t_n]\!]$, whereas the old descendant of t is now the corresponding descendant of t_i, so it only remains to reduce the piled $\tau[\![t_1, \ldots, t_{i-1}, \Box, t_{i+1}, \ldots, t_n]\!]$ as previously in the reduction to obtain exactly the same situation as before, and the new reduction can proceed in the mimicking of the old reduction (see Figure 3). Note how these postponed reductions provoke no m-collapsings.
- We inserted $\tau[\![t_1, \ldots, t_{i-1}, \Box, t_{i+1}, \ldots, t_n]\!]$ below all the $r_1, \ldots, r_k$, but actually pure descendants of t may be absorbed in the old reduction only by part of

the descendants of these special subterms.
But we can get rid of these superfluous occurrences of material acting, as hinted previously, with the rules that in the initial reduction made $\tau[\![t_1, \ldots, t_{i-1}, \Box, t_{i+1}, \ldots, t_n]\!]$ collapse into $\Box$: they are applied to all of these extra descendants when the piled material is not needed any more. This means that these 'deleting sequences' must be applied

i) When in the sequel of the original reduction the descendant of an r_p will not absorb a pure descendant of t any more.
ii) When the descendant of an r_p absorbs another descendant of an r_q (Fig. 4).

Again, it is immediate to see these deleting sequences produce no m-collapsings.

This way we have obtained a new reduction that is infinite since all of the infinite rules acting on the top of the start term are still present. Note that left-linearity, via Proposition 2.5, was essential to be able to mimic the old reduction.
This new reduction has the number of special subterms of the start with a pure descendant that m-collapses in the reduction itself diminished by one: indeed, t is no more present, and as remarked no new m-collapsings are introduced modifying the original reductions.

Therefore, repeating this 'pile and delete' process leads, ultimately, to an infinite reduction without m-collapsings.

But this is impossible: this reduction has an infinite number of reduction rules acting on the top of the starting term, as seen before (alternatively, note this also follows from the fact the 'pile and delete' process does not increase the rank of the term), but since there are no m-collapsings the principal subterms of the starting term can be substituted by fresh variables without preventing the application of these infinite number of reduction rule (Proposition 2.5). Hence we obtain an infinite reduction of a term of an unique colour, contradicting the fact $\mathcal{A}$ and $\mathcal{B}$ are SN. $\Box$

4 Remarks

Note how the previous proof relies upon modularity of CR ([Toy87]) to show that only SN has to be proved: this is not necessary, however, since as noticed in the introduction CR + SN = UN$^{\rightarrow}$ + SN, and therefore the modularity of UN$^{\rightarrow}$ for left-linear TRSs proved in [Mar93] could be used as well, thus giving a proof completely based on the 'pile and delete' technique.

In fact, in [Mar93] it was noticed that the 'pile and delete' technique does not need the full power of UN$^{\rightarrow}$, but it can be applied under the weaker assumption of ***consistency with respect to reduction*** (briefly CON$^{\rightarrow}$), that is satisfied if a term cannot be rewritten to two different variables. This is true since the 'pile and delete' technique essentially relies upon Proposition 2.6, that still holds if CON$^{\rightarrow}$ is required in place of UN$^{\rightarrow}$. Hence, it was shown in [Mar93] that exactly the same proof used for the modularity of UN$^{\rightarrow}$ proved that CON$^{\rightarrow}$ is modular for left-linear TRSs.

The same observation is pertinent here: (the proof of) Theorem 3.1 *extends* the result in [TKB89] showing that

Theorem 4.1 *Termination is a modular property for left-linear and consistent with respect to reduction TRSs.*

Observe that, using the aforementioned modularity of $CON^{\rightarrow}$, it even holds the stronger result that SN + $CON^{\rightarrow}$ is a modular property for left-linear TRSs.

An easy consequence of the above theorem is the following. Recall that a TRS is said *non-erasing* if for every its rule $l \rightarrow r$ the variables in r are the same as in l. It is immediate to see that if a TRS is non-erasing then it is consistent w.r.t. reduction. Hence by Theorem 4.1 we obtain right away

Corollary 4.2 *Termination is a modular property for left-linear and non-erasing TRSs.*

Acknowledgments: Thanks are due to Jan Willem Klop for his continuous support and encouragement in getting a better and shorter proof. Moreover, I want to thank Aart Middeldorp for careful reading an earlier version of this paper and for having implicitly suggested me that the proof here given was not only a great simplification of [TKB89] but also an extension: indeed, I heard from him about the existence of the draft [SSP94], and that title gave me the suggestion to weaken $UN^{\rightarrow}$ in $CON^{\rightarrow}$ (incidentally, we note the basic character of the pile and delete technique developed in [Mar93], since in [SSP94] an analogous technique is employed). Thanks also to Vincent van Oostrom and Enno Ohlebusch.

References

[Klo92] J.W. Klop. Term rewriting systems. In S. Abramsky, Dov M. Gabbay, and T.S.E. Maibaum, editors, *Handbook of Logic in Computer Science*, volume 2, chapter 1, pages 1–116. Clarendon Press, Oxford, 1992.

[Mar93] M. Marchiori. Modularity of $UN^{\rightarrow}$ for left-linear term rewriting systems. Draft, March 1993. Extended and revised version: Technical Report CS-R9433, CWI, Amsterdam.

[Mid90] A. Middeldorp. *Modular Properties of Term Rewriting Systems.* PhD thesis, Vrije Universiteit, Amsterdam, November 1990.

[SSP94] M. Schmidt-Schauß and S. Panitz. Modular termination of consistent and left-linear TRSs. Draft, May 1994.

[TKB89] Y. Toyama, J.W. Klop, and H.P. Barendregt. Termination for the direct sum of left-linear term rewriting systems. In N. Dershowitz, editor, *Proceedings of the 3rd RTA*, vol. 355 of *LNCS*, pp. 477–491, Springer, 1989. Extended version: Technical Report CS-R8923, CWI, Amsterdam.

[TKB94] Y. Toyama, J.W. Klop, and H.P. Barendregt. Termination for direct sums of left-linear complete term rewriting systems. *JACM*, 1994. To appear.

[Toy87] Y. Toyama. On the Church-Rosser property for the direct sum of term rewriting systems. *JACM*, 1(34):128–143, 1987.

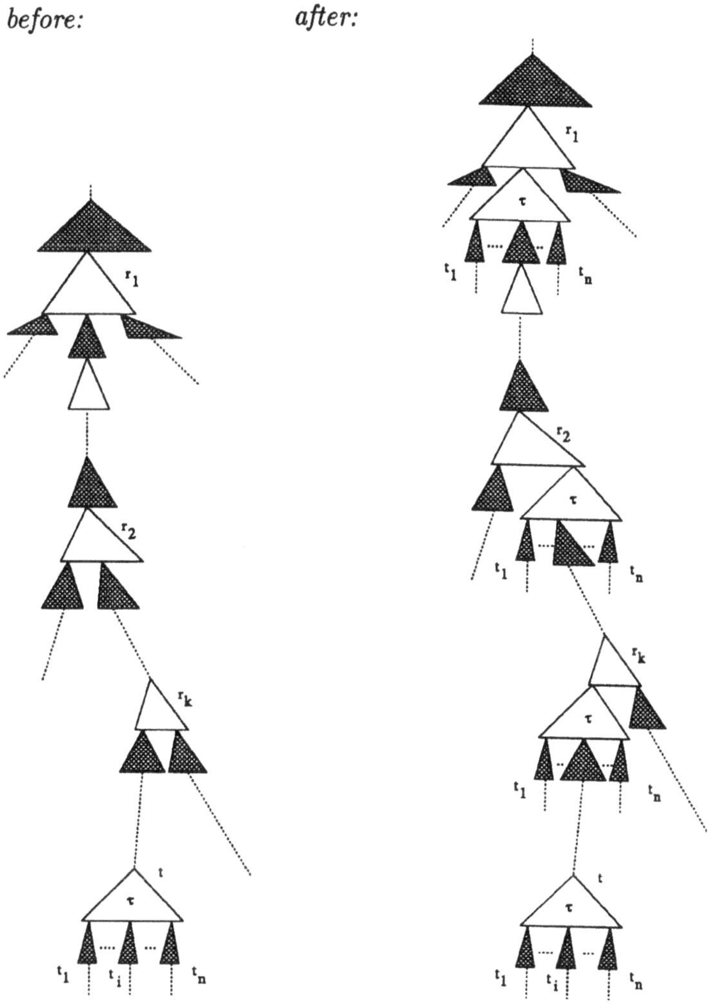

Fig. 1. The 'pile' process.

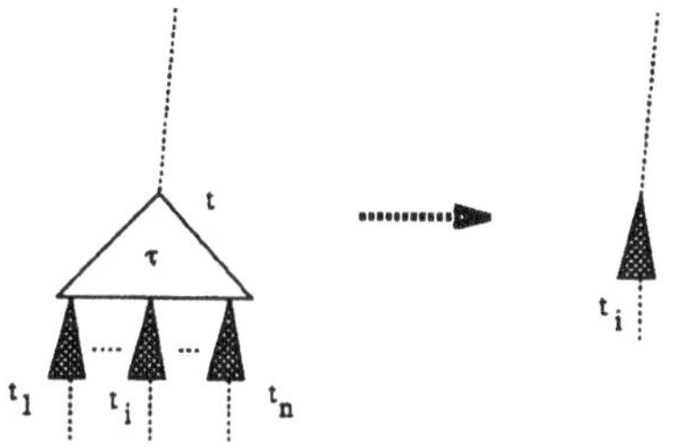

Fig. 2. The 'delete' process.

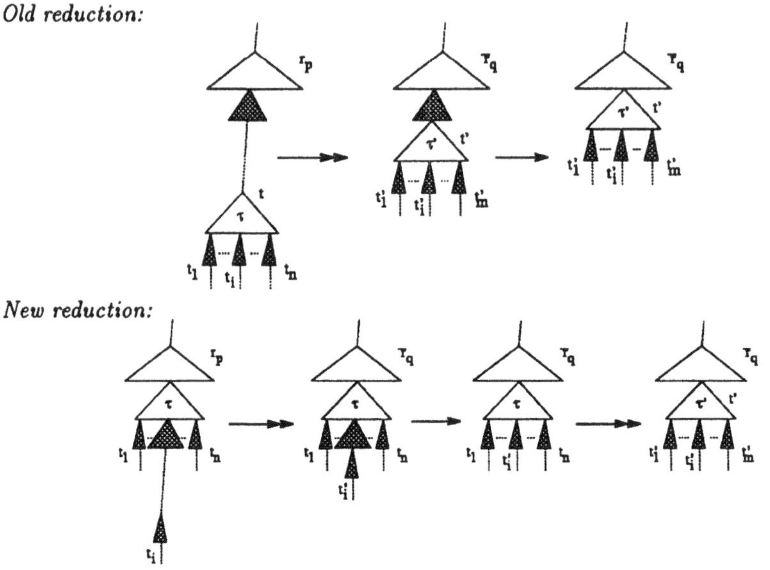

Fig. 3. One case of mimicking.

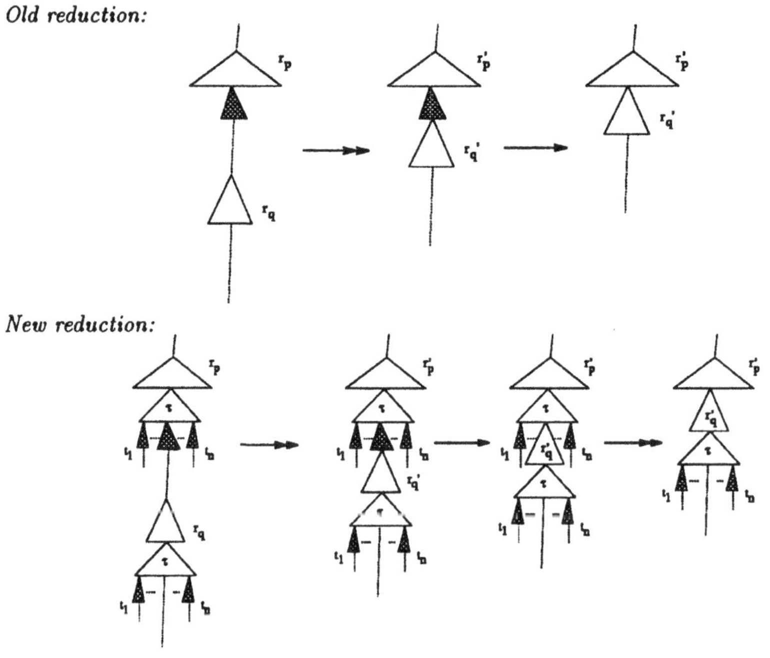

Fig. 4. An application of a deleting sequence.

Automatic Termination Proofs With Transformation Orderings

Joachim Steinbach

Institut für Informatik
Technische Universität München
80290 München (Germany)
steinbac@informatik.tu-muenchen.de

Abstract. Transformation orderings are a powerful tool for proving termination of term rewriting systems. However, it is rather hard to establish their applicability to a given rule system. We introduce an algorithm which automatically generates transformation orderings for many non-trivial systems including Hercules & hydra and sorting algorithms.

1 Introduction

Term rewriting systems (TRSs) are based on directed equations (called *rules*) which may be used as non-deterministic functional programs. Termination of TRSs is both a fundamental and an undecidable property ([HL78]).

Example 1 (Hercules & Hydra). In the ancient Greek myths, Hercules had to destroy a hydra, which was a snake with nine heads. For each head Hercules cut off, the hydra grew two new heads. Nevertheless, according to the Greek myths, Hercules was the winner of this fight! But he used unfair tactics!

We consider a fair modification of this problem which is 'indeed' terminating. A hydra will be represented as a tree the leaves of which are the heads. After cutting off a head, the following two cases are possible: (i) Assume that Hercules cuts off a head which is a direct son of the root. Then this head is removed and nothing else happens. Otherwise, (ii) $n > 0$ copies of the subtree – the root of which is the direct father of the lost head – are added:

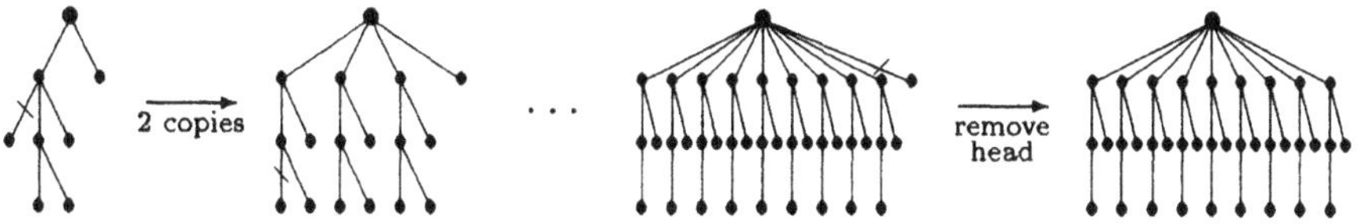

For a formal termination proof, a tree is represented as a term. Since we have to consider arbitrary trees, a tree will be represented as a list:

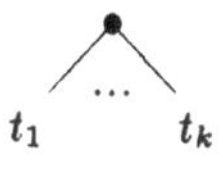

$$\begin{cases} \mathsf{f}(t_1.(t_2.\cdots.(t_{k-1}.t_k)\cdots)) & \text{if } k > 1 \\ \mathsf{f}(t_1) & \text{if } k = 1 \\ \mathsf{nil} & \text{if } k = 0 \end{cases}$$

The following TRS models the fight between Hercules and a hydra:

$$\begin{array}{rclcrcl} \mathsf{f}(\mathsf{nil}.y) & \to & \mathsf{f}(y) & \quad & \mathsf{copy}(0, y, z) & \to & \mathsf{f}(z) \\ \mathsf{f}(\mathsf{f}(\mathsf{nil}.y).z) & \to & \mathsf{copy}(\mathsf{n}, y, z) & & \mathsf{copy}(\mathsf{s}(x), y, z) & \to & \mathsf{copy}(x, y, \mathsf{f}(y).z) \end{array}$$

The example of a fight between Hercules and a hydra as given in the figure above illustrates that the reported outcome of the fight is not obvious! ◇

There exist various techniques based on so-called term orderings which together are able to cope with many TRSs occurring in practice. These *term orderings* are *stable* ($s \succ t \hookrightarrow s\sigma \succ t\sigma$ for any substitution σ) reduction orderings. *Reduction orderings* are *well-founded* (there are no infinite descending chains) and *compatible* with the structure of terms ($s \succ t \hookrightarrow \mathsf{f}(\cdots s \cdots) \succ \mathsf{f}(\cdots t \cdots)$ for any operator f). *Simplification orderings* are reduction orderings which have the *subterm property* ($\mathsf{f}(\cdots t \cdots) \succ t$). However, this property does not allow to prove the termination of, for example, a rule of the form $\mathsf{f}(\mathsf{f}(x)) \to \mathsf{f}(\mathsf{g}(\mathsf{f}(x)))$ since the left-hand side is homeomorphically embedded in the right-hand side. On the other hand, there exist orderings which are able to prove the termination of such a rule: *transformation orderings*. These orderings (recently called *completion orderings*) have been developed by Bellegarde & Lescanne (see, for example, [BL90]) from a work of Bachmair & Dershowitz ([BD86]). The main characteristics of this method are two rewriting relations represented by TRSs $\mathcal{T}$ and $\mathcal{S}$, one for transforming terms and the other for ensuring the well-foundedness of the ordering. The framework of transformation orderings is general enough to subsume *all* other term orderings.

The central problem with transformation orderings is to find adequate TRSs $\mathcal{S}$ and $\mathcal{T}$ such that the termination of a given TRS can be proved. In this paper, we introduce a heuristic-driven algorithm which automatically generates $\mathcal{S}$ and $\mathcal{T}$ for many TRSs (see Section 3). Furthermore, the practical relevance of this technique is illustrated in Section 4 by applying it to several examples.

2 The Transformation Orderings

This section deals with the formal definition of transformation orderings as described in [BL90]. For more details, we refer to [BL90].

We assume familiarity with the definitions used in term rewriting and especially in the area of termination of TRSs (see, for example, [DJ90], [DJ91] and [Der87]).

$\mathcal{T}(\mathcal{F},\mathcal{X})$ denotes the set of (first-order) *terms* over *operators* $\mathcal{F}$ and *variables* $\mathcal{X}$. An *atom* is either a constant or a variable. A *trivial term* is one containing not more than one operator. The *set of variables* of a term t is denoted by $\mathcal{V}ar(t)$.

A TRS $\mathcal{R}$ of *rules* $l \to r$ over $\mathcal{T}(\mathcal{F},\mathcal{X})$ is *left-linear* if the left-hand side of each rule of $\mathcal{R}$ is linear. It is called *variable-preserving* if $\mathcal{V}ar(l) = \mathcal{V}ar(r)$ for each rule $l \to r \in \mathcal{R}$. If $\to_{\mathcal{R}}$ is well-founded, then $\mathcal{R}(t)$ usually is the *set of normal forms* of a term t. Here, we use $\mathcal{R}(t)$ to denote the *smallest* (w.r.t. any total ordering on $\mathcal{T}(\mathcal{F},\mathcal{X})$) normal form t' of t. In particular, $\mathcal{R}(t)$ stands for *the unique* normal form whenever $\to_{\mathcal{R}}$ is confluent.

$\rightarrow_S$ *locally cooperates* with $\rightarrow_T$ iff ${}_T\!\leftarrow \cdot \rightarrow_S \subseteq \xrightarrow{*}_T \cdot \rightarrow_S \cdot \xrightarrow{*}_{S\cup T} \cdot {}_T\!\xleftarrow{*}$. The relation $\rightarrow_S$ *cooperates* with $\rightarrow_T$ iff ${}_T\!\xleftarrow{*} \cdot \rightarrow_S \cdot \xrightarrow{*}_{S\cup T} \subseteq \xrightarrow{*}_T \cdot \rightarrow_S \cdot \xrightarrow{*}_{S\cup T} \cdot {}_T\!\xleftarrow{*}$. As shown in [BL90], $\rightarrow_S$ cooperates with $\rightarrow_T$ if $\rightarrow_S$ locally cooperates with $\rightarrow_T$, $\rightarrow_{S\cup T}$ is well-founded and $\rightarrow_T$ is locally confluent.

Definition 1 (Transformation Ordering, [BL90]). Let $\rightarrow_S$ and $\rightarrow_T$ be two rewriting relations such that $\rightarrow_S$ cooperates with $\rightarrow_T$, $\rightarrow_{S\cup T}$ is included in some term ordering $\succ_{\mathsf{T}}$ and $\rightarrow_T$ is confluent. Then the *transformation ordering* TO on terms s and t is defined as

$$s \succ_{\mathsf{TO}} t \iff \begin{array}{l} \mathcal{T}(s) = \mathcal{T}(t) \ \wedge \ s \succ_{\mathsf{T}} t \quad \text{or} \\ \mathcal{T}(s) \neq \mathcal{T}(t) \ \wedge \ \mathcal{T}(s) \rightarrow_S \cdot \xrightarrow{*}_{S\cup T} \mathcal{T}(t) \end{array}$$ ◇

Cooperation of $\mathcal{S}$ and $\mathcal{T}$ is crucial for a transformation ordering to be a term ordering ([BL90]). In order to use the TO as a term ordering, there are three properties to check: local cooperation, confluence and termination. The last two properties can be proved by classical methods. For testing local cooperation the following theorem provides a means for a restricted class of transformation systems $\mathcal{T}$ (see also [Ges91]). Note that a *critical pair* $\langle p, q \rangle$ between a rule of $\mathcal{S}$ and a rule of $\mathcal{T}$ such that $p \ {}_T\!\leftarrow \cdot \rightarrow_S q$ is said to be *cooperative* iff $p \xrightarrow{*}_T \cdot \rightarrow_S \cdot \xrightarrow{*}_{S\cup T} \cdot {}_T\!\xleftarrow{*} q$.

Theorem 2 ([BL90]). *Let $\mathcal{T}$ be left-linear and variable-preserving. Then $\mathcal{S}$ locally cooperates with $\mathcal{T}$ iff all critical pairs between $\mathcal{S}$ and $\mathcal{T}$ are cooperative.* ◇

Example 2 (Hercules & Hydra Revisited). Let $\mathcal{R}$ be the TRS of Example 1, $\mathcal{T} = \{\mathsf{nil}.y \rightarrow y\,,\ \mathsf{copy}(\mathsf{n}, y, z) \rightarrow \mathsf{H}(y, z)\}$ and $\mathcal{S} = \{\mathsf{f}(\mathsf{f}(x).y) \rightarrow \mathsf{H}(x, y)\,,\ \mathsf{copy}(0, y, z) \rightarrow \mathsf{f}(z)\,,\ \mathsf{copy}(\mathsf{s}(x), y, z) \rightarrow \mathsf{copy}(x, y, \mathsf{f}(y).z)\}$ where H is a new binary operator. Then (i) $\rightarrow_{S\cup T} \subseteq \succ_{\mathsf{RPOS}}$ based on the precedence $\mathsf{copy} \succ \mathsf{f} \succ . \succ \mathsf{H}$ and the status function $\tau(\mathsf{copy}) = \text{left}$[1], (ii) $\mathcal{T}$ is confluent and $\mathcal{S}$ locally cooperates with $\mathcal{T}$ because there exist no critical pairs (see Theorem 2) and (iii) $l \succ_{\mathsf{TO}} r$ for each $l \rightarrow r \in \mathcal{R}$, e.g. $\mathsf{f}(\mathsf{nil}.y) \succ_{\mathsf{TO}} \mathsf{f}(y)$ since $\mathcal{T}(\mathsf{f}(\mathsf{nil}.y)) = \mathsf{f}(y) = \mathcal{T}(\mathsf{f}(y)) \wedge \mathsf{f}(\mathsf{nil}.y) \succ_{\mathsf{RPOS}} \mathsf{f}(y)$ and $\mathsf{f}(\mathsf{f}(\mathsf{nil}.y).z) \succ_{\mathsf{TO}} \mathsf{copy}(\mathsf{n}, y, z)$ since $\mathcal{T}(\mathsf{f}(\mathsf{f}(\mathsf{nil}.y).z)) = \mathsf{f}(\mathsf{f}(y).z) \rightarrow_S \mathsf{H}(y, z) = \mathcal{T}(\mathsf{copy}(\mathsf{n}, y, z))$. ◇

3 The $\mathcal{S}$-$\mathcal{T}$-Generation Technique

In this section, we are going to present a new technique for *automatically generating* a transformation ordering (i.e. for automatically generating $\mathcal{S}$ and $\mathcal{T}$) such that the termination of a TRS is ensured. In [BL90], a completion-like algorithm is suggested for establishing cooperation of $\mathcal{S}$ and $\mathcal{T}$ if this property does not already hold. Cooperation is achieved by adding new rules to $\mathcal{S}$. According to this algorithm, a confluent (and terminating) transformation system $\mathcal{T}$ must be given initially and cannot be changed any more. From a practical point of view,

[1] $\succ_{\mathsf{RPOS}}$ stands for the *recursive path ordering with status* (see, for example, [Der87]).

it is pretty necessary to generate the *whole* TO, i.e. not only $\mathcal{S}$ *but also* $\mathcal{T}$. In this paper, we introduce an algorithm doing this job. Although the technique is (and can only be) a heuristic and not a decision procedure, it proved good in practical applications.

3.1 Motivation

Before giving precise definitions and a formal description of our method, let us elucidate the technique informally. Suppose we must prove termination of a TRS $\mathcal{R}$ with a TO. The parameters $\mathcal{S}$ and $\mathcal{T}$ of an adequate TO are to be obtained solely from $\mathcal{R}$. The basic ideas of our technique are (i) to put orientable rules into the $\mathcal{S}$-part of the TO and (ii) to use the $\mathcal{T}$-part to replace subterms (of right-hand sides of rules in $\mathcal{R}$) that are 'too large' by 'smaller' ones based on *new* operators. We illustrate these general features by a pathological example.

Example 3. Let $\mathcal{R}$ be the TRS consisting of the rules[2]

$$\begin{aligned} 1:&\ \mathsf{ghg}x \rightarrow \mathsf{g}x \\ 2:&\ \mathsf{gg}x \rightarrow \mathsf{ghg}x \\ 3:&\ \mathsf{hh}x \rightarrow \mathsf{hf}(\mathsf{h}x, x) \end{aligned}$$

1. In order to find an appropriate TO for proving termination of $\mathcal{R}$ we utilize the basic idea for showing the TO to be stronger than all other term orderings: $\mathcal{T} := \emptyset$ and $\mathcal{S} := \mathcal{R}$ (then $\mathcal{T}(l) \neq \mathcal{T}(r) \ \wedge \ \mathcal{T}(l) \rightarrow_{\mathcal{S}} \mathcal{T}(r)$ for all $l \rightarrow r \in \mathcal{R}$ and $\mathcal{S}$ terminates iff $\mathcal{R}$ terminates). Thus, each rule of $\mathcal{R}$ which can be oriented w.r.t. (an extension of) the basic ordering $\succ_{\mathsf{T}}$ will be part of $\mathcal{S}$ and removed from $\mathcal{R}$. Here, we use the RPOS as $\succ_{\mathsf{T}}$:

$\mathcal{R}$: 2 3	$\mathcal{S}$: 1	$\mathcal{T}$:	RPOS: $\succ = \emptyset$

The ordering $\succ_{\mathsf{T}}$ should be extendable in an incremental way. For instance, the RPOS possesses this property (e.g., it can be started with an empty precedence). Moreover, the RPOS has the favourable property that there exist simple algorithms for automatically generating its parameters.

2. The remaining rules of $\mathcal{R}$ (i.e. each side of them) will be transformed into 'easier' ones by applying the (terminating) rules of the current $\mathcal{S}$. If these 'easier' rules can be oriented w.r.t. (an extension of) $\succ_{\mathsf{T}}$, they will be removed from $\mathcal{R}$ and added to $\mathcal{S}$ (see step 1.). Furthermore, each rule of $\mathcal{S}$ which has been part of the reduction process must be moved from $\mathcal{S}$ to $\mathcal{T}$ because these rules can be applied to both sides of a rule, especially to the right-hand side (see the definition of the TO):

$\mathcal{R}$: 3	$\mathcal{S}$: $\mathsf{gg}x \rightarrow \mathsf{g}x$	$\mathcal{T}$: 1	RPOS: $\succ = \emptyset$

3. If $\mathcal{R}$ is not yet empty – i.e. there exists at least one rule $l \rightarrow r$ (the simplification of) which cannot be oriented w.r.t. an extension of the current $\succ_{\mathsf{T}}$ – we *project* subterm(s) of r into smaller (w.r.t. an extension of $\succ_{\mathsf{T}}$) ones.

[2] For simplicity, the unary operators g and h are represented without parentheses.

For that purpose we extract those subterms of r which are 'critical', i.e. for which there is no greater (w.r.t. $\succ_{\mathcal{T}}$) subterm in l. For each such subterm r', we introduce a new rule $r' \rightarrow \mathsf{H}(x_1, \ldots, x_n)$ and include it in $\mathcal{T}$ where H is a new operator and $\{x_1, \ldots, x_n\} = \mathcal{V}ar(r')$. With this extended $\mathcal{T}$ we proceed with step 2. by applying $\mathcal{T}$ to each side of the rules of $\mathcal{R}$ in order to produce $\mathcal{T}$-normal forms:

$\mathcal{R}$:	$\mathcal{S}$: $\mathsf{gg}x \rightarrow \mathsf{g}x$ $\mathsf{hh}x \rightarrow \mathsf{hH}(x, x)$	$\mathcal{T}$: 1 $\mathsf{f}(\mathsf{h}x, y) \rightarrow \mathsf{H}(x, y)$	RPOS: $\mathsf{h} \succ \mathsf{H}$ $\mathsf{f} \succ \mathsf{H}$

Note that the set of critical subterms of the right-hand side of a rule may include *non-linear* terms (in the example, we want to project $\mathsf{f}(\mathsf{h}x, x)$[3] which is the subterm of the right-hand side of the third rule at position 1). Since these terms could become left-hand sides of rules in $\mathcal{T}$, we have to *generalize* them because $\mathcal{T}$ must contain left-linear rules only (see Theorem 2).

4. With the help of the previous step each rule of $\mathcal{R}$ can be oriented. Thus, $\mathcal{R}$ becomes empty and the required properties must be checked: confluence, left-linearity and variable-preservation of $\mathcal{T}$, cooperation of $\mathcal{S}$ with $\mathcal{T}$ and termination of $\mathcal{R}$ w.r.t. the generated TO. In our example, $\mathcal{T}$ is confluent[4], left-linear and variable-preserving. However, $\mathcal{S}$ does not cooperate with $\mathcal{T}$ since the critical pair $\langle \mathsf{H}(\mathsf{h}x, y), \mathsf{f}(\mathsf{hH}(x, x), y) \rangle$ is not cooperative. The application of the algorithm of [BL90] for extending $\mathcal{S}$ to cooperate with a given $\mathcal{T}$ could sometimes be helpful: If $\mathcal{S}$ is extended by $\mathsf{H}(\mathsf{h}x, y) \rightarrow \mathsf{H}(\mathsf{H}(x, x), y)$, the cooperation will be satisfied. Finally, $\mathcal{R}$ is terminating w.r.t. the TO based on $\mathcal{S}$ and $\mathcal{T}$, i.e. $\forall l \rightarrow r \in \mathcal{R}$: $l \succ_{\mathsf{TO}} r$. ◇

3.2 The Algorithm

In this subsection the algorithm for generating $\mathcal{S}$ and $\mathcal{T}$ is described. For simplicity, some macros for testing the required properties of $\mathcal{S}$ and $\mathcal{T}$ (see 4. of Example 3) are used:

- *s-confl(T)*: This function returns $\mathcal{T}$ if $\mathcal{T}$ is confluent. Otherwise, it tries to complete $\mathcal{T}$ by applying a *strong* version of the Knuth-Bendix algorithm ([KB67]). More precisely, we allow only a *fixed* number of critical pairs to be produced. If the completion procedure successfully terminates, *s-confl(T)* provides the completed $\mathcal{T}$ (by simultaneously setting $\mathcal{T}$ to its completed counterpart) and otherwise, *nil* is returned.
- *s-coop(S,T)*: Here, $\mathcal{S}$ will be extended for potentially ensuring cooperation of $\mathcal{S}$ with $\mathcal{T}$ (see [BL90]) if this property does not hold. The same restrictions as in *s-confl(T)* concerning the number of critical pairs and the output also apply here.

  Additionally, we will use a new technique whenever the cooperation cannot be achieved by the completion-like method of [BL90]. This approach adds

[3] It is also possible to choose the whole right-hand side of the third rule.

[4] A non-confluent TRS can possibly be transformed into a confluent one by adding new rules according to the completion algorithm of [KB67].

rules to $\mathcal{T}$ (instead of $\mathcal{S}$) and can be applied when the following critical pair is given together with some constraints:

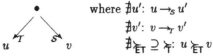

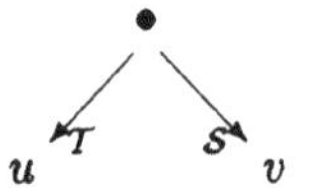

where $\nexists u'\colon u \rightarrow_{\mathcal{S}} u'$
$\nexists v'\colon v \rightarrow_{\mathcal{T}} v'$
$\nexists \succ_{\mathsf{ET}} \supseteq \succ_{\mathsf{T}}\colon u \succ_{\mathsf{ET}} v$

Let $\mathcal{T} = \mathcal{T} \cup \{l \rightarrow r\}$ with $v|_p = l\sigma$. If $\mathcal{S}$ contains such a rule $l \rightarrow r$ which is left-linear and variable-preserving, then it is a good candidate for joining u and v since it does not extend the theory of $\mathcal{R}$. Thus, in case of non-cooperation, we check all possible partitions of $\mathcal{S}$ into $\mathcal{T}$ (see Example 15).

- *Simplify*($\mathcal{R}$) reduces the complexity of a TRS $\mathcal{R}$, i.e. it simplifies the structures of the terms of the rules. Currently, we optionally use the following simplification: Assume that $\mathcal{R}$ contains a rule of the form $t[r]_u \rightarrow t[s]_u$ such that $\nexists p\colon s|_p = r$, $\mathcal{V}ar(r) \supseteq \mathcal{V}ar(s)$ and $r \notin \mathcal{X}$. Thus, $\mathcal{R}$ can be simplified by replacing this rule by $r \rightarrow s$. A transformation ordering proving the termination of the simplified $\mathcal{R}$ also proves the termination of $\mathcal{R}$.[5]
- $lv(\mathcal{P})$ is the left-linear and variable-preserving part of a TRS $\mathcal{P}$.

In addition to these functions, the operation $\mathcal{P}roj(l \rightarrow r, \succ_{\mathsf{T}})$ computes appropriate projections for r as indicated in 3. of the previous example. Simply speaking, $n > 0$ subterms $r_1, \ldots, r_n$ of r will be replaced by smaller terms by (i) including rules of the form $r_i \rightarrow \mathsf{H}_i(x_{i_1}, \ldots, x_{i_m})$ into $\mathcal{T}$[6] where $\mathcal{V}ar(r_i) = \{x_{i_1}, \ldots, r_{i_m}\}$ and $\mathsf{H}_i \notin \mathcal{F}$ and then by (ii) replacing $l \rightarrow r$ by $\mathcal{T}(l) \rightarrow \mathcal{T}(r)$. Since the idea of projections is the most essential one for generating $\mathcal{S}$ and $\mathcal{T}$, it will be separately treated in the following subsection.

Algorithm 3 ($\mathcal{S}$-$\mathcal{T}$-Generation Technique). *Let $\mathcal{R}$ be a non-empty TRS for which a termination proof is needed. Furthermore, let $\succ_{\mathsf{T}}$ be a term ordering and $\rho(\succ_{\mathsf{T}})$ a not necessarily proper extension of it. We start with empty sets $\mathcal{S}$ and $\mathcal{T}$. The following procedure generates $\mathcal{S}$ and $\mathcal{T}$.*

$\mathcal{R}' := \mathcal{R}$ (1)
[optional] *Simplify*($\mathcal{R}'$) (2)
<u>REPEAT</u>
  <u>REPEAT</u>
    $\mathcal{S}^{\mathcal{T}} := lv(\mathcal{S}) \cup \mathcal{T}$ (3)
    $\mathcal{S}' := \{\mathcal{S}^{\mathcal{T}}(l) \rightarrow \mathcal{S}^{\mathcal{T}}(r) \mid l \rightarrow r \in \mathcal{R}', \mathcal{S}^{\mathcal{T}}(l)\ \rho(\succ_{\mathsf{T}})\ \mathcal{S}^{\mathcal{T}}(r)\}$ (4)
    $\mathcal{T} := \mathcal{T} \cup \{l \rightarrow r \in lv(\mathcal{S}) \mid l \rightarrow r$ was used to compute u or v (5)
      for $u \rightarrow v \in \mathcal{S}'\}$
    $\mathcal{R}' := \mathcal{R}' \setminus \{l \rightarrow r \in \mathcal{R}' \mid \mathcal{S}^{\mathcal{T}}(l) \rightarrow \mathcal{S}^{\mathcal{T}}(r) \in \mathcal{S}'\}$ (6)
    $\mathcal{S} := (\mathcal{S} \setminus \mathcal{T}) \cup \mathcal{S}'$ (7)
  <u>UNTIL</u> $\mathcal{R}' = \emptyset$ or $\mathcal{S}' = \emptyset$ (8)
  <u>IF</u> $l \rightarrow r \in \mathcal{R}'$ <u>THEN</u> $\mathcal{S} := \mathcal{S} \cup \mathcal{P}roj(l \rightarrow r, \succ_{\mathsf{T}})$ (9)
<u>UNTIL</u> $\mathcal{R}' = \emptyset$ (10)

---

[5] Another simplification can be the application of any TO where $\mathcal{R}$ is transformed into $\mathcal{S}, \mathcal{T}$ such that the termination of $\mathcal{S} \cup \mathcal{T}$ is proved by another TO.

[6] First, these rules will be added to $\mathcal{S}$ in step (9) and then moved into $\mathcal{T}$ in step (5).

IF *s-confl(T)* $\wedge$ $lv(\mathcal{T})=\mathcal{T}$ $\wedge$ *s-coop(S,T)* $\wedge$ $\forall l \rightarrow r \in \mathcal{R}$: $l \succ_{\mathsf{TO}} r$ (11)
THEN RETURN $\mathcal{S}$, $\mathcal{T}$ and $\succ_{\mathsf{T}}$
ELSE RETURN fail ◇

Whenever one of the conditions in step (11) cannot be verified, we backtrack to step (9) and use another set of projections for the previous rule. In case of no other set existing, we backtrack to the previous rule. The execution of steps (3)–(7) for the first time corresponds to the application of (an extension of) the basic term ordering $\succ_{\mathsf{T}}$ to $\mathcal{R}'$ since $\mathcal{S}^{\mathcal{T}} = \emptyset$ (because $\mathcal{S} = \emptyset$ and $\mathcal{T} = \emptyset$) which implies $\mathcal{S}^{\mathcal{T}}(t) = t$ and $\mathcal{T}$ in step (5) to be empty. Note that $\mathcal{R}'$ contains non-orientable rules of $\mathcal{R}$ while $\mathcal{S}^{\mathcal{T}}$ consists of all rules which are potentially applicable for simplifying both sides of rules of $\mathcal{R}'$. All rules of $\mathcal{R}'$ which can be oriented after reducing them w.r.t. rules of $\mathcal{S}^{\mathcal{T}}$ are part of $\mathcal{S}'$ and $\mathcal{S}$. The basic ordering $\succ_{\mathsf{T}}$ is eventually extended in step (4) and set to $\succ_{\mathsf{T}}$.

Theorem 4. *Algorithm 3 always terminates. If it does not fail, $\mathcal{R}$ is proved to be terminating with the help of the* TO *defined by $\mathcal{S}, \mathcal{T}$ and $\succ_{\mathsf{T}}$.* ◇

Proof. The correctness is obviously satisfied by checking the last condition in step (11). The algorithm terminates since (i) every outermost repeat-loop decreases $\mathcal{R}'$ by at least one rule (because of step (9)) and (ii) step (11) terminates because *s-confl(T)* and *s-coop(S,T)* are 'finite' versions of completion procedures for confluence and cooperation. □

3.3 Projections

The technique of the TO extends each term ordering. In Algorithm 3 we use a classical term ordering (like the RPOS) for proving termination and if this ordering fails to orient an equation (or a rule), it will be improved by a TO with appropriate TRSs $\mathcal{S}$ and $\mathcal{T}$. Inability to orient a given rule $s \rightarrow t$ with the basic term ordering used in the TO is the precondition for computing a set of *critical terms* (see 3. of Example 3). Such a set contains all those subterms $l_1, \ldots, l_n$ of the right-hand side t which are responsible for failing in orienting $s \rightarrow t$. The set of *projections* for $s \rightarrow t$ consists of a set of rewriting rules, one for each critical term l_i, and maps these terms into smaller terms. In step (9) of Algorithm 3 these projections are added to $\mathcal{S}$ but in step (5) they will become a part of $\mathcal{T}$ because they are designed to reduce a right-hand side of a rule in $\mathcal{R}'$.

Definition 5 (Set of Projections). Let $\{\mathsf{H}_1, \ldots, \mathsf{H}_n\} \not\subseteq \mathcal{F}$ be a set of new operators and $s \rightarrow t$ be a rewriting rule over $\mathcal{T}(\mathcal{F},\mathcal{X})$ where t is non-trivial. Furthermore, let $\succ_{\mathsf{T}}$ be a term ordering such that $s \not\succ_{\mathsf{T}} t$. Then

$$\mathcal{M} \;=\; \{l_1 \rightarrow r_1, \ldots, l_n \rightarrow r_n\}$$

is a *set of projections* $\mathcal{P}roj(s \rightarrow t, \succ_{\mathsf{T}})$ if

$$\mathcal{M}(s) \succ_{\mathsf{ET}} \mathcal{M}(t) \quad \text{with} \quad \succ_{\mathsf{ET}} \supseteq \succ_{\mathsf{T}}$$

and the following properties hold for all $i \in [1, n]$:

(i) $\exists u_i \in \mathcal{P}os(t)$: $t|_{u_i} \geq l_i$, l_i linear and non-trivial[7]
(ii) $r_i = \mathsf{H}_i(s_{i_1}, \ldots, s_{i_m})$ with $\mathcal{V}ar(r_i) = \mathcal{V}ar(l_i)$
(iii) $l_i \not\succ_{\mathsf{ET}} r_i$ ◇

Example 4. Let $t = \mathsf{f}(x, \mathsf{g}(\mathsf{a}, \mathsf{h}(x)))$ be the right-hand side of a rule $s \rightarrow t$ which is not orientable w.r.t. a given term ordering $\succ_{\mathsf{T}}$.

- The simplest set of projections is the one which needs no generalization (i.e. $l_i = t|_{u_i}$ in (i)) and takes the set $\mathcal{V}ar(l_i)$ as the arguments of H_i in (ii). Then $\mathcal{P}roj(s \rightarrow t, \succ_{\mathsf{T}}) = \{\mathsf{g}(\mathsf{a}, \mathsf{h}(x)) \rightarrow \mathsf{H}_1(x)\}$ is the maximal set.
- Using generalizations of $t|_{u_i}$, the above set for t can be extended by the projections $\mathsf{g}(x, \mathsf{h}(y)) \rightarrow \mathsf{H}_2(x, y)$, $\mathsf{g}(\mathsf{a}, y) \rightarrow \mathsf{H}_3(y)$, $\mathsf{f}(x, \mathsf{g}(\mathsf{a}, \mathsf{h}(y))) \rightarrow \mathsf{H}_4(x, y)$, $\mathsf{f}(x, \mathsf{g}(y, \mathsf{h}(z))) \rightarrow \mathsf{H}_5(x, y, z)$, $\mathsf{f}(x, \mathsf{g}(\mathsf{a}, y)) \rightarrow \mathsf{H}_6(x, y)$ and $\mathsf{f}(x, \mathsf{g}(y, z)) \rightarrow \mathsf{H}_7(x, y, z)$.
- In case of employing only arbitrary s_{i_j} as arguments of H_i, $\mathsf{g}(\mathsf{a}, \mathsf{h}(x)) \rightarrow \mathsf{H}_8(\mathsf{a}, x)$, $\mathsf{g}(\mathsf{a}, \mathsf{h}(x)) \rightarrow \mathsf{H}_9(x, \mathsf{f}(x, \mathsf{a}))$, $\mathsf{g}(\mathsf{a}, \mathsf{h}(x)) \rightarrow \mathsf{H}_{10}(\mathsf{H}_{11}(x))$ are three examples of the infinitely many possible projections.
- $\mathsf{f}(x, \mathsf{g}(y, z)) \rightarrow \mathsf{H}_{12}(x, \mathsf{g}(y, z))$ utilizes $t|_{u_i} > l_i$ *and* $s_{i_j} \notin \mathcal{X}$. ◇

For the termination proof of Algorithm 3, it is important that each rule of $\mathcal{R}'$ can be projected into a rule which can be shown to be terminating with the help of (an extension of) $\succ_{\mathsf{T}}$. This property is satisfied since the set of projections always exists for any rule $s \rightarrow t$ and any $\succ_{\mathsf{T}}$ if t is chosen as the critical term and the right-hand side of the projection is set to $\mathsf{H}(x_1, \ldots, x_m)$ with $\{x_1, \ldots, x_m\} = \mathcal{V}ar(t)$, since t is a non-trivial term (i.e. it contains at least *two* operators) and the right-hand side of the projection of t contains *only one* operator which is *new*.

The main problem concerning the projection process is its non-determinism which swells up the search space drastically. For instance, (i) the set of critical terms is not unique. (ii) The set of critical terms may include non-linear terms (see 3. of Example 3). Since $\mathcal{T}$ must contain left-linear rules only (see Theorem 2) we eventually have to generalize them. Note that there are different ways to generalize. (iii) The right-hand side of a projection rule can be choosen (nearly) arbitrarily (see Example 4).

Thus, in the sequel of this subsection, we deal with two solutions for restricting the search space. First of all, as $\mathcal{P}roj(s \rightarrow t, \succ_{\mathsf{T}})$ we choose a *minimal* set of projections whose left-hand sides are non-overlapping.

Secondly, we only use (generalizations of) subterms of l_i as arguments of H_i, i.e. $l_i \rightarrow \mathsf{H}_i(t_{i_1}, \ldots, t_{i_m})$ with $\forall j \in [1, i_m]: \exists p_j \in \mathcal{P}os(l_i): \exists \sigma_j: t_j\sigma_j = l_i|_{p_j}$. This guide-line will be very useful if the t_j's are chosen as those subterms of l_i which overlap with left-hand sides of rules contained in $\mathcal{S}$: Let $\mathcal{R}$ be

$$s \rightarrow t \text{ such that } \nexists \succ_{\mathsf{ET}} \supseteq \succ_{\mathsf{T}}: s \succ_{\mathsf{ET}} t$$
$$v_1 \rightarrow w_1, \ldots, v_m \rightarrow w_m$$

[7] $\geq$ stands for the *subsumption ordering* defined by $s \geq t$ iff $\exists \sigma$: $s = t\sigma$.

For simplicity, we assume that only one critical term t' in t exists where a projection is necessary. Furthermore, w.l.o.g. let t' be left-linear. Suppose that $\exists i \in [1,m]$: $\exists u \in \mathcal{P}os(t')$: $\exists \sigma$: $t'|_u\sigma = v_i\sigma$. Then the following projection rule will be established:

$$t' \rightarrow \mathsf{H}(x_1, ..., x_j, t'|_u, x_k, ..., x_l)$$

where $\mathcal{V}ar(t') = \{x_1, ..., x_l\}$ and $\mathcal{V}ar(t'|_u) = \{x_{j+1}, ..., x_{k-1}\}$[8]. It is not difficult to generalize this concept to arbitrary numbers of critical terms (t') and overlappings ($v_i\sigma$). The reason for integrating $t'|_u$ on the right-hand side of the projection rule is the following:

$\mathcal{T}$: $t' \rightarrow \mathsf{H}(x_1, ..., x_j, t'|_u, x_k, ..., x_l)$
$\mathcal{S}$: $v_i \rightarrow w_i$
$\vdots$

$$\begin{array}{ccc} & t'\sigma & \\ \swarrow_{\mathcal{T}} & & \searrow_{\mathcal{S}} \\ \mathsf{H}(x_1, ..., x_j, t'|_u\sigma, x_k, ..., x_l) & & t'[w_i\sigma]_u \\ \downarrow_{\mathcal{S}} & & \\ \mathsf{H}(x_1, ..., x_j, w_i\sigma, x_k, ..., x_l) & & \end{array}$$

The rule $t'[w_i\sigma]_u \rightarrow \mathsf{H}(x_1, ..., x_j, w_i\sigma, x_k, ..., x_l)$ added to $\mathcal{T}$ ensures cooperation. Its termination proof is obvious since H is a new operator (being smaller than the operators of $\mathcal{F}$) and the set $\{x_1, ..., x_l\}$ is contained in $\mathcal{V}ar(t'[w_i\sigma]_u)$. In Examples 16, 17 and 18, this idea will be applied.

3.4 Examples

Example 5 (Intervals of Natural Numbers). We consider a TRS specifying intervals of natural numbers, i.e. the binary function $\ltimes$ will be defined as $\mathsf{s}^m(0) \ltimes \mathsf{s}^n(0) = \mathsf{s}^m(0).(\mathsf{s}^{m+1}(0).(\mathsf{s}^{m+2}(0).(\cdots(\mathsf{s}^n(0).\mathsf{nil})\cdots)))$ where s represents the successor on natural numbers:

$$\begin{array}{rrcl} 1: & 0 \ltimes 0 & \rightarrow & 0.\mathsf{nil} \\ 2: & 0 \ltimes \mathsf{s}(y) & \rightarrow & 0.(\mathsf{s}(0) \ltimes \mathsf{s}(y)) \\ 3: & \mathsf{s}(x) \ltimes 0 & \rightarrow & \mathsf{nil} \\ 4: & \mathsf{s}(x) \ltimes \mathsf{s}(y) & \rightarrow & \mathsf{h}(x \ltimes y) \end{array} \qquad \begin{array}{rrcl} 5: & \mathsf{h}(\mathsf{nil}) & \rightarrow & \mathsf{nil} \\ 6: & \mathsf{h}(x.y) & \rightarrow & \mathsf{s}(x).\mathsf{h}(y) \end{array}$$

Assume that the RPOS with the precedence $\ltimes \succ \mathsf{h} \succ . \succ \mathsf{s} \succ \mathsf{nil}$ is given. Algorithm 3 performs the following steps:

(4) $\mathcal{S}' = \{1, 3, 4, 5, 6\}$
(6) $\mathcal{R}' = \{2\}$
(7) $\mathcal{S} = \mathcal{S}'$
(3) $\mathcal{S}^{\mathcal{T}} = \{1, 4, 5, 6\}$
(4) $\mathcal{S}' = \{0 \ltimes \mathsf{s}(y) \rightarrow 0.\mathsf{h}(0 \ltimes y)\}$
(5) $\mathcal{T} = \{4\}$
(6) $\mathcal{R}' = \emptyset$
(7) $\mathcal{S} = \{1, 3, 5, 6\} \cup \mathcal{S}'$
(11) $\mathcal{S}, \mathcal{T}$ have the desired properties. ◇

Example 6 (Greatest Common Divisor). Let $\mathcal{R}$ be a TRS including the rules

$$\begin{array}{rrcl} 1: & \mathsf{gcd}(x, 0) & \rightarrow & x \\ 2: & \mathsf{gcd}(0, y) & \rightarrow & y \\ 3: & \mathsf{gcd}(\mathsf{s}(x), \mathsf{s}(y)) & \rightarrow & \mathsf{if}(x < y, \mathsf{gcd}(\mathsf{s}(x), y - x), \mathsf{gcd}(x - y, \mathsf{s}(y))) \end{array}$$

[8] Obviously, $t'|_u$ can also be generalized.

representing the greatest common divisor of two natural numbers. We apply our algorithm to $\mathcal{R}$ for generating an appropriate TO.

(4) $\mathcal{S}' = \{1,2\}$
(6) $\mathcal{R}' = \{3\}$
(7) $\mathcal{S} = \mathcal{S}'$
(3) $\mathcal{S}^{\mathcal{T}} = \mathcal{S}$
(4) $\mathcal{S}' = \emptyset$

(9) $\mathcal{S} = \{1,2\} \cup \{\mathsf{gcd}(\mathsf{s}(x), y - z) \rightarrow \mathsf{H}_1(x,y,z),\ \mathsf{gcd}(x - y, \mathsf{s}(z)) \rightarrow \mathsf{H}_2(x,y,z)\}$
(3) $\mathcal{S}^{\mathcal{T}} = \mathcal{S}$
(4) $\mathcal{S}' = \{\mathsf{gcd}(\mathsf{s}(x), \mathsf{s}(y)) \rightarrow \mathsf{if}(x < y, \mathsf{H}_1(x,y,x), \mathsf{H}_2(x,y,y))\}$
(5) $\mathcal{T} = \mathcal{S} \setminus \{1,2\}$
(6) $\mathcal{R}' = \emptyset$
(11) $\mathcal{S}, \mathcal{T}$ have the demanded properties.

Note that the smallest (w.r.t. the size) critical subterms of the right-hand side are $\mathsf{gcd}(\mathsf{s}(x), y - x)$ and $\mathsf{gcd}(x - y, \mathsf{s}(y))$. Thus, after generalizing, $\mathcal{S}$ consists of the projections of these terms. In order to prove the termination of the final $\mathcal{S} \cup \mathcal{T}$, we extend the precedence for the RPOS by $\mathsf{gcd} \succ \mathsf{H}_1, \mathsf{H}_2, \mathsf{if}, <$. ◇

3.5 Detailed Discussion of the $\mathcal{S}$-$\mathcal{T}$-Generation Technique

This subsection contains a few remarks on Algorithm 3 (and an implementation of it) and discusses some open questions.

1. The tests in step (11) are necessary since $\mathcal{T}$ is dynamically constructed in step (5)[9]. Especially, after completing $\mathcal{T}$ one has to check its variable-preservation and left-linearity once again. However, there is no need to test termination of $\mathcal{S} \cup \mathcal{T}$ since this property holds by construction of $\succ_{\mathcal{T}}$ (containing $\xrightarrow{+}_{\mathcal{S}\cup\mathcal{T}}$).
2. Algorithm 3 is designed to prove the termination of a given set of *rules*. It is not difficult (but more time consuming) to generalize this technique such that a given set of *equations* can be transformed into a provably terminating set of *rules*.
3. A problem concerning the basic ordering is that there sometimes exists the possibility of selecting different sets of incomparable rules w.r.t. $\succ_{\mathcal{T}}$. We use backtracking to find an appropriate set (see Example 12). The backtracking concept is also applied to extract adequate critical term(s) in step (9).
4. When simplifying a rule $l \rightarrow r$ of $\mathcal{R}'$ in step (2), sometimes a proper subterm of l and another left-hand side of $\mathcal{R}'$ could overlap whereas this is not possible with l. Therefore, this simplification is used by need (see Example 13).
5. Another subtle point of our algorithm, requiring careful treatment, is the use of the operation $\mathcal{S}^{\mathcal{T}}(t)$ for choosing a term from the set of normal forms derived from a term t.
   In step (4) of our approach, *sets* of normal forms of terms will be computed. It is obvious that we may replace *none* of these operations by the computation of *the* normal form since the system $\mathcal{S}^{\mathcal{T}}$ is not necessarily confluent (at the

[9] The last test of step (11) is performed although we strongly conjecture that confluence of $\mathcal{T}$ and cooperation of $\mathcal{S}$ and $\mathcal{T}$ imply the condition '$\forall l \rightarrow r \in \mathcal{R}$: $l \succ_{\mathsf{TO}} r$'.

time of its application). We assume that the normal forms are often unique (since the system in most cases contains very few rules, only). However, we need a good heuristic for choosing a convenient normal form (because these normal forms are important for step (11)). Such a heuristic is always applied to a rule $l \to r$ (which has to be transformed into a rule $\mathcal{S}^{\mathcal{T}}(l) \to \mathcal{S}^{\mathcal{T}}(r)$) and should possess (at least) the following properties: (i) The chosen normal form of l must be greater than that of r w.r.t. $\succ_{\mathcal{T}}$. (ii) The term $\mathcal{S}^{\mathcal{T}}(l)$ *should* not overlap with left-hand sides of $\mathcal{S}$ and $\mathcal{T}$.

6. Property (ii) is also important for generalizing (left-hand sides of) projection rules. Such a generalization is necessary since the projection rules are part of $\mathcal{T}$ which must be left-linear in order to apply Theorem 2. Obviously, the more general the left-hand sides of $\mathcal{T}$-rules are the more possibilities of overlappings (w.r.t. $\mathcal{S}$ and $\mathcal{T}$) exist and the more difficult it will be to guarantee the properties required in step (11). Thus, the generalization process should compute the greatest (w.r.t. $>$) generalization of a projection rule which leads to left-linearity. Note that the greatest generalization is identical to the replacement of all non-linear variables by new and unique ones.
7. An obvious extension of our algorithm concerns step (11). Only here, we eventually extend $\mathcal{T}$ and $\mathcal{S}$ for achieving confluence and cooperation. Obviously, it is possible to complete $\mathcal{T}$ (for confluence) as well as $\mathcal{S}$ (for cooperation) whenever extending these sets by *one* new rule (instead of completing them only just in step (11)).

4 Evaluation and Perspectives

We have implemented a prototpye of Algorithm 3 and tested it with a series of examples. Some of the examples which can be oriented using Algorithm 3 but *not* with an RPOS, are listed here. At the end of this section, an example will be given the termination of which cannot be shown with the presented algorithm. Furthermore, statistics will illustrate the power of the technique.

The transformation orderings presented in the following examples are *automatically* generated *exclusively* using Algorithm 3. The basic ordering used in the TO is the RPOS including the precedence $\succ$ and the status function τ on $\mathcal{F}$.

Example 7. $\mathsf{f}(\mathsf{f}(x)) \to \mathsf{f}(\mathsf{g}(\mathsf{f}(x)))$
$\mathcal{T} = \{\mathsf{g}(\mathsf{f}(x)) \to \mathsf{H}(x)\}$, $\mathcal{S} = \{\mathsf{f}(\mathsf{f}(x)) \to \mathsf{f}(\mathsf{H}(x)), \mathsf{H}(\mathsf{f}(x)) \to \mathsf{H}(\mathsf{H}(x))\}$, $\mathsf{f} \succ \mathsf{H}$
The termination of the similar rule $\mathsf{f}(\mathsf{g}(x)) \to \mathsf{f}(\mathsf{h}(\mathsf{g}(x)))$ can also be automatically proved with Algorithm 3. ◇

Example 8 ([Der87]). 1: $\mathsf{f}(\mathsf{a}) \to \mathsf{f}(\mathsf{b})$ 2: $\mathsf{g}(\mathsf{b}) \to \mathsf{g}(\mathsf{a})$
$\mathcal{T} = \{\mathsf{g}(\mathsf{a}) \to \mathsf{H}\}$, $\mathcal{S} = \{1\} \cup \{\mathsf{g}(\mathsf{b}) \to \mathsf{H}\}$, $\mathsf{a} \succ \mathsf{b}$, $\mathsf{g} \succ \mathsf{H}$ ◇

Example 9 (L. Bachmair). 1: $\mathsf{g}(x,y) \to \mathsf{h}(x,y)$ 2: $\mathsf{h}(\mathsf{f}(x),y) \to \mathsf{f}(\mathsf{g}(x,y))$
$\mathcal{T} = \{1\}$, $\mathcal{S} = \{\mathsf{h}(\mathsf{f}(x),y) \to \mathsf{f}(\mathsf{h}(x,y))\}$, $\mathsf{g} \succ \mathsf{h} \succ \mathsf{f}$ ◇

Example 10 (A. Middeldorp).

$$\begin{array}{llll} 1{:}\ \mathsf{h}(x,x,y) \to \mathsf{g}(x) & & 3{:}\ \mathsf{i}(x) \to \mathsf{f}(x,x) \\ 2{:}\ \mathsf{g}(\mathsf{a}) \to \mathsf{h}(\mathsf{a},\mathsf{b},\mathsf{a}) & & 4{:}\ \mathsf{f}(x,y) \to x \end{array}$$

$\mathcal{T} = \{\mathsf{h}(\mathsf{a},\mathsf{b},\mathsf{a}) \to \mathsf{H}\}$, $\mathcal{S} = \{1,3,4\} \cup \{\mathsf{g}(\mathsf{a}) \to \mathsf{H}\}$, $\mathsf{h} \succ \mathsf{g} \succ \mathsf{H}$, $\mathsf{i} \succ \mathsf{f}$ ◇

Example 11 (Perfect Numbers). The following TRS represents a specification of the perfect numbers. A *perfect number* is a number whose proper divisors sum up to itself (e.g. $28 = 1+2+4+7+14$).

$$\begin{array}{rrcl} 1{:} & \mathsf{perfectp}(0) & \to & \mathsf{false} \\ 2{:} & \mathsf{perfectp}(\mathsf{s}(x)) & \to & \mathsf{f}(x,\mathsf{s}(0),\mathsf{s}(x),\mathsf{s}(x)) \\ 3{:} & \mathsf{f}(0,y,0,u) & \to & \mathsf{true} \\ 4{:} & \mathsf{f}(0,y,\mathsf{s}(z),u) & \to & \mathsf{false} \\ 5{:} & \mathsf{f}(\mathsf{s}(x),0,z,u) & \to & \mathsf{f}(x,u,z-\mathsf{s}(x),u) \\ 6{:} & \mathsf{f}(\mathsf{s}(x),\mathsf{s}(y),z,u) & \to & \mathsf{if}(x \le y, \mathsf{f}(\mathsf{s}(x),y-x,z,u), \mathsf{f}(x,u,z,u)) \end{array}$$

$\mathcal{T} = \{\mathsf{f}(\mathsf{s}(x),y-z,u,v) \to \mathsf{H}(x,y,z,u,v)\}$
$\mathcal{S} = \{1,2,3,4,5\} \cup \{\mathsf{f}(\mathsf{s}(x),\mathsf{s}(y),z,u) \to \mathsf{if}(x \le y, \mathsf{H}(x,y,x,z,u), \mathsf{f}(x,u,z,u))\}$

We apply the RPOS based on $\mathsf{perfectp} \succ \mathsf{f},\mathsf{s},0$ and $\mathsf{f} \succ \mathsf{true},\mathsf{false},-,\mathsf{if},\le,\mathsf{H}$ and $\tau(\mathsf{f}) = \text{left}$. ◇

Example 12. Consider the following TRS:

$$\begin{array}{rrcl} 1{:} & \mathsf{f}(x,y,z) & \to & \mathsf{g}(x \le y, x, y, z) \\ 2{:} & \mathsf{g}(\mathsf{true},x,y,z) & \to & z \\ 3{:} & \mathsf{g}(\mathsf{false},x,y,z) & \to & \mathsf{f}(\mathsf{f}(\mathsf{p}(x),y,z), \mathsf{f}(\mathsf{p}(y),z,x), \mathsf{f}(\mathsf{p}(z),x,y)) \end{array} \qquad \begin{array}{rrcl} 4{:} & \mathsf{p}(\mathsf{s}(x)) & \to & x \\ 5{:} & \mathsf{p}(0) & \to & 0 \end{array}$$

Using the RPOS, we are able to orient rule 1 in the desired direction by including the relations $\mathsf{f}\succ\mathsf{g}$ and $\mathsf{f}\succ\le$. This decision implies inability of proving termination of rule 3. Thus, $\mathcal{T}$ must contain a subterm of the right-hand side of this rule which leads to non-cooperation of $\mathcal{S}$ and $\mathcal{T}$. However, by orienting rule 3 with the RPOS based on $\mathsf{g}\succ\mathsf{f}$ and $\mathsf{g}\succ\mathsf{p}$, Algorithm 3 will be successful. It generates the systems $\mathcal{T} = \{\mathsf{g}(x \le y,z,u,v) \to \mathsf{H}(x,y,z,u,v)\}$ and $\mathcal{S} = \{2,3,4,5\} \cup \{\mathsf{f}(x,y,z) \to \mathsf{H}(x,y,x,y,z)\}$ by adding $\mathsf{f}\succ\mathsf{H}$, $\mathsf{g}\succ\mathsf{H}$. ◇

Example 13 (Hercules & Hydra Revisited). The transformation ordering given in Example 2 for proving the termination of the TRS of Example 1 is the one that is automatically generated with Algorithm 3 by using the *Simplify* operation in step (2) applied to the rule $\mathsf{f}(\mathsf{nil}.y) \to \mathsf{f}(y)$. ◇

Example 14 (L. Bachmair).

$$\begin{array}{ll} 1{:}\ \mathsf{f}(\mathsf{h}(x)) \to \mathsf{f}(\mathsf{i}(x)) & 3{:}\ \mathsf{h}(\mathsf{a}) \to \mathsf{b} \\ 2{:}\ \mathsf{g}(\mathsf{i}(x)) \to \mathsf{g}(\mathsf{h}(x)) & 4{:}\ \mathsf{i}(\mathsf{a}) \to \mathsf{b} \end{array}$$

Algorithm 3 generates $\mathcal{T} = \{\mathsf{g}(\mathsf{h}(x)) \to \mathsf{H}(x)\}$ and $\mathcal{S} = \{1,3,4\} \cup \{\mathsf{g}(\mathsf{i}(x)) \to \mathsf{H}(x)\}$. $\mathcal{S}$ does not cooperate with $\mathcal{T}$ since the non-cooperative critical pair $\langle \mathsf{H}(\mathsf{a}), \mathsf{g}(\mathsf{b}) \rangle$ exists by overlapping $\mathsf{g}(\mathsf{h}(x))$ and $\mathsf{h}(\mathsf{a})$. However, by adding the rule $\mathsf{H}(\mathsf{a}) \to \mathsf{g}(\mathsf{b})$ to $\mathcal{S}$ in step (11), the required properties are satisfied (if $\mathsf{h}\succ\mathsf{i}$, $\mathsf{a}\succ\mathsf{g}\succ\mathsf{H}$, $\mathsf{a}\succ\mathsf{b}$). ◇

Example 15 (Cartesian Category Combinators, [BL90]). Let $\circ$ be the composition of combinators and p be the pairing, let i be the identity combinator and f and s are respectively the first and the second projection combinator.

$$
\begin{array}{lrcl}
1: & (x \circ y) \circ z & \to & x \circ (y \circ z) \\
2: & \mathsf{p}(x,y) \circ z & \to & \mathsf{p}(x \circ z, y \circ z) \\
3: & x \circ \mathsf{i} & \to & x \\
4: & \mathsf{i} \circ y & \to & y
\end{array}
\qquad
\begin{array}{lrcl}
5: & \mathsf{f} \circ \mathsf{p}(x,y) & \to & x \\
6: & \mathsf{s} \circ \mathsf{p}(x,y) & \to & y \\
7: & \mathsf{b} \circ \mathsf{p}(\mathsf{b} \circ \mathsf{p}(x,y), z) & \to & \mathsf{b} \circ \mathsf{p}(x, \mathsf{b} \circ \mathsf{p}(y,z))
\end{array}
$$

$\mathcal{T} = \{\mathsf{b} \circ \mathsf{p}(x,y) \to \mathsf{H}(x,y)\} \cup \{2\}$

$\mathcal{S} = \{1,3,4,5,6\} \cup \{\mathsf{H}(\mathsf{H}(x,y),z) \to \mathsf{H}(x,\mathsf{H}(y,z))\,,\ \mathsf{H}(x,y) \circ z \to \mathsf{H}(x \circ z, y \circ z)\}$

The RPOS based on $\circ \succ \mathsf{p}, \mathsf{H}$ and $\tau(\circ) = \tau(\mathsf{H}) =$ left proves the termination of $\mathcal{S} \cup \mathcal{T}$. During the first phase of Algorithm 3, rule 2 belongs to $\mathcal{S}$. However, then the critical pair $\langle \mathsf{H}(x,y) \circ z, \mathsf{b} \circ (\mathsf{p}(x,y) \circ z)\rangle$ between rule 1 (contained in $\mathcal{S}$) and the rule $\mathsf{b} \circ \mathsf{p}(x,y) \to \mathsf{H}(x,y)$ of $\mathcal{T}$ is not cooperative. Thus, according to the mentioned improvement of the cooperation-completion procedure of [BL90] (see function *s-coop* in Subsection 3.2), rule 2 is moved from $\mathcal{S}$ to $\mathcal{T}$ which ensures cooperation. ◇

Example 16 (Purging). The function $\mathsf{pu}(x)$ removes all but the leftmost occurrences of each number in a list x whereas $\mathsf{rem}(x,l)$ removes all occurrences of the number x in the list l.

$$
\begin{array}{lrcl}
1: & \mathsf{pu}(\mathsf{nil}) & \to & \mathsf{nil} \\
2: & \mathsf{pu}(x.l) & \to & x.\mathsf{pu}(\mathsf{rem}(x,l))
\end{array}
\qquad
\begin{array}{lrcl}
3: & \mathsf{rem}(x,\mathsf{nil}) & \to & \mathsf{nil} \\
4: & \mathsf{rem}(x,y.l) & \to & \mathsf{if}(x{=}y, \mathsf{rem}(x,l), y.\mathsf{rem}(x,l))
\end{array}
$$

$\mathcal{T} = \{\mathsf{pu}(\mathsf{rem}(x,l)) \to \mathsf{H}(\mathsf{rem}(x,l))\,,\ \mathsf{pu}(\mathsf{if}(x{=}y,u,v)) \to \mathsf{H}(\mathsf{if}(x{=}y,u,v))\,,\ \mathsf{pu}(\mathsf{nil}) \to \mathsf{H}(\mathsf{nil})\}$, $\mathcal{S} = \{3,4\} \cup \{\mathsf{H}(\mathsf{nil}) \to \mathsf{nil}\,,\ \mathsf{pu}(x.l) \to x.\mathsf{H}(\mathsf{rem}(x,l))\}$, $\mathsf{pu} \succ \mathsf{rem}, \mathsf{H}$ and $\mathsf{rem} \succ \mathsf{if}, ., =$. ◇

Example 17 (Egyptian Fractions, M. Gardner). The ancient Egyptians had a peculiarly hobbled approach to fractions. With the sole exception of 2/3, for which there was a special hieroglyph, they had symbols only for unit fractions, i.e. fractions that are the reciprocals of positive integers. To manipulate fractions with numerators higher than 1 the Egyptians expressed such fractions as sums of distinct unit fractions. For example, instead of writing 5/6 they wrote $1/2+1/3$. It is not obvious that every proper fraction can be expressed as the sum of unit fractions if a repetition is forbidden. One proof of this fact is the existence of Fibonacci's algorithm. Call the proper fraction a/b. The first term of the expansion is the largest unit fraction not greater than a/b. Now subtract the unit fraction from a/b to obtain another proper fraction. The second term of the expansion is the largest unit fraction not greater than this remainder. Continue in this manner. It can be proved that the process always terminates. Hence the algorithm always works.

One way of specifying Fibonacci's algorithm with a TRS is given below. A sum of unit fractions will be represented as a list of its denominators. For example, the sum $1/2+1/4$ has the representation $\mathsf{s}(\mathsf{s}(0))\ .\ (\mathsf{s}(\mathsf{s}(\mathsf{s}(\mathsf{s}(0))))\ .\ \mathsf{nil})$. Note that the division operator / is used as a constructor.

$$\begin{array}{rrcl}
1: & (x/y) \bowtie (u/v) & \rightarrow & ((x \star v) - (y \star u))/(y \star v) \\
2: & \mathsf{h}(\mathsf{s}(0), y, z) & \rightarrow & \mathsf{s}(0) \\
3: & \mathsf{h}(\mathsf{s}(\mathsf{s}(x)), \mathsf{s}(0), z) & \rightarrow & \mathsf{s}(\mathsf{h}(\mathsf{s}(x), z, z)) \\
4: & \mathsf{h}(\mathsf{s}(\mathsf{s}(x)), \mathsf{s}(\mathsf{s}(y)), z) & \rightarrow & \mathsf{h}(\mathsf{s}(x), \mathsf{s}(y), z) \\
5: & \mathsf{div}(x, y) & \rightarrow & \mathsf{h}(x, y, y) \\
6: & \mathsf{egypt}(0/y) & \rightarrow & \mathsf{nil} \\
7: & \mathsf{egypt}(\mathsf{s}(x)/y) & \rightarrow & \mathsf{div}(y, \mathsf{s}(x)).\mathsf{egypt}((\mathsf{s}(x)/y) \bowtie (\mathsf{s}(0)/\mathsf{div}(y, \mathsf{s}(x))))
\end{array}$$

$\mathcal{T} = \{1\} \cup \{\mathsf{egypt}(x \bowtie y) \rightarrow \mathsf{H}(x \bowtie y), \mathsf{egypt}(((x_1 \star x_2) - (x_3 \star x_4))/(x_5 \star x_6)) \rightarrow \mathsf{H}(((x_1 \star x_2) - (x_3 \star x_4))/(x_5 \star x_6))\}$, $\mathcal{S} = \{2, 3, 4, 5, 6\} \cup \{\mathsf{egypt}(\mathsf{s}(x)/y) \rightarrow \mathsf{div}(y, \mathsf{s}(x)).\mathsf{H}((\mathsf{s}(x)/y) \bowtie (\mathsf{s}(0)/\mathsf{div}(y, \mathsf{s}(x))))\}$, $\mathsf{div} \succ \mathsf{h} \succ \mathsf{s}$ and $\bowtie \succ /, -, \star$ and $\mathsf{egypt} \succ \bowtie, 0, ., \mathsf{div}, \mathsf{H}, \mathsf{nil}$ and $\tau(\mathsf{h}) = $ left. ◇

Example 18 (Quick-Sort, B. Gramlich). Consider the following specification:

$$\begin{array}{rlcl}
1: & \mathsf{qsort}(\mathsf{nil}) & \rightarrow & \mathsf{nil} \\
3: & \mathsf{low}(x, \mathsf{nil}) & \rightarrow & \mathsf{nil} \\
5: & \mathsf{high}(x, \mathsf{nil}) & \rightarrow & \mathsf{nil}
\end{array}
\qquad
\begin{array}{rlcl}
2: & \mathsf{qsort}(x.y) & \rightarrow & \mathsf{qsort}(\mathsf{low}(x, y)) \circ (x.\mathsf{qsort}(\mathsf{high}(x, y))) \\
4: & \mathsf{low}(x, y.z) & \rightarrow & \mathsf{if}(y \leq x, y.\mathsf{low}(x, z), \mathsf{low}(x, z)) \\
6: & \mathsf{high}(x, y.z) & \rightarrow & \mathsf{if}(y \leq x, \mathsf{high}(x, z), y.\mathsf{high}(x, z))
\end{array}$$

$$\begin{array}{rrcl}
\mathcal{T} = \{1\} \cup \{ & \mathsf{qsort}(\mathsf{low}(x, y)) \circ (z.\mathsf{qsort}(\mathsf{high}(u, v))) & \rightarrow & \mathsf{H}(\mathsf{low}(x, y), z, \mathsf{high}(u, v)) \\
& \mathsf{qsort}(\mathsf{low}(x, y)) \circ (z.\mathsf{nil}) & \rightarrow & \mathsf{H}(\mathsf{low}(x, y), z, \mathsf{nil}) \\
& \mathsf{qsort}(\mathsf{low}(x, y)) \circ (z.\mathsf{qsort}(\Delta_2)) & \rightarrow & \mathsf{H}(\mathsf{low}(x, y), z, \Delta_2) \\
& \mathsf{qsort}(\Delta_1) \circ (x.\mathsf{qsort}(\mathsf{high}(y, z))) & \rightarrow & \mathsf{H}(\Delta_1, x, \mathsf{high}(y, z)) \\
& \mathsf{qsort}(\Delta_1) \circ (x.\mathsf{nil}) & \rightarrow & \mathsf{H}(\Delta_1, x, \mathsf{nil}) \\
& \mathsf{qsort}(\Delta_1) \circ (x.\mathsf{qsort}(\Delta_2)) & \rightarrow & \mathsf{H}(\Delta_1, x, \Delta_2) \\
& \mathsf{nil} \circ (x.\mathsf{qsort}(\mathsf{high}(y, z))) & \rightarrow & \mathsf{H}(\mathsf{nil}, x, \mathsf{high}(y, z)) \\
& \mathsf{nil} \circ (x.\mathsf{nil}) & \rightarrow & \mathsf{H}(\mathsf{nil}, x, \mathsf{nil}) \\
& \mathsf{nil} \circ (x.\mathsf{qsort}(\Delta_2)) & \rightarrow & \mathsf{H}(\mathsf{nil}, x, \Delta_2) \ \}
\end{array}$$

$\mathcal{S} = \{3, 4, 5, 6\} \cup \{\mathsf{qsort}(x.y) \rightarrow \mathsf{H}(\mathsf{low}(x, y), x, \mathsf{high}(x, y))\}$
where $\Delta_1 = \mathsf{if}(x_1 \leq x_2, x_3.x_4, x_5)$, $\Delta_2 = \mathsf{if}(x_1 \leq x_2, x_3, x_4.x_5)$, $\mathsf{low} \succ \mathsf{if}, \leq, .$ and $\mathsf{high} \succ \mathsf{if}, \leq, .$ and $\mathsf{qsort} \succ \circ \succ \mathsf{H}, \mathsf{low}, \mathsf{high}$.
The generalized version r_2 of the right-hand side of rule 2 is projected onto the term $\mathsf{H}(\mathsf{low}(x, y), z, \mathsf{high}(u, v))$ using those subterms of r_2 which can be unified with rules of $\mathcal{S}$ (see the comments of Subsection 3.3). Most of the remaining rules of $\mathcal{T}$ are created similarly. Note that rule 1 is moved from $\mathcal{S}$ to $\mathcal{T}$. ◇

The only reasons for the technique to fail (see the following example) are either (i) $\mathcal{T}$ could not be completed to a finite and confluent TRS or (ii) the cooperation of $\mathcal{S}$ with $\mathcal{T}$ could not be achieved. Both are closely connected to the generation of $\mathcal{S}$-rules and $\mathcal{T}$-rules.

Example 19 (Factorial Function). Consider the following (unusual) specification:

$$1: \mathsf{fac}(\mathsf{s}(x)) \rightarrow \mathsf{fac}(\mathsf{p}(\mathsf{s}(x))) \star \mathsf{s}(x) \qquad \begin{array}{rrcl} 2: & \mathsf{p}(\mathsf{s}(0)) & \rightarrow & 0 \\ 3: & \mathsf{p}(\mathsf{s}(\mathsf{s}(x))) & \rightarrow & \mathsf{s}(\mathsf{p}(\mathsf{s}(x))) \end{array}$$

Due to Algorithm 3, we have to choose $\mathcal{T}$ as $\mathsf{p}(\mathsf{s}(x)) \rightarrow \mathsf{H}(x)$ or $\mathsf{fac}(\mathsf{p}(\mathsf{s}(x))) \rightarrow \mathsf{H}(x)$ or $\mathsf{fac}(\mathsf{p}(\mathsf{s}(x))) \star \mathsf{s}(y) \rightarrow \mathsf{H}(x, y)$. However, all these possibilities lead to

non-cooperation of $\mathcal{S}$ and $\mathcal{T}$. Our method can sometimes be further improved by integrating semantics (for example, in the form of lemmata of the inductive theory) of the used operators. Note that p stands for predecessor whereas s is the abbreviation for successor. Thus, $\mathsf{p}(\mathsf{s}(x)) = x$ holds using the semantics of p and s. If $\mathcal{T}$ is represented by the rule $\mathsf{p}(\mathsf{s}(x)) \rightarrow x$ and $\mathcal{S}$ to be $\mathsf{fac}(\mathsf{s}(x)) \rightarrow \mathsf{fac}(x)\star\mathsf{s}(x)$ both systems have the demanded properties.
We made the experience that the use of the semantics of the operators belonging to $\mathcal{T}$ facilitates the generation of an appropriate TO and leads to final versions of $\mathcal{S}$ and $\mathcal{T}$ which are structurally easier than without semantics. Obviously, Algorithm 3 extended by this improvement does not represent an automatic technique. ◇

We have applied Algorithm 3 to 60 TRSs (which are *not* orientable with simplification orderings) randomly chosen from the literature. For 47 TRSs (78.3 %) the method succeeds.

References

[BD86] Leo Bachmair and Nachum Dershowitz. Commutation, transformation and termination. In J.H. Siekmann, editor, *8th CADE*, volume 230 of *LNCS*, pages 52–60, Oxford (England), July 1986.

[BL90] Françoise Bellegarde and Pierre Lescanne. Termination by completion. *Applicable Algebra in Engineering, Communication and Computing (AAECC)*, 1:79–96, 1990.

[Der87] Nachum Dershowitz. Termination of rewriting. *JSC*, 3:69–116, February/April 1987.

[DJ90] Nachum Dershowitz and Jean-Pierre Jouannaud. *Rewrite systems*, volume B of *Handbook of Theoretical Computer Science*, chapter 6, pages 243–320. Elsevier Science Publisher B.V., 1990.

[DJ91] Nachum Dershowitz and Jean-Pierre Jouannaud. Notations for rewriting. *EATCS*, 43:162–172, February 1991.

[Ges91] Alfons Geser. Relative termination. Ulmer Informatik-Berichte, Fachbereich Informatik, Universität Ulm, Ulm (Germany), 1991.

[HL78] Gérard Huet and Dallas S. Lankford. On the uniform halting problem for term rewriting systems. Rapport Laboria 283, INRIA, Rocquencourt (France), March 1978.

[KB67] Donald E. Knuth and Peter B. Bendix. Simple word problems in universal algebras. In J. Leech, editor, *Conference on Computational Problems in Abstract Algebra*, pages 263–297, Oxford (England), August/September 1967. Pergamon Press. published in 1970.

Acknowledgement This research was supported by the Deutsche Forschungsgemeinschaft (SFB314-D4/Kaiserslautern, Je112/3-2/München). I would like to thank Reinhold Letz, Maximilian Moser and four anonymous reviewers for their constructive criticism and their helpful suggestions on the early version of this paper.

A Termination Ordering for Higher Order Rewrite Systems

Olav Lysne* and Javier Piris**

Laboratoire de Recherche en Informatique
Bat. 490 - 91405 Orsay Cedex
FRANCE

Abstract. We present an extension of the recursive path ordering for the purpose of showing termination of higher order rewrite systems. Keeping close to the general path ordering of Dershowitz and Hoot, we demonstrate sufficient properties of the termination functions for our method to apply. Thereby we describe a class of different orderings. Finally we compare our method to previously published extensions of the recursive path ordering into the higher order setting.

1 Introduction

During the last two decades a lot of work has been done on techniques for first order rewriting. There are still many open problems for the first order case, but even so we have in the last few years seen an increasing interest in extending the obtained results to a higher order framework. This is partly because the higher order logics offer increased expressive power, and partly due to a wish to extend the applicative programming paradigm by adding a rewriting relation to β-reduction. The focus of this research is on the combination of λ-terms and its formal theory, the typed lambda calculus [1, 2, 8], with algebraic terms and term rewriting [6].

There exist several different formalisms on the integration of typed lambda calculus and rewrite systems, and on the study of the interaction between braic rewriting and β-reduction. The earliest one seems to be the work on *combinatory reduction systems* due to Klop [12]. In [3] Breazu-Tannen combines the rules for λ-terms with arbitrary first order rewrite rules, and modular properties of the combined system are proved. More recently Jouannaud and Okada [10] introduced the *Algebraic Functional Language* allowing to define higher order functional constants by rewrite rules which follow a general schema. Also here modularity results are given, in particular with respect to termination aspects. Another approach is given in the work of Nipkow [16] where it is presented a rewriting relation modulo β and η-conversion. The decidability of validating an equation between higher order terms in the theory induced by the λ-conversions and a set of particular equations is considered, and

* E-mail: Olav.Lysne@ifi.uio.no. On leave from Dep. of Informatics, University of Oslo, P.O. box 1080 Blindern, 0316 Oslo, Norway. Supported by the Norwegian Research Council.

** E-mail: jpiris@dsic.upv.es. On leave from U.P.V., D.S.I.C., Camino de Vera S/N, P.O. box 22012, 46071 Valencia, Spain. Partially supported by CICYT under grant TIC 92-0793-C02-02.

partly solved by showing that local confluence is decidable, whenever all left hand sides of the rewrite rules are restricted to so called *patterns*.

The work of Nipkow leans on the decidability of unification of patterns. Miller discovered that the unification procedure for second order terms due to Huet [9] terminates for patterns, yielding a decision procedure [15]. Furthermore he showed that when two patterns are unifiable there exists a most general unifier.

We shall here consider the problem of termination of the rewrite relation due to Nipkow, by extending the recursive path ordering into his notion of higher order rewriting. In section 2 we give the basic notions needed to understand this paper. Section 3 describes a first approach to an ordering, whereas sections 4 and 5 present an extension of this ordering wrt. flexible terms. Section 6 is devoted to an extension of our method, and in section 7 we comment on related work.

2 Basic notions

The set $\mathcal{T}$ of *types* is generated from a set $\mathcal{B}$ of *base types* (e.g. *nat*, *bool*, *int*...) and the binary constructor $\longrightarrow$. The constructor $\longrightarrow$ is right associative, thus $\alpha_1 \longrightarrow \alpha_2 \longrightarrow \alpha_3$ means $\alpha_1 \longrightarrow (\alpha_2 \longrightarrow \alpha_3)$.

The *simply typed λ-terms* are built from λ-binders, a set of typed variables $\mathcal{V} = \bigcup_{\alpha \in \mathcal{T}} \mathcal{V}_\alpha$ and a set of constants $\mathcal{F} = \bigcup_{\alpha \in \mathcal{T}} \mathcal{F}_\alpha$, where $\mathcal{V}_\alpha \cap \mathcal{V}_{\alpha'} = \mathcal{F}_\alpha \cap \mathcal{F}_{\alpha'} = \emptyset$ if $\alpha \neq \alpha'$. Lowercase letters in the middle of the alphabet (s, t, u, v) denote simply typed λ-terms (*terms* for short), and letters at the end of the alphabet (x, y, z) denote variables. An inductive definition of terms is the following:

$$\frac{x \in \mathcal{V}_\alpha}{x : \alpha} \qquad \frac{c \in \mathcal{F}_\alpha}{c : \alpha} \qquad \frac{s : \alpha \longrightarrow \alpha' \quad t : \alpha}{(s\ t) : \alpha'} \qquad \frac{x : \alpha \quad s : \alpha'}{(\lambda x.s) : \alpha \longrightarrow \alpha'}$$

To enhance readability we shall use lowercase letters to denote bound variables, whereas capital letters always denote free variables. Thereby the set of bound and the set of free variables are always kept disjoint. A term which contains no free variables is a *ground term*. A term is said to be *basic* if it is of base type.

We use the following conventions for notation: Lowercase letters at the end of the alphabet denote bound variables, whereas capital letters always denote free variables. Thereby the set of bound and the set of free variables are always kept disjoint. We let $\mathcal{F}$ denote the set of *constants* (function symbols), and single constants are denoted by the lowercase letters f, g and h. *Atoms* may be either constants, bound or free variables, and are denoted by a and b.

By $\lambda\overline{x_n}.t$ or simply $\lambda\overline{x}.t$ we mean $\lambda x_1 \ldots \lambda x_n.t$, and that t is not a λ-abstraction. Furthermore we shall write $a(t_1, \ldots, t_n)$ meaning $((\ldots((a\ t_1)\ t_2)\ldots)\ t_n)$.

Mappings from variables to terms are written $\{x_n \mapsto t_n\}$. *Substitutions* are type preserving mappings from free variables to terms and are denoted by lowercase Greek letters. Mappings are treated as (postfix) operators to terms, which simultaneously replace the occurrences of variables with the corresponding terms.

Two terms t and u are α-equivalent, written $t =_\alpha u$ if t can be reached from u by a bijective mapping which renames bound variables. A *β-reduction* is the transformation of a term $(\lambda x.s)\ t$ into $s\{x \mapsto t\}$. A *β-normal form* is a term which cannot be β-reduced. It is well known that β-reduction terminates for simply typed λ-terms, thus every term has a β-normal form. An *η-reduction* is the transformation of a term $\lambda x.(t\ x)$ into t whenever x does not occur in t. An *η-long* form is a β-normal form $\lambda\overline{x}.a(t_1,\ldots,t_n)$ where $a(t_1,\ldots,t_n)$ is of base type and every t_i is η-long. In this way every occurrence of an atom a of type $\alpha_1 \longrightarrow \ldots \longrightarrow \alpha_n \longrightarrow \beta$ in a η-long form, will be at the head of a basic subterm of the form $a(t_1,\ldots,t_n)$. It is well known that every term may be mechanically transformed into a β-normal η-long form. Whenever we speak of a β-normal form we implicitly assume that it is η-long as well, unless stated otherwise. The β-normal (η-long) form of t is denoted $t\downarrow$. The *set of external λ-binders* in a β-normal form is the set containing all the outermost λ-binders in the term down to the first atom. For brevity we allow ourselves to write variables in β-normal forms as x and X instead of $\lambda\overline{y}.x(\overline{y})$ and $\lambda\overline{y}.X(\overline{y})$ whenever the outermost λ-binders are not important to the discussion. A β-normal form $\lambda\overline{x}.a(t_1,\ldots,t_n)$ is *flexible* if a is a free variable, and *rigid* otherwise. A β-normal form t is called a *pattern* if all arguments of each free variable in t are distinct bound variables.

We shall in this paper use notions of positions and subterms that are standard for first order algebraic terms, and nonstandard for λ-calculus. We shall view a term in β-normal form as an ordered tree. This by first regarding the occurring λ-binders as nodes with a single subtree, and second by viewing subterms on the form $a(t_1,\ldots,t_n)$ as a tree with a as root, and with the n subtrees $t_1,\ldots,t_n$. Notice that in our notion of term tree the "application" nodes of the λ-term have disappeared, giving a more restrictive notion of subterm. A *position* in a term can be viewed as a finite sequence of natural numbers, pointing out a path from the root of this tree. The place in the term where the path ends is the actual position. By $\mathcal{P}os(t)$ we denote the set of positions in t. The subterm of t at position p, written $t|_p$, is the subterm of t which has its root symbol at position p. The result of replacing the subterm of t at position p by the term u is written $t[u]_p$. A position in a term is said to be a *flexible position* if it occurs within a flexible subterm.

A *higher order rewrite rule* is an ordered pair of β-normal forms, written $t \rightarrow u$, such that t and u are of the same type, and such that all free variables in u occurs free in t as well. In this paper we only consider rewrite rules in which the left hand sides are basic patterns, and the right hand sides are arbitrary terms of base type. A *rewrite system* is a set of rewrite rules. A rewrite system R induces a rewrite relation $\rightarrow_R$ on the set of β-normal forms:

$$s \rightarrow_R t \Leftrightarrow \exists(l \rightarrow r) \in R, p \in \mathcal{P}os(s), \sigma\ .\ s|_p =_\alpha l\sigma \wedge t =_\alpha s[r\sigma]_p\downarrow$$

A sequence of rewrite steps is written $t_1 \rightarrow_R t_2 \rightarrow_R t_3 \ldots$, and consists only of terms in β-normal forms. It has been shown that local confluence of this rewrite relation is decidable [16], and that given convergence this relation may be used to decide $\alpha \cup \beta \cup \eta \cup R$-equivalence.

3 RPO for higher order rewriting

In order to prove that a term rewriting system is terminating it is not sufficient to show that every rewrite rule is contained in some well founded ordering. It is also necessary to show that every possible rewrite step is contained in the well founded ordering. This is usually implied by the ordering being *monotone* wrt. application of rewrite rules. In our higher order setting monotonicity is a more complex notion than in the fist order setting. This is because it will have to take into account the β-reductions that take place as a part of each rewrite step. We shall establish what monotonicity amounts to for the rewrite relation of Nipkow, and begin by presenting some basic facts on the rewriting relation and the β-normal forms we consider:

Lemma 1. *Let u be in β-normal η-long form. Then for any substitution σ containing only β-normal η-long forms we reach $u\sigma\downarrow$ from $u\sigma$ by β-reductions only.*

Thereby we need not worry about the effect of η-reduction in the following.

Lemma 2. *If u is of base type then $t\downarrow[u\downarrow]_p$ is in β-normal form.*

Lemma 3. *If there exists an infinite rewriting derivation for a set R of higher order rewrite rules, then there also exists one which only contains ground terms.*

Definition 4. An ordering $\succ$ on higher order rewrite rules is

- *stable wrt. substitutions* if $t\downarrow\succ u\downarrow \Rightarrow (t\downarrow\sigma)\downarrow\succ(u\downarrow\sigma)\downarrow$ for all basic patterns t, basic terms u and ground substitutions σ.
- *stable wrt. context application* if $t\downarrow\succ u\downarrow\Rightarrow v\downarrow[t\downarrow]_p\succ v\downarrow[u\downarrow]_p$ for all positions p, ground contexts v and basic ground terms t and u, such that t is an instance of a basic pattern.
- *monotone* if it is stable wrt. substitutions and context application.

Theorem 5. *Let R be a higher order rewrite system, and let $>$ be a well founded and monotone term ordering. If $l > r$ for every rewrite rule $l \to r \in R$ then there exist no infinite rewrite sequence wrt. R.*

Let us now present a basic fact on the effect of first applying a substitution to a term and then finding the β normal form.

Lemma 6. *Let $\mathcal{A}(s_1,\ldots,s_n)$ be a term in β-normal form, such that $\mathcal{A}$ is either a lambda-binder, constant or a bound variable. Then we have $\mathcal{A}(s_1,\ldots,s_n)\sigma\downarrow=\mathcal{A}(s_1\sigma\downarrow,\ldots,s_n\sigma\downarrow)$.*

Obviously if $\mathcal{A}(s_1,\ldots,s_n)$ is a term then $\mathcal{A}$ cannot be a bound variable. The above lemma is however stated like it is in order to be used in induction on the complexity of the terms. This is necessary because a term may have subterms that has bound variables on top.

Now we are ready to present a first approach to termination orderings for higher order terms. We begin by giving a way to interpret β-normal higher order terms as first order ones. These first order terms shall in turn be compared by a slightly modified recursive path ordering [4].

Definition 7. Let $\mathcal{M}$ interpret higher order terms as first order terms by

1. replacing all bound variables by a (one and the same) fresh constant c,
2. replacing all corresponding λ-binders λx with the new unary constant λ and
3. replacing all free variables with new constants, conserving the identity relation between the variables.

Notice that the structure of the terms are kept by $\mathcal{M}$. This means that we may use $\mathcal{M}$ as a means to view the terms as both first order and higher order, maintaining the obvious one to one correspondence between the nodes. In the following we therefore allow ourselves to switch focus between the higher order view and the first order view of expressions like $\mathcal{M}(t)$ without further ado. Each time it will be clear from the context what our view is.

We shall now define a recursive path ordering for β-normal higher order terms, staying close to the definition of the general path order of Dershowitz and Hoot [5]. We start by introducing *termination functions.* These are mappings from the set of terms to some partially ordered set. Examples of such functions are the following:

1. A function that returns the outermost constant symbol to be compared using a precedence.
2. A function that extracts an immediate subterm at a specified position, to be compared recursively in the induced term ordering.
3. A function that extracts an immediate subterm given by a specified selection criterion. (e.g. the i'th largest according to an ordering).

In [5] there are several additional (up to seven) classes of termination functions, opening for the definition of a wide range of different path orderings. At a later stage we shall also introduce an additional branch of termination functions, but for the time being we shall only be concerned with those mentioned above.

Definition 8. Let $\tau_0, \ldots, \tau_k$ be termination functions, $>_{\tau_0}, \ldots, >_{\tau_k}$ the corresponding partial orderings and $(>_{\tau_0} \ldots >_{\tau_k})_{lex}$ their lexicographic combination. The *recursive path ordering for mapped higher order terms*, denoted $\succ_{rpo}$ is defined as it appears in the following inference system:

$$\frac{\begin{array}{c} s_i \succsim_{rpo} t \quad \text{for some } i \quad 1 \leq i \leq m \\ f \textit{ is not mapped from a free variable} \end{array}}{f(s_1, \ldots, s_m) \succ_{rpo} t} \quad \textbf{subterm}$$

$$\frac{\begin{array}{c} s = f(s_1, \ldots, s_m), \quad t = g(t_1, \ldots, t_n) \\ f(s_1, \ldots, s_m) \succ_{rpo} t_1, \ldots, t_n \\ \langle \tau_0(s), \ldots, \tau_k(s) \rangle \; (>_{\tau_0} \cdots >_{\tau_k})_{lex} \; \langle \tau_0(t), \ldots, \tau_k(t) \rangle \\ f \textit{ and } g \textit{ are not mapped from free variables} \end{array}}{s \succ_{rpo} t} \quad \textbf{recpath}$$

Finally we define the ordering $\succ_{\mathcal{M}}$ on higher order patterns to be such that $t \succ_{\mathcal{M}} u$ iff $\mathcal{M}(t) \succ_{rpo} \mathcal{M}(u)$ holds for the first order interpretation of t and u.

For the time being we fix the termination function τ_0 such that it returns the outermost function symbol to be compared using a precedence. Furthermore $\tau_1, \ldots, \tau_k$ is restricted to be functions that return immediate subterms, such that we have either the lexicographic or the multiset path ordering. Notice that the inference rules above represent a restriction of the general path ordering considered by Dershowitz and Hoot [5], thus inherits the well foundedness property whenever the termination functions are used as we have indicated. See also [7] for a general and elegant proof of well foundedness. The reason for the additional restrictions we have made on the leading symbols in the inference rules is that without these restrictions it is easy to construct counterexamples to stability wrt. substitutions. This may be done by considering substitutions where higher order variables are instantiated by λ-terms which disregards all of its input arguments (e.g. $\lambda x.zero$).

The following lemma implies that the ordering is *sound* in the sense that any rewrite system contained in the ordering actually terminates:

Lemma 9. The relation $\succ_{\mathcal{M}}$ *is monotone.*

Proof. Stability wrt. substitutions: We show that $t\sigma\downarrow \succ_{\mathcal{M}} u\sigma\downarrow$ by cases of $t \succ_{\mathcal{M}} u$ using induction on the depth of t and u.

subterm: Here $t = f(t_1, \ldots, t_n)$ and $t_i \succsim_{\mathcal{M}} u$ for some i. By induction we get $t_i\sigma\downarrow \succsim_{\mathcal{M}} u\sigma\downarrow$, and according to lemma 6 we have $t\sigma\downarrow = f(t_1\sigma\downarrow, \ldots, t_n\sigma\downarrow)$, because from the inference rule we know that f is not mapped from a free variable. Thus by the **subterm** inference $t\sigma\downarrow \succ_{\mathcal{M}} u\sigma\downarrow$.

recpath: Now $t = f(t_1, \ldots, t_n)$ and $u = g(u_1, \ldots, u_m)$. If f has precedence over g then we have $f(t_1, \ldots, t_n) \succ_{\mathcal{M}} u_1, \ldots, u_m$, and by induction $f(t_1, \ldots, t_n)\sigma\downarrow \succ_{\mathcal{M}} u_1\sigma\downarrow, \ldots, u_m\sigma\downarrow$. From lemma 6 we have $u\sigma\downarrow = g(u_1\sigma\downarrow, \ldots, u_m\sigma\downarrow)$ thus the **recpath** rule gives $t\sigma\downarrow \succ_{\mathcal{M}} u\sigma\downarrow$. Otherwise f and g have the same precedence and $t_1, \ldots, t_n$ is either lexicographically or multiset greater than $u_1, \ldots, u_m$. By induction we easily get that $t_1\sigma\downarrow, \ldots, t_n\sigma\downarrow$ is respectively lexicographically or multiset greater than $u_1\sigma\downarrow, \ldots, u_m\sigma\downarrow$. In the same way as above lemma 6 and the **recpath** rule finishes the proof.

Stability wrt. context application: Due to lemma 2 This is immediate from the corresponding proof for the first order case. □

Example 1. This example is fetched from [16]. The following is a higher order rewrite system for computing the negation normal form of formulae in classical first order logic. We have two base types, *term* and *form* of terms and formulae, and here are the profiles of the constants:

$$\neg : form \longrightarrow form$$
$$\_ \wedge \_, \_ \vee \_ : form \longrightarrow form \longrightarrow form$$
$$\forall, \exists : (term \longrightarrow form) \longrightarrow form$$

Terms like $\forall(\lambda x.P(x))$ should be interpreted as $\forall x.P(x)$. Following Nipkow we find it sufficient to present half of the rewrite system. The other half is the dual.

$$\neg(\neg(\lambda x.P(x))) \to \lambda x.P(x)$$
$$\neg(\lambda x.P(x) \wedge \lambda x.Q(x)) \to (\neg(\lambda x.P(x)) \vee \neg(\lambda x.Q(x)))$$
$$\neg(\forall(\lambda x.P(x))) \to \exists(\lambda x.\neg(P(x)))$$

This rewrite system is shown to be confluent in [16], and termination may easily be proven by $\succ_{\mathcal{M}}$ with the precedence $\neg$ greater than everything.

4 Handling higher order variables

In the previous section we had to restrict the comparison of flexible terms due to the fact that for these terms it is hard to control stability wrt. substitutions. The effect of this restriction is that we may only prove termination of rules that have patterns in the right hand side as well as the left hand side. This excludes most of the higher order programs that one would like to write, e.g. the definition of the *map* function:

$$map(\lambda x.F(x), \epsilon) \rightarrow \epsilon$$
$$map(\lambda x.F(x), X :: Y) \rightarrow F(X) :: map(\lambda x.F(x), Y)$$

The strengthening of the ordering proceeds in two steps. The first step is to relieve the first termination function τ_0 from being a precedence on the outermost function symbols. The second step is to extend the underlying inference system.

Let us concentrate on τ_0 first. We leave open the exact definition of it, and instead we indicate what the requirements are on τ_0 in order for the proof of monotonicity of our extended ordering to go through.

Definition 10. Let τ_0 be a termination function taking a higher order term as input. Furthermore let $\succ_{\mathcal{M}^{\tau_0}}$ be the corresponding recursive path ordering on higher order terms. We say that τ_0 is *monotone* if it is a mapping into a well founded set of values such that

1. $\tau_0(t) > \tau_0(u) \Rightarrow \tau_0(t\sigma\downarrow) > \tau_0(u\sigma\downarrow)$ for every pair of terms t and u both in β-normal form, and every ground substitution σ.
2. $\tau_0(t) = \tau_0(u) \Rightarrow \tau_0(t\sigma\downarrow) = \tau_0(u\sigma\downarrow)$ for every pair of terms t and u both in β-normal form, and every ground substitution σ.
3. $t \succ_{\mathcal{M}^{\tau_0}} u \Rightarrow \tau_0(v[t\sigma\downarrow]_p) \geq \tau_0(v[u\sigma\downarrow]_p)$ for every ground term v in β-normal form, basic pattern t and basic term u both in β-normal form, position p not at the top, and ground substitution σ.

The first two of the above restrictions ensures that $\succ_{\mathcal{M}^{\tau_0}}$ is stable wrt. substitutions, whereas the last is there for stability wrt. context application. The above restrictions on τ_0 do not suffice to guarantee that τ_0 falls into one of the categories of termination functions of Dershowitz and Hoot. This because according to these categories a termination function that maps terms into a well founded set must either be a precedence on the outermost function symbols, or a homomorphism. Neither of the two are implied by our notion of monotonicity, and in a later section we shall indeed propose an instance of τ_0 that is monotone but no homomorphism. Still the proof of well foundedness of their ordering (which is stated to be akin to Kamin and Lévy [11]) trivially extends to ours. The proof is only based on the facts that the ordering contains the subterm relation, and that each of the termination functions $\tau_0, \ldots, \tau_k$ induces a well founded ordering. Both of these facts are kept by monotonicity of τ_0.

The next definitions establish the basic concepts in our proposal for the treatment of flexible subterms. First we take into account the paths from the top of a higher order ground term down to the subterms that are prone to change due to β-reduction.

Definition 11. Let t be a higher order ground term. The set of *critical positions* in t consists of the positions in t that lie on a path from the outermost λ-binders down to an occurrence of a corresponding bound variable, excluding the λ-binders and the bound variables themselves.

Example 2. The set of symbols occurring in critical positions in the higher order term $\lambda xy.f(g(x), h(h', y))$ is $\{f, g, h\}$.

Now we prepare for the control of the β-reduction by indicating how the τ_0 function may be used to handle free variables.

Definition 12. Let X be a higher order free variable, t be a basic pattern containing X and τ_0 be a termination function. We say that $\tau_0(t)$ *dominates* X, iff $\tau_0(t\sigma\downarrow) > \tau_0(X\sigma\downarrow|_p)$ for all σ, and all positions p that are critical in $X\sigma\downarrow$.

Informally the idea is that when $\tau_0(t)$ dominates a free variable X, then the comparison of $t\sigma\downarrow$ with $X(s_1, \ldots, s_n)\sigma\downarrow$ in the ordering shall be reduced to the comparison of $t\sigma\downarrow$ with the terms $s_1\sigma\downarrow, \ldots, s_n\sigma\downarrow$. This is obtained by recursive applications of the **recpath** rule along the critical positions in $X\sigma$. At this point in the development we encounter the reason why we need to be able to maintain also the higher order view on expressions like $\mathcal{M}(t)$. This is because the information needed to compute critical positions is lost in the first order version of t.

Definition 13. A higher order variable is *simple* if all of its arguments are basic.

Now a lemma on the effect of β-reduction on simple variables:

Lemma 14. Let X be a simple higher order variable and let $u_1, \ldots, u_n$ be terms of base type. Then for all σ we have that $X(u_1, \ldots, u_n)\sigma\downarrow$ may be obtained from $X\sigma\downarrow$ by replacing all occurrences of variables bound by outermost lambdas by $u_i\sigma$ for some i, and removing the outermost lambda-binders.

In order to carry through the idea sketched above, we need to require another property of τ_0, namely that it is in a way insensitive to β-reduction:

Definition 15. A termination function τ is said to be *stable wrt. β-reduction* if whenever $\tau(t) > \tau(G\sigma|_p)$ for G a simple variable and p a critical position in $G\sigma$ then $\tau(t) > \tau(G(v_1, \ldots, v_n)\sigma\downarrow|_{p'})$ where p' is obtained from p by removing the n first 1's.

The definition of β-stability above is rather technical, but the last part regarding the computation of p' from p is only there to remedy the fact that β-reduction will remove all outermost λ-binders from $G\sigma\downarrow$ in $G(v_1, \ldots, v_n)\sigma\downarrow$.

At this point we extend $\succ_{rpo}$ into $\succ_{erpo}$ by changing the name of the actual relation in **subterm** and **recpath** rules, and adding the following inference rule for flexible subterms:

$$\frac{\tau_0(s) \text{ dominates } G \qquad s \succ_{erpo} t_1, \ldots, t_n \qquad \textit{G is simple and occurs in s without free variables above it}}{s \succ_{erpo} G(t_1, \ldots, t_n)} \quad \textbf{flexrecpath}$$

Example 3. Let us study the problem we started this section with, namely the rule

$$map(\lambda x.F(x), X :: Y) \rightarrow F(X) :: map(\lambda x.F(x), Y)$$

Let us assume that $\tau_0(map(\lambda x.F(x), X :: Y)) > \tau_0(F(X) :: map(\lambda x.F(x), Y))$, and that $\tau_0(map(\lambda x.F(x), X :: Y)) \geq \tau_0(map(\lambda x.F(x), Y))$. Then we would be able to prove that this higher order rewrite rule is contained in $\succ_{\mathcal{M}^{\tau_0}}$ with $\succ_{erpo}$ at the bottom, granted that $\tau_0(map(\lambda x.F(x), X :: Y))$ dominates F.

We are now ready to prove that $\succ_{\mathcal{M}^{\tau_0}}$ is monotone.

Lemma 16. *Let τ_0 be a monotone and β-stable termination function. Furthermore let $\tau_1, \ldots, \tau_k$ correspond to lexicographic or multiset comparison of immediate subterms. Furthermore let $\succ_{\mathcal{M}^{\tau_0}}$ be the corresponding ordering on higher order terms based on $\succ_{erpo}$. Then $\succ_{\mathcal{M}^{\tau_0}}$ is monotone.*

Proof. Stability wrt. substitutions: The proof is constructed in the same manner as the proof of lemma 9:

subterm: Identical to the corresponding case in the proof of lemma 9.

recpath: Clear from the corresponding case of lemma 9 by replacing the comparison of outermost function symbols by a monotone termination function τ_0.

flexrecpath: Then we are in the situation $s \succ_{\mathcal{M}^{\tau_0}} G(t_1, \ldots, t_n)$, where s is a pattern, $\tau_0(s)$ dominates G and G occurs in s in such a way that it cannot disappear by β-reduction. We must prove $s\sigma\downarrow \succ_{\mathcal{M}^{\tau_0}} G(t_1, \ldots, t_n)\sigma\downarrow$.

Consider the topmost position in $G(t_1, \ldots, t_n)\sigma\downarrow$. From lemma 14 we know that $G(t_1, \ldots, t_n)\sigma\downarrow$ may be obtained from $G\sigma\downarrow$ by replacing all occurrences of variables bound by outermost lambdas by $t_i\sigma$ for some i, and removing the outermost lambda-binders. We therefore have one of three cases:

1. The topmost position stems from a non-critical position in $G\sigma$. Due to lemma 14 and the definition of critical positions, we get that $G(t_1, \ldots, t_n)\sigma\downarrow$ is a subterm of $G\sigma\downarrow$, and by the occurrence of G in s this case is finished by the **subterm** rule.
2. The topmost position stems from a critical position in $G\sigma$. Since $\tau_0(s)$ dominates G we get that $\tau_0(s\sigma\downarrow) > \tau_0(G\sigma\downarrow|_p)$ where p is the topmost position in $G\sigma$ which is not a λ-binder. Since τ_0 is stable wrt. β-reduction we must have that $\tau_0(s\sigma\downarrow) > \tau_0(G(t_1, \ldots, t_n)\sigma\downarrow)$. This means that we may use the **recpath** inference and transform the proof obligation into that of proving $s\sigma\downarrow \succ_{\mathcal{M}^{\tau_0}} u$ for all immediate subterms u of $G(t_1, \ldots, t_n)\sigma\downarrow$. The proof follows by induction on the depth of $G(t_1, \ldots, t_n)\sigma\downarrow$.
3. $G(t_1, \ldots, t_n)\sigma\downarrow$ equals $t_i\sigma$ for some i such that $1 \leq i \leq n$. In that case we must prove that $s\sigma \succ_{\mathcal{M}^{\tau_0}} t_i\sigma$ The inference rule **flexrecpath** gives us that $s \succ_{\mathcal{M}^{\tau_0}} t_i$, thus this follows by induction.

Stability wrt. context application: Since τ_0 is monotone this follows by a straightforward adaptation of corresponding proofs from the first order case. □

Theorem 17. *Let τ_0 be a monotone and β-stable termination function, and let $\succ_{\mathcal{M}^{\tau_0}}$ be the corresponding recursive path ordering on higher order terms based on $\succ_{erpo}$. Then any higher order rewrite system which is contained in $\succ_{\mathcal{M}^{\tau_0}}$ is terminating.*

Proof. By analysis of the proof of lemma 16 we see that every ground rewrite step of R is reducing by $\succ_{\mathcal{M}^{\tau_0}}$ using the inference rules of $\succ_{rpo}$ only. We have already shown that $\succ_{\mathcal{M}^{\tau_0}}$ with $\succ_{rpo}$ at the bottom is well founded. □

The intended application of our ordering is within a setting where τ_0 is known to be β-stable, and where domination is easily decidable. An example of such a termination function is given in the next section.

5 A monotone and β-stable termination function

So let us now present an example of a termination function which has the desired properties. Let us for the sake of simplicity first restrict ourselves to ground terms. The idea is as follows: Let a denote a constant symbol or a bound variable. If a has a higher order argument, we make sure that $\tau_0(a(\ldots,\lambda x.t,\ldots)) > \tau_0(\lambda x.t|_p)$ for all p that are critical in $\lambda x.t$. Following this idea we may assume that $\tau_0(a(\ldots,X,\ldots))$ dominates the higher order variable X.

We start by assigning a positive integer value, $val(\mathcal{A})$ to each symbol $\mathcal{A} \in \mathcal{F} \cup \{\lambda, c\}$. As a means to control the critical positions in a term we define a function $\mathcal{O}$:

Definition 18. Let $\mathcal{O}(t)$ be $max(\tau_0(t|_p))$ for all critical positions p in t.

Definition 19. We hereby fix τ_0 such that $\tau_0(\mathcal{A}(t_1,\ldots,t_m)) = val(\mathcal{A}) + max(\mathcal{O}(t_i))$ for $1 \leq i \leq m$.

The ordering assigned to τ_0 is simply the *greater than* relation on natural numbers. Notice that the function τ_0 is not a homomorphism. (A mapping function ϕ is a homomorphism if $\phi(\mathcal{A}(t_1,\ldots,t_n)) = \mathcal{A}_\phi(\phi(t_1),\ldots,\phi(t_n))$ for some function $\mathcal{A}_\phi$.) This because it is impossible to compute $\tau_0(\mathcal{A}(t_1,\ldots t_m))$ from $\tau_0(t_1),\ldots,\tau_0(t_m)$. Also note that τ_0 needs the higher order version of the term. For terms with free variables, it is straightforward to extend the ordering. The most conservative way of doing this is by assuming that if t_i is a flexible subterm, then neither $\mathcal{O}(t_i)$ nor $\tau_0(t_i)$ is known.

In order to be able to use this version of τ_0 in the method we have described, we must prove that is is monotone and β-stable. First we need to demonstrate a simple technical property.

Lemma 20. Let X be a free variable, $t_1,\ldots,t_n$ be arbitrary terms, and σ be a substitution. Consider the term $X(t_1,\ldots,t_n)\sigma$. Then the term introduced for the X-occurrence by σ will not contain variables bound in any of $t_1,\ldots,t_n$ and vice versa.

Proof. We have that $X(t_1,\ldots,t_n)\sigma = ((\ldots(X\sigma\ \ t_1\sigma)\ldots)\ \ t_n\sigma)$, thus the scoping rules of λ-binders give the rest. □

Lemma 21. Domination is stable wrt. β-reduction when τ_0 is as defined above.

Proof. Obvious from lemma 20. □

Lemma 22. The termination function τ_0 is monotone.

Proof. Regarding the three points stated in definition 10 the proofs that τ_0 fulfills the requirements of points 1. and 2. are trivial. To enhance the reading of the proof of point 3. we restate the proof obligation:

$t \succ_{\mathcal{M}^{\tau_0}} u \Rightarrow \tau_0(v[t\sigma\downarrow]_p) \geq \tau_0(v[u\sigma\downarrow]_p)$ *for every ground term v in β-normal form, basic pattern t and basic term u both in β-normal form, position p not at the top, and ground substitution σ.*

Let us assume that we have a counterexample to the above. Without loss of generality we may assume that the term assigned to v in the counterexample is minimal, in the sense that no subterm of v provides another counterexample. According to the definition of τ_0 this would mean that there exists an immediate subterm v' of v, such that $\mathcal{O}(v'[t\sigma\downarrow]_{p'}) < \mathcal{O}(v'[u\sigma\downarrow]_{p'})$. Also v' must be the smallest term satisfying this latter inequality.

Therefore we need consider the τ_0-values of the critical positions in $v'[t\sigma\downarrow]_{p'}$ and in $v'[u\sigma\downarrow]_{p'}$. Since v provided the smallest counterexample, the τ_0 values cannot have increased above position p', thus critical positions at and below p' are the only ones we need to study.

In order to abstract away the context v' of the above, we let $\lambda x_1 \ldots \lambda x_j$ be the outermost λ-binders of v'. The counterexample may now be rewritten as

$$\mathcal{O}(\lambda x_1 \ldots \lambda x_j \;\; t\sigma\downarrow) < \mathcal{O}(\lambda x_1 \ldots \lambda x_j \;\; u\sigma\downarrow)$$

Let $\mathcal{O}'$ be the slight modification of $\mathcal{O}$ that only considers the at most j first outermost λ-binders in the computation of critical positions. We shall prove that

$$\mathcal{O}'(\lambda x_1 \ldots \lambda x_j \;\; t\sigma\downarrow) \geq \mathcal{O}'(\lambda x_1 \ldots \lambda x_j \;\; u\sigma\downarrow)$$

Notice that for t and u of base type $\mathcal{O}$ and $\mathcal{O}'$ coincide in the above expression, thus this actually suffices to finish this subproof. We make the proof by cases of $\mathcal{M}(t) \succ_{erpo} \mathcal{M}(u)$ by induction on the depth of t and u. In the following $\mathcal{A}$ and $\mathcal{B}$ denotes elements in $F \cup \{c, \lambda\}$.

subterm Here we know that $t = \mathcal{A}(\overline{v})$, and that $\mathcal{M}(v_i) \succ_{erpo} \mathcal{M}(u)$ for some index i. By induction it follows that $\mathcal{O}'(\lambda x_1 \ldots \lambda x_j \;\; v_i\sigma\downarrow) \geq \mathcal{O}'(\lambda x_1 \ldots \lambda x_j \;\; u\sigma\downarrow)$ which clearly implies $\mathcal{O}'(\lambda x_1 \ldots \lambda x_j \;\; \mathcal{A}(\overline{v})\sigma\downarrow) \geq \mathcal{O}'(\lambda x_1 \ldots \lambda x_j \;\; u\sigma\downarrow)$.

recpath Let $t = \mathcal{A}(\overline{v})$ and $u = \mathcal{B}(\overline{s})$. The inference rule gives that $\mathcal{M}(t) \succ_{erpo} \mathcal{M}(s_i)$ for all i. By induction $\mathcal{O}'(\lambda x_1 \ldots \lambda x_j \;\; \mathcal{A}(\overline{v})\sigma\downarrow) \geq \mathcal{O}'(\lambda x_1 \ldots \lambda x_j \;\; s_i\sigma\downarrow)$ thus the position with the biggest τ_0-value in $s\sigma\downarrow$ does not have a bigger τ_0 value than the corresponding one in $t\sigma\downarrow$. We have $\tau_0(\mathcal{A}(\overline{v})) \geq \tau_0(\mathcal{B}(\overline{s}))$ from the inference rule and therefore $\mathcal{O}'(\lambda x_1 \ldots \lambda x_j \;\; \mathcal{A}(\overline{v})\sigma\downarrow) \geq \mathcal{O}'(\lambda x_1 \ldots \lambda x_j \;\; \mathcal{B}(\overline{s})\sigma\downarrow)$.

flexrecpath Now $t = \mathcal{A}(\overline{v}), u = G(\overline{s})$ and $\tau_0(\mathcal{A}(\overline{v}))$ dominates G, and G occurs in $\mathcal{A}(\overline{v})$ in a non-flexible position. Let us study the position of interest in $G(\overline{s})\sigma\downarrow$ with the biggest τ_0-value. According to lemma 14 the subterm of this position falls into one of three classes:

1. It is a subterm of $G\sigma\downarrow$. In that case it is also a subterm of $\mathcal{A}(\overline{v})\sigma\downarrow$, thus the τ_0 value of the subterm cannot be greater than $\mathcal{O}'(\lambda x_1 \ldots \lambda x_j \;\; \mathcal{A}(\overline{v})\sigma\downarrow)$.
2. It is a subterm of $s_i\sigma\downarrow$ for some i. In this case the inference rule **flexrecpath** states that $\mathcal{M}(\mathcal{A}(\overline{v})) \succ_{erpo} \mathcal{M}(s_i)$. By induction $\mathcal{O}'(\lambda x_1 \ldots \lambda x_j \;\; \mathcal{A}(\overline{v})\sigma\downarrow) \geq \mathcal{O}'(\lambda x_1 \ldots \lambda x_j \;\; s_i\sigma\downarrow)$ thus this case is unproblematic.

3. The position of the subterm corresponds to a critical position in $G\sigma\downarrow$. In that case the β-stability and the fact that $\tau_0(\mathcal{A}(\overline{v}))$ dominates G guarantees that the τ_0-value is less than $\tau_0(\mathcal{A}(\overline{v}))$.

□

It should now be clear what it takes for a variable to be dominated by a τ_0-expression:

Lemma 23. *The expression $\tau_0(\mathcal{A}(t_1,\ldots,t_n))$ dominates a simple higher order variable F, whenever $t_i = \lambda\overline{y}.F(\overline{x})$ for some i, $1 \leq i \leq n$, and F is not a free variable.*

Example 4. In order to prove termination of the rewrite system from example 3 we needed to establish the following three facts:

- $\tau_0(map(\lambda x.F(x), X :: Y)) > \tau_0(F(X) :: map(\lambda x.F(x), Y))$. This reduces to requiring $max(\mathcal{O}(\lambda x.F(x)),\ val(map)) > val(::)$. Here it suffices to require that $val(map) > val(::)$.
- $\tau_0(map(\lambda x.F(x), X :: Y)) \geq \tau_0(map(\lambda x.F(x), Y))$. By simple inspection we get that the two expressions have identical values.
- $\tau_0(map(\lambda x.F(x), X :: Y))$ dominates F, which is clear from lemma 23.

Example 5. Let us take a look at another higher order function that is common in functional programming, namely the function *iter* which iterates the application of a function on the elements of a list. Here we use *iter* to define a function *sum* which computes the sum of the elements in a list of natural numbers:

$$iter(\lambda x.\lambda y.F(x,y), X, []) \rightarrow X \qquad (1)$$

$$iter(\lambda x.\lambda y.F(x,y), X, Y :: L) \rightarrow iter(\lambda x.\lambda y.F(x,y), F(X,Y), L) \qquad (2)$$

$$sum(L) \rightarrow iter(\lambda x.\lambda y.(x+y), 0, L) \qquad (3)$$

The first rule is trivially contained in our ordering, the second follows by letting $\tau_1,\ldots,\tau_k$ give the lexicographic ordering of subterms from right to left. The third rule is handled by letting $val(sum)$ be greater than $val(\lambda)$ and greater than the sum of $val(+)$ and $val(iter)$.

6 Handling non-simple variables

The key to the approach we have described above is hidden in lemma 14. This is because in some cases this lemma allows us to handle the comparison of an expression $\tau_0(f(u_i,\ldots,u_n))$ with a higher order variable G. These are the cases where one may utilize the **recpath** rule through all the critical positions of $G\sigma$. This will let the term $f(u_i,\ldots,u_n)$ "survive" down to a subterm over which we have control.

This scheme works for simple variables G, because then any substitution σ will leave $t_i\sigma\downarrow$ unchanged in $G(t_i,\ldots,t_n)\sigma\downarrow$, and therefore controllable. If any of t_i are of higher order, however, this pleasant state of affairs is altered.

Let us now study what happens if t_i is second order. Then we face the same problem as we faced with $G\sigma$ previously, namely that we must control the ability of an expression $\tau_0(f(u_i,\ldots,u_n))$ to survive through the critical positions of $t_i\sigma\downarrow$ as well. This problem may of course also be solved in the same way, by altering the

flexrecpath rule such that it makes sure that $\tau_0(f(u_i,\ldots,u_n))$ "dominates" each of $t_i,\ldots,t_n$. In that case $\tau_0(f(u_i,\ldots,u_n))$ will survive down to a term which must be a subterm of a bound variable in $G\sigma$. If we allow t_i to be of arbitrary order, then this subterm of $G\sigma$ may also be of higher order, thus $\tau_0(f(u_i,\ldots,u_n))$ must survive down through this subterm in the same way, and so on. The scheme we have presented here may therefore be iterated in order to create an ordering where **flexrecpath** can handle flexible subterms with non-simple variables on top.

7 Related work

To our knowledge there have been two previous attempts to extend the recursive path ordering into higher order rewriting. The first is a short paper by Loría-Sáenz and Steinbach [14]. This paper focuses on the more general relation being the union between R reduction and β-reduction. It is clear that this relation subsumes the rewrite relation of Nipkow. Rewrite rules consists of β-normal η-long forms, with the additional restriction that variables in the left hand sides can only occur as leaf nodes in the term tree. This means that free higher order variables in the left hand side must be written as F rather than $\lambda\overline{x}.F(\overline{x})$.

From [14], however, it is not clear which restrictions have to be made for their ordering to work, and below we present a counterexample to the main result of the paper. Let us project their ordering into our setting. Their version of τ_0, denoted τ_0', maps every term of form $a(t_1,\ldots,t_n)$ into a term $a(u_1,\ldots,u_m)$. Here $u_1,\ldots,u_m$ is the subset of $t_1,\ldots,t_n$ consisting of higher order terms. The ordering assigned to τ_0' is simply the recursive path ordering itself. We shall see that this scheme has to be restricted in a way not indicated in the paper in order to be correct: Consider the higher order rewrite system

$$g(F,Y)\rightarrow F(Y) \tag{4}$$
$$f(\lambda x.h(g'(f')))\rightarrow g(\lambda z.f(\lambda x.h(z)),g'(f')) \tag{5}$$

When we let f' be greater in the precedence than all other constant symbols, this set of rewrite rules is contained in the ordering in [14]. For rule (4) we get that $\tau_0'(g(F,Y)) = g(F)$ is greater than $\tau_0'(F(Y)) = F$ by the subterm property. For rule (5) we get that $\tau_0'(f(\lambda x.h(g'(f')))) = f(\lambda x.h(g'(f')))$ is greater than both $\tau_0'(g(\lambda z.f(\lambda x.h(z)),g'(f'))) = g(\lambda z.f(\lambda x.h(z)))$ and $\tau_0'(f(\lambda x.h(z))) = f(\lambda x.h(z))$ because neither of the two latter contain the most significant constant symbol f'. For both of the rewrite rules the rest of the reasoning needed in order to show that it is contained in the ordering is straightforward. But this set is not terminating since it gives rise to the infinite sequence

$$g(\lambda z.f(\lambda x.h(z)),g'(f'))\rightarrow_{(4)} f(\lambda x.h(g'(f')))\rightarrow_{(5)} g(\lambda z.f(\lambda x.h(z)),g'(f'))\ldots$$

What seems like an immediate remedy, is to restrict the ability of any constant symbol to be greater than bound variables and lambda-binders. In the paper there is such a restriction on the type of the constant f', but notice that in the above example f' always occurs within the context of a constant symbol g', thus may be given an arbitrary complex type, simply by adding arguments to it in the example.

In his thesis [13], Loría-Sáenz describes a restricted version of the ordering presented in [14] which is correct. In the thesis all higher order terms are regarded as first order terms by viewing lambda binders as unary constants and bound and free variables as constants of the proper arity. There is of course a restriction saying that free variables are not comparable to anything. In addition the bound variables are restricted in such a way that they are incomparable to every other constant symbol except for that they may be equivalent to another bound variable of equivalent type. The standard recursive path ordering is then extended by extending the subterm relation. This is roughly done by extending the **subterm** inference rule such that it considers a term obtained by applying t_i to u be a subterm of $f(t_1, \ldots, t_i, \ldots, t_n)$ whenever u is a basic subterm of any of $t_1 \ldots t_{n-1}, t_{n+1} \ldots t_n$.

Let us compare the ordering of Loría-Sáenz to our ordering parameterized by the termination function τ_0 that we have described. We are able to show termination of rules like $f(F, zero) \to F(succ(zero))$ by letting $f > succ$ and $f > zero$. The ordering of Loría-Sáenz is not able to cope with this case. This because the only way it can handle higher order variables in the right hand side, is by the extended subterm relation. With this subterm notion there are two subterms of the left hand side which contain F, namely $F(zero)$ and F itself. Neither of these can be greater than $F(succ(zero))$. It is also fairly easy to see that if t is a subterm of u in the extended sense of [13], then we are able to show that $t \succ_{\mathcal{M}^{\tau_0}} u$. Thus we conjecture that our approach parameterized by the version of τ_0 that we have given is a proper extension of the approach in [13].

Finally we mention that the method of "termination by interpretation in a well-founded monotone algebra" described by Zantema in [18] has been extended to a higher order setting by van de Pol [17]. Here we do not make an explicit comparison with our method, since it is clear that the methods inherit the differences that are present in first order case.

8 Conclusion

We have extended the recursive path ordering so that it may be used to prove termination of higher order rewrite systems. Our method is parameterized by a function which collects information from the subterms of each occurring constant symbol. We have demonstrated what properties this function must entail in order for the soundness of the ordering to be maintained. This means that our approach really defines a family of termination orderings. We believe that further work on the development of different versions of τ_0 will provide a wide platform for proving termination of many different classes of higher order rewrite systems.

There exist some previous attempts to utilize the recursive path ordering in a higher order setting, and we have presented a counterexample to the results presented in [14]. Regarding the ordering put forward in [13] we have demonstrated that our approach is stronger when it is parameterized by the particular extraction function we have presented in this paper.

Acknowledgments: We are grateful to Hubert Comon and Jean-Pierre Jouannaud, both for initiating this research and for numerous discussions. Special thanks are due to Jean-Pierre Jouannaud for commenting on previous versions of this paper.

References

1. H. P. Barendregt. *The Lambda Calculus, its Syntax and Semantics.* North Holland, Amsterdam, 2nd ed., 1984.
2. H. P. Barendregt. *Handbook of Logic in Computer Science*, chapter Typed lambda calculi. Oxford Univ. Press, 1993. eds. Abramsky et al.
3. V. Breazu-Tannen. Combining algebra and higher-order types. In *Proceedings 3rd IEEE Symposium on Logic in Computer Science, Edinburgh (UK)*, July 1988.
4. N. Dershowitz. Orderings for term-rewriting systems. *Theoretical Computer Science*, 17:279–301, 1982.
5. N. Dershowitz and C. Hoot. Topics in termination. In *Proceedings 5th Conference on Rewriting Techniques and Applications, Montreal (Canada)*, volume 690 of *Lecture Notes in Computer Science*, pages 198–212. Springer-Verlag, 1993.
6. N. Dershowitz and J.-P. Jouannaud. Rewrite systems. In J. van Leeuwen, editor, *Handbook of Theoretical Computer Science*, volume B, chapter 6. Elsevier, Amsterdam, 1990.
7. M. C. F. Ferreira and H. Zantema. Well-foundedness of term orderings. In *Proceedings 4th International Workshop on Conditional Term Rewriting Systems, Jesuralem (Israel)*, 1994. To be published by Springer Verlag.
8. R. Hindley and J. Seldin. *Introduction to Combinators and λ-calculus.* Cambridge University Press, 1986.
9. G. Huet. A unification algorithm for typed λ-calculus. *Theoretical Computer Science*, 1(1):27–57, June 1975.
10. J.-P. Jouannaud and M. Okada. A computation model for executable higer-order algebraic specification languages. In *Proceedings 6th IEEE Symposium on Logic in Computer Science*, pages 350–361, 1991.
11. S. Kamin and J.-J. Lévy. Two generalizations of the recursive path ordering. Unpublished Note, Department of Computer Science, University of Illinois, Urbana, IL, 1980.
12. J. W. Klop. *Combinatory Reduction Systems.* Mathematical Centre Tracts 127, Mathematisch Centrum,Amsterdam, 1980.
13. C. Loría-Sáenz. *A Theoretical Framework for Reasoning about Program Construction based on Extensions of Rewrite Systems.* PhD thesis, Fachbereich Informatik der Universität Kaiserslautern, 1993.
14. C. Loría-Sáenz and J. Steinbach. Termination of combined (rewrite and λ-calculus) systems. In *Proceedings 3rd International Workshop on Conditional Term Rewriting Systems, Pont-a-Mousson (France)*, volume 656 of *Lecture Notes in Computer Science*, pages 143–147. Springer-Verlag, 1992.
15. D. Miller. A logic programming language with lambda-abstraction function variables, and simple unification. *Journal of Logic and Computation*, 1(4):497–536, 1991.
16. T. Nipkow. Higher order critical pairs. In *Proceedings 6th IEEE Symposium on Logic in Computer Science*, pages 342–349, 1991.
17. J. van de Pol. Termination proofs for higher-order rewrite systems. In *First International Workshop on Higher-Order Algebra, Logic and Term Rewriting*, volume 816 of *Lecture Notes in Computer Science*, pages 305–325. Springer-Verlag, 1993.
18. H. Zantema. Termination of term rewriting by interpretation. In *Proceedings 3rd International Workshop on Conditional Term Rewriting Systems, Pont-a-Mousson (France)*, volume 656 of *Lecture Notes in Computer Science*, pages 155–167. Springer-Verlag, 1992.

A complete characterization of termination of $0^p\,1^q \;\rightarrow\; 1^r\,0^s$

Hans ZANTEMA[1] and Alfons GESER[2]

[1] Universiteit Utrecht, P.O. Box 80089, NL-3508 Utrecht, The Netherlands,
E-mail: hansz@cs.ruu.nl

[2] Lehrstuhl für Programmiersysteme, Universität Passau, D-94030 Passau, Germany,
E-mail: geser@fmi.uni-passau.de

Abstract. We characterize termination of one-rule string rewriting systems of the form $0^p\,1^q \rightarrow 1^r\,0^s$ for every choice of positive integers p, q, r, and s. For the simply terminating cases, we give the precise complexity of derivation lengths.

Keywords: string rewriting, term rewriting, termination, simple termination, transformation ordering, dummy elimination, derivation length

1 Introduction

A term rewriting system R terminates, if every R-derivation $t_1 \rightarrow_R t_2 \rightarrow_R \cdots$ is finite. Much of the success of term rewriting is due to the availability of powerful termination criteria. String rewriting is a special case of term rewriting where function symbols are of arity 1, and may be taken as characters.

Termination of term rewriting systems is known to be undecidable, even for the special case of string rewriting systems [12], and even for left-linear, one-rule term rewriting systems [5]. The question whether termination is decidable for one-rule string rewriting systems is still open. In this paper we give a decision procedure for a non-trivial subclass, namely one-rule string rewriting systems Z of the form $0^p\,1^q \rightarrow 1^r\,0^s$.

Theorem 1. *Let p, q, r, s denote positive integer numbers. Then the one-rule string rewriting system $0^p\,1^q \rightarrow 1^r\,0^s$ terminates if, and only if,*

1. *$p \geq s$ or $q \geq r$ (simple termination), or*
2. *$p < s < 2p$ and $q < r$ and $q \nmid r$ or $q < r < 2q$ and $p < s$ and $p \nmid s$ (non-simple termination).*

Here $x|y$ denotes that x is a divisor of y. Moreover we show that the worst-case derivation length of a simply terminating system is either

1. linear ($p > s$ or $q > r$), or
2. quadratic ($p = s$, $q = r$), or
3. exponential ($p = s$, $q < r$, or $p < s$, $q = r$),

in the size of the initial term.

The main goal of the paper is not only to prove this theorem, but also to show how a very difficult termination proof can be given in a purely transformational style. We give such a proof for the non-simply terminating case. Instead of giving a complicated recursively defined ordering as is quite usual in termination proofs, we transform the system a number of times. For every transformation the termination of the original system follows from termination of the transformed system, which is equivalent to saying that the transformation preserves non-termination. This preservation follows from theorems that are generally applicable. These theorems follow the underlying ideas of *transformation ordering* [2, 3] and *dummy elimination* [9].

The paper is organized as follows. First we treat the case of simple termination in section 3. Here the termination proof is routine and we extend our attention to derivation lengths, on which we obtain sharp bounds. In section 4 we deal with the non-terminating cases.

The remainder and the main part of the paper is devoted to the difficult, non-simply terminating case $p < s < 2p$, $q < r$, $q \not| r$; the other non-simply terminating case is obtained by symmetry. In section 5, we describe how the system is transformed a number of times. In one step a fresh symbol $\Box$ is introduced in a right hand side of a rule, whose purpose is to stand there as a proof for the absence of information flow. This step is called *dummy introduction*, and is treated in detail in section 6. We employ an impoverished form of transformation order to prove preservation of non-termination. The next step is *dummy elimination*, the symbol $\Box$ is removed again by splitting the rule $l \to r_1 \Box r_2$ into two rules $l \to r_1$, $l \to r_2$. In section 8 the representation of strings over 0 and 1 is changed by describing such a string by $0^m\, 1^n$ packages. In this representation termination of the final system is proved by a lexicographical argument. We conclude by comparing related work.

2 Basic Notions

We assume that the reader is familiar with term rewriting, and in particular with termination proofs. A survey on termination of rewriting is [7]. For surveys on string rewriting see [13] and [4].

A binary relation $\to \subseteq S \times S$ on a set S is said to terminate, if there is no infinite $\to$-derivation $t_1 \to t_2 \to \cdots$.

For $\to$ a binary relation, $\to^{-1}$ and $\leftarrow$ denote the inverse relation: $s \leftarrow t$ holds if $t \to s$. Likewise $\to^{+}$ and $\to^{*}$ denote the transitive, transitive-reflexive closure of $\to$, respectively. The composition of two relations, $\to$ and $\hookrightarrow$, is denoted by $\to\hookrightarrow$, where $s \to\hookrightarrow t$ means that there is u such that $s \to u$ and $u \hookrightarrow t$. The expression $\to / \hookrightarrow$ abbreviates for $\hookrightarrow^{*}\to\hookrightarrow^{*}$, and $\leftrightarrow$ for the symmetric closure $\to \cup \leftarrow$ of $\to$.

A binary relation on terms is called an *order* if it is irreflexive and transitive, and a *quasiorder* if it is reflexive and transitive. A quasiorder defines an order $>$ by $s > t$, if s t and not t s, and an equivalence relation $\sim$ by $s \sim t$, if

s t and t s. By abuse, we call a quasiorder terminating, or wellfounded, if $>$ is terminating.

3 Simple Termination

If we try to apply well-known simplification orders like the recursive path order [6] to a one-rule rewrite system of the form $0^p\,1^q \to 1^r\,0^s$, we find that recursive path order can handle the case $p \geq s$ for arbitrary q, r, using the precedence $0 > 1$. The same is done by polynomial interpretation $[0](x) = (r+1)x, [1](x) = x + 1$ [14]. So in this case obviously Z is simply terminating. Moreover, since the interpretation is linear, the derivation length $D(n)$ is at most exponential in n [16]. Here $D(n)$ is defined to be the maximal number of steps in a reduction starting with a string of length n. Below we show that there are systems having linear, quadratic, and exponential derivation lengths.

Now it is easy to see that exchanging 0 by 1 and reversing strings gives only a renamed copy of the problem. By this symmetry argument the case $q \geq r$, with arbitrary p, s is simply terminating as well.

Proposition 2. *Z is simply terminating if $p \geq s$ or $q \geq r$.*

It is a surprising fact that in spite of the symmetry neither recursive path order nor one-level polynomial interpretations are able to handle the case $p < s, q = r$. Both techniques imply ω-termination as introduced in [19]. In [19] it is proved that the system $0\,1 \to 1\,0\,0$ is not ω-terminating; the same holds for the more general case $p < s, q = r$.

3.1 Linear and Quadratic Derivation Lengths

If the number of 0 symbols strictly decreases, then obviously the length of a derivation is bounded by the number of 0 symbols in the initial term. The complexity of derivation lengths is thus linear in this case.

Proposition 3. *If $p > s$ or $q > r$ then $D(n) = O(n)$.*

If the number of 0 symbols remains constant, i.e. if $p = s$ holds, then substrings 0^p behave exactly like single characters. So it suffices to treat the case $p = s = 1$.

Now consider the case $p = s = 1 = q = r$. The rewrite system $0\,1 \to 1\,0$ amounts to swap adjacent symbols if the former symbol is smaller (w.r.t. $0 < 1$) than the latter. This is nothing but the well-known *bubblesort* algorithm. Bubblesort has quadratic worst-case complexity.

Proposition 4. *If $p = s$ and $q = r$ then $D(n) = O(n^2)$.*

Indeed $0^n\,1^n \to_Z^{n^2} 1^n\,0^n$, which proves that a term of size linear in n yields a reduction of length quadratic in n.

3.2 Exponential Derivation Lengths

The considerations of the previous subsection leave the case $p = s = 1$, $q < r$. We claim that here for every choice of $q < r$, the derivation lengths are indeed exponential in the worst case. For instance, consider the case $q = 1$, $r = 2$. The rewrite system is $\mathtt{0\,1} \to \mathtt{1\,1\,0}$. A worst-case initial term is $\mathtt{0}^n\,\mathtt{1}$, for it has the normal form $\mathtt{1}^{2^n}\,\mathtt{0}^n$, and every rewrite step contributes only 1 to the length of a term. The derivation has therefore length $2^n - 1$.

Things are however much less easy in the general case; many terms initiate only derivations of polynomial length. The following example is typical. Let $q = 2$, $r = 3$, so Z is $\mathtt{0\,1\,1} \to \mathtt{1\,1\,1\,0}$. The term $\mathtt{0}^n\,\mathtt{1\,1}$ rewrites in only n steps to its normal form $\mathtt{1\,1(1\,0)}^n$. The shortness is caused by the fact that $\mathtt{1}$ symbols are not used up completely since 3 is not divisible by 2. The following pattern is free of such losses, as an easy induction on k shows.

Proposition 5. *For Z the system $\mathtt{0\,1}^q \to \mathtt{1}^r\,\mathtt{0}$, the following derivation holds.*

$$\mathtt{0}^k\,\mathtt{1}^{q^k} \to_Z^* \mathtt{1}^{r^k}\,\mathtt{0}^k$$

Its derivation length is $\frac{r^k - q^k}{r-q}$. Note however that this length may be not exponential in the size of the initial term. For, by an easy calculation, the asymptotic derivation length is described by

$$d(\mathtt{0}^k\,\mathtt{1}^{q^k}) = \begin{cases} O(r^{n-2}), & \text{if } q = 1, \\ O(n^{\log_q r}), & \text{else} \end{cases}$$

where n is the size $k + q^k$ of the initial term. This is an exponential function in n for $q = 1$ but a polynomial function for $q > 1$.

To finally achieve a worst-case pattern we choose a fixed number k which is large enough to lead to (at least) a duplication of the exponent of $\mathtt{1}$. Thus we can simulate the behaviour of case $q = 1$, $r = 2$ by a macro step.

Proposition 6. *Let Z denote the system $\mathtt{0\,1}^q \to \mathtt{1}^r\,\mathtt{0}$, where $1 < q < r$, and let $k = \lceil \frac{1}{\log_2 r - \log_2 q} \rceil$. Then for every $n > 0$, the term $\mathtt{0}^{nk}\,\mathtt{1}^{q^k}$ initiates a Z-derivation of a length exponential in n.*

Proof. By definition, k is the smallest number such that $r^k \geq 2q^k$ holds. Let c denote the (constant) length of the derivation $\mathtt{0}^k\,\mathtt{1}^{q^k} \to_Z^c \mathtt{1}^{r^k}\,\mathtt{0}^k$. By an easy induction on m one shows that $\mathtt{0}^k\,\mathtt{1}^{mq^k} \to_Z^{cm} \mathtt{1}^{mr^k}\,\mathtt{0}^k$ holds. We prove by induction on n that every term of the form $\mathtt{0}^{nk}\,\mathtt{1}^{mq^k}$ has a derivation length $\geq c(2^n - 1)m$. The base case $n = 1$ is given. For $n > 1$, there is a derivation starting with

$$\mathtt{0}^{nk}\,\mathtt{1}^{mq^k} = \mathtt{0}^{(n-1)k}\,\mathtt{0}^k\,\mathtt{1}^{mq^k} \to_Z^{cm} \mathtt{0}^{(n-1)k}\,\mathtt{1}^{mr^k}\,\mathtt{0}^k$$

$$\underset{r^k \geq 2q^k}{=} \boxed{\mathtt{0}^{(n-1)k}\,\mathtt{1}^{2mq^k}}\,\mathtt{1}^{mr^k - 2mq^k}\,\mathtt{0}^k\ ,$$

where the boxed substring $\mathtt{0}^{(n-1)k}\,\mathtt{1}^{2mq^k}$ by inductive hypothesis still makes for at least $c(2^{(n-1)} - 1)2m = c(2^n - 2)m$ steps. Altogether there are $cm + c(2^n -$

$2)m = c(2^n - 1)m$ steps. This finishes the proof of the intermediate lemma. The claim is immediate by $m = 1$. □

Note that since k is fixed, the initial term indeed has size linear in n. Thus we have exponential derivation lengths, in the size of the initial term.

4 Non-termination

As we claim "if and only if" in our main theorem, we should be able to prove non-termination for the following cases:

1. $2p \leq s$, $2q \leq r$,
2. $p < s < 2p$, $q < r$, $q|r$,
3. $q < r < 2q$, $p < s$, $p|s$.

To do it we employ the well-known fact that every looping derivation extends to an infinite derivation. For a string rewriting relation $\rightarrow$, a proper derivation $t \rightarrow^+ u$ is called *looping*, if $u = vtw$ holds for some v, w. Indeed we have:

Lemma 7. *For $2p \leq s$, $2q \leq r$, the system Z has a looping derivation.*

Proof. The following derivation is looping. (Redexes are underlined; the re-occurrence of the initial string, $0^p\,1^{2q}$, is enclosed by a frame box.)

$$0^p\,1^{2q} = \underline{0^p\,1^q}\,1^q \rightarrow_Z 1^r\,0^s\,1^q = 1^r\,0^{s-p}\,\underline{0^p\,1^q}$$
$$\rightarrow_Z 1^r\,0^{s-p}\,1^r\,0^s = 1^r\,0^{s-2p}\boxed{0^p\,1^{2q}}1^{r-2q}\,0^s \qquad \square$$

To prove that the other case is looping as well, is more difficult. First we establish a lemma saying that a certain suffix can be reached. We use the notation $t \rightarrow \ldots u$ to express the fact that t admits a derivation to a string which has suffix u.

Lemma 8. *If $p < s < 2p$, $1 = q < r$, then $0^{pk}\,1^{m+1} \rightarrow_Z^* \ldots 0^{sk}$ holds for all nonnegative k, m.*

Proof. By induction on k and m lexicographically. If $k = 0$ then the claim is trivial. So let $k > 0$. Here we get

$$0^{pk}\,1^{m+1} = 0^{p(k-1)}\,\underline{0^p\,1}\,1^m \rightarrow_Z 0^{p(k-1)}\,1^r\,0^s\,1^m \underset{IH,(k-1,r)}{\rightarrow_Z^*} \ldots 0^{s(k-1)}\,0^s\,1^m = \ldots 0^{sk}\,1^m\,.$$

If $m = 0$ then we are done. Else, the derivation finishes by

$$\ldots 0^{sk}\,1^m \underset{s>p}{=} \ldots 0^{pk}\,1^m \underset{IH,(k,m-1)}{\rightarrow_Z^*} \ldots 0^{sk}\,. \qquad \square$$

This lemma is used to prove our claim:

Lemma 9. *If $p < s < 2p$, $q < r$, $q|r$, or symmetrically, $q < r < 2q$, $p < s$, $p|s$, then Z has a looping derivation.*

Proof. Again, we may assume $p < s < 2p$, $1 = q < r$. Then the following derivation is looping.

$$0^{p^2}\,1^2 = 0^{p^2-p}\,\underline{0^p\,1}\,1 \to_Z 0^{p^2-p}\,1^r\,0^s\,1 = 0^{p^2-p}\,1^r\,0^{s-p}\,\underline{0^p\,1} \to_Z 0^{p^2-p}\,1^r\,0^{s-p}\,1^r\,0^s$$

$$\underset{\text{lemma 8}}{\to_Z^*} \ldots 0^{s(p-1)}\,0^{s-p}\,1^r\,0^s = \ldots 0^{(s-1)p}\,1^r\,0^s \underset{\substack{s-1\geq p\\ r\geq 2}}{=} \ldots \boxed{0^{p^2}\,1^2}\,1^{r-2}\,0^s$$

□

5 The Proof Architecture for the Complex Case

Finally, we are left with the case $p < s < 2p$, $q < r$, $q \not| r$, and its symmetric counterpart. Here, as we are going to show below, Z is terminating again, but no longer simply terminating for Z has a self-embedding derivation.

Given a string rewriting relation $\to$, a proper derivation $t \to^+ u$ is called *self-embedding*, if t is a *subsequence* of u, i.e. if $t = t_1 \ldots t_m$ and $u = u_0 t_1 u_1 \ldots t_m u_m$ holds for suitable (potentially empty) strings t_i and u_i, $0 \leq i \leq m$. It is well known that then $\to$ may still terminate, but termination of $\to$ is not simple.

Lemma 10. *If $p < s < 2p$, $q < r$, then Z has a self-embedding derivation.*

Proof. One easily proves by induction on n that $0^p\,1^{qn} \to_Z^n (1^r\,0^{s-p})^n\,0^p$ holds for all n. This derivation is self-embedding if we can choose integers n_1 and n_2 such that $0 < n = n_1 + n_2$ and 0^p is a subsequence of $(1^r\,0^{s-p})^{n_1}$ and 1^{qn} is a subsequence of $(1^r\,0^{s-p})^{n_2}$. An easy calculation shows that this is the case whenever $n_1 \geq \lceil \frac{p}{s-p} \rceil$ and $n_2 \geq \lceil \frac{qn_1}{r-q} \rceil$ hold; such n_1 and n_2 can effectively be computed. □

As we will show in the remainder of the paper, Z is terminating if moreover $q \not| r$.

The rewrite system Z, whose termination we want to prove, is transformed to a rewrite system C in such a way that termination of C entails termination of Z. In the same way, C is transformed to another system, B, next B to S, and, finally, S to R.

$$Z \mapsto C \mapsto B \mapsto S \mapsto R$$

Let us briefly explain how the rewrite systems look like, and by which intuition we justify the steps.

5.1 Step 1: Encompassment by Left and Right Contexts

The first transformation step is easy: We consider the rule in every pair of left and right context each of length 1. As the alphabet is $\{0, 1\}$, we get the four-rule rewrite system $C = \{(1), (2), (3), (4)\}$, as follows.

(1) $\quad 0\,0^p\,1^q\,0 \to 0\,1^r\,0^s\,0$

(2) $\quad 0\,0^p\,1^q\,1 \to 0\,1^r\,0^s\,1$

(3) $\quad 1\,0^p\,1^q\,0 \to 1\,1^r\,0^s\,0$

(4) $\quad 1\,0^p\,1^q\,1 \to 1\,1^r\,0^s\,1$

It is obvious that for every Z-derivation, starting with term t, there is a corresponding C-derivation of the same length, starting e.g. with $\mathtt{0}\,t\,\mathtt{0}$. For this reason, as soon as we have proved termination of $\rightarrow_C$, termination of $\rightarrow_Z$ follows. The purpose of this transformation is simply to split up Z into four cases which may be treated each in a different way.

5.2 Step 2: Dummy Introduction

The main idea for this step is that there is no "information flow" between the left half, $\mathtt{0}\,\mathtt{1}^r$, and the right half, $\mathtt{0}^s\,\mathtt{1}$, of the right hand side, $\mathtt{0}\,\mathtt{1}^r\,\mathtt{0}^s\,\mathtt{1}$, of this rule. More precisely, there is no redex that needs a proper part of both the left and right half. This fact allows one to introduce a "barrier", or "block", $\square$, between the two, without changing the applicability of rewrite steps. Thus the termination proof of rewrite system C reduces to that of the rewrite system $B = \{(1), (2'), (3), (4)\}$, where rule (2) has been replaced by the following rule.

(2') $$\mathtt{0}\,\mathtt{0}^p\,\mathtt{1}^q\,\mathtt{1} \rightarrow \mathtt{0}\,\mathtt{1}^{r'}\,\square\,\mathtt{0}^{s'}\,\mathtt{1}$$

In section 6 we prove that this step indeed preserves non-termination, provided that $p \not| s$, $q \not| r$, $s' > s - s \bmod p$, and $r' > r - r \bmod q$ hold.

5.3 Step 3: Dummy Elimination

The introduction of the $\square$ symbol enables a further step which splits rule (2') into two rules

(21) $$\mathtt{0}\,\mathtt{0}^p\,\mathtt{1}^q\,\mathtt{1} \rightarrow \mathtt{0}\,\mathtt{1}^{r'}$$

(22) $$\mathtt{0}\,\mathtt{0}^p\,\mathtt{1}^q\,\mathtt{1} \rightarrow \mathtt{0}^{s'}\,\mathtt{1}$$

This transforms B towards a system $S = \{(1), (21), (22), (3), (4)\}$. In section 7 we prove that this step is non-termination preserving.

5.4 Step 4: Relative Termination

In an arbitrary string consider the number of nonempty *packages* of zeroes (separated by ones). In system S each rule decreases the number of packages, rule (3) even strictly. Let R denote $S \setminus \{(3)\}$. Obviously any S-derivation contains only finitely many (3)-steps. Hence termination of S follows from termination of R. This can also be stated as "$\rightarrow_{(3)}$ terminates *relative to* $\rightarrow_R$"; relative termination is studied in [10]. Termination of R for $s' = p + 1$ is proven in section 8.

6 Dummy Introduction

The idea that leads to this step can be expressed informally as follows. Call a string d *dead* in context (v, w) if every derivation starting from a string of the form $lvdwr$ will only take steps in the part strictly left or strictly right of d in the string. In particular never a nonempty part of d appears in a redex. (But the definition makes sense even when d is empty.) When one replaces the dead part by a new symbol, no derivation (and, particularly, no *infinite* derivation) disappears. So the introduction of this new symbol may be used as a non-termination preserving transformation step. By construction, the new symbol only appears at right hand sides of the new rewrite system; such a symbol is called a *dummy symbol* in [9].

Technically, dummies are introduced by rewrite steps using rules of the form

$$vdw \longrightarrow v\Box w$$

where d is dead for C in context (v, w). These rules are collected in an additional rewrite system T. In the next subsection we present abstract criteria for which termination of C can be concluded from termination of B.

6.1 A New Abstract Commutation Criterion

The commutation property is first expressed by a local criterion on abstract reduction systems.

Lemma 11. *Let $\longrightarrow_B$, $\longrightarrow_T$, and $\longrightarrow_C$ be binary relations on a given set. If*

1. *$\longrightarrow_B$ terminates,*
2. *$\longrightarrow_C \subseteq \longrightarrow_B^+ \longleftarrow_T^*$,*
3. *$\longleftarrow_T \longrightarrow_C \subseteq \longrightarrow_C^+ \longleftarrow_T^*$*

then $\longrightarrow_C$ terminates.

Proof. We have

$$\longrightarrow_{T^{-1}} \longrightarrow_C \;=\; \longleftarrow_T \longrightarrow_C \;\subseteq\; \longrightarrow_C^+ \longleftarrow_T^* \;=\; \longrightarrow_C^+ \longrightarrow_{T^{-1}}^* \;\subseteq\; \longrightarrow_C^+ \longrightarrow_{C \cup T^{-1}}^* \;=\; \longrightarrow_C \longrightarrow_{C \cup T^{-1}}^* \;.$$

Since $\longrightarrow_C \longrightarrow_C \subseteq \longrightarrow_C \longrightarrow_{C \cup T^{-1}}^*$ we obtain

$$\longrightarrow_{C \cup T^{-1}} \longrightarrow_C \;\subseteq\; \longrightarrow_C \longrightarrow_{C \cup T^{-1}}^* \;;$$

straightforward induction yields $\longrightarrow_{C \cup T^{-1}}^* \longrightarrow_C \subseteq \longrightarrow_C \longrightarrow_{C \cup T^{-1}}^*$. Using this property and premise 2, we obtain by the following diagram that for every element t issuing an infinite $\longrightarrow_C$-derivation there exists an element t' again issuing an infinite $\longrightarrow_C$-derivation such that $t \longrightarrow_B^+ t'$.

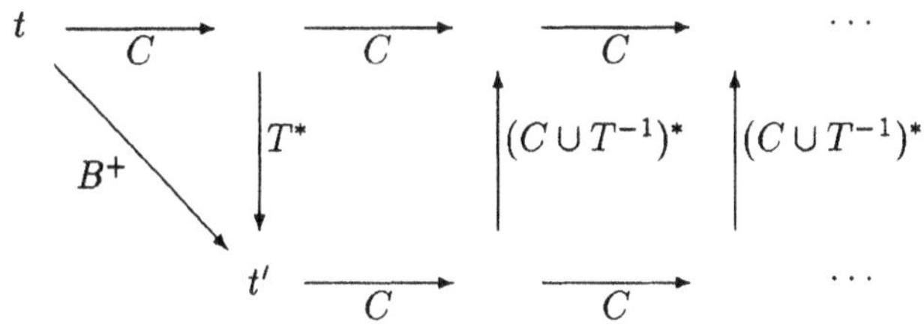

Repeating the argument shows that the existence of t having an infinite $\to_C$-derivation leads to an infinite $\to_B$-derivation of t, contradicting the termination of $\to_B$. □

6.2 Application to Term Rewriting Systems

We recall a few technical notions concerning critical pair criteria. A *unifier* of a pair of terms (s,t) is a substitution σ such that $s\sigma = t\sigma$. Unifiers need not exist. If they do, then there are most general unifiers, unique up to bijective renaming of variables.

Two rules $l \to r$ and $l' \to r'$ (with their variables renamed apart) are said to form a *critical pair* $(c(r')\sigma, r\sigma)$, if $l = c(s)$ can be split into a (left) context, $c(\_)$, and a subterm, s, such that s is not a variable and (s, l') has a most general unifier σ. If R, S are rewrite systems, the set $CP(R,S)$ of (R,S)-critical pairs is the set of all critical pairs (s,t) of a rule from R with a rule from S, together with all pairs (s,t) where (t,s) is a critical pair of a rule from S with one from R. Intuitively, the set $CP(R,S)$ represents all forking derivations $s \leftarrow_R u \to_S t$ where the redexes overlap.

A rewrite rule $l \to r$ is called *non-erasing* if each variable in l also appears in r, and *left-linear* if each variable occurs at most once in l. A rewrite system is non-erasing, left-linear, respectively, if each of its rules is so.

By a straightforward critical pair analysis (as e.g. in [3]), lemma 11 yields the following result for term rewriting systems.

Theorem 12. *Let B, T, and C be term rewriting systems. If*

1. *$\to_B$ terminates,*
2. *$C \subseteq \to_B^+ \leftarrow_T^*$,*
3. *$CP(T,C) \subseteq \to_C^+ \leftarrow_T^*$,*
4. *T left-linear and non-erasing,*
5. *C left-linear,*

then $\to_C$ terminates.

We stipulate that our theorem is applicable not only in string rewriting, but in proper term rewriting as well, as in the following example.

Example 1. Let C be given by the rule

$$f(h(x)) \to h(f(g(h(x),x))) \ .$$

Since C is self-embedding all methods for simple termination fail. Let k be a new binary function symbol, and let T be the system $g(h(x), y) \to k(x, y)$. Choose B to be $f(h(x)) \to h(f(k(x,x)))$. Now all conditions are satisfied: there are no critical pairs and $\to_B$ terminates by recursive path order with precedence $f > h > k$. So theorem 12 applies, by which $\to_C$ terminates. □

6.3 Termination by Completion

Let us now apply theorem 12 to do the dummy introduction step, $C \mapsto B$. Here B was obtained from C by replacing the rule $0\,0^p\,1^q\,1 \to 0\,1^r\,0^s\,1$ by $0\,0^p\,1^q\,1 \to 0\,1^{r'}\,\Box\,0^{s'}\,1$. For the dead part d we have $d = 1^{r-r'}\,0^{s-s'}$ which may be empty. The notion of dead part preassumes $r \geq r'$, $s \geq s'$, but the construction below works as well without this restriction. The precise definition of r' and s' is postponed.

Recall that for string rewriting systems, conditions 4 and 5 are always satisfied. With $r'' =_{\text{def}} r' - q\lfloor \frac{r}{q} \rfloor = r' - r + r \bmod q$, choose T as

$$T \underset{\text{def}}{=} \left\{ \; 0\,1^{iq+r \bmod q}\,0^{jp+s \bmod p}\,1 \;\to\; 0\,1^{iq+r''}\,\Box\,0^{jp+s''}\,1 \;\middle|\; i \in \left\{0, \dots, \lfloor \tfrac{r}{q} \rfloor\right\},\; j \in \left\{0, \dots, \lfloor \tfrac{s}{p} \rfloor\right\} \right\};$$

then condition 2 is satisfied. In order to check condition 3 we consider a typical critical pair. Other critical pairs are either similar or trivial — overlapping in a pair of contexts. The overlapping region of the peak string is enclosed by a frame box, otherwise redexes are underlined.

$$\begin{array}{ccc}
0\,0^{p-1}\,\boxed{0\,1^q\,1}\,1^{r-q-1}\,0^s\,1 & \xrightarrow[C]{} & 0\,1^r\,0^{s-1}\,\underline{0\,1^{r-q}\,0^s\,1} \\
\downarrow T & & \downarrow T \\
\underline{0\,0^p\,1^q\,1}\,1^{r'-q-1}\,\Box\,0^{s'}\,1 & \xrightarrow[C]{} & 0\,1^r\,0^s\,1^{r'-q}\,\Box\,0^{s'}\,1
\end{array}$$

The $\to_T$ arrow at the right column is required by the critical pair condition. There is such a step, provided that $r' \geq q$ holds. T may so be determined step by step during the test of critical pairs, similar to the "termination by completion" method [3].

The number r'' should be positive, otherwise some critical pairs would not close. In other words, our construction works only if $r' > r - r \bmod q$. Of course, we might choose $r' = r$ but we will see that $r' = r - r \bmod q + 1$ suits our purposes better. Moreover observe that the requirement $q \not| r$ indeed turns out essential. For, if $r \bmod q = 0$, then the T-rule for $i = 0 = j$, which is $0\,0^{s \bmod p}\,1 \to 0\,1^{r''}\,\Box\,0^{s''}\,1$, causes the critical pair

$$1\,0^{p-s \bmod p-1}\,0\,1^{r''}\,\Box\,0^{s''}\,1\,1^{q-1}\,0 \leftarrow_T 1\,0^{p-s \bmod p-1}\,\boxed{0\,0^{s \bmod p}\,1}\,1^{q-1}\,0 \to_C 1\,1^r\,0^s\,0$$

which does not close as the string $1\,1^r\,0^s\,0$ is not able to develop a □ symbol, and C steps cannot get rid of □.

We conclude: If $p \not| s$, $q \not| r$, and $r' > r - r \bmod q$, $s' > s - s \bmod p$, then T satisfies the critical pair criterion, $CP(T,C) \subseteq \rightarrow_C \leftarrow_T$. Hence theorem 12 can be applied and termination of $\rightarrow_B$ implies termination of $\rightarrow_C$.

7 Dummy Elimination

Let $\Box$ be a symbol which only occurs at right hand sides of a string rewriting system. This symbol $\Box$ can never be removed by any rewrite rule and will act as a separator between parts of the string. Intuitively an infinite derivation can be localized between these separators, hence a rule $l \rightarrow r_1 \Box r_2 \Box \cdots \Box r_n$ may be split into n rules $l \rightarrow r_1, \ldots, l \rightarrow r_n$ whose termination can be easier to prove. In this section we formalize this idea.

Definition 13. For each string of the form $s = r_1 \Box r_2 \cdots \Box r_n$ where $r_i \in (\mathcal{A} \setminus \{\Box\})^*$ for all $i \in \{1, \ldots, n\}$, let $\mathcal{E}(s) =_{\text{def}} \{r_1, \ldots, r_n\}$.

Lemma 14. *Let B be a string rewriting system on the alphabet $\mathcal{A}$ where the symbol $\Box \in \mathcal{A}$ does not occur on left hand sides of B. Let*

$$S \underset{\text{def}}{=} \{l \rightarrow u \mid (l \rightarrow r) \in B \wedge u \in \mathcal{E}(r)\} \ .$$

Then $\rightarrow_B$ terminates if $\rightarrow_S$ terminates.

Proof. In the definition of $\mathcal{E}(s)$ it does not make any difference whether $\mathcal{E}(s)$ is considered as a set or as a multiset. Here we consider $\mathcal{E}(s)$ as a multiset in order to apply well-foundedness of the multiset order.

Let B and S be as specified in the claim and suppose $\rightarrow_S$ is terminating. Define order $>$ on strings on $\mathcal{A} \setminus \{\Box\}$ by $v > w$ if there exist q, q' such that $v \rightarrow_S^+ qwq'$. Clearly $>$ is an order. Assume $v_1 > v_2 > v_3 > \cdots$ with $v_i \rightarrow_S^+ q_i v_{i+1} q_i'$, then

$$v_1 \rightarrow_S^+ q_1 v_2 q_1' \rightarrow_S^+ q_1 q_2 v_3 q_2' q_1' \rightarrow_S^+ q_1 q_2 q_3 v_4 q_3' q_2' q_1' \rightarrow_S^+ \cdots$$

contradicting termination of $\rightarrow_S$, hence $>$ is well-founded.

We claim that $s \rightarrow_B t$ implies $\mathcal{E}(s) >^{mult} \mathcal{E}(t)$. Suppose $s \rightarrow_B t$ using rule $l \rightarrow r$ in B, which means that s is of the form $s = s_1 l s_2$. Let us first assume that s_1, s_2 do not contain $\Box$, whence $\mathcal{E}(s) = \{s_1 l s_2\}$. If r does not contain $\Box$, then $\mathcal{E}(t) = \{s_1 r s_2\}$, and the claim follows by $s_1 l s_2 \rightarrow_S s_1 r s_2$, as rule $l \rightarrow r$ is also in S. Else, suppose that $r = r_1 \Box r_2 \Box \cdots \Box r_n$ with $n > 1$. Then $\mathcal{E}(t) = \{s_1 r_1, r_2, r_3, \ldots, r_{n-1}, r_n s_2\}$. Now by defintion $s_1 l s_2$ is greater than every element of $\mathcal{E}(t)$, hence again the claim follows. By closure under multiset union, this reasoning carries over to the case where s_1 or s_2 contain dummy symbols and the claim has been proved. Since $>^{mult}$ is wellfounded, termination of $\rightarrow_B$ follows. $\Box$

In [9] a general dummy elimination theorem is proved for term rewriting instead of string rewriting. Our lemma can also be proved using that theorem.

8 Finish of the Proof

It remains to prove termination of R consisting of the rules

$$
\begin{array}{ll}
(1) & 0\,0^p\,1^q\,0 \to 0\,1^r\,0^s\,0 \\
(4) & 1\,0^p\,1^q\,1 \to 1\,1^r\,0^s\,1 \\
(21) & 0\,0^p\,1^q\,1 \to 0\,1^{r'} \\
(22) & 0\,0^p\,1^q\,1 \to 0^{s'}\,1
\end{array}
$$

In this system we still have some freedom in choosing r' and s'; the validity of dummy introduction only required $r' > r - r \bmod q$ and $s' > s - s \bmod p$. Here we require $p < s < 2p$, hence we may choose $s' = p + 1$ and replace s' in rule (22) by $p + 1$.

Now we switch the representation of a string, $0^{m_1}\,1^{n_1} \ldots 0^{m_k}\,1^{n_k}$, to a sequence of pairs of non-negative integers, $(m_1, n_1) \ldots (m_k, n_k)$, where for uniqueness we require that except possibly m_1, n_k, all numbers are positive. Now R can be presented in the form

$$
\begin{array}{ll}
(1) & (m+p, q)(m', z) \to (m, r)(m' + s, z) \\
(4) & (z, n)(p, n' + q) \to (z, n + r)(s, n') \\
(21) & (m+p, n+q) \to (m, n + r' - 1) \\
(22) & (m+p, n+q) \to (m+p, n)
\end{array}
$$

where $z \geq 0$ and $m, m', n, n' > 0$.

Choose any well-founded order on non-negative integers for which p s and $n + p$ n for all n, for example $n \sqsupset n' \iff f(n) \geq f(n')$ for

$$
f(n) \underset{\text{def}}{=} \begin{cases} n + p, & \text{if } p | n, \\ n, & \text{else} \end{cases}
$$

Now we see that by an R-reduction step of type (1), (4) and (21) of the sequence $(m_1, n_1) \ldots (m_k, n_k)$ the string $(m_1, \ldots, m_k)$ lexicographically decreases according to , while it remains the same by a step of type (22). Hence any R-reduction contains only finitely many steps of type (1), (4) and (21). Since the rule (22) is clearly terminating, we conclude that $\to_R$ terminates.

It can be shown that this proof works only with the choice $s' = p + 1$, whence the requirement $s < 2p$ is essential.

Related Work

Our work on this subject began with proving termination of the one-rule string rewriting system, sometimes called "Zantema's problem",

$$0011 \to 111000$$

which corresponds to the case $p = 2 = q$, $r = 3 = s$, of this paper. To our knowledge, there is a proof sketch by Nachum Dershowitz and Charles Hoot [8], and a detailed proof including a treatment of derivation lengths by Elias Tahhan-Bittar [17]. Dershowitz/Hoot's line of argument is by minimal counterexample, and by forward closures. Tahhan-Bittar uses the notion of "inner redex" and shows termination by the fact that all inner redexes terminate. Our notion of dead part corresponds to his "strongly irreducible" strings. He was able to extend his termination result to prove a sharp upper bound for the lengths of derivation.

Theorem 15 [17]. *If $p = 2 = q$, $r = 3 = s$ then $D(n) = 2n - 6$.*

A completely different approach is currently investigated by Jan-Willem Klop (personal communication). He uses a reasoning by cases, visualized at rectangular figures where 0 characters are represented by upwards arrows, and 1 characters by rightbound arrows. Rewrite steps are understood as commuting diagrams. Another proof by case analysis has been given by Robert MacNaughton [15].

The notion of "transformation ordering" and "termination by completion" have been coined by Françoise Bellegarde and Pierre Lescanne [2, 3].

Theorem 16 Transformation order, [2, 3]. *Let B, C, and T be term rewriting systems. If*

1. *$\to_B \cup \to_T$ terminates,*
2. *$\to_T$ is confluent,*
3. *T is non-erasing and left-linear, and*
4. *$CP(T, B)$ is* cooperative, *i.e. it satisfies $CP(T, B) \subseteq (\to_B / \to_T)^+ \leftarrow_T^*$,*

then even

$$> \underset{\text{def}}{=} (\to_B / \leftrightarrow_T)^+ \cup \to_T^+$$

terminates. Moreover, every rewrite system C that satisfies $C \subseteq >$, is a terminating rewrite system.

In fact, an earlier attack to our problem has shown that the transformation ordering is quite useful, too, to describe the dummy introduction step [11]. On closer observation however we found that termination of the transformer system is an unnatural requirement: T steps did not occur as C rules, and the critical pair criteria worked as well without assuming any normal forms. We felt that we could do without it.

The result is our theorem 12. It is a criterion similar to the quasi-commutation criterion of Leo Bachmair and Nachum Dershowitz. Quasi-commutation is the property $\to_T \to_B \subseteq \to_B \to_{B\cup T}^*$, which, provided that $\to_B$ terminates, is equivalent to $\to_T \to_B \subseteq \to_B^+ \to_T^*$.

Theorem 17 [1]. *Let B, T be term rewriting systems. If*

1. *$\to_B$ terminates,*
2. *$CP(T^{-1}, B)$ is quasi-commuting, i.e. $CP(T^{-1}, B) \subseteq \to_B^+ \to_T^*$,*
3. *B is left-linear and non-erasing,*

4. T is right-linear,

then $\to_B \to_T^*$ *terminates.*

Comparing the abstract versions, we find we can simulate their version, replacing T by T^{-1} and setting $\to_C = \to_B \leftarrow_T^*$. But there is no such simulation vice versa. As a counterexample choose $a \to_B b, b \to_B c, a \to_C c, b \to_T c$. Here $\to_B$ terminates, and so $\to_C$ by our criteria, but $\to_B \leftarrow_T^*$ does not terminate. All the same, we do not claim that our improvement over Bachmair/Dershowitz' criterion is essential. By accident, our criterion serves dummy introduction better, where critical pairs with C are more comfortable to handle than critical pairs with B.

9 Conclusions

We gave a complete and precise characterization when a one-rule string rewriting system Z of the form $0^p\,1^q \to 1^r\,0^s$ terminates, where p, q, r, s are positive integers. For the simply terminating cases we gave sharp upper bounds for the complexity of derivation lengths.

We attacked the difficult, non-simply terminating case, $p < s < 2p$, $q < r$, $q \nmid r$, by a series of transformation steps, each preserving non-termination. We demonstrated how to design a termination proof and how to split it into small steps each of which can be supported by standard methods. For the dummy introduction, we used an impoverished form of transformation order. Dummy elimination is about to become a standard method. Another standard method, *semantic labelling* [18] turned out not to support the dummy introduction step, but a twin-labelling (first label as usual, then label the reversed strings) looks promising.

Of course, we would like to have an estimate of the derivation length in the non-simple termination case, too. We expect that derivation lengths are linear, as in the case $p = 2 = q$, $r = 3 = s$. On close observation of the termination proof, we get that each of the transformation steps, except the dummy elimination step, preserves the length of derivations. Dummy elimination, however, gives only an exponential upper bound.

We want to thank John Tromp and Elias Tahhan-Bittar for fruitful discussions on this topic.

References

1. Leo Bachmair and Nachum Dershowitz. Commutation, transformation, and termination. In *8th Int. Conf. Automated Deduction*, pages 5–20. Springer LNCS 230, 1985.
2. Françoise Bellegarde and Pierre Lescanne. Transformation ordering. In *2nd TAPSOFT*, pages 69–80. Springer LNCS 249, 1987.
3. Françoise Bellegarde and Pierre Lescanne. Termination by completion. *Applicable Algebra in Engineering, Communication, and Computing*, 1:79–96, 1990.

4. Ronald Book and Friedrich Otto. *String-rewriting systems*. Texts and Monographs in Computer Science. Springer, New York, 1993.
5. Max Dauchet. Simulation of turing machines by a left-linear rewrite rule. In *Proc. 3rd Int. Conf. Rewriting Techniques and Applications*, pages 109–120. LNCS 355, 1989.
6. Nachum Dershowitz. Orderings for term rewriting systems. *Theoretical Computer Science*, 17(3):279–301, March 1982.
7. Nachum Dershowitz. Termination of rewriting. *J. Symbolic Computation*, 3(1&2):69–115, Feb./April 1987. Corrigendum: 4, 3, Dec. 1987, 409-410.
8. Nachum Dershowitz and Charles Hoot. Topics in termination. In Claude Kirchner, editor, *5th Int. Conf. Rewriting Techniques and Applications*, pages 198–212. Springer LNCS 690, 1993.
9. Maria C. F. Ferreira and Hans Zantema. Dummy elimination: Making termination easier. Technical Report UU-CS-1994-47, Utrecht University, October 1994. Available via `ftp.cs.ruu.nl/pub/RUU/CS/techreps/CS-1994`.
10. Alfons Geser. *Relative termination*. Dissertation, Fakultät für Mathematik und Informatik, Universität Passau, Germany, 1990. Also available as: Report 91-03, Ulmer Informatik-Berichte, Universität Ulm, 1991.
11. Alfons Geser. A solution to Zantema's problem. In Rudolf Berghammer and Gunther Schmidt, editors, *Proc. coll. "Programmiersprachen und Grundlagen der Programmierung"*, pages 88–96, Bericht Nr. 9309, Univ. der Bundeswehr, Neubiberg, Germany, December 1993. Also appeared as report MIP-9314, Universität Passau, Germany, December 1993.
12. Gérard Huet and Dallas Lankford. On the uniform halting problem for term rewriting systems. Technical Report 283, INRIA, 1978.
13. Matthias Jantzen. *Confluent string rewriting*, volume 14 of *EATCS Monographs on Theoretical Computer Science*. Springer, Berlin, 1988.
14. Dallas S. Lankford. On proving term rewriting systems are noetherian. Technical Report MTP-3, Louisiana Technical University, Math. Dept., Ruston, LA, 1979.
15. Robert McNaughton. The uniform halting problem for one-rule Semi-Thue Systems. Technical Report 94-18, Dept. of Computer Science, Rensselaer Polytechnic Institute, Troy, NY, August 1994.
16. Vincent Meeussen and Hans Zantema. Derivation lengths in term rewriting from interpretations in the naturals. In H. A. Wijshoff, editor, *Computing Science in the Netherlands*, pages 249–260, November 1993. Also appeared as report RUU-CS-92-43, Utrecht University.
17. Elias Tahhan-Bittar. Non-erasing, right-linear orthogonal term rewrite systems, application to Zantema's problem. Technical Report RR2202, INRIA, France, 1994.
18. Hans Zantema. Termination of rewriting by semantic labelling. Technical Report RUU-CS-92-38, Utrecht University, December 1992. Extended and revised version appeared as RUU-CS-93-24, July 1993, accepted for special issue on term rewriting of Fundamenta Informaticae.
19. Hans Zantema. Termination of term rewriting: interpretation and type elimination. *J. Symbolic Computation*, 17:23–50, 1994.

On Narrowing, Refutation Proofs and Constraints

Robert Nieuwenhuis*

Technical University of Catalonia
Pau Gargallo 5, 08028 Barcelona, Spain.
E-mail: roberto@lsi.upc.es.

Abstract. We develop a proof technique for dealing with narrowing and refutational theorem proving in a uniform way, clarifying the exact relationship between the existing results in both fields and allowing us to obtain several new results. Refinements of narrowing (basic, LSE, etc.) are instances of the technique, but are also defined here for arbitrary (possibly ordering and/or equality constrained or not yet convergent or saturated) Horn clauses, and shown compatible with simplification and other redundancy notions. By narrowing modulo equational theories like AC, compact representations of solutions, expressed by AC-equality constraints, can be obtained. Computing AC-unifiers is only needed at the end if one wants to "uncompress" such a constraint into its (doubly exponentially many) concrete substitutions.

1 Introduction

Answer computation in some logic is at the heart of many applications in computer science, such as (functional) logic programming, automated theorem proving/discovering or deductive data bases. Well-known simple examples of such mechanisms are Prolog's SLD-resolution, where the accumulated unifiers are kept as answers, or E-unification procedures for equational (or more general) theories E in which, given a goal $s = t$, answers σ are computed such that $E \models s\sigma = t\sigma$.

Narrowing was originally devised as an efficient E-unification procedure using a convergent (confluent and terminating) set of rewrite rules R for E ([Fay79, Hul80]). A rewrite rule $l \rightarrow r$ can be applied to narrow a term s into $s\sigma[r\sigma]_p$, denoted $s \rightsquigarrow s\sigma[r\sigma]_p$, if σ is the mgu of $s|_p$ and l. Assume we have two new symbols *goal* and *true*, and in R the additional rule $goal(x, x) \rightarrow true$. Then (naive unconditional) narrowing is complete in the sense that for each (irreducible wrt. R) solution σ of $s = t$, there is a narrowing derivation $goal(s, t) \rightsquigarrow^* true$

* This work was presented in a preliminary stage at the Unif'94 workshop in june 1994 in Val d'Ajol, France. I wish to thank Albert Rubio and Christopher Lynch for several useful discussions. The author is partially supported by the ESPRIT Basic Research working group 6028, CCL.

in which the variables of the goal $s = t$ are instantiated with some accumulated solution θ more general than σ. To find all such solutions θ, the possible derivations can be explored by building a (possibly infinite, but finitely branching) narrowing *tree*. Many extensions (to e.g. conditional TRS's) and optimizations of narrowing have been proposed during the last years (see e.g. [RKL85, BGM88, Höl89, NRS89, BKW92, BW94]).

In the literature, most completeness proofs of narrowing strategies are based on lifting arguments applied to rewrite proofs ([Hul80]). For example, to prove the completeness of *basic* narrowing (no steps needed on subterms introduced by unifiers of previous steps), one easily shows that for each (irreducible) solution σ every rewrite proof $goal(s\sigma, t\sigma) \rightarrow^* true$ can be *lifted*, i.e. each rewrite step has a corresponding basic narrowing step in a narrowing derivation $goal(s,t) \rightsquigarrow^* true$ in which some solution θ more general than σ is computed. For a rewrite strategy with unique rewrite proofs one may obtain a corresponding narrowing strategy which is *optimal* in the sense that each solution will be computed exactly once. This is in essence the idea behind LSE-narrowing ([BKW92, BW94]), in which leftmost-innermost rewrite proofs, with a priority ordering on the rewrite rules, are lifted.

However, the proof technique of lifting rewrite proofs has severe limitations. For more efficient strategies, more general clauses without restrictions on the occurrences of variables, more general simplification and redundancy notions, or with constrained clauses, it quickly becomes extremely complicated or impossible (see [CAR93, Cha94] for the case of constrained equations). Here we follow an alternative approach, based on the well-known fact that in refutation theorem proving, each refutation proof provides *one* answer, like in SLD-resolution. The main difficulty thereby lies in proving that *all* answers can be obtained, even if one considers arbitrary Horn clauses with equality, more restrictive strategies, or constrained clauses. Our completeness proofs are based on the *model generation* framework with its *abstract redundancy notions* for detecting redundant clauses and inferences during the theorem proving process, defined by Bachmair and Ganzinger in [BG94]. As clauses may have both equality and non-equality predicates, the technique uniformly covers E-unification-like methods and Prolog-like resolution strategies, and we believe that it can be applied to all known major narrowing strategies parameterized by a total (on ground terms) reduction ordering.

Symbolic constraints à la [KKR90] are very useful in our framework. On one hand, ordering and equality constraints prune the search tree in several ways ([NR92b]). Furthermore, equality constraints have also turned out to be ideally suited for expressing basicness restrictions [NR92a], and more "operational" formalisms like clauses with a set of *marked* positions produced by earlier substitutions can be avoided. Moreover, the concept of equality constraints in which equality is interpreted *modulo* an equational theory like associativity and commutativity (AC), allows us to express sets of solutions in a compact way. In this context, computing AC-unifiers is completely unnecessary during the deduction process. In fact, AC-unification can be seen in this framework as the operation of "uncompressing" a compact set of solutions, expressed by a constraint, into its (doubly exponentially many) concrete substitutions.

This paper is structured as follows. Section 2 gives basic definitions and notation. In section 3 we revisit model generation for constrained clauses in order to more carefully analyse the permissible kind of initial constrained clauses. This is needed in section 4 to prove that a narrowing derivation can be formulated as a refutation proof of such admissible constrained clauses, which leads to our main completeness results. After this, in section 5 we relate some existing results from narrowing and refutational theorem proving; for example, it becomes clear that the "natural" notion (from the refutation point of view) of basic normal narrowing is indeed complete, and that left-to-right narrowing strategies generalize to the *redex orderings* of [BGLS92]. Section 6 is on narrowing modulo equational theories like AC.

2 Clauses with Equality and Constraints

For more details on basic notions see [DJ90] (rewriting), [BG94] (first-order clauses and model generation) and [KKR90, NR92b, NR94] (constraints and constrained deduction).

Here an *equation* is a multiset of terms $\{s, t\}$, which will be written in the form $s \simeq t$. A (first-order) Horn clause C is a pair of (finite) multisets of equations Γ (the *antecedent*) and Δ (the *succedent*), denoted by $\Gamma \rightarrow \Delta$, where Δ consists of exactly one equation (then C is a *program* clause) or Δ is empty (*goal* clause). We sometimes also write C like $\neg e_1 \vee \ldots \vee \neg e_n \vee e'_1 \vee \ldots \vee e'_n$ if Γ is $\{e_1, \ldots, e_n\}$ and Δ is $\{e'_1, \ldots, e'_n\}$.

Ordering and equality constraints are quantifier-free first-order formulae built over the binary predicate symbols $\succ$ and $=$ relating terms in $\mathcal{T}(\mathcal{F}, \mathcal{X})$, where $=$ is interpreted as equality of ground terms in $\mathcal{T}(\mathcal{F})$ (sometimes modulo an equational theory like AC), and $\succ$ as a (possibly AC-compatible) reduction ordering total on ground terms. A ground substitution σ *satisfies* a constraint T (is a *solution* of T) if $T\sigma$ evaluates to true (denoted $T\sigma \equiv true$) in the given interpretation. We extend $\succ$ to an ordering on ground equations (in fact, to their *occurrences* in clauses) and to clauses by using standard multiset extensions (see [BG94]). We will ambiguously use $\succ$ to denote all these (total) orderings on ground terms, equations and clauses. A pair consisting of a clause C and a constraint T is a *constrained clause*, written $C\,[\![T]\!]$.

An interpretation I is a congruence on ground terms. The congruence on $\mathcal{T}(\mathcal{F})$ generated by a set of equations (or rewrite rules) E (which is an interpretation) will be denoted by E^*. I satisfies a ground clause $\Gamma \rightarrow \Delta$, denoted $I \models \Gamma \rightarrow \Delta$, if $I \not\supseteq \Gamma$ or else $I \cap \Delta \neq \emptyset$. An interpretation I satisfies (is a model of) a constrained clause $C\,[\![T]\!]$, denoted $I \models C\,[\![T]\!]$, if it satisfies every ground instance of $C\,[\![T]\!]$, i.e. every $C\sigma$ such that σ is ground and $T\sigma$ is true. Therefore, clauses with unsatisfiable constraints are tautologies, and $\Gamma \rightarrow \Delta\,[\![T]\!]$ is an *empty clause* denoting an inconsistence iff $\Gamma \cup \Delta = \emptyset$ and T is satisfiable. I satisfies a set of clauses S, denoted by $I \models S$, if it satisfies every clause in S. A clause C

follows from a set of clauses S (and is called a *logical consequence* of S), denoted $S \models C$, if $I \models C$ whenever $I \models S$. For dealing with non-equality predicates, atoms A can be expressed by equations $A \simeq true$ (sometimes denoted simply by A), where *true* is a special symbol (minimal in $\succ$).

3 Model Generation for Narrowing

Bachmair and Ganzinger's model generation method [BG94] defines a way to build from a set S of clauses a (convergent) set of ground rewrite rules R_S. Then it is shown that if S is a *saturated* set of clauses (i.e. closed up to redundant inferences under a given superposition inference system), then either the empty clause is in S or else S has a model, namely R_S^*, the congruence generated by R_S, thus proving the refutational completeness of saturation. In [NR92b] this method is applied to constrained clauses, and in [NR94] to constrained clauses modulo equational theories like AC.

Here we give an adapted version of it for the case of constrained Horn clauses. The essential modification wrt. [NR92b] for our completeness proofs of narrowing is that most notions are parameterized by R, a set of ground rewrite rules (e.g. a clause is redundant if all its *irreducible* (by R) instances follow from R and other smaller irreducible instances). This parameterization by R allows us to more carefully analyse the kind of initial constrained clauses for which the methods are complete. It is known that arbitrary initial constraints cannot be dealt with:

Example 1. Let S be the set of clauses $\{\ a \simeq c,\ P(x)\,[\![x \succ b]\!],\ \neg P(c)\ \}$ and suppose $a \succ b \succ c$. Then S is inconsistent, since we have $P(a)$ as an instance of $P(x)\,[\![x \succ b]\!]$ contradicting $\neg P(c)$ and $a \simeq c$. However, the empty clause is not in S and no superposition inferences exist: the only possible paramodulation inferences are either below the variable position x in $P(x)$ or involve the small side of $a \simeq c$ (applied to $P(c)$) or lead to a tautology (when unifying $P(x)$ and $P(c)$, getting a clause with the unsatisfiable constraint $[\![c \succ b]\!]$; note that this inference can also be seen as a resolution step). □

Definition 1. Let R be a set of ground rewrite rules.

1. A ground substitution σ is *irreducible* wrt. R if $x\sigma$ is irreducible wrt. R, for every variable x in the domain of σ.
2. A clause $C\sigma$ is a *ground instance with* σ of a constrained clause $C\,[\![T]\!]$ if σ is a ground substitution s.t. $T\sigma \equiv true$.
3. Similarly, if π is an inference with premises $C_1\,[\![T_1]\!], \ldots C_n\,[\![T_n]\!]$ and conclusion $C\,[\![T]\!]$ and σ is a ground substitution σ satisfying T, then the inference $\pi\sigma$ with premises $C_1\sigma, \ldots, C_n\sigma$ and conclusion $D\sigma$ is a *ground instance with* σ of π.
4. The set of ground instances of a constrained clause $C\,[\![T]\!]$ with substitutions that are irreducible wrt. R is denoted by $Ir_R(C\,[\![T]\!])$. Similarly, we write $Ir_R(\pi)$ for inferences π and $Ir_R(S)$ for the irreducible instances of clauses of a set S.

5. A set of constrained clauses S is *pure* wrt. R if $Ir_R(S) \cup R \models S$.
6. If C is a ground clause and S is a set of clauses, then we denote by $S^{\prec C}$ (resp. $S^{\preceq C}$) the set of ground instances of clauses in S that are smaller (resp. smaller or equal) wrt. $\succ$ than C.
7. A constrained clause $D\,[\![T]\!]$ is *redundant* wrt. R in a set of constrained clauses S if $Ir_R(S)^{\preceq C} \cup R \models C$ for all C in $Ir_R(D\,[\![T]\!])$.
8. An inference π is *redundant* wrt. R in a set S if $Ir_R(S)^{\prec C} \cup R \models D$ for all inferences in $Ir_R(\pi)$ with maximal premise C and conclusion D.

Pure sets of *initial* clauses (for a certain R) are the ones for which completeness is preserved. Note that the set of the previous example is not pure if $a \to c$ is in R: then the reducible instance $P(a)$ of $P(x)\,[\![x \succ b]\!]$ does not follow from R and the irreducible instances, since $P(c)$ is not an instance of $P(x)\,[\![x \succ b]\!]$.

Now we define the set of rewrite rules R_S for a set S by induction. Each instance C may *generate* a rule depending on the set R_C of rules generated by smaller instances:

Definition 2. Let S be a set of constrained clauses and let C be a ground instance $\Gamma \to s \simeq t$ with σ of a clause in S. Then C *generates* a rule $s \to t$ if:

1. $R_C^* \not\models C$,
2. $s \succ t$ and $s \succ t'$ for all terms t' occurring in Γ and
3. both s and σ are irreducible by R_C.

where R_C and R_S are the sets of rules generated by $S^{\prec C}$ and S respectively.

Definition 3. The inference system $\mathcal{I}$ consists of the following two inference rules for constrained clauses, where OC denotes the ordering constraint stating that the inferences take place with the maximal equations of each premise, and, in the case of superposition, with the maximal terms and where $D|_p$ is not a variable:

superposition:
$$\frac{C \vee s \simeq t\,[\![T]\!] \qquad D\,[\![T']\!]}{C \vee D[t]_p\,[\![T \wedge T' \wedge OC \wedge D|_p = s]\!]}$$

equality resolution:
$$\frac{C \vee s \not\simeq t\,[\![T]\!]}{C\,[\![T \wedge OC \wedge s = t]\!]}$$

Definition 4. A set of constrained clauses S is *saturated* if all inferences of $\mathcal{I}$ with premises in S are redundant wrt. R_S in S.

Lemma 5. *Let S be a saturated set of constrained clauses. Then either the empty clause is in S or else $R_S^* \models Ir_{R_S}(S)$.*

Proof. The proof is rather similar to [NR92b]. Here we only sketch the main idea. Suppose there is a minimal (non-empty) counter example C in $Ir_{R_S}(S)$, such that $R_S^* \not\models C$. If the maximal equation $s \simeq t$ of C is negative and $s \succ t$, then, since $R_S^* \not\models C$ we have $R_S \models s \simeq t$, so s must be reducible by a rule $l\theta \to r\theta$ of the (convergent) system R_S. But then there is an inference by superposition, with the clause $D \vee l \simeq r$ of S that has generated the rule, on the clause of which C is an instance. This inference is redundant since S is saturated. From the definition of redundancy it follows that then there must be a smaller clause C' in $Ir_{R_S}(S)$ such that $R_S^* \not\models C'$, contradicting the minimality of C. Similar arguments apply in the remaining cases: if the maximal equation of C is a negative equation $s \simeq s$ (inference by equality resolution) or a positive equation $s \simeq t$ (superposition).

□

Corollary 6. *Let S be a saturated set of constrained clauses that is pure wrt. R_S. Then either the empty clause is in S or else $R_S^* \models S$, i.e. S is consistent.*

Now, since we have the previous result for saturated sets, our aim will be to compute such sets, by means of *theorem proving derivations*. The following definition of such derivations is still parameterized by R, so from the definition it is not clear how to compute derivations if R is not known, because e.g. the redundancy of clauses depends on R. Later on this becomes clear, and sufficient conditions for redundancy will be given.

Definition 7. Let R be a ground rewrite system.

1. A *theorem proving derivation* wrt. R is a sequence of sets of constrained clauses $S_0, S_1, \ldots$, such that each S_{i+1} is obtained from S_i by adding a logical consequence of S_i or by removing a clause $C\,[\![T]\!]$ that is redundant wrt. R in $S_i \setminus C\,[\![T]\!]$.
2. The set S_∞ of *persistent* clauses in $S_0, S_1, \ldots$ is defined as $\cup_j(\cap_{k \geq j} S_k)$.
3. A theorem proving derivation wrt. R is *fair* if every inference of $\mathcal{I}$ with persisting premises is redundant wrt. R in some S_j.

Lemma 8. *Let $S_0, S_1, \ldots$ be a theorem proving derivation wrt. some R s.t. S_0 is pure wrt. R. Then S_∞ is also pure wrt. R.*

Proof. We first prove $R \cup Ir_R(S_\infty) \models Ir_R(S_0)$, by showing something more general, namely $R \cup Ir_R(S_\infty) \models Ir_R(S_k)$ for all k. Let C be the smallest counter example, i.e. C is the minimal clause of all $Ir_R(S_k)$ such that $R \cup Ir_R(S_\infty) \not\models C$.

C is not an instance of any clause in S_∞, so there is some j such that C is an instance of some clause in S_j but not of any clause in S_{j+1}. This means $R \cup Ir_R(S_{j+1})^{\prec C} \models C$, which contradicts the minimality of C.

Now we have proved $R \cup Ir_R(S_\infty) \models Ir_R(S_0)$. By pureness of S_0 (i.e. $R \cup Ir_R(S_0) \models S_0$), this implies $R \cup Ir_R(S_\infty) \models S_0$ which in turn, by soundness of the derivation (only consequences are added) implies $R \cup Ir_R(S_\infty) \models S_\infty$, i.e. S_∞ is pure wrt. R. □

If we instantiate R by the set R_{S_∞} for some derivation $S_0, S_1, \ldots$, then we get the following theorem. Again, note that R_{S_∞} is not known in advance, and that therefore in practice sufficient conditions for pureness and redundancy have to be used:

Theorem 9. *If $S_0, S_1, \ldots$ is a fair theorem proving derivation wrt. R_{S_∞} and S_0 is pure wrt. R_{S_∞}, then (i) S_∞ is saturated, (ii) S_∞ is pure wrt. R_{S_∞}, and (iii) either the empty clause is in S_∞ or else $R^*_{S_\infty} \models S_\infty$, hence S_∞ in consistent.*

Proof. (i) follows from fairness and the fact that inferences that are redundant in some S_j remain redundant in S_∞ (see [NR92b] for details); (ii) is an instance of the previous lemma where R is R_{S_∞} and (iii) follows from (i) and (ii) by Corollary 6. □

Before applying these results to narrowing, let us see how theorem proving derivations can be computed in practice, i.e. how a practical theorem prover can be an instance of the framework. For the notions of redundancy, we have the following sufficient conditions:

Lemma 10. *A constrained clause $C[\![T]\!]$ is redundant (wrt. every R) in a set of constrained clauses S if $un(S)^{\preceq C\sigma} \models C\sigma$ for all its ground instances $C\sigma$, where $un(S)$ denotes the subset of unconstrained clauses of S.*

This lemma is powerful enough to allow simplification by rewriting and subsumption with unconstrained clauses, deletion of tautologies, etc. For instance, a simplification step of a clause C into D by rewriting (demodulation) fits into the framework by adding the new clause D and then removing C, which has become redundant. This allows us in the next section to prove the completeness of narrowing techniques with simplification steps and other redundancy proofs.

In fact, the previous lemma can be sharpened (see [NR92b]), e.g. by replacing $un(S)$ by the subset of S with constraints that do not lower bound any variables. If also this condition fails one can *propagate* information from the constraint to the clause part until it holds (ordering constraints generated by the inferences can be discarded without propagation to the clause part). For redundancy of inferences similar sufficient conditions exist (see [NR92b] for details). When applying these sufficient conditions for redundancy, one can speak about fair derivations in general (wrt. *any* R, so the concrete R can be omitted).

For pureness of initial clauses, obviously all sets S of clauses without constraints are pure for all R: we have $R \cup Ir_R(S) \models S$, and again this can be sharpened by allowing constraints that do not lower bound any variables.

Finally, let us say some words about finite saturated sets. If in a theorem proving derivation $S_0, S_1, \ldots$ a subset S of the initial clauses S_0 is a finite saturated set (all inferences between clauses in S are redundant in S), then these inferences are also redundant in the superset S_0 of S, i.e. no inferences between clauses of S have to be considered in the derivation. This will be applied below for narrowing in the theory defined by S.

4 Narrowing and Constraints

In the following, let S be an arbitrary consistent set of Horn clauses, and let $\Gamma \rightarrow$ be a goal clause. Our aim is to find all *irreducible* solutions σ such that $S \models \Gamma\sigma$. Here "irreducible" refers to the following:

Definition 11. Let S be an arbitrary consistent set of Horn clauses. Then a ground substitution σ is called *irreducible* for S if for each x in the domain of σ, the term $x\sigma$ is minimal wrt. $\succ$ in its congruence class[2] of the unique minimal (*initial*) Herbrand model of S.

Note that this notion of irreducibility coincides with the usual notion in the particular case where S is a convergent conditional TRS, which is irreducibility wrt. conditional rewriting with S. We also have the following not too surprising fact, which implies that for saturated sets S it also coincides with irreducibility wrt. the convergent set of rewrite rules R_S defined as in the previous section:

Lemma 12. *Let S be a pure (wrt. R_S) consistent saturated set of Horn clauses. Then R_S^*, the congruence generated by R_S, is the (unique) minimal model of S.*

Proof. By Corollary 6, we have $R_S^* \models S$, and by construction of R_S we have $S \models R_S$. This implies that R_S^* is indeed the minimal Herbrand model of S. □

For finding all solutions, we will proceed as follows. As in e.g. Prolog, we could start a theorem proving derivation from an initial set containing S plus the goal clause $\Gamma \rightarrow$; then, the accumulated unifiers of each refutation proof (i.e. each proof of the empty clause) provide a solution. However, we cannot do exactly this because we would not be able to obtain all solutions. For example, our redundancy notions would allow to subsume all clauses as soon as the first empty clause appears. Furthermore, we don't want to accumulate unifiers, because sometimes (e.g. in the AC case, see section 6) it is preferable not to compute eagerly all solutions of the unification problems. Also, for technical reasons we prefer to keep the whole set of clauses (including the goal clause) consistent.

Therefore we will add an answer literal to each goal clause $\Gamma \rightarrow$, which becomes $answer(x_1, \ldots, x_n), \Gamma \rightarrow$, where $\{x_1, \ldots, x_n\} = \mathcal{V}ars(\Gamma)$. The new predicate *answer* does not appear in the program, so $S \cup \{\, answer(x_1, \ldots, x_n), \Gamma \rightarrow \,\}$ is consistent. Furthermore, this answer literal is not taken into account wrt. the ordering $\succ$ on equations and clauses (for ordering purposes, assume it is the minimal constant *true*). The following definition tells us how answers can be read off from clauses of the form $answer(x_1, \ldots, x_n) \rightarrow [\![T]\!]$ (also see the examples in the next sections):

Definition 13. Let $S_0, S_1, \ldots$ be a theorem proving derivation. A ground answer σ *is computed* in the derivation if there is some clause $answer(x_1, \ldots, x_n) \rightarrow [\![T]\!]$ in some S_i, and σ is a solution of T restricted to the domain $\{x_1, \ldots, x_n\}$.

Regarding the previous definition, note that in fact all solutions of only the accumulated equality constraint part $\mathcal{E}$ of T are correct solutions, but completeness is preserved even if we require σ to be solution of the whole constraint T, including the ordering constraint part.

---

[2] Remind that we require $\succ$ to be (extendable to) a total ordering on ground terms. This is the case for all general-purpose orderings like LPO, RPO, polynomial orderings, etc.

Theorem 14. *Let $S_0, S_1, \ldots$ be a fair theorem proving derivation wrt. R_{S_∞} s.t. S_0 is a consistent set of Horn clauses that is pure wrt. R_{S_∞}.*

For every goal clause $\mathrm{answer}(x_1, \ldots, x_n), \Gamma \to$ *in S_0, every irreducible for S_0 answer σ such that $S_0 \models \Gamma\sigma$ is computed in the derivation.*

Proof. We derive a contradiction from the existence of a minimal ground (wrt. $\succ$) instance C with an irreducible σ of a clause $answer(x_1, \ldots, x_n), \Gamma \to [\![T]\!]$ in $\cup_j S_j$ such that $S_0 \models \Gamma\sigma$ and σ is not computed. Note that irreducibility for S_0 of σ is equivalent by Lemma 12 to irreducibility wrt. R_{S_∞}. There are several cases to be considered:

1. If C is not an instance of S_∞, then there is some i such that C is an instance of S_{i-1} but not of S_i, i.e. C is redundant in S_i and therefore $irred_{R_{S_\infty}}(S_i)^{\prec C} \cup R_{S_\infty} \models C$.

This implies that $irred_{R_{S_\infty}}(S_i)^{\prec C} \cup R_{S_\infty} \cup \Gamma\sigma \cup \{answer(x_1, \ldots, x_n)\sigma\}$ is inconsistent, so in particular R^* is not a model of it, if R is the rewrite system $R_{S_\infty} \cup \{answer(x_1, \ldots, x_n)\sigma \to true\}$. But we know (applying the results of the previous section) that $R^*_{S_\infty} \models irred_{R_{S_\infty}}(S_i)^{\prec C} \cup R_{S_\infty} \cup \Gamma\sigma$, and that R is convergent because σ is irreducible wrt. R_{S_∞}. This means that there must be some goal clause D in $irred_{R_{S_\infty}}(S_i)^{\prec C}$ such that $R^* \not\models D$ (note that adding the rewrite rule $answer(x_1, \ldots, x_n)\sigma \to true$ can only affect the satisfaction of clauses with answer literals, i.e. goal clauses).

Let D be of the form $answer(x_1, \ldots, x_n)\sigma', \Gamma'\sigma' \to$, where σ' is irreducible wrt. R_{S_∞}. From $R^* \not\models D$ we have (i) $R \models answer(x_1, \ldots, x_n)\sigma' \simeq true$ and (ii) $R \models \Gamma'\sigma'$. The only rewrite rule of R applicable for (i) is $answer(x_1, \ldots, x_n)\sigma \to true$, because σ' is irreducible wrt. R_{S_∞}, and therefore $answer(x_1, \ldots, x_n)\sigma$ is $answer(x_1, \ldots, x_n)\sigma'$. We also have $R^*_{S_\infty} \models \Gamma'\sigma'$, because the rule $answer(x_1, \ldots, x_n)\sigma \to true$ is not applicable in a rewrite proof of $\Gamma'\sigma'$.

So if Γ' is empty, this contradicts the fact that σ is not computed, and if Γ' is not empty, then $C \succ D$ contradicts the minimality of C.

2. If C is an instance of S_∞ then let $s\sigma \simeq t\sigma$ be the maximal equation in $\Gamma\sigma$. Since $S_0 \models \Gamma\sigma$, also $R^*_{S_\infty} \models \Gamma\sigma$ and therefore $R^*_{S_\infty} \models s\sigma \simeq t\sigma$.

1. If $s\sigma \succ t\sigma$, then $s\sigma$ must be reducible by some rule $l\sigma \to r\sigma$ in the convergent set of rules R_{S_∞}, and $l\sigma \to r\sigma$ is generated by an instance C' of a clause $\Gamma' \to l \simeq r [\![T']\!]$ in S_∞ (note that we can use σ for both clauses since they can be supposed not to share any variables). Since σ is irreducible by R_{S_∞}, then there exists an inference by superposition between the clauses of which C and C' are instances, with a conclusion with an instance D of the form $\Gamma'\sigma, answer(x_1, \ldots, x_n)\sigma, \Gamma\sigma[r\sigma] \to$ such that $R^*_{S_\infty} \models \Gamma'\sigma, \Gamma\sigma[r\sigma]$. By fairness of the derivation, this inference is redundant in some S_i. This leads, as in the previous point 1., to a contradiction (either σ is generated or C is not minimal).
2. Finally, if $s\sigma$ is the same term as $t\sigma$, then there exists an inference by equality resolution that is redundant in some S_i, leading to a contradiction as in the previous case.

□

5 Relationship between Narrowing and Refutation

The previous theorem has several implications. On one hand, as it states the completeness of narrowing techniques for arbitrary sets S of Horn clauses, a particular case of it applies when S is a saturated set. This is the case if S is a convergent set of conditional rewrite rules, but there are also saturated sets that do not fall under any existing notion of convergent TRS. If S is saturated, then no new inferences have to be computed between the clauses of S, and the theorem proving derivation amounts to the construction of a narrowing tree. The notion of fair theorem proving derivation does not impose any concrete strategy wrt. the order in which the narrowing tree is built. Inferences by superposition correspond to narrowing steps with other clauses, and inferences by equality resolution correspond to steps with reflexivity of equality, like in usual narrowing. Completeness is also preserved if in a clause C an arbitrary negative literal is *selected* and no inferences involving other literals of C are computed (see e.g. [BG94]). Simplification steps by rewriting correspond to normalization of the goal. Also other redundancy proofs apply, like constrained and clausal rewriting, case analysis, etc. [NN93].

If S is not a saturated set, then in the theorem proving derivation a saturated set for S is computed, while building at the same time the narrowing tree for the goal. In the particular case of (unconstrained) pure equations, this corresponds to narrowing combined with unfailing Knuth-Bendix completion.

Another implication of the theorem is that constraints can be inherited between clauses, which allows us to cut the search space by exploiting information from the meta-level (the ordering and unifiability restrictions). This information is kept and inherited to restrict future inferences: clauses with unsatisfiable constraints are tautologies, hence redundant. This leads to basicness restrictions (no inferences are needed on subterms introduced by the unifiers of previous inferences) because the accumulated unification problems are kept in the equality constraints (no propagation is needed) and future superpositions can take place only on non-variable subterms of the clause part.

In the remaining part of this section, our aim is to have a closer look, from the viewpoint of our proof technique which deals uniformly with narrowing and refutation proofs, to the results that have been independently obtained in both fields and to clarify the relationship between them. Here, because of the lack of space, we cannot analyse all cases, so let us only mention two of them: basic normal strategies and the relationship between redex orderings and left-to-right narrowing.

5.1 Basic Normal Narrowing

Example 2. Let R be the following program (which is in fact a set of two rewrite rules): $\{\ f(a) \simeq a,\ g(x) \simeq x\ \}$, and consider the goal $g(f(y)) \simeq y$. The only irreducible answer is here $\{y \mapsto a\}$.

In *normal narrowing* one originally used to view a rewrite (or normalization) step as a particular case of a narrowing step where the unification can be done by matching, in which case one does not need to consider alternative choices of narrowing steps (i.e. one has don't care non-determinism in normalization steps and don't know non-determinism in other narrowing steps). In our example, $g(f(y)) \simeq y$ normalizes into $f(y) \simeq y$, which narrows into $a \simeq a$ finding the answer $\{y \mapsto a\}$.

If one extends this viewpoint to *basic normal* narrowing, after the first (normalization) narrowing step obtaining $f(y) \simeq y$, the subterm $f(y)$ is blocked by the basicness restriction since it is part of the unifier (formulated with constraints, starting with $answer(y), g(f(y)) \simeq y \rightarrow [\![true]\!]$, we would obtain the clause $answer(y), x \simeq y \rightarrow [\![x = f(y)]\!]$). Then no further inferences can be made and the solution is not found. This illustrates the classical incompleteness result for (naive) basic normal narrowing.

From the viewpoint of refutational theorem proving, this is not surprising: normalization is in fact simplification, that is, a redundancy proof. (If it were an inference, one could not remove the simplified clause, i.e. one could not get don't care non-determinism.) For obtaining completeness of basic normal narrowing in the sense mentioned above, the clause $answer(y), x \simeq y \rightarrow [\![x = f(y)]\!]$, together with $g(x) \simeq x$, should make $answer(y), g(f(y)) \simeq y \rightarrow [\![true]\!]$ redundant. This is not the case: the instance of $answer(y), g(f(y)) \simeq y \rightarrow$ with the irreducible substitution $\{y \mapsto a\}$ does not follow from irreducible instances of $g(x) \simeq x$ and $answer(y), x \simeq y \rightarrow [\![x = f(y)]\!]$ (for which if y is a we get $\{y \mapsto a, x \mapsto f(a)\}$, which is reducible).

On the other hand, a normalization step, i.e. a classical simplification step adding the clause $answer(y), f(y) \simeq y \rightarrow [\![true]\!]$, does indeed make the clause $answer(y), g(f(y)) \simeq y \rightarrow [\![true]\!]$ redundant in the presence of $g(x) \simeq x$, so the natural and complete notion of basic normal narrowing from the refutational point of view is to have basic inferences and redundancy proofs.

5.2 Left-to-right Narrowing vs. Redex Orderings

In [BGLS92] the concept of *redex orderings* is introduced to further restrict the search space in refutation proofs for full first-order clauses with equality without loosing completeness. The idea is that inferences are only needed at positions which are minimal wrt. an arbitrary redex ordering which orders independent positions of the clause. A particular case of redex ordering is the one which orders positions in a leftmost-innermost way.

The completeness proof of this technique is also based on the model generation technique and is compatible with the one we apply here. Only our proof of lemma 5 has to be refined: instead of considering that a term s must be reducible by *some* rule of R_S, we pick the particular rule that reduces s in a minimal position in the redex ordering, which leads to the result that inferences on such minimal positions suffice. This implies that a particular case of our results given here is the completeness of narrowing strategies like left-to-right narrowing which are based on some particular redex ordering. Arbitrary redex orderings correspond to the selection orderings of ***basic selection narrowing*** [BGM88].

LSE-narrowing [BKW92, BW94] is a refined version of left-to-right narrowing in which in case of several rules reducing at the minimal position of the redex ordering, the minimal rule is chosen wrt. an ordering on the rewrite rules (e.g. by numbering them). This is also straightforwardly covered by our results by again considering in the lemma the rule (this time the one generated by the clause with minimal number) that reduces s in a minimal position in the redex ordering.

6 Narrowing Modulo Equational Theories

In some cases special treatments for some equational subset E of the axioms are preferable, since they generate many slightly different permuted versions of clauses, and for efficiency reasons it is many times better to treat all these clauses together as a single one representing the whole class, by working with special E-matching and E-unification algorithms (see [DJ90] for the basic notions and further references). A special attention has always been devoted to the case where E includes axioms of associativity (A) and commutativity (C), as they occur very frequency in practical specifications, and are well-suited for being built in due to their permutative nature. The traditional way of working modulo AC is with inferences in which one conclusion is added for each σ in $cU_{AC}(s|_p, s')$, a minimal complete set of AC-unifiers of $s|_p$ and s':

$$\frac{s' \simeq t' \qquad s \simeq t}{(s[t']_p \simeq t)\sigma} \quad \forall\, \sigma \in cU_{AC}(s|_p, s')$$

One of the drawbacks of building-in AC in theorem proving or narrowing is the complexity of AC-unification: there may be doubly exponentially many AC-unifiers for two terms (and therefore as many conclusions in an inference) whereas the decision problem of AC-unifiability is "only" NP-complete.

However, now that constraints have become available, these problems can be avoided. In the following inference system, the predicates $=$ and $\succ$ of the constraints are interpreted as AC-equality and an AC-compatible total ordering on terms (e.g. the ordering [RN93]) respectively.

Definition 15. The inference system $\mathcal{I}_{AC}$ consists of the inference rules of $\mathcal{I}$ plus the following two inference rules where f denotes an AC-function symbol, x and x' are new variables, and OC denotes the ordering constraint stating that the inferences take place with the maximal terms of the maximal equations of each premise and where $D|_p$ is headed with f:

AC-superposition:

$$\frac{C \vee s \simeq t\,[\![T]\!] \qquad D\,[\![T']\!]}{C \vee D[f(t,x)]_p\,[\![T \wedge T' \wedge OC \wedge D|_p = f(s,x)]\!]}$$

AC-top-superposition:

$$\frac{C \vee s \simeq t\,[\![T]\!] \qquad D \vee s' \simeq t'\,[\![T']\!]}{C \vee D \vee f(t',x') \simeq f(t,x)\,[\![T \wedge T' \wedge OC \wedge f(s,x) = f(s',x')]\!]}$$

Equality constraints turn out to be extremely useful in this context: instead of AC-unifying the terms, the unification problem is kept in an equality constraint. Apart from the fact that we obtain basic strategies in this way, dealing with a constrained clause $C\,[\![s = t]\!]$ is also much more efficient than having n clauses $C_1, \ldots, C_n$, one for each AC-unifier of s and t, since many inferences are computed at once, and each inference generates *one single conclusion* with an additional equality $s = t$ in its constraint. Such a constrained clause can be proved redundant by means of efficient incomplete methods detecting cases of non-AC-unifiability of its constraint.

In the context of narrowing, AC-equality constraints have the additional advantage that they can be used to express sets of solutions in a compact way.

Example 3. Suppose $a \succ b \succ c \succ d$ and the program R contains one single clause (in fact, a rewrite rule) $f(a,b) \simeq c$ where f is an AC symbol, and we have the goal $f(x,y) \simeq f(c,d)$.

If we do not want to work modulo AC, then we need to do narrowing while running the saturation procedure (like unfailing Knuth-Bendix completion), because no (ground) confluent system exists for $R \cup$ AC. Then, apart from the trivial solutions where one of x and y is c and the other one is d, we will obtain, in a rather inefficient procedure, many other irreducible solutions: $\{x \mapsto f(a,b), y \mapsto d\}$, $\{x \mapsto f(b,d), y \mapsto a\}$, $\{x \mapsto f(d,b), y \mapsto a\}$, $\{x \mapsto f(d,a), y \mapsto b\}$, etc.

If we work modulo AC and with constraints, the set R is a convergent rewrite system, i.e. R is saturated, so no inferences between program clauses are needed, and, starting with $answer(x,y), f(x,y) \simeq f(c,d) \rightarrow [\![true]\!]$, we obtain in one inference by equality resolution the answer clause: $answer(x,y) \rightarrow [\![f(x,y) = f(c,d)]\!]$ which represents the first two solutions. Furthermore, by one single inference of AC-superposition, we obtain the clause $answer(x,y), f(c,z) \simeq f(c,d) \rightarrow [\![f(f(a,b),z) = f(x,y)]\!]$ which in turn by equality resolution produces the answer clause with a (simplified) constraint $answer(x,y) \rightarrow [\![f(f(a,b),d) = f(x,y)]\!]$ representing in a compact way all other irreducible solutions. □

Computing AC-unifiers is needed only at the end if one wants to "uncompress" such a constraint into its (doubly exponentially many) concrete substitutions. In [NR94], by means of the model generation method, we proved the completeness of such a basic strategy for the AC-case combined with ordering constraints (see also [Vig94], and see [Rub94] for a detailed analysis of constrained theorem proving modulo equational theories). Indeed, when working modulo AC and considering the inference system $\mathcal{I}_{AC}$, Theorem 14 holds too.

7 Conclusions and Further Work

We have given a uniform completeness proof covering both refutational theorem proving and narrowing, and have analysed the relationship between some results obtained in both fields. We have proved several new results for answer computation with arbitrary Horn clauses (e.g. no restrictions on occurrences of variables) with constraints and with powerful abstract redundancy notions.

We hope that this work will help to avoid situations in which the same work is done twice (once for narrowing and once for refutational theorem proving), as it has happened with results on basic strategies or redex orderings.

In this paper we have focussed on Horn clauses. However, most techniques of model generation applied here can be extended to work for the non-Horn case. Also a natural semantics for answer computation in such general logic programs with equality exists, namely the *perfect model semantics* given in [BG91], where it is shown that the model generated by R_S for saturated (under an inference system similar to the one used here) sets S coincides with the unique perfect model for S under the given term ordering. It would be interesting to work out practical narrowing techniques which avoid the problem of *floundering* that appears in the theoretical method outlined in [BG91].

References

[BG91] Leo Bachmair and Harald Ganzinger. Perfect model semantics for logic programs with equality. In Koichi Furukawa, editor, *Logic Programming, Proceedings of the Eighth International Conference*, pages 645–659, Paris, France, June 24–28, 1991. The MIT Press.

[BG94] Leo Bachmair and Harald Ganzinger. Rewrite-based equational theorem proving with selection and simplification. *Journal of Logic and Computation*, 4(3):1–31, 1994.

[BGLS92] Leo Bachmair, Harald Ganzinger, Christopher Lynch, and Wayne Snyder. Basic paramodulation and superposition. In Deepak Kapur, editor, *11th International Conference on Automated Deduction*, LNAI 607, pages 462–476, Saratoga Springs, New York, USA, June 15–18, 1992. Springer-Verlag. To appear in Information and Computation.

[BGM88] P. Bosco, E Giovanetti, and C. Moiso. Narrowing vs. sld-resolution. *Theoretical Computer Science*, 2(59):3–23, 1988.

[BKW92] Alexander Bockmayr, Stefan Krischer, and Andreas Werner. An optimal narrowing strategy for general canonical systems. In M. Rusinowitch and J.-L. Rémy, editors, *The Third International Workshop on Conditional Term Rewriting Systems*, LNCS 656, Pont-à-Mousson, France, July 8–10, 1992. Springer-Verlag.

[BW94] Alexander Bockmayr and Andreas Werner. LSE narrowing for decreasing conditional term rewrite systems. In N. Dershowitz, editor, *The fourth International Workshop on Conditional Term Rewriting Systems*, Jerusalem, July 11–15, 1994.

[CAR93] J. Chabin, S. Anantharaman, and P. Réty. E-unification via constrained rewriting. In *7th Workshop on Unification*, Boston, USA, June 13–14, 1993. Boston University.

[Cha94] Jacques Chabin. Unification generale par surreduction ordonnee contrainte et surreduction dirigee. Thèse de Doctorat, Université d'Orleans, France, January 1994.

[DJ90] Nachum Dershowitz and Jean-Pierre Jouannaud. Rewrite systems. In Jan van Leeuwen, editor, *Handbook of Theoretical Computer Science*, volume B: Formal Models and Semantics, chapter 6, pages 244–320. Elsevier Science Publishers B.V., Amsterdam, New York, Oxford, Tokyo, 1990.

[Fay79] M. Fay. First-order unification in an equational theory. In *Proceedings of the Fourth Workshop on Automated Deduction*, pages 161–167, Austin, TX, February 1979.

[Höl89] Steffen Hölldobler. *Foundations of equational logic programming.* LNCS 353. Springer-Verlag, 1989.

[Hul80] J. M. Hullot. Canonical forms and unification. In *Proc. 4th International Conference on Automated Deduction*, LNCS 87, Les Arcs, 1980.

[KKR90] Claude Kirchner, Hélène Kirchner, and Michaël Rusinowitch. Deduction with symbolic constraints. *Revue Française d'Intelligence Artificielle*, 4(3):9–52, 1990.

[NN93] Pilar Nivela and Robert Nieuwenhuis. Practical results on the saturation of full first-order clauses: Experiments with the saturate system. (system description). In C. Kirchner, editor, *5th International Conference on Rewriting Techniques and Applications*, LNCS 690, Montreal, Canada, June 16–18, 1993. Springer-Verlag.

[NR92a] Robert Nieuwenhuis and Albert Rubio. Basic superposition is complete. In B. Krieg-Brückner, editor, *European Symposium on Programming*, LNCS 582, pages 371–390, Rennes, France, February 26–28, 1992. Springer-Verlag.

[NR92b] Robert Nieuwenhuis and Albert Rubio. Theorem proving with ordering constrained clauses. In Deepak Kapur, editor, *11th CADE*, LNAI 607, pages 477–491, Saratoga Springs, New York, 1992. Extended version to appear in Journal of Symbolic Computation.

[NR94] Robert Nieuwenhuis and Albert Rubio. AC-Superposition with constraints: No AC-unifiers needed. In Allan Bundy, editor, *12th International Conference on Automated Deduction*, LNAI, Nancy, France, June 1994. Springer-Verlag.

[NRS89] W. Nutt, P. Réty, and G Smolka. Basic narrowing revisited. *Journal of Symbolic Computation*, 7:295–317, 1989.

[RKL85] P. Rety, C. Kirchner, and P Lescanne. NARROWER: a new algorithm for unification and its application to logic programming. In Jean-Pierre Jouannaud, editor, *Rewriting Techniques and Applications, 1st International Conference*, LNCS 202, Dijon, France, May 20–22, 1985. Springer-Verlag.

[RN93] Albert Rubio and Robert Nieuwenhuis. A precedence-based total AC-compatible ordering. In C. Kirchner, editor, *5th International Conference on Rewriting Techniques and Applications*, LNCS 690, pages 371–388, Montreal, Canada, June 16–18, 1993. Extended version to appear in Theoretical Computer Science.

[Rub94] Albert Rubio. Automated deduction with ordering and equality constrained clauses. PhD. Thesis, Technical University of Catalonia, Barcelona, Spain, 1994.

[Vig94] Laurent Vigneron. Associative Commutative Deduction with constraints. In Allan Bundy, editor, *12th International Conference on Automated Deduction*, LNAI, Nancy, France, June 1994. Springer-Verlag.

Completion for Multiple Reduction Orderings

Masahito Kurihara[1], Hisashi Kondo[2] and Azuma Ohuchi[1]

[1] Department of Information Engineering
Hokkaido University, Sapporo, 060 Japan
[kurihara|ohuchi]@huie.hokudai.ac.jp
[2] Department of Systems Engineering
Ibaraki University, Hitachi, 316 Japan
kondo@lily.dse.ibaraki.ac.jp

Abstract. We present a completion procedure (called MKB) which works with multiple reduction orderings. Given equations and a set of reduction orderings, the procedure simulates a computation performed by the parallel processes each of which executes the standard Knuth-Bendix completion procedure (KB) with one of the given orderings. To gain efficiency, however, we develop new inference rules working on objects called nodes, which are data structure consisting of a pair $s : t$ of terms associated with the information to show which processes contain the rule $s \to t$ (or $t \to s$) and which processes contain the equation $s \leftrightarrow t$. The idea is based on the observation that some of the inferences made in the processes are closely related, so we can design inference rules that simulate multiple KB inferences in several processes all in a single operation. Our experiments show that MKB is significantly more efficient than the naive simulation of parallel execution of KB procedures, when the number of reduction orderings is large enough.

1 Introduction

Given equations and a reduction ordering, the Knuth-Bendix completion procedure (KB) [9] tries to compute a complete (convergent) set of rewrite rules. As a result, it may either succeed (with a finite, convergent set of rules), or fail (because of unorientable equations), or loop (in a diverging process trying to generate an infinite set of rules). The practical interest of the completion processes is limited by the possibility of the failure and divergence. The success of the procedure heavily depends on the choice of the reduction ordering. Thus the simplest (but often effective) way of trying to hopefully avoid the failure or divergence is to change the orderings. Actually, in many existing implementations the user can interactively change (or extend) the orderings. Note, however, that this kind of implementation necessarily requires that the users have knowledge of appropriate class of reduction orderings and intuition which is hopefully correct. From the viewpoint of interface with software designers and/or AI researchers who are not familiar with termination proof techniques, automatic change of (or search for) the orderings is desired. However, the problem is that since the completion process can diverge (and we can never decide the divergence in general), it is inappropriate to search for a correct ordering by just sequentially scanning possible orderings. This means that we have to consider, more or less, parallel execution of the completion procedures each working with one of possible orderings. However, naive implementation would result in serious inefficiency.

In this paper, we present a *single* completion procedure (called MKB) which works with *multiple* reduction orderings. Basically, given equations and a set of reduction orderings, the procedure simulates a computation performed by the parallel processes each of which executes KB with one of the given orderings. To gain efficiency, however, we develop new inference rules working on objects called nodes, which are data structure consisting of a pair $s : t$ of terms associated with the information to show which processes contain the rule $s \to t$ (or $t \to s$) and which processes contain the equation $s \leftrightarrow t$. The idea is based on the observation that some of the inferences made in the processes are closely related, so we can design inference rules that simulate multiple KB inferences in several processes all in a single operation. Our experiments show that MKB is significantly more efficient than the naive simulation of parallel execution of KB procedures, when the number of reduction orderings is large enough. In Section 2 we review the standard completion very briefly. Then we present MKB as an inference system in Section 3. A possible MKB completion procedure is presented in Section 4. Section 5 summarizes our work.

2 Standard Completion

We assume that the reader is familiar with the general idea of term rewriting systems. The reader may consult the surveys by Dershowitz and Jouannaud[5], Klop[8], Huet and Oppen[6], Avenhaus and Madlener[1], and Plaisted[11]. In this section we briefly review the standard completion techniques, based on [2, 3, 4].

A set R of rewrite rules is *convergent* (or *complete*) if it is terminating and confluent. The system is *(inter)reduced* if for all $l \to r$ in R, r is irreducible with R and l is irreducible with $R - \{l \to r\}$. A convergent, reduced system is called *canonical*. Let $\succ$ be a reduction ordering (i.e., a well-founded, strict partial ordering on terms such that $s \succ t$ implies $C[s\sigma] \succ C[t\sigma]$ for all contexts $C[\,]$ and substitutions σ). Given a set E of equations and a reduction ordering $\succ$, the standard completion procedure tries to generate a convergent (canonical) set R of rewrite rules which is contained in $\succ$ and which induces the same equational theory as E (if rules of R are regarded as equations). The standard completion is defined in terms of the inference system KB that consists of the following six inference rules:

DELETE:	$(E \cup \{s \leftrightarrow s\}; R\}) \vdash (E; R)$	
COMPOSE:	$(E; R \cup \{s \to t\}) \vdash (E; R \cup \{s \to u\})$	if $t \to_R u$
SIMPLIFY:	$(E \cup \{s \leftrightarrow t\}; R) \vdash (E \cup \{s \leftrightarrow u\}; R)$	if $t \to_R u$
ORIENT:	$(E \cup \{s \leftrightarrow t\}; R) \vdash (E; R \cup \{s \to t\})$	if $s \succ t$
COLLAPSE:	$(E; R \cup \{t \to s\}) \vdash (E \cup \{u \leftrightarrow s\}; R)$	if $l \to r \in R$, $t \to_{\{l \to r\}} u$, and $t \rhd l$
DEDUCE:	$(E; R) \vdash (E \cup \{s \leftrightarrow t\}; R)$	if $s \leftarrow_R u \to_R t$

where $\rhd$ is a well-founded ordering on terms. In this paper, we use as $\rhd$ the *encompassment ordering*: $s \rhd t$ if a subterm of s is an instance of t, but not vice versa.

In practice we assume that the symbol $\cup$ used in the left-hand sides of the inference rules denotes disjoint union. We write $(E; R) \vdash_{\mathrm{KB}^{\succ}} (E'; R')$ if the latter may be obtained from the former by one application of rule in KB. We usually leave $\succ$ implicit and write $\vdash_{\mathrm{KB}}$. Moreover, the subscript KB will be left out, if it is understood.

A *completion procedure* is any program that takes as input a finite set E_0 of equations and a reduction ordering $\succ$, and computes a sequence of inferences from $(E_0; R_0)$ with $R_0 = \emptyset$. The results of a possibly infinite completion sequence $(E_0; R_0) \vdash (E_1; R_1) \vdash \cdots$ are the sets $E_\infty = \bigcup_{i\geq 0}\bigcap_{j\geq i} E_j$ and $R_\infty = \bigcup_{i\geq 0}\bigcap_{j\geq i} R_j$ of *persisting* equations and rules. For a finite sequence ending with $(E_n; R_n)$, we let $(E_\infty; R_\infty) = (E_n; R_n)$. A completion sequence is *successful* if E_∞ is empty and R_∞ is convergent.

The rules in KB are evidently *sound*, in that the class of provable theorems is unchanged by an inference step, i.e., $\leftrightarrow^*_{E\cup R} = \leftrightarrow^*_{E'\cup R'}$ whenever $(E; R) \vdash (E'; R')$. Moreover, $R' \subseteq \succ$ whenever $R \subseteq \succ$. Consequently, the result R_∞ of any successful completion sequence is terminating and presents the same equational theory as E_0.

Since the rule DEDUCE can lead to infinitely long chains of inference, fairness conditions aim to minimize applications of that rule, while ensuring that it is not completely ignored. Let $CP(R)$ denote the set of all critical pairs between rules in R. In particular, $CP(R_\infty)$ denotes the set of all *persistent critical pairs*. A completion sequence $(E_0; R_0) \vdash (E_1; R_1) \vdash \cdots$ in KB is *fair* if (1) all persistent critical pairs are generated ($CP(R_\infty) \subseteq \bigcup_{i\geq 0} E_i$) and (2) no equation persists ($E_\infty = \emptyset$). An n-step KB completion sequence *fails* at step n if no fair sequence has it as a prefix; in that case the completion procedure fails. Assuming the procedure never discriminates against any critical pair or simplifiable rule or equation, the only possible reason for failure is that all equations are unorientable. A completion procedure is *correct* if it generates only fair or failing sequences. The main result in Huet[7] and Bachmair, Dershowitz, and Hsiang[2] is that if a completion sequence is fair, then the limit rewrite system R_∞ is convergent. Therefore, if a correct completion procedure does not fail, then it generates a convergent system in the limit. Moreover, if neither COMPOSE nor COLLAPSE is applicable to $(\emptyset, R_\infty)$, then R_∞ is even canonical.

3 Completion for Multiple Reduction Orderings

3.1 Inference Rules

Let us define a completion procedure MKB for multiple reduction orderings. Let $O = \{\succ_1, \ldots, \succ_n\}$ be a finite set of reduction orderings and $I = \{1, \ldots, n\}$ its index set. (Actually, we may allow O to be a multiset in order to avoid the problem of deciding the identity of orderings.) Given O and an initial set of equations, MKB simulates a computation performed by the parallel processes $\{P_1, \ldots, P_n\}$, where the process P_i executes KB for the reduction ordering $\succ_i$. Thus we may regard I as indices of the parallel processes. However, naive implementation could result in serious inefficiency, because many inferences made by several processes are related so closely (or even essentially the same) that they could cause much waste of computation time. We can be smarter by exploiting this observation to design inference rules that simulate the related KB inferences all in a single operation. The following inference system MKB is based on this idea.

MKB is an inference system that works on objects called *nodes*. A *node* is a tuple $\langle s : t, L_1, L_2, L_3\rangle$, where $s : t$ is an ordered pair of terms s and t, and L_1, L_2, L_3 are subsets of I. The pair $s : t$ is called a *datum* and L_1, L_2, L_3 are called *labels* of the node. This node is identified with the node $\langle t : s, L_2, L_1, L_3\rangle$. Intuitively, L_1 (L_2) denotes the set of indices of processes (executing the KB) in which the current set of rules, R, contains a rule $s \to t$ ($t \to s$); L_3 denotes the set of indices of processes in which the current set of equations, E, contains an equation $s \leftrightarrow t$ (or $t \leftrightarrow s$). MKB consists of the following inference rules:

DELETE: $N \cup \{\langle s : s, \ldots, \ldots, L\rangle\} \vdash N \quad \text{if } L \neq \emptyset$

REWRITE-1: $N \cup \{\langle s : t, L_1, L_2, L_3\rangle\} \vdash N \cup \left\{\begin{array}{l}\langle s : t, L_1 \backslash L, L_2, L_3 \backslash L\rangle, \\ \langle s : u, L \cap L_1, \emptyset, L \cap L_3\rangle\end{array}\right\}$
if $\langle l : r, L, \ldots, \ldots\rangle \in N$, $t \to_{\{l \to r\}} u$, $L \cap (L_1 \cup L_3) \neq \emptyset$, and either $t \doteq l$ or $L \cap L_2 = \emptyset$

REWRITE-2: $N \cup \{\langle s : t, L_1, L_2, L_3\rangle\} \vdash N \cup \left\{\begin{array}{l}\langle s : t, L_1 \backslash L, L_2 \backslash L, L_3 \backslash L\rangle, \\ \langle s : u, L \cap L_1, \emptyset, L \cap (L_2 \cup L_3)\rangle\end{array}\right\}$
if $\langle l : r, L, \ldots, \ldots\rangle \in N$, $t \to_{\{l \to r\}} u$, $t \rhd l$ and $L \cap L_2 \neq \emptyset$

ORIENT: $N \cup \{\langle s : t, L_1, L_2, L_3 \cup L\rangle\} \vdash N \cup \{\langle s : t, L_1 \cup L, L_2, L_3\rangle\}$
if $L \neq \emptyset$ and $s \succ_i t$ for all $i \in L$

DEDUCE: $N \vdash N \cup \{\langle s : t, \emptyset, \emptyset, L \cap L'\rangle\}$
if $\langle l : r, L, \ldots, \ldots\rangle \in N$, $\langle l' : r', L', \ldots, \ldots\rangle \in N$, $L \cap L' \neq \emptyset$, and $s \leftarrow_{\{l \to r\}} u \to_{\{l' \to r'\}} t$

where N denotes a finite set of nodes such that if $i \in L_1$ then $s \succ_i t$ and if $i \in L_2$ then $t \succ_i s$, whenever $\langle s : t, L_1, L_2, L_3\rangle \in N$. The equation $t \doteq l$ in the condition of REWRITE-1 denotes that t is an instance of l and vice versa; in other words, t and l are syntactically the same up to renaming variables. Note that with $t \to_{\{l \to r\}} u$ and $\rhd$ being the encompassment ordering, the negation of $t \rhd l \wedge L \cap L_2 \neq \emptyset$, appearing in REWRITE-2, is equivalent to $t \doteq l \vee L \cap L_2 = \emptyset$, appearing in REWRITE-1.

DELETE rule of MKB removes a node with a trivial equation $s \leftrightarrow s$. It simulates, in a single operation, applications of DELETE operation (of KB) performed in all processes P_i with $i \in L$.

REWRITE-1 and REWRITE-2 rewrite a term t to u by the rule $l \to r$. The result is the modification of the labels of the node $\langle s : t, \ldots\rangle$ and the creation of a node $\langle s : u, \ldots\rangle$. REWRITE-1 simulates COMPOSE operation in the processes P_i, $i \in L \cap L_1$, and SIMPLIFY operation in the processes P_i, $i \in L \cap L_3$. All these KB operations in the processes P_i, $i \in L \cap (L_1 \cup L_3)$, are simulated by this single MKB operation. Moreover, REWRITE-2 additionally simulates COLLAPSE operation in the processes P_i, $i \in L \cap L_2$. To see this, consider the following two nodes:

$$\langle l : r, L, \ldots, \ldots\rangle \tag{1}$$

$$\langle s : t, L_1, L_2, L_3\rangle \tag{2}$$

We assume that $t \to_{\{l \to r\}} u$. In our interpretation, the current set of rules, R, in every process P_i, $i \in L \cap L_1$, contains both $l \to r$ and $s \to t$. Therefore, COMPOSE may be applied on the two rules. As a result, the rule $s \to t$ is replaced by a new rule $s \to u$. This could be simulated by the modification of the node (2) to

$$\langle s : t, L_1 \backslash L, L_2, L_3\rangle$$

and the creation of the node

$$\langle s : u, L \cap L_1, \emptyset, \emptyset\rangle.$$

Similarly, in every process P_i, $i \in L \cap L_3$, R contains $l \to r$ and E contains $s \leftrightarrow t$. Therefore, SIMPLIFY operation would result in the replacement of the equation $s \leftrightarrow t$ by a new equation $s \leftrightarrow u$. This could be simulated by the modification of the node (2) to

$$\langle s : t, L_1, L_2, L_3 \backslash L\rangle$$

and the creation of the node

$$\langle s : u, \emptyset, \emptyset, L \cap L_3\rangle.$$

It follows that the combination of COMPOSE and SIMPLIFY could be simulated by REWRITE-1, which modifies the node (2) to

$$\langle s : t, L_1 \backslash L, L_2, L_3 \backslash L\rangle$$

and creates the node

$$\langle s : u, L \cap L_1, \emptyset, L \cap L_3\rangle.$$

To make this inference really effective, we naturally require that $L \cup (L_1 \cap L_3) \neq \emptyset$.

If $t \rhd l$ holds and $L \cap L_2$ is not empty, we could combine more. In this case, every process P_i, $i \in L \cap L_2$, contains $l \to r$ and $t \to s$ in R. Application of COLLAPSE would result in the removal of $t \to s$ and the creation of $u \leftrightarrow s$. This could be simulated by the modification of the node (2) to

$$\langle s : t, L_1, L_2 \backslash L, L_3\rangle$$

and the creation of the node

$$\langle s : u, \emptyset, \emptyset, L \cap L_2\rangle.$$

It follows that the combination of COMPOSE, SIMPLIFY, and COLLAPSE could be simulated by REWRITE-2. Note that if labels are implemented as bit vectors (in which the ith bit is 1 if and only if i belongs to the label), then their union, intersection, and difference can be computed very quickly. We refer to the nodes $\langle s : t, L_1, \ldots\rangle$, $\langle s : t, L_1 \backslash L, \ldots\rangle$, and $\langle s : u, L \cap L_1, \ldots\rangle$ in these inference rules as *original*, *updated*, and *created* nodes, respectively. In actual implementation, we could directly modify the labels of an original node in order to put the updated node on the same memory location as is occupied by the original node.

ORIENT orients an equation $s \leftrightarrow t$ to $s \rightarrow t$ in processes P_i with $s \succ_i t$. This is achieved by suitable modification of labels of the node. In practice we let L to be the maximal subset of the third label of the node such that $s \succ_i t$ for all $i \in L$. The maximal label may be trivially obtained by scanning the indices in the third label one by one to see if $s \succ_i t$, but for some classes of reduction orderings this may be obtained more quickly by other means. For example, if O is a set of recursive path orderings and the equation is $f(x) \rightarrow g(x)$, then L is the intersection of the third label and the set of indices corresponding to orderings containing the precedence $f \succ g$. A practically most effective case is when O is a set of simplification orderings and the right-hand side of the rule is homeomorphically embedded in the left-hand side; then L is identical with the third label, because then the left-hand side is greater than the right-hand side in every simplification ordering.

DEDUCE creates a node for equational consequences derived from two rules. Of course, only critical pairs need to be considered. Every process P_i, $i \in L \cap L'$, contains rules $l \rightarrow r$ and $l' \rightarrow r'$ in R. Therefore, the critical pair $s \leftrightarrow t$ (if it exists) can be deduced in all these processes. The node $\langle s : t, \ldots\rangle$ created here is called a *critical* node.

Note that we have assumed that we never distinguish a node $\langle s : t, L_1, L_2, L_3\rangle$ from $\langle t : s, L_2, L_1, L_3\rangle$. This implies that some inference rules implicitly specify the symmetric cases. For example, ORIENT rule implicitly specify the following case:

$$\text{ORIENT}': N \cup \{\langle s : t, L_1, L_2, L_3 \cup L\rangle\} \vdash N \cup \{\langle s : t, L_1, L_2 \cup L, L_3\rangle\}$$
$$\text{if } L \neq \emptyset \text{ and } t \succ_i s \text{ for all } i \in L.$$

We write $N \vdash_{\mathrm{MKB}} N'$ if the latter may be obtained from the former by one application of rule in MKB. An MKB completion procedure is any program that takes as input a finite set E_0 of equations and a set O of reduction orderings, and computes a sequence of inferences from the initial set of nodes, $N_0 = \{\langle s : t, \emptyset, \emptyset, I\rangle \mid s \leftrightarrow t \in E_0\}$, where $I = \{1, \ldots, |O|\}$.

3.2 Soundness and Completeness

Let us see the relationships between MKB and KB.

Definition 1. Let $N = \{\langle s_j : t_j, L_1^j, L_2^j, L_3^j\rangle \mid 1 \leq j \leq m\}$ be a set of nodes and $i \in I$ be an index. The *projection* $E[N, i]$ of N *onto equations* of process i is a set of equations defined by

$$E[N, i] = \{s_j \leftrightarrow t_j \mid i \in L_3^j, 1 \leq j \leq m\}.$$

Similarly, The projection $R[N, i]$ of N *onto rules* of process i is a set of rules defined by

$$R[N, i] = \{s_j \rightarrow t_j \mid i \in L_1^j, 1 \leq j \leq m\} \cup \{t_j \rightarrow s_j \mid i \in L_2^j, 1 \leq j \leq m\}.$$

In the following proposition, $\vdash_{\mathrm{KB}}^{=}$ denotes the reflexive closure of $\vdash_{\mathrm{KB}}$. In other words, $\vdash_{\mathrm{KB}}^{=}$ means either $\vdash_{\mathrm{KB}}$ or $=$. Moreover, $\vdash_{\mathrm{KB}}$ is an abbreviation for $\vdash_{\mathrm{KB}^{\succ_i}}$. The proposition formally states that MKB actually simulates KB.

Proposition 2 (Soundness). *If $N \vdash_{\mathrm{MKB}} N'$ then for all $i \in I$,*

$$(E[N, i]; R[N, i]) \vdash_{\mathrm{KB}}^{=} (E[N', i]; R[N', i]),$$

where the strict part, $\vdash_{\mathrm{KB}}$, holds for at least one i.

Usefulness of MKB comes from the observation that the strict part, $\vdash_{\mathrm{KB}}$, often holds for many i's. The following proposition shows that MKB is as powerful as KB in its ability of inference.

Proposition 3 (Completeness). *If $(E[N, i]; R[N, i]) \vdash_{\mathrm{KB}} (E'; R')$ then there exists a set N' of nodes such that $E' = E[N', i]$, $R' = R[N', i]$, and $N \vdash_{\mathrm{MKB}} N'$.*

3.3 Optional Rules

It is possible to add the following optional inference rules to MKB. They do not correspond to any inference rules of KB, but can affect efficiency of the completion procedure.

REMOVE: $N \cup \{\langle s : t, \emptyset, \emptyset, \emptyset\rangle\} \vdash N$

MERGE: $N \cup \left\{ \begin{array}{l} \langle s : t, L_1, L_2, L_3\rangle, \\ \langle s : t, L_1', L_2', L_3'\rangle \end{array} \right\} \vdash N \cup \{\langle s : t, L_1 \cup L_1', L_2 \cup L_2', L_3 \cup L_3'\rangle\}$

We will abuse the notation and write $N \vdash_{\mathrm{MKB}} N'$ if N' is obtained from N by one application of rule in either the original MKB rules or the two optional rules. REMOVE removes a node if its projections onto equations and rules are empty for all processes. MERGE merges two nodes into a single one if they have the same datum $s : t$. Note that the optional operations make the size of the current node set smaller, without affecting projections.

Proposition 4. *If $N \vdash_{\mathrm{MKB}} N'$ by applying* REMOVE *or* MERGE, *then*

$$(E[N, i]; R[N, i]) = (E[N', i]; R[N', i])$$

for all $i \in I$.

This proposition implies that when the optional inference rules are included in MKB, we have to revise the previous proposition on soundness as follows.

Proposition 5 (Soundness). *If $N \vdash_{\mathrm{MKB}} N'$ then for all $i \in I$,*

$$(E[N,i]; R[N,i]) \vdash^{=}_{\mathrm{KB}} (E[N',i]; R[N',i]),$$

where the strict part, $\vdash_{\mathrm{KB}}$, holds for at least one i if the employed rule is not optional.

On the other hand, the completeness result need not be revised. The optional rules are helpful in saving the memory space. Moreover, MERGE operation can make the reasoning process more efficient, because a single inference on the new node can replace the corresponding two inferences on the two old nodes. This saving can be significant when the number of nodes with the same datum increases rapidly. Note, however, that naive implementation of MERGE can make the program slower, because it requires the search for the same datum in the node database.

3.4 Fairness

Given a completion sequence $N_0 \vdash_{\mathrm{MKB}} N_1 \vdash_{\mathrm{MKB}} \cdots$ in MKB, the set of persisting nodes is defined by $N_\infty = \bigcup_{i \geq 0} \bigcap_{j \geq i} N_j$. For a finite sequence ending with N_n, we let $N_\infty = N_n$. The sequence is *successful* if there exists an index $i \in I$ such that $E[N_\infty, i]$ is empty and $R[N_\infty, i]$ is convergent. The sequence is *fair for* $i \in I$ if all persistent critical nodes are generated ($CN[N_\infty, i] \subseteq \bigcup_{i \geq 0} N_i$) and no equation persists ($E[N_\infty, i] = \emptyset$) for process i, where $CN[N,i]$ denotes the set of all critical nodes between nodes $\langle l : r, L_1, L_2, L_3 \rangle \in N$ and $\langle l' : r', L'_1, L'_2, L'_3 \rangle \in N$ with $i \in (L_1 \cup L_2) \cap (L'_1 \cup L'_2)$. The sequence *fails for* $i \in I$ if no MKB sequence which is fair for i has it as a prefix. The sequence *fails* if a prefix of it fails for every $i \in I$.

Fairness for i only ensures fair creation of (and selection from) critical pairs in process i. For MKB to be really useful, we need a stronger notion of fairness which ensures that every non-failing process is fair. Then we would have a greater possibility of success. This motivates the following definition.

Definition 6. A completion sequence in MKB is *fair* if it satisfies the following conditions:

- It is fair for some $i \in I$.
- If it is infinite, then it is either fair or failing for every $i \in I$.

Let S be a completion sequence $N_0 \vdash_{\mathrm{MKB}} N_1 \vdash_{\mathrm{MKB}} \cdots$ in MKB. By projecting each node set onto equations and rules of process $i \in I$, we have a sequence $(E_0; R_0) \vdash^{=}_{\mathrm{KB}} (E_1; R_1) \vdash^{=}_{\mathrm{KB}} \cdots$, where $E_j = E[N_j, i]$ and $R_j = R[N_j, i]$ for $j = 0, 1, \ldots$. By removing all equivalent steps $(E_j; R_j) = (E_{j+1}; R_{j+1})$ we have a proper completion sequence $(E_{k_0}; R_{k_0}) \vdash_{\mathrm{KB}} (E_{k_1}; R_{k_1}) \vdash_{\mathrm{KB}} \cdots$ in KB, where $0 \leq k_0 < k_1 < \cdots$. We denote this sequence by $KB[S, i]$. Then we can easily verify the following relationships between completion sequences in MKB and in KB.

```
procedure mkb(E,O);
begin
  NE := {⟨s : t, ∅, ∅, I⟩ | s ↔ t ∈ E} where I = {1,...,|O|};
  NR := ∅;
  while success(NE, NR) = false do
    if NE = ∅ then return(fail)
    else
      n := choose(NE);
      NE := merge(NE\{n}, delete(rewrite({n}, NR)));
      if n ≠ ⟨...,∅,∅,∅⟩ then
        n := orient(n);
        if n ≠ ⟨...,∅,∅,...⟩ then
          NE := merge(NE, delete(rewrite(NR, {n})));
          NE := merge(NE, cp(n, NR))
        end;
        NR := merge(NR, {n})
      end
    end
  end;
  return(R[NR, success(NE, NR)])
end mkb.
```

Fig. 1. A completion procedure for multiple reduction orderings

Proposition 7. *Let S be a completion sequence in MKB.*

1. *If S fails for $i \in I$, then $KB[S, i]$ fails.*
2. *If S fails, then $KB[S, i]$ fails for all $i \in I$.*
3. *If S is fair for $i \in I$, then $KB[S, i]$ is fair.*
4. *If S is fair, then $KB[S, i]$ is fair for some $i \in I$. Moreover, if S is infinite, then every $KB[S, i]$, $i \in I$, is either fair or failing.*

An MKB completion procedure is *correct* if it generates only fair or failing MKB sequences. If a correct MKB completion procedure generates a non-failing MKB completion sequence S, then there exists an index $i \in I$ such that the KB completion sequence $KB[S, i]$ is fair, thus the limit rewrite system $R[N_\infty, i]$ is convergent.

4 Completion Procedure

4.1 Completion Procedure

A possible MKB completion procedure, named *mkb*, is given in Fig. 1. It accepts as input a set E of equations and a set O of reduction orderings, and return as output a convergent set of rewrite rules if it successfully halts. The procedure is based on the open/closed lists algorithm which is well-known in the literature of search techniques for artificial intelligence; the sets NE and NR of nodes play a role of the open and closed lists, respectively. Initially, NR is empty and NE consists of all the initial nodes. The union $NE \cup NR$ of the current set of nodes defines a set N_j of nodes in an MKB completion sequence. Although we have seen the inference rules as working on a set of nodes, we can naturally see them as working on a single node or two nodes as well. More precisely, let us call DELETE, ORIENT, and REMOVE the ***single-node operations***, and REWRITE-1, REWRITE-2, DEDUCE, and MERGE the ***double-node operations***. Then the former is applied to a single node, while the latter to a pair of nodes. We assert that in the computation of *mkb* every node in NR has been fully considered for application of single-node operations, and that every pair of nodes in NR has been fully considered for double-node operations. This implies that all we have to do is applying single-node operations to nodes in NE and double-node operations on pairs of nodes of which at least one is from NE.

The procedure *success*(NE, NR) checks if the completion process has succeeded. More precisely, it is successful if there exists an index $i \in I$ such that i is not contained in any labels of NE nodes and any L_3 labels of NR nodes. Then $E[NE \cup NR, i]$ is the empty set of equations and $R[NR, i]$ is a convergent set of rules contained in $\succ_i$. We assume that the procedure returns such an index i if it is successful, and returns **false** otherwise. Actually, we need not scan all the nodes every time *success* is invoked; this decision could be made more efficiently, if we introduce integer variables c_i $(i = 1, \ldots, |I|)$ for counting the occurrences of the index i in the labels as above, increasing and/or decreasing them every time the labels are updated.

The procedure *choose*(NE) selects a node from NE. We assume that this selection is *fair* in that every node in NE will be eventually selected if *mkb* does not fail. It may be a heuristically good strategy to select a node with the "smallest" datum (in a sense, say, measured by its size). However, do not forget to meet the fairness requirement.

The procedure *merge*(N, N') computes the union of N and N'. If the optional MERGE operation is employed, the operation is applied to a suitable pair of nodes of which at least one is from N'.

The procedure *delete*(N) applies DELETE and optional REMOVE to N and returns the set of the remaining nodes. Note that REMOVE is also implemented implicitly by the second **if** statement.

The procedure *rewrite*(N, N') repeatedly applies REWRITE-1 and REWRITE-2 (zero or more times) to $N \cup N'$, rewriting the data of N by the rules of N', until no more rewriting is possible. It returns the set of nodes "created" in this process. We assume that this process contains mutation operations in which the labels of original nodes of N are directly modified in order to make the updated nodes in the same memory location as the original node. This means that the procedure implicitly executes the assignment such as

$$N := N \backslash \{\text{original nodes}\} \cup \{\text{updated nodes}\}.$$

The procedure *orient*(n) repeatedly applies ORIENT (zero or more times) to the node n until no more application is possible, and returns the resultant node n which is the original object but whose labels may have been modified.

The procedure $cp(n, N)$ applies DEDUCE to $\{n\} \cup N$, and returns a set of all critical nodes between n and a node from $\{n\} \cup N$.

Note that the assertion made in the beginning of this section together with the fair selection by *choose* ensures the correctness of the procedure.

Proposition 8 (Correctness). *Mkb is a correct MKB completion procedure.*

4.2 Example

Let us illustrate the completion procedure by a very simple problem. Consider the equational system consisting of the two equations $(x+y)+z = x+(y+z)$ and $f(x)+f(y) = f(x+y)$. Let $\succ_1$ and $\succ_2$ be the lexicographic path orderings induced by the precedence $+ \succ f$ and $f \succ +$, respectively. It is known [1] that if we use $\succ_1$ in the standard completion then the procedure will loop, generating the infinite canonical system

$$\begin{aligned} R_\infty = \{ & \ (x+y)+z \to x+(y+z), \\ & f(x)+f(y) \to f(x+y) \ \ \} \\ \cup\{ & \ f^n(x+y)+z \to f^n(x)+(f^n(y)+z) \mid n = 1, 2, \ldots\}, \end{aligned}$$

while the use of $\succ_2$ would lead almost immediately to the finite canonical system

$$R_\infty = \{(x+y)+z \to x+(y+z),\ f(x+y) \to f(x)+f(y)\}.$$

Let us apply the *mkb* procedure to this problem with $O = \{\succ_1, \succ_2\}$ and $I = \{1, 2\}$. The initial sets of nodes are

$$NE = \begin{cases} \langle (x+y)+z : x+(y+z), \emptyset, \emptyset, \{1,2\}\rangle & (1) \\ \langle f(x)+f(y) : f(x+y), \emptyset, \emptyset, \{1,2\}\rangle & (2) \end{cases}$$

and $NR = \emptyset$. We assume that the node (1) is selected by *choose*. Then by *orient* the node is modified to

$$\langle (x+y)+z : x+(y+z), \{1,2\}, \emptyset, \emptyset\rangle \qquad (1')$$

and a critical node

$$\langle (w+x)+(y+z) : (w+(x+y))+z, \emptyset, \emptyset, \{1,2\}\rangle \qquad (3)$$

between itself is created. After $(1')$ is moved from NE into NR, we have $NE = \{(2), (3)\}$ and $NR = \{(1')\}$.

Suppose the node (2) is selected then. It is oriented to

$$\langle f(x)+f(y) : f(x+y), \{1\}, \{2\}, \emptyset\rangle. \qquad (2')$$

Two critical nodes

$$\langle f(x+y)+z : f(x)+(f(y)+z), \emptyset, \emptyset, \{1\}\rangle \qquad (4)$$

$$\langle f(x+(y+z)) : f(x+y)+f(z), \emptyset, \emptyset, \{2\}\rangle \tag{5}$$

are created between (1′) and (2′), and (2′) is moved into NR. At this point, we have $NE = \{(3), (4), (5)\}$ and $NR = \{(1'), (2')\}$.

By fairness the nodes (3) and (5) will be eventually selected. To make the story short, let us suppose that they are selected successively from now. Then by *rewrite* both are reduced to a node with trivial equation and thus DELETEd. All the intermediate nodes (including the original nodes), which contain only empty labels, are REMOVEd.

Now we have $NE = \{(4)\}$ and $NR = \{(1'), (2')\}$. Since the third labels of (1′) and (2′) are empty and since the index 2 is not contained in any labels of (4), the *success* procedure reports the success of completion by returning 2, and the procedure is finished. This means that the completion has succeeded under the reduction ordering $\succ_2$, yielding the finite canonical set $R[NR, 2]$ of rules described before.

Although this simple example shows almost nothing but *mkb*'s ability of simulating parallel execution of KB, its effect on performance should be clear if we think of the extension of this example with more equations on a larger set of function symbols. For example, let O be a larger set of lexicographic path orderings and I the corresponding set of indices. Nevertheless, by *orient* we can expect to have a node

$$\langle (x+y)+z : x+(y+z), I, \emptyset, \emptyset\rangle \tag{1'}$$

at almost the same cost as before, because $(x+y)+z$ is greater than $x+(y+z)$ in every ordering in O and an appropriate implementation of *orient* could determine and exploit this fact very quickly. Similarly, we would have a node

$$\langle f(x)+f(y) : f(x+y), I_1, I_2, \emptyset\rangle \tag{2'}$$

where I_1 and I_2, both being subsets of I, correspond to the sets of all the orderings (in O) containing $+ \succ f$ and $f \succ +$, respectively. Then by *cp* we would have the following critical nodes:

$$\langle f(x+y)+z : f(x)+(f(y)+z), \emptyset, \emptyset, I_1\rangle \tag{4}$$

$$\langle f(x+(y+z)) : f(x+y)+f(z), \emptyset, \emptyset, I_2\rangle \tag{5}$$

This simulates the creation of critical pairs in all processes P_i, $i \in I_1 \cup I_2$, at virtually the same cost as the simple example.

4.3 Experiments

We have implemented the *mkb* procedure in Lisp and made some experiments on sample problems taken from Steinbach and Kühler [12]. *Mkb* was compared with a standard completion procedure *kb* and also with a procedure (named *pkb*) that simply simulates parallel execution of *kb*. *Kb* is implemented in a framework proposed by Lescanne[10] and *pkb* is implemented on the multitasking facility of Lucid Common Lisp. The results show that because of the overhead for node

Table 1. Experimental results

n	1	5	10	20	40	120
kb	1					
pkb	1.1	8.4	17	35	72	99
mkb	1.2	7.1	9.2	13	17	18

(a) Execution time

n	1	5	10	20	40	120
kb	1					
pkb	1	5.8	13.6	24	38	102
mkb	1.4	4.5	4.5	6.6	5.5	4.1

(b) The number of equations/nodes

manipulation ***mkb*** is slightly slower than ***pkb*** when the number of reduction orderings, n, is relatively small, but when n is large enough, ***mkb*** is significantly faster than ***pkb***.

To show how the efficiency depends on n, we present the results on the following problem (Example 3.14 of [12]).

$$\begin{array}{ll} s(s(x)) & = x \\ f(0, y) & = y \\ f(s(x), y) & = s(f(x, y)) \\ f(f(g(x, y), 0), 0) & = g(x, y) \\ g(0, y) & = y \\ g(s(x), y) & = f(g(x, y), 0) \\ h(0) & = s(0) \end{array}$$

The problem is to complete a system of these *equations*. (In [12] the problem is to complete a system of rewrite rules defined by orienting these equations from left to right.) The completion succeeds if we specify as a reduction ordering a recursive path ordering induced by a precedence satisfying $g \succ f \succ s, g \succ 0, h \succ s$. The result is the following canonical system of rewrite rules.

$$\begin{array}{ll} s(s(x)) & \rightarrow x \\ f(0, y) & \rightarrow y \\ f(s(x), y) & \rightarrow s(f(x, y)) \\ g(0, y) & \rightarrow y \\ g(s(x), y) & \rightarrow f(g(x, y), 0) \\ h(0) & \rightarrow s(0) \\ f(f(x, 0), 0) & \rightarrow x \end{array}$$

The statistics are given in Table 1. We have considered $5! = 120$ recursive path orderings induced by all total precedences on the set $\{f, g, h, s, 0\}$. From them we have randomly selected n $(= 1, 5, 10, 20, 40, 120)$ distinct orderings except that the total ordering $g \succ f \succ h \succ s \succ 0$ has been always included to ensure success. The table (a) shows the average execution time. The table (b) is included for evaluating the consumed memory space. It shows the average number of equations (or nodes for ***mkb***) required in the computation, i.e., the number of initial equations (nodes) plus the number of equations (nodes) created in the processes. (For ***pkb***, the number of initial equations is n times greater than those for ***kb***.) In both tables the entries for ***kb*** are normalized to 1. We see that ***mkb*** is

Table 2. Experimental results for ten problems ($n = 40$)

No.	13	14	19	20	21	22	23	29	30	31
pkb	38	72	70	48	1.6	45	15	48	93	2
mkb	8.4	17	16	8.7	1.2	30	6.1	8.8	8.5	1.8

(a) Execution time

No.	13	14	19	20	21	22	23	29	30	31
pkb	38	38	52	51	8.6	49	24	47	65	24
mkb	3.8	5.5	6.5	12	3.4	62	3.3	4.4	4.7	3.8

(b) The number of equations/nodes

less efficient than *pkb* when n is less than about 5, but when n is greater than 10, it is significantly more efficient.

The Table 2 shows the results for other problems. From [12] we selected ten problems which contained five or more function symbols and were solved with recursive path orderings with left-to-right status. (The problem number xx denotes Example 3.xx of [12].) The number of reduction orderings to consider was fixed to $n = 40$. We see that in most problems (except problem 22) the number of nodes for *mkb* was less than that of equations for *pkb*, and that in all the problems *mkb* was faster than *pkb* on average.

5 Conclusion

We have presented a completion procedure MKB for multiple reduction orderings. Basically, MKB simulates a parallel execution of KB procedures. Formally, MKB is defined in an abstract framework as an inference system which works on a set of nodes consisting of a pair $s : t$ of terms associated with three labels to show which processes contain the rule $s \rightarrow t$ (or $t \rightarrow s$) and which processes contain the equation $s \leftrightarrow t$. This makes it possible to simulate multiple KB inferences in several processes all in a single operation. We have also discussed the soundness, completeness, and fairness. We have proposed a possible correct implementation and made some experiments to show that MKB is significantly more efficient than the naive simulation of parallel execution of KB procedures, when the number of the reduction orderings is large enough.

Acknowledgment

This work is partially supported by the Grants-in-Aid for Scientific Research, No.04650298, the Education Ministry of Japan; and also by the donation from Toshiba Corporation.

References

1. Avenhaus J. and Madlener, K., Term rewriting and equational reasoning, Banerji, R.B. ed., *Formal Techniques in Artificial Intelligence: A Sourcebook*, North-Holland, 1-44, 1990.
2. Bachmair, L., Dershowitz, N., and Hsiang, J., Orderings for equational proofs, *Proc. Symp. on Logic in Computer Science*, 346 357, 1986.
3. Bachmair, L., *Canonical Equational Proofs*, Birkhäuser, 1991.
4. Dershowitz, N., Completion and its applications, Aït-Kaci, H. and Nivat, M. eds., *Resolution of Equations in Algebraic Structures*, Vol.2: Rewriting Techniques, Academic Press, 31 85, 1989.
5. Dershowitz, N. and Jouannaud, J.-P., Rewrite systems, van Leeuwen, J. ed., *Handbook of Theoretical Computer Science*, vol. B, North-Holland, 243 320, 1990.
6. Huet, G. and Oppen, D. C., Equations and rewrite rules: a survey, Book, R. ed., *Formal Language Theory: Perspectives and Open Problems*, Academic Press, 349 405, 1980.
7. Huet, G., A complete proof of correctness of the Knuth and Bendix completion algorithm, *J. Comput. Syst. Sci.* 23, 11 21, 1981.
8. Klop, J.W., Term rewriting systems, Abramsky, S., et al. eds., *Handbook of Logic in Computer Science*, vol.*II*, Oxford Univ. Press, 1 116, 1992.
9. Knuth, D.E. and Bendix, P.B., Simple word problems in universal algebras, Leech, J. ed., *Computational Problems in Abstract Algebra*, Pargamon Press, 263 297, 1970.
10. Lescanne, P., Completion procedures as transition rules + control, *Proc. TAPSOFT (vol. 1)*, Lect. Notes in Comput. Sci. 351, 28 41, 1989.
11. Plaisted, D. A., Equational reasoning and term rewriting systems, Gabbay, D. M. et al. eds., *Handbook of Logic in Artificial Intelligence and Logic Programming*, Vol. 1, Oxford Univ. Press, 274 367, 1993.
12. Steinbach, J. and Kühler, U., Check your ordering termination proofs and open problems, SEKI report, SR-90-25 (SFB), 1990.

Towards an efficient construction of test sets for deciding ground reducibility

Klaus Schmid and Roland Fettig

Universität Kaiserslautern, FB Informatik, Postfach 3049,
67653 Kaiserslautern, Germany,
email: {schmid,fettig}@informatik.uni-kl.de

Abstract. We propose a method for constructing test sets for deciding whether a term is ground reducible w.r.t. an arbitrary, many-sorted, unconditional term rewriting system. Our approach is based on a suitable characterization of such test sets using a certain notion of transnormality. It generates very small test sets and shows some promise to be an important step towards a practicable implementation.

1 Introduction

Ground reducibility tests have widely been discussed in the literature since the decidability of ground reducibility was shown by Plaisted [13] in a general setting and by Kapur, Narendran, and Zhang [8] for many-sorted term rewriting systems. Testing for ground reducibility has many applications, if one considers initial model semantics of term rewriting systems. For example, the notions of sufficient completeness [8] and inductive validity [7, 9] are closely related to the ground reducibility problem. Recently, the question whether a rewrite system can be replaced by a left-linear one while preserving the set of its irreducible ground terms has been shown to be decidable [14, 12, 6]. Again, the key to this result is the notion of ground reducibility.

There are several competing approaches to testing ground reducibility. One can distinguish two current main streams. The first method constructs a certain kind of finite automaton for each term to be tested such that the term is ground reducible if and only if the language accepted by the automaton is empty. This idea was introduced by Bogaert and Tison [1] and further improved by Comon and Jacquemard [3].

We will not discuss this approach in detail, because our paper is a contribution to the second one, which is based on finite test sets. They are distinguished by the following property: a term is ground reducible if and only if all its test set instances are reducible. A characterization of test sets for left-linear rewrite systems as well as a method for the generation of test sets was first given by Jouannaud and Kounalis in the preliminary version of [7]. Kapur, Narendran, and Zhang gave a different characterization of test sets [9]. Their generation method produces very small test sets. In [2] Bündgen invented a procedure, which is also able to handle a certain class of non-left-linear rewrite systems. Kounalis proposed a method for the general case [11], but his approach generates fairly large test sets even for linear rewrite systems and is applicable only to one-sorted signatures. In [6] Hofbauer and Huber developed a characterization of test sets for arbitrary many-sorted rewrite systems. However, they only proved a theoretical computability result, which is not suitable for an implementation due to its computational complexity.

In this paper we will propose an efficient method for computing small ground reducibility test sets for arbitrary many-sorted term rewriting systems. Our approach achieves the following goals: It extends techniques for left-linear rewrite systems such that additional overhead is necessary only in the presence of non-linear left hand sides. Compared to other approaches in the literature it produces the smallest test sets. The computational complexity of the generation method allows a practicable implementation. Our approach incorporates ideas of Kapur et al. [9] for the linear part, ideas of Bündgen [2] for some pruning mechanisms, ideas of Kounalis [11] for the notion of transnormality and ideas of Hofbauer and Huber [6] for considering problem positions.

This paper is organized as follows. After introducing some basic notions and notations in Sect. 2 we describe our method for the linear case in Sect. 3. This method combines the concept of extensibility of Kapur et al. with pruning mechanisms given by Bündgen and forms the basis of the general case. In Sect. 4 we characterize the problems caused by non-linear left hand sides and propose to handle them by additional expansions. Stimulated by the notion of problem positions of Hofbauer and Huber we describe constraints to identify the positions at which these expansions are necessary. The notion of independent transnormality allows us in Sect. 5 to formulate an extensibility criterion which leads to our main theorem about the characterization of test sets. The remaining problem is the classification of terms as being independently transnormal. In Sect. 6 we prove two results related to this problem and give an outline of our method for constructing test sets. The test sets generated by our method can further be reduced in size using the knowledge that they are already test sets. This is described in Sect. 7. Finally, we summarize our work in Sect. 8 and give an outlook on our future plans.

2 Notations

As far as possible we will use the notations introduced in [5]. In this paper we will use *many-sorted signatures*, i.e. we will distinguish between different, unrelated sorts. The different sorts are called $s_1, \ldots, s_n$. Let x_{s_i} be a variable of sort s_i and f_{s_i} a function symbol with target sort s_i. $sort(t)$ denotes the sort of the term t, i.e. $sort(x_{s_i}) = s_i$ and $sort(f_{s_i}(t_1, \ldots, t_m)) = s_i$.

Positions are represented by sequences of natural numbers. Λ denotes the empty sequence. The concatenation of two positions p and q is denoted by $p.q$. We write $p \geq q$ ($p > q$) if there exists a position r such that $p = q.r$ ($r \neq \Lambda$); further, r is denoted by $p \setminus q$. p^n stands for $\underbrace{p. \ldots .p}_{n \text{ times}}$ and $p^0 = \Lambda$. $|p|$ is the size of a position p ($|\Lambda| = 0$; $|p.q| = |p| + |q|$).

Let $t|_p$ be the subterm of t at position p and $t[s]_p$ the term resulting from substituting this subterm by the term s. $\mathcal{P}os(t)$ is the set of all positions in t. Similar $\mathcal{VP}os(t)$ ($\mathcal{NVP}os(t)$) is the set of all (non-linear) variable positions in t.

For the purposes of this paper it is sufficient to identify the *depth of a term* $depth(t)$ with the size of its largest position (w.r.t. $>$). $t|_p$ is called a *proper subterm* of t if $p \neq \Lambda$ and it is called a *principal subterm* if $|p| = 1$. Additionally, we define $\mathcal{V}ar(t) = \{t|_p \mid p \in \mathcal{VP}os(t)\}$ and $\mathcal{NV}ar(t) = \{t|_p \mid p \in \mathcal{NVP}os(t)\}$.

Throughout the paper we will denote the current rewrite system by R. The set of the left hand sides of the rules in R is denoted by $LHS(R)$. In Sect. 3 we will assume R to be linear. In the sections after that it can also be non-linear.

The set of all substitutions is denoted by $\Omega(\mathcal{X})$. A substitution τ is called a *variable-renaming* if $x\tau \in \mathcal{X}$ for all $x \in \mathcal{X}$ and τ is bijective.

We will define substitutions also using sets. Let T be a set of terms, then σ is called a *T-substitution*, if for each $x \in \mathcal{D}om(\sigma)$ there exist $t \in T$ and a variable-renaming τ such that $x\sigma \equiv t\tau$ and for all $x, y \in \mathcal{D}om(\sigma)$ with $x \not\equiv y$ holds $\mathcal{V}ar(x\sigma) \cap \mathcal{V}ar(y\sigma) = \emptyset$. t is a *T-instance* of s if there is a T-substitution σ with $t \equiv s\sigma$ and $\mathcal{V}ar(s) \subseteq \mathcal{D}om(\sigma)$.[1] A *ground instance* is a $\mathcal{G}(\mathcal{F})$-instance and a *ground substitution* is a $\mathcal{G}(\mathcal{F})$-substitution. $\Omega_{\mathrm{g}}(\mathcal{X})$denotes the set of all ground substitutions. Substitutions can be restricted to a subset of their domain denoted by $\sigma|_{\mathcal{V}}$: $\tau = \sigma|_{\mathcal{V}}$ iff $\mathcal{D}om(\tau) = \mathcal{D}om(\sigma) \cap \mathcal{V}$ and $x\tau \equiv x\sigma$ for all $x \in \mathcal{D}om(\tau)$.

A term t is said to be *reducible (at position $p \in \mathcal{P}os(t)$)* if there exist $l \in LHS(R)$ and $\sigma \in \Omega(\mathcal{X})$ such that $l\sigma \equiv t|_p$. Otherwise t is called *irreducible (at p)*. Similarly, t is called *ground reducible* if each ground instance of t is reducible. $IRG(R)$ denotes the set of all irreducible ground terms.

3 Test Sets for Linear Term-rewriting Systems

The basic idea of the test set approach is to construct a finite set of terms T, such that a term t is ground reducible iff every T-instance of t is reducible.

Definition 1. A finite set $T \subset \mathcal{T}(\mathcal{F}, \mathcal{X})$ is a *test set (for R)* if the following holds for all $t \in \mathcal{T}(\mathcal{F}, \mathcal{X})$: each T-instance of t is reducible iff t is ground reducible.

In this paper we will focus on test sets which comply with the following two conditions:[2]

1. Every irreducible ground instance $t\tau$ is an instance of a T-instance $t\sigma$.
2. For all terms t, T-instances $t\sigma$ holds: $t\sigma$ is reducible iff $t\sigma$ is ground reducible.

The following definition formalizes condition 1:

Definition 2. A set $T \subseteq \mathcal{T}(\mathcal{F}, \mathcal{X})$ is called *complete (w.r.t. R)*, if for all $s \in IRG(R)$ there exist a term $t \in T$ and a ground substitution σ such that $s \equiv t\sigma$.

As we will see, completeness can easily be achieved. It is much more difficult to guarantee compliance with condition 2. The problem is simplified if no term in T is ground reducible:

Definition 3 (cf. [9, Definition 3.5]). A set $T \subseteq \mathcal{T}(\mathcal{F}, \mathcal{X})$ is called *minimal (w.r.t. R)* if no $t \in T$ is ground reducible.

[1] This definition is unequivocal modulo variable-renaming. Therefore, whenever we speak of a T-instance we will mean an unique representative. Besides this, the application of a T-substitution to a term does not necessarily result in a T-instance.

[2] Most test set approaches use test sets which satisfy these constraints. However, Kapur et al. described in [9, p. 91] a class of test sets which does not comply with condition 1. Yet, we do not know of any test set which does not satisfy condition 2.

If T is minimal then no T-instance can be ground reducible merely because a subterm introduced by the T-substitution is ground reducible. Thereby it is guaranteed that at least one reducible ground instance $(t\tau)\sigma$ of a T-instance $t\tau$ is reducible at a position $p \in \mathcal{P}os(t)$. Under these conditions it is sufficient to demand that the terms in T are 'big enough', to guarantee that a T-instance $t\tau$ is not ground reducible iff it is irreducible.[3] This idea of a term being 'big enough' is formalized by the following definition of a linearly expanded term:

Definition 4. A term $t \in \mathcal{T}(\mathcal{F}, \mathcal{X})$ is called *linearly extensible at position* p *w.r.t.* $l \in \mathcal{T}(\mathcal{F}, \mathcal{X})$, if t and l are unifiable, p is a variable position in t and p is a non-variable position in l.

Additionally, t is called *linearly expanded w.r.t.* l if t is not linearly extensible at any position $q \in \mathcal{P}os(l)$ w.r.t. l and t is said to be *linearly expanded w.r.t.* $S \subseteq \mathcal{T}(\mathcal{F}, \mathcal{X})$ if t is linearly expanded w.r.t. each $l \in S$. A set $T \subset \mathcal{T}(\mathcal{F}, \mathcal{X})$ is called *linearly expanded w.r.t.* $l \in \mathcal{T}(\mathcal{F}, \mathcal{X})$ $(S \subseteq \mathcal{T}(\mathcal{F}, \mathcal{X}))$ if each $t \in T$ is linearly expanded w.r.t. l (S).

With this definition the following theorem can be formulated:

Theorem 5. *If R is a linear rewrite system and $T \subseteq \mathcal{T}(\mathcal{F}, \mathcal{X})$ is linear, complete and minimal w.r.t. R and linearly expanded w.r.t. each proper subterm of any $l \in LHS(R)$, then T is a test set (for R).*

Proof. This theorem is a corollary to Theorem 15 about the non-linear case. □

A similar definition of test sets (so-called *standard test sets*) is given in in [9]. But Kapur et al. make two additional requirements: for any two distinct terms t_1 and t_2 in T, t_1 and t_2 are not unifiable, and T is linearly expanded w.r.t. $LHS(R)$. As we will see, none of these conditions is really necessary for test sets.

Now, we want to show how a test set, as characterized in Theorem 5, can be effectively constructed. The first step is the generation of a set of linearly expanded terms. This can be done using the following procedure:

Procedure 6. The *candidate set* $\mathcal{C}$ for a rewrite system R is given by:

1. set $\mathcal{C} := \{x\}$ initially (or $\mathcal{C} := \{x_{s_1}, \ldots, x_{s_n}\}$ if the signature is many-sorted)
2. while there exists a $t \in \mathcal{C}$ which is linearly extensible at a (variable) position p w.r.t. a proper subterm of a term $l \in LHS(R)$:

$$\mathcal{C} := \mathcal{C} \setminus \{t\} \cup \{t[s]_p \mid t[s]_p \text{ is irreducible and } sort(s) = sort(t|_p),$$
$$s \equiv f(x_1, \ldots, x_n),\ n \text{ is the arity of } f,\ x_1, \ldots, x_n \notin \mathcal{V}ar(t)\}^4$$

This construction method is similar to the method described by Kapur et al. in [9]. Bündgen described an alternative procedure in [2].

Lemma 7. *Procedure 6 terminates on each finite input and the resulting set T is complete and linearly expanded w.r.t. each proper subterm of a term $l \in LHS(R)$.*

Proof. Procedure 6 terminates, because no extensible position exists, which is larger than $\max_{l \in LHS(R)} \max_{p \in \mathcal{P}os(l)} |p|$. The fact that Procedure 6 starts with variables and that these variables are replaced by all function symbols obviously guarantees that the resulting set $\mathcal{C}$ is complete w.r.t. R. □

---

[3] This argument is only valid for left linear rewriting systems, as non-linear left hand sides can relate subterms of any depth. In the next section we will return to this point.

[4] Variants, i.e. terms which are equal modulo variable-renaming, are omitted.

Now, the only obstacle left is to extract a minimal subset from the output of Procedure 6. For this an algorithm called *Closure* is used. It is based on ideas from Bündgen's *Closure*-algorithm ([2]). However, the requirements of our algorithm are less restrictive. Central to both approaches is the following observation:

If $t \equiv f(t_1, \ldots, t_n)$ is linearly expanded w.r.t. $LHS(R)$, irreducible at Λ and no t_i $(1 \le i \le n)$ is ground reducible then t is not ground reducible.

The successful application of the *Closure*-algorithm to a set of terms T requires compliance with the following two prerequisites:

1. T is linearly expanded w.r.t. $LHS(R)$. This can be easily accomplished by exchanging 'linearly extensible w.r.t. a proper subterm' by 'linearly extensible w.r.t. a subterm (including the term itself)' in the condition of step 2 of Procedure 6. This leads to a similar modification of Lemma 7.
2. For all $t \equiv f(t_1, \ldots, t_n) \in T, i \in \{1, \ldots, n\}$ holds: if t_i and $s \in T$ are unifiable with most general unifier σ then $t_i\sigma \equiv s$ (that is σ is a match).
   This can be ensured by a proper selection of the term t and the extensible position p in step 2 of Procedure 6. One such strategy is to select a smallest term (w.r.t. depth) and in this term the left-most linearly extensible variable position of minimal size. A similar strategy is used in [9].

Procedure 8. The *Closure* $\mathcal{C}^*$ of a set $T \subseteq \mathcal{T}(\mathcal{F}, \mathcal{X})$ is defined as:

$$\mathcal{C}_0 := T \cap \mathcal{G}(\mathcal{F})$$

$$\mathcal{C}_{i+1} := \mathcal{C}_i \cup \{t \in T \mid t \equiv f(t_1, \ldots, t_n) \text{ and for all } t_j \ (1 \le j \le n) \text{ either } t_j \in \mathcal{G}(\mathcal{F}) \text{ or there exists a term } s \in \mathcal{C}_i \text{ and } s \text{ is an instance of } t_j\}$$

$$\mathcal{C}^* := \lim_{i \to \infty} \mathcal{C}_i$$

Since $\mathcal{C}^* \subseteq T$ and $\mathcal{C}_i = \mathcal{C}_{i+1} \rightsquigarrow \mathcal{C}_i = \mathcal{C}^*$, the procedure stops for each finite input T. This procedure can be used for *pruning* the set T by extracting a minimal subset $\mathcal{C}^*$ of T:

Theorem 9. *Let $T \subseteq \mathcal{T}(\mathcal{F}, \mathcal{X})$ be complete, such that for each $f(t_1, \ldots, t_n) \in T$, $i \in \{1, \ldots, n\}$ we have: if t_i and $s \in T$ are unifiable with mgu σ then $t_i\sigma \equiv s$. Let $\mathcal{C}^*$ be the Closure of T:*

1. *Each $t \in T \setminus \mathcal{C}^*$ is ground reducible. This holds for arbitrary R.*
2. *Additionally, if T is linearly expanded w.r.t. every subterm of each $l \in LHS(R)$ then no $t \in \mathcal{C}^*$ is ground reducible. This only holds if R is linear.*

Proof. 1. Assume there is a term $t \in T \setminus \mathcal{C}^*$ such that t has an irreducible ground instance $\hat{t}$, then choose $t \equiv f(t_1, \ldots, t_n)$ such that $\hat{t}$ has minimal depth among all irreducible ground instances of terms in $T \setminus \mathcal{C}^*$. As $t \notin \mathcal{C}^*$, there must be a t_i $(1 \le i \le n)$ such that t_i is no ground term and t_i is no instance of any $s \in \mathcal{C}^*$. As t_i has an irreducible ground instance $\hat{t}_i$ and T is complete, there must exist a term $s \in T \setminus \mathcal{C}^*$ which unifies with t_i (and therefore matches t_i). However, then s would have a smaller ground instance than t. *Contradiction.*

2. This can be proved for each $\mathcal{C}_i$ by induction over i, if R is linear. □

Thus $\mathcal{C}^*$ is minimal and complete w.r.t. R and linearly expanded w.r.t. every proper subterm of any $l \in LHS(R)$. Therefore, $\mathcal{C}^*$ is a test set. Hence, it is possible to generate a test set for any linear rewrite system R by pruning the result of Procedure 6 (using the *Closure*). Due to part 1 of Theorem 9 the *Closure* can also be applied to the intermediate results of step 2 of Procedure 6.

4 Problems Caused By Non-linear Rewrite Systems

The set of terms generated by the method described in the previous section is usually not a test set if R is non-linear. This is shown in the following example:

Example 1. $\mathcal{F} = \{a, f\}$ $\quad R = \{f(x,x) \to \dots\}$
The *candidate set* for R is $C = \{a,\ f(x,y)\}$. Pruning C using the *Closure* results in $C^* = C$. But C^* is not a test set because $f(x,y)$ is ground reducible. This can be easily shown by expanding $f(x,y)$ in positions 1 and 2 simultaneously: we get $C = \{a,\ f(a, f(x,y)),\ f(f(x,y),a),\ f(f(u,v),f(x,y))\}$ ($f(a,a)$ is reducible).

If we apply the *Closure*-algorithm to this set, which is possible due to part 1 of Theorem 9, we get $C^* = \{a\}$. This shows that $f(x,y)$ is ground reducible.

Example 1 indicates how we can tackle the problems introduced by non-linearity: we will make additional expansions. In order to apply this method we need to answer two questions: *Which positions shall be considered for expansion?* and *When can we stop making expansions?*

The new aspect introduced by non-linear rules is that they can relate subterms of any depth. This relation between subterms can be regarded as a constraint superimposed on a linear rule. We can make this idea more explicit by transforming the rewrite system R from Example 1 into the following notation:

$$R = \{1 = 2 : f(x,y) \to \dots\}$$

with the semantics, that a term t is reducible if it is an instance of $f(x,y)$ *and* $t|_1 \equiv t|_2$. In this way we can transfer any non-linear rewrite system into a linear one with additional *constraints*. A rule can have more than one constraint, if it contains more than one non-linear variable.

We can distinguish two types of problems caused by non-linear rules with respect to the method described in Sect. 3:

1. A term t, which is included in the *Closure* C^*, can be ground reducible even if all preconditions mentioned for Procedure 8 are satisfied (cf. Example 1).
2. The criterion of linear extensibility is not sufficient for guaranteeing that a complete and minimal set of terms is a test set.

These two problems need to be handled in order to extend the method introduced in Sect. 3 to non-linear rewrite systems. The first problem is due to the fact that a non-linear rule, which unifies with an irreducible term $t \in C^*$ (at Λ), can reduce an instance of t if and only if the subterms at the positions of the non-linear variables are instantiated to identical terms. If they are instantiated differently, we say the corresponding *constraints are eliminated*. Consequently, only instantiations of variables of t at positions, which are larger than a position of a non-linear variable of l, must be treated differently from the linear case. Therefore, the necessary information for dealing with the first problem is contained in a set of those constraints which belong to rules which unify with t. In the following definitions we will represent a constraint by the set of positions it relates.

Definition 10 (Root Constraints). Let t be a term.

1. Let l be a term with $\mathcal{V}ar(l) \cap \mathcal{V}ar(t) = \emptyset$

$$RCon_l(t) := \begin{cases} \{p \in \mathcal{NVP}os(l) \mid p \notin \mathcal{P}os(t) \text{ or } t|_p \notin \mathcal{G}(\mathcal{F})\} & l,t \text{ are unifiable} \\ \emptyset & \text{otherwise} \end{cases}$$

2. Given a set S of terms with $\mathcal{V}ar(l) \cap \mathcal{V}ar(t) = \emptyset$ for each $l \in S$

$$RCon_S(t) := \{RCon_l(t) \mid l \in S\} \setminus \{\emptyset\}$$

Positions, which correspond to ground terms in t are not included in $RCon_l(t)$, because ground terms can not be further instantiated and our method for eliminating constraints is based on instantiation. Additionally, it is not necessary in $RCon_l(t)$ to distinguish between the variable positions which belong to different constraints of l, because our method is based on proving that there are instantiations for a *single* position, such that the rule is no longer applicable.

If $RCon_{LHS(R)}(t) = \emptyset$ holds for a term t, then its ground-reducibility can be deduced from the ground-reducibility of its principal subterms (cf. *Closure*). Therefore, we want to decrease the size of $RCon_{LHS(R)}(t)$ as far as possible. This can be done by expanding t at positions larger than the positions contained in $RCon_{LHS(R)}(t)$. However, $RCon_l(t) = RCon_l(t\sigma)$ may hold for some $\sigma \in \Omega(\mathcal{X})$. This may even be the case for an infinite sequence of terms, where each term in the sequence is generated by expanding its predecessor. In Sect. 5 we will give other criteria for handling these terms.

A similar set can be given for Problem 2 above. The main difference is that in this case the reduction does not take place at Λ, but at some position above (that is in the context the term is embedded in). Therefore we do not consider the whole left hand side of the rule, but each proper subterm of it:

Definition 11 (Partial Constraints). Let t be a term.

1. Let l be a term, $q \in \mathcal{P}os(l)$ and $\mathcal{V}ar(l) \cap \mathcal{V}ar(t) = \emptyset$

$$PCon_{l,q}(t) := \begin{cases} \{\{p \in \mathbb{N}^* \mid l|_{q.p} \equiv x,\ p \notin \mathcal{P}os(t) \text{ or } t|_p \notin \mathcal{G}(\mathcal{F})\} \\ \qquad \mid x \in \mathcal{NV}ar(l)\} \setminus \{\emptyset\} & l|_q, t \text{ are unifiable} \\ \emptyset & \text{otherwise} \end{cases}$$

2. Given a set S of terms with $\mathcal{V}ar(l) \cap \mathcal{V}ar(t) = \emptyset$ for each $l \in S$

$$PCon'_S(t) := \bigcup_{l \in S} \bigcup_{q \in \mathcal{P}os(l) \setminus \{\Lambda\}} PCon_{l,q}(t)$$

$$PCon_S(t) := \begin{cases} PCon'_S(t) \cup \{\{\Lambda\}\} & \text{there exist } l \in S,\ \sigma \in \Omega(\mathcal{X}),\ u \in \mathcal{NVP}os(l), \\ & q \in \mathcal{P}os(l\sigma), \text{ such that } q \geq u \text{ and } (l\sigma)|_q \equiv t \\ PCon'_S(t) & \text{otherwise} \end{cases}$$

$RCon_l(t)$ and $PCon_{l,\Lambda}(t)$ contain the same positions. However, positions in $PCon_{l,\Lambda}(t)$, which belong to the same rule but not to the same non-linear variable, are in different sets. This grouping by variable is also the main difference to the problem positions $P_R(t)$ introduced in [6]. In our approach this is necessary because we demand only *one* transnormal position per constraint (cf. Sect. 5) and at least one transnormal position for each non-linear variable is needed.

The property of being non-linearly expanded (Definition 14) is tightly connected with the set $PCon_{LHS(R)}(t)$ ($t \in T$). This will turn out to be a necessary additional criterion for test sets for non-linear rewrite systems (Theorem 15).

The definition of $PCon'_S(r)$ is sufficient if we are dealing with one-sorted signatures. With many-sorted signatures it is possible for a rule l to match a context of r and the position of r in this context is larger than the position of a non-linear variable of l. Such a situation is illustrated in Fig. 1 (l matches the context t at position q, and the position p_1 of r in t is larger than the position $q.v$ of the non-linear variable of l). Moreover, no part of l needs to unify with r. In this case $PCon'_{l,q}(r)$ may be empty for each $q \in \mathcal{P}os(l) \setminus \{\Lambda\}$, but r can be influenced in some test instances by this non-linear variable. Therefore we deliberately add the set $\{\Lambda\}$. If $PCon'_{LHS(R)}(r)$ is not empty, then this additional set will have no effect on the operation of the procedure we will outline in Sect. 6.

In this section we developed the notions to identify the positions where the terms need to be expanded: It is only helpful to expand a variable x in t ($t|_p \equiv x$) if there exists $V \in RCon_{LHS(R)}(t) \cup PCon_{LHS(R)}(t)$ and $q \in V$ such that $p \geq q$.

5 Independent Transnormality

The central question of this section is: *When can we stop the expansion?* Before we can answer this question, we want to turn to an only seemingly unrelated fact: terms in test sets for non-linear rewrite-system may need an infinite set of irreducible ground instances. This illustrates the following example:

Example 2. $\mathcal{F} = \{a, g, h, f\}$ (Let $g^i(s)$ denote the term $\underbrace{g(\cdots g(}_{i\text{-times}} s) \cdots)$.)
Let $LHS(R_n) = \{f(x,x), g^{n+1}(x), g(h(x)), g(f(x,y)), h(h(x)), h(f(x,y))\}$.

Note that $h(g(x))$ is linearly expanded and has exactly n irreducible ground instances (namely $h(g^i(a))$ $(1 \leq i \leq n)$).
Now there exists a context c_n for each $n \in \mathbb{N}$ such that the test instance, which results from substituting the variable x in c_n by $h(g(x))$, is irreducible but ground reducible: $c_0 = f(h(g(a)), x)$; $c_{i+1} = f(c_i\{x \leftarrow h(g^{i+1}(a))\}, c_i)$
The test instance $t_i \equiv c_i\{x \leftarrow h(g(x))\}$ is irreducible and ground reducible.

Therefore, it is not sufficient for a term in a test set to possess only a finite set of ground instances, because in this case a test instance containing this term may be irreducible but ground reducible. Consequently, whenever $PCon_S(t) \neq \emptyset$ we will require that t has an infinite set of irreducible ground instances.

Kounalis already suggested that in the non-linear case the terms in the test set may need infinitely many irreducible ground instances. In [10, p. 224] he called a term with this property transnormal. In [11] he rectified his older version and refined the notion of transnormal terms.

The concept of transnormal terms is also the key to handling those terms, for which expansion generates an infinite sequence of non-empty constraint sets. This is illustrated in Fig. 1. There t denotes a test instance of a context c and s, r are terms from a test set ($t|_{p_2} \equiv s$ and $t|_{p_1} \equiv r$). l unifies with t at q; u, $v \in \mathcal{NVP}os(l)$, $l|_u \equiv l|_v$.

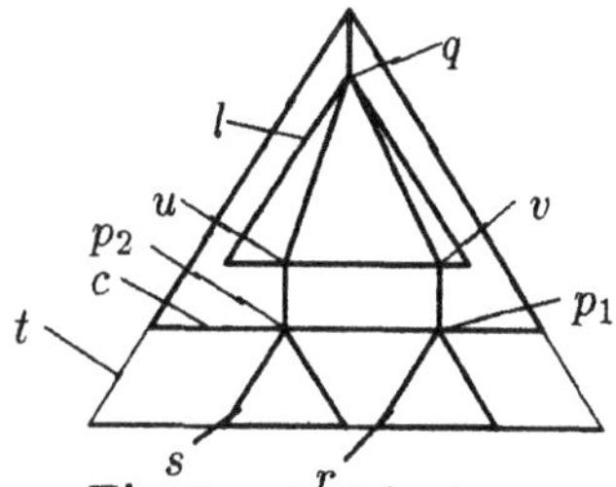

Fig. 1. a test instance

If r has infinitely many irreducible ground instances and $t|_{q.u} \not\equiv t|_{q.v}$ holds[5], then it is impossible that l reduces each ground instance of t at q.[6] Moreover, for every in-

[5] $t|_{q.u} \equiv t|_{q.v}$ is possible if c is non-linear.

[6] Positions q, p_1, and p_2 are measured relatively to t, whereas u and v are relative to l.

stance of s there can be at most one instance of r such that the resulting ground term (the other variables being arbitrarily instantiated) is reducible by l at q. Therefore, there are infinitely many instances of r left, such that the resulting term is irreducible. This idea is central to the proofs of Theorems 15 and 18.

The situation becomes much more complex if l overlaps with s and/or r. With the positions used in Fig. 1 this means that $p_1 \backslash q \in \mathcal{P}os(l)$ and/or $p_2 \backslash q \in \mathcal{P}os(l)$. If $p_1 \setminus q \in \mathcal{P}os(l)$, it is no longer sufficient that r is transnormal. Instead the subterm $r|_{(q.v)\backslash p_1}$ needs to be transnormal within the term r.[7] This refined concept of transnormality (the refinement goes much further than the one presented in [11]) is described in the following definition of *independent transnormality*:

Definition 12. Let R be a rewrite system; $t \in \mathcal{T}(\mathcal{F}, \mathcal{X})$, $\Sigma \subseteq \Omega_{\mathbf{g}}(\mathcal{X})$, $U \subseteq \mathbb{N}^*$. t is called *independently transnormal with Σ at U (w.r.t. R)*, if:

- $\Sigma \neq \emptyset$
- there exists $\tau \in \Omega(\mathcal{X})$ with $t\tau$ linear, $U \subseteq \mathcal{P}os(t\tau)$ and $\mathcal{V}ar((t\tau)|_u) \neq \emptyset$ for all $u \in U$ and additionally for all $\sigma \in \Sigma$ holds:
  - $\mathcal{D}om(\sigma) = \bigcup_{u \in U} \mathcal{V}ar((t\tau)|_u)$
  - f.a. $u \in U$ $\{\hat{\sigma}|_{\mathcal{V}ar((t\tau)|_u)} \mid \hat{\sigma} \in \Sigma, \hat{\sigma}|_{\mathcal{V}ar(t\tau)\backslash\mathcal{V}ar((t\tau)|_u)} = \sigma|_{\mathcal{V}ar(t\tau)\backslash\mathcal{V}ar((t\tau)|_u)}\}$ is infinite
- there exists $\eta \in \Omega_{\mathbf{g}}(\mathcal{X})$ with $\mathcal{D}om(\eta) = \mathcal{V}ar(t\tau)\backslash\bigcup_{u \in U}\mathcal{V}ar((t\tau)|_u)$ such that: for all $\sigma \in \Sigma$ holds: $((t\tau)\sigma)\eta \in \mathcal{G}(\mathcal{F})$ is irreducible.

t is called *independently transnormal at U (w.r.t. R)*, if there exists $\Sigma \subseteq \Omega_{\mathbf{g}}(\mathcal{X})$, such that t is independently transnormal with Σ at U (w.r.t. R).

We will use *transnormal* and *independently transnormal* synonymously.

The substitution τ in Definition 12 is necessary, because we want to allow that a term t is independently transnormal at positions that are not in $\mathcal{P}os(t)$ (cf. Theorem 18). The substitution η is used for constructing a ground instance, as we want to guarantee that for every σ the term $(t\tau)\sigma$ is not ground reducible.

Our characterization is similar to the typical terms defined in [6]. Indeed, each U-typical term is independently transnormal at U (but not vice versa).

Corollary 13. *Let a term t be an instance of s. If t is independently transnormal at U with Σ, then s is independently transnormal at U with Σ.*

Proof. There exists a substitution ρ with $s\rho \equiv t$. Simply use $\tau \circ \rho$ instead of τ. □

Using the concept of independent transnormality we can answer the central question of this section: if V is a constraint and there exists a transnormal position u, such that $u \geq v$ for any $v \in V$ then V can be eliminated. Consequently, we stop making expansions if every (partial) constraint is eliminated.

Definition 14. Let R be a rewrite system and t a linear term such that for all $l \in LHS(R)$ holds $\mathcal{V}ar(l) \cap \mathcal{V}ar(t) = \emptyset$.
t is called *non-linearly expanded (w.r.t. R)* if there exists U, such that t is independently transnormal at U (w.r.t. R) and for each $V \in PCons_S(t)$ there exists $v \in V$, $u \in U$, such that $v \leq u$.[8]

[7] This is not identical to the term $r|_{(q.v)\backslash p_1}$ being transnormal, because the context r may allow for additional reductions.

[8] Note, that every ground term is non-linearly expanded.

$T \subseteq \mathcal{T}(\mathcal{F}, \mathcal{X})$ is *non-linearly expanded* if each $t \in T$ is non-linearly expanded. A term t (or a set T) is *non-linearly extensible* if it is not non-linearly expanded.

With this definition we can give sufficient conditions for characterizing test sets for arbitrary rewrite systems:

Theorem 15. *If R is a rewrite system and $T \subseteq \mathcal{T}(\mathcal{F}, \mathcal{X})$ is linear, complete and non-linearly expanded w.r.t. R*[9] *and linearly expanded w.r.t. each proper subterm of any $l \in LHS(R)$, then T is a test set (for R).*

Proof. The proof of this theorem is based on the argumentation we made above with respect to the situation depicted in Fig. 1. There we have a situation, which does not contribute directly to the set $PCon_S(r)$, because l does not overlap with r. However, $PCon_S(r)$ will at least contain $\{\Lambda\}$, either because a non-linear variable in l unifies with r or because this set is added to $PCon'_S(r)$ to give $PCon_S(r)$. Consequently, we have $V = \{\Lambda\}$ and any transnormal position of r will do. A similar argument can be given for s, too.

If l overlaps with r, then there is a position $\hat{q}$, such that $PCon_{l,\hat{q}}(r) \neq \emptyset$. Further, there exists $V \in PCon_{l,\hat{q}}(r)$ with $(q.v) \setminus p_1 \in V$. In this case r must be transnormal with Σ at U, such that there exists $\hat{u} \in U$ with $\hat{u} \geq (q.v) \setminus p_1$ or $\hat{u} \geq \hat{v}$, $\hat{v}$ being a different position in V. If we delete each substitution σ from Σ, for which holds $t\sigma|_{q.v} \equiv t\sigma|_{q.u}$, then r is still transnormal with the resulting set Σ', as for each instance of $t|_{q.v}$ there is at most one substitution $\sigma|_{\mathcal{V}ar(r|_{(q.v)\setminus p_1})}$ such that $t\sigma|_{q.v} \equiv t\sigma|_{q.u}$ holds. Therefore, even if l reduces some instances of t at q (even infinitely many) we still have ground instances of t, which are irreducible by l at q. This way we can exclude each position of reduction in turn, always leaving infinitely many irreducible ground instances. □

Note that for linear rewrite systems Theorem 15 is equivalent to Theorem 5. Therefore, we get the same small test sets for linear rewrite systems as we got with the criteria of Sect. 3. This is contrary to the method described by Kounalis [11] which generates large test sets even for linear rewrite systems.

Interpreting the test set criteria given by Hofbauer and Huber in [6] in our terms, one observes that they demand the existence of one transnormal position (i.e. a position in U) per constraint position ($v \in V$). Requiring instead only one transnormal position per constraint in Definition 14 makes it possible to construct smaller test sets. This is demonstrated in the following example:

Example 3. $LHS(R) = \{g(g(f(x,x))),\ f(g(x), g(y)),\ f(f(x,y),z),\ f(x, f(y,z))\}$ $T = \{a,\ g(a),\ f(a,a),\ g(g(a)),\ g(g(g(x))),\ g(g(f(x,y))),\ g(f(x,y)),\ f(a,g(x)),$ $f(g(x),a)\}$ is a test set by Theorem 15, but would not be one if we would request one transnormal position for each constraint position, because $t \equiv g(f(x,y))$ is independently transnormal at $\{1.1\}$ and $\{1.2\}$, but not at $\{1.1, 1.2\}$. ($PCon_{LHS(R)}(t) = \{\{1.1,\ 1.2\}, \{\Lambda\}\}$)

As Example 3 shows, the set of independently transnormal positions is not uniquely determined. Accordingly, the procedure we will outline in the following section is not deterministic, i.e. at some points arbitrary choices are possible.

[9] We need not demand that T is minimal, because this can be easily deduced from the fact that the terms in T are non-linearly expanded.

6 Construction of Test Sets

In this section we will show how test sets for non-linear rewriting systems can be effectively constructed. For this we need a method for computing the set of independently transnormal positions of a term. The procedure we will describe here is similar to the *Closure* described in Sect. 3. We will start with those terms which we can directly classify as being independently transnormal. Such terms are instances of one of their subterms, such that replacing this subterm by the whole term results in an instance which is not ground reducible.

Theorem 16. *Let t be a linear term with an irreducible ground instance $\hat{t}$. p is a position in $\mathcal{P}os(t)$, $p \neq \Lambda$. No $l \in LHS(R)$ is unifiable with $t|_q$ for any $q < p$.*

If t is an instance of $t|_p$, then t is independently transnormal at $\{p^n\}$ (for arbitrary $n \in \mathbb{N}$).[10]

Proof. $\hat{t}$ is an instance of $t|_p$. Therefore, $\hat{t}[\hat{t}]_p$ is an irreducible ground instance of t. Each term in the sequence $t_0 = \hat{t}$; $t_{i+1} = t_0[t_i]_p$ is an irreducible ground instance of t, too. Further, for all $i, j \in \mathbb{N}$ $i, j \geq n$ holds $t_i[x]_{p^n} \equiv t_j[x]_{p^n}$ (x:new variable). Choose η, τ such that $t(\eta \circ \tau) = t_n[x]_{p^n}$ and $\Sigma = \{\{x \leftarrow t_i|_{p^n}\} \mid i \geq n\}$. □

Theorem 16 is not sufficient for computing the set of independently transnormal positions of any term, because it can only be applied to a very limited class of terms. Transnormality of the other terms will be deduced from the transnormality of their principal subterms. For this we need an additional definition:

Definition 17. The *disjoint union* of the sets $U_1, \ldots, U_n \subseteq \mathbb{N}^*$ is defined by:

$$\bigoplus_{i=1}^{n} U_i := \{i.u_i \mid i \in \{1, \ldots, n\},\ u_i \in U_i\}$$

Now we can describe the propagation of transnormal positions:

Theorem 18. *Let R be a rewrite system and $t \equiv f(t_1, \ldots, t_n)$ a linear term, which is linearly expanded w.r.t. $LHS(R)$ and irreducible at Λ. Let t_i be independently transnormal at U_i for each $i \in \{1, \ldots, n\}$.*

If for every $l \in LHS(R)$ holds: $RCon_l(t) \neq \emptyset$ implies that there exist $u \in \bigoplus_{i=1}^{n} U_i, v \in RCon_l(t)$ with $v \leq u$ and for each $u \in \bigoplus_{i=1}^{n} U_i$ holds: $|u| \geq \max_{l \in S} \max_{p \in \mathcal{NVP}os(l)} |p|$[11]*, then t is independently transnormal at $\bigoplus_{i=1}^{n} U_i$.*

Proof. The basic idea of this proof is identical to that of Theorem 15. While the term t corresponds to the test instance to be tested for ground reducibility, the terms t_i correspond to the terms from the test set. Because the terms t_i are known to be principal subterms of t, this proof is simpler than that of theorem 15. □

Using Theorems 16 and 18, and Corollary 13 we can show that the set T from Example 3 is indeed a test set.

Example 4 (Example 3 continued). Obviously, T is complete and linearly expanded w.r.t. each proper subterm of each $l \in LHS(R)$. For demonstrating that T is a test set, we need to show that the terms in this set are non-linearly expanded; i.e. they are independently transnormal at a set U, such that for each $V \in PCon_{LHS(R)}(t)$ there exist positions $v \in V, u \in U$ with $u \geq v$.

---

[10] This must not be confused with $\{p^n \mid n \in \mathbb{N}\}$. No term can be independently transnormal in the latter set.

[11] This requirement is needed in the proof, as the case that a transnormal position is smaller than a non-linear variable position can not be handled. However, this is no real restriction, as we can make each transnormal position arbitrarily large.

$g(g(a))$ is not (ground) reducible and an instance of $g(g(x)) \equiv g(g(g(x)))|_1$. Therefore, $g(g(g(x)))$ is not ground reducible, too.[12]
$g(g(g(x)))$ is an instance of $g(g(g(x)))|_1$ and no $l \in LHS(R)$ unifies at Λ.

$\overset{\text{Theorem 16}}{\leadsto}$ $g(g(g(x)))$ is transnormal at $\{1.1.1\}$ and $PCon_{LHS(R)}(g(g(g(x)))) = \{\{\Lambda\}\}$, therefore $g(g(g(x)))$ is non-linearly expanded.

$\overset{\text{Corollary 13}}{\leadsto}$ $g(x)$ is transnormal at $\{1.1.1\}$.

$\overset{\text{Theorem 18}}{\leadsto}$ $f(g(x), a)$ is transnormal at $\{1.1.1.1\}$
$(PCon_{LHS(R)}(f(g(x), a)) = \{\{\Lambda\}, \{1\}\} \surd)$ $(f(a, g(x))$ analogously)

$\overset{\text{Theorem 18}}{\leadsto}$ $g(f(x, y))$ is transnormal at $\{1.1.1.1.1\}$ (or at $\{1.2.1.1.1\}$)
$(PCon_{LHS(R)}(g(f(x, y))) = \{\{\Lambda\}, \{1.1, 1.2\}\} \surd)$

$\overset{\text{Theorem 18}}{\leadsto}$ $g(g(f(x, y)))$ is transnormal at $\{1.1.1.1.1.1\}$ (or at $\{1.1.2.1.1.1\}$); note, that $RCon_{LHS(R)}(g(g(f(x, y)))) = \{\{1.1.1, 1.1.2\}\}$.
$(PCon_{LHS(R)}(g(g(f(x, y)))) = \{\{\Lambda\}\} \surd)$

Finally, we outline the whole test set generation in the form of an algorithm:

Procedure 19 (test set construction in the general case).

```
Compute C* by linear expansion as described in section 3 (Procedures 6 and 8);
Compute transn. positions of all terms in C* using theo. 16 and 18, and cor. 13;
E := {t ∈ C* | t is non-linearly extensible};
WHILE E ≠ ∅
    Choose a term t ∈ E and a position p ∈ Var(t) adequately[13];
    Expand t at p as shown in step 2 of procedure 6 generating the set C;
    Compute the Closure C* of C using procedure 8;[14]
    Compute transn. positions of all terms in C* using theo. 16 and 18, and cor. 13;
    E := {t ∈ C* | t is non-linearly extensible};
```

In Example 4 no expansions besides those necessary for ensuring linear expansion are made. Thus, the while-loop in Procedure 19 is not entered. Numerous examples we computed suggest that this is a typical case. Yet, few examples need a small number of additional expansions. This experimental evidence as well as some theoretical considerations lead us to the conjecture that there exists a small finite bound (dependent on R) on the number of additional expansions needed. However, we could not prove this conjecture, yet. So we can not deny the possibility that there exists an infinite sequence of terms generated by expansion such that Procedure 19 can not classify any term in this sequence as being ground reducible or as being independently transnormal at the necessary positions.

However, this poses no principal problem for our approach, as ground reducibility and independent transnormality are decidable.[15] Both decision meth-

[12] This is a special case of Definition 12 and Theorem 18 with $U = \emptyset$.

[13] No position may be infinitely long deferred from expansion and the prerequisites of the *Closure*-algorithm must always be satisfied, e.g. $|p|$ minimal in $\{|q| \mid$ there exist $t \in E$, $q \in VPos(t)$, $V \in PCon_{LHS(R)}(t) \cup RCon_{LHS(R)}(t)$, $q' \in V$, $q \geq q'\}$.

[14] The *Closure*-algorithm can be used for eliminating ground reducible terms, because part 1 of Theorem 9 even holds if R is non-linear.

[15] Independent transnormality is decidable, because U-typicality is decidable ([6]).

ods are based on the construction of all ground instances of a term up to a certain bound. Consequently, we can extend Procedure 19 into a procedure which can easily be shown to be terminating:

Let us call the set E, which results after the first three instructions of Procedure 19, E_{init}. If, in the first instruction in the while-loop, we always select the smallest position in all the terms in E, then Procedure 19 constructs instances of those terms in E_{init}, which can not be classified as being ground reducible or non-linearly expanded by our theorems. This instances are built such that all the variables in the terms are at the same depth. With the help of the decidability results mentioned above, all terms in E_{init} can be either classified as being ground reducible or as being independently transnormal. The latter ones are either non-linearly expanded or a finite set of instances can be generated in which each term is non-linearly expanded.

However, such a modification is only of theoretical interest, because the bounds needed in the known decision procedures are extremely large. Consequently, we plan to implement our procedure without any additions of this kind.

7 Optimized Test Sets

By comparing our test set criteria (cf. Theorem 15) with our procedure for generating test sets (cf. Procedure 19), one can see that some expansions of the terms are not required by the test set criteria. They are only necessary for constructing the test set. These are the linear expansions w.r.t. the terms $l \in LHS(R)$ (cf. the first prerequisite of the *Closure*-algorithm), and some expansions necessary for ensuring that there is a transnormal position for each constraint in $RCon_{LHS(R)}(t)$ so that Theorem 18 can be applied.

Things are different if we already have a test set T for R, which was constructed according to the procedure outlined above. In this case we know the transnormal positions of the terms in T. This information can then be used for constructing a better test set T'. The basic instrument for this is Corollary 13. It can be used for transferring information about the transnormal positions from the test set terms to the newly constructed terms:

Procedure 20 (Optimization of test sets).

1. Construct a set T', which is linearly expanded w.r.t. every proper subterm of each term $l \in LHS(R)$, using Procedure 6.
2. Transfer sets of transnormal positions from terms in T to the terms in T' using Corollary 13.
   If there are several instances $t \in T$ of $t' \in T'$ then choose the most adequate instance. If there is no instance, eliminate t' because it is ground reducible.
3. If there is a term in T', which is non-linearly extensible then choose and expand a term $t' \in T'$ and a position $p \in \mathcal{P}os(t')$ as outlined in Procedure 19. (Follow the expansions made for constructing T). Proceed with step 2.

Typically, optimized test sets are much smaller than the original ones. Experiments showed, that the fact, that the requirement of only one transnormal position per constraint, is very useful especially for optimizing test sets.

Usually, the depth of the terms in the optimized set is by one smaller than the depth of the terms in the original set. Therefore, T' is smaller than T approximately by a factor of $|\mathcal{F}|$. This cut-down is important, because the total complexity of a ground reducibility test is given by $|T|^n$, where T is the test set used and n the number of different variables in the term to be tested. Using this approach we can optimize the test set given in Example 4 and get $T' = \{a, g(a), g(g(x)), g(f(x,y)), f(x,y)\}$, thereby reducing the size of the test set from 9 to 5 elements.

8 Conclusion

In this paper we gave a new characterization of test sets for non-linear rewrite systems even with many-sorted signatures. Additionally, we proposed a procedure for efficiently generating such test sets. This procedure is supplemented by an optimization procedure. Multiple experiments with this approach are very encouraging.

An interesting aspect of our method is that for linear rewrite systems it gives the same results as the best method (measured by the size of the test sets) we know of ([9]). This is due to the fact that our general criteria become equivalent to the criteria for the linear case, if R is linear. Therefore the general method can be used for the linear case without incurring any additional overhead. Moreover, these results can be further improved by using our optimization procedure. To the contrary, Kounalis' approach gives huge test sets even for linear rewrite systems. Additionally, his method is restricted to one-sorted signatures. In [11] no complete test set generated with his method is given. Therefore we used his method for generating a test set for a non-linear rewrite system. This lead to a test set with 43 elements, whereas our approach gave a test set with 6 elements (3 elements with optimization). This is equivalent to a more than 100 times (1000 times with optimization) smaller number of test instances to be tested for a term with 3 variables. We used our approach on every partial test set constructed in [11], too, with similar results.

The method proposed in [2] gives a correct test set only if $IRG(R)$ is regular, that is, if there exists a linear rewrite system with $IRG(R') = IRG(R)$. Usually this is the case if either $IRG(R)$ is finite or R is already linear.

The only method which gave a comparable result to our approach is the procedure given in [4]. The result given there in the appendix is identical to the result Procedure 19 produces on the same input. (The result in [4] is stated as a normal form grammar, but can be translated into a test set.) However, our optimization results in a reduction of the original test set from 13 to 5 elements.

Hofbauer and Huber only give a characterization of test sets in [6], but no practical method for constructing them. Consequently, all test sets they give as examples are hand-crafted. We compared each test set they give in Sect. 3 of their paper with the results of our method (with optimization). The test sets were identical with two exceptions: 1) In T_4' (p. 101) the terms $f(a,x)$ and $f(x,a)$ can be substituted by $f(x,y)$. This is due to the less restrictive requirements of our approach. 2) In T_4'' (p. 104) $f(x,x')$ can be replaced by x_f. Because this would be allowed by their approach, too, we believe that this is simply an oversight.

Further outlook: a proof is still lacking, which shows that Procedure 19 always terminates (without the extensions we described above) and which gives a small bound on the size of the resulting test sets, as can be expected from experimental evidence. We are optimistic that our characterization of test sets is also sufficient for deciding idleness of variables, thereby leading to a practical procedure for the linearization of rewriting systems, which is still lacking (cf. [6] for details).

An important project for the future is the implementation of our method. This is, however, not as straightforward as it may seem, because the different choices which can be made (e.g. which position to expand, which instance to use for propagating transnormality, etc.) influence the quality of the resulting test set. In the linear case, good choices can even lead to test sets superior (w.r.t. to the size) to those generated by the approach described in [9]. This even holds without using the optimization outlined in Sect. 7.

References

1. B. Bogaert and S. Tison. Equality and disequality constraints on direct subterms in tree automata. In *Proc.* 9th *STACS*, volume 577 of *LNCS*, pages 161–171. Springer, 1992.
2. R. Bündgen. *Term completion versus algebraic completion.* Ph.D. thesis, Fakultät für Informatik, Eberhard-Karls-Universität, Tübingen, Germany, May 1991.
3. H. Comon and F. Jacquemard. Ground reducibility and automata with disequality constraints. In *Proc.* 11th *STACS*, LNCS 775, pages 151–162. Springer, 1994.
4. H. Comon and J.-L. Remy. How to characterize the language of ground normal forms. Rapports de Recherche 676, INRIA, 1987.
5. N. Dershowitz and J.-P. Jouannaud. Notations for rewriting. *Bulletin EATCS, No.43*, pages 162–172, Feb 1991.
6. D. Hofbauer and M. Huber. Linearizing term rewriting systems using test sets. *Journal of Symbolic Computation*, 17:91–129, 1994.
7. J. Jouannaud and E. Kounalis. Automatic proofs by induction in theories without constructors. *Information and Computation*, 82:1–33, July 1989. An earlier version of this paper appeared in *Bulletin EATCS, No.27, 1985*, pages 49–55.
8. D. Kapur, P. Narendran, and H. Zhang. On sufficient completeness and related properties of term rewriting systems. *Acta Informatica*, 24:395–415, 1987.
9. D. Kapur, P. Narendran, and H. Zhang. Automating inductionless induction using test sets. *Journal of Symbolic Computation*, 11:83–111, 1991.
10. E. Kounalis. Testing for inductive (co)-reducibility. In *CAAP'90, Copenhagen, Denmark*, volume 431 of *LNCS*, pages 221–238, 1990.
11. E. Kounalis. Testing for the ground (co)-reducibility property in term-rewriting systems. *Theoretical Computer Science, Elsevier*, 106:87–117, 1992.
12. G. Kucherov and M. Tajine. Decidability of regularity and related properties of ground normal form languages. In 3rd *CTRS*, LNCS 656, pages 272–286, 1992.
13. D. Plaisted. Semantic confluence tests and completion methods. *Information and Control*, 65:182–215, 1985.
14. S. Vágvölgyi and R. Gilleron. For a rewriting system it is decidable whether the set of irreducible ground terms is recognizable. *Bulletin EATCS, No.48*, pages 197–200, 1992.

Term Rewriting in Contemporary Resolution Theorem Proving

Mark E. Stickel

Artificial Intelligence Center
SRI International
Menlo Park, Californa 94025

The use of term rewriting methods for efficient equality reasoning in resolution theorem proving predates even the Knuth and Bendix paper that provided the starting point for the term rewriting field. The resolution operation for nonequational reasoning over first-order predicate calculus formulas was augmented by the demodulation and paramodulation operations, which closely resemble Knuth and Bendix's reduction and superposition operations.

Since then, research on term rewriting has contributed a great deal to validate and extend methods employed in resolution theorem proving. For example, the completeness of combinations of resolution, paramodulation, and demodulation without functional reflexive rules, stronger ordering restrictions on resolution and paramodulation operations, and simplification orderings such as recursive path ordering have all added to the theory and practice of resolution theorem proving.

Some fundamental problems of great practical importance remain. Paramodulation and demodulation are often used in conjunction with the set of support strategy (to make resolution more goal-directed) although their combination is incomplete. Efficient, complete, goal-directed equality reasoning in resolution systems is a good target for further research.

Resolution systems are being employed in query answering and program synthesis applications. Efficient equality reasoning is important for their performance, but completeness of consequence-finding (as opposed to refutational completeness) is often an unresolved issue.

More generally, insufficient research has been done on how to integrate various technologies such as goal-directedness, equality reasoning, and answer or program construction. For another example, a lot of research has been done on unification for equational theories, sorted unification, and indexing terms for efficient retrieval, but research on their combinations has been sparse, which impedes their acceptance and use in resolution theorem-proving programs.

$\delta \circ ! \epsilon = 1$

Optimizing optimal λ-calculus implementations

Andrea Asperti*

Dipartimento di Matematica
P.zza di Porta S.Donato 5, Bologna, Italy
asperti@cs.unibo.it

Abstract. In [As94], a correspondence between Lamping-Gonthier's operators for Optimal Reduction of the λ-calculus [Lam90, GAL92a] and the operations associated with the comonad "!" of Linear Logic was established. In this paper, we put this analogy at work, adding new rewriting rules directly suggested by the categorical equations of the comonad. These rules produce an impressive improvement of the performance of the reduction system, and provide a first step towards the solution of the well known and crucial problem of accumulation of control operators.

1 Introduction

Fifteen years ago, Lévy [Le78] proposed a complex notion of *redex family* to formalize the intuitive idea of optimal sharing in the λ-calculus (see also [Le80, AL93b]). As a main consequence, the length of the *family reduction* would provide a lower bound to the *intrinsic complexity* of λ-term reduction, *in any possible implementation.*

Lévy's thesis raised two different kinds of problems:

- to get some evidence that his abstract notion of sharing could be feasible; actually, no "traditional" implementation of λ-calculus (supercombinators, environment machines, etc...) is able to share all redexes in Levy's families (see [Fie90] for a discussion).
- to understand if the lower bound to the complexity of the reduction provided by the lenght of the family reduction could be actually reached (or at least approached, say, up to some polynomial order).

Note that a positive answer to the first problem is an *essential prerequisite* for a positive answer to the second one. However, it does not imply it, for at least two reasons:

- the computational overhead introduced for handling sharing;
- the complexity of beta-reduction, that in lambda calculus (and all the more reason, in family reduction) is not an atomic operation (due to the meta-operation of substitution).

* Partially supported by the ESPRIT Basic Research Project 6454 - CONFER.

The first problem has been solved by Lamping [Lam90], using a complex graph reduction technique. The second problem is still open: no relation has been established by Lamping between the complexity of his reduction technique and the lenght of the family reduction (it is not even clear if there is *any* relation at all, due to the "cost of doing a substitution" [Lam89]). So, the very abstract problem of establishing a concrete mesure for the *intrinsic* complexity of λ-term reduction is still far from being solved.

This discussion should clarify an important point. Since Lamping's seminal work, it is customary to call "optimal", in the literature, any implementation which is able to share all redexes in Levy's families. This has very little to do with its actual performance, and it should not be surprising that among "optimal implementations", there are some which are "more optimal" (or we should say more efficient, maybe) than others. For instance, the "simplified" implementation proposed in [GAL92a], which is based on a restricted set of rewriting rules, is much less efficient than Lamping's original one: there are examples where the reduction grows exponentially in [GAL92a] (w.r.t. the lenght of the family reduction), while it remains linear in [Lam90].

Remark. The previous comment is not meant at all to dimimish [GAL92a], whose theoretical relevance is out of question. By pointing out the relations between optimal reductions, Linear Logic [Gi86] and the Geometry of Interaction [Gi88a, Gi88b], it provided an essential brackthrough in the topic, suggesting new and absolutely innovative perspectives (see [GAL92b, As94, AL93b, ADLR94]).

The analogy between Optimal Reductions and Linear Logic was further developed in [As94], from a categorical perspective. In particular, the two *control operators* of Lamping and Gonthier (i.e. the "croissant" and the "square bracket") were related to the two natural transformations $\epsilon :! \rightarrow I$ and $\delta :! \rightarrow !!$ associated with the comonad "!" of linear logic, and the optimal graph reduction rules in [GAL92a] were explained as a sort of "local" implementation of naturality laws (see also [ADLR94] for the relations with geometry of interaction and dynamic algebras [DR93, Re92]).

On the contrary, and somewhat surprisingly, among the reduction rules of [GAL92a] there was no counterpart at all for the *three comonad equations* involving δ and ϵ. In [As94], we already explained that this absence was the main cause for the well known and crucial problem of accumulation of control operators [GAL92a] (that is the explosion of the book-keeping work required for the correct handling of optimal sharing). However, adding these rules to the rewriting system seemed problematic (in some cases, it generated deadlocks, that is "wrong" configurations where interacting nodes can never be eliminated).

In this paper, we solve this problem, by proposing a "safe" use of these rules. Their introduction in the rewriting system produces a really amazing improvement of its performance.

The paper is organized as follows. In section 2, we shall start with providing some examples of reduction, comparing the mutual performances of a standard (environment-based) implementation such as Caml Light [LM92], and our optimal compiler (with and without the new rules suggested by the comonad equations). Section 3 contains the definition of sharing graphs, and the associated rewriting rules. In section 4 we define the translation of λ-terms in sharing graphs following [As94]. Then, we provide a brief explanation of the relation between optimal reductions and the categorical interpretation of Linear Logic (section 5). The problem of accumulation of control operators is discussed in section 6. Finally, in section 7, we explain our solution, and the *safe* use of the comonad equations.

The paper requires some knowledge of optimal reductions.

2 Examples and Benchmarks

A prototype implementation of the optimal compiler described in [As94] has been jointly developed by the author and Cosimo Laneve (INRIA-Sophia Antipolis), who helped in implementing the readback procedure. The compiler is available by anonymous ftp, at `ftp.cs.unibo.it`, in the directory /pub/asperti, under the name `opt1.tar.Z` (compressed tar format).

At present, the prototype merely works for pure λ-calculus (with some syntactic sugar), but in the near future we plan to add new primitive types and delta rules, following the the main idea of Interaction Systems [AL93a, AL94]. The current implementation pursues reduction up to weak head normal forms, according to a lazy strategy (leftmost outermost reduction). The prototype is written in standard C, which is also the target language.

In this section, we show the practical relevance of the optimization rules suggested by the comonad equations, by comparing, on a few standard examples, the relative performances of the rewriting system with and without these rules. Moreover, in order to give the gist of optimality and its actual power, we shall compare the performance of our system with a different, fully developed, and largely diffused implementation: CamlLight. CamlLight [LM92] is a small, portable implementation of the ML language (about 100K for the runtime system, and another 100K of bytecode for the compiler) developed at INRIA-Rocquencourt. In spite of its limited dimension (versions for Macintosh and IBM PC are also available), the performance of CamlLight is quite good for a byte-coded implementation: five to ten times slower than SML-NJ.

Our first example is a "primitive recursive" version of the factorial function (on Church integers).

```
let Succ = \n.\x.\y.(x (n x y));;
let Mult = \n.\m.\x.(n (m x));;
let Pair = \x.\y.\z.(z x y);;
let Fst = \x.\y.x;;
let Snd = \x.\y.y;;
let Nextfact = \p.let n1 = (p Fst) in
                  let n2 = (Succ (p Snd)) in
                      (Pair (Mult n1 n2) n2);;
let Fact = \n.(n Nextfact (Pair one zero) Fst);;
```

Since the evaluation stops at weak head normal forms, we have to supply enough arguments (identities, for instance) to get an interesting computation. The result of the test is in Figure 1.

Input	Optimal Implementation As94 (GAL92)	Optimal Implementation new version	CamlLight
(fact one I I)	0.00 s 561 interact.	0.00 s 273 interact.	0.00 s
	20 family reductions		
(fact three I I)	0.04 s. 3375 interact.	0.01 s 983 interact.	0.00 s
	45 family reductions		
(fact five I I)	0.23 s. 17268 interact.	0.04 s 2380 interact.	0.02 s
	74 family reductions		
(fact seven I I)	explodes	0.13 s 5625 interact.	0.17 s
	107 family reductions		
(fact nine I I)		0.26 s 12402 interact.	10.60 s
	144 family reductions		
(fact ten I I)		0.38 s 17916 interact.	explodes
	164 family reductions		
(fact twenty I I)		5.58 s 307839 interact.	
	425 family reductions		

Fig. 1. Factorial

The first column of the table is the input term. The two following columns respectively refer to the optimal implementation without the "comonad" rules (old version), and with them (new version). Note that the performance of our old version (the one described in [As94]) is similar to the performance of the reduction system considered in [GAL92a] (the formal, syntactical correspondence between these translations has been recently settled in [AL95]). The last column refers to CamlLight. Each entry of the table contains the user time required for reducing the input by each system (on a Spark Station ELC). In the case of optimal systems, we also give the total number of interactions (i.e. of elementary operations) executed during the computation, and the lenght of the family reduction. The latter datum is the number of redex-families (i.e. *optimally shared* β-reductions) required for normalizing the term. This is the same as the total number of interactions between application and lambda nodes in the graph. Note that this is obviously the same for the two optimal implementations.

The improvement provided by the "comonad" rules is amazing. The old version explodes for input seven, CamlLight for input ten, while in the new version we compute the factorial of twenty in less than six seconds.

Unfortunately, things are not always that good (w.r.t. CamlLight). Our next example is fibonacci, defined as follows:

```
let Add = \m.\n.\x.\y.(m x (n x y));;
let Nextfibo = \p.let n1 = (p Fst) in
                  let n2 = (p Snd) in
                  (Pair (Add n1 n2) n1);;
let Fibo = \n.(n Nextfibo (Pair zero one) Fst);;
```

The result of the test is in Figure 2. Again, there is a substantial improvement w.r.t. the old version, but in this case CamlLight is still more efficient.

Input	Optimal Implementation As94 (GAL92)	Optimal Implementation new version	CamlLight
(fibo one I I)	0.01 s. 513 interact. 20 family reductions	0.00 s. 248 interact. 20 family reductions	0.00 s.
(fibo four I I)	0.06 s. 3152 interact. 55 family reductions	0.02 s. 868 interact. 55 family reductions	0.00 s.
(fibo seven I I)	0.19 s. 12749 interact. 98 family reductions	0.05 s. 2209 interact. 98 family reductions	0.01 s.
(fibo ten I I)	0.74 s. 52654 interact. 173 family reductions	0.19 s. 6456 interact. 173 family reductions	0.02 s.
(fibo thirteen I I)	3.08 s. 230548 interact. 390 family reductions	0.71 s. 24487 interact. 390 family reductions	0.04 s.
(fibo sixteen I I)	explodes 1177 family reductions	2.94 s. 95199 interact. 1177 family reductions	0.12 s.
(fibo nineteen I I)	4404 family reductions	13.11 s 385911 interact. 4404 family reductions	0.44 s.

Fig. 2. Fibonacci

Note that the total number of family reductions (fr) grows quadratically in the case of the factorial function ($fr(n) = (n^2 + 21n + 18)/2$), while it grows as fibonacci for the fibonacci function ($fr(n) = fib(n) + 11n + 8$). In the first case, the optimal compiler is able to profit of the big amount of sharing, obtaining very good performances on big input values, while in the second case the computation is already intrinsically exponential, and CamlLight takes advantage of its simpler abstract reduction model, avoiding book-kepping operations that in this case are useless. Note in particular that the execution time in the CamlLight system is *linear* in the number of family reductions (i.e. optimal, in the strongest sense) in the case of fibonacci, so this term is surely *one of the best examples we could choose for CamlLight.*

Another important point is that, even if the computation of the new version of the optimal compiler works very good for the factorial function, still it is not linear in the number of family reductions (as it is also the case for fibonacci). As we remarked in the introduction, there is no theoretical evidence that such a linear behaviour can be reasonably expected, in general, due to some plausible intrinsic cost of "doing a substitution" (even if, in graph reduction, this operation actually looks *atomic*!). Unfortunately, as in the case of Lamping, we have been unable so far to provide a bound for the amount of bookkeeping work required for each family reduction. We just hope that our cleaner implementation will help to find a solution.

After having considered an example particularly favorable to CamlLight (and other standard implementation models), let us see an example particularly suited to Lamping's technique.

Input	Optimal Implementation As94-GAL92	 new version	CamlLight
(g one)	0.00 s. 124 interact.	0.00 s 78 interact.	0.00 s
	7 family reductions		
(g four)	0.00 s. 608 interact.	0.00 s 221 interact.	0.00 s
	16 family reductions		
(g seven)	0.04 s. 3052 interact.	0.00 s 386 interact.	0.00 s
	25 family reductions		
(g ten)	0.43 21176 interact.	0.01 s 551 interact.	0.03 s
	34 family reductions		
(g thirteen)	explodes	0.02 s 831 interact.	0.20 s
	49 family reductions		
(g sixteen)		0.02 s 973 interact.	1.60 s
	58 family reductions		
(g nineteen)		0.03 s 1106 interact.	12.65 s
	67 family reductions		

Fig. 3. g = λ n.(n two I I)

This is provided by the term $g = \lambda n.(n\ two\ I\ I)$, where *two* is a Church Integer, and I is the identity. In this case (see Figure 3) the (new version of the) optimal compiler works linearly w.r.t. its input, while CamlLight is just exponential!

3 Sharing Graphs

In [AL93a], it has been clarified that all redexes in a same family define a unique common *path* in the initial term of the derivation. In a sense, these paths are the "physical" informations that *must be shared* in optimal reduction systems. Lamping's idea was to define a set of local rewriting rules, on a suitable graph representation of λ-terms, that avoided duplications of these paths without preventing the reduction of the term to its normal form.

The graph we shall consider here are build up by means of the following indexed nodes:

In the reduction rules below, we shall use $\blacktriangleright^n$ as either $\sqsupset^n$ or $\subset^n$, and

as any of

Rewriting in sharing graphs is then definied by the following rules.
Effacement rules (E-rules):

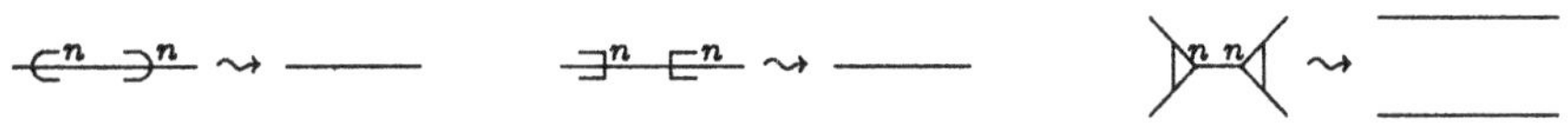

Principal Propagation rules (P-rules); $0 \leq n < m$:

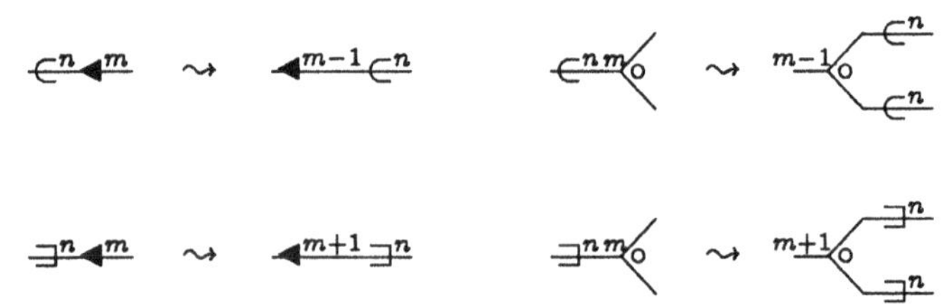

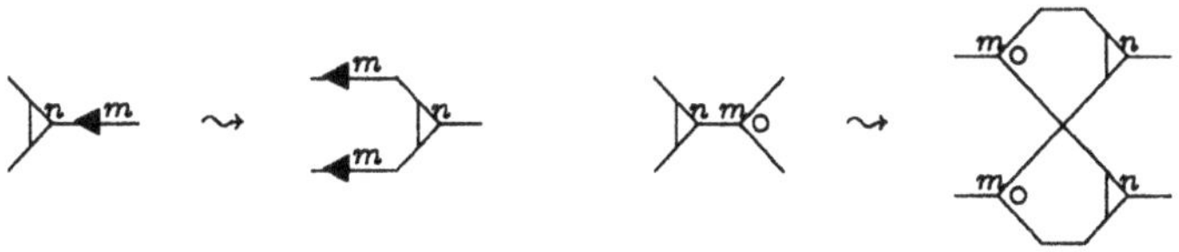

The rewriting system is completed by the the following β-rules ($n \geq 0$):

4 The graph representation of λ-terms

A λ-term N with n free variables will be represented with a graph with $n+1$ entries (free edges): n for the free variables (the inputs), and one for the "root" of the term (the output).

The translation is inductively defined by the following rules (inputs are upwards, while the output is downwards):

$$[M] \quad = \quad [M]_0$$

where:

$$[x]_n \quad = \quad \psi^n$$

$[MN]_n \quad =$

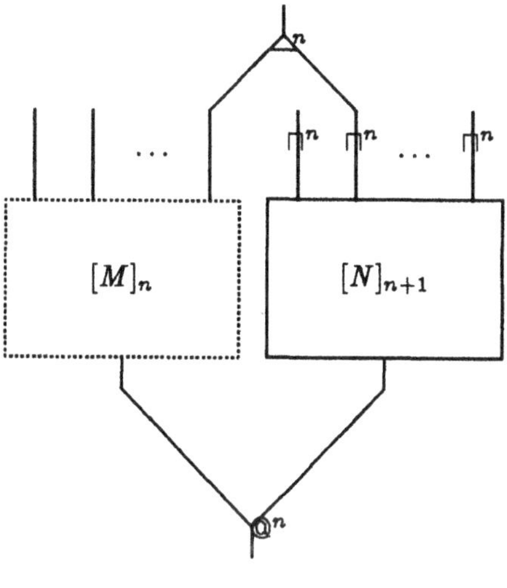

(all variables common to M and N are shared.)

$$[\lambda x.M]_n \;=$$

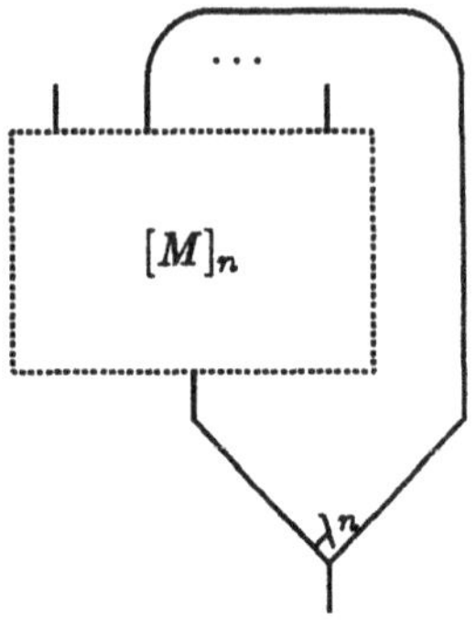

when $x \in FV(M)$, and

$$[\lambda x.M]_n \;=$$

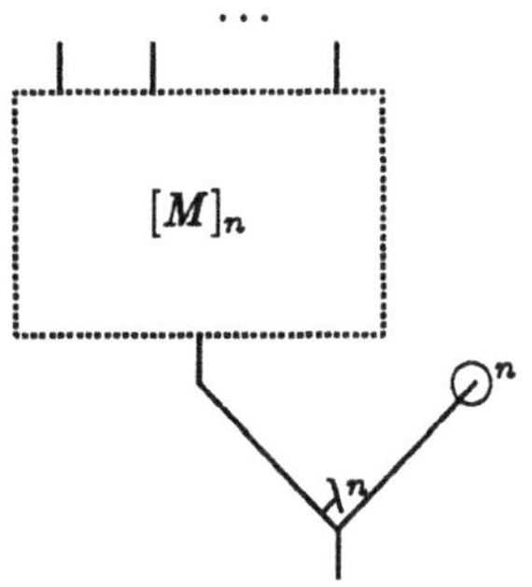

when $x \notin FV(M)$.

5 Optimal Reductions and Linear Logic

The correspondence between the previous graph representation and the categorical interpretation of λ-calculus, passing through Linear Logic, is essentially based on the following remarks [As94]:

1. a node with index n is an operator inside n applications of the comonad "!";
2. the two control nodes *bracket* and *croissant* are respectively associated with the two natural transformations $\delta :! \rightarrow !!$ and $\epsilon :! \rightarrow I$ of the comonad "!";
3. the two structural nodes *fan* and *garbage* are respectively associated with the two comonoid operations $\Delta :! \rightarrow ! \otimes !$ and $E :! \rightarrow 1$.

Reduction can be then (roughly) expressed by *linear* β-reduction, together with the four rewriting rules suggested by naturality, namely:

$$\epsilon \circ !(f) \rightarrow f \circ \epsilon$$

$$\delta \circ !(f) \rightarrow !!(f) \circ \delta$$

$$E \circ !(f) \rightarrow E$$

$$\Delta \circ !(f) \rightarrow !(f) \otimes !(f) \circ \Delta$$

In the previous rules, the box !(-) is considered as a single, global entity. In particular, the reduction

$$\Delta \circ !(f) \rightarrow !(f) \otimes !(f) \circ \Delta$$

amounts to physically duplicate the box !(-).

In order to get optimal reduction, we are forced to proceed to this duplication in a sort of "lazy" way. In other words, the information Δ at the output of the box must be propagated to the inputs travelling *inside* the box, and this propagation must be stopped at suitable positions.

In this way, part of the box may be shared by different terms, and we are forced to implement *all* the reductions expressing the naturality of transformations over boxes following the same local strategy as for Δ.

The idea for performing this local implementation of a naturality law is that of broadcasting the information from the output door to the input doors of the box, following the connected structure of the net.

The information to be propagated is at a lower level with respect to the term it must traverse. The propagation stops when it reaches a region at its own level. During the propagation, each information operates on the net according to its meaning. In particular, every link traversed by ϵ must be shifted down by one level, every link traversed by δ must be lifted up by one level, every link traversed by Δ must be duplicated (and every link passed by E should be erased, but we shall not consider this case). The same rules take also into account the case of a mutual crossing of different informations during their propagation.

Propagation is essentially a complete broadcasting. That is, if some information representing a natural transformation at level n enters a link of level $m > n$ at same of its edges, it is propagated to all the other edges of the link.

During its propagation, the information may be duplicated. When two instances of a same information meet face to face, they are erased.

In order to achieve optimal reduction, it is enough (and, in some cases, necessary) to limit the propagation to the case when the information enters a link at its *principal port* (see [Laf90, AL93a] for the terminology). Following this analogy, we easily derive the set of rewriting rules in section 3.

6 Accumulation of control operators

If you try to reduce the λ-term (two I) according to the previous system, you will end up with the first graph in figure 4, while one would obviously expect to get the second configuration, as a result.

Actually, these two graphs are *equivalent* w.r.t. the standard semantics for sharing graphs, provided by means of *contexts* [Lam90, GAL92a]. However, the first representation contains a lot of redundant control information. This accumulation of control operators is the main source of inefficency of the optimal graph reduction technique (for instance, it is responsible for the exponential explosion in the reduction of the function g of Section 2).

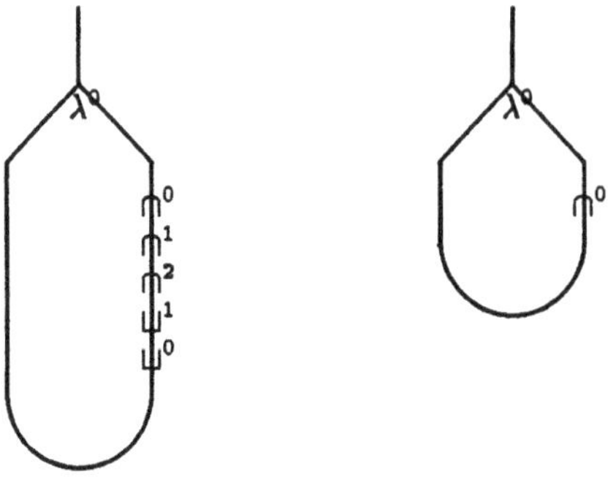

Fig. 4. Two representations for the identity

A similar problem already arises when working with a global notion of box (or at the level of the categorical semantics, if you prefer). Actually, in our analogy, the row of control operators in the first graph above would correspond to the following sequence of terms

$$\delta \circ !(\delta) \circ !!(\epsilon) \circ !(\epsilon) \circ \epsilon$$

At the categorical level, the problem is solved by using the comonad equations associated with these natural transformations, namely

$$\delta \circ ! \epsilon = 1$$

$$\delta \circ \epsilon ! = 1$$

$$! \delta \circ \delta = \delta ! \circ \delta$$

By a repeated used of the first equation, you may actually prove that

$$\delta \circ !(\delta) \circ !!(\epsilon) \circ !(\epsilon) \circ \epsilon = \epsilon.$$

So, the natural idea would be to add the following rewriting rule to our system:

Unfortunately, due to the locality of graph rewriting, this rule gives problems, in general. Consider for instance the term

$$\lambda x.I(I(x\ P))\ I$$

where I is the identity and P is any term. If you reduce this term according to a leftmost outermost strategy in the enriched system, after four beta reductions you are left with the following row of control operators leading to P:

$$\in^0 \; \in^1 \; \ni^0 \; \in^1 \; \sqsubset^0 \; P$$

Now, the correct final configuration would be:

$$\in^0 \; \in^1 \; \in^2 \; P$$

but pursuing a leftmost outermost reduction you would get instead the following deadlock situation:

$$\in^0 \; \sqsubset^0 \; \in^2 \; P$$

Remark. A different way to look at the same problem is by noticing that all "comonad equations" are inconsistent with the Dynamic Algebra of the Geometry of Interaction [Re92, DR93] (a problem that was pointed out to us by Laurent Regnier).

Let us recall that a Dynamic Algebra (LS) is an equational theory of *monomials* defined by the following items and equational axioms:

- an associative operation with a 0 and a 1;
- an involution u^* satisfying $0^* = 0$, $1^* = 1$ and $(PQ)^* = Q^* P^*$ for any P and Q in LS;
- a morphism ! for composition, 0, 1 and $*$;
- two *multiplicative coefficients* p and q satisfying the *annihilation axioms*:

  $$p^*p = q^*q =, 1 \qquad\qquad p^*q = q^*p = 0$$

- four *exponential constants* r, s, t, d satisfying the *annihilation axioms*:

  $$r^*r = s^*s = d^*d = t^*t = 1 \qquad\qquad s^*r = 0$$

  and the *commutation axioms*

  $$!(P)r = r!(P) \qquad !(P)s = s!(P) \qquad !(P)t = t!^2(P) \qquad !(P)d = dP.$$

  where P is any monomial of LS.

Note that there is an obvious analogy between the categorical operations δ and ϵ and the dynamic algebra operators t and d (see [ADLR94] for the formal relations between dynamic algebras and optimal reductions).
Here is Regnier's proof of the inconsistency of $td = 1$.

Suppose that $td = 1$. Since $t^*t = 1$, we have $d = t^*$, and by involution $d^* = t$. So, $dd^* = t^*t = 1$. Then, since $dd = d!(d)$ by the commutation axioms of d, we also have $d = ddd^* = !(d)dd^* = !(d)$.
Let now $u = !(r^*)ds$. Then, $u = dr^*s = 0$, but also $u = !(r^*)!(d)s = !(r^*)(!(!(d))s = !(r^*!(d))s$, and by the commutation axioms for s and r^*, $u = s!(r^*!(d)) = s!(!(d)r^*)$. Summing up, $u = s!(!(d)r^*) = 0$, that implies

$$1 = !(d^*)s^*s!(!(d)r^*)!(r) = 0$$

Similarly, you can prove the inconsistency of the other "comonad equations".

7 Safe operators

Our solution is to limit the application of the above rules to particularly *safe* positions in the graph. A safe position is a place where control operators have a "global status". This means that they can be safely regarded as their categorical counterpart, in that no residual of their information is still (or yet) propagating inside the graph (you may look at a safe operator as a sort of "fresh" node). Precisely,

1. all operators are safe in the initial graph;
2. an operator becomes unsafe when it starts its broadcasting (i.e. it interacts with another operator of higher index);
3. an (unsafe) operator that, during its broadcasting, reaches another *safe* operator of equal or lower level becomes safe again (since it has completed its broadcasting inside a box, reaching an extremity).

Remark. Note that a safe operator is eventually a "in"-form, generalizing Lamping's distinction between fan-in and fan-out.

From the implementative point of view, it is sufficient to add a tag to each node, expressing its "safeness". The tag is modified in the obvious way, according to the above mentioned rules.

It is possible to prove that a residual of a safe operator (created by its broadcasting) may be only annihilated by another residual of the *same* operator (the residual relation is defined in the obvious way). Then, we immediately get the following result:

Theorem 1. *The three "comonad" rules*

$$\in^{n+1}\ \exists^{n} \quad \rightsquigarrow \quad \text{———}$$

$$\in^{n}\ \exists^{n} \quad \rightsquigarrow \quad \text{———}$$

$$\exists^{n+1}\ \exists^{n} \quad \rightsquigarrow \quad \exists^{n}\ \exists^{n}$$

are correct between safe *operators.*

Proof. An easy consequence of the following considerations:

1. all pairs of control operators in the left hand side of a comonad equation may be always propagated together (i.e., they can be seen as a unique operator);
2. the total control effect of this "compound" operator on the nodes they are propagated through is equal to the control effect of the right hand side of the rule (i.e. it is null for the two first equations, and it amounts to a double level shift for the third equation);

3. if the innermost form of (a residual of) the pair is annihilated, then the outermost form can be annihilated at the next step. Indeed, the two operators are residuals of a "compound" safe operator, and they can be only annihilated by other residuals of the same "compound" configuration.

Note that the third rule fails for unsafe operators.

Remark. "Safe elimination" rules are not entirely new. Actually, there are some analogies between our safe bracket and Lamping's unrestricted bracket (see rules VII.b, VII.d and III.b in [Lam90]). Lamping only needs a single kind of "safe" square at level 0 due to the "duality" of his translation scheme (see [AL95] for a formal explanation of this "duality"). Similarly, according to the translation in [GAL92a], an operator is safe *if and only if* it is at level 1.

8 Conclusions

In this paper, we proposed an essential improvement of the optimal graph reduction technique, based on a *safe* use of the comonad equations. Although "safe elimination" rules are not entirely new [Lam90], our approach to this problem is completely original. Moreover, the correct understanding (and generalization) of the notion of "safeness" to all control operators has some important applications in other aspects of the implementation (such as garbage collection, for instance).

As in the case of Lamping [Lam90], we have not been able yet to quantify the complexity of the reduction in terms of the lenght of the the family reduction (it looks *really* difficult). We just hope that our cleaner theoretical approach will help to get some more intuition on this problem.

A lot of work is still left in studying (and understanding) other possible optimizations. There are some "pragmatic" rules in Lamping's system [Lam90] that need further study, such as the "rapid (or auxiliary [As94]) propagation" of brackets (rules IV, VII and IX), that can be useful to put in evidence more "comonads reductions", and the elimination/reorganization of fan nodes (rules VIII) which can be profitably rephrased in terms of safe-operators. However, according to our experimental results, all these rules seem merely produce a linear (even if in some cases impressive) speed-up (while they substantially complicate the rewriting system: by some of these rules, a control operator may now *enter* inside the path of a virtual redex).

On the contrary, some polynomial improvement could be reasonably expected by using different representations (such as the bus notation [GAL92a]), or by collapsing long sequences of (safe) control operators into a single one.

In conclusion, the big problems are still open: is it possible to have a reduction technique that works linearly (polynomially) in the number of family reductions? Otherwise, what is the real intrinsic complexity of λ-term reduction?

References

[As94] A. Asperti. *Linear Logic, Comonads, and Optimal Reductions.* To appear in Fundamenta Informaticae, Special Issue devoted to Categories in Computer Science. 1994.

[AL93a] A. Asperti, C. Laneve. *Optimal Reductions in Interaction Systems.* Proc. of the 4th Joint Conference on the Theory and Practice of Software Development, TAPSOFT'93, Orsay (France). April 1993.

[AL93b] A. Asperti, C. Laneve. *Paths, Computations and Labels in the λ-calculus.* To appear in Theoretical Computer Science, Special issue devoted to RTA'93, Montreal. June 1993.

[AL94] A. Asperti, C. Laneve. *Interaction Systems I: the theory of optimal reductions.* To appear in Mathematical Structures in Computer Science. 1994.

[AL95] A. Asperti, C. Laneve. *Relating λ-calculus translations in sharing graphs.* Proc. of the Second International Conference on Typed Lambda Calculi and Applications, TLCA'95, Edinburgh, Scotland. 1995.

[ADLR94] A. Asperti, V. Danos, C. Laneve, L. Regnier . *Paths in the λ-calculus: three years of communications without understandings.* Proceedings of LICS'94. Paris. 1994.

[DR93] V. Danos, L. Regnier. *Local and asynchronous beta-reduction.* Proc. of the 8th Annual Symposium on Logic in Computer Science (LICS 93), Montreal. 1993.

[Fie90] J. Field. *On laziness and optimality in lambda interpreters: tools for specification and analysis.* Proc. of the 17th ACM Symposium on Principles of Programmining Languages (POPL 90). 1990.

[Gi86] J. Y. Girard. *Linear Logic.* Theoretical Computer Science, 50. 1986.

[Gi88a] J. Y. Girard. *Geometry of Interaction I: Interpretation of system F.* In Ferro, Bonotto, Valentini and Zanardo eds., Logic Colloquium '88, North Holland. 1988.

[Gi88b] J. Y. Girard. *Geometry of Interaction II: Deadlock-free algorithms.* Proc. of the International Conference on Computer Logic, (COLOG 88), P.Martin Löf and G.Mints eds., Springer Verlag. 1988.

[GAL92a] G. Gonthier, M. Abadi, J.J. Lévy. *The geometry of optimal lambda reduction.* Proc. of the 19th Symposium on Principles of Programming Languages (POPL 92). 1992.

[GAL92b] G. Gonthier, M. Abadi, J.J. Lévy. *Linear Logic without boxes.* Proc. of the 7th Annual Symposium on Logic in Computer Science (LICS'92). 1992.

[Laf90] Y. Lafont. *Interaction Nets.* Proc. of the 17th Symposium on Principles of Programming Languages (POPL 90). San Francisco. 1990.

[Lam89] J. Lamping. *An algorithm for optimal lambda calculus reductions.* Xerox PARC Internal Report. 1989.

[Lam90] J. Lamping. *An algorithm for optimal lambda calculus reductions.* Proc. of the 17th Symposium on Principles of Programming Languages (POPL 90), San Francisco. 1990.

[Le78] J.J.Levy. *Réductions correctes et optimales dans le lambda-calcul.* Thèse de doctorat d'état, Université de Paris VII. 1978.

[Le80] J. J. Lévy. *Optimal reductions in the lambda-calculus.* In J.P. Seldin and J.R. Hindley, editors, *To H.B. Curry, Essays on Combinatory Logic, Lambda Calculus and Formalism*, pages 159 – 191. Academic Press. 1980.

[LM92] X. Leroy, M.Mauny. *The Caml Light system, release 0.5. Documentation and user's manual.* INRIA Technicl Report. September 1992.

[Re92] L. Regnier. *Lambda Calcul et Réseaux.* Thèse de doctorat, Université Paris VII. 1992.

Substitution Tree Indexing

Peter Graf*

Max-Planck-Institut für Informatik
Im Stadtwald
66123 Saarbrücken, Germany
email: graf@mpi-sb.mpg.de

Abstract. Sophisticated maintenance and retrieval of first-order predicate calculus terms is a major key to efficient automated reasoning. We present a new indexing technique, which accelerates the speed of the basic retrieval operations such as finding complementary literals in resolution theorem proving or finding critical pairs during completion. Subsumption and reduction are also supported. Moreover, the new technique not only provides maintenance and efficient retrieval of terms but also of idempotent substitutions. Substitution trees achieve maximal search speed paired with minimal memory requirements in various experiments and outperform traditional techniques such as path indexing, discrimination tree indexing, and abstraction trees by combining their advantages and adding some new features.

1 Introduction

Most automated reasoning systems accumulate enormous amounts of data such as terms, substitutions, clauses, formulae [7], or rewrite rules. These data must be stored and accessed in various ways. As in standard database technology, *indexing* is the key to *efficiently* retrieving data from large databases. However, the structure of logical data and the queries to a logical database [2] are much more complicated than in relational databases.

Typical queries that arise in the context of rewriting [6] or resolution theorem proving [11] are: Given a database D containing rewrite rules/literals and a query term t, find all left hand sides of rewrite rules/literals in D that are unifiable with, instances of, or more general than t. Most deduction systems currently in use or under development employ sophisticated term indexing techniques to achieve efficient retrieval. Hitherto, the most often applied strategies have been path indexing [12, 9, 4], discrimination tree indexing [9, 3, 1], and abstraction tree indexing [10].

In this paper *substitution tree indexing* is presented as a new indexing technique, which combines the advantages of discrimination and abstraction tree indexing. Memory requirement and retrieval times being the main criteria for judging an indexing technique, this paper will show that substitution tree indexing is superior to the known strategies in these points in the average case.

* This work was supported by the German Science Foundation (DFG).

A substitution tree can represent any set of idempotent substitutions. In the simplest case all these substitutions have identical domains and consist of a single assignment, which implies that the substitution tree can be used as a term index as well. Figure 1 shows an index for the three substitutions $\{u \mapsto f(a,b)\}$, $\{u \mapsto f(y,b)\}$, and $\{u \mapsto f(b,z)\}$ representing a term index for the terms $f(a,b)$, $f(y,b)$, and $f(b,z)$. As the name indicates, the labels of substitution tree nodes are substitutions. Each branch in the tree represents a binding chain for variables. Consequently, the substitutions of a branch from the root down to a particular node can be composed and yield an instance of the root node's substitution. Consider the substitution $\tau = \{u \mapsto f(a,b)\}$, which is represented by the chain of substitutions $\tau_0 = \{u \mapsto f(x_1,x_2)\}$, $\tau_1 = \{x_2 \mapsto b\}$, and $\tau_2 = \{x_1 \mapsto a\}$. The original substitution τ can be reconstructed by simply applying the substitution $\tau_0\tau_1\tau_2$ to u. The result of this application is $\tau = \{u \mapsto u(\tau_0\tau_1\tau_2)\} = \{u \mapsto f(x_1,x_2)(\tau_1\tau_2)\} = \{u \mapsto f(x_1,b)\tau_2\} = \{u \mapsto f(a,b)\}$.

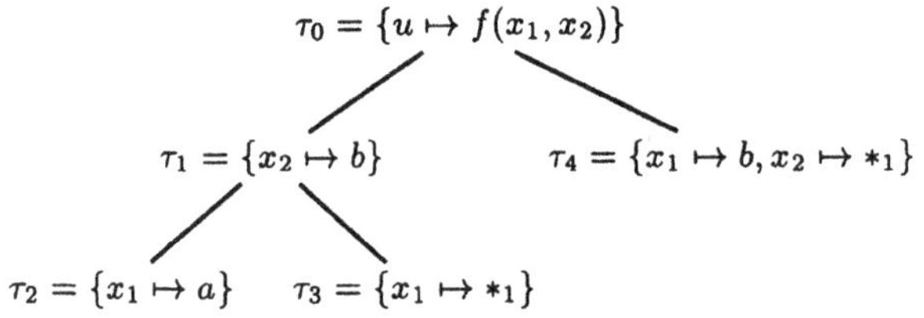

Fig. 1. Substitution tree

Substitutions are *normalized* before being inserted into the index. Normalization renames all variables in the codomain to so-called *indicator variables*, which are denoted by $*_i$. The substitutions represented by the index in Fig. 1 therefore are $\{u \mapsto f(a,b)\}$, $\{u \mapsto f(*_1,b)\}$, and $\{u \mapsto f(b,*_1)\}$. The renaming is done for two main reasons: There is more structure sharing in the index if the substitutions are normalized and, when searching for instances in the index, indicator variables must not be instantiated.

Retrieval in substitution trees is based on a backtracking algorithm in addition to an ordinary representation of substitutions as lists of variable-term pairs. The retrieval also needs a backtrackable variable binding mechanism, similar to the one used in PROLOG.

As an example, we describe the search for substitutions *compatible* with $\{u \mapsto f(a,x)\}$, i.e. substitutions τ are sought such that $u\tau$ is unifiable with $f(a,x)$. We begin by binding the variable u to the term $f(a,x)$ and start the retrieval: The substitution tree is traversed by testing at each node marked with the substitution $\tau = \{x_1 \mapsto t_1, \ldots, x_n \mapsto t_n\}$ whether under the current bindings all x_i are unifiable with their appropriate t_i. At the root node in our example we unify the terms $f(a,x)$ and $f(x_1,x_2)$, which yields the two bindings $x_1 \mapsto a$ and $x \mapsto x_2$. Then the son marked with τ_1 is considered. The variable x_2 is unified with b, because x_2 has not been bound yet. The resulting binding is $x_2 \mapsto b$

and the leaf node τ_2 is the next node to be investigated. As x_1 is bound to a, the unification problem is trivial and therefore the substitution represented by this leaf node is compatible with $\{u \mapsto f(a,x)\}$. After backtracking node τ_3 is found to represent another solution, because the variable $*_1$ is unifiable with a. Backtracking deletes the bindings of $*_1$ and x_2 and then proceeds with node τ_4. Obviously, retrieval can be stopped at this point, because the binding of x_1 is not unifiable with b.

Substitution trees provide high search speed paired with low memory requirements. Additionally, they work not only for term sets but also for sets of substitutions. There is no need for the substitutions to have identical domains. The structure of the index is very simple; only a tree of substitutions has to be maintained.

Some technical definitions as well as proofs and a detailed discussion of further operations like the merge of trees, which have to be omitted due to lack of space, can be found in [5].

2 Preliminaries

The standard notions for first order logic are used. $\mathbf{F}_n$ is the set of n-ary *function symbols*, $\mathbf{V}$ denotes the set of *variable* symbols and $\mathbf{V}^* \subset \mathbf{V}$ is the set of *indicator variables*. The variables that occur in a term or a set of terms are denoted by $\mathsf{VAR}(t)$. In our examples the symbols $u, v, w, x, y, z \in \mathbf{V}$ and $*_i \in \mathbf{V}^*$ are used for variables. The symbols f, g, h denote function symbols and a, b, c denote constants, i.e. $a, b, c \in \mathbf{F}_0$.

The set $\mathsf{DOM}(\sigma) := \{x \in \mathbf{V} \mid x\sigma \neq x\}$ is called *domain* of the substitution σ, the set $\mathsf{COD}(\sigma) := \{x\sigma \mid x \in \mathsf{DOM}(\sigma)\}$ the *codomain* of σ, and $\mathsf{IM}(\sigma) := \mathsf{VAR}(\mathsf{COD}(\sigma))$ is the set of variables *introduced* by σ. The *composition* $\sigma\tau$ of substitutions $\sigma = \{x_1 \mapsto s_1, \ldots, x_n \mapsto s_n\}$ and $\tau = \{y_1 \mapsto t_1, \ldots, y_m \mapsto t_m\}$ is defined as $x(\sigma\tau) := (x\sigma)\tau$ for all x. The *join* of the substitutions σ and τ is defined as $\sigma \bullet \tau := \{x_1 \mapsto s_1\tau, \ldots, x_n \mapsto s_n\tau\} \cup \{y_i \mapsto t_i \mid y_i \in \mathsf{DOM}(\tau)\backslash\mathsf{IM}(\sigma)\}$. For $\sigma = \{z \mapsto g(x)\}$ and $\tau = \{x \mapsto a, y \mapsto c\}$ we have $\sigma\tau = \{z \mapsto g(a), x \mapsto a, y \mapsto c\}$ and $\sigma \bullet \tau = \{z \mapsto g(a), y \mapsto c\}$.

A *position* in a term is a finite sequence of natural numbers. We write sequences in square brackets and denote the concatenation of sequences by @. The subterm of a term t at position p is denoted by t/p and $t/[] = t$, because the empty sequence represents the top position of a term. Note that $[]@p = p@[] = p$. The set of all positions of a term is recursively defined by $\mathsf{O}(f(t_1, \ldots, t_n)) := \{[]\} \cup \bigcup_{t_i} \{[i]@p \mid p \in \mathsf{O}(t_i)\}$. Constants and variables have the position $\{[]\}$ only. For example, $\mathsf{O}(h(a, g(b), x)) = \{[], [1], [2], [2,1], [3]\}$. The definition of the normalization of terms is based on the lexicographical extension $<^*$ of the natural ordering $<$ on natural numbers. For example, $[1,1] <^* [1,2]$, $[1,1] <^* [2]$, and $[1,2] <^* [1,2,1]$.

Definition 1 (Normalization of Terms and Substitutions). Let t be a term and $F = \{p \mid p \in \mathsf{O}(t),\ t/p \in \mathbf{V},\ \forall q \in \mathsf{O}(t), q <^* p.\ t/q \neq t/p\}$ the set of first occurrences of variables in t. If $p_1, \ldots, p_m \in F$ and $m = |F|$ and $p_i <^* p_j$ for $1 \leq i < j \leq m$ then the substitution $\sigma = \{t/p_1 \mapsto *_1, \ldots, t/p_m \mapsto *_m\}$ is called *normalization* and $\bar{t} := t\sigma$ is a *normalized term.*

Moreover, let $\sigma = \{x_1 \mapsto t_1, \ldots, x_n \mapsto t_n\}$ be a substitution and $f_n \in F_n$ an n-ary function symbol. Additionally, let $<$ be a fixed total ordering on variables and $x_1 < \ldots < x_n$. We call $\overline{\sigma} := \{x_1 \mapsto \overline{f_n(t_1, \ldots, t_n)}/1, \ldots, x_n \mapsto \overline{f_n(t_1, \ldots, t_n)}/n\}$ a *normalized substitution.*

The condition $m = |F|$ in Def. 1 ensures that *all* variables in the term are renamed. For example, consider $\overline{h(x,x,y)} = \overline{h(z,z,x)} = h(*_1, *_1, *_2)$. For $\sigma = \{x \mapsto f(u,v), y \mapsto f(a,v)\}$ and $x < y$ we have $\overline{\sigma} = \{x \mapsto f(*_1, *_2), y \mapsto f(a, *_2)\}$. Note that $\overline{\sigma} = \{x \mapsto f(*_2, *_1), y \mapsto f(a, *_1)\}$ if $y < x$.

Most definitions in this paper are based on sets of conditional equations. We read the equations top down. Each condition that holds in a defining equation is assumed to occur negated in all equations below.

3 Indexing Techniques Related to Substitution Trees

Discrimination Tree Indexing. A discrimination tree (DT) is a single tree representing the structure of all indexed terms. Pointers to these terms are stored at the leaf nodes of the tree. Each path from the root to a leaf of the discrimination tree corresponds to a set of terms such that all terms in the set are equal modulo variable renaming. All these terms are represented by the unique normalized form of the terms. In Fig. 2 a discrimination tree is depicted. To answer a query one has to traverse the tree using a backtracking algorithm.

Discrimination trees are deterministic, i.e. there is only one position at which to insert a new entry. Therefore, insertion of entries is very fast. The memory requirement depends on the sharing of common prefixes of the indexed terms. In the example there are 3 terms which end in $g(*_1)$ and the whole tree consists of 14 nodes.

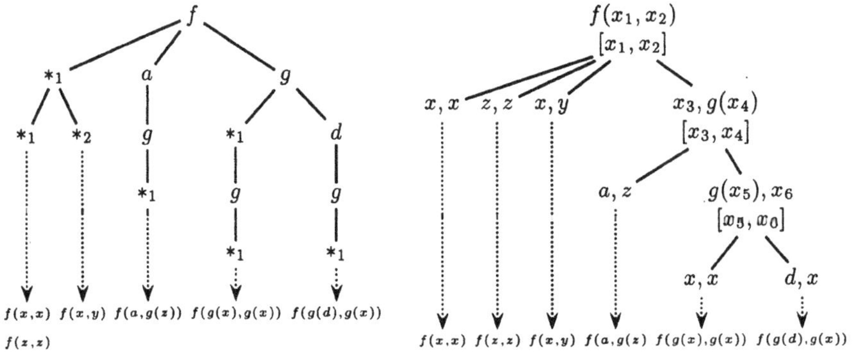

Fig. 2. Discrimination and abstraction tree

Abstraction Tree Indexing. Abstraction tree indexing exploits the lattice structure of terms [10]. An abstraction tree (AT) is based on the usual instance relation on terms which forms a partial ordering.

The nodes of an abstraction tree are labeled with termlists such that the free variables of the termlists at node N and the termlists of N's subnodes form the domain and codomain of a set of matchers. If all matchers from the root to a leaf of the abstraction tree are applied to the termlist in the root of the tree then the resulting term is the one which is represented by this path. Abstraction trees are not deterministic. Their structure depends on the order in which terms are inserted and on the insertion heuristic, which has been used.

Due to a better configuration of the terms the abstraction tree in our example consists of 9 nodes only. However, abstraction trees contain lots of variable renamings like $\{x_1 \mapsto x_3\}$ and $\{x_4 \mapsto x_6\}$ that are not necessary. Additionally, variables of indexed terms may occur in leaf nodes of abstraction trees only. This implies that an algorithm looking for instances of a query term cannot exploit the fact that a variable in an indexed term must not be instantiated at inner nodes of the tree. The abstraction tree contains 16 assignments.

4 Substitution Tree Indexing

Substitution trees (ST) were developed to increase the performance of indexing. The main difference compared to abstraction trees lies in the representation of variables of indexed substitutions. Additionally, variable renamings are avoided. To this end the structure of the nodes is simplified: Substitutions are stored in contrast to termlists and lists of variables. Variables of indexed terms are represented by indicator variables and may occur at arbitrary positions in the substitution tree. Figure 3 shows our standard term set. We only need 3 auxiliary variables and the whole tree contains just 9 assignments in contrast to the 16 assignments of the abstraction tree in Fig. 2. However, the main advantages of abstraction trees are preserved. Essentially, substitution trees unify the advantages of abstraction and discrimination trees and result in a very promising indexing method.

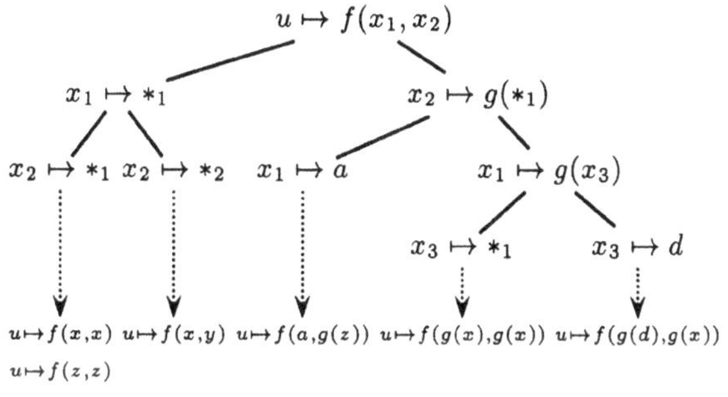

Fig. 3. Substitution tree

In our example we stored 6 substitutions in the substitution tree. The domains of all these substitutions are identical, but this is not necessary. Substitution trees may also contain substitutions with different domains. We use a backtracking algorithm to find substitutions in the tree with specific properties. All retrieval algorithms are based on backtrackable variable bindings and algorithms for unification and matching that take variable bindings into account. Insertion of a substitution into the index is a complex operation. The deletion of entries is much easier and even complex deletion operations, like the deletion of all compatible substitutions in a substitution tree, can easily be accomplished.

Definition 2 (Substitution Tree). A *substitution tree* is a tuple (τ, Σ) where τ is a substitution and Σ is an ordered set of substitution trees. The following conditions hold:

- A substitution tree is either a *leaf* $(\tau, \emptyset)$ or $|\Sigma| \geq 2$.
- For every path (τ_1, Σ_1), ..., $(\tau_n, \emptyset)$ starting from the root of a substitution tree we have $\mathsf{IM}(\tau_1 \bullet \ldots \bullet \tau_n) \subset \mathbf{V}^*$.
- For every path (τ_1, Σ_1), ..., (τ_i, Σ_i) from the root to any node of a tree we have $\mathsf{DOM}(\tau_i) \cap \bigcup_{1 \leq j < i} \mathsf{DOM}(\tau_j) = \emptyset$.

If (τ_1, Σ_1), ..., (τ_n, Σ_n) is a path from the root of the tree to node (τ_n, Σ_n) and $x \in (\bigcup_{1 \leq i \leq n} \mathsf{IM}(\tau_i)) \setminus (\bigcup_{1 \leq i \leq n} \mathsf{DOM}(\tau_i))$ then the variable x is called *open* at node (τ_n, Σ_n). Variables that are not open at a node N are called *closed* at N. The *empty* tree is denoted by $\emptyset$.

The second condition in Def. 2 implies that all non-indicator variables are closed at leaf nodes of substitution trees.

4.1 Retrieval

The retrieval algorithm checks the root of a substitution tree for some special conditions and proceeds with the subtrees if the conditions are fulfilled. If subtrees of the current tree do not exist, the entry of the index represented by this leaf is retrieved. Four different retrieval tasks are supported: Find more general substitutions, compatible substitutions, instances, and variant substitutions.

Definition 3 (Test Functions for Retrieval). Let τ and ρ be substitutions. The *test functions* are defined as

$$\mathcal{G}(\tau, \rho) := \{\sigma \mid \forall x_i \in \mathsf{DOM}(\tau).\ x_i\tau\rho\sigma = x_i\rho\} \quad (1)$$

$$\mathcal{U}(\tau, \rho) := \{\sigma \mid \forall x_i \in \mathsf{DOM}(\tau).\ x_i\tau\rho\sigma = x_i\rho\sigma \ \wedge\ \sigma \text{ is most general}\} \quad (2)$$

$$\mathcal{I}(\tau, \rho) := \{\sigma \mid \sigma \in \mathcal{U}(\tau, \rho) \ \wedge\ \mathsf{DOM}(\sigma) \cap \mathbf{V}^* = \emptyset\} \quad (3)$$

$$\mathcal{V}(\tau, \rho) := \{\sigma \mid \sigma \in \mathcal{G}(\tau, \rho) \ \wedge\ \mathsf{DOM}(\sigma) \cap \mathbf{V}^* = \emptyset\} \quad (4)$$

The functions $\mathcal{G}$, $\mathcal{I}$, $\mathcal{U}$, and $\mathcal{V}$ compute matchers and unifiers at nodes of a substitution tree and are used as parameters for the search.

Definition 4 (Retrieval Function search). Let (τ, Σ) be a substitution tree, ρ a substitution, and $\mathcal{X}$ one of the test functions $\mathcal{G}$, $\mathcal{U}$, $\mathcal{I}$, or $\mathcal{V}$. The retrieval function `search` computes a set of leaves and is defined as

$$\texttt{search}(\emptyset, \rho, \mathcal{X}) := \emptyset \tag{5}$$

$$\texttt{search}((\tau, \emptyset), \rho, \mathcal{X}) := \{(\tau, \emptyset)\} \quad \text{if } \exists\sigma \in \mathcal{X}(\tau, \rho) \tag{6}$$

$$\texttt{search}((\tau, \Sigma), \rho, \mathcal{X}) := \bigcup_{N' \in \Sigma} \texttt{search}(N', \rho\sigma, \mathcal{X}) \quad \text{if } \exists\sigma \in \mathcal{X}(\tau, \rho) \tag{7}$$

$$\texttt{search}((\tau, \Sigma), \rho, \mathcal{X}) := \emptyset \quad \text{if } \mathcal{X}(\tau, \rho) = \emptyset \tag{8}$$

Note that for the search of variants the query substitution has to be normalized.

Implementing Standard Retrieval. A stack of variable bindings is maintained. The function `unify`(*N,STACK,BINDINGS*), for example, implements the test function $\mathcal{U}$ by checking for each assignment $x_i \mapsto t_i$ of N's substitution $\tau = \{\dots, x_i \mapsto t_i, \dots\}$ whether x_i is unifiable with t_i. The bindings of variables in the unifier are pushed on the *STACK* and counted in *BINDINGS*. Obviously, this unification has to consider variable bindings in the terms to be unified. Additionally, the function `backtrack`(*STACK, BINDINGS*) resets the *STACK* by popping *BINDINGS* bindings from it. In order to extract the unifiers one can evaluate the *STACK* after a successful unification at a leaf node and store the unifier in – a substitution tree!

A retrieval algorithm based on these functions and a sequence of stacks resulting from the search for substitutions stored in the substitution tree in Fig. 3 and compatible with $\{u \mapsto f(a, y)\}$ is depicted in Fig. 4. Originally, the stack is empty. Before we start the retrieval algorithm, all variables in the domain of the query substitution are bound to their corresponding codomain, i.e. the bindings are pushed on the stack (compare stack "Init"). The recursive retrieval algorithm is started on the root node. In case it succeeds in testing the current substitution, the modified stack is marked with "Success". If the corresponding node in the tree additionally is a leaf node, "**Success**" is written boldface. In the example three leaves are tested successfully. They correspond to the substitutions $\{u \mapsto f(x, x)\}$, $\{u \mapsto f(y, y)\}$,$\{u \mapsto f(x, y)\}$, and $\{u \mapsto f(a, g(z))\}$.

4.2 Insertion

The insertion of new entries into a substitution tree is difficult. We introduce the central notion of the *most specific common generalization*, which is needed if the insertion of new entries modifies inner nodes of the tree.

Definition 5 (Most Specific Common Generalization). Let τ_1 and τ_2 be two substitutions. If there exist substitutions μ, σ_1, and σ_2 such that $\mu\sigma_1 = \tau_1$ and $\mu\sigma_2 = \tau_2$ and there are no substitutions λ, σ_1', and σ_2' such that $\mu\lambda\sigma_1' = \tau_1$ and $\mu\lambda\sigma_2' = \tau_2$ then $\mathsf{mscg}(\tau_1, \tau_2) := (\mu, \sigma_1, \sigma_2)$. The substitution μ is called the *most specific common generalization* (mscg) for τ_1 and τ_2. The substitutions σ_1 and σ_2 are called *specializations*.

```
function search;
input N, STK, X;
output HITS;
begin
  HITS = ∅;
  if X(N, STK, BINDINGS) then
    if N is a leaf node then
      HITS = HITS ∪{N};
    else forall subnodes N' of N do
      HITS =
          HITS ∪ search(N', STK, X);
  backtrack(STK, BINDINGS);
  return HITS;
end;
```

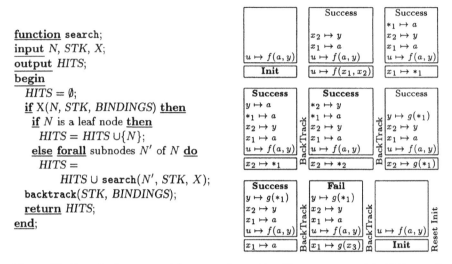

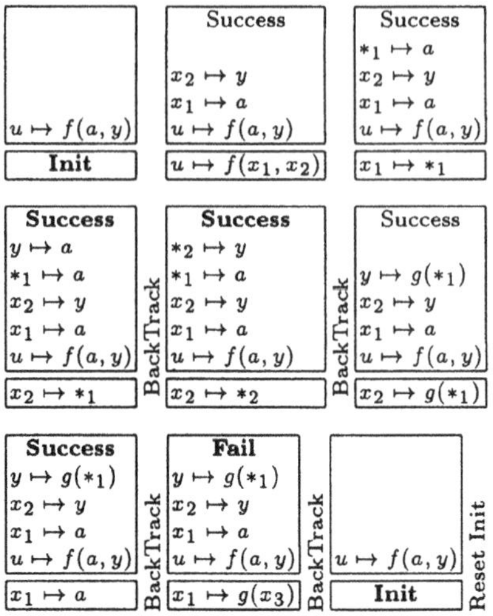

Fig. 4. Bindings during retrieval for query $\{u \mapsto f(a,y)\}$ using search(N, STK,unify)

Suppose $\tau = \{x \mapsto g(b), y \mapsto a\}$ and $\rho = \{x \mapsto g(a), y \mapsto b\}$. Then $\mathsf{mscg}(\tau,\rho) = (\{x \mapsto g(x_1)\}, \{x_1 \mapsto b, y \mapsto a\}, \{x_1 \mapsto a, y \mapsto b\})$. The auxiliary variable x_1 represents the different parts of τ and ρ in the mscg.

Generally speaking, the insertion process is very similar to finding variant entries in a substitution tree. The only difference is, that only a single path in the tree is considered. A heuristic select that has to cope with three different situations is used for descending into a substitution tree: First, the heuristic has to select a variant subnode of the current node for descending if such a variant exists. Second, the heuristic selects a non-variant subnode which will yield a non-empty mscg if a variant could not be found. Third, if neither a variant nor a subnode that yields a non-empty mscg could be found the heuristic has to select the empty tree for insertion. In this case our insertion function will create a new leaf node.

Definition 6 (Insertion Function insert). The function for inserting a substitution ρ into a substitution tree N computes a modified tree.

$$\mathtt{insert}(N,\rho) := \mathtt{ins}(N,\overline{\rho},\mathsf{DOM}(\rho)) \tag{9}$$

$$\mathtt{ins}(\emptyset,\rho,OV) := (\rho|_{OV},\emptyset) \tag{10}$$

$$\mathtt{ins}((\tau,\emptyset),\rho,OV) := (\tau,\emptyset) \quad \text{if } \exists\sigma \in \mathcal{V}(\tau,\rho) \tag{11}$$

$$\mathtt{ins}((\tau,\Sigma \uplus \{N\}),\rho,OV) := (\tau,\Sigma \cup \mathtt{ins}(N,\rho\sigma,\mathsf{IM}(\tau) \cup OV \backslash \mathsf{DOM}(\tau))) \quad \text{if } \exists\sigma \in \mathcal{V}(\tau,\rho) \wedge N = \mathsf{select}((\tau,\Sigma \uplus \{N\})) \tag{12}$$

$$\mathtt{ins}((\tau,\Sigma),\rho,OV) := (\tau,\Sigma \cup (\rho\sigma|_{\mathrm{IM}(\tau)\cup\mathrm{OV}\backslash\mathrm{DOM}(\tau)},\emptyset)) \quad \text{if } \exists\sigma \in \mathcal{V}(\tau,\rho) \wedge \emptyset = \mathsf{select}((\tau,\Sigma)) \tag{13}$$

$$\mathtt{ins}((\tau,\Sigma),\rho,OV) := (\mu,\{(\sigma_1,\Sigma),(\sigma_2 \cup \rho|_{OV\backslash\mathrm{DOM}(\tau)},\emptyset)\}) \quad \text{if } \neg\exists\sigma \in \mathcal{V}(\tau,\rho) \wedge \mathsf{mscg}(\tau,\rho) = (\mu,\sigma_1,\sigma_2) \tag{14}$$

A substitution ρ is inserted into a tree N by insert(N, ρ), which causes the normalization of ρ. In OV the set of open variables is maintained. A new leaf node is created in case the substitution is inserted into an empty tree (10). An existing leaf node is returned if ρ corresponds to a variant substitution which has been inserted already (11). At inner nodes we have two possibilities. First, if a variant node is found, a subnode of the current node is selected for continued insertion by the heuristic (12). Depending on the chosen subtree, different insertions can be performed (13). Second, a new inner node and a new leaf are created in case the substitution in the tree is not a variant of ρ. In this case the set of open variables OV is needed in (14) for completely describing the inserted substitution at the new leaf node.

As the index is used as a means for accessing data, it should be possible to store additional information at the leaf nodes of the tree. Therefore, in a real implementation it is reasonable that insert also returns the found or created leaf node, so that the user can perform some additional operations.

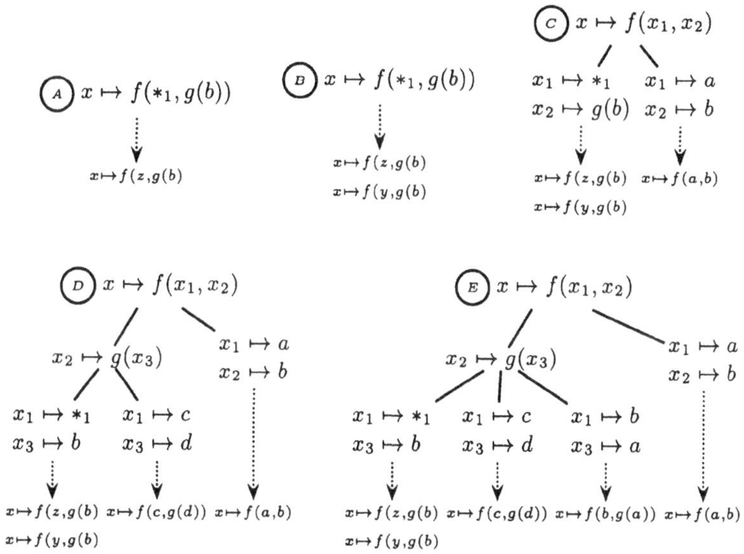

Fig. 5. Insertion sequence for substitutions with identical domains

Consider the example insertion sequence in Fig. 5: Insertion of the substitution $\{x \mapsto f(z, g(b))\}$ into an empty tree yields tree A marked with $\{x \mapsto f(*_1, g(b))\}$. Insertion of $\{x \mapsto f(y, g(b))\}$ into tree A yields tree B, which has an identical tree structure. Tree B is changed to tree C by inserting $\{x \mapsto f(a, b)\}$. As $f(*_1, g(b))$ is not a variant of $f(a, b)$, a new root and a new leaf node are added to tree B. Insertion of $\{x \mapsto f(c, g(d))\}$ to tree C employs the heuristic mentioned in (12) which selects the node marked with $\{x_1 \mapsto *_1, x_2 \mapsto g(b)\}$. In the following step a new leaf node and a new inner node marked with $\{x_2 \mapsto g(x_3)\}$

are created. Eventually, $\{x \mapsto f(b, g(a))\}$ is inserted into tree D, which yields tree E. Here the heuristic cannot find an appropriate subnode and therefore an additional leaf is added in (13).

So far, all nodes in our substitution trees have been marked with non-empty substitutions. Insertion of the substitutions $\{x \mapsto a\}$ and $\{y \mapsto b\}$ into an empty tree, however, will yield a root marked with the empty substitution. Additionally, if a part of a substitution that has already been inserted is to be added, the new leaf node will also be empty. Inner nodes other than the root are never marked with empty substitutions. During the retrieval, tests of $\mathcal{G}$, $\mathcal{U}$, $\mathcal{I}$, or $\mathcal{V}$ on empty substitutions are always successful with no bindings being established.

In the following we show how the mscg is derived in detail. The mscg for terms is considered first.

Definition 7 (Most Specific Common Generalizations for Terms mscg). The function mscg for terms is defined. The meta variables X, Y, and Z are supposed to match sequences of terms.

$$\begin{aligned}
\mathtt{mscg}(t,t,\sigma_1,\sigma_2) &:= (t,\sigma_1,\sigma_2) && (15)\\
\mathtt{mscg}(x_i,t,\sigma_1,\sigma_2) &:= (x_i,\sigma_1,\sigma_2 \cup \{x_i \mapsto t\}) \text{ if } x_i \neq t && (16)\\
\mathtt{mscg}(f(X),f(Y),\sigma_1,\sigma_2) &:= (f(Z),\sigma_1',\sigma_2') && (17)\\
&\quad \text{if } X \neq Y \wedge \langle X,Y,\sigma_1,\sigma_2,[]\rangle \downarrow_{\mathtt{Arg}} = \langle [],[],\sigma_1',\sigma_2',Z\rangle\\
\mathtt{mscg}(f(X),g(Y),\sigma_1,\sigma_2) &:= (y,\sigma_1,\sigma_2) \text{ if } y\sigma_1 = f(X) \wedge y\sigma_2 = g(Y) && (18)\\
\mathtt{mscg}(f(X),g(Y),\sigma_1,\sigma_2) &:= (x_j,\sigma_1 \cup \{x_j \mapsto f(X)\},\sigma_2 \cup \{x_j \mapsto g(Y)\}) && (19)\\
&\quad \text{if } \neg\exists y.\ y\sigma_1 = f(X) \wedge y\sigma_2 = g(Y)
\end{aligned}$$

$$\mathtt{Arg}: \frac{\langle [t_x]@X,[t_y]@Y,\sigma_1,\sigma_2,Z\rangle}{\langle X,Y,\sigma_1',\sigma_2',Z@[t_z]\rangle} \text{ if } \mathtt{mscg}(t_x,t_y,\sigma_1,\sigma_2) = (t_z,\sigma_1',\sigma_2')$$

The mscg of two identical terms is the term itself (15). In case the first argument[1] is a non-indicator variable, this variable can be used in the specialization for term t (16). The transition Arg is used to process the arguments of identical top symbols in (17) by computing the normal form with respect to Arg. The reuse of assignments found in the specializations is described in (18). Therefore non-linear mscg's can be created. If (18) is omitted only linear generalizations will be produced. The effect on the substitution trees is minimal, because most generalizations are changed by further insertions anyway. Finally, new non-indicator variables are introduced in case none of the other equations could be applied (19). For example, $\mathtt{mscg}(h(a,*_2,x_1,*_1,*_1), h(a,*_2,c,b,b), \emptyset, \emptyset) = (h(a,*_2,x_1,x_2,x_2), \{x_2 \mapsto *_1\}, \{x_1 \mapsto c, x_2 \mapsto b\})$.

Definition 8 (Transition System MSCG). The mscg of two substitutions τ and ρ can be computed by the set MSCG of transition rules. The symbol $\uplus$ denotes the disjoint set union.

[1] Later the first argument will correspond to a domain variable in a substitution τ stored in a substitution tree.

$$\texttt{Bind:}\ \frac{\langle \tau \uplus \{x_i \mapsto t\}, \rho, \mu, \sigma_1, \sigma_2\rangle}{\langle \tau, \rho, \mu, \sigma_1 \cup \{x_i \mapsto t\}, \sigma_2\rangle}\ \text{if}\ x_i \notin \mathsf{DOM}(\rho)$$

$$\texttt{Freeze:}\ \frac{\langle \tau \uplus \{x_i \mapsto t\}, \rho, \mu, \sigma_1, \sigma_2\rangle}{\langle \tau, \rho, \mu \cup \{x_i \mapsto t\}, \sigma_1, \sigma_2\rangle}\ \text{if}\ x_i\rho = t$$

$$\texttt{Divide:}\ \frac{\langle \tau \uplus \{x_i \mapsto f(X)\}, \rho, \mu, \sigma_1, \sigma_2\rangle}{\langle \tau, \rho, \mu, \sigma_1 \cup \{x_i \mapsto f(X)\}, \sigma_2 \cup \{x_i \mapsto x_i\rho\}\rangle}\ \text{if}\ x_i\rho = g(Y)$$

$$\texttt{Mix:}\ \frac{\langle \tau \uplus \{x_i \mapsto f(X)\}, \rho, \mu, \sigma_1, \sigma_2\rangle}{\langle \tau, \rho, \mu \cup \{x_i \mapsto f(Z)\}, \sigma_1', \sigma_2'\rangle}\ \text{if}\ \begin{array}{l} x_i\rho = f(Y)\ \wedge\ X \neq Y\ \wedge \\ \langle X, Y, \sigma_1, \sigma_2, []\rangle \downarrow_{\texttt{Arg}}= \\ \langle [], \sigma_1', \sigma_2', Z\rangle \end{array}$$

In particular, we have

$$\mathsf{mscg}(\tau, \rho) = (\mu, \sigma_1, \sigma_2)\ \text{iff}\ \langle \tau, \rho, \emptyset, \emptyset, \emptyset\rangle \downarrow_{\mathsf{MSCG}} = \langle \emptyset, \rho, \mu, \sigma_1, \sigma_2\rangle$$

The transition system MSCG considers every assignment in τ. We have seen that during insertion variable bindings are established by the function $\mathcal{V}$. These current bindings are stored in ρ. The transition Bind handles assignments in τ that map variables which do not occur in the substitution that is being inserted. Freeze detects assignments where the binding of x_i in ρ is identical to the binding to represent the substitution to be inserted. In case the terms under consideration do not even have the same top symbol, Divide completely splits the information into the specializations. Finally, Mix initiates the computation of mscg's on the term level.

For example, for $\tau = \{y \mapsto a, x_1 \mapsto a, x_2 \mapsto b, x_3 \mapsto g(c)\}$ and $\rho = \{x_1 \mapsto a, x_2 \mapsto c, x_3 \mapsto g(a)\}$ we have $\langle \tau, \rho, \emptyset, \emptyset, \emptyset\rangle \downarrow_{\mathsf{MSCG}} = \langle \emptyset, \rho, \{x_1 \mapsto a, x_3 \mapsto g(x_4)\}, \{y \mapsto a, x_2 \mapsto b, x_4 \mapsto c\}, \{x_2 \mapsto c, x_4 \mapsto a\}\rangle$.

In equation (19) we have to create new auxiliary variables x_j. Such a non-indicator variable does not really have to be *new*. For example, in tree D in Fig. 5 we introduced x_3. Obviously, the variable x_3 could be used again if we had to create another mscg in the right subtree of the root. Generally speaking, let $N_0 = (\tau_0, \Sigma_0)$, ..., $N_i = (\tau_i, \Sigma_i)$ be a path in a substitution tree and N_i a node which has to be extended. Then the set of non-indicator variables that can be reused is the set of all non-indicator variables in the tree minus the domain variables on the path from N_0 to N_i minus the domain variables that occur in the subtree N_i. In various tests variable reuse was able to reduce the number of different non-indicator variables by a factor of 40 in the average case.

In our implementation we used a very simple *first-fit* insertion heuristic: We choose the first variant son for descending. If such a son does not exist the first non-variant son that produces a non-empty mscg is selected. Using a more complex insertion heuristic implies the traversal of the whole tree rating all possible insertion positions. Finally, the substitution is inserted at the best position. Several heuristics that minimize the size of the index, the number of auxiliary variables x_i in the index, the number of sons for inner nodes of the tree, or the depth of indicator variables $*_i$ in the tree have been tried. Experiments showed that insertion was *very* slow. However, size and retrieval times did not change significantly. Due to this experience, using a complex insertion heuristic cannot be recommended.

4.3 Deletion

Like the insertion function, deletion has to cope with additional information that is stored at the leaf nodes of the tree. To this end, deletion uses a predicate Δ in order to determine whether a specific leaf node *really* has to be deleted. Assume the user stores different pointers at a single leaf node of a substitution tree. If a leaf contains just the pointer which was searched for, the whole leaf node has to be deleted. If the leaf contains more than this pointer, the pointer has to be deleted from the list of pointers stored at the leaf node, but the leaf itself is not deleted. Moreover, Δ is true for all inner nodes of the tree.

Definition 9 (Deletion Function `delete`). Let (τ, Σ) be a substitution tree and ρ a substitution to be deleted. The function `delete` computes a modified tree by removing variants of ρ according to the predicate Δ. Note that ρ need not be normalized.

$$\texttt{delete}(\emptyset, \rho, \Delta) := \emptyset \tag{20}$$

$$\texttt{delete}((\tau, \Sigma), \rho, \Delta) := (\tau, \Sigma) \quad \text{if } \neg\exists\sigma \in \mathcal{V}(\tau, \rho) \vee \neg\Delta((\tau, \Sigma)) \tag{21}$$

$$\texttt{delete}((\tau, \Sigma), \rho, \Delta) := \texttt{repair}(\tau, \bigcup_{N_i \in \Sigma} \texttt{delete}(N_i, \rho\sigma, \Delta)) \quad \text{if } \exists\sigma \in \mathcal{V}(\tau, \rho) \wedge \Delta((\tau, \Sigma)) \tag{22}$$

$$\texttt{repair}(\tau, \emptyset) := \emptyset \tag{23}$$

$$\texttt{repair}(\tau, \{(\tau', \Sigma')\}) := (\tau \bullet \tau', \Sigma') \tag{24}$$

$$\texttt{repair}(\tau, \Sigma) := (\tau, \Sigma) \quad \text{if } |\Sigma| \geq 2 \tag{25}$$

The deletion of substitutions from a substitution tree is divided into two phases. In the first phase the substitution is searched for and the corresponding leaf node of the tree is possibly deleted. In a second phase the substitution tree is reconstructed, so that each inner node has at least two subtrees. The tree remains unchanged if either the represented substitution is not a variant of the substitution that has to be deleted or if Δ does not cause the deletion the node (maybe after having performed some side effects on a leaf node). Otherwise, we continue the deletion in all subtrees of the current node. Obviously, Fig. 5 also represents a deletion sequence if the trees are read in reverse order.

A great advantage of substitution trees is that the deletion function can easily be modified so it will remove instances, generalizations, or unifiable entries from the index. We simply have to change all occurrences of $\mathcal{V}(\tau, \rho)$ to $\mathcal{I}(\tau, \rho)$, $\mathcal{G}(\tau, \rho)$, or $\mathcal{U}(\tau, \rho)$, respectively. Additionally, $\Delta(N)$ has to be true for all nodes N.

5 Experiments

Term Sets. For the experiments special term sets were used. Some of them were introduced in [9]. These sets were taken from typical OTTER [8] applications. As the sets are paired, there is a set of positive literals and a set of negative literals in each pair. The sets EC+ and EC− consist of 500 terms each and are derived from a theorem in equivalential calculus. The sets CL+ and CL− have 1000 members and are derived from a theorem in combinatory logic. The sets BO+ and BO− are derived from a theorem in the relational formulation of Boolean algebra and consist of 6000 terms each.

The other term sets were produced randomly. Each of these sets contains 10000 terms. Three different function symbols with varying arities and three different constants have been used. The terms contain at most three different variables with possibly multiple occurrences. The set WIDE contains function symbols with an arity of at least 3. In the other sets the maximal arity of function symbols is 2. The maximal depth of all terms is 3 except for the set DEEP where the maximal term depth is 6. Terms in LIN contain each variable at most once and the set GND contains no variables at all. All other sets contain linear as well as non-linear terms.

Memory Requirements, Insertion, and Deletion. In the experiments substitution trees are smallest. In the average case a substitution tree consumes 70% of the memory occupied by the worst of the three techniques. This fact is illustrated in Fig. 7: For each of the three indexing techniques we have three bars. The white bar represents the average behavior. For example, discrimination tree indexing is most greedy in memory consumption. The gray bar shows the best result of the experiments and the black bar the worst behavior. Consider the bars for abstraction tree indexing: In the average case, the trees need 97% of the memory needed by the most greedy method. However, there was an experiment where abstraction trees occupied just 53% of the memory occupied by the worst technique. Nevertheless, the black bar tells us that abstraction trees have at least once been the indexing technique consuming most memory.

Discrimination trees are created most quickly. The other two indexing techniques are slower at adding entries to the index, because these indexes are non-deterministic and a position for insertion has to be found using a rather complex algorithm. In contrast to substitution trees, the abstraction tree technique does not require normalization of the entries to be inserted and therefore abstraction trees are faster at inserting new entries.

In spite of slow insertion, substitution trees are equally appropriate for dynamic data, because they perform best for deletion. The wide range of the times for deletion in substitution trees is striking (6% to 100%). Deletion in substitution trees performs extremely well for the deep terms in EC−, CL+, and DEEP.

Retrieval Times. All experiments were run on a Sun SPARCstation SLC computer with 16 MBytes of RAM. The indexing problems are described as follows: EC+− corresponds to storing the set EC+ in an index and starting a retrieval

Task	Memory [KBytes]			Insertion [Seconds]			Deletion [Seconds]			General. [Seconds]			Instances [Seconds]			Unif. Terms [Seconds]		
	DT	AT	ST	DT	AT	ST	DT	AT	ST	DT	AT	ST	DT	AT	ST	DT	AT	ST
EC++	108	153	**88**	**0.3**	**0.3**	**0.3**	0.4	0.2	**0.1**	0.6	0.4	**0.3**	6.5	6.0	**1.0**	27.9	32.1	**11.8**
EC+−										1.7	0.8	**0.4**	3.8	6.0	**0.8**	28.6	60.6	**21.8**
EC−+										0.4	**0.1**	**0.1**	42.8	14.9	**0.3**	99.9	89.2	**5.8**
EC−−	539	483	**252**	1.0	**0.9**	1.3	1.6	1.2	**0.2**	2.1	0.7	**0.6**	76.4	18.6	**1.4**	308.8	211.1	**46.1**
CL++	612	505	**318**	**0.8**	0.9	1.3	1.6	1.0	**0.1**	3.4	**0.8**	**0.8**	17.9	4.6	**1.7**	50.2	12.7	**6.8**
CL+−										4.0	**0.2**	0.3	3.2	3.5	**0.1**	42.7	17.6	**4.8**
CL−+										1.3	**0.1**	**0.1**	100.0	11.9	**6.7**	309.4	22.5	**16.9**
CL−−	2113	1116	**797**	**2.4**	**2.4**	3.2	4.1	3.1	**1.5**	5.6	1.7	**1.5**	5.2	5.1	**3.1**	19.1	7.3	**6.8**
BO++	685	780	**610**	**1.1**	1.6	2.4	**1.9**	2.2	2.7	**3.3**	3.5	3.4	6.1	7.5	**5.2**	11.1	10.0	**9.2**
BO+−										3.6	3.1	**3.0**	2.6	4.2	**2.3**	**4.9**	5.5	5.3
BO−+										1.7	0.9	**0.8**	18.7	2.4	**1.8**	23.9	2.4	**2.3**
BO−−	958	1090	**858**	**1.7**	2.3	3.6	2.7	**2.1**	3.0	3.6	2.3	**2.1**	3.7	3.5	**2.6**	4.0	3.5	**3.3**
AVG	969	1278	**935**	**1.7**	6.4	4.2	**2.7**	6.6	4.7	13.5	28.0	**13.3**	60.8	50.6	**32.2**	100.2	86.5	**61.5**
WIDE	13402	9312	**8056**	**9.0**	13.6	17.7	19.2	16.9	**13.3**	27.9	**16.2**	18.0	550.3	60.0	**46.7**	672.7	**84.3**	110.3
GND	846	1015	**831**	**1.5**	2.6	4.0	**2.4**	**2.4**	4.0	4.6	4.1	**3.7**	5.0	5.5	**4.3**	4.6	5.1	**4.5**
LIN	1020	1265	**950**	**1.8**	6.0	4.3	**2.7**	6.2	4.6	13.0	25.5	**12.8**	49.0	44.3	**30.0**	74.7	71.1	**52.2**
DEEP	6191	4312	**3766**	**5.3**	11.3	9.3	9.9	11.0	**7.6**	22.7	28.2	**14.8**	643.1	70.4	**52.8**	736.6	142.3	**138.6**

Fig. 6. Experiments with different indexing techniques

for each member of EC−. For the randomly created terms the index set and the query set are identical. McCune reports similar tests using discrimination tree indexing and path indexing in [9]. We do not report experiments with path indexing, because substitution tree indexing was much faster in all experiments. In the average case substitution tree indexing is also the fastest of the methods examined. Discrimination trees are slowest. Abstraction trees seem to work well on "wide" terms. There are only three experiments in which substitution tree indexing is not the fastest technique for the retrieval of more general entries. Due to the introduction of indicator variables, the search for instances using substitution trees takes just 1% of the time of discrimination or abstraction trees in case of the sets EC+ and EC−, which contain many variables. Substitution trees provide the fastest retrieval of instances in all examples. In all but two experiments substitution trees find unifiable entries most quickly.

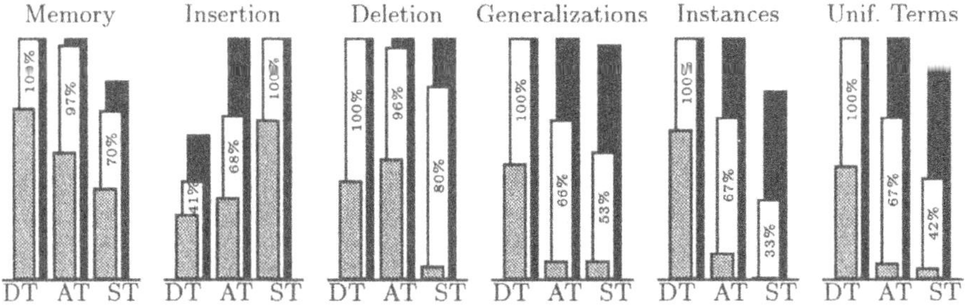

Fig. 7. Average performance of different indexing techniques

Implementation. Substitution trees, abstraction trees, discrimination trees, and (extended) path indexing are implemented in C and are available via anonymous ftp. They are part of "A Collection of Indexing Data Structures (ACID)", which is a library for efficient data structures and algorithms for theorem provers. Some of the techniques support subterm retrieval. Our implementations do not depend on term data structures and can easily be embedded into other software. For more information contact *acid@mpi-sb.mpg.de.*

6 Conclusion

A new data structure for indexing substitutions was presented. Substitution trees are based on a simple data structure and perform very well on a variety of tasks. They are stable for large sets of entries and memory requirements are low. In our experiments retrieval using substitution trees was faster than with abstraction and discrimination trees. The disadvantage of relatively slow insertion is compensated by a very fast and powerful deletion procedure.

Acknowledgments. I thank Hans Jürgen Ohlbach, Penny Anderson, and the referees for their comments on earlier versions of this paper.

References

1. L. Bachmair, T. Chen, and I.V. Ramakrishnan. Associative-commutative discrimination nets. In *Proceedings TAPSOFT '93, LNCS 668*, pages 61–74. Springer Verlag, 1993.
2. R. Butler and R. Overbeek. Formula databases for high-performance resolution/paramodulation systems. *Journal of Automated Reasoning*, 12:139–156, 1994.
3. J. Christian. Flatterms, discrimination nets, and fast term rewriting. *Journal of Automated Reasoning*, 10(1):95–113, February 1993.
4. P. Graf. Extended path-indexing. In *12th Conference on Automated Deduction*, pages 514–528. Springer LNAI 814, 1994.
5. P. Graf. Substitution tree indexing. Technical Report MPI-I-94-251, Max-Planck-Institut für Informatik, Saarbrücken, Germany, 1994. Full version of this paper.
6. D. Knuth and P. Bendix. *Simple Word Problems in Universal Algebras.* Computational Problems in Abstract Algebras. Ed. J. Leech, Pergamon Press, 1970.
7. E. Lusk and R. Overbeek. Data structures and control architectures for the implementation of theorem proving programs. In *5th International Conference on Automated Deduction*, pages 232–249. Springer Verlag, 1980.
8. W. McCune. Otter 2.0. In *10th International Conference on Automated Deduction*, pages 663–664. Springer Verlag, 1990.
9. W. McCune. Experiments with discrimination-tree indexing and path-indexing for term retrieval. *Journal of Automated Reasoning*, 9(2):147–167, October 1992.
10. H.J. Ohlbach. Abstraction tree indexing for terms. In *Proceedings of the 9th European Conference on Artificial Intelligence*, pages 479–484. Pitman Publishing, London, August 1990.
11. J.A. Robinson. A machine-oriented logic based on the resolution principle. *Journal of the ACM*, 12(1):23–41, 1965.
12. M. Stickel. The path-indexing method for indexing terms. Technical Note 473, Artificial Intelligence Center, SRI International, Menlo Park, CA, October 1989.

Concurrent Garbage Collection for Concurrent Rewriting

Ilies Alouini

INRIA-Lorraine & CRIN
615, rue du Jardin Botanique, BP 101
54602 Villers-lès-Nancy Cedex FRANCE
E-mail: Ilies.Alouini@loria.fr

Abstract. We describe an algorithm that achieves garbage collection when performing concurrent rewriting. We show how this algorithm follows the implementation model of concurrent graph rewriting. This model has been studied and directly implemented on MIMD machines where nodes of the graph are distributed over a set of processors. A distinguishing feature of our algorithm is that it collects garbage concurrently with the rewriting process. Furthermore, our garbage collection algorithm never blocks the process of rewriting; in particular it does not involve synchronisation primitives. In contrast to a classical garbage collection algorithm reclaiming unused blocks of memory, the presented algorithm collects active nodes of the graph (i.e. nodes are viewed as processes). Finally, we present different results of experimentations based on our implementation (RECO) of concurrent rewriting using the concurrent garbage collection algorithm and show that significant speed-ups can be obtained when computing normal forms of terms.

Keywords: Concurrent rewriting, Graph rewriting, MIMD architectures, Concurrent garbage collection algorithms.

1 Introduction

Rewriting provides a powerful model of computation. As Meseguer showed in [Mes92] rewriting logic is used as a general model of concurrency. Concurrent graph rewriting [KV92, Vir92] provides an operational method to implement efficiently rewriting on parallel machines. Following [GKM87] an implementation has been defined [KV92, Alo93] on MIMD machines using both a message passing mechanism between processors and a distributed matching algorithm. The system RECO [Alo93], concurrently computing a normal form of a term, has been built using a fine-grained parallelism on terms. This normalisation process is non-deterministic since several rewrite rules can be applied arbitrarily on different processors. Among the advantages of the implementation is its large scalability. In contrast to other implementations of parallel term rewriting [DL90] reducing disjoint subterms in parallel, the model of concurrent rewriting allows the concurrent reduction of overlapping redexes, and thus does not need any synchronisation process.

The lack of a concurrent garbage collector when performing concurrent rewriting has a dramatic effect on the efficiency of the implemented model. In contrast with standard garbage collection where only memory is collected, we are faced here with the problem of collecting processes producing themselves new processor cells to be collected. We are in a situation where "garbage produces garbage", and then an efficient collector is needed in order to avoid the implementation being overloaded by unnecessary computations. So the need for a garbage collector becomes in this situation extremely important.

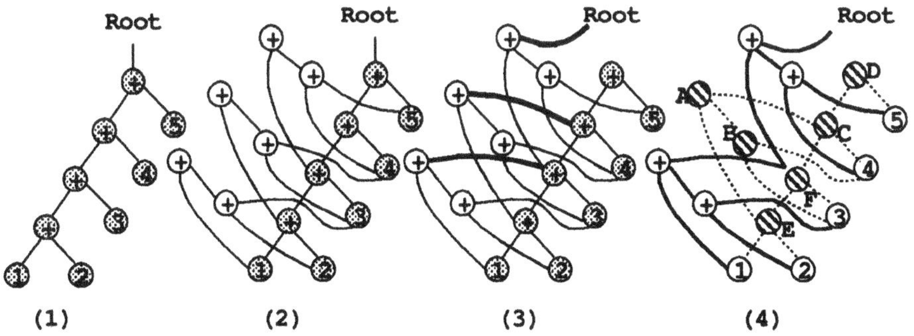

Fig. 1. Concurrent Rewriting steps

In the computational process schematised in figure 1 the program manipulates acyclic directed graphs to produce the final result. It shows three concurrent applications of the associative rewrite rule $(x + y) + z \rightarrow x + (y + z)$. As explained in details in [KV92], given a term represented as a direct acyclic graph (1), right-hand sides of rules are created (2), then redirection to the newly created right-hand sides are done (step 3). Step (4) distinguish nodes that are reachable from the root. All other nodes (A,B,C,D,E,F) are garbage. We note here that the garbage node C performs one rewriting step and then not only consumes memory but also processor resources. The garbage collector should detect them and allocate them to the free memory.

Our main goal in this paper is to build a concurrent garbage collector algorithm which can be embedded into the implemented model of concurrent rewriting.

The highlights of the presented concurrent garbage collector algorithm are:

(1) It uses local information in a node to detect garbage.
(2) It uses address references.
(3) It provides a concurrent garbage collector without synchronisation.
(4) It increases the efficiency of concurrent graph rewriting and thus it is an indispensable complement to the implementation of concurrent rewriting.
(5) It uses information on the implemented model of concurrent rewriting as much as possible.

The paper is organised as follows: In section 2 we review briefly the basic aspects of the implemented model of concurrent rewriting on distributed memory machines. In section 3 we describe related work and show why known garbage collectors are not adaptable to the implemented model of concurrent rewriting. In section 4 we describe the detection method of garbage nodes and we discuss the problem of copies of references. In section 5 we show how the garbage collection algorithm improves the efficiency of the concurrent rewriting model. Remarkable speed-ups for computing normal forms of terms are possible. All experiments are based on our system RECO that we are developing in Nancy. Finally, conclusions are given in section 6.

For details on the model of concurrent graph rewriting and the implemented model the reader can refer to [Alo93, KV92]. The reader can also refer to [DJ90] of definitions on terms and rewriting.

2 A brief description of the implemented concurrent rewriting model

Our system RECO (implementation of concurrent rewriting) uniformly implements the model of concurrent graph rewriting on distributed memory machines. We do not recall here the technical aspects about this model of implementation. The reader can refer to [Alo93] for a complete description. The implementation (RECO) is written in a so-called single program multiple data (SPMD) style. All processors apply the same program to different data. Initially, the term to be reduced is distributed on a given number of processors. All processes apply rewrite rules concurrently until a normal form of the term is obtained (if it exists).

We present an outline of the basic operations needed to implement concurrent graph rewriting. Two basic operations are implemented, matching and replacement. A distributed matching algorithm is used [KV92, Vir92]. It is a bottom-up algorithm efficient and incremental with respect to replacement. Its message-passing implementation avoids all the locking and synchronisation problems. The term represented as a jungle[1] is distributed over the processors and has no more a notion of global state. Jungles are based on objects (nodes and edges) and pointers (local and distant pointers). The basic operations needed for performing a replacement are implemented using message passing. To solve the lack of global state, each node stores a data structure called photo (part of the graph below it). The photo, denoted in prefix notation (we assume that we memorise the arity of each symbol of the rewriting system in a global table), is used to calculate addresses of variables substitution and then rewriting is done with those photos which remains coherent with the matching information. The two main characteristics of the implemented model are:

(1) No explicit parallelisation directives are to be given, the same program may run on a sequential or a parallel machine.

(2) No synchronisation is needed.

[1] special acyclic directed hyper-graph.

RECO offers two mechanisms for right-hand side rewrite rule creation. The first, called local rewriting, creates the right side locally in a processor. Figure 2 shows an application of the rule $F(x) + F(y) \rightarrow F(x + y)$ when a term is distributed over three processors P1,P2 and P3. Replacement (Fig 2) is done in three phases. Phase 1 creates the right-hand side of the rule, phase 2 computes addresses of variables substitution (using the N_1 photo equal to $N_1 = N_1N_2N_3N_4N_5$) by sending connect messages to node N_3 and N_4. In phase 3, each node receiving a connect message (for example node N_3) creates the ascendant pointer and sends match messages which contain its photo and its matching information. The second mechanism, not detailed here, offers the possibility to create right-hand sides of rewrites rules on a distant processor. Thus, load balancing is possible when creating nodes and then can be done at the same time when applying rules. This is an important feature of our implemented model since we have not to "move" subgraphs from a processor to another one.

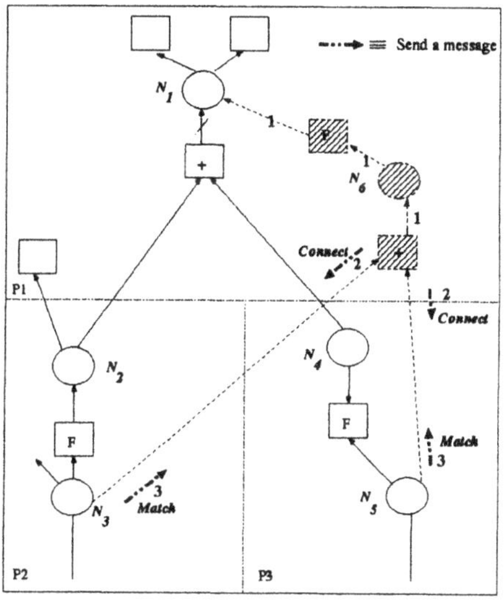

Fig. 2. A concurrent reduction step

3 Related work and motivations

Many garbage collectors have been proposed in the literature but most of them work in a sequential environment and they reclaim only memory blocks. The published parallel garbage collection algorithms contain simplifying assumptions that cannot be met in practice in a multiprocessor system. Moreover, few of them apply to active nodes (i.e. node as a dynamic entity). In our case, we shall construct an effective garbage collection for graph rewriting on multiprocessors.

For the first step, let us briefly describe the basic techniques for detecting garbage nodes in a graph.

There are two basic techniques for detecting garbage. The first technique visits all nodes of the graph. During the marking phase, every node reachable from the root of the computational graph is marked. At the end of the marking phase, all unmarked nodes are garbage. Finally, all marked nodes are reset. For the second technique, each node has a reference counter field which maintains the number of references to that node. When the reference count of a node becomes zero, the nodes can be removed from memory. The reader can refer to [PK82, BA84] for a survey of garbage collection algorithms.

When the graph is distributed over several processors, Ashoke Deb [Deb86] proposed a parallel garbage collection algorithm with multiple collectors running in parallel. Each collector works on a part of the graph and synchronisation is needed between collectors. Besides that, simplifying hypotheses are introduced, for example assuming that there is no delay in transferring messages from one collector to another cannot be verified in practice. Recent papers evoke the use of local garbage collectors with a logical global collector [WK91, Pua94] performing garbage collection in distributed systems of active objects. The local ones need not be synchronised with each other but still here we need a central garbage collector. To our knowledge major existing parallel garbage algorithms for graphs fail to fit into the implemented model of concurrent rewriting. In fact, major existing garbage collectors require covering the whole graph to detect garbage nodes or need synchronisation with a global garbage collector in a distributed environment. This overhead of marking nodes or testing reachability of nodes is not suitable when the graph is distributed over a set of processors. The main purpose of this paper is to have a method that represents an efficient garbage collector for the implemented model. Any solution introducing a supervisor or a global snapshot[WK91, Pua94] of the system state is inadequate.

We recall that the implemented model introduces additional data structures called photos to solve the lack of global state. Each node stores a photo (see section 2) which is a part of a graph below a node. Thus, a node N is referred to by its antecedents (because of sharing graphs, a node may have several antecedents) and in all photos that referred to it. The technique of photos detailed above has the disadvantage of requiring a complex method for garbage. To delete a node, we must ensure that N is not referred to in any photo above. The last condition introduces a complex method for detecting garbage nodes.

In this paper, we use a variant of the reference counting technique adapted to our model of computation where the graph is distributed over a set of processors.

4 Garbage collector

4.1 Overview of the proposed garbage collector

The garbage collector we are proposing is composed only of a set of local garbage collectors integrated into each rewriting process. Each one running on a processor detects and reclaims garbage nodes by using local information in nodes (photos and lists of address references). A key point of our approach is that we can use photos and a mechanism for sending messages to detect such nodes. This allows a local detection process. The technique presented is derived from the reference counting algorithm.

The detection phase is performed using a local difference of photos in a node N. At each rewriting step, newly referenced or dereferenced nodes in the photo of N are computed by comparing the newest photo to the oldest one. Each time a node N refers to a node N' it sends an increment message to N'. In the opposite case, N sends a decrement message to N'. Each node N stores a list of couples (address, processor identifier) of nodes that refer to it (increment list). In the same way, each node N stores a list of tuples of nodes that dereference it (decrement list). The core of the collection algorithm is quite simple: if the two lists above are equal then the node N is deleted from memory. Nevertheless, the condition above is not sufficient. We have to resolve two problems essentially. The first is the synchronisation of increment and decrement messages. The second is the problem of copies of references in a distributed environment. All these aspects are detailed in section 4.3.

4.2 Detection of garbage nodes

The most important part of the algorithm is to detect a set of nodes that may get disconnected. These nodes are called garbage nodes. The four requirements that the detection process should verify are:

(i) It is incremental i.e. it is possible to collect garbage without having to detect all of it.
(ii) It should be done locally in a node.
(iii) It should not lead to a big overhead.
(iv) It is unreasonable to have hypotheses on the order of arrival of messages.

Requirement (i) is important since concurrent garbage collection becomes possible. Requirement (ii) is a constraint that avoids to cover all nodes of a graph; the deletion decision of a node is made locally in a node and then no synchronisation is needed. Requirement (iii) involves a small degradation of the performance of concurrent rewriting. Requirement (iv) means that the order of the arrival of messages has no effect on the detection process . The basic idea of the detection process is to compare old and new photos in nodes to detect garbage. First we introduce some notation and afterwards we present a simple example and we formalise the idea below.

We consider R a labelled term rewriting system, $left(l_i)$ denotes the left-hand side of a rule l_i. We consider the value $hls = Max_{1 \leq i \leq n} h_i$ where h_i is the height of $left(l_i)$ and n the number of rules in R. Photos at nodes N_i are subgraphs of height hls denoted P_i. In the following, photos are denoted in prefix notation. We define the sets Δ_{N_i} and Δ'_{N_i} for each node N_i as follows:

$$\Delta_{N_i} = \{N_j \mid N_j \in P_i \text{ and } N_j \notin P'_i\}$$
$$\Delta'_{N_i} = \{N_j \mid N_j \in P'_i \text{ and } N_j \notin P_i\}$$

where the old photo is P_i and the new one is P_i'. To facilitate the comprehension of the detection method, we present it using the classical reference counting technique. Each node N has a reference count field which maintains the number of references to N. The first step of detection of possible deleted nodes is to compute the sets Δ_{N_i} and Δ'_{N_i}. The second step is to send increment messages (respectively decrement messages) to nodes $\in \Delta'_{N_i}$ (respectively $\in \Delta_{N_i}$) indicating an increase (respectively a decrease) of their references. Intuitively, by computing Δ_{N_i} and Δ'_{N_i}, we locally detect nodes that may be deleted. When a reference count of a node becomes zero, that node is deleted as garbage. But in our case, the last condition does not imply a deletion of a node. As explained, in the next section we have to resolve two problems: synchronisation of increment and decrement messages and copies of references in a distributed environment. In this paragraph we are only interested in the detection method of nodes that may be deleted. The condition of deletion of nodes is discussed later.

We recall that each node N_i contains a photo P_i of height *hls*, so a node N_i refers to all nodes of P_i. We have here two kinds of references. The first induced by the structure of the graph, each node refers to its sons and the second type of references is induced by the additional data-structure (photo) in a node. Naturally the first kind of reference is a special case of the second.

Differences of photos can be done concurrently at different nodes without any constraint, we satisfy then requirement (i) and (ii) (i.e. *hls* is generally a small value). We note that because of sharing subgraphs, nodes in photos P_i and P_i' represent the variable substitution and are then already referenced.

As mentioned in section 2, a photo in a node N is computed at each rewriting step indicating a transformation of the subgraph below N. We recall that a bottom-up message called "match" (containing the newest photo) is received at N. Receiving a match message at a node $(N_i, Proc_i)$, releases the following operation:

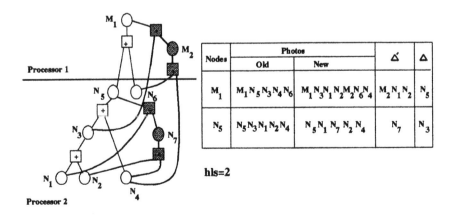

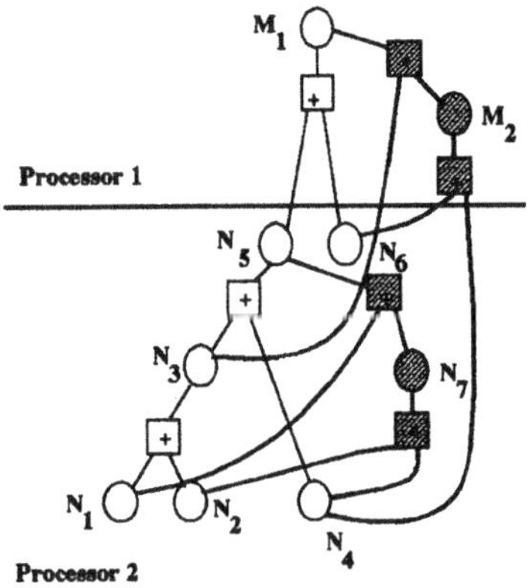

Nodes	Photos		Δ'	Δ
	Old	New		
M_1	$M_1N_5N_3N_4N_6$	$M_1N_3N_1N_2M_2N_6N_4$	$M_2N_1N_2$	N_5
N_5	$N_5N_3N_1N_2N_4$	$N_5N_1N_7N_2N_4$	N_7	N_3

hls=2

Fig. 3. Local differences of photos in node M_1 and N_5

DETECTGARBAGE:

$$\textbf{Compute}(\Delta_{N_i}, \Delta'_{N_i})$$
$$\textbf{Send}(INCREMENT_{(N_i,Proc_i)\rightarrow\Delta'_{N_i}})$$
$$\textbf{Send}(DECREMENT_{(N_i,Proc_i)\rightarrow\Delta_{N_i}})$$

Comment: The primitive Send($INCREMENT_{(N_i,Proc_i)\rightarrow\Delta'_{N_i}}$) send an increment message from node N_i in processor $Proc_i$ to each node of the set Δ'_{N_i}. We suppose that the structure of the message contains its origin (node address,node processor).

Given the term $t = +(+(+(x, y), z), w)$, the above figure 3 illustrates the technique of detection. It shows two concurrent rewriting steps (associative rule) at nodes M_1 and N_5. We suppose M_1 is on a processor 1 and all others nodes are on processor 2. For example, node M_2 is referenced, so we must increment its reference counter. Computing Δ'_{M1} decrements N_5 reference counter.

Unfortunately, reference counters fails to take account synchronisation of increment and decrement messages (figure 4). To solve this problem, reference addresses are introduced. We store for each node N_i the set of all addresses N_j referencing N_i.

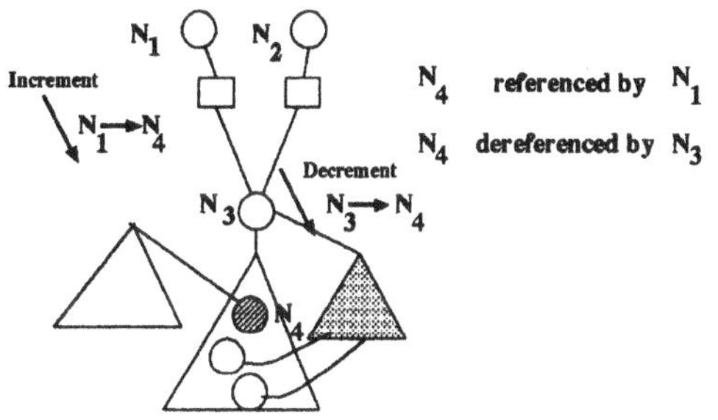

Fig. 4. Why reference counter fails ?

It is clear that using reference counter abstracts the origin of the reference. In figure 4, after receiving an increment message from N_1 and a decrement message from N_3, figure shows that N_4 counter reference is equal to zero but cannot be deleted (i.e. referenced by N_2). With reference addresses, we can distinguish between these two messages. We recall that we do not assume any hypotheses on the order of arrival of messages (N_3, N_4 and N_1 can be in different processors). So the idea is to correspond each incrementation and decrementation by its addresses reference.

Two fields are added to each node N. One list denoted $L_{INC}(N)$ for storing addresses of nodes referencing N and another list denoted $L_{DEC}(N)$ to store addresses of nodes dereferencing N. Therefore, receiving an increment or decrement message in a node N adds to $L_{INC}(N)$ and $L_{DEC}(N)$ origin addresses of such messages. We need here to store increment and decrement messages to solve the problem of copies of references (see section 4.3). The problem of synchronisation of messages still remains but the identification of origin of increment and decrement messages and is a step towards solving it.

The advantages of the presented technique is the use of pre-defined data structures in the implemented model of concurrent graph rewriting (photos) that resolve not only determine the addresses of variable node substitution when the graph is distributed but also allows the detection of nodes that may be deleted from memory. It is clear that we satisfy requirements (i), (ii) and (iv). Requirement (iii) is satisfied because the lists $L_{INC}(N)$ and $L_{DEC}(N)$ are of small size (maximum depth of variables in left-hand sides of rewrite rules is in general a small value). The garbage detection method is easily generalised to take into account of several references for a node N by the same node N'.

4.3 The problem of copies of references

The major problems which are faced are :
(a) Synchronisation of increment and decrement messages.
(b) Copies of node references.

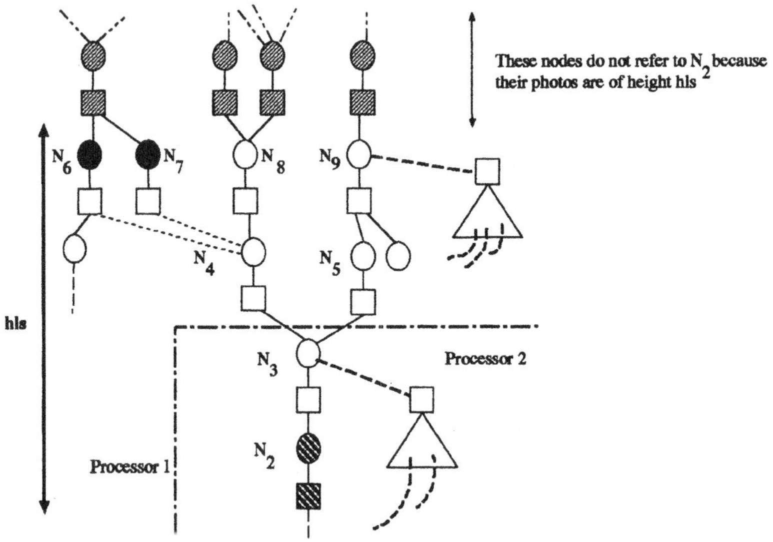

Fig. 5. The problem of copies of references

We analyse the subgraph shown by the double arrow in the figure 5. The height of this graph is *hls*. We note that only nodes of this subgraph reference N_2. As illustrated in figure 5, computing Δ and Δ' at nodes N_3,N_4,N_5,N_8 and N_9 and sending corresponding messages to N_2 detect it as a possible garbage node. Nevertheless, rewriting on a distant processor of the graph may connect N_4. In this case we have a copy of the reference N_4 in node N_6 and N_7 (black nodes). How can we decide locally in node N_2 about the copy of its reference ?. This problem introduces a complex method since we assume that there are no hypotheses on the order of arrival of increment or decrement messages. Also, any solution that takes into account message arrival time is not suitable. To satisfy requirement (ii), any call to synchronisation primitives is not admitted.

The idea to solve the problem is to use message decrement to anticipate the messages increment that will arrive at node N. This is possible since messages match are bottom-up and the height photos are limited to ***hls***. In the above example, (figure 5) N_4 send decrement messages to N_2 containing antecedents of node N_4 (Note here that necessary N_4 figures in N_6 and N_7 photos). The added field in the structure of node N is a list of incrementing messages waiting for in a node N denoted $L_{INCWAIT}(N)$. Two cases are discussed:

(1) *Simplest one: No concurrent rewriting occurs in subgraphs of height hls.*

If the three lists $L_{INC}(N)$, $L_{DEC}(N)$ and $L_{INCWAIT}(N)$ of node N detailed above are equal, we can delete the node N from memory. We must show that the proposed solution does not delete a node as being garbage when it is not (correctness), and every garbage node is deleted (liveness). We do not provide any formal proof for the correctness of this algorithm. We only give some arguments that help us to prove the correctness. The argument is: (N references M) $\Rightarrow$ ($\exists$ R, $\exists$ message DEC / Send($DEC_{R \rightarrow M}(L)$) where N, M and R are nodes and $N \in L$. Send($DEC_{R \rightarrow M}(L)$) denotes a decrement message from R to M containing the list L of addresses of antecedent nodes of R. Note here that N is an antecedent node of M.

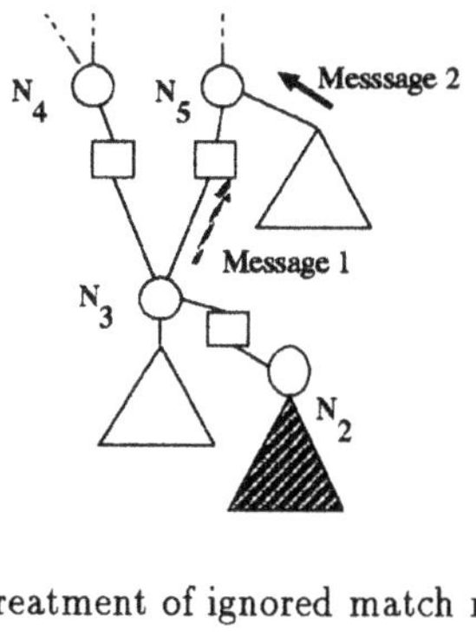

Fig. 6. Treatment of ignored match messages

The deletion condition of a node N is simple if the three lists $L_{INC}(N)$, $L_{DEC}(N)$ and $L_{INCWAIT}(N)$ are equal then node N is collected from memory.

(2) *Concurrent rewriting occurs in subgraphs of height hls.*

Figure 6 analyses the node N_2 deletion. Applying concurrently another rule at node N_5 implies ignoring match messages received from N_3. We introduce a mechanism by sending ignored messages to nodes of the N_5 photo that we never reference. Nodes in dark triangle (term t) can not be referenced by node N_5 because we ignored messages match (new photos) issued from t. This is why we have to treat the ignored message, denoted message 1, shown in figure 6. This problem is solved by adding a list of ignored messages $L_{IGNO}(N)$ at each node N of term t (for example node N_2). The condition of deletion detailed in (1) is modified to take into account the ignored match messages (see section 4.4). Finally, all nodes in the photos of a deleted node from memory must receive a decrement message ('cascade' decrement).

4.4 The core algorithm

We first present the basic operations needed to gather a node N. We briefly describe these fundamental actions. Only, the principal parts of the algorithm are given. All specials cases and details are omitted due to lack of space.

The algorithm is decomposed into three parts:

(1) Initialisation part: At the beginning, we initialise the lists of node N to empty lists.

(2) Computing part (a): Here is the main operation that compute the differences between photos received in match message.

(3) Testing equality part (b): At each increment or decrement message received, we test if the three lists of node N are equal. The last condition allows us to physically delete this active node from memory.

Notations

$\text{message}_{(N_i,Proc_i)\rightarrow(N_j,Proc_j)}$: Send message from the node N_i in processor $Proc_i$ to node $(N_j, Proc_j)$.
valid(message): True if message treated, False if message ignored.
message(L): denotes a message containing the list L.
$L'_{INCWAIT}(N_1) = Diff(L_{INCWAIT}(N_1), L_{IGNO}(N_1))$ where $Diff$ computes the difference between $L_{INCWAIT}(N_1)$ and $L_{IGNO}(N_1)$.
pho(message) denotes the photo of the message $Match$.
$\Delta'^{=hls}_{N_1}$ denotes nodes of Δ'_{N_1} of height equal to hls in the photo of N_1.
Add-list($L, (N, Proc)$): adds the couple $(N, Proc)$ to the list L.

Outline of the messages treatment algorithm at node $(N_1, Proc_1)$

Initialisation:
$L_{INC}(N_1) = L_{DEC}(N_1) = L_{INCWAIT}(N_1) = L_{IGNO}(N_1) = \oslash$ at node $(N_1, Proc_1)$.
Receive($message$)
case type($message$)
  $Connect_{(N_i,Proc_i)\rightarrow(N_1,Proc_1)}$:Add-list($L_{INCWAIT}(N_1), (N_i, Proc_i)$);
Comment: An increment message from $(N_i, Proc_i)$ will arrive to node N_1
  $Match_{(N_i,Proc_i)\rightarrow(N_1,Proc_1)}$ and valid($message$) (a)
    Compute($\Delta_{N_1}, \Delta'_{N_1}$);
    Send(Increment(Δ_{N_1}));
    Send(Decrement($antecedents(N_1)$), $\Delta'^{<hls}_{N_1}$);
    Send(Decrement($\oslash$), $\Delta'^{=hls}_{N_1}$);
Comment: Basic operations to detect garbage nodes.

$Match_{(N_i,Proc_i)\rightarrow(N_1,Proc_1)}$ **and not** valid($message$):
  Send($Igno(pho(message))$);
Comment: Sending ignored messages to nodes not referenced before by $(N_1, Proc_1)$
$Increment_{(N_i,Proc_i)\rightarrow(N_1,Proc_1)}$: (b)
  Add-list($L_{INC}(N_1), (N_i, Proc_i)$);
  if $L'_{INCWAIT}(N_1) = L_{DEC}(N_1) = L_{INC}(N_1)$ **then** delete(N_1);
Comment: Test of deletion condition
$Decrement_{(N_i,Proc_i)\rightarrow(N_1,Proc_1)}$: (b)
  Add-list-list($L_{INCWAIT}(N_1), list(message)$);
  Add-list($L_{DEC}(N_1), (N_i, Proc_i)$);
  if $L'_{INCWAIT}(N_1) = L_{DEC}(N_1) = L_{INC}(N_1)$ **then** delete(N_1);
Comment: Test of deletion condition
endcase

5 Experimental results and performance

In this section we present and comment on actual timings of our implementation of concurrent rewriting with the concurrent garbage algorithm introduced in this paper. Our prototype runs on a cluster of SUN4 Sparc stations. The network use an ETHERNET protocol. The bandwidth supported is at most 1MB/s ($\sim$ 0.6 measured in practice) and the latency time for sending a message is 4 ms. Although the cluster was relatively unloaded, a certain amount of interference by other users' jobs could not be avoided. The prototype RECO (implementation of COncurrent REwriting) and the concurrent garbage algorithm are written in C using the library PVM (parallel virtual machine) [BDGG94] for messages passing primitives.

We recall that the implementation of concurrent rewriting is totally asynchronous. All primitives for receiving messages are non-blocked calls. Processes running on each processor never wait for messages, and processing times do not include waiting times for messages. We note that the normalisation process is non-deterministic. This leads to different possible executions for computing a normal form of a term concurrently. Statistics presented are mean values for different executions for a given term.

Times showed in table 1 and 2 are times to compute normal forms of terms. User time includes rewriting time, time for packing and unpacking messages, etc. System time include time for systems unix calls memory allocation, etc.

The results are based on the following rewrite system R, and tests computing normal form of terms have been done with different sizes of terms using the concurrent garbage collection algorithm.

$$R = \begin{cases} 1 : F(x) + F(y) & \rightarrow F(x+y) \\ 2 : x + (y+z) & \rightarrow (x+y)+z \\ 3 : F(0) & \rightarrow 0 \\ 4 : 0 + x & \rightarrow x \\ 5 : x + 0 & \rightarrow x \end{cases}$$

The structure of terms used in the tests have a large amount of overlapping redexes. To test our implementation of the concurrent garbage algorithm, we compare the number of total rules using or not the garbage collection algorithm. The times and the number of total rules to compute the normal form of a term are given in Table 1 and Table 2. The terms are built as follows: $t_0 = F(x) + F(y)$; $t_i = t_{i-1} + t_{i-1}$, $i \geq 1$. The scheme of t_i normal forms are: $F((((var + var) + var) + var) + \ldots)$ where var is a variable name. We note that these results are sensitive to interference from other users and are therefore of a qualitative nature.

Number of processors	1	2		3			4			
Number of nodes	2559	1478	1179	833	1095	1000	560	693	824	716
Number of deleted nodes	1785	1103	780	577	838	745	400	533	568	524
Number of applied rules	1408	752	736	448	703	607	272	400	528	419
User time	35,8	15,9	13,5	10,9	14,2	11,8	8,0	7,7	7,8	9,3
System time	6,0	10,28	3,3	6,9	6,7	6,1	7,2	6,4	2,8	3,7

Table 1: Computing term normal form (1151 nodes, garbage on): general statistics

Number of processors	1	2		3			4			
Number of nodes	2559	1504	2279	833	1095	1000	560	813	1423	1172
Number of applied rules	1408	764	1847	448	951	1416	272	519	1125	871
User time	12,5	11,0	12,3	9,7	10,1	10,8	14,6	13,4	15,3	13,6
System time	9,7	15,6	8,9	12,4	8,2	9,3	21,5	21,48	12,06	15,35

Table 2: Computing term normal form (1151 nodes, garbage off): general statistics

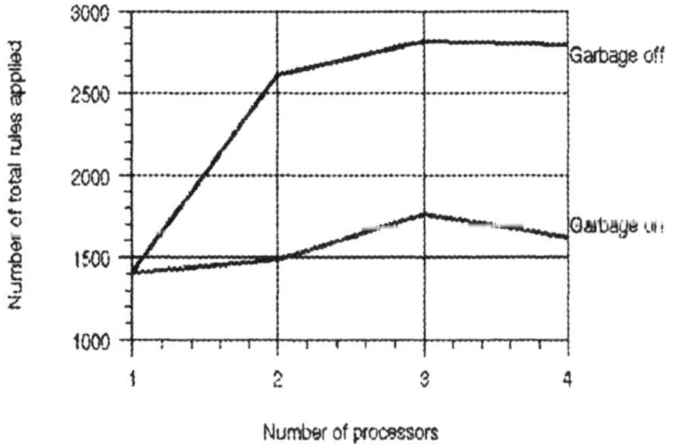

Fig. 7. Comparison of total applied rules

From figure 7, we remark that if we use the garbage collector, the total number of applied rules in all processors is approximately equal to those applied in a sequential execution. It is clear, that big terms have greater speed-ups (up to 3.8 for term having 2303 nodes on four processors) (figure 8) than smaller ones. Although, increasing the number of processors do not lead a better speed-up (figure 8). This is due to the fact that initial granularity (i.e. the number of initial nodes in a processor) produces more communication than rewriting processing. Moreover, the network used is based on a sequential bus (ETHERNET) where all the communications are concentrated. On real MIMD machines with thousands of processors, this bottleneck does not exist and then significant speed-ups remains when we increase the number of processors.

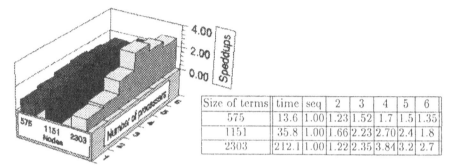

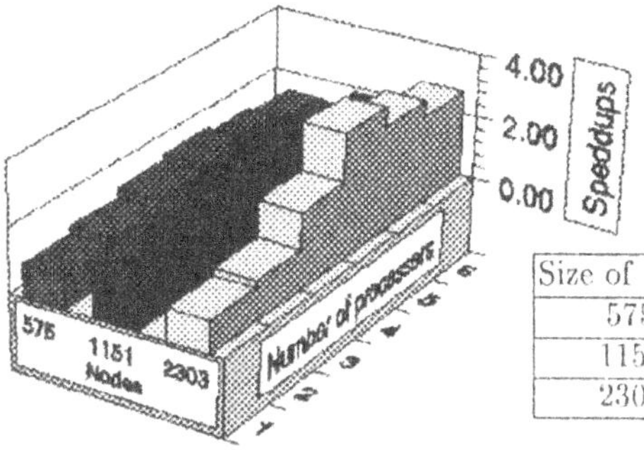

Fig. 8. Speed-ups w.r.t. times

Size of terms	time	seq	2	3	4	5	6
575	13.6	1.00	1.23	1.52	1.7	1.5	1.35
1151	35.8	1.00	1.66	2.23	2.70	2.4	1.8
2303	212.1	1.00	1.22	2.35	3.84	3.2	2.7

Table 3: Time speed-ups w.r.t. sequential

To summarise, experimental results using the concurrent garbage algorithm show that we obtain good speed-ups for big terms with a large amount of overlapping redexes without using any load balancing algorithm. We believe that using a load balancing algorithm in conjunction with a high bandwidth network between processors will ameliorate the speed-ups.

6 Conclusions

What we have shown in this paper is that a concurrent garbage collection algorithm can be successfully used to increase the efficiency of the implemented model of concurrent rewriting by significantly decreasing the number of total rules applied in processors. This algorithm operates concurrently with the process of rewriting and never involves synchronisation primitives. Each garbage collector runs on an independent processor. Speed-ups up to a factor of 3.8 on four processors shows that computing normal forms for terms coupled with the concurrent garbage collector provides good speed-ups. Our implementation of concurrent rewriting on a cluster of workstations and the presented concurrent garbage collector seems to be a efficient normalisation process of terms at a fine grained parallelisation. Finally, we believe that the techniques and ideas introduced in this paper are practical issues and a contribution towards solving the question of how to have an efficient normalisation process of terms in concurrent or distributed deduction.

Acknowledgements

I am grateful to Claude Kirchner and the referees for helpful comments on the paper. I would also like to thank Patrick Viry and Christopher Lynch for reading earlier drafts.

References

[Alo93] Ilies Alouini. Réécriture concurrente: Etude et implantation. Rapport de DEA, Université de Nancy 1, September 1993.

[BA84] M. Ben-Ari. Algorithms for on-the fly garbage collection. In *Proc. ACM symposium on Prog. Lang and Sys*, volume 6, pages 333–444, 1984.

[BDGG94] A. Beguelin, J.J Dongarra, and A Geist G. Parallel virtual machine user's guide and reference manual. Technical report, Oak Ridge National Laboratory, May 1994.

[Deb86] A. Deb. Parallel garbage collection for graph machines. In *Proc. of a Workshop Santa Fe, New Mexico, USA*, volume 279 of *Lecture Notes in Computer Science*, pages 252–264. Springer-Verlag, September 1986.

[DJ90] N. Dershowitz and J.-P. Jouannaud. Rewrite Systems. In J. van Leeuwen, editor, *Handbook of Theoretical Computer Science*, chapter 6, pages 244–320. Elsevier Science Publishers B. V. (North-Holland), 1990.

[DL90] N. Dershowitz and N. Lindenstrauss. An abstract concurrent machine for rewriting. In *Proc. of Algebraic and logic programming*, volume 463 of *Lecture Notes in Computer Science*, pages 318–331. Springer-Verlag, October 1990.

[GKM87] J. A Goguen, C Kirchner, and J Meseguer. Concurrent term rewriting as a model of computation. In *Proceedings of Graph Reduction Workshop R.Keller et J. Fasel*, volume 279 of *Lecture Notes in Computer Science*, pages 53–93, Santa Fe (NM USA), 1987. Springer-Verlag.

[KV92] Claude Kirchner and P. Viry. Implementing parallel rewriting. In B. Fronhöfer and G. Wrightson, editors, *Parallelization in inference systems*, number 590 in Lecture Notes in Computer Science, pages 123–138. Springer-Verlag, 1992.

[Mes92] J. Meseguer. Conditional rewriting logic as a unified model of concurrency. *Theoretical Computer Science*, 96(1):73–155, 1992.

[PK82] P.Hudak and R. M. Keller. Term graph rewriting. In *Proc. ACM symposium on LISP and Functionnel Programming*, volume 201 of *Lecture Notes in Computer Science*. Springer-Verlag, 1982.

[Pua94] I. Puaut. A distributed garbage collector for active objects. In *Proc. of PARLE'94*, volume 817 of *Lecture Notes in Computer Science*, pages 539–552. Springer-Verlag, July 1994.

[Vir92] Patrick Viry. *La réécriture concurrente*. Thèse de Doctorat d'Université, Université de Nancy 1, October 1992.

[WK91] D. M. Washabaugh and D. Kafura. Distributed garbage collection of active objects. In *Proc. of 11th Int Conf on Distributed Computing Systems*, pages 369–376, May 1991.

Lazy Rewriting and Eager Machinery

J.F.Th. Kamperman (jasper@cwi.nl)
H.R. Walters (pum@cwi.nl)

CWI

Abstract. We[1] define *Lazy Term Rewriting Systems* and show that they can be realized by local adaptations of an *eager* implementation of conventional term rewriting systems. The overhead of lazy evaluation is only incurred when lazy evaluation is actually performed.
Our method is modelled by a transformation of term rewriting systems, which concisely expresses the intricate interaction between pattern matching and lazy evaluation. The method easily extends to term *graph* rewriting. We consider only left-linear, confluent term rewriting systems, but we do not require orthogonality.

1 Introduction

It is well-known that outermost rewriting strategies often have a better termination behaviour than innermost rewriting strategies [O'D77, HL79]. Innermost strategies (also called *eager evaluation*) can be implemented much more efficiently, however. We propose to solve this dilemma by transforming a term rewriting system (TRS) in such a way that the termination behaviour of innermost rewriting is improved. At the core of our transformation are established ideas of Ingermann [Ing61] and Plotkin [Plo75].

Figure 1 illustrates the bad termination behaviour of innermost rewriting.

$$\begin{aligned} inf(x) &\to cons(x, inf(s(x))) && (1)\\ nth(0, cons(x, y)) &\to x && (2)\\ nth(s(x), cons(y, z)) &\to nth(x, z) && (3)\end{aligned}$$

Figure 1

There, the term $nth(0, inf(0))$ has an infinite reduction sequence $inf(0) \stackrel{(1)}{\to} cons(0, inf(s(0))))$ $\stackrel{(1)}{\to} cons(0, cons(s(0), inf(s(s(0))))) \stackrel{(1)}{\to} \ldots$ This can be avoided by applying rule (1) only once, before applying rule (2): $nth(0, inf(0)) \stackrel{(1)}{\to} nth(0, cons(0, inf(\ s(0)))) \stackrel{(2)}{\to} 0$. By postponing some reductions, outermost rewriting may succeed in avoiding them altogether. 'Optimal avoidance' is studied, amongst others, in [O'D77, HL79, Fie90, Mar92]. However, this work crucially depends on orthogonality (no overlap of patterns) of TRSs. Our method applies to arbitrary TRSs. There seems to be no sensible definition of optimality for non-orthogonal TRSs. Furthermore, there are terms which do not terminate under any strategy, such as $cons(0, inf(0))$.

[1] Partial support received from the European Communities under ESPRIT project 5399 (Compiler Generation for Parallel Machines – COMPARE) and from the Foundation for Computer Science Reasearch in the Netherlands (SION) under project 612-17-418, "Generic Tools for Program Analysis and Optimization".

Therefore, rather than simulating a pure outermost strategy, our transformation simulates a variant of *lazy* evaluation, which is used to implement *lazy functional programming languages* [PvE93]. We will briefly discuss this. During lazy evaluation, non-outermost redexes are contracted in order to establish that the outermost function symbol will never become part of a redex. The resulting term is said to be in Weak Head Normal Form (WHNF). E.g., the term $cons(0, inf(0)))$ in the example above is in WHNF. For this term, the (implicit) output routine first produces $cons(0,$ on output, before recursively reducing the argument to WHNF. The term $cons(0, inf(0))$ still leads to an infinite computation, but 'useful' output is produced during this computation. Because rewriting of outermost redexes is expensive, it is usually avoided as much as possible. Arguments that can be rewritten eagerly without affecting termination behaviour, are called *strict*. Strictness analysis (initiated by Mycroft, [Myc80]) attempts to identify these arguments statically.

$$
\begin{aligned}
inf(x) &\rightarrow cons(x, thunk(x)) && (1)\\
inst(thunk(x)) &\rightarrow inf(s(x)) && (2)\\
nth(0, cons(x, y)) &\rightarrow x && (3)\\
nth(s(x), cons(y, z)) &\rightarrow nth(x, inst(z)) && (4)
\end{aligned}
$$

Figure 2

Now consider the TRS in Figure 2, which differs only slightly from the one in Figure 1. The term $nth(0, inf(0))$ still rewrites to 0, but there are no infinite reduction sequences. This example illustrates that only minor changes are needed to achieve the desired effect, and that these changes can be made local to "lazy positions" (cf. the second argument of $cons$). To some extent, this explains the success of strictness analysis. In many cases, only a few positions need a lazy treatment in order to preserve termination.

The example also demonstrates the common observation that a good implementation of a lazy language spends most time in "eager mode". Given the locality of the changes above, it is worthwhile to investigate how an *eager* implementation can be adapted to do some *lazy* evaluation, rather than adapting a *lazy* implementation to do (a lot of) *eager* evaluation.

We use laziness annotations to indicate argument positions where rewriting should be postponed if possible. These annotations could be provided by the programmer or by a strictness analyzer. In the latter case, all arguments that are not found to be strict, will get the annotation *lazy*, and the reductions performed by our implementation will correspond closely to the reductions performed by an implementation of a lazy functional programming language using the same strictness analyzer.

Even though Figure 2 is a simplified version of the result of our transformation, applied to the TRS of Figure 1 (with only the second argument of $cons$ annotated with *lazy*), it exhibits a peculiarity of our scheme. The term $inf(0)$ rewrites to the normal form $cons(0, thunk(0))$, which is not a term in the original signature. However, the term $thunk(0)$ (called a "thunk" after Ingermann [Ing61]) represents a (possibly infinite) term in the original system, which can be further approximated by repeatedly replacing terms $thunk(x)$ by the normal form of $inst(thunk(x))$. Our lazy normal forms (LNFs) generalize WHNF, and the approximation process corresponds to what is done by the output routine of an implementation of a lazy functional language. We do not assume such an output routine, because leaving the thunk in place offers the possibility of preventing uninteresting work, and yields a larger class of terminating systems.

We give definitions and notations pertaining to term (graph) rewriting in Section 2, our definition of *lazy* term rewriting in Section 3, a complete version of the transformation sketched above in Section 4, and some remarks on a realistic implementation in Section 5. We end with a discussion of related work and conclusions.

2 Term (graph) rewriting

We mostly repeat definitions and results from [Klo92] and [DJ90].

A Term Rewriting System (TRS) is a pair (Σ, R) of a *signature* Σ and a set of *rewrite rules* R. The signature Σ consists of:

- A countably infinite set *Var* of *variables* $x_1, x_2, \ldots$
- A non-empty set of *function symbols* $F, G, \ldots$, each with an *arity*, which is the number of arguments it requires. Function symbols with arity 0 are called *constant symbols*.

The set of terms over Σ is the smallest set $Ter(\Sigma)$ such that

- $x_1, x_2, \ldots \in Ter(\Sigma)$,
- if F is an n-ary function symbol and $t_1, \ldots t_n \in Ter(\Sigma)$ $(n \geq 0)$, then $F(t_1, \ldots, t_n) \in Ter(\Sigma)$. The t_i $(i = 1, \ldots, n)$ are called the *arguments*.

Terms in which no variable occurs more than once are called *linear*.

A path in a term is represented as a sequence of positive integers. By $t|_p$, we denote the *subterm* of t at path p. For example, if $t = push(0, pop(push(y, z)))$, then $t|_{2.1}$ is the first subterm of t's second subterm, which is $push(y, z)$. We will say $p \in t$ if the path p is defined in t, i.e., p leads to a subterm of t. The empty path (denoting the root) is written ε. We will call a set of paths P *prefix-reduced* if there are no pairs $p, p' \in P$ such that p is a prefix of p'. We will call a set S of subterms of s *prefix-reduced with respect to* s if there is a *prefix-reduced* set of paths $\{p_1, \ldots, p_n\}$ such that $S = \{s|_{p_1}, \ldots, s|_{p_n}\}$. We write $t[s]_p$ for the term resulting from the replacement of $t|_p$ by s in t.

A *substitution* is a map $\sigma : Ter(\Sigma) \to Ter(\Sigma)$ that satisfies $\sigma(F(t_1, \ldots, t_n)) = F(\sigma(t_1), \ldots, \sigma(t_n))$ for every function symbol F. By convention, we write t^σ instead of $\sigma(t)$.

A *rewrite rule* is a pair (t, s) of terms $\in Ter(\Sigma)$. It will be written as $t \to s$. Often a rewrite rule will get a name, e.g. r, and we write $r : t \to s$. Two conditions are imposed:

- the LHS (left-hand side) t is not a variable,
- the variables in the RHS (right-hand side) s already occur in t.

A rewrite rule $r : t \to s$ determines a relation: the set of *rewrites* $t^\sigma \to_r s^\sigma$ for all substitutions σ. The LHS t^σ is called a *redex* (from "reducible expression"), and the RHS s^σ is called the *contractum*. Allowing replacement inside other terms, $\to_r$, the *one-step rewrite relation* generated by r, is defined by:

$$u|_p = t^\sigma \ \Rightarrow \ u \to_r u[s^\sigma]_p$$

We call the relation $\rightarrow_R = \bigcup_{r \in R} \rightarrow_r$ the *rewrite relation* defined by R. Usually, the subscript R is omitted if it is clear from the context. Concatenating rewrite steps we have (possibly infinite) *rewrite sequences* $t_0 \rightarrow t_1 \rightarrow \ldots$ or *rewrites* for short. We write $\xrightarrow{+}$ and $\xrightarrow{*}$ for the transitive and reflexive-transitive closures of $\rightarrow$. If $t_0 \xrightarrow{+} t_n$, we call t_n a *reduct* of t_0. A term $t \in Ter(\Sigma)$ is said to be in *normal form* if there is no s such that $t \xrightarrow{+}_R s$. A function symbol F is called a *constructor* symbol if there is no LHS which has F as its outermost symbol. It is understood that R does not contain rewrite rules that are equal up to a bijective renaming of variables.

A TRS is called *left-linear* if all left-hand sides are linear. A TRS is called *confluent* if, for all terms t_1, t_2, t_3, we have that $t_1 \xrightarrow{*} t_2$ and $t_1 \xrightarrow{*} t_3$ implies that there exists a term t_4 such that $t_2 \xrightarrow{*} t_4$ and $t_3 \xrightarrow{*} t_4$. A TRS is called *terminating* if there are no infinite rewrite sequences. In the sequel, we will only consider left-linear, confluent TRSs. However, we will not require TRSs to be terminating. Note that it is undecidable whether a TRS is confluent or terminating.

In general, a term can contain many redexes. In an implementation of a TRS, a rewriting *strategy* determines which of the many possible rewrite sequences is chosen. Confluence guarantees unique normal forms. A redex is *needed* if it is contracted in every sequence leading to a normal form, and a strategy is called *optimal* if it only contracts needed redexes.

In this article, we will assume the existence of an implementation of the *leftmost-innermost strategy* (LI: the leftmost-innermost redex takes precedence). By a transformation, we will simulate lazy evaluation.

A typical implementation of an LI strategy for TRSs is given in [Heu88], where the rules are compiled into Lisp functions (one function for every defined function symbol in the TRS). The body of such a function consists of pattern matching code that determines which code is used for instantiation of the RHS. The former code is produced by a pattern matching compiler, the latter code is typically a number of nested function calls, with references to terms as arguments. On many architectures, this type of recursive code performs badly, which leads to several alternatives [TAL90, KW93, Bak94].

Term *graph* rewriting [BvEJ+87], where terms and rules are replaced by graphs, can be seen as a restriction of rewriting with infinite terms [KKSdV93]. An implementation of term rewriting can be turned into an implementation of graph rewriting by taking care that the sharing expressed by graphs is retained. Note that, in general, this is not easy.

3 Lazy term rewriting

We define *lazy term rewriting* as term rewriting with a restriction on the (one-step) rewrite relation. First, we define *lazy* signatures, which make a distinction between *eager* argument positions and *lazy* argument positions.

The choice to annotate the *arguments* rather than the *function symbols* themselves is not only motivated by compatibility with lazy functional languages, but has two additional advantages. First, if functions are annotated, we must expect thunks at every argument position, thus losing the locality of our transformation. Second, for functions such as `if(Bool,Exp,Exp)`, it is more natural to annotate an argument position than to annotate all function symbols that may occur there. Unfortunately, not all TRSs can be made terminating by only annotating arguments (cf. the rule $inf(x) = inf(x)$).

A lazy signature includes a predicate Λ on function symbols and natural numbers, where $\Lambda(F, i) = true$ means that the ith argument position of F $(0 \leq i \leq arity(F))$ is lazy, and $\Lambda(F, i) = false$ means that it is eager. As an abbreviation, we write $F(!, ?)$ for a function F of arity 2, the first argument of which is eager and the second argument of which is lazy.

This notion is easily extended to paths in terms.

Definition

- *For all terms t, ε is an* eager path *in t.*
- *If p is an eager path in t and* $t|_p = F(t_1, \ldots, t_n)$ *with* $\neg\Lambda(F, i)$ *for some* i $(1 \leq i \leq n)$ *then p.i is eager.*
- *All other paths are* lazy.

In other words, a path is *eager* precisely if it passes through eager arguments only. A lazy path p in t is called *directly lazy* if $p = p'.i$ with i a one-element path, and i lazy in $t|_{p'}$. For example, given the signature *cons*(!,?), bin(!,!)} and the terms $t_1 = cons(x, cons(y, z))$ and $t_2 = \text{bin}(cons(x, y), cons(x, z))$, the paths 1 in t_1 and $1, 1.1, 2, 2.1$ in t_2 are eager; $2, 2.1, 2.2$ in t_1 and $1.2, 2.2$ in t_2 are lazy, of which only 2.1 in t_1 is not also directly lazy. With $Lazy(t)$, we will denote the prefix reduced set of lazy paths in t, and a subterm at a lazy path will be called a *lazy subterm*.

With $\zeta(t)$, we will denote the term obtained by replacing every lazy subterm of t with the unique constant ζ. For any normal form n, $\zeta(n)$ is exactly the part we are interested in. We will say that terms t_1 and t_2 are ζ-equal, or equal *up to* ζ, when $\zeta(t_1) = \zeta(t_2)$.

Ideally, we would like to rewrite a lazy subterm at path p only if this is *needed* to establish a *needed* redex at an eager prefix e of p. Then, the termination behaviour of lazy rewriting would be at least as good as the termination behaviour of rewriting only needed redexes.

If there are overlapping LHSs, however, the notion of needed redex cannot be defined. Therefore, we give a weaker definition, which only requires that a redex at an eager prefix of p can be established by replacing lazy subterms. The ideal of *needed* rewriting can be approximated by demanding a particular relation between the lazy subterms and their replacements. We will not try to achieve this, because most interesting relations seem to be either undecidable or hard to implement or have such a large bias towards a particular strategy that they are unnatural as a restriction on the rewrite relation. Instead, we try to make the restriction on the rewrite relation as weak as possible, by considering only LHSs and outermost lazy positions. The rewrite *strategy* is expected to approximate the ideal by avoiding as much rewrites at lazy paths as reasonably possible. The transformation presented in section 4 implements such a strategy.

We will first present our definition informally, using Figure 3 as illustration. Let the lazy path p consist of an eager path e, a lazy path l, and a path m, which may be either eager or lazy. We allow rewriting at $p = e.l.m$ in t only if at the eager prefix e, a redex g^τ can be established by replacing some lazy subterms of t, such that the nonvariable part of g (shown as a triangle labeled with g) overlaps with l. The endpoints of lazy paths where rewriting is allowed, are indicated by a dotted triangle. The actual rewrite at $e.l.m$ is indicated by an arrow annotated with "LR".

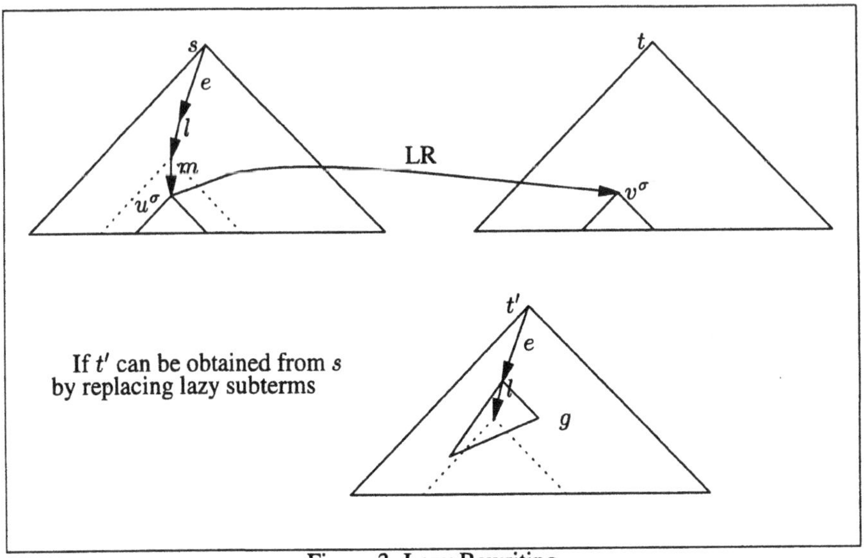

Figure 3: Lazy Rewriting

Formally, this is described by the following definition:

Definition *s rewrites* lazily *to t, written* $s \stackrel{LR}{\rightarrow} t$, *if* $\exists u \rightarrow v \in R, \sigma, p$ *such that*

- $s|_p = u^\sigma$
- $t = s[v^\sigma]_p$
- – *p is eager in s, or*
  – $p = e.l.m$, *where e is eager in s, $e.l$ is lazy in s, and*
  $\exists p_1, \ldots, p_n \in Lazy(s), r_1, \ldots, r_n, t' \in Ter(\Sigma), g \rightarrow h \in R, \tau$
  such that $t' = s[r_1]_{p_1} \cdots [r_n]_{p_n}$, $t'|_e = g^\tau$ *and* $g|_l \notin Var$.

This restriction of the one-step rewrite relation yields an extended class of normal forms. We will call these *lazy normal forms (LNF)*. For instance, given the TRS of Figure 1, if $\Lambda(cons, 2) = true$, then $cons(0, inf(0))$ is an LNF which is not a normal form. If $\Lambda(f, i)$ is *true* for all f, i LNF coincides with WHNF. If t is an LNF, we call $\zeta(t)$ a ζ-LNF.

Because $\stackrel{LR}{\rightarrow}$ is a restriction of $\rightarrow$, it follows easily that termination is preserved. We write $\stackrel{LR*}{\rightarrow}$ for the reflexive-transitive closure of $\stackrel{LR}{\rightarrow}$. We have that lazy rewriting is both correct and complete in the following sense:

Theorem 1 *(Completeness) For all normal forms s of t, there is a ζ-equal LNF s'.*

Proof sketch s' can be constructed from the rewriting sequence $t_1 \rightarrow t_2 \ldots \rightarrow s$ by directly performing the rewrites that are allowed, and maintaining a residual map [O'D77] of the paths where rewriting is not allowed. When a path where rewriting is forbidden is mapped to a context which is either eager, or may turn into a redex by replacing lazy subterms, the suspended rewrite is also performed. Thus, all non-performed rewrites pertain either to a term that is deleted, or are mapped (by the residual map) to a lazy subterm in s'. Therefore, $\zeta(s') = \zeta(s)$ ■

Theorem 2 *(Correctness) If t is an LNF, then for all normal forms s of t,* $\zeta(s) = \zeta(t)$.
Proof From the definition, it follows that there are no redexes at eager paths, and no lazy path leading to a redex has an eager prefix which may become a redex by replacing lazy subterms. Therefore, all eager paths in an LNF are stable ■

Corollary 1 *If there is a unique normal form t of s, then all LNFs of s are ζ-equal to t.*

From the fact that an arbitrary number of "irrelevant" rewrite steps can in general be performed before the rewrite that turns a term into an LNF, it follows that confluence is not preserved. However, given the fact that we are only really interested in $\zeta(n)$ for any LNF n, it is fair to consider only *ζ-confluence*:

Definition *A TRS R is* ζ-confluent *iff for every* t_1, t_2 *and* t_3, *if* $t_1 \rightarrow t_2$ *and* $t_1 \rightarrow t_3$ *then there are terms* t_4, t_5, *such that* $t_2 \overset{*}{\rightarrow} t_4, t_3 \overset{*}{\rightarrow} t_5$ *and* $\zeta(t_4) = \zeta(t_5)$.

Theorem 3 *Lazy rewriting preserves ζ-confluence.*

Proof Suppose R is ζ-confluent, and let t_1, t_2 and t_3 be such that $t_1 \overset{LR*}{\rightarrow} t_2$ and $t_1 \overset{LR*}{\rightarrow} t_3$. Then there are ζ-equal terms t_4, t_5, and rewriting sequences $s_1 : t_2 \overset{*}{\rightarrow} t_4$ and $s_2 : t_3 \overset{*}{\rightarrow} t_5$. If at some term t' from the sequence s_1, a rewrite at a (lazy) path p is forbidden by lazy rewriting, then no redex can later occur at an eager path above p. Therefore, at all eager paths above p, t' has the same function symbol as t_4. We can thus skip the forbidden rewrite and all rewrites that occur below p, because they only affect subterms that do not make a difference from the viewpoint of ζ-equality. Repeating our reasoning for all other forbidden rewrites, we arrive at a term that is ζ-equal to t_4. Similarly for s_2 ■

Of course, ζ-confluence implies uniqueness up to ζ of LNFs.

4 A transformation to achieve laziness

Here, we will specify a transformation $\mathcal{L}$ from TRSs to TRSs and a transformation $\mathcal{T}$ from terms to terms, such that when $\mathcal{T}(t)$ is rewritten by an innermost strategy in $\mathcal{L}(R)$ to a normal form n, then $\zeta(n)$ is the ζ-LNF of t with respect to R. The transformed system avoids rewriting lazy subterms to a large extent (*optimal* avoidance cannot be defined for general TRSs). Basically, the transformation $\mathcal{T}$ replaces all lazy subterms of an input term by irreducible encodings (*thunks*), and $\mathcal{L}$ supplies rules for "unthunking" both input thunks and thunks that encode right-hand sides. Furthermore, $\mathcal{L}$ ensures that

- Lazy subterms of right-hand sides are thunked. Thunks are irreducible because they are built from a constructor symbol and (already normalized) subterms of the left-hand side, so reduction at lazy paths is blocked temporarily.
- When a subterm (matched to a variable) is moved from a lazy *lhs* position into an eager *rhs* position, it is unthunked, so thunks only occur at lazy positions.

- A lazy argument is unthunked before a match overlapping with it is rejected.

We start with some definitions. A *thunk* is a term with a special function symbol δ at the top, a name of a pattern (p) as first argument, and a tuple of terms (denoted by $\mathbf{vec}_n(t_1, \ldots, t_n)$) as second argument:

$$\delta(p, \mathbf{vec}_n(t_1, \ldots, t_n))$$

Given a rule $s \to t$, we call a variable *migrating* if it occurs at a *directly* lazy position in s and at some eager position in t. Because we want to keep the effect of our transformation local, rules must be added that 'unthunk' migrating variables.

4.1 The transformation $\mathcal{L}$

$\mathcal{L}$ takes a TRS (Σ, R), and transforms it into a system $(\Sigma \cup N \cup A, RG \cup RI \cup R')$. In the transformed system,

- N is a countably infinite set of function symbols that do not occur in Σ (they are used in thunks as names of patterns). There is a set $T \subset N$ of "tokens", such that for every function symbol f in Σ, we have a unique $t_f \in T$,
- A is a set of "administrative" function symbols

$$\{\delta(!,!), \delta?(!), \mathbf{inst}(!), \pi(!,!), \mathbf{true}\} \cup_{m,n \in Nat} \{\mathbf{vec}_{mn}(l_{m1}, \ldots, l_{mn})\},$$

  where δ will be used as the top symbol of a thunk, $\delta?$ is a predicate that recognizes thunks, a function $\mathbf{vec}_{mn}(l_{m1} \ldots l_{mn})$ is used to "pack" n variables in a thunk (m encodes the laziness annotations: the l_{mi} are either ! or ?. Mostly, m will be omitted). Finally, π is a projection function that makes implementation of *graph rewriting* easy, which will be discussed in Section 5.1.
- RG contains the general rules defining the projection π and the thunk-recognizer $\delta?$:

$$\pi(x, y) \to y$$
$$\delta?(\delta(x, y)) \to \mathbf{true}$$

- RI contains the rules describing selective unthunking of input terms. For every f with arity n, of which k are eager positions (with indices $e_1, \ldots, e_k$), RI contains the rules (with $c_{f_i} \in N$):

$$\mathbf{inst}(\delta(t_f, \mathbf{vec}_n(x_1, \ldots, x_n))) \to c_{f_1}(\delta?(x_{e_1}), x_1, \ldots, x_n) \quad (1)$$
$$c_{f_1}(\mathbf{true}, x_1, \ldots, x_n) \to c_{f_2}(\delta?(x_{e_2}), x_1, \ldots, \mathbf{inst}(x_{e_1}), \ldots, x_n) \quad (2)$$
$$c_{f_1}(\delta?(y), x_{e_1}, \ldots, x_n) \to c_{f_2}(\delta?(x_{e_2}), x_1, \ldots, x_{e_1}, \ldots, x_n) \quad (3)$$
$$\ldots$$
$$c_{f_k}(\mathbf{true}, x_1, \ldots, x_n) \to f(x_1, \ldots, inst(x_{e_k}), \ldots, x_n) \quad (4)$$
$$c_{f_k}(\delta?(y), x_1, \ldots, x_n) \to f(x_1, \ldots, x_{e_k}, \ldots, x_n) \quad (5)$$

  Here, (1) starts the instantiation of a delayed term with function symbol f, (2,4) handle the case that an argument (x_{e_1} and x_{e_k}, respectively) is still thunked and (3,5) handle the case that an argument is already unthunked. Note that the distinction between thunked and unthunked arguments relies on the partial function $\delta?$ being evaluated eagerly.

- The rules in R' are obtained by applying the three transformations below (RHS for thunk introduction, LR for left-right unthunking and LS for left-hand side matching) to R as follows: RHS until fixpoint, LR once for every equation in the fixpoint, LS once for every equation in the result of LR.

RHS (Thunk Introduction) This transformation is applicable to all rules $r : s \to t$ where t contains a directly lazy path p, such that $t|_p$ is neither a variable, nor a subterm already occurring in s, nor a thunk. Let $\{t_1, \ldots, t_n\}$ be the set of terms occurring in both s and $t|_p$, and prefix-reduced with respect to t, then r is *replaced* by two rules (i unique in N):

$$s \to t[\delta(i, \mathbf{vec}_n(t_1, \ldots, t_n))]_p$$
$$\mathbf{inst}(\delta(i, \mathbf{vec}_n(x_1, \ldots, x_n))) \to \pi(\delta(i, \mathbf{vec}_n(x_1, \ldots, x_n)), t[t_i/x_i])$$

Here $\Lambda(\mathbf{vec}_n, i) =?$ if and only if t_i is a variable occuring at a directly lazy path in s.

LR (Migrating Thunk Elimination) This transformation applies to rules $r : s \to t$ containing *migrating* variables. Supposing $\{t_1, \ldots, t_n\}$ is a set of subterms which occur both in s and t, and which is prefix-reduced with respect to t, and let $e_1, \ldots, e_k$ be the indices of the *migrating* variables, then r is replaced by the following rules, similar in form and intent to the rules in RI:

$$s \to c_{i_1}(\delta?(x_{e_1}), t_1, \ldots, t_n)$$
$$c_{i_1}(\mathbf{true}, x_1, \ldots, x_n) \to c_{i_2}(\delta?(x_{e_2}), x_1, \ldots, \mathbf{inst}(x_{e_1}), \ldots, x_n)$$
$$c_{i_1}(\delta?(y), x_1, \ldots, x_n) \to c_{i_2}(\delta?(x_{e_2}), x_1, \ldots, x_{e_1}, \ldots, x_n)$$
$$\ldots$$
$$c_{i_k}(\mathbf{true}, x_1, \ldots, x_n) \to c_{i_k+1}(x_1, \ldots, \mathbf{inst}(x_{e_k}), \ldots, x_n)$$
$$c_{i_k}(\delta?(y), x_1, \ldots, x_n) \to c_{i_k+1}(x_1, \ldots, x_{e_k}, \ldots, x_n)$$
$$c_{i_k+1}(x_1, \ldots, x_n) \to t[t_i/x_i]$$

LS (Matching Thunk Elimination) This transformation is applicable to rules $r : s \to t$ if s contains nonvariable lazy positions. For every element $i = \{i_1, \ldots, i_n\}$ in the prefix-reduced powerset of lazy paths in s, add a rule (all n_j and v_j fresh):

$$s[\delta(n_1, v_1)]_{i_1} \ldots [\delta(n_n, v_n)]_{i_n}$$
$$\to s[\pi(\delta(n_1, v_1), \mathbf{inst}(n_1, v_1))]_{i_1}[\delta(n_2, v_2)]_{i_2} \ldots [\delta(n_n, v_n)]_{i_n}$$

4.2 The transformation $\mathcal{T}$

$\mathcal{T}$ thunks all non-variable lazy subterms of the original input term, by the token of their outermost function symbol and their thunked arguments.

$$\mathcal{T}[f(t_1, \ldots, t_n)] = f(t'_1, \ldots, t'_n) \text{ (where } t'_i = \mathcal{T}[t_i] \text{ iff } \neg\Lambda(f, i)\text{, otherwise } t'_i = \mathcal{T}_l[t_i])$$
$$\mathcal{T}_l[f(t_1, \ldots, t_n)] = \delta(t_f, \mathbf{vec}_n(\mathcal{T}_l[t_1], \ldots, \mathcal{T}_l[t_n]))$$
$$\mathcal{T}[x] = \mathcal{T}_l[x] = x$$

4.3 Correctness and completeness of the transformation

First, we remark that the transformation itself terminates, because every application of RHS replaces one (non-thunked) lazy argument by a thunk, and LR and LHS terminate trivially.

Theorem 4 *(Correctness of $\mathcal{L}$ and $\mathcal{T}$) Given a TRS R and a term t, every step in an innermost rewriting of $\mathcal{T}(t)$ in $\mathcal{L}(R)$ is either an administrative step (checking if an argument is a thunk), or it corresponds to a legal step in $\stackrel{LR}{\rightarrow}_R$.*

Proof Note that for all terms t, $\mathcal{T}(t)$ has only R-redexes above lazy positions, because all lazy subterms are thunked by $\mathcal{T}_l$. By RHS, all rules have been transformed into rules that put normal forms at lazy paths, and LR preserves this property. The only redexes at lazy paths are $\mathcal{L}(R)$-redexes, introduced by LS, but the conditions for application of LS imply that there is a nonvariable R-pattern overlapping with the hole in which the redex is introduced, so the condition for lazy rewriting is fulfilled ■

Theorem 5 *(Completeness of $\mathcal{L}$ and $\mathcal{T}$) Given a ζ-confluent TRS R and a term t, any normal form t_n of $\mathcal{T}(t)$ with respect to $\mathcal{L}(R)$ is ζ-equal to any LNF t_l of t.*

Proof Because of correctness, we have that $t \stackrel{LR}{\twoheadrightarrow} t'_n$, where t'_n is obtained from t_n by replacing thunks with the terms they encode (so $\zeta(t'_n) = \zeta(t_n)$). Because t_l is a LNF of t, we have that $t \stackrel{LR}{\twoheadrightarrow} t_l$. Lazy rewriting preserves ζ-confluence, so there must be a term t'_l with $\zeta(t_l) = \zeta(t'_l)$ such that $t'_n \stackrel{LR}{\twoheadrightarrow} t'_l$. Because t_n only differs from t'_n by having thunks at lazy paths, and LS introduces rules that remove any thunk which blocks matching of an LHS at an eager path, all the rewrites in the sequence $t'_n \stackrel{LR}{\twoheadrightarrow} t'_l$ occur at lazy paths. Therefore, $\zeta(t'_l) = \zeta(t'_n)$, so $\zeta(t_l) = \zeta(t_n)$ ■

5 From transformation to implementation

The transformation in Section 4 is useful both as a tool for experimentation, and as a concise model of an implementation of lazy rewriting. To obtain an implementation that competes with special-purpose lazy implementations such as TIM ([FW87]) or the Spineless Tagless G-machine (STG, [JS89]), some details have to be changed.

First, in order to prevent multiple reductions of the same term, the TRS should be interpreted as a *graph* rewriting system. We give details on this in Section 5.1.

Second, some glaring inefficiency is caused by the LS transformation. This can be overcome by simulating the effect of LS in the pattern-matching code, which is explained in Section 5.2.

5.1 Graph rewriting by adding sharing

By the following modifications, the advantages of graph rewriting are incorporated:

- In the implementation of $\mathcal{T}$, sharing should be retained.

- The function $\pi(!, !)$ is implemented such, that it overwrites its first argument (always a thunk) with the LNF of its second argument (always the LNF corresponding with the thunk). Note that this requires a fixed node size, or some other means to avoid overwriting smaller with bigger nodes.
- If a subterm occurs both in LHS and RHS of some rule, no copy should be made. Then it follows from the construction of the transformed system, that thunks are never duplicated, so every thunk is only evaluated once.
- For cyclic graphs, the code that is generated for the construction of a right-hand side must be modified slightly. Without loss of generality, we consider a prototypical RHS $x : f(\ldots, x, \ldots)$. For this RHS, the compiler should emit code corresponding to $\mathbf{inst}(T)$, where T is a thunk for $f(\ldots, T, \ldots)$. Note that this requires that the "address" of a node under construction is available during the construction.

5.2 Optimizations

When implemented naively, our transformation has a large impact on the number of equations. A worst-case analysis shows that the maximal number of additional equations is

$$3 + n.r + n.2^l + 2s$$

where n is the number of rules, r is the maximal number of nonvariable lazy positions in a RHS, l is the maximal number of nonvariable lazy positions in a LHS, and s is the number of lazy positions in the signature. It should be noted that, measured in function symbols, the rules added by RHS are compensated for by a reduction in size of the original rule, and s is generally small compared to n.

Thus, the only dangerous term is the exponential term in l, caused by the powerset construction in transformation LS. We will illustrate both the problem and its solution with an example. Assuming we have a signature $\{a, b, i(!), t(?, ?)\}$ and a rule $i(t(a, b)) \to a$, then LS adds the rules

$$i(t(\delta(p, \mathbf{vec}_O), b)) \to i(t(\pi(\delta(p, \mathbf{vec}_O), \mathbf{inst}(\delta(p, \mathbf{vec}_O))), b))$$
$$i(t(\delta(p, \mathbf{vec}_O), \delta(p', \mathbf{vec}_O))) \to i(t(\pi(\delta(p, \mathbf{vec}_O), \mathbf{inst}(\delta(p, \mathbf{vec}_O))), \delta(p', \mathbf{vec}_O)))$$
$$i(t(a, \delta(p, \mathbf{vec}_O))) \to i(t(a, \pi(\delta(p, \mathbf{vec}_O), \mathbf{inst}(\delta(p, \mathbf{vec}_O)))))$$

When a term $i(t(x, y))$ is rewritten, where both x and y are thunks which will instantiate to a and b respectively, this leads to the following inefficiencies:

- i and t are matched 3 times (2 times to discover the thunks, and the last time to find the original match),
- the function symbol t is copied 2 times, because the subterm from the LHS cannot be reused.

This can be repaired by changing the pattern matching code to instantiate the thunks, such that the rules introduced by LS are no longer needed (even though they give a nice model of what is happening). In the pseudo code below, italics indicate the modifications that remove the need for the LS transformation:

```
case x of
i(y): case y of
   t(z1,z2): case z1 of
      a: label1: case z2 of
                 b: continue(a) /* matched ! */
                 thunk: inst(z2); goto label1
                 otherwise: return(x) /* normal form */
      thunk: label2: case z2 of
         b: inst(z1)
           case z1 of
           a) continue(a) /* matched ! */
           otherwise: return(x) /* normal form */
         thunk: inst(z2); goto label2
         otherwise: return(x) /* normal form */
      otherwise: return(x) /* normal form */
   otherwise: ...
otherwise: ...
```

This pattern matching code is bigger than the code for the single rule in the original system, but it is somewhat smaller than the code that corresponds to the transformed system. The increase in time with respect to eager matching is linear in the number of unevaluated thunks encountered during matching, because every unevaluated thunk causes a jump to the state from which the thunk was discovered, and a repeated atomic match operation on the instantiation of the thunk. This means, that laziness is only paid for if it is actually used!

The implementation can be further improved by implementing δ as a tag-bit, $\delta?$ as a bit-test, and **inst** and $\mathbf{vec}_n$ as built-in functions. Finally, the effect of the LR transformation can be achieved by generating slightly different code for right-hand sides.

6 Related work

A very early related paper is [Plo75], which gives simulations of call-by-name by call-by-value (eager evaluation), and vice versa, in the context of the λ-calculus. Call-by-name evaluation differs from lazy evaluation (or call-by-need): Thunks are not overwritten with the result of evaluation, but evaluated on every use (which is essential in a language with side-effects).

In the context of functional programs, [Amt93] developed an algorithm to transform call-by-name programs into call-by-value equivalents. In [SW94], dataflow analysis is done in order to minimize thunkification in this context.

In [OLT94], a continuation passing style (cps) transformation of call-by-need into call-by-value equivalents is given. To their knowledge, it is the first. Apart from the fact that they consider a particular λ-calculus, whereas we consider general TRSs, our transformation differs mainly by completely integrating pattern matching of algebraic datatypes in the transformation. It is unclear how much can be gained by taking pattern matching into account in a transformation for a lazy functional implementation. An abstract approach to strictness analysis of algebraic datatypes is investigated in [Ben93].

The effect of our transformations of rewrite systems is somewhat similar in spirit to the use of evaluation transformers in [Bur91]. Not only in theory, but also in practice, our technique does not rely on properties of built-in algebraic datatypes such as lists or trees. In [BM92], some of the techniques in [Bur91] are formulated in the context of continuation passing transformations.

Another approach to obtain better termination properties are the sequential strategies investigated by [HL79, O'D77]. In this approach, only *needed* redexes are rewritten, i.e., redexes that would be rewritten in any reduction to a normal form. Unfortunately, *neededness* is only well-defined in TRSs that do not have overlapping redexes. This restriction is hard to live with in practice.

To our knowledge, only the Clean [PvE93] and the OBJ3 systems support laziness annotations. Clean supports the annotation of *strict* arguments, OBJ3 [GWM$^+$92] features annotations for the evaluation order of arguments which are somewhat more explicit than ours. It appears that a similar transformation can implement OBJ's annotations.

A rule occuring in the context of an E-unification algorithm, presented in [MMR86], is called "lazy rewriting" in [Klo92]. It might be interesting to investigate if our technique of implementing lazy rewriting on eager machinery is useful in that context.

In CAML (Categorial ML, [CH90]) there are lazy constructors, which can be used to achieve similar effects as our transformation does. However, the transformation of the program must then be carried out manually for the most part (only equivalents of **inst**, δ? and δ are supplied by the implementation).

It is obvious, that our final implementation of lazy term rewriting is similar to the implementation of modern lazy functional languages. As far as we know, these implementations are completely lazy by nature, but are optimized to perform as much eager evaluation as possible.

Therefore, it is appropriate to provide a discussion of the cost of basic datastructures and actions in our scheme, compared with the cost in those implementations. It should be noted that it is *extremely* difficult ([JS89]) to assess the effect of different design choices on performance, so we will only give a qualitative discussion.

- Only a little structure (δ, a thunk constant and a vector containing references to subterms from the left-hand side) occurs below a lazy position in any rhs after the transformation. This is comparable to the frames used in TIM [FW87], or the closures in the STG. Similarly to the latter, our scheme only uses space for the subterms from the LHS that may actually be used later. In the ABC machine [PvE93], complete graphs are built for lazy arguments, which is a drawback compared to all other implementations.
- No runtime cost is incurred when all arguments in the original TRS are annotated eager. Even when all arguments are found to be strict, TIM and STG do a function call to obtain the tag of a constructor term (this is the reason they are called "tagless"), whereas our implementation only needs to dereference a pointer.
- There is no need for the dreaded indirection nodes ([O'D77, JL92], because δ fullfills this role; every term (input or rhs) is evaluated exactly once, either by immediate innermost rewriting, or later, by overwriting the δ node. In [JL92], the indirection nodes are also transformed away, but some very complicated analysis is needed to

arrive at this result. In the ABC machine, the indirection nodes are indispensable.

- In the rules added by transformation LR, testing if a lazy argument is thunked, is done by rewriting. Even if this is replaced by a bit-test implementation, a subsequent call of **inst** must be done. This is less efficient than the "tagless" reduction which is done in both TIM and STG.
- Unthunking is only done if all eager pattern matching was succesful. Because the order of pattern matching and its effects on evaluation of subterms are fixed in the semantics of lazy functional languages, this cannot be done in the other implementations. Usually, the interaction between pattern matching of algebraic datatypes and lazy evaluation is not incorporated in strictness analysis.

Taking into account these points, we expect our scheme to perform better than ABC, TIM or the STG, when there is a small number of lazy arguments.

It is clear, that a strictness analyzer can provide laziness annotations (by annotating all arguments that are not found to be strict). However, strictness analyzers being very conservative beasts, this will lead to far too many annotated arguments. So, how much work is involved in finding laziness annotations manually? It is well-known, that even with *lazy* functional programming languages, a thorough understanding of a program is required to make sure that it terminates. In our experience, this level of understanding is adequate to provide complete laziness annotations. Therefore, we hold that programmer-provided laziness annotations are a suitable way of achieving lazy evaluation.

7 Conclusions and acknowledgements

We have defined *lazy rewriting* and have generalized the notion of *Weak Head Normal Form* to the less operational notion of *Lazy Normal Form*.

We have modeled lazy rewriting by a transformation of term rewriting systems, which avoids rewriting of lazy subterms to a large extent (optimal avoidance cannot be defined for general TRSs), and completely integrates pattern matching of algebraic datatypes. When all arguments are annotated, the transformed system computes WHNFs.

We derive an efficient implementation on already efficient eager machinery from this model. Our method compares favourably to existing methods.

Our notion of Lazy Normal Forms (LNFs) could also be helpful in an implementation of *abstract rewriting*, as described in [BEØ93], or in the context of *theorem proving*.

We would like to thank John Field for his very insightful comments on an earlier version of this paper, and Jan Heering for his meticulous reading of a later version.

References

[Amt93] Torben Amtoft. Minimal thunkification. In *Third International Workshop on Static Analysis, Padova, Italy*, volume 724 of *Lecture Notes in Computer Science*, pages 218–229. Springer-Verlag, 1993.

[Bak94] Henry G. Baker. Cons should not cons its arguments, part II: Cheney on the M.T.A. Draft Memorandum, January 1994.

[Ben93] P.N. Benton. Strictness properties of lazy algebraic datatypes. In *Third International Workshop on Static Analysis, Padova, Italy*, volume 724 of *Lecture Notes in Computer Science*, pages 206–217. Springer-Verlag, 1993.

[BEØ93] Didier Bert, Rachid Echahed, and Bjarte M. Østvold. Abstract rewriting. In *Third International Workshop on Static Analysis, Padova, Italy*, volume 724 of *Lecture Notes in Computer Science*, pages 178–192. Springer-Verlag, 1993.

[BM92] Geoffrey Burn and Daniel Le Métayer. Cps-translation and the correctness of optimising compilers. Technical Report DoC92/20, Imperial College, Department of Computing, 1992.

[Bur91] Geoffrey Burn. *Lazy Functional Languages: Abstract Interpretation and Compilation*. Pitman, 1991.

[BvEJ+87] H.P. Barendregt, M.C.J.D. van Eekelen, J.R.W.Glauert, J.R. Kennaway, M.J. Plasmeijer, and M.R. Sleep. Term graph rewriting. In J.W. de Bakker, A.J. Nijman, and vol. II P.C. Treleaven, editors, *Proceedings PARLE'87 Conference*, volume 259 of *Lecture Notes in Computer Science*, pages 141–158. Springer Verlag, 1987.

[CH90] Guy Cousineau and Gérard Huet. The CAML primer. Technical report, Inria, 1990. Version 2.6.1, available by ftp from ftp.inria.fr.

[DJ90] N. Dershowitz and J.-P Jouannaud. Rewrite systems. In J. van Leeuwen, editor, *Handbook of Theoretical Computer Science, Vol B.*, pages 243–320. Elsevier Science Publishers, 1990.

[Fie90] J. Field. On Laziness and Optimality in Lambda Interpreters: Tools for Specification and Analysis. In *Proc. ACM Conference on Principles of Programming Languages, San Francisco*, 1990.

[FW87] Jon Fairbairn and Stuart Wray. Tim: A simple, lazy abstract machine to execute supercombinators. In Gilles Kahn, editor, *Functional Programming Languages and Computer Architecture*, volume 274 of *Lecture Notes in Computer Science*, pages 34–45. Springer-Verlag, 1987.

[GWM+92] J.A. Goguen, T. Winkler, J. Meseguer, K. Futatsugi, and J.P. Jouannaud. Introducing OBJ. In J.A. Goguen, D. Coleman, and R. Gallimore, editors, *Applications of Algebraic Specification Using OBJ*. Cambridge University Press, 1992.

[Heu88] Thierry Heuillard. Compiling conditional rewriting systems. In S. Kaplan and J.P. Jouannaud, editors, *Proceedins of the First International Workshop on Conditional Term Rewriting Systems*, volume 308 of *Lecture Notes in Computer Science*, pages 111–128. Springer-Verlag, 1988.

[HL79] G. Huet and J.-J. Lévy. Call-by-need computations in non-ambiguous linear term rewriting systems. Technical Report 359, INRIA, 1979. Also appeared as: *Computations in Orthogonal Rewriting Systems part I and II*, in: 'Computational Logic; essays in honour of Alan Robinson' (eds. J.-L. Lassez and G. Plotkin), MIT Press, Cambridge, MA, 1991, 395-443.

[Ing61] P.Z. Ingermann. Thunks – a way of compiling procedure statements with some comments on procedure declarations. *Communications of the ACM*, 4(1):55–58, 1961.

[JL92] Simon L Peyton Jones and David Lester. *Implementing Functional Languages – A Tutorial*. Prentice Hall, 1992.

[JS89] Simon L Peyton Jones and Jon Salkild. The Spineless Tagless G-machine. In *Functional Programming and Computer Architecture*, pages 184–201. ACM, 1989.

[KKSdV93] J.R. Kennaway, J.W. Klop, M.R. Sleep, and F.J. de Vries. The adequacy of term graph rewriting for simulating term rewriting. In Ronan Sleep, Rinus Plasmeijer, and Marko van Eekelen, editors, *Term Graph Rewriting: Theory and Practice*. John Wiley & Sons Ltd, 1993.

[Klo92] J.W. Klop. Term rewriting systems. In S. Abramsky, D. Gabbay, and T. Maibaum, editors, *Handbook of Logic in Computer Science, Volume 2.*, pages 1–116. Oxford University Press, 1992.

[KW93] J.F.Th. Kamperman and H.R. Walters. ARM – Abstract Rewriting Machine. In H.A. Wijshoff, editor, *Computing Science in the Netherlands*, pages 193–204, 1993.

[Mar92] Luc Maranget. *La stratégie paresseuse*. PhD thesis, L'Université Paris VII, July 1992.

[MMR86] A. Martelli, C. Moiso, and C.F. Rossi. An algorithm for unification in equational theories. In *Proceedings of the Symposium on Logic Programming*, pages 180–186. IEEE Computer Society, 1986.

[Myc80] Alan Mycroft. The theory and practice of transforming call-by-need into call-by-value. In B. Robinet, editor, *International Symposium on Programming*, volume 83 of *Lecture Notes in Computer Science*. Springer-Verlag, 1980.

[O'D77] M.J. O'Donnell. *Computing in Systems Described by Equations*, volume 58 of *Lecture Notes in Computer Science*. Springer-Verlag, 1977.

[OLT94] Chris Okasaki, Peter Lee, and David Tarditi. Call-by-need and continuation-passing style. *Lisp and Symbolic Computation*, 7:57–82, 1994.

[Plo75] G. D. Plotkin. Call-by-name, call-by-value and the λ-calculus. *Theoretical Computer Science*, 1(1):125–159, 1975.

[PvE93] M J. Plasmeijer and M C J D. van Eekelen. *Functional Programming and Parallel Graph Rewriting*. Addison Wesley, 1993.

[SW94] Paul Steckler and Mitchell Wand. Selective thunkification. In *First International Static Analysis Symposium*, Namur, Belgium, 28-30 September 1994. also available by ftp as sas94.ps.Z from ftp.ccs.neu.edu:/pub/people/steck.

[TAL90] David Tarditi, Anurag Acharya, and Peter Lee. No assembly required: Compiling Standard ML to C. Technical Report CMU-CS-90-187, School of Computer Science, Carnegie Mellon University, november 1990.

A Rewrite Mechanism for Logic Programs with Negation

Siva Anantharaman and Gilles Richard

LIFO, Dépt. d'Informatique, Université d'Orléans
B.P. 6759, 45067 Orléans Cedex 02 (Fr.)
e-mail: {siva, richard}@lifo.univ-orleans.fr

Abstract. Pure logic programs can be interpreted as rewrite programs, executable with a version of the Knuth-Bendix completion procedure called linear completion. The main advantage here is in avoiding many of the loops inherent in the resolution approach : for most productive loops, linear completion yields a finite set of answers and a finite set of rewrite rules (involving just one predicate), from which all the remaining answers can be deduced. And this 'program synthesizing' aspect can be easily combined with other loop avoiding techniques using 'marked' literals and substitutions. It is thus natural to ask how much of the rewrite mechanism carries through for deducing negative information from pure programs, and more generally for any normal logic program. In this paper we show that such an extension can be built in a natural way, with ideas from the Clark completion for normal logic programs, and the domain of constrained rewriting. The correction and completeness of this extended mechanism is proved w.r.t. the 3-valued declarative semantics of Künen for normal programs. We also point out how the semantics of a normal program can in a certain sense be 'parametrized', in terms of the 'meta-reduction' rule set of our approach.

Keywords : logic program - negation - rewrite system - linear completion - constraints

1 Introduction

Logic programming by linear completion seems to have been first investigated in [7]. Predicates are encoded as boolean valued functions and the clauses are turned into rewrite rules. The mechanism to execute the 'rewrite program' thus obtained, is a version of the Knuth-Bendix completion procedure, called *linear completion*. For any pure logic program P, and any query $Q(x)$?, in the rewrite mechanism described in [7], [2], a substitution $x \leftarrow x_0$ is an answer for $Q(x)$?, w.r.t. P iff $Q(x_0)$ can be reduced to a certain (new) symbol '$\top$'. But the mechanism of [7], [2] is limited to pure (i.e. negation-free) logic programs and to queries for 'positive information', which are only part of our concern in logic programming. For instance, for the pure logic program $P : p(a). \, , \; q(b).$, the atom $p(b)$ is 'intended false' (that is the literal $\neg p(b)$ is 'intended true'). Now P gets translated (in the above approach) into the rewrite program: $p(a) \rightarrow \top, \quad q(b) \rightarrow \top$, which a priori can in no way reduce $\neg p(b)$ to '$\top$'. Our aim in this paper is therefore to extend the above rewrite mechanism to cover logic programming with negation : that is, deal with negative information as well as negative formulæ. As in [2], we proceed in two steps :

- first we define a translation function ψ, which takes as argument a logic program (possibly with negation) P, and produces a rewrite program $\psi(P)$, which in general will be constrained.

- next we define an inference system to describe the operational mechanism of an extended linear completion in presence of negation.

The initial hints for this refinement come from the Clark completion mechanism ([11]); rewrite rules under constraints appear for this reason. For instance, in our translation of the above 2-clause program, will also figure the following two constrained rewrite rules :

$$p(x) \rightarrow \bot\, [\![x \neq a]\!], \qquad q(x) \rightarrow \bot\, [\![x \neq b]\!]$$

where '$\neq$' means 'syntactic disequality'. We will also need to extend the initial signature with a few further symbols : in particular, a new unary predicate '$\supset$' and a new constant '$\bot$', will be introduced, with the respective intended semantics 'implies', and 'false'.

An inference step called 'Resolve-Neg' ('Resolution under Negation') will be playing a major role in our inference mechanism. Such steps will in general yield constrained rewrite rules, among which some will have a few 'box'-arguments ; these symbolize variables (implicitly) quantified existentially, in the logical sense. The usual rewrite steps on such rules (like simplification) will therefore be meaningful only when these 'boxes' are eliminated or sufficiently instantiated. This is the role of our 'Constraint Manipulating' inference step.

These are the only two inference rules which get added in our approach, to the linear completion mechanism of [2]. Negative queries, with variables, can then be easily treated, in a manner which is 'incremental from the logic programming viewpoint'. The solutions are obtained under the form of constraints. The loop avoiding facility, offered by simplification in the pure case, is preserved. The constraints which appear, are in general conjunctions and/or disjunctions of 'equalities' or 'disequalities' (w.r.t. the variables of the query) in the Clark Equational Theory ('CET'). This paper is organized as follows. In Section 2, we briefly recall the basic results on the rewrite approach for pure logic programs and positive queries, from [2] ; a second subsection will then present the extensions needed for treating negative queries. In section 3, we show that no additional inference-steps are needed, for considering normal logic programs by the rewrite approach. We show in particular that 'rewrite-semantics' can be defined for normal logic programs, in complete conformity with their 3-valued Fitting/Künen declarative semantics. The notion of 'rewrite synthesis' for normal programs is also presented, and its role in loop avoiding is studied briefly. Section 4 points out how these rewrite semantics can be in a sense parametrized, by the choice of the *meta-reduction rules* of the rewrite mechanism, and how by doing so, one can go *beyond* Künen's semantics, if so desired. The concluding section presents some related works, and a few directions for future work.

2 The Rewrite mechanism for Pure Logic Programs

We begin with an example which shows how rewriting can be a tool for 'synthesizing' logic programs. Consider the following classical program for addition over integers :

$$plus(0,x,x).\,, \qquad plus(s(x),y,s(z)) :- plus(x,y,z).$$

If we consider the query : $plus(x, s(0), y)$?, it is well-known that a resolution-based interpreter will 'loop forever', giving the infinite set of ground answers :

$$plus(0, s(0), s(0)), plus(s(0), s(0), s(s(0))), \ldots$$

Such an interpreter cannot give here a finite representation of the infinite set of solutions. Let us now view the above program as being equivalent to the following 'rewrite program' :

$$plus(0, x, x) \rightarrow \top, \qquad plus(s(x), y, s(z)) \rightarrow plus(x, y, z)$$

For doing this, the new symbol '$\top$' (with intended semantics 'true') is added to the signature. For treating the above query, a new predicate Ans is also added, with arity equal to the number of free variables in this query ; and the query is transformed into the 'goal rule' $plus(x, s(0), y) \rightarrow Ans(x, y)$. The linear completion mechanism gives rise then to the following search tree :

$$plus(x, s(0), y) \longrightarrow Ans(x, y)$$

overlap / overlap

$$Ans(0, s(0)) \rightarrow \top \qquad plus(x1, s(0), y1) \longrightarrow Ans(s(x1), s(y1)$$

simplify

$$Ans(s(x1), s(y1)) \rightarrow Ans(x1, y1)$$

No inference rule is now applicable, and the last rewrite rule obtained (which symbolizes the logical-equivalence $Ans(s(x_1), s(y_1)) \leftrightarrow Ans(x_1, y_1)$), together with the solution obtained in the first branch, gives a nice way of representing the whole set of ground solutions. The essential step rendering this possible in the rewrite approach is the notion of simplification, which is absent in logic programming. The point here is that, whatever be the number of predicates in the initial program, the 'synthesized rewrite program' obtained by the rewrite mechanism at the end, will contain *only* the predicate Ans, and is thus much simpler to manipulate.

We briefly describe now, in formal terms, the linear completion mechanism in the pure case, essentially as presented in [2]; this is necessary for a clear understanding of our approach, presented subsequently as its extension to normal programs.

2.1 Linear Completion for Pure Logic Programs

Suppose given an alphabet containing variables, function symbols and predicate symbols. Terms, atomic formulæ or atoms, substitutions and instances are defined in the usual way. In general, $\overline{t}$ will denote a list of terms $t_1, \ldots, t_n$, and $\overline{x}$, a list of variables. A list of atoms $A_1, \cdots, A_n$ will be denoted with an overline as $\overline{A}$ (and this will mean their conjunction). A *simplification ordering* on terms is an irreflexive and transitive relation on these, which is stable under substitutions, and 'monotonic' (i.e. stable under contexts) , and is such that any term is strictly bigger than any of its proper subterms; such a notion is easily extended to any first order signature. Such orderings are very easily built over any specified precedence ordering on the symbols. We assume given one such ordering '$\succ$'.

A *clause* is a pair $(head, body)$ where $head$ is an atom and $body$ is a set of atoms. If the body is empty, we have a $fact$. A clause is written under the form : $head :- body$. ($head$. for a fact). Consider any logic program P (i.e. a set of clauses), with no negation.

According to the 'translation mechanism' ϕ of [2], any fact A of P is transformed into the rewrite rule $A \rightarrow \top$. Any clause $A :- \overline{B}$ in P is (in general) transformed into the rewrite rule $A, \overline{B} \rightarrow \overline{B}$, *except* when the ordering '$\succ$' is such that $A \succ \overline{B}$ *and no other head in P unifies with A* : in such a case the clause gets transformed more simply into $A \rightarrow \overline{B}$.

This translation introduces a new constant '$\top$', which will be subsequently assumed minimal w.r.t. the ordering $\succ$. For instance, the *plus* program above gets transformed 'more simply', but the following program :

$$plus(0,x,x).,\quad suc(x,s(x)).,\quad plus(s(x),y,z) :- suc(z_1,z), plus(x,y,z_1).$$

is translated into the rewrite program :

$$plus(0,x,x) \rightarrow \top,\qquad suc(x,s(x)) \rightarrow \top$$
$$plus(s(x),y,z),\ suc(z_1,z),\ plus(x,y,z_1) \ \rightarrow\ suc(z_1,z), plus(x,y,z_1)$$

To describe the operational semantics for the rewrite program $\phi(P)$ (as in [2]), one needs to define what a query will be in this context ; for this a special predicate Ans (assumed 'new') is introduced (with arity equal to the 'number of free variables in the query'); the ordering $\succ$ is extended such that for every symbol s other than '$\top$', we have $s \succ Ans \succ \top$.

Definition 1. i) A query rule is a rule of the form : $\overline{Q} \rightarrow Ans(\overline{x})$ where $\overline{x}$ denotes the set of free variables appearing in $\overline{Q}$.

ii) A state of computation is a triple $(P; G; M)$ where : P denotes a rewrite program, G denotes a rule (the current 'goal rule' to process), and M denotes a set of rules which are the ancestors of the current goal. M is the 'stored memory', containing the goal rules used to simplify the current goal rule.

iii) An initial state of a rewrite program P is a triple $(P; \overline{Q} \rightarrow Ans(\overline{x}); \emptyset)$.

The objective now is that if σ is a substitution which is an answer for a query $\overline{Q}$ w.r.t. a pure logic program P, then the inference mechanism we look for, must be able to lead from an initial state $(P;\ \overline{Q} \rightarrow Ans(\overline{x});\ \emptyset)$ to a 'final state' of the form $(P; \sigma(Ans(\overline{x})) \rightarrow \top; M')$. The following inference system given in [2], called *linear completion* that we will denote by LC, does the job for queries without negation.

Rule 1) **Answer** : Stop if the current state is of the form $(P; Ans(\sigma(\overline{x})) \rightarrow \top; M)$

Rule 2) **Delete** : Suppress the goal rules of the form '$X \leftrightarrow X$'.

Rule 3) **Orient** : Orient the goal rules w.r.t. the ordering '$\succ$'.

Rule 4) **Simplify** : Simplify the current goal rule in G, with the rules in the rewrite program P, and with the rules in the current 'memory' M.

Rule 5) **Overlap** : From any chosen goal rule, generate a new goal rule by 'overlapping' it with the rewrite rules in P. And store the chosen goal rule in M.

Besides, at any stage of computation, any current goal rule $\overline{L} \rightarrow \overline{R}$ is also 'reduced' into a normal form using the following two *meta*-reduction rules :

$$(X, \top) \rightarrow X\,,\qquad (X, X) \rightarrow X$$

Here, and for the rest of the paper, variables denoted by capitals are assumed to be meta-variables. This set of rules will be called the meta-reduction system and denoted MR in the following. We write : $(P;\ Q;\ M)\ \vdash_{LC}\ (P;\ Q';\ M')$ if the latter state derives from the former, by the application of one or more steps of the linear completion inference mechanism.

Definition 2. If $(P\,;\,\overline{Q} \rightarrow Ans(\overline{x})\,;\,\emptyset) \vdash_{LC} (P;\,Ans(\sigma(\overline{x})) \rightarrow \top\,;\,M')$ then we write : $P \vdash_{LC} \sigma(\overline{Q})$, and say that σ is a computed answer to the query $\overline{Q}$.

Let P be a logic program, without negation, and $\mathcal{B}_P$ the Herbrand base associated to P (i.e., the set of ground atoms built with the symbols appearing in P) ; the smallest Herbrand model of P (included in $\mathcal{B}_P$) is defined as the canonical declarative semantics for P. This set denoted by M_P is the set of ground atomic logical consequences of P.

Let now $\phi(P)$ be the rewrite program deduced from P, and set $\phi(P)^* = \phi(P) \cup MR$. This latter defines over the set $\mathcal{B}_P \cup \{\top\}$ a congruence relation denoted $\simeq$. The 'declarative rewrite semantics' of $\phi(P)$ is the set of ground atoms equivalent to $\top$: $M_{\phi(P)} = \{B \in \mathcal{B}_P \mid \phi(P)^* \models B \simeq \top\}$. The operational semantics of $\phi(P)$ is defined naturally as :

$$op(\phi(P)) = \{B \in \mathcal{B}_P \mid \phi(P) \vdash_{LC} B\}$$

We have then the following result, proved in [2] :

Theorem 3. *For any pure logic program P, and any (positive) query we have :*

i) $M_P = M_{\phi(P)} = op(\phi(P))$.

ii) Any computed answer is also logically correct.

iii) Any logical answer is 'equivalent' to a computed answer.

The above notion of 'equivalence' between answers is upto the rewrite program given by the set of all rewrite-rules obtained as the terminal leaves of the LC-search tree. (It should be pointed out here that our formulation of the 'Answer' rule is different from that of [2]; the reasons for this will become clear in Section 3.4).

We will see in the following subsection what kind of extension is needed on the rewrite mechanism LC, in order to extract negative information, even if we consider only pure logic programs.

2.2 Negative Information from Pure Logic Programs

We observed above that a ground atom in $\mathcal{B}_P$ is 'true' in the intended semantics of P, iff it can be 'rewritten to $\top$' by the rewrite mechanism above. On the other hand, the 'intended' negative information for any pure logic program P, is defined to be its finite-failure set $FF(P)$. As we pointed out earlier, this negative information is not obtainable that easily by the rewrite mechanism. It does not suffice to introduce a new symbol '$\bot$' (symbolizing 'false', and assumed *minimal* w.r.t. the given ordering '$\succ$'), and expect the elements of $FF(P)$ to rewrite to '$\bot$' : at least some rules rewriting to '$\bot$' must appear in the rewrite program. Such rules are rather naturally suggested by the Clark completion of the logic program P (cf. [11]) : e.g. any predicate appearing only in the bodies of the clauses of P but not in any head, is deemed 'false'. We will retain this suggestion. And to trace down the additional inference rules needed in the rewrite approach for recovering negative information, we propose first to look at a few simple programs.

Example 1. Consider the following three programs :

$(P_1)\quad p(x) :- q(x), r(x).,\quad q(b).$

$(P_2)\quad p(x) :- q(x), r(x).,\quad q(b).,\quad r(a).$

$(P_3)\quad p(a).,\quad p(x) :- q(x).$

For each of these $p(b)$ is intended to be 'false'. Now, we may a priori transform P_1 into the rewrite program :

$$p(x) \to q(x), r(x), \qquad q(b) \to \top, \qquad r(x) \to \bot$$

which allows us to rewrite $p(b)$ to '$\bot$' as follows. Start with the query rule $p(b) \to Ans$, 'overlap' with the first rewrite rule, and then simplify by the last rewrite rule, to get $q(b), \bot \to Ans$; and finally use a 'meta-reduction' rule $(X, \bot) \to \bot$ (which says simply that 'false' is absorbing w.r.t. conjunction in Boolean algebra).

The second program P_2 will get a priori transformed into the rewrite program :

$$p(x) \to q(x), r(x), \qquad q(b) \to \top, \qquad r(a) \to \top$$

from which we *cannot* deduce now that $p(b)$ rewrites to $\bot$. We need to be more precise here in 'copying' the Clark completion step, and add two more rewrite rules :

$$r(x) \to \bot\ [\![x \neq a]\!], \quad q(x) \to \bot\ [\![x \neq b]\!]$$

These are no longer simple rewrite rules, but are *constrained* (cf. [9]). We can deduce then that $p(b)$ can now be rewritten to $\bot$, by 'overlap' and 'simplify'.

The last program P_3 can be transformed now into the rewrite program :

$$p(a) \to \top, \qquad p(x), q(x) \to q(x), \qquad q(x) \to \bot$$

from which however *we have no way of deducing that* $p(b)$ rewrites to $\bot$; since starting with the query rule, 'overlapping' and 'simplifying' will lead here to the tautology $\bot \leftrightarrow \bot$, which is non-informative. □

These simple examples bring out three essential facts : in order to extract negative information, via the the rewrite approach, even from a pure logic program,

- one needs more than just the 'base rules rewriting to $\bot$' in the manner of Clark,
- 'false' being absorbing for conjunction, rewriting to '$\bot$' may too often lead to the non-informative tautology $\bot \leftrightarrow \bot$. If an atom B is 'intended negative' it should thus be a better idea to try rewriting $\neg B$ to '$\top$', rather than B to '$\bot$'.
- and finally something more powerful than the 'overlap' step is needed.

Before introducing such an inference step, let us adopt a few conventions on the syntax of our mechanism. We first introduce a binary symbol '$\supset$' with intended logical meaning 'implies'; so the term '$(B \supset \bot)$' will have the intended logical meaning $\neg B$. With this convention, we can now formulate our transformation function ψ, which takes as input any pure logic program P and gives a constrained rewrite program. The transformation $\psi(P)$ is defined as follows :

i) Each negated atom $\neg A$ is translated into $A \supset \bot$. Otherwise the method for transforming clauses into rewrite rules remains the same as in Section 2.1.

ii) For each predicate $B(\bar{t})$ appearing in the body of a clause and *not defined elsewhere* in the program P, we add a rule : $B(\bar{x}) \to \bot$.

iii) For each predicate $C(\bar{t})$ appearing in the body of a clause, and *defined elsewhere only with facts* of the form $C(\overline{t_1}).,\dots,C(\overline{t_p}).$, we add a constrained rewrite rule :

$$C(\bar{x}) \to \bot\ [\![\bar{x} \neq \overline{t_1}, \dots, \bar{x} \neq \overline{t_p}]\!]$$

The rules of the rewrite program $\psi(P)$ which rewrite to $\top$ (resp. $\bot$) will be called 'fact rules' (resp. 'negative fact rules') ; the label *base rewrite rules* will cover both.

Example 2-a. (from Chan ([5])). The program :

$p(x) :- m(x, y), n(y).\,, \quad n(1).\,, \quad m(a, 1).\,, \quad m(b, 2).$

whose ground semantics is : $p(a), \neg p(b), \neg p(1), \neg p(2)$, gets translated in our approach into the following rewrite program :

$$p(x), m(x,y), n(y) \to m(x,y), n(y),\ n(1) \to \top,\ m(a,1) \to \top,\ m(b,2) \to \top,$$
$$n(x) \to \bot[\![x \neq 1]\!],\qquad m(x,y) \to \bot[\![(x,y) \neq (a,1), (x,y) \neq (b,2)]\!]$$

From now on, we will be assuming well-known the notions of 'equality' and 'dis-equality' constraints. The term 'constraint' will mean in general a conjunction of such elementary constraints. A constrained literal is a pair (B, C), where B is a literal, and C is a constraint. *Notation* : $B[\![\mathrm{C}]\!]$, or $[\![\mathrm{C}]\!]B$. A substitution σ is often seen as an equality constraint on its variables, expressed in solved form. In general, we will write $[\![\overline{x} \neq \overline{t}]\!]$ instead of $[\![\overline{x} = \overline{t}]\!] \supset \bot$. A constraint appearing in a conjunction is naturally seen as a constraint on the variables of the literals in this conjunction.

The inference step we have in mind, labeled *Resolve-Neg* , with which we want to replace the overlap step for recovering negative information, will function as follows on program P_3 in Example 1 : starting from the query rule $(p(b) \supset \bot) \to Ans$, it will infer the following rewrite rule :

$$(\top[\![b = a]\!] \supset \bot), (q(x)[\![x = b]\!] \supset \bot) \to Ans$$

To the left of this rewrite rule are two constrained terms, respectively obtained with 'top-resolving the heads' of the first two rewrite rules with $p(b)$; the constraints are given by the substitutions for this resolution. These two contrained terms will both get simplified then to $(\bot \supset \bot)$, and we will eventually 'meta-reduce' the above rule to $Ans \to \top$.

The general form of the Resolve-Neg rule as given below, is necessarily more complicated, since the body of a clause in a program P can contain variables not appearing in its head.

Rule 6) Resolve-Neg : the symbol B below stands for an atom, and $\overline{C}$ stands for a conjunction of literals.

$$\frac{(P\,;\ ((B,\overline{C}) \supset \bot), \overline{L} \to \overline{R}\,;\ M)}{(P\,;\ (\prod_i(\tilde{C}_i, \overline{C}\ [\![\sigma_i]\!]) \supset \bot), \overline{L} \leftrightarrow \overline{R}\,;\ M')}$$

where the product $\prod$ means a conjunction, and is taken over all indices i, for which there exist rules $A_i, \overline{C_i} \to \overline{C_i}$, (or of the form $A_i \to \overline{C_i}$) in $\psi(P)$, and σ_i is the mgu of A_i and B. Each $\tilde{C}_i$ is practically the same as $\overline{C_i}$ except that the (possible) extra-variables figuring in $\overline{C_i}$ but *not* in A_i are replaced by 'boxes' (or 'holes'), *different extra variables being represented by different boxes*. The constraint $[\![\sigma_i]\!]$ expresses the unifier σ_i on the variable of B in 'solved form'. And finally, M' is the set $M \cup \{((B,\overline{C}) \supset \bot), \overline{L} \to \overline{R}\}$.

It is assumed tacitly here that if no atom A on the lhs of the rules (of our rewrite program) unifies with B, the 'empty product' will then stand for $\top$; that is to say, the goal rule derived in that case by 'Resolve-Neg' will be $\overline{L} \leftrightarrow \overline{R}$.

These *Resolve-Neg* inference steps will also be *applied, more generally*, to goal rules with possible 'holes' or 'box'-variables. In practice the 'holes' or 'box-variables' (symbolizing the existentially quantified variables) will be represented by dashes. For instance, for the above program of Chan (Example 2-a), and the negative query : $\neg p(b)$?, we start with the query rule $(p(b) \supset \bot) \to Ans$; overlap will become inapplicable ; and Resolve-Neg gives the goal rule : $((m(x,-), n(-))[\![x = b]\!] \supset \bot) \to Ans$.

Our objective is to allow for queries with variables and with negation, and find all the answers for such queries ; for instance, again in Example 2-a, for the query rule

$(p(x) \supset \bot) \rightarrow Ans(x)$, we want to find the set of answers $x = b, x = 1, x = 2$, say under the form $[x \neq a]$. For doing this, we will express the notion of 'answer' for a query with variables, in terms of constraints, as follows. The basic idea is the same as earlier : an answer to a query $\overline{Q}$ should be proved equivalent to $\top$ (that is, if $\sigma(Ans(\overline{x})) \rightarrow \top$, then σ is an answer to $\overline{Q}$). Starting from a rewrite program $\psi(P)$, and a query rule $Q(\overline{x}) \rightarrow Ans(\overline{x})$, if we derive a goal rule of the form $Ans(\overline{x}) \rightarrow \top$ $[\![C]\!]$, then we say that C is an answer to the query. Obviously, such a definition also covers the earlier case of pure programs and positive queries, by viewing the substitution obtained there as an 'equality' constraint.

We now complete our rewrite mechanism, by adding one final rule to our inference mechanism. This rule is meant to manipulate a constraint appearing under negation in any goal rule (possibly with box-variables). It has 'two branches' which are complementary; one leads to an 'answer' in terms of the negated constraint, while the other 'ripples' the constraint away from negation. Both branches have to be traversed in general, for completeness.

Rule 7) Manipulation of a constraint under negation : the symbol σ in these rules stands for an equality constraint (and M' is the set M augmented with the addition of the current goal-rule).

$\mathcal{NC}$**-extract** :

$$\frac{(P\,;\ (\overline{D}\ [\![\sigma \wedge \mathbf{C}]\!] \supset \bot), \overline{L} \rightarrow \overline{R}\,;\ M)}{(P\,;\ \overline{L}\ [\![\neg\sigma]\!] \leftrightarrow \overline{R}\,;\ M')}$$

This branch is logically justified, because when the constraint σ evaluates to 'false', the 'parenthesized' part of the goal rule in the numerator evaluates to 'true'.

$\mathcal{C}$**-ripple** :

$$\frac{(P\,;\ (\overline{D}\ [\![\sigma \wedge \mathbf{C}]\!] \supset \bot), \overline{L} \rightarrow \overline{R}\,;\ M)}{(P\,;\ (\sigma(\overline{D}) \supset \bot)\ [\![\sigma \wedge \mathbf{C}]\!], \overline{\sigma L} \leftrightarrow \overline{\sigma R}\,;\ M')}$$

This is the case naturally complementary to the previous one.

Remark 1. This is the step where we get rid of the 'hole'-variables in goal rules. For if σ in the initial goal rule is of the form $[\![(\overline{x}, -) = (\overline{s}, \overline{t})]\!]$, then the $\mathcal{NC}$-extract branch gives the goal rule $[\![\overline{x} \neq \overline{s}]\!], \overline{L} \leftrightarrow \overline{R}$; while the goal rule generated by $\mathcal{C}$-ripple studies the case where the 'hole' variable can be instantiated to $\overline{t}$ (and $\overline{x}$ to $\overline{s}$). Here is an example.

Example 2-b. Consider again the program of Chan in Example 2-a. Starting with the query rule $(p(x) \supset \bot) \rightarrow Ans(x)$, our mechanism will function as in Figure 1.

In the two $\mathcal{C}$-ripple steps to the right in Figure 1, we have instantiated from their respective constraints, assumed 'true' along these branches (e.g. for the first $\mathcal{C}$-ripple branch, x in $Ans(x)$ has been instantiated to b, and the box '$-_2$' to 2). The two final 'simplify'-steps use the base (fact or negative fact) rules in the rewrite program. □

For treating negative information, we need to extend the meta-reduction system MR to a set MR_{ext}, with three additional rules manipulating the symbols '$\supset$', '$\bot$' :

$$(X, \bot) \rightarrow \bot, \qquad ((X \supset \bot) \supset \bot) \rightarrow X, \qquad (\top \supset \bot) \rightarrow \bot$$

The inference system $LC \cup \{\text{Rules } 6, 7\}$ will be denoted by LC_{ext}, and referred to as 'extended linear completion'.

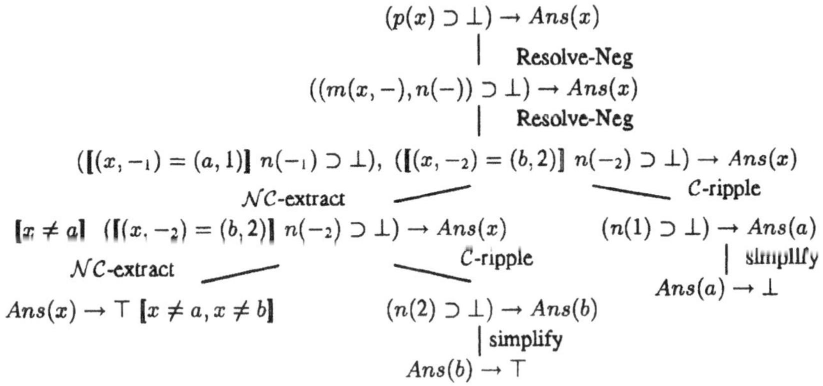
$(p(x) \supset \bot) \rightarrow Ans(x)$
Resolve-Neg
$((m(x,-), n(-)) \supset \bot) \rightarrow Ans(x)$
Resolve-Neg
$([\![(x,-_1) = (a,1)]\!]\ n(-_1) \supset \bot),\ ([\![(x,-_2) = (b,2)]\!]\ n(-_2) \supset \bot) \rightarrow Ans(x)$
$\mathcal{NC}$-extract
$\mathcal{C}$-ripple
$[\![x \neq a]\!]\ ([\![(x,-_2) = (b,2)]\!]\ n(-_2) \supset \bot) \rightarrow Ans(x)$
$(n(1) \supset \bot) \rightarrow Ans(a)$
$\mathcal{NC}$-extract
$\mathcal{C}$-ripple
simplify
$Ans(a) \rightarrow \bot$
$Ans(x) \rightarrow \top\ [\![x \neq a, x \neq b]\!]$
$(n(2) \supset \bot) \rightarrow Ans(b)$
simplify
$Ans(b) \rightarrow \top$

Fig. 1. Search Tree for Chan's Program : Example 2-b.

3 Linear Completion for Normal Logic Programs

A *normal* logic program is a set of clauses, where negation is allowed only in the body of clauses. It turns out that the rewrite approach presented above for extracting negative information from pure programs, needs no additional sophistication for treating normal programs. We keep unchanged our translation function ψ as well as the operational mechanism LC_{ext}, and the meta-reduction set MR_{ext}. Examples 3-a below shows for instance, that the 'program synthesizing' aspect of our extended mechanism is well-preserved in the case of normal programs. Other interesting - but more complicated - examples can be found in [1].

Example 3-a. For the following normal program P, specifying even integers :

$$even(0).,\qquad even(s(x)) :- \neg even(x).$$

the associated rewrite program $\psi(P)$ is :

$$even(0) \rightarrow \top,\qquad even(s(x)) \rightarrow (even(x) \supset \bot).$$

The query $even(x) \rightarrow Ans(x)$, gives rise here to the search tree of Figure 2.

Note that the rewrite rules obtained at the end can generate *all* the ground solutions of the query, from the goal rules which rewrite to $\top$, via overlaps. □

Remark 2. i) As is now apparent, the operational mechanism of LC_{ext} on a negated literal, can be resumed as follows : apply 'Resolve-Neg', then on the literals that this might bring in, continue with 'Resolve-Neg', so as to replace all these by equality constraints ; subsequently get rid of these equality constraints with '$\mathcal{NC}$-extract' and '$\mathcal{C}$-ripple' ; finally end up with overlap or simplification steps.

ii) Variables which are 'existential' in the logical sense might appear in two ways in our constraints : either as a 'box' when they are 'extra' variables appearing only in the body of a clause, or as those not appearing in the current query rule but brought in by unification. When a constraint is negated, as in the case of $\mathcal{NC}$-extract, these variables get implicitly quantified *universally*. For instance, this is the case of the variable y_2 in the disequality constraint on the final goal rules for Example 3-a. This gives also the full justification of our Remark 1, on the elimination of the 'box' variables.

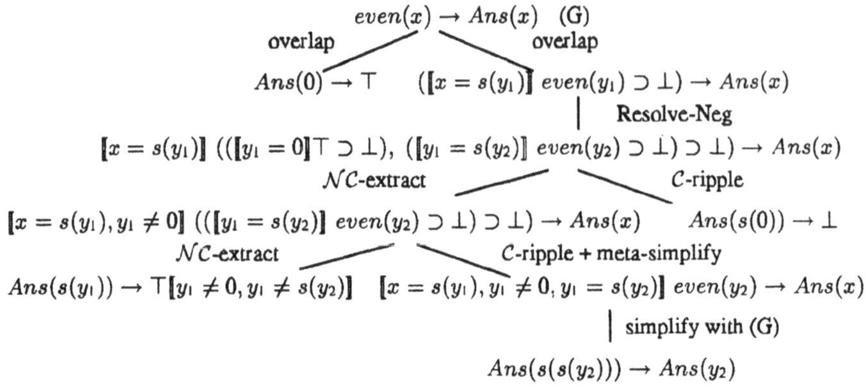

Fig. 2. Search Tree for program 'even' : Example 3-a.

3.1 Semantics

Recall, to begin with, that for the first two programs in Example 1 of Section 2.2, the atom $p(b)$ (intended false), could be rewritten to 'false'; and it is easy to see that, in these two cases, we can also rewrite $(p(b) \supset \bot)$ to 'true'. However, for the third program in this Example 1 we *cannot* rewrite $p(b)$ to 'false' (despite our added inference rules), although we can rewrite $(p(b) \supset \bot)$ to 'true'. Thus, the following definition appears now as natural for our operational semantics : we take into account only the things we can rewrite to '$\top$'.

Definition 4. The operational semantics of $\psi(P)$ consists of the following two subsets of the Herbrand base $\mathcal{B}_P$ associated to P :

1) $op^+(\psi(P)) = \{B \in \mathcal{B}_P \mid \psi(P) \vdash_{LCext} B\}$
2) $op^-(\psi(P)) = \{B \in \mathcal{B}_P \mid \psi(P) \vdash_{LCext} (B \supset \bot)\}$

Our aim is to compare these operational (rewrite) semantics of the rewrite program $\psi(P)$, with a declarative one. The simplest way to define declarative semantics for $\psi(P)$ seems to be to adopt a logic programming point of view. Let us recall briefly how such semantics are usually built in logic programming. According to Clark (cf. [11]) a normal program is best seen as a shorthand for its 'completion' $Comp(P)$. This latter consists of axioms, one for each predicate symbol appearing in P and of the form $\forall \overline{x}, p(\overline{x}) \leftrightarrow D(\overline{x})$, where $D(\overline{x})$ is the so-called 'completed definition' of p, obtained as follows :

- each clause $p(\overline{t}) :- \overline{B}$. is replaced with : $p(\overline{x}) :- \tilde{B}$, where $\tilde{B} = \exists \overline{y}\ [\overline{x} = \overline{t},\ \overline{B}]$, where $\overline{x}$ are new variables and $\overline{y}$ are the variables appearing in $\overline{t}$ and $\overline{B}$.
- if the predicate p is defined by the clauses $p :- \overline{D_i}$, then the expression for D looked for is the disjunction over all the $\tilde{D}_i$.

In addition, for every predicate p of P *not* appearing in any of the heads, one adds a formula : $\neg p(\overline{x})$ to $Comp(P)$, to which is also added the theory of Clark Equality.

'Declarative semantics' for a logic program P attempt to divide the Herbrand base $\mathcal{B}_P$ into subclasses of atoms 'true' or 'false' w.r.t. P. If we adopt the closed world assumption ("whatever is not provable as true must be false"), we may be inconsistent. The interpretation of Fitting ([8]), based on Kleene's three-valued logic came in precisely for this reason. In this logic, we have 3 truth values, $\top$, $\bot$, and u, with 'natural' truth-tables for $\neg$, $\wedge$ and $\vee$. Using such a logic, it is possible to rebuild the semantics presented in section 2 in terms of a new 'immediate consequence' operator Φ_P introduced by Fitting ([8]) Φ_P is a mapping on the set of all three-valued interpretations of P : if I is a three-valued interpretation of P, then the truth value of any predicate $p(\bar{t})$ w.r.t. $\Phi_P(I)$, is defined as the truth value of the completed definition of p w.r.t. I.

The three-valued Herbrand models of $Comp(P)$ are the fixed points of Φ_P and the least Herbrand model is then $\Phi_P \uparrow \alpha$, where α denotes the closure ordinal of Φ_P. If, from a theoretical point of view, this set reflects our program semantics because it models all Herbrand consequences of $Comp(P)$, from a computability point of view, $\Phi_P \uparrow \omega$ is what is really reachable. Thus $\Phi_P \uparrow \omega$ is the natural definition of 'things computable as true or false' from P. It has been proved by Künen ([10]) that this is just the set of three-valued logical consequences of $Comp(P)$. We consider this set as the declarative semantics of P as well as that of our translated $\psi(P)$. We will be denoting below by '$\models_3$' the three-valued logical consequence relation, and the symbol '$\cong$' will mean "*has the same truth value as*".

3.2 Correctness of LC_{ext} w.r.t. the 3-valued Künen Semantics

We proceed to give now our definition of a 'correct' answer to a given query $\overline{Q}$, w.r.t. the declarative semantics described above. Recall that we start from the query rule $\overline{Q} \to Ans(\overline{x})$ and try to get $Ans(\overline{x}) \to \top[\![\mathrm{C}]\!]$ which is then interpreted as : if the value $\overline{v}$ given to the variables $\overline{x}$ satisfies the constraint C, then $\overline{Q}(\overline{v})$ is true. C is what we call a computed answer. For any goal $\overline{Q}$, we denote by $\overline{Q^*}$ the naturally associated first order conjunctive formula. A natural definition of the 'correctness' of a computed answer C is then : $Comp(P) \models_3 \mathrm{C} \wedge \overline{Q^*} \cong \mathrm{C}$.

In the setup of Fitting, the truth value of an equality or disequality constraint C is never 'u', so the congruence $\mathrm{C} \wedge \overline{Q^*} \cong \mathrm{C}$ says simply that : $\mathrm{C} \Rightarrow \overline{Q^*}$. The soundness of our inference system w.r.t. the Fitting semantics means then that a computed answer is such a 'correct' answer. For proving this, we have to look at the intermediary rules generated during the LC_{ext} mechanism. These are rules looking like : $\overline{L} \to \overline{R}\ [\![\mathrm{C}]\!]$, which is a notation for : $\overline{L},\ \mathrm{C} \leftrightarrow \overline{R},\ \mathrm{C}$. We first show that the rules in $\psi(P)$ are 3-valued consequences of $Comp(P)$.

Lemma 5. *If $\overline{L} \to \overline{R}\ [\![\mathrm{C}]\!]$ belongs to $\psi(P)$, then $Comp(P) \models_3 \overline{L^*} \wedge \mathrm{C} \cong \overline{R^*} \wedge \mathrm{C}$*

Proof. It suffices to consider only the two kinds of rules not already appearing explicitly in $Comp(P)$. i) $A, \overline{C} \to \overline{C}$: Since $A : - \overline{C^*}$ belongs to P, it is clear here that $A \wedge \overline{C^*} \cong \overline{C^*}$; ii) $A \to \bot\ [\![\mathrm{C}]\!]$: this follows from the 'completed definition' of A. □

The next lemma says that the meta-reduction rules of MR_{ext} are valid in the 3-valued context, and its proof is straightforward.

Lemma 6. *If $X \to Y$ belongs to MR_{ext} then $X^* \cong Y^*$ is a valid 3-valued formula.*

Theorem 7. *If* $\psi(P) \vdash_{LC_{ext}} \overline{L} \rightarrow \overline{R}\,[\![\mathrm{C}]\!]$, *then we have :*
$$Comp(P) \cup \{Ans(\overline{x}) \cong \overline{Q^*}\} \models_3 \overline{L^*} \wedge \mathrm{C} \cong \overline{R^*} \wedge \mathrm{C}$$

Proof. The proof is by induction on the length of the derivation $\psi(P) \vdash_{LC_{ext}} \overline{L} \rightarrow \overline{R}[\![\mathrm{C}]\!]$. For details see the full version ([1]). □

Remark 3. The fact that the sets op^+ and op^- defined in the preceding subsection (for our operational semantics) are disjoint, is now immediate, since we know that the Fitting/Künen semantics is consistent.

3.3 Completeness of LC_{ext} w.r.t. the 3-valued Künen Semantics

The completeness here is stated naturally with respect to answers which are correct under the three-valued declarative semantics. Consider any ground goal $\overline{Q} = Q_1, \cdots, Q_n$.

Theorem 8. *If the truth value of* $\overline{Q^*}$ *in* $\Phi_P \uparrow \omega$ *is true then for every* i, $\overline{Q_i} \in op^+(\psi(P))$. *If the truth value of* $\overline{Q^*}$ *in* $\Phi_P \uparrow \omega$ *is* $false$, *then for some* i *we have* $\overline{Q}_i \in op^-(\psi(P))$.

Proof. We know that the truth value of $\overline{Q^*}$ in $\Phi_P \uparrow \omega$ is obtained for a finite power k of Φ_P. The reasoning goes by induction on k. The basic idea is as follows : if the truth value' of $\overline{Q}$ is 'true' in $\Phi \uparrow \omega$, then an 'overlap' or 'simplify' step will allow us go down to from $\Phi \uparrow k$ to $\Phi \uparrow (k-1)$; while if the truth value of $\overline{Q}$ is 'false' a 'Resolve-Neg' step will allow us to do the same. For details see the full version. □

Proposition 9. *For any* pure *logic program* P, *we have* : $FF(P) = op^-(\psi(P))$.

Proof. This follows from the correctness and completeness of our approach : when P is pure, $FF(P)$ is exactly the set of atoms which are $false$ in $\Phi(P) \uparrow \omega$; cf. [8]). □

3.4 Rewrite Synthesis for Normal Logic Programs

We proceed now to formulate an analogue of Theorem 1 for our extended linear completion. For doing this we first 'synthesize' our rewrite-program $\psi(P)$, w.r.t. a given query rule, as follows. We assume given a fair and complete strategy for selecting the literals in our goal rules, on which our inference steps are to be applied (e.g. the head-first strategy, as in the pure case of [2]). We can then associate in a natural way an LC_{ext}-search tree with any query rule $Q(\overline{x}) \rightarrow Ans(\overline{x})$. Call *terminal rules* of such a search tree the rules generated by the LC_{ext}-mechanism, to which *no further inference step is applicable.* The 'synthesized' rewrite program associated to the given query Q?, is then the set of rewrite rules obtained by adding to MR_{ext} all such terminal rules. We will also refer to it as the 'synthesized rewrite program' w.r.t. P and Q?. The congruence relation that this program defines on $\mathcal{B}_P \cup \{\top, \bot\}$ will be denoted by '$\approx$'. We will not be assuming explicitly here, that such a synthesized rewrite program is necessarily *finite*, nor that the LC_{ext}-search tree is finite.

Thus, for the program $even(0).$, $even(s(x)) \;:-\; \neg even(x).$, our 'synthesized rewrite program' for the query $even(x) \rightarrow Ans(x)$, contains (besides the meta reduction rules) the two terminal rules : $Ans(0) \rightarrow \top$, $Ans(s(s(x)) \rightarrow Ans(x)$. If we consider the substitution $\sigma : x \leftarrow s(s(0))$, which is a logical answer, we see that it is 'equivalent' to the LC_{ext}-calculated $\theta : x \leftarrow 0$, modulo this synthesized rewrite program. We are going to show that the general situation is similar. We begin with a lemma (harder to enounce than to prove).

Lemma 10. (Lifting lemma) *Let P be any normal program, $\psi(P)$ its associated rewrite program, $Q(\overline{x}) \to Ans(\overline{x})$ any given query rule, and $\mathcal{A}$ the associated search tree. Let σ be any ground substitution, such that $Q(\sigma\overline{x})$ is in $op^{+}(\psi(P))$; and consider a given path, on the LC_{ext}-search tree $\mathcal{A}'$, leading from $Q(\sigma\overline{x}) \to Ans$ to $Ans \to \top$. Then for every node i along this path on $\mathcal{A}'$, there corresponds a node η_i on the tree $\mathcal{A}$, and a substitution τ_i, such that one of the following conditions holds :*

i) every (sub-)goal rule $G'_i[\![\mathrm{C}'_i]\!]$ at the node i on $\mathcal{A}'$ can be seen as a τ_i-instance of a (sub-)goal rule $G_i[\![\mathrm{C}_i]\!]$ at the node η_i on $\mathcal{A}$; and the substitution σ satisfies $[\![\tau_i \wedge \mathrm{C}_i]\!]$

ii) or, the goal-rule at the node η_i is simplifiable by the goal rule at an earlier node η_j, for some $j < i$, on $\mathcal{A}$.

Proof. The proof is in fact straightforward. We begin by observing that every initial 'simplify' branch on the tree $\mathcal{A}'$ lifts in a natural way to an initial 'overlap' or a 'simplify' branch on the tree $\mathcal{A}$, while every other (initial) LC_{ext} step on $\mathcal{A}'$ lifts (also in a natural way) to an (initial) LC_{ext} step of exactly the same type on $\mathcal{A}$. For instance, consider an overlap step on the initial goal rule $G'_0 = Q(\sigma\overline{x}) \to Ans$, with a rule $l \to r$. To simplify, assume that there is only one variable x in $\overline{x}$, and $l :- r$ is a clause in P. So, at some position $u \in G'_0$, we have unifiability of $G'_0\,|_u$ with l, say under an mgu θ_0 ; and the new goal G'_1, generated at node 1 on $\mathcal{A}'$, is of the form $G'_1 = G'_0[u \leftarrow \theta_0(r)]$. But then the subterm at position u of $Q(x)$ is also unifiable with l, with an mgu θ. So we conclude that there exists τ_1 such that $\theta_0 \circ \sigma = \tau_1 \circ \theta$; since σ is ground, this means precisely $\sigma = \tau_1 \wedge \mathrm{C}_1$, where C_1 is the equality constraint corresponding to θ. This corresponds to assertion i). The reasoning is similar for the other initial steps on $\mathcal{A}'$.

The reasoning is just as similar, also for the subsequent nodes $i > 0$ with their (non initial) goal rules on $\mathcal{A}'$, until we reach an i_p, where assertion ii) may become true. In such a case, suffices to set $\eta_i = \eta_{i_p}$, and $\tau_i = \tau_{i_p}$, for all $i > i_p$. □

Theorem 11. *Let P be any normal program, $Q(\overline{x})?$ any given query, and $\mathcal{R}$ the synthesized rewrite program w.r.t. P and $Q(\overline{x})?$. Suppose σ is a ground substitution which is a 'logical' answer for the given query. Then $\sigma(\overline{x})$ can be rewritten to $\top$ with the rewrite rules of $\mathcal{R}$.*

Proof. Let $\mathcal{A}$ be the LC_{ext}-search tree for the query rule $Q(\overline{x}) \to Ans(\overline{x})$. Due to the logical correctness and completeness of our semantics, the hypothesis on σ means that $Q(\sigma\overline{x})$ can be rewritten to $\top$ by the mechanism LC_{ext}. Fix then a path on the LC_{ext}-search tree $\mathcal{A}'$ leading from $Q(\sigma\overline{x}) \to Ans$ to $Ans \to \top$; and let n be its length.

Case $n = 1$: This means that a base rule has simplified $Q(\sigma\overline{x})$, in which case this step lifts into a 'simplify' or 'overlap' step with the same base rule at the root of $\mathcal{A}$; this step will obviously yield an element of $\mathcal{R}$ which rewrites $\sigma(\overline{x})$ to $\top$.

Case $n > 1$: We have two sub-cases to study.

Sub-Case (i) : we assume here that for *every* node along the path on $\mathcal{A}'$ assertion i) of the lifting lemma holds.

Here we go along the given path upto the $(n-1)$-th node ; let η be the node on $\mathcal{A}$ which 'corresponds' to this node, in the sense of the lifting lemma. Now, the goal rule at the $(n-1)$-th node on $\mathcal{A}'$ leads to $Ans \to \top$ in one step, and this LC_{ext} (or meta) step lifts to a similar step below η on $\mathcal{A}$. The subset of the rules of $\mathcal{R}$ *below this node* η on $\mathcal{A}$ suffices then to normalize $\sigma(\overline{x})$ to $\top$.

Sub-Case (ii) : where for some intermediary node along the given path on $\mathcal{A}'$, assertion ii) of the lifting lemma holds.

Let $1 < m < n$ be the least integer such that for the m-th node on the path, we have assertion ii) of the lifting lemma. We then go along the given path upto this m-th node. With the notation of the lemma, the goal rule at η_m on $\mathcal{A}$ is simplifiable by the goal rule at an earlier node η_p, for a $p < m$; so the goal rule G_{η_p} at η_p matches into the goal rule at η_m on $\mathcal{A}$, and combining this match with τ_m gives us a 'new' substitution w.r.t. which every (sub-)goal at the node m on $\mathcal{A}'$ can be seen as an instance of some (sub-)goal at the node p on $\mathcal{A}$. But then we can lift the LC_{ext}-step on $\mathcal{A}'$, from the node m to the node $m+1$ along our given path, as a step on the goal rule G_{η_p} at η_p on $\mathcal{A}$; this would be in general along some other branch on $\mathcal{A}$, starting at the node η_p. Repeating the same kind of construction w.r.t. all the subsequent nodes from $m+1$ to n along our path, we finally obtain a finite set of branches on the search tree $\mathcal{A}$, such that the set of terminal rules below them suffices to normalize $\sigma(\overline{x})$ to $\top$. □

Remark 4. Since we admit negative queries, for which our 'answers' are given in terms of constraints, the above analogue of Theorem 1 cannot have the same kind of formulation as given there. To see this, we may go back to Example 2-b. For the substitution $\sigma : x \leftarrow 1$ here, the LC_{ext}-proof tree for $(p(1) \supset \bot)$ 'corresponds' to the left-most branch of the tree in Figure 1 (this falls into Sub-case (i) of the above proof). The synthesized rewrite program here consists of (MR_{ext} and) the 3 terminal rules therein; the left-most terminal rule rewrites $\sigma(x)$ to $\top$. We have *no* way here for building a substitution 'more general than σ and solution for the generic query'. □

4 Meta-Reduction system versus Semantics

In this section, we point out briefly how the set MR_{ext} plays an important role in determining our operational semantics, and how by modifying this set, we can also modify these semantics. We begin with a simple example.

Example 4. Consider the following normal program P :

$$r :- p(x), \neg p(x). , \qquad p(a). , \qquad p(s(x)) :- p(x).$$

With respect to the corresponding 'immediate consequence' operator Φ_P of Fitting (for this program P), consider the respective powers upto the ordinals ω and $\omega + 1$. The former gives the semantics for Künen. It is easily seen here that r is '$undefined$' in $\Phi_P \uparrow \omega$, and is '$false$' in $\Phi_P \uparrow \omega + 1$. And if we apply our above rewrite-mechanism LC_{ext} on the query rule $r \rightarrow Ans$, it is immediate that we get no information on r, so r is also 'rewrite-undecidable' ; this is not a surprise, since our operational semantics, as they stand, are those of Künen.

However, suppose now that we *modify our meta-reduction system*, by adding the meta rule : $X, (X \supset \bot) \rightarrow \bot$ (which is a valid formula in two-valued logic). Then, for the same query, the rewrite mechanism with this enlarged meta-reduction system gives immediately $Ans \rightarrow \bot$. As a matter of fact, if we start with the opposite query $(r \supset \bot) \rightarrow Ans$, a Resolve-Neg step followed by a simplifying step (using this new meta-reduction rule), will lead to $Ans \rightarrow \top$; that is r is in $op^{-}(\psi(P))$. □

The extra meta rule that we have added 'contradicts' the truth-value-table of Kleene's 3-valued logic only when the truth-value of X is 'u'. So (even if enlarging the meta-reduction system with this extra rule will lead to semantics no longer 'correct' w.r.t. those of Künen), such an addition can be seen as a way of 'refining' the undecidable fragment of Fitting's semantics. Let us show however, that adding the new meta rule $X, (X \supset \bot) \to \bot$ to MR_{ext}, does *not* necessarily eliminate the 'undecidable'-part ; i.e. does not necessarily lead to a 2-valued semantics.

Example 5. Consider the following program (without negation) :

$$p :- q(x)., \qquad q(x) :- q(f(x)).$$

Here p and q are undecidable w.r.t. Fitting/Künen. The associated rewrite program is :

$$p, q(x) \to q(x), \qquad q(x), q(f(x)) \to q(f(x))$$

Let us consider now the two respective queries $p \to Ans$, and $(p \supset \bot) \to Ans$. With or without the meta rule added above, we get the following search trees, leading in either case to undecidabilty :

$$p \to Ans$$
| overlap
$$q(x), Ans \to q(x)$$
| overlap
$$q(x), q(f(x)) \to q(f(x)), Ans$$
| simplify
$$q(f(x)), q(x) \to q(f(x))$$
| simplify
$$q(f(x)) \leftrightarrow q(f(x))$$

$$(p \supset \bot) \to Ans$$
| Resolve-Neg
$$(q(-) \supset \bot) \to Ans$$
| Resolve-Neg
$$(q(f(-)) \supset \bot) \to Ans$$
| simplify
$$Ans \leftrightarrow Ans$$

(The reader can find in [1], some more details on the refinement that can be obtained on the 'undecidable' fragment of Fitting's semantics, via such additional meta rules). We conclude this section with a remark. The reader has probably noticed that certain leaves of some of our LC_{ext}-search trees (w.r.t. our 'standard' meta reduction system) are occasionally of the form $Ans(x_0) \to \bot$, with x_0 ground. But our operational semantics make no explicit mention or use of such leaves. Actually, it is not difficult to prove that such leaves lead to elements of $op^-(\psi(P))$. Our Example 1 shows however that such a conclusion is, in general, only a one way implication.

5 Conclusion

The difficulty of handling non-ground negative literals is a major difficulty for all known operational mechanisms in Logic Programming. The classical SLDNF, based on 'negation as failure' (NAF) can only be used as a test, and never 'constructs' answers. So several extensions of NAF were conceived to tackle these problems. Two such extension for instance, are Chan's 'constructive negation' ([5]) incorporating the so-called SLD-CNF mechanism, and the recent work of [4]. Besides the complicated nature of their search trees (in particular, due to the presence of disjunction in the formulas of [4]), neither of these mechanisms addresses the loop problem for non-ground queries with an infinite number of solutions (for instance for the query $even(x)$? w.r.t. the program of Example 3-a).

The mechanism that we have developed in this paper in order to treat negation, has several interesting features. It can in a sense be considered constructive, and it does at least as well as the above mentioned mechanisms when the set of solutions is finite. When this set is infinite, we get a finite set of solutions and in general a synthesized rewrite program (simpler than the intitial logic program), from which the other solutions are easily derived. Besides this program synthesis aspect, we have also shown how one can go *beyond* the usual operational semantics of normal logic programs, by enlarging the meta-reduction rule set.

Projects for future work at the LIFO (Orléans, Fr.) include : i) the enhancement of LC_{ext} with 'strategies for marking literals', (e.g. [13], [3]), as a first step towards building 'concurrency' into LC_{ext}, ii) the extension of the approach of this paper to deal with Constraint Logic Programming ([12]), as a necessary step for incorporating more powerful notions of equality than CET, into our mechanism.

References

1. **Anantharaman S., Richard G.**. "A Rewrite Mechanism for Logic Programs with Negation", Research Report $n°94 - 13$, LIFO, Université d'Orléans (Fr.), 1994.
2. **Bonacina M., Hsiang J.** "On rewrite programs: semantics and relationship with Prolog", Journal of Logic Programming, $n°14$, pp. 155-180, 1992.
3. **Bronsard F., Lakshman T.K., Reddy Uday S.** "A Framework of Directionality for Proving Termination of Logic programs", Proc. of ICLP-92, MIT Press (ed. K. Apt), pp. 321-335, 1992.
4. **Bottoni A., Levi G.** "Computing in the Completion" Proc. of GULP'93, , pp. 375-389 Orsay (Fr.), 1993.
5. **Chan D.** "Constructive Negation Based on the Completed Database", Proceedings of the 5th ICLP, Seattle (USA) The MIT press, Cambridge, Mass. pp. 111-125, 1988.
6. **Clark K.L.** "Negation as failure", Logic and data bases Plenum, Eds. Gallaire and Minker, 1978.
7. **Dershowitz N., Josephson N.A.** "Logic programming by completion", Proceedings of the 2nd ICLP, Uppsala, Sweden, 1984.
8. **Fitting M.** "A Kripke-Kleene semantics for logic programs", J. Logic Programming, vol 4, pp. 295-312, 1985.
9. **Kirchner C., Kirchner H., Rusinowitch M.** "Deduction with symbolic constraints", Revue d'Intelligence Artificielle, vol 4 ($n°3$), pp. 9-52, 1990, (Special Issue on Automated Deduction).
10. **Künen K.** "Negation in logic programming", J. of Logic Programming, vol 4, pp. 289-308, 1987.
11. **Lloyd J. W.** "Foundations of Logic Programming", Symbolic Computation series, Springer Verlag, 1984.
12. **Maher M. J.** "A Logic Programming View of CLP", Proceedings of 10th ICLP, Budapest, pp. 737-753, 1993
13. **Swift T, Warren D.S.** "Compiling OLDT Evaluation : Background and Overview", Technical Report 92/04, SUNY at Stony Brook (NY).

Level-Confluence of Conditional Rewrite Systems with Extra Variables in Right-Hand Sides

Taro Suzuki   Aart Middeldorp   Tetsuo Ida

Institute of Information Sciences and Electronics
University of Tsukuba, Tsukuba 305, Japan

{taro,ami,ida}@softlab.is.tsukuba.ac.jp

ABSTRACT

Level-confluence is an important property of conditional term rewriting systems that allow extra variables in the rewrite rule because it guarantees the completeness of narrowing for such systems. In this paper we present a syntactic condition ensuring level-confluence for orthogonal, not necessarily terminating, conditional term rewriting systems that have extra variables in the right-hand sides of the rewrite rules. To this end we generalize the parallel moves lemma. Our result bears practical significance since the class of systems that fall within its scope can be viewed as a computational model for functional logic programming languages with local definitions, such as let-expressions and where-constructs.

1. Introduction

There is a growing interest in combining the functional and logic programming paradigms in a single language, see Hanus [12] for a recent overview of the field. The underlying computational mechanism of most of these integrated languages is (conditional) narrowing. Examples of such languages include BABEL [18] and K-LEAF [8]. In order to ensure the desirable completeness of narrowing strategies, restrictions have to be imposed on the programs, which for the purpose of this paper are viewed as conditional term rewriting systems, written in these languages. In this paper we are concerned with the level-confluence restriction, a key property (Giovannetti and Moiso [9], Middeldorp and Hamoen [17]) for ensuring the completeness of narrowing in the presence of so-called extra variables. Very few techniques are available for establishing level-confluence of conditional systems, this in contrast to the confluence property for which several sufficient criteria are known, e.g. [2, 3, 6, 7, 11, 16, 19, 20, 21]. We only know of an early paper by Bergstra and Klop. In [3] they show that orthogonal normal conditional systems are level-confluent. (Actually they show confluence—Giovannetti and Moiso [9] remark that the proof yields level-confluence.) Bergstra and Klop restrict the use of extra variables to the conditional part of the rewrite rules. Several authors remarked that it makes good sense to lift this restriction, since it enables a more natural and efficient way of writing programs in a functional logic language. For example, the Haskell program

```
divide 0     (y+1)           = (0, 0)
divide (x+1) (y+1) | x < y  = (0, x+1)
                   | x >= y = (q+1, r)
                   where (q, r) = divide (x - y) (y+1)
```

corresponds to the conditional term rewriting system

$$\begin{cases} div(0,S(x)) & \rightarrow & (0,0) & & \\ div(S(x),S(y)) & \rightarrow & (0,S(x)) & \Leftarrow & x < y = true \\ div(S(x),S(y)) & \rightarrow & (S(q),r) & \Leftarrow & x \geqslant y = true, \\ & & & & div(x-y,S(y)) = (q,r) \end{cases}$$

which has extra variables q and r in the right-hand side of the last rewrite rule.

The criterion—orthogonality together with normality—of Bergstra and Klop [3] is no longer sufficient when extra variables are permitted in right-hand sides. For instance, the orthogonal normal system

$$\begin{cases} a & \rightarrow & f(x) & \Leftarrow & g(x) = true \\ g(b) & \rightarrow & true & & \\ g(c) & \rightarrow & true & & \end{cases}$$

from [14] is not confluent, let alone level-confluent, since the term a can be rewritten to the different normal forms $f(b)$ and $f(c)$. In this paper we present a useful syntactic condition for level-confluence in the presence of extra variables in right-hand sides of rewrite rules.

The remainder of the paper is organized as follows. In the next section we recapitulate the basics of conditional term rewriting. In Section 3 we introduce and motivate our syntactic criterion. In Section 4 we prove that our criterion indeed implies level-confluence. In the next section we extend our result to the larger class of join conditional systems. In Section 6 we relate our result to the recent work of Avenhaus and Loría-Sáenz [2] and Hanus [13]. In the final section we discuss further extensions of our result.

2. Preliminaries

We assume the reader is familiar with term rewriting. (See [5] and [15] for extensive surveys.) In this preliminary section we recall only some less common definitions and introduce the basic facts concerning conditional term rewriting.

The set of function symbols $\mathcal{F}$ of a term rewriting system (TRS for short) $(\mathcal{F},\mathcal{R})$ is partitioned into disjoint sets $\mathcal{F}_\mathcal{D}$ and $\mathcal{F}_\mathcal{C}$ as follows: a function symbol f belongs to $\mathcal{F}_\mathcal{D}$ if there is a rewrite rule $l \rightarrow r$ in $\mathcal{R}$ such that $l = f(t_1,\dots,t_n)$ for some terms $t_1,\dots,t_n$, otherwise $f \in \mathcal{F}_\mathcal{C}$. Function symbols in $\mathcal{F}_\mathcal{C}$ are called *constructors*, those in $\mathcal{F}_\mathcal{D}$ *defined* symbols. A term built from constructors and variables is called a *data* term. A left-linear TRS without critical pairs is called *orthogonal.*

The rules of a conditional TRS (CTRS for short) have the form $l \to r \Leftarrow c$. Here the conditional part c is a (possibly empty) sequence $s_1 = t_1, \ldots, s_n = t_n$ of equations. At present we only require that l is not a variable. A rewrite rule without conditions will be written as $l \to r$. Depending on the interpretation of the equality sign in the conditions of the rewrite rules, different rewrite relations can be associated with a given CTRS. In this paper we are mainly concerned with what we will call *oriented* CTRSs. The rewrite relation $\to_{\mathcal{R}}$ associated with an oriented CTRS $\mathcal{R}$ is obtained by interpreting the equality signs in the conditional part of a rewrite rule as reachability ($\to^*$). Formally, $\to_{\mathcal{R}}$ is the smallest (w.r.t. inclusion) rewrite relation $\to$ with the property that $l\sigma \to r\sigma$ whenever there exist a rewrite rule $l \to r \Leftarrow c$ in $\mathcal{R}$ and a substitution σ such that $s\sigma \to^* t\sigma$ for every equation $s = t$ in c. The existence of $\to_{\mathcal{R}}$ is easily proved. For every oriented CTRS $\mathcal{R}$ we inductively define TRSs[1] $\mathcal{R}_n$ ($n \geqslant 0$) as follows:

$$
\begin{aligned}
\mathcal{R}_0 &= \varnothing, \\
\mathcal{R}_{n+1} &= \{\, l\sigma \to r\sigma \mid l \to r \Leftarrow c \in \mathcal{R} \text{ and } s\sigma \to^*_{\mathcal{R}_n} t\sigma \text{ for all } s = t \text{ in } c \,\}.
\end{aligned}
$$

In the sequel we write $\mathcal{R}_n \vdash c\sigma$ instead of $s\sigma \to^*_{\mathcal{R}_n} t\sigma$ for every $s = t$ in c. Observe that $\mathcal{R}_n \subseteq \mathcal{R}_{n+1}$ for all $n \geqslant 0$. We have $s \to_{\mathcal{R}} t$ if and only if $s \to_{\mathcal{R}_n} t$ for some $n \geqslant 0$. The minimum such n is called the *depth* of $s \to t$. The depth of a reduction $s \to^*_{\mathcal{R}} t$ is the minimum n such that $s \to^*_{\mathcal{R}_n} t$. The depth of a 'valley' $s \downarrow_{\mathcal{R}} t$ is similarly defined. We abbreviate $\to_{\mathcal{R}_n}$ to $\to_n$. The same applies to the derived relations of $\to_{\mathcal{R}_n}$.

The TRS obtained from a CTRS $\mathcal{R}$ by dropping the conditions in rewrite rules is called the *underlying* TRS of $\mathcal{R}$ and denoted by $\mathcal{R}_u$. Concepts like orthogonality and data term are defined for CTRSs via the underlying TRS. Following [17], we classify rewrite rules $l \to r \Leftarrow c$ of CTRSs according to the distribution of variables among l, r, and c, as follows:

type	requirement
1	$\mathcal{V}ar(r) \cup \mathcal{V}ar(c) \subseteq \mathcal{V}ar(l)$
2	$\mathcal{V}ar(r) \subseteq \mathcal{V}ar(l)$
3	$\mathcal{V}ar(r) \subseteq \mathcal{V}ar(l) \cup \mathcal{V}ar(c)$
4	*no restrictions*

An n-CTRS contains only rules of type n. An *extra* variable x in a rewrite rule $l \to r \Leftarrow c$ satisfies $x \in (\mathcal{V}ar(r) \cup \mathcal{V}ar(c)) - \mathcal{V}ar(l)$. So a 1-CTRS contains no extra variables, a 2-CTRS may only contain extra variables in the conditions, and a 3-CTRS may also have extra variables in the right-hand sides provided these occur in the corresponding conditional part. Most of the literature on

[1] If $\mathcal{R}$ contains rewrite rules that have extra variables in their right-hand sides, the TRSs $\mathcal{R}_n$ may violate the usual restriction $\mathcal{V}ar(r) \subseteq \mathcal{V}ar(l)$ imposed on (unconditional) rewrite rules. This doesn't cause us any concern.

conditional term rewriting is concerned with 1 and 2-CTRSs. We are concerned with *level-confluence* of 3-CTRSs in this paper. An (oriented) CTRS $\mathcal{R}$ is called *level-confluent* if every TRS $\mathcal{R}_n$ ($n \geqslant 0$) is confluent.

A *normal* CTRS $\mathcal{R}$ is an oriented CTRS satisfying the additional restriction that every right-hand side of an equation in the conditions of the rewrite rules is a ground $\mathcal{R}_u$-normal form.

3. Syntactic Restrictions

In the introduction we saw that 3-CTRSs are not confluent in general, even if they are orthogonal and normal. In this section we present syntactic conditions that ensure (level-)confluence. The first consideration is that we have to severely restrict the many possible terms substituted for extra variables in right-hand sides of the rules.

DEFINITION 3.1. An oriented CTRS $\mathcal{R}$ is called ***properly oriented*** if every rewrite rule $l \to r \Leftarrow s_1 = t_1, \ldots, s_n = t_n$ with $\mathcal{V}ar(r) \nsubseteq \mathcal{V}ar(l)$ in $\mathcal{R}$ satisfies the following property:

$$\mathcal{V}ar(s_i) \subseteq \mathcal{V}ar(l) \cup \bigcup_{j=1}^{i-1} \mathcal{V}ar(t_j)$$

for all $i \in \{1, \ldots, n\}$.

A properly oriented oriented CTRS is simply called a properly oriented CTRS. Clearly every 2-CTRS is properly oriented. For 3-CTRS proper orientedness guarantees that the value of extra variables in the right-hand side is determined by the values of the variables in the left-hand side. So extra variables in a properly oriented CTRS are not really 'extra'. The following example illustrates this point.

EXAMPLE 3.2. Consider the properly oriented 3-CTRS

$$\mathcal{R} = \left\{ \begin{array}{lcll} f(x) & \to & g(x,y,z) & \Leftarrow \; h(a,x) = i(y), \;\; h(a,y) = i(z) \\ h(a,a) & \to & i(b) & \\ h(a,b) & \to & i(c) & \\ h(b,b) & \to & i(d) & \end{array} \right.$$

Suppose we rewrite the term $f(a)$ by the first rewrite rule. In this rule y and z are extra variables. The value of y is determined by the condition $h(a,x) = i(y)$ since a is substituted for x and $h(a,a)$ reduces to $i(b)$. So y is bound to b. This determines the value of the extra variable z as $h(a,b)$ reduces to $i(c)$. Hence the term $f(a)$ rewrites only to $g(a,b,c)$.

Since we didn't impose any restrictions on the right-hand side of the conditions so-far, properly oriented orthogonal CTRSs are in general not normal.

Bergstra and Klop [3] showed that orthogonal oriented 2-CTRSs are in general not confluent. Hence it is necessary to further restrict the class of properly oriented 3-CTRSs, before we can conclude level-confluence. In order to get a better understanding of such a restriction, we first present a number of counterexamples against the level-confluence of properly oriented 3-CTRSs.

COUNTEREXAMPLE 3.3. Consider the properly oriented orthogonal 3-CTRS

$$\mathcal{R} = \begin{cases} f(x) & \to & g(y) & \Leftarrow & h(x,a) = i(y,y) \\ h(x,y) & \to & i(x,f(y)) & & \\ a & \to & f(a) & & \end{cases}$$

and the term $f(f(a))$. Because $h(a,a) \to i(a,f(a)) \to i(f(a),f(a))$, $f(a)$ rewrites to $g(f(a))$, and hence $f(f(a)) \to f(g(f(a)))$. We can also rewrite $f(f(a))$ to $g(f(a))$ because $h(f(a),a) \to i(f(a),f(a))$. Both steps have depth 2. We claim that $f(g(f(a)))$ and $g(f(a))$ don't have a common reduct in $\mathcal{R}_2$. Suppose to the contrary that $f(g(f(a))) \downarrow_2 g(f(a))$. This is only possible if there exist terms t_1, t_2, and t_3 such that $f(a) \to_2^* t_1$, $h(g(t_1),a) \to_1^* i(t_2,t_2)$, $f(a) \to_2^* t_3$ and $t_2 \to_2^* t_3$. The sequence $h(g(t_1),a) \to_1^* i(t_2,t_2)$ must have the following form: $h(g(t_1),a) \to_1^* h(g(t_4),f^n(a)) \to_1 i(g(t_4),f^{n+1}(a)) \to_1^* i(t_2,t_2)$ for some term t_4 and $n \geqslant 0$. However, since all $\mathcal{R}_1$-reducts of $f^{n+1}(a)$ are of the form $f^m(a)$ for some $m \geqslant n+1$, the common $\mathcal{R}_1$-reduct t_2 of $g(t_4)$ and $f^{n+1}(a)$ doesn't exist. We conclude that $\mathcal{R}_2$ is not confluent.

COUNTEREXAMPLE 3.4. Consider the properly oriented orthogonal 3-CTRS

$$\mathcal{R} = \begin{cases} f(x) & \to & g(y) & \Leftarrow & x = h(y), \quad i(x) = y \\ i(x) & \to & a & & \\ a & \to & b & \Leftarrow & c = d \\ c & \to & d & & \end{cases}$$

and the term $f(h(a))$ with the two $\mathcal{R}_2$ reducts $f(h(b))$ and $g(a)$. These two terms have a common reduct $g(b)$, but the only sequence from $f(h(b))$ to $g(b)$ has depth 3 because the instantiated second condition of the first rule is $i(h(b)) = b$ which requires depth 2: $i(h(b)) \to_1 a \to_2 b$.

COUNTEREXAMPLE 3.5. Finally, consider the following properly oriented orthogonal 3-CTRS:

$$\mathcal{R} = \begin{cases} f(x) & \to & y & \Leftarrow & x = g(y) \\ g(a) & \to & h(b) & & \end{cases}$$

We can rewrite the term $f(g(a))$ both to $f(h(b))$ and a. These two reducts are not joinable since they are (different) normal forms.

Based on the above findings, we introduce the following restriction.

DEFINITION 3.6. A CTRS $\mathcal{R}$ is called *right-stable* if every rewrite rule $l \to r \Leftarrow s_1 = t_1, \ldots, s_n = t_n$ in $\mathcal{R}$ satisfies the following conditions:

$$(\mathcal{V}ar(l) \cup \bigcup_{j=1}^{i-1} \mathcal{V}ar(s_j = t_j) \cup \mathcal{V}ar(s_i)) \cap \mathcal{V}ar(t_i) = \emptyset$$

and t_i is either a linear data term or a ground $\mathcal{R}_u$-normal form, for all $i \in \{1, \ldots, n\}$.

4. Level-Confluence

In this section we show that orthogonal properly oriented right-stable 3-CTRSs are level-confluent. It is not difficult to see that every normal 2-CTRS is right-stable. Hence our class of CTRSs properly extends the class of orthogonal normal 2-CTRSs (III_n systems in the terminology of [3]) of Bergstra and Klop. They showed that orthogonal normal 2-CTRSs satisfy the so-called parallel moves lemma. Hence these systems are confluent. Giovannetti and Moiso [9] observed that the confluence proof in [3] actually reveals level-confluence. Let us briefly recapitulate the result of Bergstra and Klop.

DEFINITION 4.1. Let $A\colon s \to_{[p, l \to r \Leftarrow c]} t$ be a rewrite step in a CTRS $\mathcal{R}$ and let $q \in \mathcal{P}os(s)$. The set $q\backslash A$ of *descendants* of q in t is defined as follows:

$$q\backslash A = \begin{cases} \{q\} & \text{if } q < p \text{ or } q \parallel p, \\ \{p{\cdot}p_3{\cdot}p_2 \mid r_{|p_3} = l_{|p_1}\} & \text{if } q = p{\cdot}p_1{\cdot}p_2 \text{ with } p_1 \in \mathcal{P}os_{\mathcal{V}}(l), \\ \varnothing & \text{otherwise.} \end{cases}$$

If $Q \subseteq \mathcal{P}os(s)$ then $Q\backslash A$ denotes the set $\bigcup_{q \in Q} q\backslash A$. The notion of descendant is extended to rewrite sequences in the obvious way.

DEFINITION 4.2. Let $\mathcal{R}$ be a CTRS. We write $s \mathrel{+\!\!\!\!\!\!\;\twoheadrightarrow}_n t$ if t can be obtained from s by contracting a set of pairwise disjoint redexes in s by $\mathcal{R}_n$. We write $s \mathrel{+\!\!\!\!\!\!\;\twoheadrightarrow} t$ if $s \mathrel{+\!\!\!\!\!\!\;\twoheadrightarrow}_n t$ for some $n \geqslant 0$. The minimum such n is called the depth of $s \mathrel{+\!\!\!\!\!\!\;\twoheadrightarrow} t$. The relation $\mathrel{+\!\!\!\!\!\!\;\twoheadrightarrow}$ is called *parallel rewriting*.

The parallel moves lemma for orthogonal normal 2-CTRSs can now be stated as follows.

LEMMA 4.3. *Let $\mathcal{R}$ be an orthogonal normal 2-CTRS. If $t \mathrel{+\!\!\!\!\!\!\;\twoheadrightarrow}_m t_1$ and $t \mathrel{+\!\!\!\!\!\!\;\twoheadrightarrow}_n t_2$ then there exists a term t_3 such that $t_1 \mathrel{+\!\!\!\!\!\!\;\twoheadrightarrow}_n t_3$ and $t_2 \mathrel{+\!\!\!\!\!\!\;\twoheadrightarrow}_m t_3$. Moreover, the redexes contracted in $t_1 \mathrel{+\!\!\!\!\!\!\;\twoheadrightarrow}_n t_3$ ($t_2 \mathrel{+\!\!\!\!\!\!\;\twoheadrightarrow}_m t_3$) are the descendants in t_1 (t_2) of the redexes contracted in $t \mathrel{+\!\!\!\!\!\!\;\twoheadrightarrow}_n t_2$ ($t \mathrel{+\!\!\!\!\!\!\;\twoheadrightarrow}_m t_1$).* □

Unfortunately, the parallel moves lemma does not hold for our class of 3-CTRSs, as shown in the following example.

EXAMPLE 4.4. Consider the properly oriented right-stable 3-CTRS

$$\mathcal{R} = \begin{cases} f(x) & \to & g(x,y) & \Leftarrow & x = i(y) \\ h(x) & \to & i(y) & \Leftarrow & x = j(y) \\ k(x) & \to & j(x) & & \\ a & \to & b & & \\ b & \to & c & & \end{cases}$$

and the term $t = f(h(k(a)))$. Because $k(a) \to_1 j(a) \to_1 j(b) \to_1 j(c)$ we have $h(k(a)) \to_2 i(c)$ and hence $t \to_2 f(i(c)) = t_1$. We also have $t \to_3 g(h(k(a)), a) = t_2$ since $k(a) \to_1 j(a)$ and thus $h(k(a)) \to_2 i(a)$. From t_1 we can only perform a single rewrite step: $t_1 \to_2 g(i(c), c)$. However, we can never reach the normal form $g(i(c), c)$ from t_2 in a parallel step because we clearly need two steps ($a \to_1 b \to_1 c$) in the second argument of t_2. The first argument of t_2 rewrites in a single $\mathcal{R}_2$-step to $i(c)$, so t_1 and t_2 do have a common reduct.

In the above example the depth of the non-parallel sequence from a to c in the sequence from t_2 to $g(i(c), c)$ is lower than the depth of the step from t to t_1. This is the key to level-confluence for orthogonal properly oriented right-stable 3-CTRSs. First we introduce a new relation on terms.

DEFINITION 4.5. Let $\mathcal{R}$ be a CTRS. We write $s \mathrel{\dashv\!\!\!+\!\!\!+\!\!\!\!\to}_n t$ if there exists a set $P \subseteq \mathcal{P}os(s)$ of pairwise disjoint positions such that for all $p \in P$ either
(1) $s_{|p}$ rewrites in a single *root* reduction step to $t_{|p}$ whose depth does not exceed n, or
(2) there exists a rewrite sequence from $s_{|p}$ to $t_{|p}$ whose depth is *less* than n.
Clearly $\mathrel{\dashv\!\!\!+\!\!\!+\!\!\!\!\to}_0$ is the identity relation and $\mathrel{\dashv\!\!\!+\!\!\!+\!\!\!\!\to}_1$ coincides with $\mathrel{+\!\!\!+\!\!\!\!\to}_1$. We also have $\mathrel{\dashv\!\!\!+\!\!\!+\!\!\!\!\to}_n \subseteq \to_n^* \subseteq \mathrel{\dashv\!\!\!+\!\!\!+\!\!\!\!\to}_{n+1}$ for all $n \geqslant 0$. The infinite union $\mathrel{\dashv\!\!\!+\!\!\!+\!\!\!\!\to}$ of the relations $\mathrel{\dashv\!\!\!+\!\!\!+\!\!\!\!\to}_n$ ($n \geqslant 0$) is called *extended parallel rewriting*. The depth of an extended parallel rewrite step $s \mathrel{\dashv\!\!\!+\!\!\!+\!\!\!\!\to} t$ is the smallest n such that $s \mathrel{\dashv\!\!\!+\!\!\!+\!\!\!\!\to}_n t$.

The following result states that in case of orthogonal properly oriented right-stable 3-CTRSs extended parallel rewriting satisfies the (important) first part of the parallel moves lemma.

THEOREM 4.6. *Let $\mathcal{R}$ be an orthogonal properly oriented right-stable 3-CTRS. If $t \mathrel{\dashv\!\!\!+\!\!\!+\!\!\!\!\to}_m t_1$ and $t \mathrel{\dashv\!\!\!+\!\!\!+\!\!\!\!\to}_n t_2$ then there exists a term t_3 such that $t_1 \mathrel{\dashv\!\!\!+\!\!\!+\!\!\!\!\to}_n t_3$ and $t_2 \mathrel{\dashv\!\!\!+\!\!\!+\!\!\!\!\to}_m t_3$.*

PROOF. We use induction on $m+n$. The case $m+n=0$ is trivial. Suppose the result holds for all m and n such that $m+n<k$ for some $k \geqslant 1$ and consider the case $m+n=k$. Before we proceed, observe that the induction hypothesis implies the validity of the diagrams in Figure 1. If $m=0$ then $t=t_1$ and we can take $t_3=t_2$. The case $n=0$ is just as simple. So we may assume that $m,n>0$. Let P (Q) be the set of positions justifying (according to Definition 4.5) $t \mathrel{\dashv\!\!\!+\!\!\!+\!\!\!\!\to}_m t_1$ ($t \mathrel{\dashv\!\!\!+\!\!\!+\!\!\!\!\to}_n t_2$). Define a set U of pairwise disjoint positions as follows: $U = \{p \in P \mid$ there is no $q \in Q$ with $q < p\} \cup \{q \in Q \mid$ there is no $p \in P$ with $p < q\}$. Since U dominates all positions in $P \cup Q$, we have $t_1 = t[(t_1)_{|u}]_{u \in U}$ and $t_2 = t[(t_2)_{|u}]_{u \in U}$.

$$m' + n' < k, \qquad \mathrel{+\!\!\!+\!\!\!\!\to} = \mathrel{\dashv\!\!+\!\!\!+\!\!\!\!\to}$$

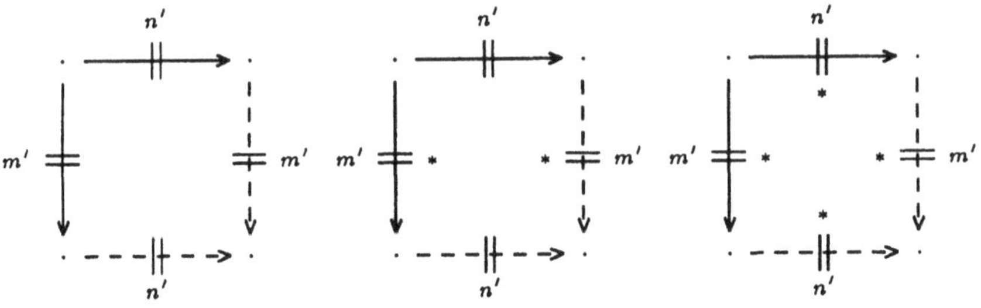

Fig. 1.

Hence for every $u \in U$ we have extended parallel rewrite steps $t_{|u} \mathrel{\dashv\!\!+\!\!\!+\!\!\!\!\to}_m (t_1)_{|u}$ and $t_{|u} \mathrel{\dashv\!\!+\!\!\!+\!\!\!\!\to}_n (t_2)_{|u}$. If for every $u \in U$ we can show the existence of a term t^u such that $(t_1)_{|u} \mathrel{+\!\!\!+\!\!\!\!\to}_n t^u$ and $(t_2)_{|u} \mathrel{+\!\!\!+\!\!\!\!\to}_m t^u$, then we can take $t_3 = t[t^u]_{u \in U}$ since $t_1 \mathrel{\dashv\!\!+\!\!\!+\!\!\!\!\to}_n t_3$ and $t_2 \mathrel{\dashv\!\!+\!\!\!+\!\!\!\!\to}_m t_3$. Let $u \in U$. Without loss of generality we assume that $u \in P$. Let P_u (Q_u) be the subset of P (Q) contributing to $t_{|u} \mathrel{\dashv\!\!+\!\!\!+\!\!\!\!\to}_m (t_1)_{|u}$ ($t_{|u} \mathrel{\dashv\!\!+\!\!\!+\!\!\!\!\to}_n (t_2)_{|u}$). We have $P_u = \{\varepsilon\}$ and $Q_u = \{q \mid u{\cdot}q \in Q\}$ by assumption. Let m_u (n_u) be the depth of the extended parallel rewrite step $t_{|u} \mathrel{\dashv\!\!+\!\!\!+\!\!\!\!\to} (t_1)_{|u}$ ($t_{|u} \mathrel{\dashv\!\!+\!\!\!+\!\!\!\!\to} (t_2)_{|u}$). We clearly have $m_u \leqslant m$ and $n_u \leqslant n$. If $m_u + n_u < k$ then we obtain a term t^u such that $(t_1)_{|u} \mathrel{+\!\!\!+\!\!\!\!\to}_{n_u} t^u$ and $(t_2)_{|u} \mathrel{+\!\!\!+\!\!\!\!\to}_{m_u} t^u$ from the induction hypothesis. Since $m_u \leqslant m$ and $n_u \leqslant n$ this implies the desired $(t_1)_{|u} \mathrel{\dashv\!\!+\!\!\!+\!\!\!\!\to}_n t^u$ and $(t_2)_{|u} \mathrel{\dashv\!\!+\!\!\!+\!\!\!\!\to}_m t^u$. So we may assume that $m_u + n_u = k$. This implies that $m_u = m$ and $n_u = n$. By definition the extended parallel rewrite step $t_{|u} \mathrel{\dashv\!\!+\!\!\!+\!\!\!\!\to}_m (t_1)_{|u}$ is either a rewrite step of depth m at root position or a rewrite sequence $t_{|u} \to^* (t_1)_{|u}$ whose depth is $m-1$. In the latter case we have $t_{|u} \mathrel{\dashv\!\!+\!\!\!+\!\!\!\!\to}^*_{m-1} (t_1)_{|u}$. The strengthened induction hypothesis (cf. Figure 1) yields a term t^u such that $(t_1)_{|u} \mathrel{\dashv\!\!+\!\!\!+\!\!\!\!\to}_n t^u$ and $(t_2)_{|u} \mathrel{\dashv\!\!+\!\!\!+\!\!\!\!\to}^*_{m-1} t^u$. Clearly $(t_2)_{|u} \mathrel{\dashv\!\!+\!\!\!+\!\!\!\!\to}^*_{m-1} t^u$ implies $(t_2)_{|u} \mathrel{\dashv\!\!+\!\!\!+\!\!\!\!\to}_m t^u$. So we may assume that $t_{|u} \mathrel{\dashv\!\!+\!\!\!+\!\!\!\!\to}_m (t_1)_{|u}$ is a rewrite step of depth m at root position. There exists a rewrite rule $l \to r \Leftarrow c$ and a substitution σ such that $t_{|u} = l\sigma$, $(t_1)_{|u} = r\sigma$, and $\mathcal{R}_{m-1} \vdash c\sigma$. Let c be the sequence of conditions $s_1 = t_1, \ldots, s_j = t_j$. For every $i \in \{0, \ldots, j\}$ let c_i be the subsequence of c consisting of the first i conditions. (So c_0 is the empty sequence.) We distinguish two cases.

(1) Suppose $Q_u = \{\varepsilon\}$. By definition the extended parallel rewrite step $t_{|u} \mathrel{\dashv\!\!+\!\!\!+\!\!\!\!\to}_n (t_2)_{|u}$ is either a rewrite step of depth n at root position or a rewrite sequence $t_{|u} \to^* (t_2)_{|u}$ whose depth is $n-1$. The latter case follows from the strengthened induction hypothesis, exactly as above. So we may assume that $t_{|u}$ rewrites in a single root $\mathcal{R}_n$-step to $(t_2)_{|u}$. By orthogonality the employed rewrite rule of $\mathcal{R}$ must be $l \to r \Leftarrow c$. Let τ be a substitution such that $t_{|u} = l\tau$, $(t_2)_{|u} = r\tau$, and $\mathcal{R}_{n-1} \vdash c\tau$. If $\mathcal{V}ar(r) \subseteq \mathcal{V}ar(l)$ then $r\sigma = r\tau$ and we simply define $t^u = r\sigma$. If the right-hand side r of the rewrite rule contains extra variables, we reason as follows. By induction on $i \in \{0, \ldots, j\}$ we show the existence of a substitution ρ_i such that

(a) $\rho_i = \sigma = \tau\ [\mathcal{V}ar(l)]$,

(b) $\mathcal{D}(\rho_i) \subseteq \mathcal{V}ar(l) \cup \mathcal{V}ar(c_i)$,
(c) $x\sigma \mathrel{\|\!\!\!\leftrightarrow}^*_{n-1} x\rho_i$ and $x\tau \mathrel{\|\!\!\!\leftrightarrow}^*_{m-1} x\rho_i$ for every variable $x \in \mathcal{V}ar(l) \cup \mathcal{V}ar(c_i)$.

The substitution $\rho_0 = \sigma\lceil_{\mathcal{V}ar(l)}$ clearly satisfies the requirements for $i = 0$. Suppose $i > 0$ and consider the i-th condition $s_i = t_i$. Because $\mathcal{R}$ is properly oriented we have $\mathcal{V}ar(s_i) \subseteq \mathcal{V}ar(l) \cup \mathcal{V}ar(c_{i-1})$. Using part (c) of the induction hypothesis, we easily obtain $s_i\sigma \mathrel{\|\!\!\!\leftrightarrow}^*_{n-1} s_i\rho_{i-1}$ and $s_i\tau \mathrel{\|\!\!\!\leftrightarrow}^*_{m-1} s_i\rho_{i-1}$. Two applications of the strengthened induction hypothesis (cf. Figure 1) yields a term s' such that $t_i\sigma \mathrel{\|\!\!\!\leftrightarrow}^*_{n-1} s'$ and $t_i\tau \mathrel{\|\!\!\!\leftrightarrow}^*_{m-1} s'$, see Figure 2.

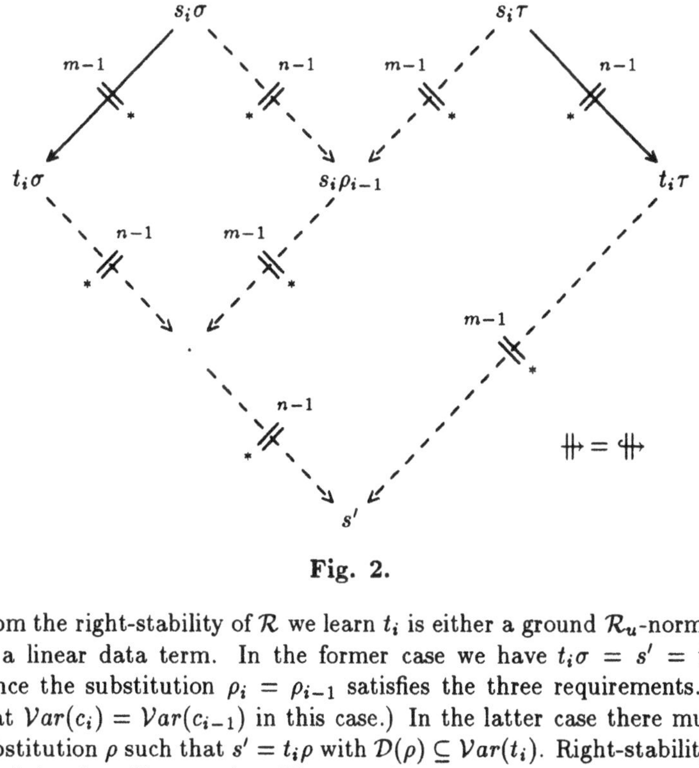

Fig. 2.

From the right-stability of $\mathcal{R}$ we learn t_i is either a ground $\mathcal{R}_u$-normal form or a linear data term. In the former case we have $t_i\sigma = s' = t_i\tau$ and hence the substitution $\rho_i = \rho_{i-1}$ satisfies the three requirements. (Note that $\mathcal{V}ar(c_i) = \mathcal{V}ar(c_{i-1})$ in this case.) In the latter case there must be a substitution ρ such that $s' = t_i\rho$ with $\mathcal{D}(\rho) \subseteq \mathcal{V}ar(t_i)$. Right-stability yields $\mathcal{V}ar(t_i) \cap (\mathcal{V}ar(l) \cup \mathcal{V}ar(c_{i-1}))) = \varnothing$. As a consequence $\rho_i = \rho_{i-1} \cup \rho$ is a well-defined substitution. It is easy to see that ρ_i satisfies the requirements (a)–(c). This concludes the induction step. Now consider the substitution ρ_j. Since $\mathcal{R}$ is a 3-CTRS we have $\mathcal{V}ar(r) \subseteq \mathcal{V}ar(l) \cup \mathcal{V}ar(c_j)$. Hence, using property (c), we obtain $(t_1)_{|u} = r\sigma \mathrel{\|\!\!\!\leftrightarrow}^*_{n-1} r\rho_j$ and $(t_2)_{|u} = r\tau \mathrel{\|\!\!\!\leftrightarrow}^*_{m-1} r\rho_j$. Since $\mathrel{\|\!\!\!\leftrightarrow}^*_{n-1} \subseteq \mathrel{\|\!\!\!\leftrightarrow}_n$ and $\mathrel{\|\!\!\!\leftrightarrow}^*_{m-1} \subseteq \mathrel{\|\!\!\!\leftrightarrow}_m$, we can define $t^u = r\rho_j$.

(2) The second case is $Q_u \neq \{\varepsilon\}$. Using the orthogonality of $\mathcal{R}$ we obtain a substitution τ such that $(t_2)_{|u} = l\tau$, $\mathcal{D}(\tau) \subseteq \mathcal{V}ar(l)$, and $x\sigma \mathrel{\|\!\!\!\leftrightarrow}_n x\tau$ for every $x \in \mathcal{V}ar(l)$. In this case there is no need to distinguish $\mathcal{V}ar(r) \subseteq \mathcal{V}ar(l)$ from $\mathcal{V}ar(r) \not\subseteq \mathcal{V}ar(l)$. By induction on $i \in \{0, \ldots, j\}$ we show the existence of a substitution ρ_i such that

(a) $\rho_i = \tau\ [\mathcal{V}ar(l)]$,

(b) $\mathcal{D}(\rho_i) \subseteq \mathcal{V}ar(l) \cup \mathcal{V}ar(c_i)$,
(c) $\mathcal{R}_{m-1} \vdash c_i\rho_i$, and
(d) $x\sigma \nparallel\!\!\rightarrow_n x\rho_i$ for every variable $x \in \mathcal{V}ar(l) \cup \mathcal{V}ar(c_i)$.

If $i = 0$ then $\rho_0 = \tau\lceil_{\mathcal{V}ar(l)}$ satisfies the requirements. Suppose $i > 0$ and consider the i-th condition $s_i = t_i$. From $\mathcal{R}_{m-1} \vdash c\sigma$ we infer that $s_i\sigma \rightarrow^*_{m-1} t_i\sigma$. Hence also $s_i\sigma \nparallel\!\!\rightarrow^*_{m-1} t_i\sigma$. Let $V = \mathcal{V}ar(s_i) - (\mathcal{V}ar(l) \cup \mathcal{V}ar(c_{i-1}))$ and define the substitution ρ as the (disjoint) union of ρ_{i-1} and $\sigma\lceil_V$. Using part (d) of the induction hypothesis we learn that $s_i\sigma \nparallel\!\!\rightarrow_n s_i\rho$. The strengthened induction hypothesis (cf. Figure 1) yields a term t' such that $s_i\rho \nparallel\!\!\rightarrow^*_{m-1} t'$ and $t_i\sigma \nparallel\!\!\rightarrow_n t'$. From the right-stability of $\mathcal{R}$ we learn that t_i is either a ground $\mathcal{R}_u$-normal form or a linear data term. In the former case we have $t_i\sigma = t_i = t'$ and hence the substitution $\rho_i = \rho$ satisfies the four requirements. In the latter case there must be a substitution ρ' such that $t' = t_i\rho'$ with $\mathcal{D}(\rho') \subseteq \mathcal{V}ar(t_i)$. Right-stability yields $\mathcal{V}ar(t_i) \cap (\mathcal{V}ar(l) \cup \mathcal{V}ar(c_{i-1})) = \varnothing$. Hence $\rho_i = \rho \cup \rho'$ is a well-defined substitution. It is not difficult to check that ρ_i satisfies the requirements (a)–(d). For instance, we have $s_i\rho_i = s_i\rho$ and $t_i\rho_i = t_i\rho'$, hence $\mathcal{R}_{m-1} \vdash (s_i = t_i)\rho'$ follows from $s_i\rho \nparallel\!\!\rightarrow^*_{m-1} t_i\rho'$. This concludes the induction step. Now consider the substitution ρ_j. We have $l\tau = l\rho_j$ by property (a) and $\mathcal{R}_{m-1} \vdash c\rho_j$ by property (c). Therefore $(t_2)_{|u} = l\tau \rightarrow_m r\rho_j$ and thus $(t_2)_{|u} \nparallel\!\!\rightarrow_m r\rho_j$. Since $\mathcal{R}$ is a 3-CTRS we have $\mathcal{V}ar(r) \subseteq \mathcal{V}ar(l) \cup \mathcal{V}ar(c_j)$. Hence, using property (d), we obtain $(t_1)_{|u} = r\sigma \nparallel\!\!\rightarrow_n r\rho_j$. Thus $r\rho_j$ is the desired term t^u.

So for every $u \in U$ we could define the desired term t^u. This concludes the proof. □

The main result of this paper is an immediate consequence of the above theorem.

COROLLARY 4.7. *Orthogonal properly oriented right-stable 3-CTRSs are level-confluent.* □

5. Join Systems

We extend the result of the previous section to join 3-CTRSs. The (only) difference between join and oriented CTRSs is that the equality signs in the conditions of the rewrite rules is interpreted differently: $\downarrow$ in the case of join CTRSs and $\rightarrow^*$ in the case of oriented CTRSs. In the following we make the explicit notational convention of writing $\mathcal{R}^j$ ($\mathcal{R}^o$) if the CTRS $\mathcal{R}$ is considered as a join (oriented) CTRS.

For an arbitrary CTRS $\mathcal{R}$, the rewrite relation associated with the oriented CTRS $\mathcal{R}^o$ is in general a proper subset of the one associated with the join variant $\mathcal{R}^j$. Consider for example the CTRS

$$\mathcal{R} = \begin{cases} a \rightarrow x \Leftarrow b = x \\ b \rightarrow d \\ c \rightarrow d \end{cases}$$

We have $a \to_{\mathcal{R}^j} c$ because $b \downarrow_{\mathcal{R}^j} c$, but $a \to_{\mathcal{R}^o} c$ does not hold because b does not rewrite to c (in $\mathcal{R}^o$). Nevertheless $a \downarrow_{\mathcal{R}^o} c$ holds since $a \to_{\mathcal{R}^o} d$ and $c \to_{\mathcal{R}^o} d$. This relationship ($\to_{\mathcal{R}^j} \subseteq \downarrow_{\mathcal{R}^o}$) holds for all orthogonal properly oriented right-stable 3-CTRSs, as will be shown below. First we prove a special case.

LEMMA 5.1. *Let $\mathcal{R}$ be an orthogonal properly oriented right-stable 3-CTRS. If $s \to_{\mathcal{R}^j_n} t$ by application of a rewrite rule $l \to r \Leftarrow c$ with substitution σ such that $s'\sigma \downarrow_{\mathcal{R}^o_{n-1}} t'\sigma$ for every equation $s' = t'$ in c then $s \downarrow_{\mathcal{R}^o_n} t$.*

PROOF. Let c be the sequence of conditions $s_1 = t_1, \ldots, s_j = t_j$. For every $i \in \{1, \ldots, j\}$ let c_i be the subsequence of c consisting of the first i conditions. By induction on $i \in \{0, \ldots, j\}$ we show the existence of a substitution τ_i such that

(a) $\tau_i = \sigma\ [\mathcal{V}ar(l)]$,
(b) $\mathcal{D}(\tau_i) \subseteq \mathcal{V}ar(l) \cup \mathcal{V}ar(c_i)$,
(c) $\mathcal{R}^o_{n-1} \vdash c_i\tau_i$, and
(d) $x\sigma \to^*_{\mathcal{R}^o_{n-1}} x\tau_i$ for every variable $x \in \mathcal{V}ar(l) \cup \mathcal{V}ar(c_i)$.

The substitution $\tau = \sigma\lceil_{\mathcal{V}ar(l)}$ clearly satisfies the requirements for $i = 0$. Suppose $i > 0$ and consider the i-th condition $s_i = t_i$. By assumption $s_i\sigma$ and $t_i\sigma$ have a common $\mathcal{R}^o_{n-1}$-reduct, say t'. Let $V = \mathcal{V}ar(s_i) - (\mathcal{V}ar(l) \cup \mathcal{V}ar(c_{i-1}))$ and define the substitution ρ as $\tau_{i-1} \cup \sigma\lceil_V$. Using part (d) of the induction hypothesis we obtain an $\mathcal{R}^o_{n-1}$-rewrite sequence from $s_i\sigma$ to $s_i\rho$. According to Corollary 4.7 $\mathcal{R}^o$ is level-confluent. This implies that the $\mathcal{R}^o_{n-1}$-reducts t' and $s_i\rho$ of $s_i\sigma$ have a common $\mathcal{R}^o_{n-1}$-reduct, say t''. From the right-stability of $\mathcal{R}$ we learn that t_i is either a ground $\mathcal{R}_u$-normal form or a linear data term. In the former case we have $t_i\sigma = t_i = t' = t''$ and hence the substitution $\tau_i = \rho$ satisfies the four requirements. In the latter case there must be a substitution ρ' such that $t'' = t_i\rho'$ with $\mathcal{D}(\rho') \subseteq \mathcal{V}ar(t_i)$. Right-stability yields $\mathcal{V}ar(t_i) \cap (\mathcal{V}ar(l) \cup \mathcal{V}ar(c_{i-1})) = \varnothing$. Hence $\tau_i = \rho \cup \rho'$ is a well-defined substitution which is easily seen to be satisfying the requirements (a)–(d). This concludes the induction step. Consider the substitution τ_j. We have an $\mathcal{R}^o_{n-1}$-rewrite sequence from $r\sigma$ to $r\tau_j$ as a consequence of property (d). From properties (a) and (c) we learn that $l\sigma = l\tau_j \to_{\mathcal{R}^o_{n-1}} r\tau_j$. Therefore $s \downarrow_{\mathcal{R}^o_{n-1}} t$. □

LEMMA 5.2. *Let $\mathcal{R}$ be an orthogonal properly oriented right-stable 3-CTRS. If $s \to_{\mathcal{R}^j_n} t$ then $s \downarrow_{\mathcal{R}^o_n} t$.*

PROOF. We use induction on n. If $n = 0$ then we have nothing to prove, so suppose $n > 0$. Let $l \to r \Leftarrow c$ be the rewrite rule and σ the substitution used in $s \to_{\mathcal{R}^j_n} t$. We have $\mathcal{R}^j_{n-1} \vdash c\sigma$. Using Corollary 4.7, we obtain $\mathcal{R}^o_{n-1} \vdash c\sigma$ from the induction hypothesis by a routine induction argument. Lemma 5.1 yields the desired $s \downarrow_{\mathcal{R}^o_n} t$. □

The main result of this section is an easy consequence of Corollary 4.7 and Lemma 5.2.

COROLLARY 5.3. *Orthogonal properly oriented right-stable join 3-CTRSs are level-confluent.* □

6. Related Work

Bertling and Ganzinger [4] defined the class of *quasi-reductive* and *deterministic* 3-CTRSs. Deterministic CTRSs are very similar to properly oriented CTRSs, the difference being that we don't impose the restrictions ($i \in \{1, \ldots, n\}$)

$$\mathcal{V}ar(s_i) \subseteq \mathcal{V}ar(l) \cup \bigcup_{j=1}^{i-1} \mathcal{V}ar(s_j \mathrel{\overset{?}{=}} t_j)$$

when the right-hand side r of the rewrite rule $l \rightarrow r \Leftarrow s_1 = t_1, \ldots, s_n = t_n$ doesn't contain extra variables. So deterministic CTRSs are a proper subclass of properly oriented CTRS. Quasi-reductivity is an (undecidable) criterion guaranteeing termination. Bertling and Ganzinger showed that every quasi-reductive deterministic 3-CTRS has a decidable rewrite relation.

In a recent paper Avenhaus and Loría-Sáenz [2] provide a simple but powerful decidable condition for quasi-reductivity. Moreover, they show that every *strongly* deterministic and quasi-reductive 3-CTRS with joinable critical pairs[2] is confluent. A strongly deterministic CTRS $\mathcal{R}$ is a deterministic one with the additional property that for every right-hand side t of the equations in the conditional parts of the rewrite rules and every normalized substitution σ, the term $t\sigma$ is a normal form. Every right-stable deterministic CTRS is strongly deterministic but not vice-versa, e.g., the 3-CTRS of Counterexample 3.3 is strongly deterministic but not right-stable. So the class of strongly deterministic quasi-reductive 3-CTRSs is incomparable with our class of orthogonal properly oriented right-stable 3-CTRSs. The essential difference however is that we do not impose the termination restriction. From a programming point of view, the termination assumption is quite severe. Strongly deterministic quasi-reductive 3-CTRSs are in general not level-confluent. For instance, the 3-CTRS

$$\mathcal{R} = \begin{cases} a & \rightarrow & b \\ a & \rightarrow & c \\ b & \rightarrow & c & \Leftarrow & d = e \\ d & \rightarrow & e \end{cases}$$

is clearly strongly deterministic. The reduction order $a \succ b \succ c \succ d \succ e$ can be used to show quasi-reductivity. We have $a \rightarrow_1 b$ and $a \rightarrow_1 c$, but the step from b to c has depth 2. Observe that $\mathcal{R}$ is properly oriented and right-stable. Hence we know that the system cannot be orthogonal, and indeed there is a critical pair between the first two rules. Finally, we would like to mention that the termination assumption makes the task of proving confluence easier.

Very recently (and independently) Hanus [13] presented a sufficient condition for the level-confluence of properly oriented orthogonal 3-CTRSs with *strict equality*. Strict equality means that a substitution σ satisfies a condition $s = t$ of a rewrite rule only if $s\sigma$ and $t\sigma$ reduce to the same ground data term. CTRSs

[2] Overlays obtained from a rewrite rule and (a renamed version of) itself don't have to be considered.

with strict equality can be viewed as a special case of normal CTRSs. The proof in [13] is however insufficient since it is based on the parallel moves lemma for orthogonal normal 2-CTRSs (Bergstra and Klop [3]), which, as we have seen in Section 4, is not valid for 3-CTRSs. Our proof method can be specialized to complete Hanus' proof. Actually, Hanus considers *almost orthogonal* ([1]) CTRSs. These are left-linear systems in which the non-overlapping restriction is relaxed by allowing trivial overlays. Although we only considered orthogonal CTRSs in this paper, our result immediately extends to almost orthogonal systems.

7. Concluding Remarks

In this final section we discuss some further extensions of our sufficient condition. First of all, Dershowitz *et al.* [7] and Gramlich [11], among others, distinguish feasible from infeasible critical pairs. A critical pair is *infeasible* if the instantiated combination of the two conditional parts of the rewrite rules that induce the critical pair is unsolvable. Infeasibility is undecidable in general, but see González–Moreno *et al.* [10] for a decidable sufficient condition. Infeasible critical pairs are harmless, so orthogonality can be strengthened by allowing infeasible critical pairs. This is important in practice since it permits systems like

$$\left\{\begin{array}{lcll} div(0,S(x)) & \rightarrow & (0,0) & \\ div(S(x),S(y)) & \rightarrow & (0,S(x)) & \Leftarrow\ x<y=true \\ div(S(x),S(y)) & \rightarrow & (S(q),r) & \Leftarrow\ x\geqslant y=true, \\ & & & div(x-y,S(y))=(q,r) \end{array}\right.$$

with disambiguating conditions. Because the condition $x < y = true$, $x \geqslant y = true$ has no solutions, the critical pair between the last two rules is infeasible. The proofs in this paper remain valid if we allow infeasible critical pairs. Secondly, satisfiability of the conditional part of a rewrite rule is independent of the order of its equations. Hence we can relax proper orientedness and right-stability by allowing any permutation of the equations in the conditions to satisfy the requirements in Definitions 3.1 and 3.6. Finally, and more in spirit with the results of this paper, we consider relaxing the proper orientedness restriction. Proper orientedness requires that extra variables in the left-hand side of some equation in the conditional part of some rewrite rule occur in the preceding equations. This requirement is not necessary, however, if the extra variable under consideration does not affect the values of extra variables that occur in the right-hand side of the rewrite rule. This leads us to the following definition.

DEFINITION 7.1. In an *extended properly oriented* CTRS the conditional part c of every rewrite rule $l \rightarrow r \Leftarrow c$ with $\mathcal{V}ar(r) \nsubseteq \mathcal{V}ar(l)$ can be written as $s_1 = t_1, \ldots, s_m = t_m, s'_1 = t'_1, \ldots, s'_n = t'_n$ such that the following two conditions are satisfied:

$$\mathcal{V}ar(s_i) \subseteq \mathcal{V}ar(l) \cup \bigcup_{j=1}^{i-1} \mathcal{V}ar(t_j)$$

for all $i \in \{1, \ldots, m\}$, and

$$\mathcal{V}ar(r) \cap \mathcal{V}ar(s'_i = t'_i) \subseteq \mathcal{V}ar(l) \cup \bigcup_{j=1}^{m} \mathcal{V}ar(t_j)$$

for all $i \in \{1, \ldots, n\}$.

The proofs in this paper can be adapted to deal with extended properly oriented CTRSs. Hence all orthogonal extended properly oriented right-stable 3-CTRSs are level-confluent. An interesting application is the class of 3-CTRSs with strict equality and so-called ***local definitions***. With respect to Definition 7.1 this means that t'_i denotes the constant *true* or *false* and s'_i the strict equation $s'_{1i} \equiv s'_{2i}$, for $i \in \{1, \ldots, n\}$. Here $\equiv$ is a function that tests whether its two arguments denote the same ground data term. The local definitions $s_1 = t_1, \ldots, s_m = t_m$ support (potentially) infinite data structures. An interesting example is the following Haskell like program:

```
filter n r []     = (-1, [])
filter n r (x:xs) = (x, xs),      if divide x n == (q, r)
                  = (y, x:ys),    otherwise
                    where (y, ys) = filter n r xs
```

The idea is that evaluating `filter n r l` filters out a natural number x such that x *mod* `n` = `r` from the list of natural numbers `l`. This program can handle (potentially) infinite lists. The function `filter` corresponds to the following orthogonal (in the sense of having only infeasible critical pairs) 3-CTRS with strict equality and local definitions:

$$\left\{\begin{array}{lcll} \mathit{filter}(n,r,\mathit{nil}) & \to & (-1,\mathit{nil}) & \\ \mathit{filter}(n,r,x:xs) & \to & (x,xs) & \Leftarrow \; \mathit{div}(x,n) \equiv (q,r) = \mathit{true} \\ \mathit{filter}(n,r,x:xs) & \to & (y,x:ys) & \Leftarrow \; \mathit{filter}(n,r,xs) = (y,ys), \\ & & & \; \mathit{div}(x,n) \equiv (q,r) = \mathit{false} \end{array}\right.$$

This system is extended properly oriented and right-stable but not properly oriented due to the presence of the extra variable q in the last rewrite rule.

Acknowledgements. We thank Michael Hanus for scrutinizing the paper.

References

1. S. Antoy and A. Middeldorp, *A Sequential Reduction Strategy*, Proc. ALP-94, Madrid, LNCS **850**, pp. 168–185, 1994.
2. J. Avenhaus and C. Loría-Sáenz, *On Conditional Rewrite Systems with Extra Variables and Deterministic Logic Programs*, Proc. LPAR-94, Kiev, LNAI **822**, 1994.
3. J.A. Bergstra and J.W. Klop, *Conditional Rewrite Rules: Confluence and Termination*, JCSS **32**(3), pp. 323–362, 1986.

4. H. Bertling and H. Ganzinger, *Completion-Time Optimization of Rewrite-Time Goal Solving*, Proc. RTA-89, LNCS **355**, pp. 45–58, 1989.
5. N. Dershowitz and J.-P. Jouannaud, *Rewrite Systems*, in: Handbook of Theoretical Computer Science, Vol. B ed. J. van Leeuwen), North-Holland, pp. 243–320, 1990.
6. N. Dershowitz and M. Okada, *A Rationale for Conditional Equational Programming*, TCS **75**, pp. 111–138, 1990.
7. N. Dershowitz, M. Okada, and G. Sivakumar, *Confluence of Conditional Rewrite Systems*, Proc. CTRS-87, Orsay, LNCS **308**, pp. 31–44, 1987.
8. E. Giovannetti, G. Levi, C. Moiso, and C. Palamidessi, *Kernel-LEAF: A Logic plus Functional Language*, JCSS **42**(2), pp. 139–185, 1991.
9. E. Giovannetti and C. Moiso, *A Completeness Result for E-Unification Algorithms based on Conditional Narrowing*, Proc. Workshop on Foundations of Logic and Functional Programming, Trento, LNCS **306**, pp. 157–167, 1986.
10. Juan Carlos Gonzälez-Moreno, M.T. Hortalá-Gonzälez, M. Rodríguez-Artalejo, *Denotational versus Declarative Semantics for Functional Programming*, Proc. CSL-92, Berne, LNCS **626**, pp. 134–148, 1992.
11. B. Gramlich, *On Termination and Confluence of Conditional Rewrite Systems*, Proc. CTRS-94, Jerusalem, LNCS, 1994. To appear.
12. M. Hanus, *The Integration of Functions into Logic Programming: From Theory to Practice*, JLP **19 & 20**, pp. 583–628, 1994.
13. M. Hanus, *On Extra Variables in (Equational) Logic Programming*, report MPI-I-94-246, Max-Planck-Institut für Informatik, 1994.
14. T. Ida and S. Okui, *Outside-In Conditional Narrowing*, IEICE Transactions on Information and Systems, **E77-D**(6), pp. 631–641, 1994.
15. J.W. Klop, *Term Rewriting Systems*, in: Handbook of Logic in Computer Science, Vol. II (eds. S. Abramsky, D. Gabbay and T. Maibaum), Oxford University Press, pp. 1–116, 1992.
16. A. Middeldorp, *Completeness of Combinations of Conditional Constructor Systems*, JSC **17**, pp. 3–21, 1994.
17. A. Middeldorp and E. Hamoen, *Completeness Results for Basic Narrowing*, AAECC **5**, pp. 213–253, 1994.
18. J.J. Moreno-Navarro and M. Rodrìguez-Artalejo, *Logic Programming with Functions and Predicates: The Language BABEL*, JLP **12**, pp. 191–223, 1992.
19. E. Ohlebusch, *Modular Properties of Constructor-Sharing Conditional Term Rewriting Systems*, Proc. CTRS-94, Jerusalem, LNCS, 1994. To appear.
20. P. Padawitz, *Generic Induction Proofs*, Proc. CTRS-92, Pont-à-Mousson, LNCS **656**, pp. 175–197, 1993.
21. Y. Toyama and M. Oyamaguchi, *Church-Rosser Property and Unique Normal Form Property of Non-Duplicating Term Rewriting Systems*, Proc. CTRS-94, Jerusalem, LNCS, 1994. To appear.

A Polynomial Algorithm Testing Partial Confluence of Basic Semi-Thue Systems

Géraud Sénizergues

LaBRI
Université de Bordeaux I
351, Cours de la Libération 33405 Talence, France ** ***

Abstract. We give a polynomial algorithm solving the problem "is S partially confluent on the rational set R ?" for finite, basic, noetherian semi-Thue systems. The algorithm is obtained by a polynomial reduction of this problem to the equivalence-problem for deterministic 2-tape finite automata, which has been shown to be polynomially decidable in [Fri-Gre82].

Keywords: semi-Thue systems; confluence; two tape finite automata.

1 INTRODUCTION

1.1 Presentation

Among all rewriting systems, those which are *confluent* and *noetherian* are considered as particularly nice because they have a decidable word-problem (provided the one-step reduction and the irreducibility property are themselves computable predicates). In some particular contexts (semi-Thue systems presenting groups [11, 8, 21], sets of words defined by a rewriting system [4, 5, 30, 2], sets of graphs defined by a rewriting system [1]), noetherian systems which only fulfill some *weak* condition of confluence, may still have nice algorithmic properties.

A rewriting-system S over some set E will be called *partially confluent* on a subset R of irreducible elements if and only if, for every $e \in E$, e is *equivalent* (mod S) with some element of R if and only if e has some *descendant* in R. (One can easily see that, in the case where R consists of a single irreducible element r, this property is equivalent to the fact that S is confluent on $[r]_{\overset{*}{\underset{S}{\longleftrightarrow}}}$).

Let us call PArtial Confluence Problem (denoted **PACP**) the following problem:

instance : A finite alphabet X, a noetherian semi-Thue system S and a deterministic finite automaton $\mathcal{R}$ over X such that $L(\mathcal{R}) \subseteq Irr(S)$.

** mailing adress:LaBRI and UER Math-info, Université Bordeaux1
351 Cours de la libération -33405- Talence Cedex.
email:ges@labri.u-bordeaux.fr; fax: (33)-56-84-66-69; tel:(33)-56-84-60-48
*** This work has been supported by the PRC Maths et Informatique and the ASMICS project

question : is S partially confluent on $L(\mathcal{R})$?
Up to now, it is known that :

(1) this problem is *undecidable* for finite, length-reducing, semi-Thue systems ([26, 34]), even for $L(\mathcal{R})$ reduced to a single word

(2) this problem is decidable in *double-exponential* time for finite, basic, length-decreasing, semi-Thue systems ([26, 34])

(3) it is decidable in *P-time* whether a finite monadic semi-Thue system S is confluent on *every word in range* (S) ([35, 15, 22], thus improving a decidability result of [29, 30])

(4) the PACP is decidable in *P-time* for finite monadic semi-Thue systems presenting *groups* and for $L(\mathcal{R}) = \{\epsilon\}$ ([21, 22])

(5) the PACP is decidable in *P-time* for finite *special* semi-Thue systems and for $L(\mathcal{R})$ reduced to a single word ([27]).

We prove here the following

Theorem 4.5 *The PArtial Confluence Problem is decidable in P-time for finite, basic, noetherian semi-Thue systems.*

This result both *improves the complexity* of result (2) and *enlarges the domain* of applicability of results (3) (4) (5). Let us mention two other results which can motivate interest in theorem 4.5:

(6) given a finite (or even rational) basic, noetherian semi-Thue system S, which is partially confluent on a rational set of irreducible words R, $[R]_{\overset{*}{\underset{S}{\longleftrightarrow}}}$ is a *deterministic context-free* language ([6, 25]) and the equality $[R]_{\overset{*}{\underset{S}{\longleftrightarrow}}} = L(\mathcal{A})$ is *decidable* for any Deterministic PushDown Automata $\mathcal{A}$ ([31, 32])

(7) the so-called *equivalence-problem for DPDA*[4] is Turing-equivalent with the PACP for finite, *left-basic*, [5] length-reducing semi-Thue systems and $L(\mathcal{R})$ reduced to a single word ([34, Theorem 5.17]).

Two other related results are :

(8) confluence of a finite, monadic semi-Thue system over a rational set R is *decidable* ([23])

(9) partial confluence of a *rational*, basic, noetherian semi-Thue system S on a rational set $R \subseteq Irr(S)$ is *decidable* ([18])

(but the time upper-bounds derived from (8)(9) are multi-exponential).

---

[4] This problem asks for every pair $(\mathcal{A}_1, \mathcal{A}_2)$ of DPDA whether $L(\mathcal{A}_1) = L(\mathcal{A}_2)$? Up to now the dedidability of this problem is unknown, see [3].

[5] See in §. 2 a precise definition of *basic* or *left-basic* systems.

1.2 Contents

We give in the preliminaries all the notation, definitions and results concerning either rewriting systems (§2.1) or 2-d.f.a's (§2.2) which are required in the proof of the main result.

We give in §3 a detailed proof of the fact that the PACP is decidable in P-time for finite, basic, noetherian semi-Thue systems, in the case where L($\mathcal{R}$) consists of a single word.

We use a criterion established in [18] which is similar to the one established in [35]. In order to test whether this criterion is satisfied, we associate to every system S and every S-irreducible word m a set of 7-tuples of words E_m and a rewriting system T_m over the set E_m; in some sense, T_m expresses a *careful* way of computing the normal form $\rho_S(\alpha m\beta)$ obtained by leftmost-reduction from the initial word $\alpha m\beta$, for every irreducible words α, β. This computation is *careful* in that it can be simulated by a 1-turn pushdown-automaton with only a *polynomial* amount of states (while the analogous 1-turn DPDA's defined in [29, 30, 26, 34] had an *exponential* amount of states in general). Moreover, in order to use the tight complexity result of [16], we use a *2 tape d.f.a* instead of a 1-turn DPDA as simulating device, which is a more or less classical trick in automata theory (see [17, Theorem 6.1 p.451]).

In §4 we sketch the extension of §3 to the general case.

2 PRELIMINARIES

2.1 Rewriting systems

Abstract rewriting systems Let E be some set and $\longrightarrow$ some binary relation over E .We shall call $\longrightarrow$ the *direct reduction*.We shall use the notations $\xrightarrow{i}$ for every integer ($i \geq 0$), and $\xrightarrow{*}, \xrightarrow{+}$ in the usual way (see [20]). By $\longleftrightarrow$, we denote the relation $\longrightarrow \cup \longleftarrow$. The three relations $\xrightarrow{*}, \xleftarrow{*}$ and $\xleftrightarrow{*}$ are respectively the *reduction*, *derivation* and the *equivalence* generated by $\longrightarrow$.

We use the notions of *confluent* relation and *noetherian* relation in their usual meaning ([20] or [14, §4, p.266-269]).

An element $e \in E$ is said *irreducible*(resp. *reducible*) modulo ($\longrightarrow$) iff there exists no (resp. some) $e' \in E$ such that $e \longrightarrow e'$. By Irr($\longrightarrow$) we denote the set of all the elements of E which are irreducible modulo ($\longrightarrow$).

Given some subset A of E, we use the following notation:

$$< A >_{\xleftarrow{*}} = \{e \in E \mid \exists a \in A, a \xleftarrow{*} e\}, \quad \Delta^*_{\longrightarrow}(A) = \{e \in E \mid \exists a \in A, a \xrightarrow{*} e\}$$

$$[A]_{\xleftrightarrow{*}} = \{e \in E \mid \exists a \in A, a \xleftrightarrow{*} e\}$$

Definition 1. The relation $\longrightarrow$ is said *partially confluent* over a subset $A \subseteq \mathrm{Irr}(\longrightarrow)$ if and only if :

$$< A >_{\overset{*}{\longleftrightarrow}} = [A]_{\overset{*}{\longleftrightarrow}}$$

Words Given an alphabet X, by $(X^*, ., \epsilon)$ we denote the free monoid generated by X (where . is the concatenation-product and ϵ is the empty word). If u denotes a word , $\tilde{u}$ denotes its mirror-image. A word u is a *factor* of the word v iff $\exists \alpha, \beta \in X^*$ such that $v = \alpha.u.\beta$. By $\mathrm{F}(u)$ we denote the set of all factors of u and given $L \subseteq X^*$, by $\mathrm{F}(L)$ we denote the set $\mathrm{F}(L) = \{u \in X^* \mid \exists v \in L, u \text{ is a factor of } v\}$. For every non-empty word $w \in X^*$, First(w) (resp. Last(w)) is the *first* or leftmost (resp. *last* or rightmost) letter of w. We set First(ϵ)=Last(ϵ)=ϵ.

Semi-Thue systems We call semi-Thue system over X every subset $S \subseteq X^* \times X^*$. By $\underset{S}{\longrightarrow}$ we denote the binary relation $\forall f, g \in X^*, f \underset{S}{\longrightarrow} g$ iff there exists $(u, v) \in S$ and $\alpha, \beta \in X^*$ such that $f = \alpha u \beta, g = \alpha v \beta$. $\underset{S}{\longrightarrow}$ is the one-step reduction generated by S. All the definitions and notation defined in the §"abstract rewriting systems" apply to the binary relation $\underset{S}{\longrightarrow}$. Let us give now additional notions, notation and results which are specific to semi-Thue systems.

We use now the notation Irr(S) for Irr($\underset{S}{\longrightarrow}$), $\Delta_S^*(A)$ for $\Delta^*_{\underset{S}{\longrightarrow}}(A)$. We set

$$\mathrm{Dom}(S) = \{u \in X^* \mid \exists v \in X^*, (u, v) \in S\}, \mathrm{Range}(S) = \{v \in X^* \mid \exists u \in X^*, (u, v) \in S\}$$

We define the *size* of S as $||S|| = \Sigma_{(u,v) \in S} |u| + |v|$

Let us fix some total ordering $\preceq$ on the set X and let us denote by the same symbol $\preceq$ the short-lex ordering induced on X^*. We call *redex* of S, every 3-tuple $(r, u, v) \in X^* \times X^* \times X^*$ such that $(u, v) \in S$. A redex (r, u, v) is said *leftmost* iff

1. no proper prefix of ru is S-reducible
2. no proper suffix u' of u is such that: $ru = r'u'$ and (r', u', v') is a redex
3. $\forall v' \in X^*, (u, v') \in S \Longrightarrow v \preceq v'$

(a *proper* prefix (resp. suffix) of u is a prefix (resp. suffix) which is *not equal* to u).
We denote by $\underset{S}{\longrightarrow}\!\!\!\!\!l$ the *leftmost reduction* generated by S, which is the binary relation [6]:

$$f \underset{S\,l}{\longrightarrow} g \Longleftrightarrow \exists r, u, v, s \in X^* \text{ such that } f = rus, g = rvs \text{ and } (r, u, v) \text{ is a leftmost redex}$$

In the case where $\underset{S}{\longrightarrow}$ is noetherian, S defines a mapping $\rho_S : X^* \longrightarrow \mathrm{Irr}(S)$ that

[6] Our definition differs slightly from that of [7]

we call the leftmost *total* reduction (mod S) and is defined by: for all $f \in X^*$,

$$\rho_S(f) \text{ is the unique irreducible word such that } f \xrightarrow[S]{*} \rho_S(f).$$

By CP(S) we denote the set of *critical pairs* of S (see [7]). By RCP(S) we denote the set of *reduced* critical pairs defined by:

$$\mathrm{RCP}(S) = \{(\rho_S(u), \rho_S(v)) \mid (u,v) \in \mathrm{CP}(S)\}$$

Definition 2. Let us consider the following conditions on the rules of a semi-Thue system S:

C1 : for every $(u,v),(u',v') \in S$ and every $r', s' \in X^*$
$v = r'u's' \Longrightarrow |s'| = |r'| = 0$
C2 : for every $(u,v),(u',v') \in S$ and every $r, s' \in X^*$
$rv = u's' \Longrightarrow |s'| = 0$ or $|s'| \geq |v|$
C3 : for every $(u,v),(u',v') \in S$ and every $r', s \in X^*$
$vs = r'u' \Longrightarrow |r'| = 0$ or $|r'| \geq |v|$
S is said *basic* (resp. *left-basic*) iff it fulfills C1, C2 and C3 (resp. C1 and C2).

These definitions appeared in [24, 13, 10, 28]. An extension to the so-called *controlled* rewriting systems is studied in ([9, 12, 33, 34]). One can notice that every *special* or *monadic* semi-Thue system is basic.

Algorithmic problem We focus in this paper on the PArtial Confluence Problem (defined in §1). In case L($\mathcal{R}$) consists of a single word, this problem is the *Confluence on a Congruence Class* problem investigated in [26, 21, 27]. One can also notice that a system S is *weakly confluent* in the sense of [22] iff it is partially confluent on the image by ρ_S of every righthand-side of S; this problem was investigated in [30, 35, 15, 22]. It follows that these two last problems are polynomially reducible to the PACP.

We use the following criterion for testing partial confluence

Lemma 2.1 (partial confluence criterion, [18]) *Let S be some noetherian semi-Thue system over X and $R \subseteq Irr(S)$. S is partially confluent on R iff* $\forall (u,v) \in RCP(S), \forall (\alpha, \alpha') \in Irr(S) \times Irr(S), \rho_S(\alpha u \alpha') \in R \Longleftrightarrow \rho_S(\alpha v \alpha') \in R$

This lemma is analogous to [35, Theorem 3.3] and can be proved in a similar way (complete proof in [18]). We refer the reader to [7, 11] for more information on rewriting systems.

2.2 2-tape deterministic finite automata

Definition 3. A deterministic n-tape finite automaton (abbreviated n-d.f.a. in the sequel) is a 5-tuple:

$$\mathcal{A} = < X, Q, \delta, q_0, F >$$

where

- X is a finite alphabet, the input-alphabet
- Q is a finite set, the set of states
- q_0 is a distinguished state , the initial state
- $F \subseteq Q$ is a set of distinguished states , the final states
- δ, the transition function, is a map from $Q \times (X \cup \{\#, \epsilon\})^n$ to Q (where $\#$ is a new letter not in X) fulfilling the restrictions:
  1. $\forall q \in Q, \forall \mathbf{u} \in (X \cup \{\#\})^n$ if $\delta(q, (\epsilon)^n)$ is defined, then
  $$\delta(q, \mathbf{u}) \text{ is defined } \Longrightarrow \mathbf{u} = \epsilon^n$$
  2. $\forall q \in Q, \forall \mathbf{u} \in (X \cup \{\#\})^n, \forall a \in X \cup \{\#\}$, if $\delta(q, (\epsilon)^i, a, (\epsilon)^j)$ (with $i + j + 1 = n$) is defined, then
  $$\delta(q, \mathbf{u}) \text{ is defined } \Longrightarrow \mathbf{u} = ((\epsilon)^i, b, (\epsilon)^j), \text{ for some } b \in X \cup \{\#\}$$

A *configuration* of $\mathcal{A}$ is any $n+1$-tuple $(q, u_1, \ldots, u_n) \in Q \times ((X \cup \{\#\})^*)^n$.By Config($\mathcal{A}$) we denote the set of all configurations of $\mathcal{A}$. The *one-step move relation* $\xrightarrow[\mathcal{A}]{}$ is the binary relation on Config($\mathcal{A}$) defined by:

$$(q, \mathbf{u}) \xrightarrow[\mathcal{A}]{} (q', \mathbf{u}') \Longleftrightarrow \exists \mathbf{a} \in (X \cup \{\#, \epsilon\})^n \text{ such that } \delta(q, \mathbf{a}) = q' \text{ and } \mathbf{u} = \mathbf{a}.\mathbf{u}'$$

The general notions introduced in §2.1 then apply on this binary relation, so that we can use the relation $\xrightarrow[\mathcal{A}]{*}$ and the notion of *irreducible* configuration (mod $\mathcal{A}$). The n-d.f.a $\mathcal{A}$ is called *loop-free* iff:

$$\forall q \in Q, \forall n \in \mathbb{N}, (q, \epsilon) \xrightarrow[\mathcal{A}]{n} (q, \epsilon) \Longrightarrow n = 0$$

From now on we suppose the n-d.f.a.'s we are dealing with are loop-free. We define the map $\tau_{\mathcal{A}}$: Config($\mathcal{A}$) $\longrightarrow$ Config($\mathcal{A}$) (the map *computed* by $\mathcal{A}$) by:

$$\forall c \in \text{Config}(\mathcal{A}), \tau_{\mathcal{A}}(c) \text{ is } \mathcal{A}\text{-irreducible and } c \xrightarrow[\mathcal{A}]{*} \tau_{\mathcal{A}}(c)$$

The language recognized by $\mathcal{A}$ is:

$$\mathrm{L}(\mathcal{A}) = \{\mathbf{u} \in X^n \mid \exists q \in F, (q_0, \mathbf{u}.(\#)^n) \xrightarrow[\mathcal{A}]{*} (q, (\epsilon)^n)\}$$

Theorem 2.2 ([16]) *The problem to determine whether two 2-d.f.a's $\mathcal{A}, \mathcal{B}$ are recognizing the same language is decidable in Polynomial time (namely in time $O((p+q)^4)$, where p, q are the sizes of the automata under comparison).*

We recall that even the more general problem of the *multiplicity equivalence* for n-tape finite automata is decidable ([19]), (but it is not known whether this last problem is in the class P, for $n \geq 3$).

Definition 4. Let us consider two sets E, F, a 2-d.f.a. $\mathcal{A}$ and three maps

$$h : E \longrightarrow F, H_1 : E \longrightarrow \text{Config}(\mathcal{A}), H_2 : \text{Config}(\mathcal{A}) \longrightarrow F.$$

We say that $\mathcal{A}$ *computes h via the encoding H_1 and the decoding H_2* iff

$$h = H_2 \circ \tau_{\mathcal{A}} \circ H_1$$

(Of course this notion can be of some use only in case the maps H_i are of a simple kind).

3 Partial confluence on a word

Let us fix some finite, basic, noetherian, semi-Thue system $S \subseteq X^* \times X^*$. To every $m \in \mathrm{Irr}(S)$ we associate a set $E_m \subseteq (X^*)^7$ and a relation $\underset{T_m}{\longrightarrow}$ on E_m [7]:

$E_m = \{(\alpha, x, \beta, \gamma, \beta', x', \alpha') \in (X^*)^7 \mid |x| \leq 1, |x'| \leq 1, \alpha x \beta \in \mathrm{Irr}(S), \beta' x' \alpha' \in \mathrm{Irr}(S), \beta \in \mathrm{F}(\mathrm{Dom}(S)), \beta' \in \mathrm{F}(\mathrm{Dom}(S))$ and $\gamma \in (\mathrm{Range}(S)+\varepsilon)\mathrm{F}(m)(\mathrm{Range}(S)+\varepsilon)\}$

We define the relation $\underset{T}{\longrightarrow}$ by describing the successors of every element $e = (\alpha, x, \beta, \gamma, \beta', x', \alpha') \in E_m$

Case 1 : $\alpha = \alpha_1 a$ where $a \in X$ and $x = \varepsilon$

$$(\alpha, x, \beta, \gamma, \beta', x', \alpha') \underset{T}{\longrightarrow} (\alpha_1, a, \beta, \gamma, \beta', x', \alpha')$$

Case 2 : Last $(\alpha x) = x$, $\alpha' = a\alpha_1'$ where $a \in X$ and $x' = \varepsilon$

$$(\alpha, x, \beta, \gamma, \beta', x', \alpha') \underset{T}{\longrightarrow} (\alpha, x, \beta, \gamma, \beta', a, \alpha_1')$$

Case 3 : Last $(\alpha x) = x$, First$(x'\alpha') = x'$ and ($|x| = 1$ and $x\beta \in \mathrm{F}(\mathrm{Dom}(S))$)

$$(\alpha, x, \beta, \gamma, \beta', x', \alpha') \underset{T}{\longrightarrow} (\alpha, \varepsilon, x\beta, \gamma, \beta', x', \alpha')$$

Case 4 : Last $(\alpha x) = x$, First$(x'\alpha') = x'$, ($|x| = 0$ or $x\beta \notin \mathrm{F}(\mathrm{Dom}(S))$) and ($|x'| = 1$ and $\beta' x' \in \mathrm{F}(\mathrm{Dom}(S))$)

$$(\alpha, x, \beta, \gamma, \beta', x', \alpha') \underset{T}{\longrightarrow} (\alpha, x, \beta, \gamma, \beta' x', \varepsilon, \alpha')$$

Case 5 : Last $(\alpha x) = x$, First$(x'\alpha') = x'$, ($|x| = 0$ or $x\beta \notin \mathrm{F}(\mathrm{Dom}(S))$), ($|x'| = 0$ or $\beta' x' \notin \mathrm{F}(\mathrm{Dom}(S))$), and $\beta\gamma\beta'$ is S-reducible

subcase 1 : $\beta = \beta_1\beta_2, \gamma = \gamma_1\gamma_2$, and $(\beta_1, \beta_2\gamma_1, u_1)$ is the leftmost redex of $\beta\gamma\beta'$

$$(\alpha, x, \beta, \gamma, \beta', x', \alpha') \underset{T}{\longrightarrow} (\alpha, x, \beta_1, u_1\gamma_2, \beta', x', \alpha')$$

subcase 2 : $\gamma = \gamma_1\gamma_2, \beta' = \beta_1'\beta_2'$ and $(\beta\gamma_1, \gamma_2\beta_1', u_2)$ is the leftmost redex of $\beta\gamma\beta'$

$$(\alpha, x, \beta, \gamma, \beta', x', \alpha') \underset{T}{\longrightarrow} (\alpha, x, \beta, \gamma_1 u_2, \beta_2', x', \alpha')$$

subcase 3 : $\beta = \beta_1\beta_2, \beta' = \beta_1'\beta_2'$, and $(\beta_1, \beta_2\gamma\beta_1', u')$ is the leftmost redex of $\beta\gamma\beta'$

$$(\alpha, x, \beta, \gamma, \beta', x', \alpha') \underset{T}{\longrightarrow} (\alpha, x, \beta_1, u', \beta_2', x', \alpha')$$

subcase 4 : $\gamma = \gamma_1 u \gamma_2$ and $(\beta\gamma_1, u, u')$ is the leftmost redex of $\beta\gamma\beta'$

$$(\alpha, x, \beta, \gamma, \beta', x', \alpha') \underset{T}{\longrightarrow} (\alpha, x, \beta, \gamma_1 u' \gamma_2, \beta', x', \alpha')$$

[7] the subscript m in T_m will be omitted in the sequel when it causes no confusion

One can check that if $e \in E$ fulfills the hypothesis of one of cases 1-5, then the 7-tuple e' which we introduce on the right-hand side of the arrow "$\underset{T}{\longrightarrow}$" is itself in E (the non-trivial checks are cases (5.1), (5.2), (5.3) and (5.4) which lean on the basicity of S and on the facts that $\gamma \in (\mathrm{Range}(S)+\varepsilon)\mathrm{F}(m)(\mathrm{Range}(S)+\varepsilon)$ and $m \in \mathrm{Irr}(S)$).

Let us define the map $\Pi : E_m \rightarrow X^*$ by

$$\Pi(\alpha, x, \beta, \gamma, \beta', x', \alpha') = \alpha x \beta \gamma \beta' x' \alpha'$$

and the map $\mathcal{O} : E_m \rightarrow \mathbb{N}^6 \times X^*$ by

$$\mathcal{O}(\alpha, x, \beta, \gamma, \beta', x', \alpha') = (|\alpha|, |\alpha'|, |x|, |x'|, |\beta|, |\beta'|, \beta\gamma\beta')$$

Lemma 3.1 $\underset{T}{\longrightarrow}$ *is confluent and noetherian.*

Proof : $\underset{T}{\longrightarrow}$ is functional, hence confluent.

Let $\sqsubseteq$ be the lexicographic ordering on $\mathbb{N}^6 \times X^*$ induced by the natural ordering on $\mathbb{N}$ and the partial ordering $\underset{S}{\overset{*}{\longleftarrow}}$ on X^*. As $\underset{S}{\longrightarrow}$ is noetherian, $\sqsubseteq$ is a well-founded partial ordering. $\forall e, e' \in E_m, e \underset{T}{\longrightarrow} e' \Rightarrow \mathcal{O}(e) \sqsupset \mathcal{O}(e')$, hence $\underset{T}{\longrightarrow}$ is noetherian. □

Let $\rho_S : X^* \rightarrow \mathrm{Irr}(S)$ be as defined in §2.

Let $\rho_T : E_m \rightarrow \mathrm{Irr}(T)$ be defined by :

$\forall e \in E_m, \rho_T(e)$ is the unique element of E_m such that

$$e \underset{T}{\overset{*}{\longrightarrow}} \rho_T(e) \text{ and } \rho_T(e) \in \mathrm{Irr}(T)$$

($\rho_T(e)$ is well-defined by lemma 3.1)

Lemma 3.2 : *Let $e_1, e_2 \in E_m$. If $e_1 \underset{T}{\overset{*}{\longrightarrow}} e_2$ then $\rho_S(\Pi(e_1)) = \rho_S(\Pi(e_2))$.*

Proof : It is sufficient to prove this lemma under the hypothesis that $e_1 \underset{T}{\longrightarrow} e_2$ (the general statement will follow by induction).

Let us suppose that $e_1 \underset{T}{\longrightarrow} e_2$. We consider all the cases occuring in the definition of $\underset{T}{\longrightarrow}$.

Case 1, 2, 3, 4 : $\Pi(e_1) = \Pi(e_2)$, hence $\rho_S(\Pi(e_1)) = \rho_S(\Pi(e_2))$

Case 5, subcase 1 :

- As ($\alpha x = \epsilon$ or $x\beta \notin \mathrm{F}(\mathrm{Dom}(S))$) and no proper prefix of $\beta\gamma_1$ is S-reducible, no proper prefix of $\alpha x \beta \gamma_1$ and longer than $\alpha x \beta$ is S-reducible

- As $\alpha x\beta \in \mathrm{Irr}(S)$, no proper prefix of $\alpha x\beta\gamma_1$ shorter than $\alpha x\beta$ is S-reducible.
- Hence $\Pi(e_1) \xrightarrow[S]{l} \Pi(e_2)$, so that $\rho_S(\Pi(e_1)) = \rho_S(\Pi(e_2))$.

Cases 5.2, 5.3, 5.4 can be treated in a similar way.□

Lemma 3.3 : *For every* $e \in E_m$, $\Pi(\rho_T(e)) = \rho_S(\Pi(e))$.

Proof : As $e \xrightarrow[T]{*} \rho_T(e)$, by lemma 3.2 we have

$$\rho_S(\Pi(\rho_T(e))) = \rho_S(\Pi(e)) \tag{1}$$

Let us show that $\Pi(\rho_T(e))$ is S-irreducible.
Let us suppose $\rho_T(e) = (\alpha, x, \beta, \gamma, \beta', x', \alpha')$ and

$$\alpha x\beta\gamma\beta' x'\alpha' = pus \tag{2}$$

where (p, u, v) is the leftmost redex of $\Pi(\rho_T(e))$.

As $\alpha x\beta, \beta' x'\alpha' \in \mathrm{Irr}(S)$, the given occurrence of u must contain at least one letter of the given occurrence of γ or contain at least one letter in both factors $\alpha x\beta, \beta' x'\alpha'$.

If (u contains $x\beta$ and $|x| = 1$) or (u contains $\beta' x'$ and $|x'| = 1$), a rule of T of type (1) (2) (3) or (4) would apply on $\rho_T(e)$, contradicting the fact that $\rho_T(e)$ is T-irreducible. If u is a factor of $\beta\gamma\beta'$, either a rule of type (1) (2) (3) or (4) can be applied on $\rho_T(e)$ or (Last(αx) $= x$,First($x'\alpha'$) $= x'$, ($|x| = 0$ or $x\beta \notin$ F(Dom(S))), ($|x'| = 0$ or $\beta' x' \notin$ F(Dom(S)))). If the second hypothesis is true, then one of cases (5.1), (5.2), (5.3), (5.4) is realized, contradicting again the fact that $\rho_T(e)$ is T-irreducible. Hence (2) is impossible and $\Pi(\rho_T(e))$ is S-irreducible. Hence

$$\rho_S(\Pi(\rho_T(e))) = \Pi(\rho_T(e)) \tag{3}$$

By (1) (3) the lemma is proved. □

Let us define now a 2-d.f.a. $\mathcal{A}_m$ and two maps H_1, H_2 such that $\mathcal{A}_m$ computes ρ_{T_m} via the encoding H_1 and the decoding H_2.

automaton $\mathcal{A}_m$: $\mathcal{A}_m = < X, Q_m, \delta, q_o, Q_{m,+} >$ where
$q_0 = [\epsilon, \epsilon, m, \epsilon, \epsilon]$
$Q_m = \{[x, \beta, \gamma, \beta', x'] \in ((X \cup \{\#\})^*)^5 \mid |x| \leq 1, |x'| \leq 1, x\beta \in \mathrm{Irr}(S), \beta' x' \in \mathrm{Irr}(S), \beta \in \mathrm{F}(\mathrm{Dom}(S)), \beta' \in \mathrm{F}(\mathrm{Dom}(S))$ and $\gamma \in (\mathrm{Range}(S)+\varepsilon)\mathrm{F}(m)(\mathrm{Range}(S)+\varepsilon)\}$
$Q_{m,+} = Q_m$.
The transition function δ mimics the rules of T. Let $q = [x, \beta, \gamma, \beta', x'] \in Q_m$. We describe below all the transitions of $\mathcal{A}_m$ starting from state q and reading a pair (a, ε) or (ε, a) where $a \in X \cup \{\#\} \cup \{\varepsilon\}$.

case 1 : $a \in X \cup \{\#\}, x = \varepsilon$

$$\delta([x, \beta, \gamma, \beta', x'], a, \varepsilon) = [a, \beta, \gamma, \beta', x']$$

case 2 : $a \in X \cup \{\#\}, | x | = 1, x' = \varepsilon$

$$\delta([x, \beta, \gamma, \beta', x'], \varepsilon, a) = [x, \beta, \gamma, \beta', a]$$

case 3 : $| x | = 1, | x' | = 1$ and $x\beta \in \mathrm{F}(\mathrm{Dom}(S))$

$$\delta([x, \beta, \gamma, \beta', x'], \varepsilon, \varepsilon) = [\varepsilon, x\beta, \gamma, \beta', x']$$

case 4 : $| x | = 1, | x' | = 1, x\beta \notin \mathrm{F}(\mathrm{Dom}(S)), \beta' x' \in \mathrm{F}(\mathrm{Dom}(S))$

$$\delta([x, \beta, \gamma, \beta', x'], \varepsilon, \varepsilon) = [x, \beta, \gamma, \beta' x', \varepsilon]$$

case 5 : $| x | = 1, | x' | = 1, x\beta \notin \mathrm{F}(\mathrm{Dom}(S)), \beta' x' \notin \mathrm{F}(\mathrm{Dom}(S)), \beta\gamma\beta'$ is S-reducible

subcase 1 : $\beta = \beta_1\beta_2, \gamma = \gamma_1\gamma_2, (\beta_1, \beta_2\gamma_1, u_1)$ is the leftmost redex of $\beta\gamma\beta'$

$$\delta([x, \beta, \gamma, \beta', x'], \varepsilon, \varepsilon) = [x, \beta_1, u_1\gamma_2, \beta', x']$$

subcase 2 : $\gamma = \gamma_1\gamma_2, \beta' = \beta'_1\beta'_2$ and $(\beta\gamma_1, \gamma_2\beta'_1, u_2)$ is the leftmost redex of $\beta\gamma\beta'$

$$\delta([x, \beta, \gamma, \beta', x'], \epsilon, \epsilon) = [x, \beta, \gamma_1 u_2, \beta'_2, x']$$

subcase 3 : $\beta = \beta_1\beta_2, \beta' = \beta'_1\beta'_2$, and $(\beta_1, \beta_2\gamma\beta'_1, u')$ is the leftmost redex of $\beta\gamma\beta'$

$$\delta([x, \beta, \gamma, \beta', x'], \epsilon, \epsilon) = [x, \beta_1, u', \beta'_2, x']$$

subcase 4 : $\gamma = \gamma_1 u \gamma_2$ and $(\beta\gamma_1, u, u')$ is the leftmost redex of $\beta\gamma\beta'$

$$\delta([x, \beta, \gamma, \beta', x'], \epsilon, \epsilon) = [x, \beta, \gamma_1 u' \gamma_2, \beta'_2, x']$$

the **encoding map**, $H_1 : E_m \to \mathrm{Config}(\mathcal{A}_m)$ is defined by
$H_1(\alpha, x, \beta, \gamma, \beta', x', \alpha') = ([x, \beta, \gamma, \beta', x'], \widetilde{\alpha}\#, \alpha'\#)$
the **decoding map**, $H_2 : \mathrm{Config}(\mathcal{A}_m) \to E_m$ is defined by
$\mathrm{Dom}(H_2) = \{([x, \beta, \gamma, \beta', x'], \alpha, \alpha') \in \mathrm{Config}(\mathcal{A}_m) | \ \widetilde{\alpha}x \in \#X^*, x'\alpha' \in X^*\#, \widetilde{\alpha}x\beta \in \mathrm{Irr}(S), \beta' x' \alpha' \in \mathrm{Irr}(S)\}$
$H_2([x, \beta, \gamma, \beta', u'], \alpha, \alpha') = (h_2(\widetilde{\alpha}), h_2(x), \beta, \gamma, \beta', h_2(x'), h_2(\alpha'))$, where
$h_2 : (X \cup \{\#\})^* \to X^*$ is the homomorphism preserving each letter of X and erasing the $\#$.

Lemma 3.4 : *$\mathcal{A}_m$ computes ρ_T via the encoding H_1 and the decoding H_2*

Proof : Let $e \in E_m : e = (\alpha, x, \beta, \gamma, \beta', x', \alpha')$ and let $c = H_1(e)$.
One can check that, for every $c, c' \in \mathrm{Config}(\mathcal{A}_m)$,

$$c \in \mathrm{Dom}(H_2) \text{ and } c \xrightarrow[\mathcal{A}_m]{} c' \Longrightarrow c' \in \mathrm{Dom}(H_2) \text{ and } H_2(c) \xrightarrow[T]{*} H_2(c') \qquad (4)$$

As $c \in \mathrm{Dom}(H_2)$, one can deduce by induction that :

$$\tau_{\mathcal{A}_m}(c) \in \mathrm{Dom}(H_2) \text{ and } H_2(c) \xrightarrow[T]{*} H_2(\tau_{\mathcal{A}_m}(c)) \qquad (5)$$

One can check that

$$\forall c' \in \mathrm{Dom}(H_2), \quad c' \in \mathrm{Irr}(\mathcal{A}_m) \Longrightarrow \mathrm{H}_2(c') \in \mathrm{Irr}(T) \qquad (6)$$

By (5),(6), $\mathrm{H}_2(\tau_{\mathcal{A}_m}(\mathrm{H}_1(e))) = \rho_T(e)$. □

Let $m \in X^*, w \in \mathrm{Irr}(S)$. We define the language $L_{m,w}$ by :

$$L_{m,w} = \{(\alpha, \alpha') \in (X^*)^2 \mid \alpha, \alpha' \in \mathrm{Irr}(S) \text{ and } \rho_S(\alpha m \alpha') = w\}$$

We define a 2-d.f.a. $\mathcal{B}_{m,w} =< X \cup \{\#\}, Q'_{m,w}, \delta'_{m,w}, q'_o, Q'_{m,w,+} >$ by:
$Q'_{m,w} = Q_m \cup \{[ux, \beta, \gamma, \beta', x'u'] \mid [x, \beta, \gamma, \beta', x'] \in Q_m, u, u' \in (X \cup \{\#\})^*, \mid x \mid = 1, \mid x' \mid = 1, x\beta \notin \mathrm{F}(\mathrm{Dom}(S)), \beta' x' \notin \mathrm{F}(\mathrm{Dom}(S)) \text{ and } ux\beta\gamma\beta' x' u' \in \mathrm{F}(\#w\#)\}$.
$q'_o = q_o$, $Q'_{m,w,+} = \{[v, \beta, \gamma, \beta', v'] \in Q'_{m,w} \mid v\beta\gamma\beta' v' = \#w\#)$.
$\delta'_{m,w}$ is defined in such a way that $\mathcal{B}_{m,w}$ *simulates* first the automaton $\mathcal{A}_m$ and then *checks* that the irreducible configuration reached by $\mathcal{A}_m$ belongs to $H_2^{-1}(\Pi^{-1}(\#w\#))$. $\delta'_{m,w} = \delta_m \cup \delta_{m,w}$ where $\delta_{m,w}$ is defined below.

We describe here all the transitions of $\delta_{m,w}$ starting from a state $q \in Q'_{m,w}$ and reading a pair (a, ε) or (ε, a) where $a \in X \cup \{\#\} \cup \{\varepsilon\}$.

case 1 : $q \in Q_m$

Then $q = [x, \beta, \gamma, \beta', x']$

subcase 1 : $\mid x \mid \neq 1$ or $\mid x' \mid \neq 1$ or $x\beta \in \mathrm{F}(\mathrm{Dom}(S))$ or $\beta' x' \in \mathrm{F}(\mathrm{Dom}(S))$ or $\beta\gamma\beta'$ is S-reducible.

Then there is no transition in $\delta_{m,w}$ starting from q.

subcase 2 :$\mid x \mid = 1, \mid x' \mid = 1, x\beta \notin \mathrm{F}(\mathrm{Dom}(S)), \beta' x' \notin \mathrm{F}(\mathrm{Dom}(S)), \beta\gamma\beta' \in \mathrm{Irr}(S), |a| = 1, ax\beta\gamma\beta' x' \in \mathrm{F}(\#w\#)$,

$$\delta'_{m,w}(q, (a, \varepsilon)) = [ax, \beta, \gamma, \beta', x']$$

subcase 3 : $x = \#, \mid x' \mid = 1, x\beta \notin \mathrm{F}(\mathrm{Dom}(S)), \beta' x' \notin \mathrm{F}(\mathrm{Dom}(S)), \beta\gamma\beta' \in \mathrm{Irr}(S), |a| = 1, x\beta\gamma\beta' x' a \in \mathrm{F}(\#w\#)$.

$$\delta'_{m,w}(q, (\varepsilon, a)) = [x, \beta, \gamma, \beta', x'a]$$

case 2 : $q \in Q'_{m,w} - Q_m$

$q = [v, \beta, \gamma, \beta', v']$

subcase 1 : $v \in X^*, a \in X \cup \{\#\}, av\beta\gamma\beta' v' \in \mathrm{F}(\#w\#)$

$$\delta'_{m,w}(q, (a, \varepsilon)) = [av, \beta, \gamma, \beta', v']$$

subcase 2 : $v \in \#X^*, a \in X \cup \{\#\}, v\beta\gamma\beta' v' a \in \mathrm{F}(\#w\#)$

$$\delta'_{m,w}(q, (\varepsilon, a)) = [v, \beta, \gamma, \beta', v'a]$$

Lemma 3.5 : *For every* $(\alpha, \alpha') \in X^* \times X^*$,
$(\alpha, \alpha') \in L_{m,w} \iff (\tilde{\alpha}, \alpha') \in \mathrm{L}(\mathcal{B}_{m,w}) \cap (\mathrm{Irr}(S) \times \mathrm{Irr}(S))$.

Proof :

1 - Let $(\alpha, \alpha') \in L_{m,w}$. Then

$e = (\alpha, \varepsilon, \varepsilon, m, \varepsilon, \varepsilon, \alpha') \in E_m$ and $H_1(e) = ([\varepsilon, \varepsilon, m, \varepsilon, \varepsilon], \tilde{\alpha}\#, \alpha'\#)$.
By hypothesis $\rho_S(\alpha m \alpha') = w$, i.e.

$$\rho_S(\Pi(e)) = w$$

hence, by lemma 3.3 :

$$\Pi(\rho_T(e)) = w \tag{7}$$

By lemma 3.4 we have :

$$H_2(\tau_{\mathcal{A}_m}(H_1(e))) = \rho_T(e) \tag{8}$$

Let $\alpha_1, \alpha'_1 \in (X \cup \{\#\})^*, x, \beta, \gamma, \beta', x', \in (X \cup \{\#\})^*$ such that

$$\tau_{\mathcal{A}_m}(H_1(e)) = ([x, \beta, \gamma, \beta', x'], \widetilde{\alpha_1}, \alpha'_1) \tag{9}$$

By (7) (8) (9) we have :

$$\alpha_1 x \beta \gamma \beta' x' \alpha'_1 = \#w\# \tag{10}$$

By (9), as $\tau_{\mathcal{A}_m}(H_1(e))$ is $\mathcal{A}_m$-irreducible we have

$$| x |= 1, | x' |= 1, x\beta \notin \mathrm{F}(\mathrm{Dom}(S)), \beta' x' \notin \mathrm{F}(\mathrm{Dom}(S)), \beta\gamma\beta' \in \mathrm{Irr}(S)$$

hence the transitions of $\delta_{m,w}$ allow a computation :

$$([x, \beta, \gamma, \beta', x'], \widetilde{\alpha_1}, \alpha'_1) \xrightarrow[\mathcal{B}_{m,w}]{*} ([\alpha_1 x, \beta, \gamma, \beta', x'\alpha'_1], \varepsilon, \varepsilon)$$

By (10), the state $[\alpha_1 x, \beta, \gamma, \beta', x'\alpha'_1]$ is terminal (i.e. belongs to $Q'_{m,w,+}$).

Hence, $(\tilde{\alpha}, \alpha') \in \mathrm{L}(\mathcal{B}_{m,w})$.

2 - Suppose $(\tilde{\alpha}, \alpha') \in \mathrm{L}(\mathcal{B}_{m,w}) \cap (\mathrm{Irr}(S) \times \mathrm{Irr}(S))$. There exists a successful computation of $\mathcal{B}_{m,w}$ on $(\tilde{\alpha}, \alpha')$:

$$([\varepsilon, \varepsilon, m, \varepsilon, \varepsilon], \tilde{\alpha}\#, \alpha'\#) \xrightarrow[\mathcal{A}_m]{*} ([x, \beta, \gamma, \beta', x'], \widetilde{\alpha_1}, \alpha'_1) \xrightarrow[\mathcal{B}_{m,w}]{*} ([\alpha_1 x, \beta, \gamma, \beta', x'\alpha'_1], \varepsilon, \varepsilon) \tag{11}$$

for some $x, x' \in X \cup \{\#\}, \beta, \gamma, \beta', \alpha_1, \alpha'_1 \in (X \cup \{\#\})^*$ such that

$$\alpha_1 x \beta \gamma \beta' x' \alpha'_1 = \#w\# \tag{12}$$

Using again (4), it follows from (11) (12) that $\alpha m \alpha' \xrightarrow[T]{*} w$.

Hence $(\alpha, \alpha') \in L_{m,w}$. □

Theorem 3.6 : *The problem "is S partially confluent on the word w ?" is solvable in P-time for finite, basic, noetherian, semi-Thue systems.*

Sketch of proof. : Let $S \subseteq X^* \times X^*$ be some finite, basic, noetherian semi-Thue system and $w \in \mathrm{Irr}(S)$. By Lemma 2.1, S is partially confluent on $\{w\}$ iff:

$$\forall (u, v) \in \mathrm{RCP}(S), \quad L_{u,w} = L_{v,w}$$

which, by lemma 3.5, is equivalent, to:

$$\forall (u,v) \in \mathrm{RCP}(S), \mathrm{L}(\mathcal{B}_{u,w}) \cap (\mathrm{Irr}(S) \times \mathrm{Irr}(S)) = \mathrm{L}(\mathcal{B}_{v,w}) \cap (\mathrm{Irr}(S) \times \mathrm{Irr}(S))$$

We use the fact that each $\mathcal{B}_{m,w}$ can be constructed in P-time and Theorem 2.2.□

4 Partial confluence on a rational set

We sketch here an adaptation of §3 to the general case. Let us fix some finite, basic, noetherian, semi-Thue system $S \subseteq X^* \times X^*$ and some rational set $R \subseteq \mathrm{Irr}(S)$ recognized by some 1-d.f.a. $\mathcal{R} =< X, Q, q_-, \theta, Q_t >$ (we suppose $\mathcal{R}$ is complete). We define two maps
$\varphi : (Q \times X \times Q)^* \to X^*$ is the homomorphism "erasing" the states i.e.:

$$\forall (q,x,q') \in Q \times X \times Q, \varphi(q,x,q') = x$$

$\sigma : X^* \to (Q \times X \times Q)^*$ sends each word $w \in X^*$ on the unique computation of $\mathcal{R}$ over this word: if $w = x_1 x_2 \cdots x_p$ and $\delta(q_-, x_1) = q_1, \cdots, \delta(q_{p-1}, x_p) = q_p$

$$\text{then } \sigma(w) = (q_-, x_1, q_1)(q_1, x_2, q_2) \cdots (q_{p-1}, x_p, q_p).$$

To every $m \in X^*$ we associate the set $\overline{E}_m = \{(\alpha, x, \beta, \gamma, \beta', x', \alpha') \in (Q \times X \times Q)^* \times (X^*)^6 \mid \alpha = \sigma(\varphi(\alpha))$ and $(\varphi(\alpha), x, \beta, \gamma, \beta', x', \alpha') \in E_m\}$.
By $\overline{\varphi}, \overline{\sigma}$ we denote the maps $\overline{\varphi} : \overline{E}_m \to E_m$, $\overline{\sigma} : E_m \to \overline{E}_m$ defined by:

$$\overline{\varphi}(\alpha,x,\beta,\gamma,\beta',x',\alpha') = (\varphi(\alpha),x,\beta,\gamma,\beta',x',\alpha'), \quad \overline{\sigma}(\alpha,x,\beta,\gamma,\beta',x',\alpha') = (\sigma(\alpha),x,\beta,\gamma,\beta',x',\alpha')$$

By $\overline{\Pi}$ we denote the map: $\overline{\Pi} = \Pi o \overline{\varphi}$. The relation $\underset{U_m}{\longrightarrow}$ is defined on $\overline{E}_m$ by:

$$\forall \overline{e_1}, \overline{e_2} \in \overline{E}_m, \overline{e_1} \underset{U_m}{\longrightarrow} \overline{e_2} \Longleftrightarrow \overline{\varphi}(\overline{e_1}) \underset{T_m}{\longrightarrow} \overline{\varphi}(\overline{e_2}).$$

One can notice that $\overline{\varphi}, \overline{\sigma}$ are bijections exchanging $\underset{T_m}{\longrightarrow}$ with $\underset{U_m}{\longrightarrow}$. The two next lemma follow easily.

Lemma 4.1 : $\underset{U_m}{\longrightarrow}$ *is confluent and noetherian.*

We define $\rho_{U_m} : \overline{E}_m \to Irr(U)$ by: $\forall \overline{e} \in \overline{E}_m, \overline{e} \overset{*}{\underset{U_m}{\longrightarrow}} \rho_{U_m}(\overline{e})$ and $\rho_{U_m}(\overline{e}) \in \mathrm{Irr}(U_m)$.

Lemma 4.2 : $\forall \overline{e} \in \overline{E}_m, \quad \rho_{U_m}(\overline{e}) = \overline{\sigma} o \rho_{T_m} o \overline{\varphi}(\overline{e})$.

We can construct in P-time a 2-d.f.a. $\overline{\mathcal{A}}_m$ which, analogously with $\mathcal{A}_m$, mimics the rules of U_m. Let $\overline{H}_1 = H_1 o \overline{\varphi}$ and $\overline{H}_2 = \overline{\sigma} o H_2$. We have

Lemma 4.3 : $\overline{\mathcal{A}}_m$ *computes* ρ_{U_m} *via the encoding* $\overline{H}_1$ *and the decoding* $\overline{H}_2$.

Let $L_{m,R} = \{(\alpha, \alpha') \in (Q \times X \times Q)^* \times X^* \mid \alpha = \sigma o \varphi(\alpha), \varphi(\alpha), \alpha' \in Irr(S)$ and $\rho_S(\varphi(\alpha) m \alpha') \in R\}$.

Lemma 4.4 : *One can construct in P-time from the system S, the word m and the 1-d.f.a.* $\mathcal{R}$*, a 2-d.f.a.* $\mathcal{B}_{m,\mathcal{R}}$ *such that:* $\forall(\alpha,\alpha') \in (Q \times X \times Q)^* \times X^*$,

$$(\alpha,\alpha') \in L_{m,R} \Longleftrightarrow (\tilde{\alpha},\alpha') \in \mathrm{L}(\mathcal{B}_{m,\mathcal{R}}).$$

Sketch of proof : We can construct such an automaton $\mathcal{B}_{m,\mathcal{R}}$ by a combination of $\overline{\mathcal{A}}_m$ with $\mathcal{R}$ and with a d.f.a. I recognizing Irr(S). $\mathcal{B}_{m,\mathcal{R}}$ can then be constructed in time polynomial with respect to $\| S \|, | m |, | Q |$. □

Theorem 4.5 : *The problem "is S partially confluent on the rational set R ?" (where* $R \subseteq Irr(S)$*), is solvable in P-time for finite, basic, noetherian semi-Thue systems.*

Sketch of proof : The conditions given in Lemma 2.1 are reduced by Lemma 4.4 to a polynomial number of equality tests and each of these tests can be done in P-time (by Theorem 2.2).□

Aknowledgements: I thank R.V. Book, J. Engelfriet, F. Otto and L. Zhang for useful discussions and an anonymous referee for pointing out several mistakes in a previous version of this work.

References

1. S. Arnborg, B. Courcelle, A. Proskurowski, and D. Seese. An algebraic theory of graph reduction. *J. ACM 40*, pages 1134–1164, 1993.
2. J.M. Autebert and L. Boasson. The equivalence of pre-NTS grammars is decidable. *Math.Systems Theory 25*, pages 61–74, 1992.
3. J. Berstel and L. Boasson. Context-free languages. In *Handbook of theoretical computer science, vol.B, Chapter 2*, pages 59–102. Elsevier, 1991.
4. L. Boasson. Grammaires à non-terminaux séparés. In *Proceedings 7th ICALP*, pages 105–118. LNCS 85, 1980.
5. L. Boasson and G. Sénizergues. NTS languages are deterministic and congruential. *JCSS 31, nr 3*, pages 332–342, 1985.
6. R.V. Book, M. Jantzen, and C. Wrathall. Monadic Thue systems. *TCS 19*, pages 231–251, 1982.
7. R.V. Book and F. Otto. *String Rewriting Systems.Texts and monographs in Computer Science.* Springer-Verlag, 1993.
8. H. Bücken. Reduction systems and small cancellation theory. In *Proceedings 4th Workshop on Automated Deduction*, pages 53–59, 1979.
9. P. Butzbach. Sur l'équivalence des grammaires simples. In *Actes des Premières journées d'Informatique Théorique,Bonascre.* ENSTA, 1973.
10. P. Butzbach. Une famille de congruences de Thue pour lesquelles l'équivalence est décidable. In *Proceedings 1rst ICALP*, pages 3–12. LNCS, 1973.
11. P. Le Chenadec. *Canonical Forms in Finitely Presented Algebras.* Research Notes in Theoretical Computer Science, Pitman Wiley & sons, 1986.
12. L. Chottin. Strict deterministic languages and controlled rewriting systems. In *Proceedings 6th ICALP*, pages 104–117. LNCS 71, Springer-Verlag, 1979.

13. Y. Cochet. Sur l'algébricité des classes de certaines congruences définies sur le monoide libre. *Thèse de l'université de Rennes*, 1971.
14. N. Dershowitz and J.P. Jouannaud. Rewrite systems. In *Handbook of theoretical computer science, vol.B, Chapter 2*, pages 243–320. Elsevier, 1991.
15. J. Engelfriet. Deciding the NTS-property of context-free grammars. *Tech. report 94-05, Leiden university*, pages 1–7, 1994. To appear in the proceedings of the conference on Important Results and Trends in Theoretical Computer Science, Graz, Austria, June 1994.
16. E.P. Friedman and S.A. Greibach. A polynomial algorithm for deciding the equivalence problem for 2-tape deterministic finite state acceptors. *SIAM J. COMPUT., vol.11, No 1*, pages 166–183, 1982.
17. S. Ginsburg and E.H. Spanier. Finite-turn Pushdown automata. *J. SIAM Control, vol.4, No 3*, pages 429–453, 1966.
18. P. Grosset-Grange. Décidabilité de la confluence partielle d'un système semi-Thuéien rationnel. *Mémoire de DEA de l'université de Bordeaux 1*, pages 1–21, 1993.
19. T. Harju and J. Karhumäki. The equivalence problem of multitape finite automata. *TCS 78*, pages 347–355, 1991.
20. G. Huet. Confluent reductions:abstract properties and applications to term rewriting systems. *JACM vol. 27 no 4*, pages 797–821, 1980.
21. K. Madlener, P. Narendran, and F. Otto. A specialized completion procedure for monadic string-rewriting systems presenting groups. In *Proceedings 18th ICALP*, pages 279–290. Springer, LNCS No 510, 1991.
22. K. Madlener, P. Narendran, F. Otto, and L. Zhang. On weakly confluent monadic string-rewriting systems. *TCS 113*, pages 119–165, 1993.
23. P. Narendran. It is decidable whether a monadic Thue system is canonical over a regular set. *Math. Systems Theory 23*, pages 245–254, 1990.
24. M. Nivat. On some families of languages related to the Dyck language. *2nd Annual Symposium on Theory of Computing*, 1970.
25. C. O'Dunlaing. Infinite regular Thue systems. *TCS 2*, pages 171–192, 1983.
26. F. Otto. On deciding the confluence of a finite string-rewriting system on a given congruence class. *JCSS 35*, pages 285–310, 1987.
27. F. Otto. The problem of deciding confluence on a given congruence class is tractable for finite special string-rewriting systems. *Math. Systems Theory 25*, pages 241–251, 1992.
28. J. Sakarovitch. Syntaxe des langages de Chomsky, essai sur le déterminisme. *Thèse de doctorat d'état de l'université Paris VII*, pages 1–175, 1979.
29. G. Sénizergues. The equivalence problem for NTS languages is decidable. In *Proceedings 6th G.I. Symposium on Theoretical Computer Science*, pages 313–323. LNCS Springer-Verlag, nr 145, 1982.
30. G. Sénizergues. The equivalence and inclusion problems for NTS languages. *J. Comput. System Sci. 31(3)*, pages 303–331, 1985.
31. G. Sénizergues. Sur la description des langages algébriques déterministes par des systèmes de réécriture confluents. *Thèse d'état,université Paris 7 et rapport LITP 88-39*, pages 1–330, 1987.
32. G. Sénizergues. Church-Rosser controlled rewriting systems and equivalence problems for deterministic context-free languages. *Information and Computation 81 (3)*, pages 265–279, 1989.

33. G. Sénizergues. A characterisation of deterministic context-free languages by means of right-congruences. *TCS 70 (2)*, pages 213–232, 1990.
34. G. Sénizergues. Some decision problems about controlled rewriting systems. *TCS 71*, pages 281–346, 1990.
35. L. Zhang. Weak confluence is tractable for finite string-rewriting systems. *preprint*, pages 1–10, 1991.

Problems in Rewriting Applied to Categorical Concepts by the Example of a Computational Comonad

Wolfgang Gehrke*

Research Institute for Symbolic Computation
Johannes Kepler University
A - 4040 Linz, AUSTRIA
Wolfgang.Gehrke@risc.uni-linz.ac.at

Abstract. We present a canonical system for comonads which can be extended to the notion of a computational comonad [BG92] where the crucial point is to find an appropriate representation. These canonical systems are checked with the help of the Larch Prover [GG91] exploiting a method by G. Huet [Hue90a] to represent typing within an untyped rewriting system. The resulting decision procedures are implemented in the programming language Elf [Pfe89] since typing is directly supported by this language. Finally we outline an incomplete attempt to solve the problem which could be used as a benchmark for rewriting tools.

1 Introduction

The starting point of this work was to provide methods for checking the commutativity of diagrams arising in category theory. Diagrams in this context are used as a visual description of equations between morphisms. To check the commutativity of a diagram amounts to check the equality of the morphisms involved. One way to support this task is to solve the uniform word problem for this category. Of course this is not always decidable.

A monoid is a very useful mathematical notion which is described equationally and which has a decidable uniform word problem. The equations can be characterized by diagrams as in Fig. 1. A monad is the categorical generalization of the this concept [ML71]. This gives evidence that there can be a canonical system for monads, too.

A recent application of monads in computer science can be found in [Mog89] where they are exploited to structure the semantics of programming languages which later was used to structure purely functional programs as in [Wad93].

In [Geh94] we reduced the uniform word problem for monads to the uniform word problem for adjunctions exploiting theorems from category theory [BW85]. But when doing the same for a comonad the resulting system could not be

* sponsored by the Austrian Science Foundation (FWF) under ESPRIT BRP 6471 MEDLAR II, and by the American National Science Foundation under Grant No. CCR-9303383

$$M \times M \times M \xrightarrow{\quad \mu \times 1 \quad} M \times M \qquad\qquad 1 \times M \xrightarrow{\quad \eta \times 1 \quad} M \times M \xleftarrow{\quad 1 \times \eta \quad} M \times 1$$

$$\downarrow{1 \times \mu} \qquad\qquad\qquad \downarrow{\mu} \qquad\qquad \cong \qquad \downarrow{\mu} \qquad \cong$$

$$M \times M \xrightarrow{\qquad\qquad\qquad} M \qquad\qquad\qquad\qquad M$$
$$\mu$$

$$associativity \qquad\qquad\qquad\qquad identity \ laws$$

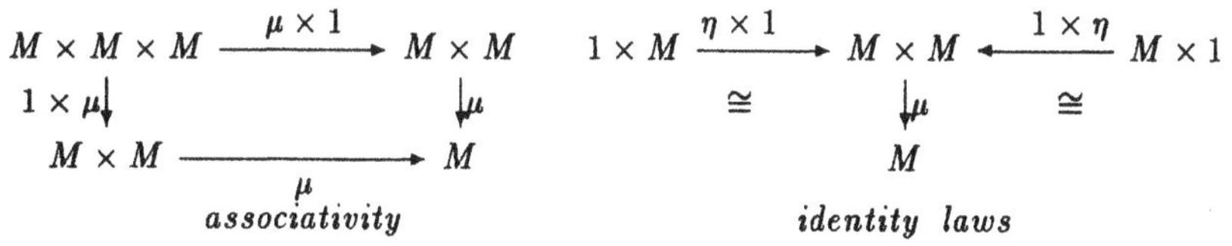

Fig. 1. laws of a monoid in form of diagrams

extended to the notion of a computational comonad [BG92] which is used to study the intensional semantics of programming languages.

In this paper we describe a canonical system for computational comonads which is based on a different canonical system for comonads. It seems that the main difficulty in handling these equational theories consists of an appropriate reformulation of the problem with the help of an equivalent theory where Cartesian closed categories are a well known instance [Hue90a]. In that case the decision problem is transferred from the language of a CCC into the language of a typed lambda calculus.

Morphisms in a category come together with a type depending on two objects – source and target. These have to be taken into account when doing rewriting. In the verification of the canonical system we will deal with types in the frame of the Larch Prover [GG91] as suggested in [Hue90a]. Finally the resulting decision procedures were implemented in the programming language Elf [Pfe89] which directly supports dependent types and they were applied to examples from [BG92].

At the end we will briefly describe another attempt to solve the same problem with the help of rewriting over a congruence.

The main contributions of this paper consist of:

- a representation of comonads which is suitable for rewriting and can be extended to the notion of a computational comonad providing a canonical system,
- the presentation of the method for encoding types within an untyped framework by using the example of a computational comonad,
- another demonstration of the usability of Elf as a tool for typed rewriting.

At the end we conclude and suggest future work.

2 Definitions

In this section the definition of a comonad [BW85] and a computational comonad [BG92] are given. Furthermore the notion of the Kleisli category is introduced. We assume throughout that the morphisms between two objects form a set.

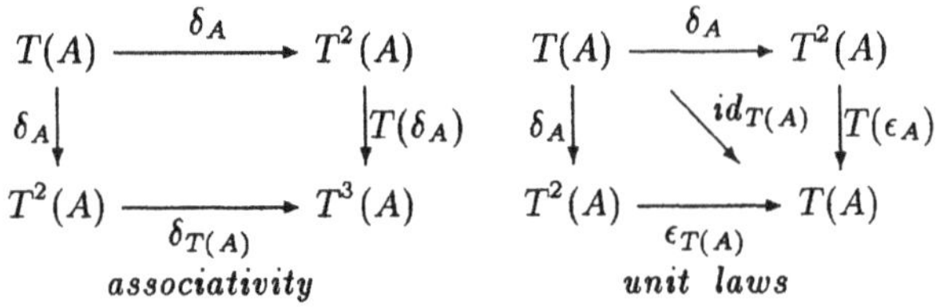

Fig. 2. comonad laws

Definition 1 (Comonad). Let C be a category. A *comonad* $CM = (T, \epsilon, \delta)$ on C is an endofunctor $T : C \to C$ with two natural transformations $\epsilon : T \to I_C$ and $\delta : T \to (T \circ T)$ where the following laws are satisfied:

(Com1)	$T(\delta_A) \circ \delta_A = \delta_{T(A)} \circ \delta_A$
(Com2)	$T(\epsilon_A) \circ \delta_A = id_{T(A)}$
(Com3)	$\epsilon_{T(A)} \circ \delta_A = id_{T(A)}$

The laws for a comonad can be visualized by diagrams as in Fig. 2 which is very similar to the diagrams of a monoid cf. Fig. 1 (a real correspondence can be better seen for monads).

Note 2. Let C be a category. A comonad $CM = (T, \epsilon, \delta)$ can be characterized by the following equational specification:

$$id_B \circ f_{A \to B} = f \quad (1) \qquad \epsilon_B \circ T(f_{A \to B}) = f \circ \epsilon_A \quad (6)$$

$$f_{A \to B} \circ id_A = f \quad (2) \qquad T(T(f_{A \to B})) \circ \delta_A = \delta_B \circ T(f) \quad (7)$$

$$(f_{A3 \to A4} \circ g_{A2 \to A3}) \circ h_{A1 \to A2} = f \circ (g \circ h) \quad (3) \qquad T(\delta_A) \circ \delta_A = \delta_{T(A)} \circ \delta_A \quad (8)$$

$$T(id_A) = id_{T(A)} \quad (4) \qquad T(\epsilon_A) \circ \delta_A = id_{T(A)} \quad (9)$$

$$T(f_{A2 \to A3} \circ g_{A1 \to A2}) = T(f) \circ T(g) \quad (5) \qquad \epsilon_{T(A)} \circ \delta_A = id_{T(A)} \quad (10)$$

Definition 3 (Kleisli category). Let $CM = (T, \epsilon, \delta)$ be a comonad on the category C. Then the corresponding *Kleisli category* K is defined by:

objects	same objects as C
morphisms	$Hom_K(A, B) = Hom_C(T(A), B)$
identity of object A	ϵ_A
composing $f \in Hom_K(A2, A3), g \in Hom_K(A1, A2)$	$(f \circ_K g) := f \circ_C T(g) \circ_C \delta_{A1}$

Remark. It can easily be verified that this construction actually is a category cf. [BW85]. We present the Kleisli category since it will be useful for characterizing comonads in another way.

Definition 4 (Computational Comonad [BG92]). Let C be a category and $CM = (T, \epsilon, \delta)$ be a comonad on C. A *computational comonad* $CCM = (T, \epsilon, \delta, \gamma)$

is a comonad having additionally one further natural transformation $\gamma : I_C \rightarrow T$ fulfilling the laws:

$$\begin{array}{ll} \textbf{(CCom1)} & \epsilon_A \circ \gamma_A = id_A \\ \textbf{(CCom2)} & \delta_A \circ \gamma_A = \gamma_{T(A)} \circ \gamma_A \end{array}$$

Note 5. Let C be a category. A computational comonad $CCM = (T, \epsilon, \delta, \gamma)$ can be characterized by an equational specification with the further equations:

$$T(f_{A\rightarrow B}) \circ \gamma_A = \gamma_B \circ f \tag{11}$$

$$\epsilon_A \circ \gamma_A = id_A \tag{12}$$

$$\delta_A \circ \gamma_A = \gamma_{T(A)} \circ \gamma_A \tag{13}$$

3 An Extensible Canonical System for a Comonad

In this section a rewriting system for comonads is presented and proved to be canonical. This particular representation of comonads is suitable to be extended to the notion of a computational comonad where a canonical system can be achieved, too.

Remark. The problem in using the original equational specification of a comonad comes from the fact that T is an endofunctor. This means that it can be iterated. But iteration requires the treatment of integer exponents which becomes difficult. Another problem is that ϵ interacts with T but δ interacts with $T \circ T$. Also this difference gives rise to complications.

Note 6. How can one take advantage from the Kleisli category which is formulated in terms of the given category? The law for multiplication in the Kleisli category suggests another auxiliary definition: $g^\star_{T(A)\rightarrow B} := T(g) \circ_C \delta_A$ (called the Kleisli star) such that $f \circ_K g = f \circ_C g^\star$. Formulating the categorical laws for the Kleisli category with this new function we get:

$$\begin{array}{rcl} in\ \ K & & in\ \ C \\ f = id_{(K)} \circ_K f & = & \epsilon \circ_C f^\star \\ f = f \circ_K id_{(K)} & = & f \circ_C \epsilon^\star \\ (f \circ_K g) \circ_K h & = & (f \circ_C g^\star) \circ_C h^\star \\ = f \circ_K (g \circ_K h) & = & f \circ_C (g \circ_C h^\star)^\star = \end{array}$$

With setting $\upsilon_A := \epsilon_A$ (called the unit of the Kleisli category) these equations can be reformulated as:

$$\begin{array}{lr} \textbf{(CKl1)} & \upsilon_B \circ f^\star_{T(A)\rightarrow B} = f \\ \textbf{(CKl2)} & \upsilon^\star_A = id_{T(A)} \\ \textbf{(CKl3)} & g^\star_{T(A2)\rightarrow A3} \circ h^\star_{T(A1)\rightarrow A2} = (g \circ h^\star)^\star \end{array}$$

Lemma 7. *A comonad can be completely characterized by* $(id, \circ, \upsilon, {}^\star)$ *on the level of morphisms assuming it is known how* T *acts on objects.*

Proof cf. [Man76]. The other components can be expressed as:
$\epsilon_A := \upsilon_A$, $\delta_A := id^\star_{T(A)}$, and $T(f_{A\rightarrow B}) := (f \circ \upsilon_A)^\star$. □

Remark. This gives a more compact way of presenting a comonad since it was previously described in terms of $(id, \circ, T, \epsilon, \delta)$. Nevertheless the action of T on the object level has to be given since it plays a role in the typing information in the rules CKl1-3. This presentation can now be used to formulate a canonical system for comonads.

Proposition 8. *Assuming it is known how $T : C \to C$ acts on objects there is the following canonical system COM for a comonad:*

$$
\begin{array}{llll}
id_B \circ f_{A\to B} \to f & (1) & (f_{A3\to A4} \circ g_{A2\to A3}) \circ h_{A1\to A2} \to f \circ (g \circ h) & (5) \\
f_{A\to B} \circ id_A \to f & (2) & f^\star_{T(A2)\to A3} \circ g^\star_{T(A1)\to A2} \to (f \circ g^\star)^\star & (6) \\
v^\star_A \to id_{T(A)} & (3) & v_{A3} \circ (f^\star_{T(A2)\to A3} \circ g_{A1\to T(A2)}) \to f \circ g & (7) \\
v_B \circ f^\star_{T(A)\to B} \to f & (4) & f^\star_{T(A3)\to A4} \circ (g^\star_{T(A2)\to A3} \circ h_{A1\to T(A2)}) \to (f \circ g^\star)^\star \circ h & (8)
\end{array}
$$

Proof. The correctness of this result was verified with the help of the Larch Prover. The check of the critical pairs can be seen in the appendix. It was done twice: firstly without taking the typing of morphisms into account and secondly including the typing. Here we only give the termination argument with the help of a polynomial interpretation $\mathcal{I}$ (cf. [Lan79]) $\mathcal{I}(f) := (\mathcal{I}_1(f), \mathcal{I}_2(f), \mathcal{I}_3(f))$:

$$
\begin{array}{ll}
\mathcal{I}_1(id) = \mathcal{I}_2(id) = \mathcal{I}_3(id) = 1 & \mathcal{I}_1(v) = \mathcal{I}_2(v) = \mathcal{I}_3(v) = 1 \\
\mathcal{I}_1(f \circ g) = \mathcal{I}_1(f) + \mathcal{I}_1(g) & \mathcal{I}_1(f^\star) = \mathcal{I}_1(f) + 2 \\
\mathcal{I}_2(f \circ g) = 2 * \mathcal{I}_2(f) * \mathcal{I}_2(g) & \mathcal{I}_2(f^\star) = \mathcal{I}_2(f) + 2 \\
\mathcal{I}_3(f \circ g) = 2 * \mathcal{I}_3(f) + \mathcal{I}_3(g) & \mathcal{I}_3(f^\star) = \mathcal{I}_3(f) + 2
\end{array}
$$

These triples of natural numbers are ordered lexicographically where the first component has the highest priority. Also this ordering is expressible in the Larch Prover and is given in the appendix. In the next section we will give more details about the typing. □

Remark. The termination argument in this case did not need any information from the typing. This is different to typed λ-calculus where mainly the types are used to prove termination. Starting from the canonical system COM we can try to extend the result to a computational comonad.

Note 9. How can the additional equations for a computational comonad be reformulated to fit into the new representation? Especially the references to T and δ have to be avoided. The key observation comes from the following equation:

$$f^\star_{T(A)\to B} \circ \gamma_A = T(f) \circ \delta_A \circ \gamma_A = T(f) \circ \gamma_{T(A)} \circ \gamma_A = \gamma_B \circ f \circ \gamma_A$$

Here both rules from the original presentation which involve T and δ have been applied leading to the single equation: $f^\star_{T(A)\to B} \circ \gamma_A = \gamma_B \circ f \circ \gamma_A$. This equation together with $v_A \circ \gamma_A = id_A$ suffices to describe a computational comonad.

Lemma 10. *A computational comonad can be completely characterized in terms of* $(id, \circ, \upsilon, {}^\star, \gamma)$ *on the level of morphisms assuming it is known how* T *acts on objects where* γ *is described by the equations:*

$$\begin{array}{ll} \textbf{(Gam1)} & \upsilon_A \circ \gamma_A = id_A \\ \textbf{(Gam2)} & f^\star_{T(A)\to B} \circ \gamma_A = \gamma_B \circ f \circ \gamma_A \end{array}$$

Proof. It has to be shown that the previous three equations for γ can be derived:

$$T(f_{A\to B}) \circ \gamma_A = (f \circ \upsilon_A)^\star \circ \gamma_A = \gamma_B \circ (f \circ \upsilon_A) \circ \gamma_A = \gamma_B \circ f$$

$$\epsilon_A \circ \gamma_A = \upsilon_A \circ \gamma_A = id_{T(A)}$$

$$\delta_A \circ \gamma_A = id^\star_{T(A)} \circ \gamma_A = \gamma_{T(A)} \circ id_{T(A)} \circ \gamma_A = \gamma_{T(A)} \circ \gamma_A$$

□

Theorem 11. *Assuming it is know how* $T : C \to C$ *acts on objects there is the following canonical system CCOM for a computational comonad extending the canonical system COM*

$$\upsilon_A \circ \gamma_A \to id_A \tag{9}$$

$$f^\star_{T(A)\to B} \circ \gamma_A \to \gamma_B \circ (f \circ \gamma_A) \tag{10}$$

$$\upsilon_B \circ (\gamma_B \circ f_{A\to B}) \to f \tag{11}$$

$$g^\star_{T(A2)\to A3} \circ (\gamma_{A2} \circ h_{A1\to A2}) \to \gamma_{A3} \circ (g \circ (\gamma_{A2} \circ h)) \tag{12}$$

□

Proof. Again the check of the critical pairs was done in the Larch Prover which can be seen in the appendix. For the typing we refer to the next section. The previous polynomial interpretation was extended by: $\mathcal{I}_1(\gamma) = \mathcal{I}_2(\gamma) = \mathcal{I}_3(\gamma) = 1$
□

Remark. At this point it should be stressed that the choice of the representation for comonads was not obvious to us. The introduction of the auxiliary $\star$ was necessary to succeed. This example could be used to test rewrite tools which allow the extension of the signature as in [KZ89] such that one can see to which extent this may be automated.

4 Encoding Types in LP

Here the method due to G. Huet (cf. [Hue90a]) is demonstrated with the concrete example of a computational comonad.

Remark. In a category it is important to check the compatibility of morphisms in order to compose them. When we assume to start with compatible morphisms the application of the three untyped rules describing the categorical axioms coincides with the application of the typed version. Since functors and natural transformations also act on the level of morphisms one also has to treat this additional information.

Problem 12. How can one check critical pairs for a rewriting system describing categorical notions taking the level of objects into account? A morphism now becomes a type depending on two objects – source and target.

Solution 13. On the level of objects usually a simple test of equality is done to check the compatibility of morphisms. Thus the unification mechanism which is present for rewriting can be exploited to perform this check, too. The typing information has to be encoded with a new function symbol representing the dependent typing. In the case of categories this looks like "$mor(f, a, b)$" where f is the untyped form of the morphism and a, b are source and target, resp.

Example 1. As an example we consider a computational comonad in the presentation which yields the canonical system. As the way to present our example we choose the specification language of the Larch Prover. The sort and variable definitions for these cases are:

untyped case

```
declare sort M       % morphisms
declare variables f, g, h: M
declare operators
   id: -> M          % identity
   *: M,M -> M       % composition
   counit: -> M      % counit
   costar: M -> M    % Kleisli
```

typed case

```
declare sort M       % typed morphisms
declare sort M'      % untyped morphisms
declare sort O       % objects
declare variables f',g',h': M'
declare variables o1, o2, o3, o4: O
declare operators
   id: -> M'         % identity
   *: M,M -> M'      % composition
   counit: -> M'     % counit
   costar: M -> M'   % Kleisli
   mor: M',O,O-> M   % explicit typing
   t': O -> O        % functor on objects
```

The following table demonstrates the translation from the untyped case into the typed case.

morphism	untyped	typed
$f_{O1\to O2}$	f	mor(f, o1, o2)
id_{O1}	id	mor(id, o1, o1)
$f_{O2\to O3} \circ g_{O1\to O2}$	f * g	mor(mor(f,o2,o3) * mor(g,o1,o2), o1, o3)
v_{O1}	counit	mor(counit, t'(o1), o1)
$f^{\star}_{T(O1)\to O2}$	costar(f)	mor(costar(mor(f,t'(o1), o2)),t'(o1),t'(o2))

Exploiting this translation the rewriting system of the computational comonad was verified again respecting the typing. This can be seen in the appendix.

Remark. All the typing information had to be hand-coded which is very error-prone. Unfortunately the Larch Prover did not support this kind of type processing. Therefore we implemented the final decision procedures for comonads and computational comonads in the logic programming language Elf which supports dependent types directly.

5 Implementation in Elf and Application

Here it is shown how the canonical system for computational comonads can be applied where examples are taken from [BG92]. The actual run of the test can be found in the appendix.

Because of the difficulty to represent dependent types in a conventional rewriting tool like the Larch Prover we implemented the rewriting in the logic programming language Elf. We think that this approach has several advantages:

- Types help to encode morphisms correctly but also represent judgments via the "propositions as types principle".
- Elf does not only give an answer substitution but also a term representing the proof which could be used for further inspection how a proof was achieved.
- Since it is a programming language several rewriting strategies can be implemented which fit a problem best (here just outermost leftmost was chosen).
- Elf allows additionally the treatment of higher order rewriting since the language supports higher order types.
- It is also possible to formulate the concept of critical overlaps in the language (unfortunately it is not possible to check the convergence of critical pairs which will be possible in a forthcoming extension of Elf).
- Elf allows to prove meta-theorems as the soundness of rewriting with respect to an axiomatic definition of the equality.

But it has to be pointed out that Elf was neither used for the completion nor for the proof of termination. This was done conveniently in the frame of the Larch Prover. On the other hand it was very easy to formulate the queries for the coming example in Elf.

An Elf program is split into a static and a dynamic part. The former is only used for type checking whereas the latter is used for proof search. In the appendix both these parts of the program are shown. The dynamic part makes already use of the definitions in the static part so that only in a few cases the type has to be made more explicit.

The sample queries which can be seen in the trace correspond to [BG92]. There a pair of functors is defined relating the category C and the Kleisli category K. We have:

$$alg : C \rightarrow K \;\; \textit{defined by} \;\; alg(f_{A \rightarrow B}) := f \circ \epsilon_A$$
$$fun : K \rightarrow C \;\; \textit{defined by} \;\; fun(f_{T(A) \rightarrow B}) := f \circ \gamma_A$$

The queries are the test whether alg and fun are indeed functors. Furthermore the following equalities were checked:

$$fun \circ alg = I_C$$
$$alg \circ fun =^e I_K \;\; \textit{where} \;\; f =^e g \iff fun(f) = fun(g)$$

The test of equality proceeds in two steps where the first one does the translation into the representation which can be used for normalization and the second one does the normalization. The knowledge about the definition of alg and fun is already coded in the translation process.

6 An Incomplete Attempt: Rewriting over a Congruence

In this section we sketch another attempt to achieve a canonical system for comonads which should be extended to computational comonads as well.

Our attempt was motivated by trying to remain within the given specification of a comonad in terms of $(id, \circ, T, \epsilon, \delta)$. Our approach was to take the equational specification of a comonad as presented above, to orient all the equations from left to right, but to work over the congruence generated by 3 for associativity and 5 for the endofunctor on composed morphisms.

Working over this congruence results in finite congruence classes which can be computed and have a canonical representation when orienting both rules from left to right. Thus the reducibility of terms is decidable cf. [Bac91]. Furthermore it is easy to find a polynomial interpretation which is decreasing on the rules and remains constant for the congruence. The hard part here would be to show confluence.

The left-linear rule method by Huet (cf. [Hue80]) was tried since all the rules are left-linear but this resulted in further critical pairs. Also the existence of a unification algorithm for a homomorphism described by 5 as given in [Vog78] but without associativity did not suffice to cover the entire congruence class.

Nevertheless the structure of the rules is rather simple since the composition occurs there only once. Besides the usual critical overlaps new ones have to be considered due to the congruence as for example:

$$T(f) \circ \epsilon_T \circ \delta \leftarrow \epsilon_T \circ T(T(f)) \circ \delta \rightarrow \epsilon_T \circ \delta \circ T(f)$$

We attempted to formalize this with rules involving exponentiation of the endofunctor as for example:

$$T^n(\epsilon_B) \circ T^{n+1}(f_{A \to B}) \to T^n(f) \circ T^n(\epsilon_A)$$
$$T^{n+2}(f_{A \to B}) \circ T^n(\delta_A) \to T^n(\delta_B) \circ T^{n+1}(f)$$

Each such rule comes together with a completed version arising from overlapping with associativity. But this really required the treatment of natural numbers and addition for the exponents which complicates the situation since AC-unification becomes necessary. On the other hand the exponents are always ground in a concrete decision problem.

Since the amount of rules was constantly growing we finally looked for a more appropriate formulation of the theory of a comonad although the structure of the rules for a computational comonad follow the same pattern. It is possible that a formalization with the ReDuX [Bün93] system which is able to handle inductive completion also in the presence of AC-operators may succeed. In the literature there several new methods are presented which try to handle infinite sets of rules arising from completion.

7 Conclusions and Future Work

The uniform word problem of computational comonads was shown to be decidable by extending an appropriate canonical system of comonads. The critical pair check had to take the presence of types into account which come from source and target of a morphism. This suggested to use the logic programming language Elf to implement typed rewriting because of its direct support of dependent types. Furthermore the user has direct influence on the strategy for rewriting. For future work we want to continue in two directions – a theoretical and a practical.

The notion of a monad as used in functional programming [Wad93] still has to be investigated. Since this requires the treatment of higher order rewriting [Nip91] Elf is still a suitable tool since it supports higher order types. This would give the right frame to reason about certain monadic functional programs.

In practice it would be very helpful to allow diagrams being a compact visual encoding of equations as input to the prover. The output could also be displayed in an appropriate form. In [FS90] one can already find a suitable graphical language used in the context of categories.

Acknowledgments

I am indebted to Frank Pfenning for his hospitality during a visit to CMU which made it possible to learn more about the programming language Elf and which allowed me fruitful discussions of these ideas. Furthermore I wish to thank Andrzej Filinski for providing me with more insight to monads. Also I am very grateful to my advisor Jochen Pfalzgraf for constant encouragement.

References

[Bac91] L. Bachmair. *Canonical Equational Proofs.* Progress in Theoretical Computer Science. Birkhäuser, 1991.

[BG92] S. Brookes and S. Geva. Computational Comonads and Intensional Semantics. In M.P. Fourman and P.T. Johnstone and A.M. Pitts, editor, *Categories in Computer Science*, number 177 in London Mathematical Society Lecture Notes, pages 1–44. Cambridge University Press, 1992.

[Bün93] R. Bündgen. Reduce the Redex → ReDuX. In C. Kirchner, editor, *Rewriting Techniques and Applications*, number 690 in Lecture Notes in Computer Science, pages 446–450. Springer-Verlag, 1993.

[BW85] M. Barr and C. Wells. *Toposes, Triples and Theories.* Number 278 in Grundlehren der mathematischen Wissenschaften. Springer-Verlag, 1985.

[Fil94] A. Filinski. Representing Monads. In *Proceedings of the 21st Annual ACM SIGPLAN-SIGACT Symposium on Principles of Programming Languages*, pages 446–457. ACM, 1994.

[FS90] P.J. Freyd and A. Sčedrov. *Categories, Allegories.* Elsevier Science Publishers, 1990.

[Geh94] W. Gehrke. Proof of the Decidability of the Uniform Word Problem for Monads Assisted by Elf. Technical Report 94-66, RISC, 1994.

[GG91] S.J. Garland and J.V. Guttag. *A Guide to LP, The Larch Prover.* MIT, 1991.

[Hue80] G. Huet. Confluent Reductions. *Journal of the Association for Computing Machinery*, 24(4):797–821, October 1980.

[Hue90a] G. Huet. Cartesian closed categories and lambda-calculus. In *[Hue90b]*, pages 7–23. Addison Wesley, 1990.

[Hue90b] G. Huet. *Logical Foundations of Functional Programming.* University of Texas at Austin Programming Series. Addison Wesley, 1990.

[KZ89] D. Kapur and H. Zhang. *RRL: Rewrite Rule Laboratory User's Manual*, revised edition, May 1989.

[Lan79] D.S. Lankford. On proving term rewriting systems are Noetherian. Technical report, Louisiana Technical University, Ruston, LA, 1979.

[Man76] E. Manes. *Algebraic Theories.* Number 26 in Graduate Texts in Mathematics. Springer-Verlag, 1976.

[ML71] S. Mac Lane. *Categories for the Working Mathematician.* Number 5 in Graduate Texts in Mathematics. Springer-Verlag, 1971.

[Mog89] E. Moggi. Computational Lambda-calculus and Monads. In *Fourth Annual Symposium on Logic in Computer Science*, pages 14–23. IEEE, June 1989.

[Nip91] T. Nipkow. Higher Order Critical Pairs. In *Sixth Annual Symposium on Logic in Computer Science*, pages 342–349. IEEE, July 1991.

[Pfe89] F. Pfenning. Elf: a language for verified meta-programming. In *Fourth Annual Symposium on Logic in Computer Science*, pages 313–322. IEEE, June 1989.

[Vog78] E. Vogel. Unifikation von Morphismen (in German). Diplomarbeit, Universität Karlsruhe, 1978.

[Wad93] P. Wadler. Monads for functional programming. In M. Broy, editor, *Program Design Calculi*, volume 118 of *NATO ASI Series F: Computer and System Sciences*, pages 233–264. Springer-Verlag, 1993.

Appendix

A LP Traces

Here the check of termination and confluence with the Larch Prover are shown. For termination the typing has not been taken into account, i.e. the termination argument does not make use of types. Afterwards the check of critical pairs is repeated in a setting which also includes types.

A.1 Termination in LP

```
Larch Prover (17 March 1993) logging on 18 January 1995 13:02:28 to
'/tmp_mnt/home/hhong/wgehrke/Data/lp/comp_comonad/termination.lplog'.
LP2: execute termination
LP2.1: declare sort Mor
LP2.2: declare variables f, g, h: Mor
LP2.3: declare operators
        id, counit: -> Mor
        costar: Mor -> Mor
        *: Mor,Mor -> Mor
..
LP2.4: set ordering polynomial 3
The ordering-method is now 'polynomial 3'.
LP2.5: register polynomial     id       1
LP2.6: register polynomial     counit   1
LP2.7: register polynomial     costar   x + 2
LP2.8: register polynomial     *        x + y, 2 * x * y, 2 * x + y
```

```
LP2.9: assert
        id * f == f
        f * id == f
        (f * g) * h == f * (g * h)
        counit * costar(f) == f
        costar(counit) == id
        costar(f) * costar(g) == costar(f * costar(g))
..
Added 6 equations named user.1, ..., user.6 to the system.
The system now contains 6 rewrite rules.
LP2.10: complete
The following equations are critical pairs between rewrite rules user.4 and
user.3.
  user.7: f * h == counit * (costar(f) * h)
The system now contains 1 equation and 6 rewrite rules.
The following equations are critical pairs between rewrite rules user.6 and
user.3.
  user.8: costar(f * costar(g)) * h == costar(f) * (costar(g) * h)
The system now contains 1 equation and 7 rewrite rules.
The system now contains 8 rewrite rules.
The system is complete.
LP2.11: declare operator gamma: -> Mor
LP2.12: register polynomial     gamma   1
LP2.13: assert counit * gamma == id
Added 1 equation named user.9 to the system.
The system now contains 9 rewrite rules.
LP2.14: assert costar(f) * gamma == gamma * (f * gamma)
Added 1 equation named user.10 to the system.
The system now contains 10 rewrite rules.
LP2.15: complete
The following equations are critical pairs between rewrite rules user.9 and
user.3.
  user.11: h == counit * (gamma * h)
The system now contains 1 equation and 10 rewrite rules.
The following equations are critical pairs between rewrite rules user.10 and
user.3.
  user.12: gamma * (f * (gamma * h)) == costar(f) * (gamma * h)
The system now contains 1 equation and 11 rewrite rules.
The system now contains 12 rewrite rules.
The system is complete.
LP2.16: display
Rewrite rules:
user.1:  id * f -> f
user.2:  f * id -> f
user.3:  (f * g) * h -> f * (g * h)
user.4:  counit * costar(f) -> f
user.5:  costar(counit) -> id
user.6:  costar(f) * costar(g) -> costar(f * costar(g))
user.7:  counit * (costar(f) * h) -> f * h
user.8:  costar(f) * (costar(g) * h) -> costar(f * costar(g)) * h
user.9:  counit * gamma -> id
user.10: costar(f) * gamma -> gamma * (f * gamma)
user.11: counit * (gamma * h) -> h
user.12: costar(f) * (gamma * h) -> gamma * (f * (gamma * h))
End of input from file
'/tmp_mnt/home/hhong/wgehrke/Data/lp/comp_comonad/termination.lp'.
LP3: quit
```

A.2 Typing in LP

```
Larch Prover (17 March 1993) logging on 18 January 1995 16:27:18 to
'/tmp_mnt/home/hhong/wgehrke/Data/lp/comp_comonad/typing.lplog'.
LP2: execute typing
LP2.1: declare sort Mor, Mor', Obj
LP2.2: declare variables f',g',h': Mor'
```

```
LP2.3: declare variables o1, o2, o3, o4: Obj
LP2.4: declare operators
        id, counit, gamma: -> Mor'
        costar: Mor -> Mor'
        *: Mor,Mor -> Mor'
        t': Obj -> Obj
        mor: Mor',Obj,Obj-> Mor
..
LP2.5: set ordering left-to-right
The ordering-method is now 'left-to-right'.
LP2.6: assert
%   category
mor(mor(id, o2, o2) * mor(f', o1, o2), o1, o2) == mor(f', o1, o2)
mor(mor(f', o1, o2) * mor(id, o1, o1), o1, o2) == mor(f', o1, o2)
mor(mor(mor(f', o3, o4) * mor(g', o2, o3), o2, o4) * mor(h', o1, o2), o1, o4)
== mor(mor(f', o3, o4) *
    mor(mor(g', o2, o3) * mor(h', o1, o2), o1, o3), o1, o4)
%   comonad
mor(mor(counit, t'(o2), o2)
  * mor(costar(mor(f', t'(o1), o2)), t'(o1), t'(o2)), t'(o1), o2)
== mor(f', t'(o1), o2)
mor(costar(mor(counit, t'(o1), o1)), t'(o1), t'(o1))
== mor(id, t'(o1), t'(o1))
mor(mor(costar(mor(f', t'(o2), o3)), t'(o2), t'(o3))
  * mor(costar(mor(g', t'(o1), o2)), t'(o1), t'(o2)), t'(o1), t'(o3))
== mor(costar(mor(mor(f', t'(o2), o3) * mor(costar(mor(g', t'(o1), o2)),
    t'(o1), t'(o2)), t'(o1), o3)), t'(o1), t'(o3))
%   completed rules
mor(mor(counit, t'(o3), o3) * mor(mor(costar(mor(f', t'(o2), o3)),
  t'(o2), t'(o3)) * mor(g', o1, t'(o2)), o1, t'(o3)), o1, o3)
== mor(mor(f', t'(o2), o3) * mor(g', o1, t'(o2)), o1, o3)
mor(mor(costar(mor(f', t'(o3), o4)), t'(o3), t'(o4))
  * mor(mor(costar(mor(g', t'(o2), o3)), t'(o2), t'(o3))
    * mor(h', o1, t'(o2)), o1, t'(o3)), o1, t'(o4))
== mor(mor(costar(mor(mor(f', t'(o3), o4)
    * mor(costar(mor(g', t'(o2), o3)), t'(o2), t'(o3)), t'(o2), o4))
        , t'(o2), t'(o4))
          * mor(h', o1, t'(o2)), o1, t'(o4))
..
Added 8 equations named user.1, ..., user.8 to the system.
The system now contains 8 rewrite rules.
LP2.7: complete
The system is not guaranteed to terminate.  If it does terminate, then it is
complete.
LP2.8: assert
%   computational comonad
mor(mor(counit, t'(o1), o1) * mor(gamma, o1, t'(o1)), o1, o1)
== mor(id, o1, o1)
mor(mor(costar(mor(f', t'(o1), o2)), t'(o1), t'(o2))
  * mor(gamma, o1, t'(o1)), o1, t'(o2))
== mor(mor(gamma, o2, t'(o2)) *
    mor(mor(f', t'(o1), o2) * mor(gamma, o1, t'(o1)), o1, o2),
        o1, t'(o2))
%   completed rules
mor(mor(counit, t'(o2), o2) *
  mor(mor(gamma, o2, t'(o2)) * mor(g', o1, o2), o1, t'(o2)), o1, o2)
== mor(g', o1, o2)
mor(mor(costar(mor(f', t'(o2), o3)), t'(o2), t'(o3))
  * mor(mor(gamma, o2, t'(o2)) * mor(g', o1, o2), o1, t'(o2)),
    o1, t'(o3))
== mor(mor(gamma, o3, t'(o3)) *
    mor(mor(f', t'(o2), o3) *
      mor(mor(gamma, o2, t'(o2)) * mor(g', o1, o2),
        o1, t'(o2)), o1, o3), o1, t'(o3))
..
Added 4 equations named user.9, ..., user.12 to the system.
```

```
The system now contains 12 rewrite rules.
LP2.9: complete
The system is not guaranteed to terminate. If it does terminate, then it is
complete.
End of input from file
'/tmp_mnt/home/hhong/wgehrke/Data/lp/comp_comonad/typing.lp'.
LP3: quit
```

B Elf Traces

All parts of the Elf program are presented to illustrate the usage of Elf for rewriting making use of dependent types. The trace shows the automated proof of examples taken from [BG92] but without printing the proof term. The current implementation of Elf is embedded into an image of the SML/NJ compiler and it is accessible through functions from the top level.

B.1 The Static Part of the Elf Program

```
%%% category
obj    : type. %name obj O
mor    : obj -> obj -> type. %name mor M
id     : {A:obj} mor A A.
*      : mor O2 O3 -> mor O1 O2 -> mor O1 O3. %infix right 10 *
%%% comonad with (T,eps,del)
T'     : obj -> obj.
T      : mor A B -> mor (T' A) (T' B).
eps    : {A:obj} mor (T' A) A.
del    : {A:obj} mor (T' A) (T' (T' A)).
%%% comonad with counit and costar via Kleisli category
counit: {A:obj} mor (T' A) A.
costar: mor (T' A) B -> mor (T' A) (T' B).
%%% missing ingredient for a computational comonad
gamma : {A:obj} mor A (T' A).
%%% for checks from paper by Brookes and Geva
alg    : mor A B -> mor (T' A) B.
fun    : mor (T' A) B -> mor A B.
```

B.2 The Dynamic Part of the Elf Program

```
%%% rules for computational comonads with the Kleisli category
rule   : mor A B -> mor A B -> type.
comon1: rule ((id B) * F) F.
comon2: rule (F * (id A)) F.
comon3: rule ((F * G) * H) (F * (G * H)).
comon4: rule ((counit B) * (costar F)) F.
comon5: rule (costar (counit A)) (id (T' A)).
comon6: rule ((costar F) * (costar G)) (costar (F * (costar G))).
comon7: rule ((counit B) * ((costar F) * G)) (F * G).
comon8: rule ((costar F) * ((costar G) * H))
 ((costar (F * (costar G))) * H).
comp1 : rule ((counit A) * (gamma A)) (id A).
comp2 : rule ((costar F) * (gamma A)) ((gamma B) * (F * (gamma A))).
comp3 : rule ((counit B) * ((gamma B) * F)) F.
comp4 : rule ((costar F) * ((gamma A) * G))
 ((gamma B) * (F * ((gamma A) * G))).
%%% rewrite relation for computational comonad
rew    : mor A B -> mor A B -> type.   % try to rewrite
step   : mor A B -> mor A B -> type.   % do at least one step
```

```
simple: step F F' <- rule F F'.
step*1: step (F * G) (F' * G) <- step F F'.
step*2: step (F * G) (F * G') <- step G G'.
step^*: step (costar F) (costar F') <- step F F'.
try   : rew F F'' <- step F F' <- rew F' F''.
fini  : rew F F.
%% dynamic equality of mors over (id, *, counit, costar, gamma)
==    : mor A B -> mor A B -> type. %name == EQ
%infix none 8 ==
moreq : F == G <- rew F H <- rew G H.
%% translation of morphisms with the help of the Kleisli category
trans : mor A B -> mor A B -> type.
tr_*  : trans (F * G) (F' * G') <- trans F F' <- trans G G'.
tr_^* : trans (costar F) (costar F') <- trans F F'.
tr_alg: trans (alg (F : mor A B)) (F' * (counit A)) <- trans F F'.
tr_fun: trans (fun (F : mor (T' A) B)) (F' * (gamma A)) <- trans F F'.
tr_T  : trans (T (F : mor A B)) (costar (F' * (counit A))) <- trans F F'.
tr_e  : trans (eps A) (counit A).
tr_d  : trans (del A) (costar (id (T' A))).
tr_0  : trans F F.
%% now morphisms over (id,*,counit,costar,gamma,T,eps,del,alg,fun)
===   : mor A B -> mor A B -> type. %name === EQU
%infix none 8 ===
cmoneq: F === G <- trans F F' <- trans G G' <- F' == G'.
```

B.3 A Sample Run

```
Standard ML of New Jersey, Version 0.93, February 15, 1993
Elf, Version 0.4, July 1, 1993, saved on Tue Jan 17 18:47:18 MET 1995
val it = () : unit
- initload ["static.elf"] ["dynamic.elf"];
  . . .
static.elf --- 1 --- static
dynamic.elf --- 2 --- dynamic
val it = () : unit
- top();
Using: dynamic.elf
Solving for: rule rew step == trans ===
?- {A} (alg (id A)) === (counit A).                          % alg on id
Solving...
solved
yes
?- {O1}{O2}{O3}{F : mor O2 O3}{G : mor O1 O2}
(alg (F * G)) === (alg F) * (costar (alg G)).                % alg on *
  Solving...
solved
yes
?- {A} (fun (counit A)) === (id A).                          % fun on id'
Solving...
solved
yes
?- {O1}{O2}{O3}{F : mor (T' O2) O3}{G : mor (T' O1) O2}
(fun (F * (costar G))) === (fun F) * (fun G).                % fun on *'
  Solving...
solved
yes
?- {A}{B}{F : mor A B} (fun (alg F)) === F.
Solving...
solved
yes
?- {A}{B}{F : mor (T' A) B} (fun (alg (fun F))) === (fun F).
Solving...
solved
yes
?-
```

Relating Two Categorical Models of Term Rewriting *

A. Corradini, F. Gadducci and U. Montanari

Università di Pisa, Dipartimento di Informatica, Corso Italia 40, 56125 Pisa, Italy
({andrea,gadducci,ugo}@di.unipi.it, Tel: +39-50-887268, Fax: +39-50-887266)

Abstract. In the last years there has been a growing interest towards categorical models for term rewriting systems (TRS's). In our opinion, very interesting are those associating to each TRS's a cat-enriched structure: a category whose hom-sets are categories. Interpreting rewriting steps as morphisms in hom-categories, these models provide rewriting systems with a concurrent semantics in a clean algebraic way.
In this paper we provide a unified presentation of two models recently proposed in literature by José Meseguer [Mes90, Mes92, MOM93] and John Stell [Ste92, Ste94], respectively, pursuing a critical analysis of both of them. More precisely, we show why they are to a certain extent unsatisfactory in providing a concurrent semantics for rewriting systems. It turns out that the derivation space of Meseguer's Rewriting Logic associated with each term (i.e., the set of coinitial computations) fails in general to form a prime algebraic domain: a condition that is generally considered as expressing a *directly implementable* model of concurrency for distributed systems (see [Win89]). On the contrary, the resulting derivation space in Stell's model is actually a prime algebraic domain, but too few computations are identified: only *disjoint concurrency* can be expressed, limiting the degree of parallelism described by the model.

Keywords: term rewriting, categorical models, concurrency, rewriting logic.

1 Introduction

Term rewriting systems (briefly, TRS's; see [DJ90]) are a simple yet powerful framework, based on the notion of *(sequence of) rewrites*: a binary, transitive relation over terms, usually generated from a finite set of rules (i.e., of pair of terms, subject to some further restriction), where each element $\langle t, s\rangle$ states the transformation from term t to term s. Despite their simplicity, TRS's propose themselves as a basic paradigm for computational devices: terms are states of an abstract machine, while rewriting rules are state-transforming functions: in this framework, computations simply are sequences of rewrites.

A very intuitive operational model for TRS's is represented by their *derivation spaces*: a class of (structured) elements representing all the possible computations

* Research partially supported by ESPRIT BRA project 6564 CONFER.

the system can perform. These structures are easily defined, but, unfortunately, the resulting operational semantics is usually too concrete: in order to obtain a description as much as possible independent from the actual implementation of the rewriting machine, we need to abstract away from irrelevant details. More precisely, if we see a term as a completely distributed structure, such that an actual rewriting machine can act simultaneously (in parallel) on different subterms, then a concurrent semantics is interested in exploiting to a maximum degree the parallelism implicit in the definition of TRS's. Usually, such a semantics can be recovered imposing an equivalence relation on computations, equating sequences of rewrites that are the same up to some conditions. Those conditions express the properties of the rewriting mechanism over the distributed system under examination: each equivalence class represents an *abstract computation*, corresponding to a family of computationally equivalent sequences of rewrites.

In the classical approach the derivation space is just a set, and the equivalence on computations relies on the notion of *residual*: given two computations d and d', then the residual $d'\backslash d$ of d' after d is intuitively "what-remains-to-be-done" of d' after the execution of d. Abstract computations are the elements of the set of "concrete" derivations, quotiented by the smallest equivalence induced by the axiom schema $d \cdot d'\backslash d = d' \cdot d\backslash d'$, where $\cdot$ indicates concatenation of rewrites. This equivalence is known as *permutation equivalence* [Lév80, Bou85], and it equates computations that are the same up to permutation of independent rewrites.

Recently, *2-categories* and *sesqui-categories* have been proposed respectively by José Meseguer [Mes90, Mes92, MOM93] and John Stell [Ste92, Ste94] as a more abstract, algebraic way to equip TRS's with a concurrent semantics. A cat-enriched structure as a 2-category or a sesqui-category is given by a category such that also each hom-set forms a category: the class of morphisms (called *cells*) of these hom-categories are closed under certain composition operators, and are subject to suitable *coherence* axioms. In the paper we show a unified, two-phase method to generate such algebraic structures starting from a TRS's. First, to each TRS $\mathcal{R}$ a *c-computad* is associated: a pair $\langle \mathbf{D}_{\mathcal{R}}, S_{\mathcal{R}} \rangle$, where $\mathbf{D}_{\mathcal{R}}$ is a category whose arrows represent terms, while $S_{\mathcal{R}}$ is a collection of cells describing the rewriting rules of $\mathcal{R}$; then from the c-computad an enriched structure is freely generated, obtaining a derivation space (the cells of the enriched structure) that is already subject to an equivalence relation, due to the coherence axioms: since there exists a correspondence between cells and computations, this way a TRS is equipped with a concurrent semantics. The resulting construction is strictly related, in the case of 2-categories, to the functorial semantics considered by Meseguer in the Appendix of [Mes90], while for sesqui-categories it can be considered as an original, algebraic presentation of Stell's proposal [Ste92].

In the second part of the paper we will consider some algebraic properties of the equivalences on derivations induced by the two models, analyzing in particular the "natural" partial order on computations. In a classical, set-theoretic context, a similar task has been tackled for λ-calculus by Laneve in [Lan94], where a new semantics is introduced (induced by the *distributive permutation equivalence*, where only permutation of *distributed* rewrites is allowed), showing that the corresponding derivation space forms a *prime algebraic domain*.

Prime algebraic domains (PAD's) are a simple, general and well-accepted model to describe the behaviour of concurrent, non-deterministic systems. Their acceptance in the concurrency field is due, besides to their historical relevance (dating back to the work of Kahn and Plotkin on *concrete domains*, and of Berry on *dI domains*) to their tight correspondence with *Prime Events Structures* [Win89]. Firstly introduced in the early eighties, Prime Event Structures (PES's) are partial orders of "events", equipped with a "conflict" relation. Such structures are very suitable for describing the behaviour of distributed and non-deterministic computational devices generating instantaneous atomic events, which can be causally related or mutually exclusive: the level of abstraction they capture is considered as *directly* reflecting that of a possible, *concurrent* implementation, where each event corresponds to a *basic* action of the underlying machine. To each PES is associated a set of *configurations*: compatible, left-closed subsets of its events. Intuitively, a configuration corresponds to a specific state of the system reached after some computation, and its events are all those generated during that specific computation. A fundamental result due to Winskel shows that the set of all configurations of a PES ordered by set inclusion (its "domain" of configurations) forms a PAD; moreover, for each PAD there is a PES such that its domain of configurations is isomorphic to the given PAD. Thus the use of PAD's or PES's for the description of computational systems is equivalent.

Coming back to term rewriting systems, in each of the two models mentioned above the collection of all abstract derivations starting from a given initial term t can be equipped easily with a *prefix* partial ordering: $d \leq d'$ iff there exists a derivation d'' such that $d \cdot d'' = d'$. In Section 5 we will show that only the prefix partial order induced by Stell's model is a PAD, while that induced by Meseguer's model fails to satisfy the "distributive property" of PAD's.

The paper has the following structure. In Section 2 we recall the basic notions of enriched category theory which are useful for our discussion. In Section 3, besides the definitions of term algebras and of term rewriting systems, we introduce some notions borrowed from Meseguer's Rewriting Logic [Mes92]. In Section 4 we present an alternative, categorical description of TRS, and we introduce in a unified presentation the two categorical models proposed by José Meseguer and John Stell. In Section 5 we analize the adequacy of such modelizations, along the line discussed in this introduction, providing a PAD semantics for sesqui-categorical models. Finally, in Section 6 we draw our conclusions about the previous results, suggesting further directions for future work.

2 Background on 2-categories and Sesqui-categories

In the following we assume that some basic definitions of category theory are known, for which we refer the reader to [ML71]. In this section we will introduce 2-categories and sesqui-categories, lifting to this enriched context some constructions of the classical category theory. Basically, a cat-enriched category simply is a category $\mathbf{C}$ such that, given any two object a, b, the hom-set $\mathbf{C}[a, b]$, i.e., the class of morphisms from a to b is a category: moreover, these *hom-categories* satisfy particular composition properties.[2] A morphism in $\mathbf{C}[a, b]$, a *cell*, is denoted as $\alpha : f \Rightarrow g : a \rightarrow b$, where $f, g : a \rightarrow b$; or graphicallly, as

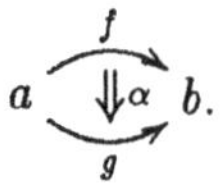

The following definition is adapted from [Ste92].

Definition 1 (2-Categories and Sesqui-Categories). Let $\mathbf{C}$ be a category such that each homset $\mathbf{C}[a, b]$ also forms a category. Moreover, let us assume that for each triple a, b, c of objects there are two composition functions $*_L$ and $*_R$ such that, given $\alpha : f \Rightarrow h : a \rightarrow b$ and $\beta : g \Rightarrow i : b \rightarrow c$, then $\beta' = f *_L \beta \in \mathbf{C}[a, c]$ and $\alpha' = \alpha *_R g \in \mathbf{C}[a, c]$. Graphically,

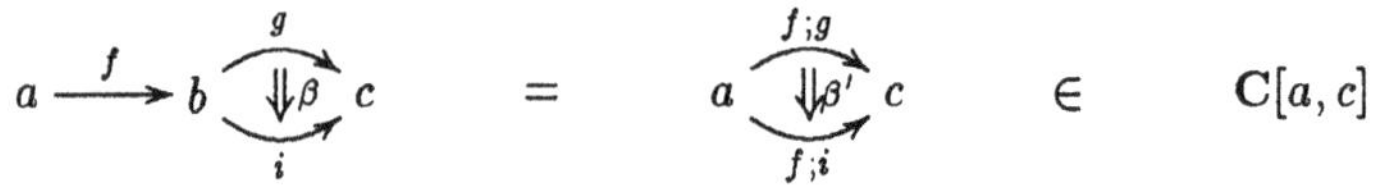

(analogously for $\alpha' = \alpha *_R g$) where ; denotes composition inside $\mathbf{C}$. Let us consider the situation denoted by the following cells:

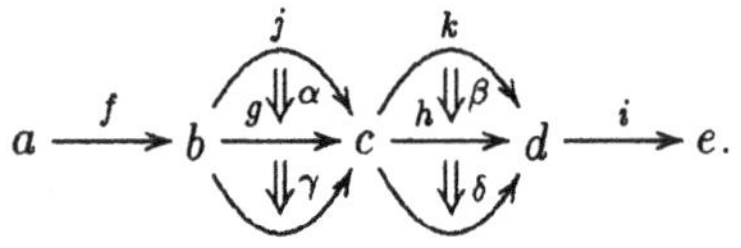

A *2-category* $\underline{\mathbf{C}}_2$ (or simply $\underline{\mathbf{C}}$) is a category $\mathbf{C}$ (called the *underlying* category) with a structure as the one above defined, such that the composition functions are subject to the following equations:

(1) $id_c *_L \beta = \beta$; (2) $(f; g) *_L \beta = f *_L (g *_L \beta)$;

(3) $g *_L id_h = id_{g;h}$; (4) $g *_L (\beta \cdot \delta) = (g *_L \beta) \cdot (g *_L \delta)$;

(5) $\alpha *_R id_c = \alpha$; (6) $\alpha *_R (h; i) = (\alpha *_R h) *_R i$;

(7) $id_g *_R h = id_{g;h}$; (8) $(\alpha \cdot \gamma) *_R h = (\alpha *_R h) \cdot (\gamma *_R h)$;

(9) $(f *_L \alpha) *_R h = f *_L (\alpha *_R h)$;

(10) $(j *_L \beta) \cdot (\alpha *_R h) = (\alpha *_R k) \cdot (g *_L \beta)$.

[2] We are quite informal here: both 2-categories and sesqui-categories are categories enriched over **Cat**, in the sense of [Kel82], but the tensor product used is different in the two cases (see [Ste94]). In the following we shall refer to both 2-categories and sesqui-categories as "cat-enriched categories", stressing that their hom-sets form categories: [Str92] contains a comprehensive introduction to this topic.

where $\cdot$ denotes composition insides hom-categories. A *sesqui-category* $\underline{\mathbf{C}}_S$ (or simply $\underline{\mathbf{C}}$: usually, there is no ambiguity) is a category $\mathbf{C}$ with a structure subject to equations (1)-(9). A *2-functor* (*sesqui-functor*, or just functor in both cases) $F : \underline{\mathbf{C}} \to \underline{\mathbf{D}}$ is a function mapping objects to objects, morphisms to morphisms and cells to cells, preserving identities and compositions of all kinds. □

Let **2-Cat** (**S-Cat**) be the category of 2-categories (sesqui-categories) and 2-functors (sesqui-functors): there exists an inclusion functor U_i : **2-Cat** $\to$ **S-Cat**, whose left adjoint F_i simply quotients the structure of a sesqui-category with respect to the axiom (10) (or *interchange* axiom). As a matter of fact, **2-Cat** is reflective inside **S-Cat**. Note also that, thanks to axiom (10), it is possible to define the notion of *horizontal* composition of cells, so that the given definition for 2-category is then equivalent to the classical one (see [KS74])

$$(j *_L \beta) \cdot (\alpha *_R h) = (\alpha *_R k) \cdot (g *_L \beta) = \alpha * \beta.$$

A category $\mathbf{C}$ is *cartesian* if it has a *terminal object* 0, such that for every object a there exists a unique arrow $!_a : a \to 0$; and *(binary) products*, i.e., a triple $\pi_{a,b} = \langle a \times b, \pi_0 : a \times b \to a, \pi_1 : a \times b \to b \rangle$ for each pair a, b of objects, such that, given the arrows $\langle f : c \to a, g : c \to b \rangle$, there is a unique arrow $\langle f, g \rangle : c \to a \times b$ satisfying $\langle f, g \rangle; \pi_0 = f$ and $\langle f, g \rangle; \pi_1 = g$. These properties are required for cells in an enriched context.

Definition 2 (Cartesian Cat-Enriched Categories). Let $\underline{\mathbf{C}}$ be a 2-category (sesqui-category) such that the underlying category $\mathbf{C}$ is cartesian. We say that $\underline{\mathbf{C}}$ has finite 2-products (sesqui-products), if for every pair $\alpha : f \Rightarrow g : c \to a$ and $\beta : h \Rightarrow k : c \to b$ of cells, there exists a unique cell $\gamma = \langle \alpha, \beta \rangle : \langle f, h \rangle \Rightarrow \langle g, k \rangle : c \to a \times b$ satisfying $\gamma *_R \pi_0 = \alpha$ and $\gamma *_R \pi_1 = \beta$. Graphically,

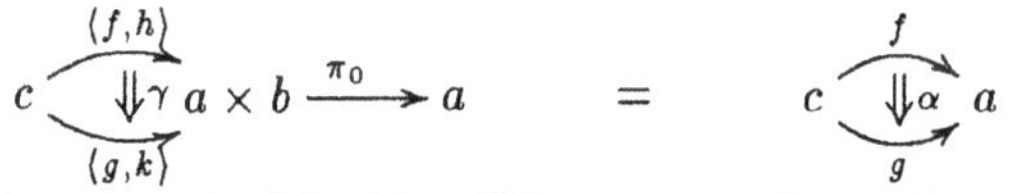

(similarly for β). $\underline{\mathbf{C}}$ has *terminal 2-object* if for every cell $\alpha : f \Rightarrow g : c \to a$ we have $\alpha *_R !_c = !_a$, where 0 is terminal in $\mathbf{C}$ and $!_a : a \to 0$, $!_c : c \to 0$. □

We define **2C-Cat** (**SC-Cat**) as the category of cartesian 2-categories (sesqui-categories) and functors preserving products and terminal object: the pair of functors $\langle U_i, F_i \rangle$ introduced above still forms an adjunction when restricted to **2C-Cat** and **SC-Cat**, respectively, if we require that functors in those categories preserve products and terminal object "on the nose".

We will be interested in a finitary presentation of an enriched structure, i.e., in a set of generators such that a cat-enriched category can be obtained freely composing cells. An appropriate structure for that is represented by *c-computads*.

Definition 3 (C-Computads). A *c-computad* is a pair $\langle \mathbf{C}, S \rangle$, where $\mathbf{C}$ is a category and S is a set of cells, each of which has two parallel arrows of $\mathbf{C}$ as *source* and *target*, respectively. Given the c-computads $\langle \mathbf{C}, S \rangle$ and $\langle \mathbf{C}', S' \rangle$, a *c-morphism* is a pair $\langle F, h \rangle$ such that $F : \mathbf{C} \to \mathbf{C}'$ is a functor, $h : S \to S'$ is a function, and for every cell $\alpha : f \Rightarrow g \in S$ we have $h(\alpha) : F(f) \Rightarrow F(g) \in S'$. □

A c-computad $\langle \mathbf{C}, S \rangle$ is cartesian if so is $\mathbf{C}$, while a c-morphism $\langle F, h \rangle$ preserves products and terminal object if F does so. Let **C-Comp** be the category of cartesian c-computads and c-morphisms preserving products and terminal object: there exists an obvious forgetful functor $U_2 : \mathbf{2C\text{-}Cat} \to \mathbf{C\text{-}Comp}$ which forgets the composition of cells, with left adjoint F_2. This adjoint composes the cells of a c-computad in all the possible ways, both horizontally and vertically, imposing further equalities in order to satisfy the axioms of a 2-category and to preserve finite products and terminal object on the underlying category. There exists also a similar adjunction pair $\langle U_s, F_s \rangle$ between **SC-Cat** and **C-Comp**, such that the following diagram commutes

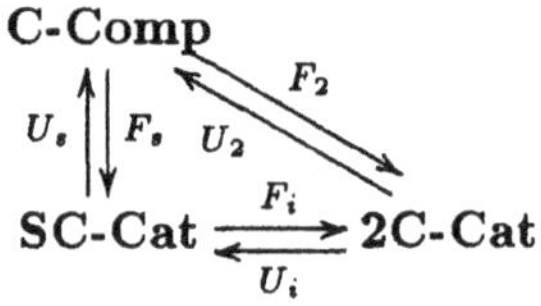

All functors preserve products and terminal object "on the nose": in particular, those originating from **C-Comp** preserve products in the underlying category.

3 Term Algebras and Rewriting Systems

We start recalling some well-known results about universal algebras (see e.g. [ADJ77]) We will introduce them in the classical, set-theoretical way, dealing in particular with the structure of terms in free algebras.

Definition 4 (The Class of Σ-algebras). Let Σ be a (one-sorted) *signature*, i.e., a ranked alphabet of operator symbols $\Sigma = \bigcup_{n \in \mathbb{N}} \Sigma_n$ (saying that f is of *arity* n for $f \in \Sigma_n$). A *Σ-algebra* is a pair $A = \langle |A|, \rho_A \rangle$ such that $|A|$ is a set (the *carrier*), and $\rho_A = \{f_A \mid f \in \Sigma\}$ is a family of functions such that for each $f \in \Sigma_n$, $f_A : |A|^n \to |A|$. Let A, B be two Σ-algebras: a *Σ-homomorphism* $\tau : A \to B$ is a function $\tau : |A| \to |B|$ preserving operators, i.e., such that for each $f \in \Sigma_n$, $\tau \circ f_A = f_B \circ \tau^n$. □

It is well-known that for each signature Σ there exists an initial algebra, often called the *word algebra* and denoted by T_Σ: the elements of the word algebra are all the terms freely generated from the constants and the operators of Σ. Given a set $X = \{x_1, \ldots, x_n\}$ of *variables*, the elements of the algebra $T_\Sigma(X)$, i.e., the algebra of terms with variables in X, are the terms of $T_{\Sigma \cup X}$, freely generated from Σ adding the elements of X as new constants.

Given a term $t = f(t_1, \ldots, t_n)$, a *subterm* of t could be simply defined as t itself, or any subterm of t_i: but since we want to distinguish among different occurrences of the same subterm in t, we need to define in a formal way the notion of "occurrence". The set $\mathcal{O}(t)$ of *occurrences* of a term t is the subset of $\mathbb{N}^*$ defined inductively in the following way

1. λ (the empty string) is an occurrence of t;

2. if $t = f(t_1, \ldots, t_n)$ and $w \in \mathcal{O}(t_i)$, then $iw \in \mathcal{O}(t)$.

In the following, as a subterm of t we mean a pair $\langle w, s\rangle$: s is the actual subterm, while w is the position of that particular subterm in t. Each $\mathcal{O}(t)$ can be equipped with a partial order relation: $w_1 \leq w_2$ iff there exists w_3 such that $w_1 w_3 = w_2$; two occurrences w_1, w_2 are *disjoint* if neither $w_1 \leq w_2$ nor $w_2 \leq w_1$; two subterms are disjoint if so are their respective occurrences.

Finally, given two finite sets X, Y of variables, a ***substitution*** is a function $\sigma : X \to T_\Sigma(Y)$, usually denoted as a set of the form $\{x_1/t_1, \ldots, x_n/t_n\}$ for $x_i\sigma = t_i$ (in postfix notation). A substitution σ extends to a function (also denoted σ) from $T_\Sigma(X)$ to $T_\Sigma(Y)$, defined inductively as $f(s_1, \ldots, s_n)\sigma = f(s_1\sigma, \ldots, s_n\sigma)$ for every operator $f \in \Sigma$.

Now we are ready to introduce *(term) rewriting systems*, which are just (labelled) sets of rules, i.e., of pairs of terms.

Definition 5 ((Term) Rewriting Systems). Let X be a set of variables. A *(term) rewriting system* $\mathcal{R}$ is a tuple (Σ, L, R), where Σ is a signature, L is a set of labels, and R is a function $R : L \to T_\Sigma(X) \times T_\Sigma(X)$, such that for all $d \in L$, if $R(d) = \langle l, r\rangle$ then $var(r) \subseteq var(l) \subseteq X$ and l is not a variable.

We write $d : l \to r \in R$ if $d \in L$ and $R(d) = \langle l, r\rangle$; sometimes, to make explicit the variables contained in a rule, we will write $d(x_1, \ldots, x_n) : l(x_1, \ldots, x_n) \to r(x_1, \ldots, x_n) \in R$ where $\{x_1, \ldots, x_n\} = var(l)$; moreover, given a substitution $\sigma = \{x_1/t_1, \ldots, x_n/t_n\}$, we will write $l(t_1, \ldots, t_n)$ for $l\sigma$. □

From the classical viewpoint, a rule $d : l \to r$ can be applied to a term t if there is a subterm $\langle w, s\rangle$ of t such that l matches s, and the result is the term t where the matched subterm is replaced by a suitable instantiation of r. Moreover, a term can be rewritten into another if there exists an appropriate chain of rules applications. This presentation makes *sequences* of rewrites the basic notion: it does not allow to reason about *how* a rewrite can be executed, i.e., to record the possible, different justifications of a derivation. To this aim, we introduce now an alternative definition of rewriting, borrowed from the seminal work by José Meseguer [Mes90, Mes92]. The idea is to take a logical viewpoint, regarding a rewriting system $\mathcal{R}$ as a theory, and any rewriting - making use of rules in $\mathcal{R}$ - as a sequent entailed by the theory. The entailment relation is defined inductively by the deduction rules of *rewriting logic*.

Definition 6 (Proof Terms and Rewriting Sequents). Let $\mathcal{R} = (\Sigma, L, R)$ be a rewriting system. Let $\Lambda = \cup_n \Lambda_n$ be the signature containing all the rules $d : l \to r \in R$ with the corresponding arity given by the number of variables in d: more precisely, for each n, $\Lambda_n = \{d \mid d(x_1, \ldots, x_n) : l(x_1, \ldots, x_n) \to r(x_1, \ldots, x_n) \in R\}$. A *proof term* α is a term of the algebra $T_\mathcal{R} = T_{\Sigma \cup \Lambda \cup \{\cdot\}}(X)$, where "$\cdot$" is a binary operator (we assume that there are no clashes of names among the various sets of operators). A *(rewriting) sequent* is a triple $\langle \alpha, t, s\rangle$ (usually written as $\alpha : t \to s$) where α is a proof term and $t, s \in T_\Sigma(X)$. □

So, sequents in rewriting logic have the form $\alpha : t \to s$, where t and s are terms and α is a proof term, encoding a justification of the rewriting of t into

s. We say that t *rewrites to* s *via* α if the sequent $\alpha : t \rightarrow s$ can be obtained by finitely many applications of certain *rules of deduction.*

Definition 7 (Rewriting Logic). Let $\mathcal{R} = (\Sigma, L, R)$ be a rewriting system. We say that $\mathcal{R}$ *entails* the *full sequent* $\alpha : s \rightarrow t$ if it can be obtained by a finite number of applications of the following rules of deduction:

(Reflexivity)

$$\frac{t \in T_\Sigma}{t : t \rightarrow t};$$

(Full Instantiation)

$$\frac{d : l \rightarrow r \in R, d \in \Lambda_n, \alpha_i : t_i \rightarrow s_i \text{ sequents for } i = 1, \ldots, n}{d(\alpha_1, \ldots, \alpha_n) : l(t_1, \ldots, t_n) \rightarrow r(s_1, \ldots, s_n)};$$

(Congruence)

$$\frac{f \in \Sigma_n, \alpha_i : t_i \rightarrow s_i \text{ sequents for } i = 1, \ldots, n}{f(\alpha_1, \ldots, \alpha_n) : f(t_1, \ldots, t_n) \rightarrow f(s_1, \ldots, s_n)};$$

(Transitivity)

$$\frac{\alpha : s \rightarrow t, \beta : t \rightarrow u}{\alpha \cdot \beta : s \rightarrow u}.$$

The rewriting system $\mathcal{R}$ entails the *flat sequent* $\alpha : s \rightarrow t$ if it can be obtained by a finite number of applications of the rules *Identity*, *Congruence*, *Transitivity*, and *Flat Instantiation*, where

(Flat Instantiation)

$$\frac{d : l \rightarrow r \in R, d \in \Lambda_n, t_i \in T_\Sigma(X) \text{ for } i = 1, \ldots, n}{d(t_1, \ldots, t_n) : l(t_1, \ldots, t_n) \rightarrow r(t_1, \ldots, t_n)}.$$

□

Let us point out the meaning of these rules. *Reflexivity* and *Transitivity* are self-explaining. *Congruence* states that disjoint rewrites can be executed in parallel: the associated proof term provides the context for the respective justifications. The most interesting rules are the two different kinds of instantiation. *Full Instantiation* allows for *nested* rewriting: two subterms matching the left-hand sides of two rules can be rewritten in parallel even if their roots are not disjoint, i.e., if one is above the other, provided that they do not overlap. On the contrary, *Flat Instantiation* does not allow such nesting: each rule can only be instantiated with elements of $T_\Sigma(X)$, and thus the rule names in a proof term generated using this rule instead of *Full Instantiation* will appear at mutually disjoint positions. It is worth stressing that on the one hand full sequents correspond exactly to the entailment relation defined by Meseguer (for the unconditional case) in [Mes92]. On the other hand, we introduced flat sequents as the syntactical counterpart of Stell's categorical model, as we will see in the next sections.

As an example, consider the rewriting system $\mathcal{V} = \{d(x) : f(x) \rightarrow g(x), d' : a \rightarrow b\}$. It entails the flat sequents $d(a) \cdot g(d')$ and $f(d') \cdot d(b)$ with source $f(a)$ and target $g(b)$: rule d has been instantiated with $\{x/a\}$ and $\{x/b\}$ respectively, while d' has been contextualized; it also entails the full sequent $d(d') : f(a) \rightarrow g(b)$, where d' is *nested inside* d. Graphically, we have the rewrites

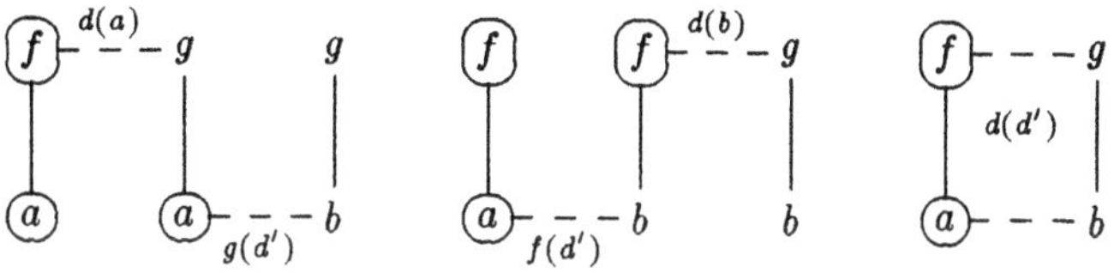

using the standard (yet suggestive, in our case) representation of terms as trees.

The complete agreement between the classical definition of derivability among terms and the one above using Rewriting Logic was already stated in [Mes92] (and it has been formally established and generalized to infinite rewriting in [CG94]). More precisely, it is not difficult to show that there exists a derivation from t to s in a TRS $\mathcal{R}$ iff $\mathcal{R}$ entails a sequent (flat or full) $\alpha : t \rightarrow s$.

The same kind of agreement holds between full and flat entailment: a rewriting system $\mathcal{R}$ entails a full sequent $\alpha : t \rightarrow s$, iff there exists a finite chain $\alpha_i : t_i \rightarrow s_i, i = 1, \ldots, n$ of flat sequents, such that $\alpha_1 \cdot \ldots \cdot \alpha_n : t \rightarrow s$. So, why to distinguish between flat and full rewrites? First of all, we must remind the reader that sequents represent an operational model for rewriting systems: each sequent can be considered as the encoding of a "concrete" computation of the machine. If we consider a term as a totally distributed structure, i.e., such that an actual rewriting machine can act separately on each occurrence of the term, then full sequents are able to describe the simultaneous execution of nested rewrites. Instead, a flat sequent can express only "disjoint" concurrency: two rewrites can be executed simultaneously only if they act on disjoint positions of a term. In other words, choosing a set of deduction rules is implicitly the same as choosing a particular implementation schema, so to say, for a rewriting system.

We already stated in the introduction the reasons why providing a concurrent semantics for a rewriting system means equipping its set of computations with a suitable equivalence relation. This is a particularly easy task in the setting of Rewriting Logic, where the elements of the space of computations are encoded by terms of the algebra $T_{\mathcal{R}}$. So, an equivalence relation can be easily expressed as an appropriate set of axioms on proof terms.

Definition 8 (Abstract Sequents). Let $\mathcal{R}$ be a rewriting system. An *abstract flat sequent* entailed by $\mathcal{R}$ is an equivalence class of flat sequents entailed by $\mathcal{R}$ modulo the following set E of axioms, which are intended to apply to the corresponding proof terms:

– *(Associativity)*

$$\frac{\alpha, \beta, \gamma \text{ proof terms}}{\alpha \cdot (\beta \cdot \gamma) = (\alpha \cdot \beta) \cdot \gamma};$$

– *(Distributivity)*

$$\frac{f \in \Sigma_n, \alpha_i, \beta_i \text{ proof terms for } i = 1, \ldots, n}{f(\alpha_1 \cdot \beta_1, \ldots, \alpha_n \cdot \beta_n) = f(\alpha_1, \ldots, \alpha_n) \cdot f(\beta_1, \ldots, \beta_n)};$$

– *(Identity)*

$$\frac{\alpha : s \to t \text{ sequent}}{s \cdot \alpha = \alpha = \alpha \cdot t}.$$

It is easy to check that proof terms equated by the above equations have the same source and target terms, thus an abstract flat sequent can be represented safely as $[\alpha]_E : s \to t$.

Similarly, an *abstract full sequent* entailed by $\mathcal{R}$ is an equivalence class $[\alpha]_{E'} : s \to t$ of full sequents entailed by $\mathcal{R}$, where $[\alpha]_{E'}$ is an equivalence class of proof terms modulo the equations in E', which is the union of the above equations and the following one:

– *(Interchange)*

$$\frac{d : l \to r \in R, d \in \Lambda_n, \alpha_i : t_i \to s_i \text{ sequents for } i = 1, \ldots, n}{d(\alpha_1, \ldots, \alpha_n) = l(\alpha_1, \ldots, \alpha_n) \cdot d(s_1, \ldots, s_n) = d(t_1, \ldots, t_n) \cdot r(\alpha_1, \ldots, \alpha_n)}.$$

□

Associativity and *Identity* need no explanation. Also *Distributivity* has an obvious meaning: to give a context to the composition of two rewrites is the same as to compose the contextualization of the single rewrites. The *Interchange* axiom is applied to full sequents only: it states that each nested rewrite can be expressed as the *sequential* composition of two simpler rewrites. Similarly to Definition 7, the set E' of axioms coincides with those given by Meseguer, while we introduced E, as we will see in the next section, as the algebraic counterpart of Stell's categorical model.

4 Categorical Models of Rewriting Systems

The aim of the previous section was to provide set-theoretic models for rewriting systems. Instead, the main concern of this section is for categorical models: we want to introduce enriched structures such that their cells are in one-to-one correspondence with the abstract sequents of a rewriting system. We first define a structure such that its set of cells is in one-to-one correspondence with the rewrite rules in $\mathcal{R}$, and, moreover, the underlying category is able to describe in a faithful way the structure of the word algebra associated to a given signature.

Given a signature Σ, we denote by $\mathbf{Th}(\Sigma)$ the associated *Lawvere theory* [Law63]: it is a category with finite products having natural numbers as objects, freely generated from the operators of Σ.

Definition 9 (Lawvere Theories). Given a signature Σ, the associated *Lawvere theory* is the category $\mathbf{Th}(\Sigma)$ with finite products such that

- its objects are underlined natural numbers: $\underline{0}$ is the terminal object and the product is defined as $\underline{n} \times \underline{m} = \underline{n+m}$;
- the arrows are generated by the following inference rules:

$$(generators)\ \frac{f \in \Sigma_n}{f_\Sigma : \underline{n} \to \underline{1}} \qquad (pairing)\ \frac{s : \underline{n} \to \underline{m}, t : \underline{n} \to \underline{m'}}{\langle s, t\rangle : \underline{n} \to \underline{m+m'}}$$

$$(identities)\ \frac{n \in \mathbb{N}}{id_{\underline{n}} : \underline{n} \to \underline{n}} \qquad (projections)\ \frac{n, m \in \mathbb{N}}{\pi_0 : \underline{n+m} \to \underline{n}, \pi_1 : \underline{n+m} \to \underline{m}}$$

$$(composition)\ \frac{s : \underline{n} \to \underline{m}, t : \underline{m} \to \underline{k}}{s; t : \underline{n} \to \underline{k}} \qquad (terminal\ arrows)\ \frac{n \in \mathbb{N}}{!_{\underline{n}} : \underline{n} \to \underline{0}}$$

and satisfying the axioms of cartesian categories. □

It can be shown that arrows from $\underline{n}$ to $\underline{1}$ are in one-to-one correspondence with Σ-terms whose variables are among $x_1,\ldots,x_n$: by $t_\Sigma : \underline{n} \to \underline{1}$ we shall denote the arrow corresponding to a term t. Similarly, an arrow $\underline{n} \to \underline{m}$ corresponds to a m-uple of Σ-terms with n variables, and arrows composition is terms substitution. Note that, since $\underline{0}$ is the terminal object, there exists a (unique) arrow $!_{\underline{m}} : \underline{m} \to \underline{0}$ for $m \in \mathbb{N}$. As for the so-called *duplicators* $\nabla_n = \langle id_{\underline{n}}, id_{\underline{n}}\rangle : \underline{n} \to \underline{2n}$, these arrows play a fundamental rôle in the one-to-one correspondence between arrows and terms. Let us consider a constant c: as a generator, the corresponding arrow is $c_\Sigma : \underline{0} \to \underline{1}$, while, when considering c as an element of $T_\Sigma(x_1, x_2)$, then the associated arrow is $!_{\underline{2}}; c_\Sigma : \underline{2} \to \underline{1}$; moreover, these arrows are unique, since for each $s : \underline{m} \to \underline{n}$, we have $s; !_{\underline{m}} = !_{\underline{n}}$.

Without going any further, the Lawvere theory $\mathbf{Th}(\Sigma)$ can be regarded simply as an alternative presentation of the signature Σ (where the basic operators are those we called *generators*), while each homset $\mathbf{Th}(\Sigma)[\underline{n}, \underline{m}]$ (i.e., the set of morphisms with source $\underline{n}$ and target $\underline{m}$) is equivalent to $[T_\Sigma(x_1, \ldots, x_n)]^m$, the set of m-tuples of terms of the free Σ-algebra with at most n variables.

Definition 10 (From Rewriting Systems to Computads). Given a rewriting system $\mathcal{R} = (\Sigma, L, R)$, the associated c-computad $Th(\mathcal{R})$ is given by the pair $\langle \mathbf{Th}(\Sigma), R_c\rangle$, where R_c is a set of cells between the arrows of $\mathbf{Th}(\Sigma)$, such that α: $s \to t \in \mathcal{R}$ iff α_c: $s_\Sigma \to t_\Sigma \in R_c$. □

After seminal studies in the late Eighties [RS87, Pow89], the correspondence between rewriting systems and c-computads has been at the basis of many works on rewriting. The usual technique consists of freely generating from a c-computad a cat-enriched structure such that its cells represent (equivalence classes of) sequences of rewrites: given a cell α_c, left-composition $s *_L \alpha_c$ instantiates the rewrite corresponding to α_c, while right-composition $\alpha_c *_R t$ inserts it in a context. The approaches by Meseguer [Mes90, Mes92] using 2-categories and by Stell [Ste92, Ste94] using sesqui-categories are instances of such a general idea. Thus they can be easily rephrased in this categorical setting by using the free functors relating c-computads and cat-enriched categories introduced after Definition 3. It is worth stressing that this presentation departs from those proposed in the original works (dramatically so for Stell's model): but it is essentially the same, in the sense that the induced equivalences on derivations are actually coincident.

Definition 11 (Space of Computations). Let $\mathcal{R}$ be a rewriting system and let $Th(\mathcal{R})$ be its associated c-computads. Then the associated *Meseguer's space of computations* $\underline{\mathbf{2\text{-}Th}(\mathcal{R})}$ is the cartesian 2-category $F_2(Th(\mathcal{R}))$, while its *Stell's space of computations* $\underline{\mathbf{S\text{-}Th}(\mathcal{R})}$ is the cartesian sesqui-category $F_s(Th(\mathcal{R}))$. □

Both these cat-enriched categories provide concurrent models for rewriting systems, since they impose an equivalence relation over cells (i.e., sequences of rewrites). *Different* equivalences, since the interchange axiom plays a crucial rôle. Let us consider the rewriting system $\mathcal{W} = \{d(x) : f(x) \to g(x,x), d' : a \to b, d''(x) : h(x) \to c\}$; the computad $Th(\mathcal{W})$ has the following set of cells

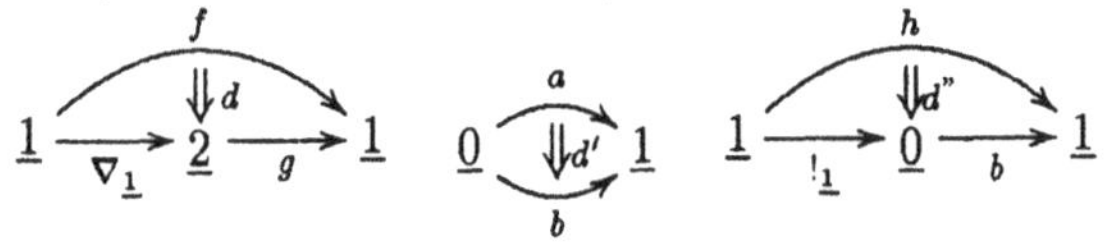

First of all, note the importance of duplicator $\nabla_{\underline{1}}$ and terminal arrow $!_{\underline{1}}$ for the correspondence between rewriting rules and cells. Note also that cells such as $(d' *_R f) \cdot (b *_L d)$ and $(a *_L d) \cdot (d' *_R (\nabla_1; g)) = (a *_L d) \cdot (\langle d', d' \rangle *_R g)$, originating from $a; f : \underline{0} \to \underline{1}$ (the morphism associated to the term $f(a)$), belong to both spaces of computations. But while in Stell's space they are different, in Meseguer's space, thanks to the interchange axiom, they are equated. The same happens to the cells $(d' *_R h) \cdot (b *_L d'')$ and $(a *_L d'') \cdot (d' *_R (!_{\underline{1}}; b)) = (a *_L d'')$ originating from the morphism $a; h$.

This suggests that sesqui-categories are suitable models of flat entailment: as an example, there is no intuitive semantical counterpart for the abstract full sequent $[d(d')]_{E'} : f(a) \to g(b,b)$ in the sesqui-category $\underline{\mathbf{S\text{-}Th}(\mathcal{W})}$. On the contrary, 2-categories are models for the full entailment relation: since the interchange axiom holds, the cell associated to that sequent is given by $(d' *_R f) \cdot (b *_L d) = (a *_L d) \cdot (d' *_R (\nabla_1; g)) = d * d'$. All this discussion is summarized by the following theorem, stating in a clear way the relationship between spaces of computations and abstract sequents.

Theorem 12 (Correspondence of Algebraic and Categorical Models). *Let $\mathcal{R}$ be a rewriting system. Then the following statements hold:*

1. *For each $n \in \mathbb{N}$ there exists a bijective function ϕ_n between the set of all abstract* flat *sequents entailed by $\mathcal{R}$ containing variables in $\{x_1, \ldots, x_n\}$ and the cells in $\underline{\mathbf{S\text{-}Th}(\mathcal{R})}[\underline{n}, \underline{1}]$, such that $\phi([\alpha]_E) : s_\Sigma \to t_\Sigma$ iff $[\alpha]_E : s \to t$.*
2. *Similarly, for each $n \in \mathbb{N}$ there exists a bijective function ψ_n between the set of all abstract* full *sequents entailed by $\mathcal{R}$ containing variables in $\{x_1, \ldots, x_n\}$ and the cells in $\underline{\mathbf{2\text{-}Th}(\mathcal{R})}[\underline{n}, \underline{1}]$, such that $\psi([\alpha]_{E'}) : s_\Sigma \to t_\Sigma$ iff $[\alpha]_{E'} : s \to t$.*

□

5 Categorical Models and Prime Algebraic Domains

In this section we aim at analyzing the algebraic properties of the categorical models we just introduced. As a start, we introduce *Prime Algebraic Domains*

(PAD's for short; see [Win89]). A PAD is simply a partial order (PO) verifying some additional properties. The use of PO's in semantics relies on the old idea that a computing machine determines an ordered space of computations: the richer structure of PAD's, however, make them especially suitable for modelling distributed systems.

Definition 13 (Prime Algebraic Domains). Let $\mathcal{D} = \langle D, \sqsubseteq \rangle$ be a PO:

1. The **least upper bound** (shortly **lub**) of a set $X \subseteq D$ is an element $\bigsqcup X$ such that $x \leq \bigsqcup X$ for all $x \in X$, and such that for all $z \in D$, $(\forall x \in X . x \leq z) \Rightarrow \bigsqcup X \leq z$. We write $x \sqcup y$ for $\bigsqcup\{x, y\}$.
2. Symmetrically, the **greatest lower bound** (shortly **glb**) of a set $X \subseteq D$ is an element $\sqcap X$ such that $\sqcap X \leq x$ for all $x \in X$, and such that for all $z \in D$, $(\forall x \in X . z \leq x) \Rightarrow z \leq \sqcap X$. We write $x \sqcap y$ for $\sqcap\{x, y\}$.
3. A **directed subset** of D is a subset $S \subseteq D$ such that for any finite subset $X \subseteq S$ there is an element $s \in S$ such that $\forall x \in X . x \leq s$.
4. An element $x \in D$ is **finite** if for all directed sets S, $x \sqsubseteq \bigsqcup S$ implies that there is some $s \in S$ such that $x \sqsubseteq s$.
5. $\mathcal{D}$ is **finitary** if for every finite element $x \in D$, the set $\{y \mid y \sqsubseteq x\}$ is finite.
6. An element $x \in D$ is **complete prime** (**prime**) if for each $X \subseteq D$ (each finite $X \subseteq D$), if $\bigsqcup X$ exists and $x \sqsubseteq \bigsqcup X$, then there exists an $y \in X$ such that $x \sqsubseteq y$.
7. $\mathcal{D}$ is **prime algebraic** if for all $x \in D$, $x = \bigsqcup\{y \sqsubseteq x \mid y \text{ is complete prime}\}$.
8. For $x, y \in D$, we write $x \uparrow y$ (and we say that x and y are **compatible**) if there exists a z such that $x \sqsubseteq z$ and $y \sqsubseteq z$. We say that $X \subseteq D$ is **pairwise compatible** if for all $x, y \in X$ we have $x \uparrow y$.
9. A finitary partial order $\langle D, \sqsubseteq \rangle$ is **distributive** if, whenever $x \uparrow y$, then we have $(x \sqcup y) \sqcap z = (x \sqcap z) \sqcup (y \sqcap z)$.
10. $\mathcal{D}$ is **finitely coherent** if it has lub's of *finite*, pairwise compatible subsets. $\mathcal{D}$ is a **coherent domain** if it has lub's of *arbitrary*, pairwise compatible subsets. □

Fact 14. *Any coherent, finitary prime algebraic domain $\mathcal{D}$ is distributive.* □

As we remarked in the introduction, whenever the derivation space of a model forms a PAD, this can be seen as an implicit confirmation of the "adequate degree of concurrency" of that model. Given term t, we define now formally its *derivation space* as the subset of cells originating from t_Σ.

Definition 15 (Derivation Spaces). Let $\mathcal{R} = \langle \Sigma, L, R \rangle$ be a rewriting system, and $t \in T_\Sigma(X)$. The *Meseguer's derivation space* $\mathcal{L}_2(t)$ associated to t is the partial order $\langle \mathcal{D}_2(t), \sqsubseteq_2 \rangle$, where $\mathcal{D}_2(t)$ is the set of coinitial cells in $\underline{\textbf{2-Th}(\mathcal{R})}$ originating from t_Σ[3], and $\sqsubseteq_2$ is the partial ordering relation defined as $\alpha \sqsubseteq_2 \beta$

[3] An equivalent definition could be given in categorical terms. Let us assume that $t_\Sigma : \underline{n} \to \underline{1}$, and let us denote $\underline{\textbf{2-Th}(\mathcal{R})}[\underline{n}, \underline{1}]$ by $\mathbf{C}$: then the objects of the comma category $(t_\Sigma \downarrow \mathbf{C})$ coincide with the elements of the derivation space $\mathcal{D}_2(t)$. The same relation holds for sesqui-categorical models.

iff there exists γ such that $\alpha \cdot \gamma = \beta$. Similarly, we define the *Stell's derivation space* $\mathcal{L}_S(t) = \langle \mathcal{D}_S(t), \sqsubseteq_S \rangle$ by simply replacing $\underline{\mathbf{2\text{-}Th}(\mathcal{R})}$ with $\underline{\mathbf{S\text{-}Th}(\mathcal{R})}$ in the above definition. □

It turns out that, in general, the derivation spaces induced by 2-categorical models fail to satisfy the distributive property of PAD's, since the ordering in which rewrites are executed influences the number of basic steps which are to be performed[4]. This is summarized by a naïve result of category theory: due to the structure of cells, in a cartesian 2-category no notion of length is definable.

As for a counterexample to distributivity, let us consider the rewriting system $\mathcal{W}$ defined in the previous section. If we look at the space of computations $\underline{\mathbf{2\text{-}Th}(\mathcal{W})}$ and the term $f(a)$, the associated derivation space $\mathcal{D}_2(f(a))$ has the following structure (where the partial ordering flows downwards):

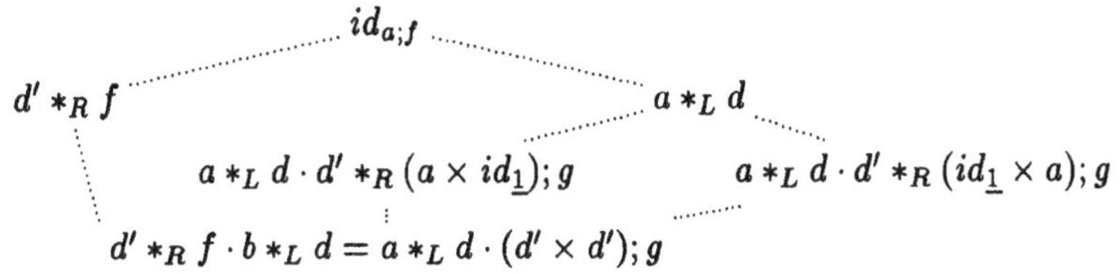

Let $x = a *_L d \cdot d' *_R (a \times id_{\underline{1}}); g$, $y = a *_L d \cdot d' *_R (id_{\underline{1}} \times a); g$, and $z = d' *_R f$: then $z = (x \sqcup y) \sqcap z \neq (x \sqcap z) \sqcup (y \sqcap z) = id_{a;f}$, hence $\mathcal{D}_2(f(a))$ is not distributive.

In categorical terms, this means that to allow together the cartesian structure and the interchange axiom does not permit to recover a PAD semantics: since the execution ordering influences in 2-categorical models the number of basic steps, it creates a causal link between different processors. Instead, in Stell's approach the interchange axiom is dropped, still preserving the cartesian structure on cells: this way, we are able to state the following theorem.

Theorem 16 (Derivation Spaces and PAD's). *Let $\mathcal{R} = \langle \Sigma, L, R \rangle$ be a rewriting system. For any $t \in T_\Sigma(X)$, the* PO *$\mathcal{L}_S(t)$ is a coherent, finitary* PAD. □

6 Conclusions

The main purpose of this paper was to analyze two categorical semantics for term rewriting systems proposed in the literature. To this aim we firstly introduced a new, unified presentation for them. Then we showed that the derivation spaces associated to Stell's model are prime algebraic domains, that is, a well accepted model for expressing the concurrency of an abstract formalism. On the contrary, this is not true in general for the derivation spaces associated to Meseguer's model (based on cartesian structure plus interchange axiom).

[4] The same result could be inferred also by the analysis of [LM92, Lan94]. In the first paper it is shown that the equivalence induced on rewrites in Meseguer's space of computations coincides with permutation equivalence. In the second, it is shown that in the case of λ-calculus (and more generally, for any TRS) permutation equivalence cannot be the basis for a prime algebraic domain semantics.

However, we view Stell's proposal as too narrow: dropping interchange means limiting the degree of concurrency expressible by the model in an unacceptable way. For instance, let us consider the rewriting system $\mathcal{V}$ (Section 3): the computations corresponding to $d(a); d'$ and $d'; d(b) : f(a) \to g(b)$ are not equated in the sesqui-category $\mathbf{S\text{-}Th}(\mathcal{V})$. On the other hand, the 2-categorical model fails to generate a śc pda when non-linear rewriting rules are involved (a rule is linear if each variable appears exactly once both in the left and right-hand side), since for such rules the execution order influences the number of basic steps to be performed. As an example, in the 2-category $\mathbf{2\text{-}Th}(\mathcal{W})$ (Section 4) the following derivations with different length are equated

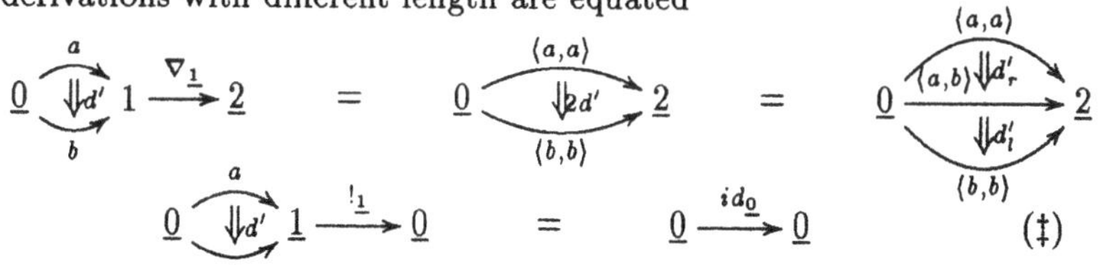

$$0 \xrightarrow[b]{a} \Downarrow\!d' \; 1 \xrightarrow{\nabla_1} 2 \quad = \quad 0 \xrightarrow[\langle b,b\rangle]{\langle a,a\rangle} \Downarrow\!d' \; 2 \quad = \quad 0 \xrightarrow[\langle b,b\rangle]{\langle a,a\rangle} {}^{\langle a,b\rangle}\!\Downarrow\!d'_r \;\; \Downarrow\!d'_l \; 2 \qquad (\dagger)$$

$$0 \xrightarrow{a} \Downarrow\!d' \; 1 \xrightarrow{!_1} 0 \quad = \quad 0 \xrightarrow{id_0} 0 \qquad (\ddagger)$$

where $2d' = d'^b *_L \nabla_1 = \langle d', d'\rangle$, $d'_l = \langle d', id_b\rangle$ and $d'_r = \langle id_a, d'\rangle$.

The cartesian structure on cells allows for implicit *garbage collection* and *duplication* (as discussed in [Cor94]): these "housekeeping" operations are performed silently, in the sense that the abstract machine corresponding to the model cannot distinguish for instance between states where garbage has been already collected, and states where it has not. The situation is well-shown by ($\dagger$) and ($\ddagger$): it seems evident that models with interchange axiom *and* implicit behaviour on these operation cannot be the basis for a PAD semantics. Stell's solution still keeps these operations silent, while dropping interchange forces the model to distinguish between derivations differing for the order of execution of nested rewrites. A current trend of work is investigating the case of preserving interchange and dealing with some kind of weak cartesianity (along the line of Jacobs [Jac93] for the semantics of linear logic), making housekeeping explicit.

Although a detailed discussion about the actual implementability of term rewriting in a way reflecting the degree of concurrency of either models goes beyond the scope of this paper, some comment is in order. Even if the computations of the 2-categorical model do not form a PAD, it does not mean that the model cannot be *implemented* on a parallel computer, as by the way has been shown by Meseguer's work on the *Rewriting Rule Machine* (see [LMR94]). However, if a *concurrent* machine (one with loosely coupled processors) is chosen as target, the results of this paper show that the 2-category model cannot be implemented *directly*, i.e., by representing operations of processors with events of a PES having the same domain of computations. Of course the 2-categorical model can be mapped to a concrete, concurrent machine, but the compilation process should be designed carefully in order to minimize hidden, expensive synchronizations and sequentializations of processors, which are unavoidable according to our results.

Acknowledgments: We wish to thank José Meseguer for the interesting discussions and his careful reviewing of the manuscript.

References

[ADJ77] J.A. Goguen, J.W. Tatcher, E.G. Wagner, J.R. Wright, *Initial Algebra Semantics and Continuous Algebras*, Journal of ACM **24** (1), 1977, pp. 68-95.

[Bou85] G. Boudol, *Computational Semantics of Term Rewriting Systems*, in Algebraic Methods in Semantics, eds. M. Nivat and J. Reynolds, Cambridge University Press, 1985.

[CG94] A. Corradini, F. Gadducci, *CPO Models for Infinite Term Rewriting*, draft.

[Cor94] A. Corradini, *Term Rewriting, in Parallel*, draft.

[DJ90] N. Dershowitz, J.P. Jouannaud, *Rewrite Systems*, Handbook of Theoretical Computer Science **B**, ed. J. van Leeuwen, North Holland, 1990, pp. 243–320.

[Jac93] B. Jacobs, *Semantics of Weakening and Contraction*, draft.

[Kel82] G.M. Kelly, *Basic Concepts of Enriched Category Theory*, London Mathematical Society, LN Series 64, 1982.

[KS74] G.M. Kelly, R.H. Street, *Review of the Elements of 2-categories*, Lecture Notes in Mathematics 420, 1974, pp. 75-103.

[Lan94] C. Laneve, *Distributive Evaluations of λ-calculus*, Fundamenta Informaticae **20** (4), 1994, pp. 333-352.

[Law63] F. W. Lawvere, *Functorial Semantics of Algebraic Theories*, Proc. National Academy of Science **50**, 1963, pp. 869-872.

[Lév80] J.J. Lévy, *Optimal Reductions in the λ-calculus*, in To H.B. Curry, Essays in Combinatory Logic, Lambda Calculus and Formalism, eds. J.P. Seldin and J.R. Hindley, Academic Press, 1980, pp. 159-191.

[LM92] C. Laneve, U. Montanari, *Axiomatizing Permutation Equivalence in the λ-calculus*, in Proc. 3rd ALP, LNCS 632, 1992, pp. 350-363.

[LMR94] P. Lincoln, J. Meseguer, L. Ricciulli, *The Rewrite Rule Machine Node Architecture and its Performance*, in Proc. CONPAR'94, LNCS 854, 1994, pp. 509-520.

[Mes90] J. Meseguer, *Functorial Semantics of Rewrite Systems*, appendix of *Rewriting as a Unified Model of Concurrency*, SRI Technical Report, CSL-93-02R, 1990.

[Mes92] J. Meseguer, *Conditional Rewriting Logic as a Unified Model of Concurrency*, in Selected Papers of 2th Workshop on Concurrency and Compositionality, Theoretical Computer Science **96**, 1992, pp. 73-155.

[ML71] S. MacLane, *Categories for the Working Mathematician*, Springer, 1971.

[MOM93] N. Martí-Oliet, J. Meseguer, *Rewriting Logic as a Logical and Semantic Framework*, SRI Technical Report, CSL-93-05, 1993.

[Pow89] A.J. Power, *An Abstract Formulation for Rewrite Systems*, in Proc. CTCS'89, LNCS 389, 1989, pp. 300-312.

[RS87] D.E. Rydeheard, J.G. Stell, *Foundations of Equational Deduction*, in Proc. CTCS'87, LNCS 283, 1987, pp. 114-339.

[Ste92] J. G. Stell, *Categorical Aspects of Unification and Rewriting*, Ph.D. Thesis, Faculty of Science, University of Manchester, 1992.

[Ste94] J. G. Stell, *Modelling Term Rewriting System by Sesqui-categories*, Technical Report TR94-02, Keele University, 1994.

[Str92] R.H. Street, *Categorical Structures*, in Handbook of Algebra, eds. M. Hazewinkel et al., Elsevier, preprint 1992.

[Win89] G. Winskel, *An Introduction to Event Structures*, Lecture Notes for the REX Summer School, LNCS 354, 1989, pp. 285-363.

Towards a Domain Theory for Termination Proofs

Stefan Kahrs

Laboratory for Foundations of Computer Science, University of Edinburgh, Edinburgh, Scotland

Abstract. We present a general framework for termination proofs for Higher-Order Rewrite Systems. The method is tailor-made for having simple proofs showing the termination of enriched λ-calculi.

1 Introduction

How does one prove the termination of a rewrite system by semantical methods? The central idea is: find a semantic interpretation $[\![\_]\!]$ of terms (pointwise for each symbol) such that $t \rightarrow_R u$ implies $[\![t]\!] > [\![u]\!]$ where $>$ is a terminating relation. Such an interpretation exists if and only if $\rightarrow_R$ is terminating. In a typed setting we would expect one terminating relation $>_\tau$ for each type τ. When we are trying to find an interpretation for a concrete rewrite system, we start by looking at the rules: if the above should succeed then in particular we need $[\![l]\!] > [\![r]\!]$ for every rule $l \rightarrow r$. However: it should be *sufficient* to look at the rules to establish a termination proof, i.e. all term-building operations should preserve the termination ordering. In other words: we try to interpret the rewrite system in the category of partially well-ordered sets (which we shall call WO), interpreting types as objects and terms as morphisms.

That is fine in principle but too abstract in practice. While any termination proof lives in this setting none can solely be expressed on this level. We need a little bit of extra structure for doing that. A simple example might tell us what this extra structure should be.

$$\begin{array}{lcr} \texttt{Length(Nil)} & \rightarrow & \texttt{0} \\ \texttt{Length(Cons(}x,y\texttt{))} & \rightarrow & \texttt{S(Length(}y\texttt{))} \end{array}$$

In contrast to an ordinary domain-theoretic interpretation we have to represent the rewrite noise somehow to make the left-hand sides bigger than the right. In the example we can interpret all constructors (`0`, `S`, `Nil`, `Cons`) in a standard way, i.e. as numbers and lists. For `Length` we could try the meaning $\lambda x.\, 2 * \#x + 1$, where $\#x$ is just the length of the list x: the interpretation of terms gives bigger values for the left-hand sides. However, this interpretation for `Length` does not preserve termination orderings, as it ignores rewriting within the list. We have to give the terms `Length(Cons(Length(Nil), Nil))` and `Length(Cons(0, Nil))` different interpretations because the former rewrites to the latter. Suppose we have a function `out` : `list` $\rightarrow$ `nat` which is order-preserving, then we can interpret

Length as $\lambda x.\, 2 * \#x + 1 + \texttt{out}(x)$ and this interpretation indeed shows the termination of our example.

The need for such out functions arises for all erasing rules, since the deleted subterm could have contained arbitrary redexes: we have to distinguish the interpretations before and after redex contraction. The Length example works smoothly as the result type of Length is numeric anyway. Let us consider another, non-numeric example:

$$\texttt{Tl}(\texttt{Cons}(x, y)) \rightarrow y$$

Using the outlined principle of how to deal with erasing rules, the interpretation of Tl should be: $\lambda y.\, \texttt{out}(\texttt{hd}(y)) + 1 + \texttt{tl}(y)$ where hd and tl are semantic "head" and "tail" functions (with e.g. $\texttt{hd}[\,] = 0, \texttt{tl}[\,] = [\,]$). But this does not quite make sense: $\texttt{tl}(y)$ is a list and we have added numbers to it! What we need is an operation $\boxplus_\tau : \omega \times \tau \rightarrow \tau$ which allows us to add numbers on top of values of arbitrary types τ. This function should be order-preserving and have the property $n \boxplus x \geq x$. With these two functions $\texttt{out}_\tau$ and $\boxplus_\tau$ we can express many semantic termination proofs. In the section on the semantics, we show how one can establish and promote the existence of these useful functions. This is one of the major contributions of this paper.

The other major contribution is to connect this semantic idea with higher-order rewriting. The essential problem here is that WO is not cartesian-closed which means that we either cannot interpret all terms as morphisms in WO or that we cannot interpret β-equivalent terms equally. Jaco van de Pol (in [7]) gave up the first objective in favour of the second, we do it the other way.

2 Preliminaries

An *Abstract Reduction System* (short: ARS) consists of a set A and a binary relation $\rightarrow$ on A. We write $\boldsymbol{A} \models P$ if the ARS $\boldsymbol{A} = (A, \rightarrow)$ has the property P. An ARS $\boldsymbol{A} = (A, \rightarrow)$ is *strongly normalising*, $\boldsymbol{A} \models \mathrm{SN}$, iff there is no non-empty relation R on A satisfying the equation $R = \rightarrow; R$, i.e. iff there are no infinite chains of $\rightarrow$-steps.

We do not have room to introduce the concepts we borrow from category theory to make this paper self-contained. The reader can find the missing definitions in most standard works on category theory, e.g. in [5]. Here is a brief summary defining some of these terms.

Given a category $\mathcal{A}$ and an object $X \in |\mathcal{A}|$, the *comma* category $\mathcal{A} \downarrow X$ has as objects pairs (Y, f), such that $Y \in |\mathcal{A}|$ and $f \in \mathcal{A}(Y, X)$, and a morphism $m \in (\mathcal{A} \downarrow X)((Y,g),(Z,h))$ is a morphism $m \in \mathcal{A}(Y,Z)$ such that $h \circ m = g$. There is a forgetful functor $U : \mathcal{A} \downarrow X \rightarrow \mathcal{A}$ defined as $U(X, f) = X$ and $U(m) = m$. Dually, $X \downarrow \mathcal{A}$ is defined as $(\mathcal{A}^{op} \downarrow X)^{op}$.

A *monad* in a category $\mathcal{A}$ is a triple $\boldsymbol{T} = (T, \mu, \eta)$ where $T : \mathcal{A} \rightarrow \mathcal{A}$ is a functor and $\mu : T(T(\_)) \rightarrow T(\_)$ and $\eta : \_ \rightarrow T(\_)$ are natural transformations satisfying $\mu \circ \eta = id$ and $\mu \circ \mu = \mu \circ T(\mu)$. Given such a monad, the category $\mathcal{A}^{\boldsymbol{T}}$ of $\boldsymbol{T}$-algebras has as objects pairs (X, f), such that $X \in |\mathcal{A}|$, $f \in \mathcal{A}(T(X), X)$,

$f \circ \mu = f \circ T(f)$ and $f \circ \eta = id$; a morphism $g \in \mathcal{A}^T((X,f),(X',f'))$ is a morphism $g \in \mathcal{A}(X,X')$ such that $g \circ f = f' \circ T(g)$. There is a forgetful functor $U : \mathcal{A}^T \to \mathcal{A}$ defined as $U(X,f) = X$ and $U(g) = g$.

Let $\mathcal{A}$ and $\mathcal{B}$ be categories and $U : \mathcal{A} \to \mathcal{B}$ be a functor. A functor $F : \mathcal{B} \to \mathcal{B}$ *can be lifted along* U if there is a functor $F' : \mathcal{A} \to \mathcal{A}$ such that $F \circ U = U \circ F'$. If $\mathcal{A} = \mathcal{B}^T$ we omit the "along U" and understand U to be the forgetful functor U of the monad. Similarly for $\mathcal{A} = \mathcal{B} \downarrow X$.

A *monoidal category* is a tuple $(\mathcal{A}, \otimes, I, a, l, r)$ where $\mathcal{A}$ is a category, $\otimes$ is a functor $\mathcal{A} \times \mathcal{A} \to \mathcal{A}$, I is an object in $\mathcal{A}$, and a, l, and r are natural isomorphisms $a : A \otimes (B \otimes C) \cong (A \otimes B) \otimes C$, $l : I \otimes A \cong A$, $r : A \otimes I \cong A$ (natural in A, B and C) such that "all coherence diagrams" commute, i.e. all diagrams only involving the isomorphisms and $\otimes$. A monoidal category is *symmetric* if there is also a natural isomorphism $s : A \otimes B \cong B \otimes A$ maintaining the property that all coherence diagrams commute. A (symmetric) monoidal category is called *closed* if the functor $\_ \otimes A$ has a right adjoint $A \Rightarrow \_$ for any $A \in |\mathcal{A}|$. We write $ap : (A \Rightarrow \_) \otimes A \to \_$ for the co-unit of the adjunction and $cur(f)$ as shorthand for $(id \Rightarrow f) \circ \eta$ where $\eta : \_ \to A \Rightarrow (\_ \otimes A)$ is the unit of the adjunction.

A *monoid* in a monoidal category $(\mathcal{A}, \otimes, I, a, l, r)$ is a triple $(X, \oplus, 0)$, where $X \in |\mathcal{A}|$, $\oplus \in \mathcal{A}(X \otimes X, X)$, and $0 \in \mathcal{A}(I, X)$, such that $\oplus \circ (0 \otimes id) \circ l^{-1} = \oplus \circ (id \otimes 0) \circ r^{-1} = id$ and $\oplus \circ (\oplus \otimes id) = \oplus \circ (id \otimes \oplus) \circ a^{-1}$. The *representation monad* of a monoid is the monad $(X \otimes \_, \mu, \eta)$, where $\mu = (\oplus \otimes id) \circ a$ and $\eta = (0 \otimes id) \circ l^{-1}$.

3 Semantics

Definition 1. A *partially well-ordered set* is an ARS $\boldsymbol{A} = (A, >_A)$ such that $\boldsymbol{A} \models$ SN and that $>_A$ is transitive. Convention: we shall write $\geq_A$ for the reflexive closure of $>_A$ on A.

Definition 2. The category WO is defined as follows:
Objects: $\boldsymbol{A} = (A, >_A) \in |\mathsf{WO}|$ if $\boldsymbol{A}$ is a partially well-ordered set.
Morphisms: a morphism $f : \boldsymbol{A} \to \boldsymbol{B}$ is a function $f : A \to B$ satisfying $\forall a, a' \in A.\ a >_A a' \Rightarrow f(a) >_B f(a')$.
Composition and identities: as in Set, the category of sets.

We want to interpret types as objects in WO. Often the rewrite system we are dealing with has type-constructors in addition to basic types, for example it may be an enriched λ-calculus with non-elementary types such as $\sigma \to \tau$, $\sigma \times \tau$, etc. To find well-ordered structures for these composite types we have to find the corresponding endofunctors on WO.

Definition 3. We define the summation functor $\mathsf{WO} \sqcup \mathsf{WO} \to \mathsf{WO}$, called categorical sum, as follows: let $\boldsymbol{A} = (A, >_A)$ and $\boldsymbol{B} = (B, >_B)$, then $\boldsymbol{A} \sqcup \boldsymbol{B} = (A \times \{0\} \cup B \times \{1\}, >_{A \sqcup B})$ such that $(a,0) >_{A \sqcup B} (a',0) \iff a >_A a'$ and $(b,1) >_{A \sqcup B} (b',1) \iff b >_B b'$; $(a,0)$ and $(b,1)$ are unrelated. On morphisms, $f \sqcup g$ is $f + g$ from Set.

The notation $\boldsymbol{A} \sqcup \boldsymbol{B}$ for the coproduct is motivated by the observation that the order type of $\alpha \sqcup \beta$ (for ordinals α and β) is just the ordinal $\alpha \cup \beta$, i.e. the maximum of the two.

Definition 4. We define two multiplication functors, categorical product $\mathsf{WO} \sqcap \mathsf{WO} \to \mathsf{WO}$ and symmetric product $\mathsf{WO} \otimes \mathsf{WO} \to \mathsf{WO}$ as follows: let $\boldsymbol{A} = (A, >_A)$ and $\boldsymbol{B} = (B, >_B)$, then

- $\boldsymbol{A} \sqcap \boldsymbol{B} = (A \times B, >_{A \sqcap B})$ with $(a,b) >_{A \sqcap B} (a',b') \iff a >_A a' \wedge b >_B b'$. On morphisms, $f \sqcap g$ is $f \times g$ from Set.
- $\boldsymbol{A} \otimes \boldsymbol{B}$ is defined just as $\boldsymbol{A} \sqcap \boldsymbol{B}$, except for the order $>_{A \otimes B}$: $(a,b) >_{A \otimes B} (a',b') \iff a >_A a' \wedge b \geq_B b' \vee a \geq_A a' \wedge b >_B b'$.

On first view, one might think that the multiplication functors differ only in minor details, but these details are quite significant. Writing bold numbers for the objects corresponding to ordinals in WO we have for example: $\mathbf{3} \sqcap \mathbf{7} \approx \mathbf{3}$ and $\mathbf{3} \otimes \mathbf{7} \approx \mathbf{9}$. Here, $\boldsymbol{A} \approx \boldsymbol{\alpha}$ is used to mean that there are morphisms $\boldsymbol{\alpha} \xrightarrow{f} \boldsymbol{A} \xrightarrow{g} \boldsymbol{\alpha}$.

Proposition 5. *$(\mathsf{WO}, \otimes, \mathbf{1})$ with $\mathbf{1} = (\{0\}, \emptyset)$ is a symmetric monoidal category.*

The categorical product does not have a neutral element — WO has no terminal object due to the Burali-Forti paradox. With a terminal object one can express constant functions (those that factor through the terminal object), but constant functions do not preserve strict orders.

Proposition 6. *The functor $\_\otimes A$ has a right adjoint $A \Rightarrow \_$, i.e. WO is monoidal closed.*

Proof. We define the functor $\_ \Rightarrow \_ : \mathsf{WO}^{op} \times \mathsf{WO} \to \mathsf{WO}$ by $(A, >_A) \Rightarrow (B, >_B) = (A \Rightarrow B, >_{A \Rightarrow B})$ with:

$$
\begin{array}{rcl}
A \Rightarrow B & = & \{f \in A \to B \mid \forall x, y \in A.\ (x >_A y \Rightarrow f(x) >_B f(y))\} \\
f >_{A \Rightarrow B} g & \iff & \forall x \in A.\ f(x) >_B g(x)
\end{array}
$$

We have to slightly amend this definition in case A is the empty set: $>_{\emptyset \Rightarrow B} = \emptyset$, the empty relation is SN. If A is non-empty, it has an element $a \in A$, and each infinite chain $f_1 >_{A \Rightarrow B} f_2 >_{A \Rightarrow B} \ldots$ can be mapped to an infinite chain $f_1(a) >_B f_2(a) >_B \ldots$ in B. On morphisms, we have as usual $(f \Rightarrow g)(h) = g \circ h \circ f$. Checking the adjunction properties is routine. □

Definition 7. An *ordinal* is a set α such that all its elements are ordinals and

- $\forall \beta \in \alpha.\ \forall \gamma \in \alpha.\ \beta \in \gamma \vee \gamma \in \beta \vee \beta = \gamma$
- $\forall \beta \in \alpha.\ \forall \gamma \in \beta.\ \gamma \in \alpha$

These are the so-called "von Neumann ordinals" [6]. In the following, I identify 0, 1, 2, etc. with their corresponding ordinal, ω is used for the ordinal corresponding to the set of natural numbers. We assume the usual operations for ordinal arithmetic (addition, multiplication, and exponentiation), see any standard book on the subject [8].

Definition 8. An ordinal α is called *indecomposable* iff $\forall \beta \in \alpha.\ \beta + \alpha = \alpha$.

Definition 9. Given two ordinals α and β, the ordinal $\alpha \# \beta$ is defined as $\sum_i \omega^i * (a_i + b_i)$ where $\alpha = \sum_i \omega^i * a_i$ and $\beta = \sum_i \omega^i * b_i$. Here, i ranges over ordinals less than $\alpha \cup \beta$, and the a_i and b_i over finite ordinals.

The function # takes the representation of two ordinals to the base ω, the so-called "Cantorian normal form", and then adds the coefficients pointwise. This form of addition is the so called "natural sum" of Hessenberg [4].

The relevance of indecomposable ordinals to this paper is that they can be understood as algebras for ordinal addition (and also natural sum), i.e. their elements are closed under ordinal addition. Indecomposable ordinals (greater than 0) are exactly those of the form ω^γ for some ordinal γ. Indecomposability of α is sufficient *and* necessary for the existence of a morphism from $\alpha \otimes \alpha$ to α, for the latter case see lemma 15 in [2].

Proposition 10. *Let α be an indecomposable ordinal. Then $(\boldsymbol{\alpha}, \#, 0)$ is a monoid in* WO, *where $0 : \mathbf{1} \to \boldsymbol{\alpha}$ maps the element of 1 to 0.*

3.1 Type-casting Ordinals

For concrete termination proofs it is useful to have "type-conversion functions" that can translate values of any type into ordinals and back.

Definition 11. Let $\mathcal{A}$ be a category and $X \in |\mathcal{A}|$. We define the category $\mathcal{A} \downdownarrows X$ as follows: $\mathcal{A} \downdownarrows X$ is the comma-category $X \downarrow \mathcal{A} \downarrow X$. Thus, an object in $\mathcal{A} \downdownarrows X$ is an object Z in $\mathcal{A}$, accompanied by two distinguished morphisms $X \xrightarrow{z} Z$ (encoding) and $Z \xrightarrow{\bar{z}} X$ (decoding), such that $\bar{z} \circ z = id_X$. A morphism from $X \xrightarrow{z} Z \xrightarrow{\bar{z}} X$ to $X \xrightarrow{y} Y \xrightarrow{\bar{y}} X$ in $\mathcal{A} \downdownarrows X$ is a morphism $f \in \mathcal{A}(Z, Y)$ such that $f \circ z = y$ and $\bar{y} \circ f = \bar{z}$.

To be precise, the category $X \downarrow \mathcal{A} \downarrow X$ should be written as either $(X \downarrow \mathcal{A}) \downarrow (X \xrightarrow{id} X)$ or as $(X \xrightarrow{id} X) \downarrow (\mathcal{A} \downarrow X)$, but both constructions result in identical categories.

In particular, we are interested in categories like $\mathsf{WO} \downdownarrows \boldsymbol{\alpha}$ where α is an ordinal. In this case, the maps z and $\bar{z}$ convert ordinals less than α into elements of Z and back.

Proposition 12. *By an abuse of notation, we write X to denote the object $X \xrightarrow{id} X \xrightarrow{id} X$ in $\mathcal{A} \downdownarrows X$. We have:*

1. *X is a null object in $\mathcal{A} \downdownarrows X$, i.e. it is initial and terminal.*
2. *For any two objects $A, B \in |\mathcal{A} \downdownarrows X|$ there is a morphism $A \xrightarrow{0} B$ uniquely defined by $0 = A \to X \to B$.*

Definition 13. We call an endofunctor $F : \mathcal{A} \to \mathcal{A}$ *retractable at* $X \in |\mathcal{A}|$ iff there is an object $X \xrightarrow{F} F(X) \xrightarrow{\bar{F}} X$ in $\mathcal{A} \downdownarrows X$.

Proposition 14. *If F is retractable at X then F can be lifted to $\mathcal{A} \downdownarrows X$.* *$F' : \mathcal{A} \downdownarrows X \rightarrow \mathcal{A} \downdownarrows X$ maps each object $X \xrightarrow{z} Z \xrightarrow{\bar{z}} X$ to* $X \xrightarrow{F} F(X) \xrightarrow{F(z)} F(Z) \xrightarrow{F(\bar{z})} F(X) \xrightarrow{\bar{F}} X$. *On morphisms, we just define $F'(f) = F(f)$.*

We can extend the notion of retractability to bifunctors in an obvious way.

Theorem 15. *Let $\mathcal{A}$ be a monoidal closed category with binary products and coproducts.*

1. *Categorical product $\sqcap$ and coproduct $\sqcup$ are retractable at any X:* $X \xrightarrow{\langle id, id\rangle} X \sqcap X \xrightarrow{\pi_1} X$ *is a retraction for the product; the coproduct is dual.*
2. *If $(X, \oplus, 0)$ is a monoid in $\mathcal{A}$ then the tensor product $\otimes$ is retractable at X.* $X \cong I \otimes X \xrightarrow{0 \otimes id} X \otimes X \xrightarrow{\oplus} X$ *is a retraction by the coherence properties of a monoid.*
3. *The exponential $\Rightarrow$ (right adjoint of $\otimes$) is retractable at monoids:* $X \xrightarrow{cur(\oplus)} X \Rightarrow X \cong (X \Rightarrow X) \otimes I \xrightarrow{id \otimes 0} (X \Rightarrow X) \otimes X \xrightarrow{ap} X$ *is a retraction; ap is the co-unit of the adjunction and cur($\oplus$) is the curried form of $\oplus : X \otimes X \rightarrow X$, also given by the adjunction.*

Taking $\mathcal{A} = \mathsf{WO}$ it follows that $\otimes$ and $\Rightarrow$ are retractable at indecomposable ordinals, $\oplus$ being the natural sum.

3.2 α-addition everywhere

Consider the representation monad $\boldsymbol{M} = (\boldsymbol{\alpha} \otimes \_, \mu, \eta)$ of the monoid $(\boldsymbol{\alpha}, \#, 0)$. $(\boldsymbol{A}, \boxplus)$ being an $\boldsymbol{M}$-algebra means in particular that (i) $\boxplus \circ \eta = id$, which is the same as saying that $0 \boxplus x = x$, and (ii) $\boxplus \circ \mu = \boxplus \circ (id \otimes \boxplus)$, which is the same as $(m \# n) \boxplus x = m \boxplus (n \boxplus x)$. Functions with these properties are just the ones we need for treating erasing rules.

Proposition 16. *Let $\mathcal{A}$ be a monoidal category and $(X, \oplus, 0)$ be a monoid in $\mathcal{A}$ and let $\boldsymbol{T}$ be the representation monad of this monoid. Then $(X, \oplus) \in |\mathcal{A}^{\boldsymbol{T}}|$.*

Although the observation in proposition 16 is rather trivial, it is important for the whole method. We can interpret "atomic" types by $\boldsymbol{\alpha}$ and leave the interpretation of composite types to functors on $\mathsf{WO}^{\boldsymbol{M}}$.

To maintain the existence of a $\boxplus$-operation with nice algebraic properties we have to make the functors we are interested in operate on $\mathsf{WO}^{\boldsymbol{M}}$.

Definition 17. Let $\mathcal{A}$ be a category and $\boldsymbol{T} = (T, \mu, \eta)$ a monad on $\mathcal{A}$. A functor $F : \mathcal{A} \rightarrow \mathcal{A}$ is called $\boldsymbol{T}$*-distributive* if there is a natural transformation δ : $T(F(U(\_))) \rightarrow F(T(U(\_)))$ such that $F(\boxplus_A) \circ \delta_A \circ \eta = id$ and $F(\boxplus_A) \circ \delta_A \circ \mu = F(\boxplus_A) \circ \delta_A \circ T(F(\boxplus_A) \circ \delta_A)$. Here, $U : \mathcal{A}^{\boldsymbol{T}} \rightarrow \mathcal{A}$ is the forgetful functor of the monad and $\boxplus_A : T(A) \rightarrow A$ is the algebra morphism on A.

Note that δ lives in $\mathcal{A}^T$. For T-distributive functors, we get a new $\boxplus'$ operation on $F(A)$ as $\boxplus' = F(\boxplus_A) \circ \delta$.

Proposition 18. *If F is T-distributive then it can be lifted to $\mathcal{A}^T$.*

We can now check whether the retractable functors we have so far, i.e. product, coproduct, symmetric multiplication, and arrow, are M-distributive.

Lemma 19. *Let $\mathcal{A}$ be a category with binary products $\_ \sqcap \_$. Let $T = (T, \mu, \eta)$ be a monad on $\mathcal{A}$. Then $\_ \sqcap X$ is T-distributive if $(X, \boxplus) \in |\mathcal{A}^T|$.*

Lemma 20. *Let $\mathcal{A}$ be a symmetric monoidal closed category with binary coproducts $\_ \sqcup \_$. Let X be a monoid in $\mathcal{A}$ and T be its representation monad. If $(Z, \boxplus) \in |\mathcal{A}^T|$ then the functor $\_ \sqcup Z$ is T-distributive.*

Proof. Because $\mathcal{A}$ is symmetric monoidal closed, the functor $X \otimes \_$ has a right adjoint which implies that it preserves colimits. Hence we have a natural isomorphism $\iota : X \otimes (\_ \sqcup Z) \cong (X \otimes \_) \sqcup (X \otimes Z)$ given by $\iota = ap \circ s \circ (id \otimes [cur(i_1 \circ s), cur(i_2 \circ s)])$ (where i_1 and i_2 are the coproduct injections and s is the symmetry isomorphism) and we get $\delta = (id \sqcup \boxplus) \circ \iota$. Checking the equations for δ is routine. □

Lemma 21. *Let $\mathcal{A}$ be a monoidal category, X a monoid in $\mathcal{A}$ and T the representation monad of X. Then the functor $\_ \otimes Y$ is T-distributive. If $\mathcal{A}$ is also monoidal closed and Y a T-algebra then $Y \Rightarrow \_$ is T-distributive.*

Theorem 22. *Let $\mathcal{A}$ be a symmetric monoidal closed category, X a monoid in $\mathcal{A}$, and T the representation monad of X. Then the functor $\_ \odot \_$ can be lifted to $\mathcal{A}^T$, where $\odot$ ranges over $\sqcup$, $\sqcap$, $\otimes$, $\Rightarrow$.*

It is similarly possible to show (though not in the available space) how the functors can be lifted to $\mathsf{WO}^M \Downarrow (\alpha, \#)$. For this, one needs that the retraction maps $\alpha \xrightarrow{F} F(\alpha) \xrightarrow{\bar{F}} \alpha$ of a functor F live in WO^M.

Putting these results together, we have the following recipe for interpreting types by partially well-ordered sets:

Theorem 23. *Let α be an indecomposable ordinal. For any object $A \in \mathsf{WO}$, which we can build from applying the functors $\sqcap$, $\sqcup$, $\otimes$, and $\Rightarrow$ in arbitrary order to α, we have:*

1. *There are morphisms $\alpha \xrightarrow{a} A \xrightarrow{\bar{a}} \alpha$ such that $\bar{a} \circ a = id_\alpha$.*
2. *For any other object B built this way, we have a morphism $A \xrightarrow{0} B$.*
3. *We have a morphism $\alpha \otimes A \xrightarrow{\boxplus} A$ such that $0 \boxplus x = x$ and $m \boxplus (n \boxplus x) = (m \# n) \boxplus x$.*
4. *The conversions $A \xrightarrow{0} B$ preserve addition, i.e. $0(n \boxplus x) = n \boxplus 0(x)$.*

Moreover, we can extend this result to other endofunctors on WO, provided they are retractable at α, they are M-distributive and their retraction maps are WO^M-homomorphisms.

4 Syntax

The previous section presented a semantic domain for the interpretation of rewrite systems that supports termination proofs. We still have to provide a connection between the syntax (Higher-Order Rewrite systems) and this semantics. Since we are not concerned here with implementability issues, we can choose Wolfram's generalisation of HRSs (see chapter 4 in [9]) as syntactic domain. HRSs are based on simply typed λ-calculus $\lambda^{\rightarrow}$. Its terms can be seen as either equivalence classes of λ-terms, the equivalence relation being $=_{\beta\eta}$, or as long β-normal forms, i.e. as canonical representatives of those classes.

The problem with HRSs is the lack of "nice" interpretations of $\lambda^{\rightarrow}$ in WO, simply because WO is not a CCC. Such an interpretation should assign the same values to $\beta\eta$-convertible terms, because HRSs rewrite modulo $\beta\eta$-conversion. Moreover, it should also interpret function types as sets of monotonic functions; this is necessary to lift termination of rewriting to its congruence closure, i.e. to rewriting on subterms. These objectives are conflicting for $\lambda^{\rightarrow}$: mapping convertible λ-terms to the same values means to interpret syntactic λ-abstraction by semantic λ-abstraction; but $\lambda^{\rightarrow}$ contains constant functions (like $\lambda x.c$) the semantic equivalent of which do not preserve strict orderings.

4.1 Term, Types, and Their Interpretations

To interpret types and terms in WO (or $\mathsf{WO}^M \downdownarrows \alpha$), we shall give some functions from sets of types or terms into $|\mathsf{WO}|$ or $\bigcup|\mathsf{WO}|$, respectively. The notation $\bigcup|\mathsf{WO}|$ is shorthand for $\bigcup\{s \mid (s,>) \in |\mathsf{WO}|\}$; similarly for $\mathsf{WO}^M \downdownarrows \alpha$, we suppress the application of the forgetful functor $U : \mathsf{WO}^M \downdownarrows \alpha \rightarrow \mathsf{Set}$. Since the domain of the mentioned functions is always a set, their graph is a set as well and so we shall not worry about foundational issues.

Definition 24. Given a set of *base types* $\mathcal{B}$ we define the set of *types over* $\mathcal{B}$, *Typ*($\mathcal{B}$), as the smallest set of words over the alphabet $\{\rightarrow, (,)\} \uplus \mathcal{B}$ satisfying:

1. $\sigma \in \mathcal{B} \Rightarrow \sigma \in Typ(\mathcal{B})$
2. $\alpha, \beta \in Typ(\mathcal{B}) \Rightarrow (\alpha \rightarrow \beta) \in Typ(\mathcal{B})$

Typ($\mathcal{B}$) comprises the types of $\lambda^{\rightarrow}$. As usual, we drop many parentheses and take $\rightarrow$ to be right-associative. Having only one type constructor for non-base types reflects the meta-level we are dealing with, i.e. $\lambda^{\rightarrow}$. This does not prevent us from giving base types an internal structure reflecting the type structure we want on the object-level.

Definition 25. Let $\mathcal{B}$ be a set of base types, let $b : \mathcal{B} \rightarrow |\mathsf{WO}|$ a function mapping base types to objects in WO. We define a map $[\![\_]\!]_b : Typ(\mathcal{B}) \rightarrow |\mathsf{WO}|$ as follows:

$$[\![\sigma \rightarrow \tau]\!]_b = [\![\sigma]\!]_b \Rightarrow [\![\tau]\!]_b$$
$$[\![\tau]\!]_b = b(\tau), \text{ if } \tau \in \mathcal{B}$$

Here, "$\Rightarrow$" is the functor from proposition 6.

Analogously, we derive from a function $b : \mathcal{B} \rightarrow |\mathsf{WO}^M \downdownarrows \alpha|$ a function for all types $[\![\_]\!]_b : \mathit{Typ}(\mathcal{B}) \rightarrow |\mathsf{WO}^M \downdownarrows \alpha|$, provided α is an indecomposable ordinal, because we can lift $\_ \Rightarrow \_$ at monoids, see theorem 23. In other words, the interpretation $[\![\sigma]\!]_b$ of a type σ in $\mathsf{WO}^M \downdownarrows \alpha$ is an object in $\mathsf{WO}^M \downdownarrows \alpha$ of the form $\alpha \xrightarrow{\sigma} ((S, >_\sigma), \boxplus_\sigma) \xrightarrow{\bar{\sigma}} \alpha$. We shall write $\bar{\sigma}$, $\boxplus_\sigma$, and $>_\sigma$ to refer to the corresponding components of $[\![\sigma]\!]_b$.

Definition 26. An HRS *signature* is a tuple $(\mathcal{B}, \mathcal{S}, \mathcal{C})$, where $\mathcal{B}$ is a set, $\mathcal{S}$ a set of *symbols*, and $\mathcal{C} : \mathcal{S} \rightarrow \mathit{Typ}(\mathcal{B})$ is a function assigning types to symbols.

Independently from particular signatures, we assume the existence of a countably infinite set of *variables*, called $\mathcal{V}$. In the following, we shall usually suppress the signature $\Sigma = (\mathcal{B}, \mathcal{S}, \mathcal{C})$ and the interpretation of base types b, i.e. we assume a fixed Σ and b unless otherwise stated.

Definition 27. A *preterm* is a λ-term with variables taken from $\mathcal{V}$ and constants taken from $\mathcal{S}$. We require abstractions to be in Church-style [1], i.e. an abstraction has the form $(\lambda x : \sigma. t)$, where x is a variable, $\sigma \in \mathit{Typ}(\mathcal{B})$, and t is a preterm. We write $\Lambda\Sigma$ for the set of all preterms. Given a preterm t, we write $t\downarrow$ for the β-normal form of t if it exists.

Preterms are just untyped λ-terms with type annotations for abstractions. They may or may not be well-typed in some type system.

Definition 28. A *context* is a finite set $\Gamma \subset \mathcal{V} \times \mathit{Typ}(\mathcal{B})$ such that $(x, \tau) \in \Gamma \wedge (x, \sigma) \in \Gamma \Rightarrow \sigma = \tau$. Convention: we write $x : \tau$ instead of (x, τ) for elements of a context. We write $x \notin \Gamma$ as shorthand for $\neg\exists\sigma.(x, \sigma) \in \Gamma$.

Definition 29. A *judgement* is a triple (Γ, t, τ), written $\Gamma \vdash t : \tau$, where Γ is a context, $t \in \Lambda\Sigma$, and $\tau \in \mathit{Typ}(\mathcal{B})$. Given a judgement $J = \Gamma \vdash t : \tau$, we write $J\downarrow$ for the judgement $\Gamma \vdash t\downarrow : \tau$ (which exists if $t\downarrow$ exists).

Definition 30. The *type theory* $\lambda^{\rightarrow}$ is the smallest set of judgements that is closed under the following rules.

$$\frac{x \notin \Gamma}{\Gamma \cup \{x : \sigma\} \vdash x : \sigma} \qquad \frac{}{\Gamma \vdash c : \mathcal{C}(c)}$$

$$\frac{\Gamma \vdash t : \sigma \rightarrow \tau \quad \Gamma \vdash u : \sigma}{\Gamma \vdash (t\, u) : \tau} \qquad \frac{\Gamma \cup \{x : \sigma\} \vdash t : \tau \quad x \notin \Gamma}{\Gamma \vdash (\lambda x : \sigma.t) : \sigma \rightarrow \tau}$$

Proposition 31. *Let $J \in \lambda^{\rightarrow}$. Then $J\downarrow$ exists and $J\downarrow \in \lambda^{\rightarrow}$.*

Proposition 32. *Let $\Gamma \vdash t : \tau$ and $\Gamma \vdash t : \tau'$ be in $\lambda^{\rightarrow}$. Then $\tau = \tau'$.*

Propositions 31 and 32 are well-known properties of $\lambda^{\rightarrow}$-Church [1].

Definition 33. A *symbol interpretation* is a function $\varrho : \mathcal{S} \to \bigcup |\mathsf{WO}^M \Downarrow \alpha|$ such that $\varrho(s) \in [\![\mathcal{C}(s)]\!]_b$. A *variable interpretation* is a function $v : \mathcal{V} \to \bigcup |\mathsf{WO}^M \Downarrow \alpha|$. If v is a variable interpretation and $x \in \mathcal{V}$ then we write $v[x \mapsto y]$ for a variable interpretation v' with $v'(x) = y$ and $v'(x') = v(x')$ if $x \neq x'$. A variable interpretation v is *consistent* w.r.t. Γ if $\forall (x, \tau) \in \Gamma.\ v(x) \in [\![\tau]\!]_b$.

An *interpretation* is a pair (ϱ, v) of a symbol interpretation ϱ and a variable interpretation v. An interpretation (ϱ, v) is called *consistent* (w.r.t. Γ) if v is.

Definition 34. Let $\rho = (\varrho, v)$ be an interpretation.We define a partial function $[\![\text{-}]\!]_\rho : \lambda^\to \rightharpoonup \bigcup |\mathsf{WO}^M \Downarrow \alpha|$ with domain $\{\Gamma \vdash t : \tau \mid \rho \text{ consistent with } \Gamma\}$ by the following equations:

$$
\begin{aligned}
[\![\Gamma \vdash c : \tau]\!]_\rho &= \varrho(c) \\
[\![\Gamma \vdash x : \tau]\!]_\rho &= v(x) \\
[\![\Gamma \vdash (f\ a) : \tau]\!]_\rho &= [\![\Gamma \vdash f : \sigma \to \tau]\!]_\rho([\![\Gamma \vdash a : \sigma]\!]_\rho) \\
&\quad \text{where } (\Gamma \vdash a : \sigma) \in \lambda^\to \\
[\![\Gamma \vdash (\lambda x : \sigma . t) : \tau]\!]_\rho &= (z \mapsto \overline{\sigma}(z) \boxplus_\tau [\![\Gamma \cup \{x : \sigma\} \vdash t : \tau]\!]_{\rho[x \mapsto z]}) \\
&\quad \text{where } z \in [\![\sigma]\!]_b
\end{aligned}
$$

To see that this definition is well-formed observe the following properties of semantic interpretation of types and judgements:

Lemma 35. *1.* $[\![\Gamma \vdash t : \tau]\!]_\rho \in [\![\tau]\!]_b$
2. Let $x \in FV(t)$, $z, z' \in [\![\sigma]\!]_b$ *and* $z >_\sigma z'$. *If* $\Gamma \vdash t : \tau$ *and if* $\rho[x \mapsto z]$ *is consistent w.r.t.* Γ *then* $[\![\Gamma \vdash t : \tau]\!]_{\rho[x \mapsto z]} >_\tau [\![\Gamma \vdash t : \tau]\!]_{\rho[x \mapsto z']}$

The chosen interpretation for λ-abstraction may look a bit peculiar, because it does not have the property that β-convertible terms have equal interpretations. Therefore, β-reduction is only of limited use for the meta-level of rewriting.

Definition 36. A *presubstitution* is a function $\theta : \mathcal{V} \to \Lambda\Sigma$. Given $t \in \Lambda\Sigma$, we write t^θ for the preterm we get by replacing all free variables in t by their image under θ, avoiding name capture by α-conversion. A *substitution* is a triple (θ, Γ, Δ), written $\theta : \Gamma \to \Delta$, if θ is a presubstitution and Γ and Δ are contexts such that $\forall (x : \tau) \in \Gamma.\ \Delta \vdash \theta(x) : \tau$.

Proposition 37. *Let* $\Gamma \vdash t : \tau$ *and* $\theta : \Gamma \to \Delta$. *Then* $\Delta \vdash t^\theta : \tau$.

This is the standard substitution lemma for $\lambda^\to$, generalised to substitutions that replace all free variables at once. It motivates the following definition:

Definition 38. Let $J = \Gamma \vdash t : \tau$ and $\vartheta = \theta : \Gamma \to \Delta$. We write J^ϑ for the judgement $\Delta \vdash t^\theta : \tau$.

Definition 39. Let $\rho = (\varrho, \upsilon)$ be an interpretation consistent w.r.t. Δ and let $\vartheta = \theta : \Gamma \to \Delta$ be a substitution. We define another interpretation $\rho \circ \vartheta$ as $(\varrho, \upsilon \circ \vartheta)$ where the variable interpretation $\upsilon \circ \vartheta$ is given by:

$$(\upsilon \circ \vartheta)(x) = \begin{cases} [\![\Delta \vdash \theta(x) : \tau]\!]_\rho, \text{ if } (x, \tau) \in \Gamma \\ \upsilon(x), \text{ otherwise} \end{cases}$$

It is easy to see that $\rho \circ \vartheta$ is consistent w.r.t. Δ.

Proposition 40. *Let $J = (\Gamma \vdash t : \tau) \in \lambda^{\to}$, $\vartheta = \theta : \Gamma \to \Delta$, and ρ be an interpretation consistent w.r.t. Δ. Then $[\![J^\vartheta]\!]_\rho = [\![J]\!]_{\rho\circ\vartheta}$.*

Proposition 40 is a typical argument often used in semantic interpretations of the λ-calculus; it does quite happily work with rather non-standard interpretations of λ-abstractions as in our case.

Lemma 41. *Let $(\Gamma \vdash t : \tau) \in \lambda^{\to}$ and $t \to_\beta t'$. Let ρ be an interpretation consistent with Γ. Then $[\![\Gamma \vdash t : \tau]\!]_\rho \geq_\tau [\![\Gamma \vdash t' : \tau]\!]_\rho$.*

4.2 Rules and Their Interpretations

The definition of HRS varies a bit in the literature. The following is another slight variation of the definitions of van de Pol or Wolfram [7, 9].

Definition 42. An *HRS-rule* is a tuple (Γ, l, r, τ) such that Γ is a context, $\tau \in \mathcal{B}$, and $\Gamma \vdash l : \tau$ and $\Gamma \vdash r : \tau$ are in $\lambda^{\to}$. Notation: we write rules as $\Gamma \vdash l \to r : \tau$.

The condition that τ is a base type does not restrict the expressive power as we can always η-expand rules by adding fresh variables to the context.

Definition 43. An HRS is a pair (Σ, R) where Σ is a signature and R a set of rules over Σ.

An HRS is associated with an ARS. The elements of this ARS are (derivable) judgements in β-normal form and the relation is given by the following notion of rule application.

Definition 44. Let (Σ, R) be an HRS. For a given a judgement $(\Gamma \vdash C : \sigma) \in \lambda^{\to}$ a *rule application* is pair $(\vartheta_l, \vartheta_r)$ of substitutions, $\vartheta_l = \theta_l : \Gamma \to \Delta$ and $\vartheta_r = \theta_r : \Gamma \to \Delta$, such that for all $(x : \sigma) \in \Gamma$ either $(*)$ $\theta_l(x) = \theta_r(x)$, or $(**)$ there is a rule $(E \vdash l \to r : \tau) \in R$ with $E = \{y_1 : \sigma_1, \ldots, y_n : \sigma_n\}$, $\sigma = \sigma_1 \to \cdots \to \sigma_n \to \tau$, and $\theta_l(x) = \lambda y_1 : \sigma_1. \cdots \lambda y_n : \sigma_n. l$ and $\theta_r(x) = \lambda y_1 : \sigma_1. \cdots \lambda y_n : \sigma_n. r$. (Notation: we shall abbreviate this as $\theta_l(x) = \boxed{l}$ and $\theta_r(x) = \boxed{r}$.) A rule application is called *proper* if for at least one $(x : \sigma) \in \Gamma$ we have $x \in \mathrm{FV}(C\!\downarrow)$ and $(**)$.

We define a relation $\to_R$ on β-normal forms of judgements in $\lambda^{\to}$ as follows: given a judgement $J = (\Gamma \vdash C : \sigma) \in \lambda^{\to}$ and a proper rule application $(\vartheta_l, \vartheta_r)$ then $J^{\vartheta_l}\!\downarrow \to_R J^{\vartheta_r}\!\downarrow$.

The above notion gives a more or less canonical definition of HRS reduction; it is slightly more general than the definitions in the literature as it supports single-step reduction with more than one rule at a time. The reason for requiring properness is the following proposition:

Proposition 45. *Let J be judgement in $\lambda^{\rightarrow}$ and $(\vartheta_l, \vartheta_r)$ be a rule application which is not proper. Then $J^{\vartheta_l}\downarrow = J^{\vartheta_r}\downarrow$.*

Therefore we need properness to give $\rightarrow_R$ a chance to be terminating. Any approach attempting to reason about termination of HRS reduction has to make similar restrictions in the definition of its reduction relation $\rightarrow_R$.

It is often convenient to assume that a rule application only instantiates one rule at one particular position in a term. We can define this as follows:

Definition 46. A rule application $(\vartheta_l, \vartheta_r)$ for a judgement J is called *linear* if $\vartheta_l(y) = \vartheta_r(y)$ for all but one y from the context of J, and if $\vartheta_l(x) = \boxed{l}$ and $\vartheta_r(x) = \boxed{r}$ then x occurs at most once in $J\downarrow$.

Lemma 47. *Let $J \rightarrow_R J'$. Then there are judgements $J_1, \ldots, J_n \in \lambda^{\rightarrow}$ such that $J = J_1 \rightarrow_R J_2 \rightarrow_R \cdots \rightarrow_R J_n = J'$ where each $J_i \rightarrow_R J_{i+1}$ by a linear rule application.*

Definition 48. Let ϱ be a symbol interpretation. A rule $\Gamma \vdash l \rightarrow r : \tau$ is called *ϱ-decreasing* if for all substitutions $\theta : \Gamma \rightarrow \Delta$ and all variable interpretations υ that are consistent with Δ it is true that $[\![\Delta \vdash l^{\theta}\downarrow : \tau]\!]_{(\varrho,\upsilon)} >_{\tau} [\![\Delta \vdash r^{\theta}\downarrow : \tau]\!]_{(\varrho,\upsilon)}$.

Theorem 49. *Let (Σ, R) be a HRS. If ϱ is a symbol interpretation such that all rules in R are ϱ-decreasing then $(\boldsymbol{\lambda}^{\rightarrow}, \rightarrow_R) \models$ SN.*

Proof. We simply prove that $J \rightarrow_R J'$ implies $[\![J]\!]_{\rho} >_{\tau} [\![J']\!]_{\rho}$ where $J = B \vdash t : \tau$ and $\rho = (\varrho, \upsilon)$ for some variable interpretation υ consistent with B. By lemma 47 it is sufficient to consider the case in which $J \rightarrow_R J'$ by a linear rule application.

Suppose we get $J \rightarrow_R J'$ by applying the proper rule application $(\vartheta_l, \vartheta_r)$ to the judgement $C = E \vdash c : \sigma$. We can assume w.l.o.g. that $c = c\downarrow$. We show $[\![E \vdash c : \sigma]\!]_{\rho\circ\vartheta_l} >_{\sigma} [\![E \vdash c : \sigma]\!]_{\rho\circ\vartheta_r}$ by induction on the term structure of c.

We only look at the most interesting case, which is when c is an application $t\, t_1 \cdots t_n$ $(n \geq 1)$ such that t is not an application; t is either a constant or a variable since we assumed c to be in β-normal form. If t is either a constant or a variable that is not instantiated to a rule by the rule application, then $t^{\theta_l} = t^{\theta_r}$ and the rule application is still proper for at least one of the judgements $E \vdash t_i : \sigma_i$. We can apply either the induction hypothesis or proposition 45 to the t_i. By monotonicity of $[\![t^{\theta_l}]\!]_{\rho}$ the result follows. If t is instantiated to a rule $\Gamma \vdash l \rightarrow r : \tau$ with $\Gamma = \{y_1 : \sigma_1, \ldots, y_k : \sigma_k\}$ then we have (if $k = n$) $c^{\theta_l}\downarrow = (t\, t_1 \cdots t_n)^{\theta_l}\downarrow = l^{\theta_l'}\downarrow$ where $\theta_l' : \Gamma \rightarrow \Delta$ is a substitution with $\theta_l'(y_i) = t_i^{\theta_l}\downarrow$. The same argument applies to $c^{\theta_r}\downarrow$, giving us another substitution $\theta_r' : \Gamma \rightarrow \Delta$. Since we assumed the rule application to be linear, it is not proper for the judgements $J_i = E \vdash t_i : \sigma_i$. Thus $t_i^{\theta_l} = t_i^{\theta_r}$ by proposition 45 and we have $\theta_l' = \theta_r'$. We conclude $[\![l^{\theta_l'}\downarrow]\!]_{\rho} >_{\tau} [\![r^{\theta_l'}\downarrow]\!]_{\rho} = [\![r^{\theta_r'}\downarrow]\!]_{\rho}$ because the rule is ϱ-decreasing.
The case $k \neq n$ is similar. □

5 Applying the method

We can apply theorem 49 to show that a given HRS is terminating. For this we need the following ingredients:

- an interpretation b for any base type
- a symbol interpretation ϱ consistent w.r.t. b and
- a proof that each rule is ϱ-decreasing.

Suppose we want to define an enriched λ-calculus $\lambda^{\rightarrow,\times,+}$ with products and coproducts over some set of elementary types.The first problem we have is that this is not quite an HRS, because we have product and coproduct types that can carry function types. The solution is to consider all types of this calculus to be base types and build $\lambda^{\rightarrow}$ on top it. We get a function b from base types to objects in $\mathsf{WO}^M \Downarrow \alpha$ by mapping each elementary type to α (any indecomposable limit ordinal will do) and each type constructor to some functor; in this case we can take the corresponding functors from theorem 23. The rewrite rules for β- and η-reduction of this calculus are the following:

$$\begin{aligned} \mathtt{AP}\,(\mathtt{LAM}\,f)\;a &\rightarrow f\;a \\ \mathtt{LAM}\,(\mathtt{AP}\,f) &\rightarrow f \\ \mathtt{FST}\,(\mathtt{PAIR}\;x\;y) &\rightarrow x \\ \mathtt{SND}\,(\mathtt{PAIR}\;x\;y) &\rightarrow y \\ \mathtt{PAIR}\,(\mathtt{FST}\;p)\,(\mathtt{SND}\;p) &\rightarrow p \end{aligned} \qquad \begin{aligned} \mathtt{CASE}\,(\mathtt{INL}\;x)\;f\;g &\rightarrow f\;x \\ \mathtt{CASE}\,(\mathtt{INR}\;x)\;f\;g &\rightarrow g\;x \\ \mathtt{CASE}\;c\;\begin{matrix}(\lambda x:\sigma.\,f\;(\mathtt{INL}\;x)) \\ (\lambda y:\tau.\,f\;(\mathtt{INR}\;y))\end{matrix} &\rightarrow f\;c \end{aligned}$$

Each rule is only a schema for infinitely many rules of the same shape for each combination of base types, so we have to look for a corresponding schematic interpretation of the schematic symbols. The easiest case are the products as they are purely first-order:

$$\begin{aligned} \varrho(\mathtt{PAIR}_{\sigma,\tau})\;x\;y &= ((1\#\bar{\tau}(y)) \boxplus_\sigma x, (1\#\bar{\sigma}(x)) \boxplus_\tau y) \\ \varrho(\mathtt{FST}_{\sigma,\tau})\;(x,y) &= x \\ \varrho(\mathtt{SND}_{\sigma,\tau})\;(x,y) &= y \end{aligned}$$

The three rules involving products are clearly ϱ-decreasing. The only problem was to define $\varrho(\mathtt{PAIR})$ in such a way that it is a morphism in $\sigma \Rightarrow \tau \Rightarrow (\sigma \sqcap \tau)$.

This was pretty simple but also very typical: to get larger values on the left-hand sides the symbols have to make some "1#_" noise and to deal with erasing rules they have to garbage-collect the erased term using the addition operator.

For function types we do essentially the same thing, but now some meta-level β-reduction can take place:

$$\begin{aligned} \varrho(\mathtt{LAM}_{\sigma,\tau})\;f &= 1 \boxplus_{\sigma\Rightarrow\tau} f \\ \varrho(\mathtt{AP}_{\sigma,\tau})\;f\;a &= f(a) \end{aligned}$$

The η-rule is trivial, the β-rule is only a little bit trickier: $[\![\mathtt{AP}\;(\mathtt{LAM}\;f^\theta)\;a^\theta]\!] = [\![\mathtt{LAM}\;f^\theta]\!]([\![a^\theta]\!]) = (1 \boxplus_{\sigma\Rightarrow\tau} [\![f^\theta]\!])([\![a^\theta]\!]) >_\tau [\![f^\theta]\!]([\![a^\theta]\!]) = [\![(f^\theta\;a^\theta)]\!] \geq_\tau [\![(f^\theta\;a^\theta)\!\downarrow]\!]$. The last step used lemma 41.

Finding the interpretation for the coproduct type is similarly simple, but again we have to be careful to make $\varrho(\mathtt{CASE})$ monotonic by collecting the garbage:

$$\begin{aligned}
\varrho(\mathtt{INL}_{\sigma,\tau})\ x &= i_1(x) \\
\varrho(\mathtt{INR}_{\sigma,\tau})\ y &= i_2(x) \\
\varrho(\mathtt{CASE}_{\sigma,\tau,\nu})\ (i_1(x))\ f\ g &= 1 \boxplus_\nu \overline{\tau \Rightarrow \nu}(g) \boxplus_\nu f(x) \\
\varrho(\mathtt{CASE}_{\sigma,\tau,\nu})\ (i_2(y))\ f\ g &= 1 \boxplus_\nu \overline{\sigma \Rightarrow \nu}(f) \boxplus_\nu g(y)
\end{aligned}$$

Showing that the case-selection rules are ϱ-decreasing is straightforward (as for the function type), but the last rule is a bit more problematic as we have meta-level β-reduction on both sides of the rule. For an arbitrary substitution $\theta(f) = \lambda x : \sigma + \tau.\, t$ we get as interpretation of the left-hand side of rule 8: if $[\![\theta(c)]\!] = i_1(c')$ then $1 \boxplus_\nu \overline{\tau \Rightarrow \nu}(g') \boxplus_\nu \overline{\sigma}(c') \boxplus_\nu [\![t[\mathtt{INL}\ y/x]]\!]_{y \mapsto c'} \geq_\nu 1 \boxplus_\nu [\![t[\mathtt{INL}\ y/x]]\!]_{y \mapsto c'} >_\nu [\![t[\mathtt{INL}\ y/x]]\!]_{y \mapsto c'} = [\![t[c^\theta/x]]\!] = [\![(f\ c)^\theta{\downarrow}]\!]$. Here we used proposition 40 to compose an interpretation with a substitution. The case of the right injection is dual.

6 Comparison to other work

Two other approaches for termination proofs for enriched λ-calculi by semantical methods are closely related to the work presented here.

Robin Gandy presented in [3] a similar solution for termination proofs of enriched λ-calculi, in particular for the given example. However, Gandy's paper appears superficially just as a bag of tricks, it is not quite clear what the underlying semantical framework is. As HRSs had not been invented yet in 1980, the difficulty of providing the syntactical framework for the method can also be felt. For example, Gandy could not treat the η-rule for coproducts with his method, but this failure was merely due to the missing foundations of the method, i.e. it was not obvious though possible. Our paper is based on the same fundamental ideas as Gandy's, but we go much deeper into the analysis of the semantics.

This paper was inspired by Jaco van de Pol's talk at HOA'93 [7] and the difficulties I encountered when trying to apply his method. Van de Pol interprets β-equivalent terms equally. This has the advantage that the method remains solely semantical, while I combine semantic interpretation with substitution. The price he pays for this elegance is a rather ugly semantic domain in which functions are split into three classes of monotonic, strictly monotonic, and "strict" functions. The latter class is particularly problematic as e.g. the identity function on higher types is not "strict", making a categorical treatment rather difficult.

While Gandy indicates how his method could be extended to other parametric types, van de Pol restricts his attention (in [7]; this may have developed since then as he has a joint paper with Schwichtenberg at TLCA'95) to atomic types. Both papers solely dealt with natural numbers as the domain for atomic types. This is tempting but problematic as addition has certain properties on ω (which one likes to exploit) that do not generalise to higher ordinals. Most prominently, the function $\#$ is the smallest strictly monotonic function in $\omega \otimes \omega \to \omega$, but it loses this property on larger ordinals.

7 Conclusion and further work

We have presented another semantic approach to termination proofs for higher-order term rewriting systems with special emphasis on enriched λ-calculi. The application of the method is fairly simple and one does not have to understand why the method works to apply it to an example. In particular, we have a number of criteria that allow us to deal with parametric types: any functor that is "retractable" at certain ordinals and "$\boldsymbol{M}$-distributive" is a candidate for the interpretation of parametric types.

However, the emphasis of this paper as the title should indicate was more to establish the abstract framework in which the method (or even different methods) can be applied rather than showing how this is done. We looked at products and coproducts, but a similar treatment for all limits and colimits (and thus co-inductive and inductive data types in ML-style) should be possible analogously. The method as presented here does not apply easily to more complicated parameterisation principles, e.g. to show termination of the calculus of constructions or System F. Functors only correpond to parametric types, not to dependent types. Thus one has to use a more sophisticated semantic construct than functors to mirror the dependency of the syntax.

Acknowledgements

People who helped developing this paper by providing feedback, advice, etc. include: Randall Dougherty, Marcello Fiore, Philippa Gardner, James McKinna, Jaco van de Pol, John Power, Alex Simpson and Judith Underwood. I would also like to thank the referees for some valuable comments and corrections.

The research reported here was supported by SERC grant GR/J07303.

References

1. H. P. Barendregt. Lambda calculi with types. In *Handbook of Logic in Computer Science, Vol.2*, pages 117–309. Oxford Science Publications, 1992.
2. M. C. F. Ferreira and H. Zantema. Total termination of term rewriting. In *Rewriting Techniques and Applications*, pages 213–227, 1993. LNCS 690.
3. R. Gandy. Proofs of strong normalization. In J. Seldin and J. Hindley, editors, *To H.B. Curry: Essays on Combinatory Logic, Lambda Calculus and Formalism*, pages 457–477. Academic Press, 1980.
4. G. Hessenberg. Grundbegriffe der Mengenlehre. *Abhandlungen der Fries'schen Schule*, pages 479–706, 1906.
5. S. MacLane. *Categories for the Working Mathematician*. Springer, 1971.
6. J. v. Neumann. Zur Einführung der transfiniten Zahlen. *Acta litterarum ac scientarum*, 1:199–208, 1923.
7. J. v. d. Pol. Termination proofs for higher-order rewrite systems. In *Higher-Order Algebra, Logic, and Term Rewriting*, pages 305–325, 1993. LNCS 816.
8. W. Sierpiński. *Cardinal and Ordinal Numbers*. Polish Scientific Publishers, 1965.
9. D. Wolfram. *The Clausal Theory of Types*. Cambridge University Press, 1993.

Higher-Order Rewrite Systems

Tobias Nipkow
Institut für Informatik
TU München
Germany

Abstract. Higher-order rewrite systems are an extension of term-rewriting to (simply typed) λ-terms. This yields a formalism with a built-in notion of variable binding (λ-abstraction) and substitution (β-reduction) which is capable of describing the manipulation of terms with bound variables. Typical examples are various λ-calculi themselves, logical formulae, proof terms and programs.
In this talk we survey the growing literature in this area with an emphasis on confluence results. In particular we examine some recent results by van Oostrom on very general confluence criteria which go beyond orthogonality and even generalize a result by Huet on parallel reduction between critical pairs. We also present a new critical pair condition for higher-order rewrite systems with arbitrary left-hand sides.

Infinitary lambda calculi and Böhm models

Richard Kennaway[1], Jan Willem Klop[2], Ronan Sleep[1] and Fer-Jan de Vries[3] *

[1] School of Information Systems, University of East Anglia, Norwich NR4 7TJ, UK
[2] Department of Software Technology, CWI, P.O. Box 94079, 1090 GB Amsterdam, The Netherlands
[3] NTT Communication Science Laboratories, Hikaridai, Seika-cho, Soraku-gun, Kyoto 619-02, Japan

Abstract. In a previous paper we have established the theory of transfinite reduction for orthogonal term rewriting systems. In this paper we perform the same task for the lambda calculus. This results in several new Böhm-like models of the lambda calculus, and new descriptions of existing models.

1 Introduction

Infinitely long rewrite sequences of possibly infinite terms are of interest for several reasons. Firstly, infinitary rewriting is a natural generalisation of finitary rewriting which extends it with the notion of computing towards a possibily infinite limit. Such limits naturally arise in the semantics of lazy functional languages, in which it is possible to write and compute with expressions which intuitively denote infinite data structures, such as a list of all the integers. If the limit of a reduction sequence still contains redexes, then it is natural to consider sequences whose length is longer than ω — in fact, sequences of any ordinal length. The question of the computational meaning of such sequences will be dealt with below. Secondly, computations with terms implemented as graphs allow the possibility of using cyclic graphs, which correspond in a natural way to infinite terms. Finite computations on cyclic graphs correspond to infinite computations on terms. Finally, the infinitary theory suggests new ways of dealing with some of the concepts that arise in the finitary theory, such as notions of undefinedness of terms. In this connection, Berarducci and Intrigila ([Ber, BI94]) have independently developed an infinitary lambda calculus and applied it to the study of consistency problems in the finitary lambda calculus.

In [KKSdV-] we developed the basic theory of transfinite reduction for orthogonal term rewrite systems. In this paper we perform the same task for the

* Email addresses: jrk@sys.uea.ac.uk, jwk@cwi.nl, mrs@sys.uea.ac.uk, and ferjan@cwi.nl. The authors were partially supported by SEMAGRAPH II (ESPRIT working group 6345). Richard Kennaway was also supported by a SERC Advanced Fellowship, and by SERC Grant no. GR/F 91582. From July 1995 Fer-Jan de Vries will be at the Hitachi Advanced Research Laboratory, Hatoyama, Saitama 350-03, Japan.

lambda calculus. In contrast to the situation for term rewriting, in lambda calculus there turn out to be several different possible domains of infinite terms which one might study. These give rise to different Böhm-like models of the calculus.

2 Basic definitions

2.1 Finitary lambda calculus

We assume familiarity with the lambda calculus, or as we shall refer to it here, the finitary lambda calculus. [Bar84] is a standard reference. The syntax is simple: there is a set Var of variables; an expression or term E is either a variable, an abstraction $\lambda x.E$ (where x is called the bound variable and E the body), or an application E_1E_2 (where E_1 is called the rator and E_2 the rand). This is the pure lambda calculus — we do not have any built-in constants nor any type system.

As customary, we identify α-equivalent terms with each other, and consider bound variables to be silently renamed when necessary to avoid name clashes.

2.2 What is an infinite term?

Drawing lambda expressions as syntax trees gives an immediate and intuitive notion of infinite terms: they are just infinite trees. Formally, we can define this set as the metric completion of the space of finite trees with a well-known (ultra-)metric. The larger the common prefix of two trees, the more similar they are, and the closer together they may be considered to be. First, some terminology. A *position* or *occurrence* is a finite string of positive integers. Given a term M and a position u, the term $M|u$, when it exists, is a subterm of M defined inductively thus:

$$\begin{aligned} M|\langle\rangle &= M \\ (\lambda x.M)|1\cdot u &= M|u \\ (MN)|1\cdot u &= M|u \\ (MN)|2\cdot u &= N|u \end{aligned}$$

$M|u$ is called the subterm of M at u, and when this is defined, u is called a position of M. The *syntactic depth* of u is its length.

Two positions u and v are *disjoint* if neither is a prefix of the other. Two redexes are disjoint if their positions are. A set of positions or redexes is disjoint if every two distinct members are.

Given two distinct terms M and N, let l be the length of the shortest position u such that $M|u$ and $N|u$ are both defined, and are either of different syntactic types or are distinct variables. Then the larger l is, the more similar are M and N. The distance between M and N is defined to be 2^{-l}. Denote this measure by $d^s(M,N)$. $d^s(M,M)$ is defined to be 0. This is the *syntactic metric*. It is easily proved that it is a metric on the set of finite terms. In fact, it is an ultrametric, i.e. $d^s(M,N) \leq max(d^s(M,P), d^s(P,N))$, although this will not be important. The completion of this metric space adds the infinite terms. We call this set Λ^s.

The above is the definition of infinite terms which we used in our study of transfinite term rewriting, but for lambda calculus the situation is a little more complicated. Consider the term $(((\ldots I)I)I)I$ where $I = \lambda x.x$. See Fig. 1. This

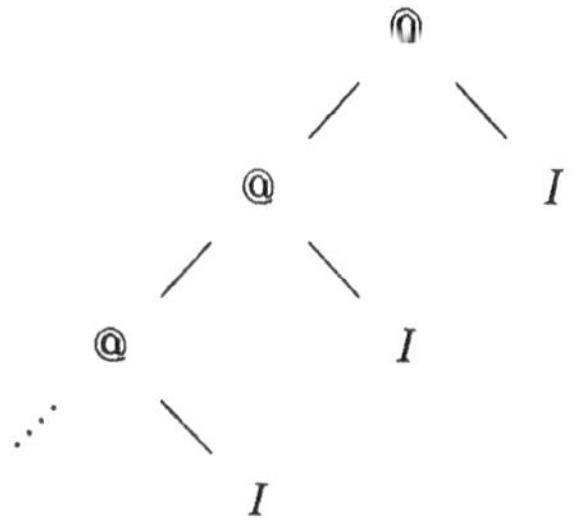

Fig. 1.

term has a combination of properties which is rather strange from the point of view of finitary lambda calculus. By the usual definition of head normal form — being of the form $\lambda x_1 \ldots \lambda x_n.yt_1 \ldots t_m$ — it is not in head normal form. By an alternative formulation, trivially equivalent in the finitary case, it is in head normal form — it has no head redex. It is also a normal form, yet it is unsolvable (that is, there are no terms $N_1, \ldots, N_n$ such that $MN_1 \ldots N_n$ reduces to I). The problem is that application is strict in its first argument, and so an infinitely left-branching chain of applications has no obvious meaning. We can say much the same for an infinite chain of abstractions $\lambda x_1.\lambda x_2.\lambda x_3.\ldots$.

Another reason for reconsidering the definition of infinite terms arises from analogy with term rewriting. In a term such as $F(x,y,z)$, the function symbol F is at syntactic depth 0. If it is curried, that is, represented as $Fxyz$, or explicitly $@(@(@(F,x),y),z)$ (as it would be if we were to translate the term rewrite system into lambda calculus), the symbol F now occurs at syntactic depth 3. We could instead consider it to be at depth zero; more generally, we can define a new measure of depth which deems the left argument of an application to be at the same depth as the application itself, and the body of an abstraction to be at the same depth as the abstraction.

Definition 1. Given a term M and a position u of M, the *applicative depth* of the subterm of M at u, if it exists, is defined by:

$$\begin{aligned} D^a(M, \langle\rangle) &= 0 \\ D^a(\lambda x.M, 1 \cdot u) &= D^a(M, u) \\ D^a(MN, 1 \cdot u) &= D^a(M, u) \\ D^a(MN, 2 \cdot u) &= 1 + D^a(N, u) \end{aligned}$$

The associated measure of distance is denoted d^a, and the space of finite and infinite terms Λ^a.

In general, we can specify for each of the three contexts $\lambda x.[]$, $[]M$, and $M[]$ whether the depth of the hole is equal to or one greater than the depth of the whole expression. Syntactic depth sets all three equal to 1. For applicative depth, the three depths are 0, 0, and 1 respectively. This suggests a general definition.

Definition 2. Given a term M a position u of M, and a string of three binary digits abc, there is an associated measure of depth D^{abc}:

$$
\begin{aligned}
D^{abc}(M, \langle\rangle) &= 0 \\
D^{abc}(\lambda x.M, 1 \cdot u) &= a + D^{abc}(M, u) \\
D^{abc}(MN, 1 \cdot u) &= b + D^{abc}(M, u) \\
D^{abc}(MN, 2 \cdot u) &= c + D^{abc}(N, u)
\end{aligned}
$$

The associated measure of distance is denoted d^{abc} and the space of finite and infinite terms Λ^{abc}.

We write Λ^∞, D, or d when we do not need to specify which space of infinite terms, measure of depth, or metric we are referring to.

We have already seen that $d^s = d^{111}$ and $d^a = d^{001}$. Some of the other measures also have an intuitive significance. d^{101} (*weakly applicative* depth, or d^w) may be associated with the lazy lambda calculus [AO93], in which abstraction is considered lazy — $\lambda x.M$ is meaningful even when M is not. Denote the corresponding set of finite and infinite terms by Λ^w. d^{000} is the discrete metric, the trivial notion in which the depth of every subterm of a term is zero. This gives the discrete metric space of finite terms, no infinite terms, and no reduction sequences converging to infinite terms — the usual finitary lambda calculus.

Many of our results will apply uniformly to all eight infinitary lambda calculi, and we will only specify the depth measure when necessary. In the final section we will find that two of them — Λ^{010} and Λ^{011} — have unsatisfactory technical properties.

Lemma 3. *Considered as a set, Λ^{abc} is the subset of Λ^{111} consisting exactly of those terms which do not contain an infinite sequence of nodes in which each node is at the same abc-depth as its parent. (Its metric and topology are not the subspace metric and topology.)* □

Both Λ^s and Λ^w contain unsolvable normal forms, such as $\lambda x_1.\lambda x_2.\lambda x_3\ldots$ In Λ^a every normal form is solvable.

2.3 What is an infinite reduction sequence?

We have spoken informally of convergent reduction sequences but not yet defined them. The obvious definition is that a reduction sequence of length ω converges if the sequence of terms converges with respect to the metric. However, this proves to be an unsatisfactory definition, for the same reasons as in [KKSdV–]. There are two problems. Firstly, a certain property which is important for attaching computational meaning to reduction sequences longer than ω fails.

Definition 4. A reduction system admitting transfinite sequences satisfies the *Compression Property* if for every reduction sequence from a term s to a term t, there is a reduction sequence from s to t of length at most ω.

A counterexample to the Compression Property is easily found in Λ^s. Let $A_n = (\lambda x.A_{n+1})(B^n(x))$ and $B = (\lambda x.y)z$. Then $A_0 \to^\omega C$ where $C = (\lambda x.C)$ (B^ω), and $C \to (\lambda x.C)(yB^\omega)$. A_0 cannot be reduced to $(\lambda x.C)(yB^\omega)$ in ω or fewer steps. (We do not know if the Compression Property holds for the above notion of convergence in Λ^a or Λ^w.)

The second difficulty with this notion of convergence is that taking the limit of a sequence loses certain information about the relationship between subterms of different terms in the sequence. Consider the term I^ω of Λ^a, and the infinite reduction sequence starting from this term which at each stage reduces the outermost redex: $I^\omega \to I^\omega \to I^\omega \to \ldots$ All the terms of this sequence are identical, so the limit is I^ω. However, each of the infinitely many redexes contained in the original term is eventually reduced, yet the limit appears to still have all of them. It is not possible to say that any redex in the limit term arises from any of the redexes in the previous terms in the sequence.

A third difficulty arises when we consider translations of term rewriting systems into the lambda calculus. Even when such a translation preserves finitary reduction, it may not preserve Cauchy convergent reduction. Consider the term rewrite rule $A(x) \to A(B(x))$. This gives a Cauchy convergent term rewrite sequence $A(C) \to A(B(C)) \to A(B(B(C)))\ldots$ If one tries to translate this by defining $A_\lambda = Y(\lambda f.\lambda x.f(Bx))$ (for some λ-term B), where Y is Church's fixed point operator $\lambda f.(\lambda x.f(xx))(\lambda x.f(xx))$, then the resulting sequence will have an accumulation point corresponding to the term $A(B^\omega)$, but will not be Cauchy convergent. The reason is that what is a single reduction step in the term rewrite system becomes a sequence of several steps in the lambda calculus, and while the first and last terms of that sequence may be very similar, the intermediate terms are not, destroying convergence.

The remedy for all these problems is the same as in [KKSdV–]: besides requiring that the sequence of terms converges, we also require that the depths of the redexes which the sequence reduces must tend to infinity.

Definition 5. A *pre-reduction* sequence of length α is a function ϕ from an ordinal α to reduction steps of Λ^∞, and a function τ from $\alpha + 1$ to terms of Λ^∞, such that if $\phi(\beta)$ is $a \to^r b$ then $a = \tau(\beta)$ and $b = \tau(\beta + 1)$. Note that in a pre-reduction sequence, there need be no relation between the term $\phi(\beta)$ and any of its predecessors when β is a limit ordinal.

A pre-reduction sequence is a *Cauchy convergent reduction sequence* if τ is continuous with respect to the usual topology on ordinals and the metric on Λ^∞.

It is a *strongly convergent reduction sequence* if it is Cauchy convergent and if, for every limit ordinal $\lambda \leq \alpha$, $\lim_{\beta\to\lambda} d_\beta = \infty$, where d_β is the depth of the redex reduces by the step $\phi(\beta)$. (The measure of depth is the one appropriate to each version of Λ^∞.)

If α is a limit ordinal, then an *open* pre-reduction sequence is defined as above, except that the domain of τ is α. If τ is continuous, the sequence is *Cauchy continuous*, and if the condition of strong convergence is satisfied at each limit ordinal less than α, it is *strongly continuous*.

When we speak of a reduction sequence, we will mean a strongly continuous reduction sequence unless otherwise stated. Different measures of depth give different notions of strong continuity and convergence.

3 Descendants and residuals

3.1 Descendants

When a reduction $M \to N$ is performed, each subterm of M gives rise to certain subterms of N — its descendants — in an intuitively obvious way. Everything works in almost exactly the same way as for finitary lambda calculus.

Definition 6. Let u be a position of t, and let there be a redex $(\lambda x.M)N$ of t at v, reduction of which gives a term t'. The set of *descendants* of u by this reduction, u/v, is defined by cases.

- If $u \not\geq v$ then $u/v = \{u\}$.
- If $u = v$ or $u = v \cdot 1$ then $u/v = \emptyset$.
- If $u = v \cdot 2 \cdot w$ then $u/v = \{v \cdot y \cdot w \mid y \text{ is a free occurrence of } x \text{ in } M\}$. If $u = v \cdot 1 \cdot w$ then $u/v = \{v \cdot w\}$.

The *trace* of u by the reduction at v, $u/\!/v$, is defined in the same way, except for the second case: if $u = v$ or $u = v \cdot 1$ then $u/\!/v = \{v\}$.

For a set of positions U, $U/v = \bigcup\{u/v \mid u \in U\}$ and $U/\!/v = \bigcup\{u/\!/v \mid u \in U\}$.

The notions of descendant and trace can be extended to reductions of arbitrary length, but first we must define the notion of the limit of an infinite sequence of sets.

Definition 7. Let $S = \{S_\beta \mid \beta < \alpha\}$ be a sequence of sets, where α is a limit ordinal. Define

$$\liminf S = \bigcup_{\beta \to \alpha} \bigcap_{\beta < \gamma < \alpha} S_\gamma \qquad \limsup S = \bigcap_{\beta \to \alpha} \bigcup_{\beta \leq \gamma \leq \alpha} S_\gamma$$

When $\liminf S = \limsup S$, write $\lim S$ or $\lim_{\beta \to \alpha} S_\beta$ for both.

Definition 8. Let U be a set of positions of t, and let S be a reduction sequence from t to t'. For a reduction sequence of the form $S \cdot r$ where r is a single step, $U/(S \cdot r) = (U/S) \cdot r$. If the length of S is a limit ordinal α then $U/S = \lim_{\beta \to \alpha} U/S_\beta$.

$U/\!/S$ is defined similarly.

Strong convergence of S ensures that the above limit exists.

Lemma 9. *Let U be a set of positions of redexes of t, and let S be a reduction from t to t'. Then there is a redex at every member of U/S.* □

Definition 10. The redexes at U/S in the preceding lemma are the *residuals* of the redexes at U.

Definition 11. Let u and v be positions in the initial and final terms respectively of a sequence S. If $v \in u//S$, we also say that u *contributes* to v (via S). If there is a redex at v, then u *contributes* to that redex if u contributes to v or $v \cdot 1$.

We do not define descendants, traces, residuals, and contribution for Cauchy convergent reductions, which is not surprising given the examples of section 2.3.

Theorem 12. *For any strongly convergent sequence $t_0 \to^{\alpha} t_{\alpha}$ and any position u of t_{α}, the set of all positions of all terms in the sequence which contribute to u is finite, and the set of all reduction steps contributing to u is finite.*

Proof. For each t_β in the sequence, we construct the set U_β of positions of t_β contributing to u, and prove that it is finite. We also show that there are only finitely many different such sets, hence their union is finite.

Suppose $U_{\beta+1}$ is finite, and $t_\beta \to t_{\beta+1}$ reduces a redex at position v. Let $w \in U_{\beta+1}$. If w and v are disjoint, or $w < v$, then w is the only position of t_β contributing to v in $t_{\beta+1}$. If $w = v$, then v, $v \cdot 1$, $v \cdot 1 \cdot 1$, and possibly $v \cdot 2$ (if the redex has the form $(\lambda x.x)N$) are the only such positions. If $w > v$, and the redex at v is $(\lambda x.M)N$, then there is a unique position in either M or N which contributes to w. In each case, the set of positions is finite, hence U_β, which is the union of those sets for all $w \in U_{\beta+1}$, is finite.

Suppose U_β is defined and finite for a limit ordinal β. By strong convergence and the finiteness of U_β, there is a final segment of $t_0 \to^{\beta} t_\beta$, say from t_γ to t_β, in which every step is at a depth more than 2 greater than the depth of every member of U. It follows that each U_δ for $\gamma \leq \delta < \beta$ is equal to U_β, and is therefore finite.

Finitely many repetitions of the above argument suffice to calculate U_β for all β, demonstrating that there are only finitely many different such sets, and all of them are finite.

Each reduction step contributing to u takes place at a prefix of a position in some U_β. By strong convergence, only finitely many steps can take place at any one position, therefore there are only finitely many such steps. □

3.2 Developments

Definition 13. A *development* of a set of redexes R of a term M is a sequence in which every step reduces some residual of some member of R by the previous steps of the sequence. It is *complete* if it is strongly convergent and the final term contains no residual of any member of R.

Not every set of redexes has a complete development. In Λ^{--1}, an example is the term $I^{\omega} = (\lambda x.x)((\lambda x.x)((\lambda x.x)(\ldots)))$. Every attempt to reduce all the redexes in this term must give a reduction sequence containing infinitely many reduction steps at the root of the term, which, by every notion of depth, is not strongly convergent. Note that the set consisting of every redex at odd syntactic depth has a complete development, as does the set consisting of every redex at even syntactic depth, but their union does not. In every other version of Λ^{∞} except 000 (the finitary calculus) the term $(\lambda x.((\lambda x.((\lambda x.(\ldots))z))z))z$ behaves in a similar manner.

Theorem 14. *Complete developments of the same set of redexes end at the same term.*

Proof. (Outline.) In the finitary case one proves this by showing that (1) it is true for a set of pairwise disjoint redexes, (2) it is true for any pair of redexes, and (3) all developments are finite. The result then follows by an application of Newman's Lemma.

In the infinitary case, (1) and (2) are still true, and indeed obvious, but (3) is of course false. The situation is complicated by the fact that a set of redexes can have a strongly convergent complete development without all its developments being strongly convergent.

One proceeds instead by picking out one particular development of the given set of redexes, analogous to the "standard" development defined in finitary rewriting, such that the set has a strongly convergent complete development if and only if its standard development is complete. Properties of the standard development then allow one to use (1) and (2) to construct a "tiling diagram" for the standard development and any other complete development, and to show that the right and bottom edges of the diagram are empty. This shows that they converge to the same limit. □

4 The truncation theorem

Some results about the finitary lambda calculus can be transferred to the infinitary setting by using finite approximations to infinite terms.

Definition 15. A $\Lambda_{\perp}$ term is a term of the version of lambda calculus obtained by adding $\perp$ as a new symbol. $\Lambda_{\perp}^{\infty}$ is defined from $\Lambda_{\perp}$ as Λ^{∞} is from Λ.

The terms of $\Lambda_{\perp}^{\infty}$ have a natural partial ordering, defined by stipulating that $\perp \leq t$ for all t, and that application and abstraction are monotonic.

A *truncation* of a term t is any term t' such that $t' \leq t$. We may also say that t' is weaker than t, or t is stronger than t'.

Theorem 16. *Let $t_0 \rightarrow^{\alpha} t_{\alpha}$ be a reduction sequence. Let s_{α} be a prefix of t_{α}, and for $\beta < \alpha$, let s_{β} be the prefix of t_{β} contributing to s_{α}. Then for any term r_0 such that $s_0 \leq r_0$ there is a reduction sequence $r_0 \rightarrow^{\leq\alpha} r_{\alpha}$ such that:*

1. *For all β, s_{β} is a prefix of r_{β}.*

2. *If $t_\beta \to t_{\beta+1}$ is performed at position u and contributes to s_α, then $r_\beta \to r_{\beta+1}$ by reduction at u.*
3. *If $t_\beta \to t_{\beta+1}$ is performed at position u and does not contribute to s_α, then $r_\beta = r_{\beta+1}$.* □

As an example of the use of this theorem, we demonstrate that Λ^∞ is conservative over the finitary calculus, for terms having finite normal forms.

Corollary 17. *If $t \to^\infty s$ and s' is a finite prefix of s, then t is reducible in finitely many steps to a term having s' as a prefix. In particular, if t is reducible to a finite term, it is reducible to that term in finitely many steps.*

Proof. From Theorems 16 and 12. □

Corollary 18. *If a finite term is reducible to a finite normal form, it is reducible to that normal form in the finitary lambda calculus.* □

5 The Compressing Lemma

One of our justifications for the interest of infinite terms and sequences is to see them as limits of finite terms and sequences. From this point of view, the computational meaning may be obscure of a sequence of length longer than ω — which performs an infinite amount of work and then doing some more work. We therefore wish to be assured that every reduction sequence of length greater than ω is equivalent to one of length no more than ω, in the sense of having the same initial and final term. This allows us to freely use sequences longer than ω without losing computational relevance.

Theorem 19. (Compressing Lemma.) *In Λ^∞, for every strongly convergent sequence there is a strongly convergent sequence with the same endpoints whose length is at most ω.*

Proof. The corresponding theorem of [KKSdV–] shows that the case of a sequence of length $\omega+1$ implies the whole theorem, and the proof is not dependent on the details of rewriting — it is valid for any abstract transfinite reduction system (as defined in [Ken92]).

Suppose we have a reduction of the form $S_{\omega+1} = s_0 \to^\omega s_\omega \to_d s_{\omega+1}$, where the final step rewrites a redex at depth d. By strong convergence of the first ω steps, the sequence must have the form $s_0 \to^* C[(\lambda x.M)N, M_1, \ldots, M_n] \to^\omega_{d+1} C[(\lambda x.M')N', M'_1, \ldots, M'_n] \to_d C[M'[x := N']]$. where the context $C[\ldots]$ is a prefix of every term of the sequence from some point onwards, and all its holes are at depth d. The reduction of $C[(\lambda x.M)N, M_1, \ldots, M_n]$ to $C[(\lambda x.M')N', M'_1, \ldots, M'_n]$ consists of an interleaving of reductions of M to M', N to N', and each M_i to M'_i of length at most ω. Conversely, any reductions of lengths at most ω starting from M, N, and each M_i can be interleaved to give a reduction of length at most ω starting from $C[(\lambda x.M)N, M_1, \ldots, M_n]$. The theorem will therefore be established if, given reductions of M to M' and N to N' of length at most ω, we can construct a reduction from $(\lambda x.M)N$ to $M'[x := N']$ of length at most ω. This can be done by first reducing $(\lambda x.M)N$ to $M[x := N]$, and then interleaving a reduction of M to M' and reductions of all the copies of N to N' in a strongly convergent way. The details are simple to work out. □

Remark. The Compressing Lemma is false for $\beta\eta$-reduction. For a counterexample, let $M = Y(\lambda f.\lambda x.I(fx))$. Then $\lambda x.Mxx \to^{\omega} \lambda x.I(I(I(...)))x \to_{\eta} I(I(I(...)))$. However, $\lambda x.Mxx$ is not reducible in ω steps or fewer to $I(I(I(...)))$.

This is not surprising. The η-rule requires testing for the absence of the bound variable in the body of the abstraction; if the abstraction is infinite, this is an infinite task, and such discontinuities are to be expected.

6 Head normal forms and Böhm trees

In the context of infinitary lambda calculus, the Böhm tree of a term can be seen as being simply its normal form with respect to transfinite reduction with respect to the β rule together with an additional rule for erasing subterms having no head normal form. More generally, we find that with each notion of depth there can naturally be associated a notion of head normal form. The classical notion of head normal form is, in the current setting, a term having no redexes at applicative depth 0. A weak head normal form is a term having no redexes at weakly applicative depth 0. A normal form is a term having no redexes at discrete depth 0 (i.e. no redexes anywhere).

Definition 20. A term of Λ^{∞} is *0-stable* if it cannot be reduced to a term having a redex at depth 0. It is *potentially* 0-stable if it can be reduced to a 0-stable term. It is *0-active* if it is not potentially 0-stable.

We shall demonstrate that for six of the eight notions of depth, the class of 0-active terms satisfies the axioms of [AKK⁺94] for a set of undefined terms. These axioms are (1) both the set and its complement are closed under reduction, and (2) the set includes all the terms which cannot be reduced to root-stable form. (A root-stable term is one which cannot be reduced to a redex, i.e. is 0-stable with respect to syntactic depth.) This immediately gives rise to a number of Böhm-like models of lambda-calculus, in which the value of a term is its unique normal form with respect to a notion of reduction which allows 0-active terms to be replaced by a symbol $\perp$.

The second of the axioms is immediate from the definition. If a term cannot be reduced to root-stable form, then it cannot be reduced to a 0-stable form, since a redex at the root of a term is at depth 0 for every notion of depth.

Half of the first axiom is immediate: the set of 0-active terms is certainly closed under reduction. It only remains to prove that the set of potentially 0-stable terms is also closed. To do this we must develop some theory of Böhm reduction.

Definition 21. *Böhm reduction* is reduction in $\Lambda_{\perp}^{\infty}$ by the β rule and the $\perp$ rule, viz. $M \to \perp$ if M is 0-active and not $\perp$. We write $\to_{\mathcal{B}}$ for Böhm reduction and $\to_{\perp}$ for reduction by the $\perp$-rule alone.

A *Böhm tree* is a normal form of $\Lambda_{\perp}^{\infty}$ with respect to Böhm reduction.

We extend the notions of 0-stability etc. to terms containing $\perp$ thus. A term of $\Lambda_{\perp}^{\infty}$ is 0-stable if it cannot be reduced to a term containing a Böhm redex or an occurrence of $\perp$ at depth 0. Potential 0-stability and 0-activeness are similarly extended.

0-stability and 0-activeness were defined in terms of reduction, but now we have defined a new notion of reduction in terms of these concepts, which in turn gives us new notions of 0-stability and 0-activeness. It is important to check that the new notions agree with the old on terms of Λ^{∞}. This turns out not to be the case for two of the eight possible notions of depth, which we regard as sufficient grounds for excluding them.

Theorem 22. *A term of Λ^{∞} is 0-stable with respect to beta reduction if and only if it is 0-stable with respect to Böhm reduction.*

Proof. We shall write $0\mathcal{B}$-stable to abbreviate 0-stable with respect to Böhm reduction. Let t be a term of Λ^{∞}. Clearly, if t is beta-reducible to a term containing a 0-redex, the same reduction sequence is a Böhm reduction to a term containing a 0-redex. Hence if t is not 0-stable, it is not $0\mathcal{B}$-stable.

Conversely, suppose t is Böhm reducible to a term s which contains either a redex or an occurrence of $\perp$ at depth 0. Omitting all the reduction steps by the $\perp$ rule in the obvious way yields a beta reduction of t to a term r which differs from s only by having 0-active subterms where s has occurrences of $\perp$. If s contains a 0-redex, then so does r. If s has an occurrence of $\perp$ at depth 0, r has a 0-active subterm at depth 0. That subterm is reducible to a term containing a 0-redex, which gives a reduction of r to a term containing a 0-redex. Hence t is not 0-stable. □

Theorem 23. *Except in Λ^{010} and Λ^{011}, a term of Λ^{∞} is potentially 0-stable with respect to beta reduction if and only if it is potentially 0-stable with respect to Böhm reduction.*

Proof. For potential 0-stability, one direction is again trivial. If t is potentially 0-stable, then it is potentially $0\mathcal{B}$-stable. Conversely, suppose that t is potentially $0\mathcal{B}$-stable. Then it is Böhm reducible to a $0\mathcal{B}$-stable term s. By the same transformation used previously, there is a beta reduction of t to a term r which differs from s only by having 0-active terms where r has $\perp$. Suppose r is not 0-stable. Then r is beta reducible to a term r' containing a 0-redex, and therefore (since the property of having a 0-redex is determined by some finite prefix) it is so reducible in finitely many steps. We shall construct a Böhm reduction of $s \to_{\mathcal{B}}^{*} s'$ which imitates the beta reduction of r.

Suppose we have reached a step $r_i \to r_{i+1}$ in the reduction of r, and the reduct of s corresponding to r_i is s_i. The condition on the depth measure says that either the depth of M in $\lambda x.M$ is 1 or the depth of M in MN is 0.

If the former is true, we take as an inductive hypothesis that r_i differs from s_i only in that r_i has 0-active subterms where s_i has occurrences of $\perp$. If the redex reduced in r_i is also present in s_i, we reduce it in s_i to obtain s_{i+1}. If the redex is in a 0-active subterm of r_i corresponding to an occurrence of $\perp$ in s_i, then we ignore it and take $s_{i+1} = s_i$. The only other possibility is where the redex is a subterm $(\lambda x.M)N$ of r_i corresponding to a subterm $\perp N'$ of s_i. But the hypothesis about the depth measure implies that $\lambda x.M$ is 0-stable, therefore by the inductive hypothesis cannot correspond to an occurrence of $\perp$ in s_i. Thus the construction can be carried out along the whole sequence, yielding a reduction of s to a term s' such that wherever s' has an occurrence of $\perp$, r' has a 0-active term. Since r' has a 0-redex, s' must have either a beta redex or an occurrence of $\perp$ at depth 0, hence s' is not $0\mathcal{B}$-stable.

Otherwise, the depth of M in $\lambda x.M$ and in MN is 0. Here the argument is similar, but now we cannot exclude the third case, where a redex $(\lambda x.M)N$ in r_i corresponds to a subterm $\perp N'$ in s_i. We deal with this by relaxing the inductive hypothesis, to require that r_i differs from s_i in having 0-active subterms where s_i has subterms of the form $\perp N_1 \dots N_n$, where $n \geq 0$. We can then carry through a similar construction, to obtain a Böhm reduction of s to s', where s' stands in that relation to r'. Since r' has a 0-redex, s' has either a beta redex or a subterm of the form $\perp N_1 \dots N_n$ at depth zero. But in the latter case, the occurrence of $\perp$ is also at depth 0, so s' is not $0\mathcal{B}$-stable. □

These theorems allow us to drop the notation $0\mathcal{B}$-stable, and to speak (potential) 0-stability and 0-activeness with respect to beta reduction or Böhm reduction interchangeably.

The two depth measures which Theorem 23 excludes are those in which the depth of M in $(\lambda x.M)N$ is 0 and in MN is 1. These appear intuitively to be unnatural. One may associate depth with strictness. These two measures regard abstraction as strict, and application as non-strict in its first argument. The rest of this section deals only with depth measures to which Theorem 23 applies.

Lemma 24. *1. For any Böhm reduction sequence $t \to_{\mathcal{B}}^{\infty} t'$, there are sequences $t' \to_{\perp}^{\infty} t''$ and $t \to_{\mathcal{B}}^{\infty} t''$, such that the latter sequence consists of alternating segments in which first a reduction is performed, no step of which is contained in any 0-active subterm, and then a reduction to normal form with respect to the $\perp$ rule is performed.*

2. For any Böhm reduction sequence $t \to_{\mathcal{B}}^{\infty} t'$, there is a sequence $t \to_{\beta}^{\infty} t'' \to_{\perp}^{\infty} t'$.

Proof. (Outline.) For the first, the proof consists in showing how any Böhm reduction sequence can be transformed step by step into the required form, by inserting $\perp$-reductions as necessary. For the second, the proof proceeds by a step-by-step transformation which postpones all $\perp$-reductions until the end. □

Theorem 25. *Böhm reduction is Church-Rosser.*

Proof. (Outline.) Given two coinitial Böhm reduction sequences, we transform them as described by Lemma 24(1). For sequences of that form, the Church-Rosser property can be proved by a tiling argument analogous to that commonly used in proving the finitary Church-Rosser property. From this the Church-Rosser property for arbitrary Böhm reductions follows. □

Theorem 26. *The set of potentially 0-stable terms is closed under reduction.*

Proof. If t is 0-active, it has the Böhm normal form $\perp$. Suppose t reduces to a non-0-active term t'. Then t' reduces to a 0-stable term, which cannot be Böhm reduced to $\perp$. But by the Church-Rosser property for Böhm reduction, this is impossible. □

Theorem 27. *Every term has exactly one Böhm normal form.*

Proof. From the previous theorem, every term has at most one Böhm normal form.

Given any term t, construct a Böhm reduction from t thus. If t is 0-active, reduce it to $\perp$. Otherwise, it is reducible to a 0-stable term. Perform such a reduction, and then repeat this construction on the maximal subterms of the term at depth 1. This generates a strongly convergent sequence whose limit contains no Böhm redexes. □

Theorem 28. *Beta reduction is Church-Rosser up to identification of 0-active terms.*

Proof. Given two beta reductions $t \rightarrow^{\infty} t_0$ and $t \rightarrow^{\infty} t_1$, Theorem 25 gives Böhm reductions $t_0 \rightarrow^{\infty}_{\mathcal{B}} t_2$ and $t_1 \rightarrow^{\infty}_{\mathcal{B}} t_2$. By Lemma 24(2), there are beta-reduct of t_0 and t_1 which are $\perp$-reducible to t_2; such reducts are identical up to identification of 0-active terms. □

We thus have a model of lambda calculus, where the objects are the Böhm normal forms, ordered according to Def. 15. The usual Böhm model is the model associated with applicative depth. The larger model described by Berarducci ([Ber]) is the one associated with syntactic depth. In this model the 0-stable terms are the root-stable terms, and the 0-active terms are the terms which Berarducci calls mute. The Böhm model for weakly applicative depth is related to Ong and Abramsky's models for lazy lambda calculus [AO93]. Discrete depth results in the trivial model (since the 0-active terms are the terms with no normal form, and identifying all of these together results in the equality of all terms). The other two depth measures which satisfy the conditions of Theorem 23 give two more models, whose relation to existing models of the lambda calculus remains to be studied.

References

[AKK+94] Z.M. Ariola, J.R. Kennaway, J.W. Klop, M.R. Sleep, and F.J. de Vries. Syntactic definitions of undefined: On defining the undefined. In *Int. Symp. on Theoretical Aspects of Computer Software, Sendai*, pages 543–554, 1994. Lecture Notes in Computer Science, vol. 789.

[AO93] S. Abramsky and L. Ong. Full abstraction in the lazy lambda calculus. *Inf. and Comp.*, 105:159–267, 1993.

[Bar84] H.P. Barendregt. *The Lambda Calculus, its Syntax and Semantics*. North-Holland, 2nd edition, 1984.

[Ber] A. Berarducci. Infinite λ-calculus and non-sensible models. Presented to the conference in honour of Roberto Magari, Siena 1994.

[BI94] A. Berarducci and B. Intrigila. Church-rosser λ-theories, infinite λ-terms and consistency problems. Dipartimento di Matematica, Università di Pisa, October 1994.

[Ken92] J.R. Kennaway. On transfinite abstract reduction systems. Technical Report CS-9205, CWI, Amsterdam, 1992.

[KKSdV-] J.R. Kennaway, J.W. Klop, M.R. Sleep, and F.J. de Vries. Transfinite reductions in orthogonal term rewriting systems. *Information and Computation*, 199–. To appear; available by ftp from ftp::/ftp.sys.uea.ac.uk/pub/kennaway/transfinite.{dvi,ps}.Z.

Proving the Genericity Lemma by Leftmost Reduction is Simple

Jan Kuper

University of Twente, Department of Computer Science
P.O.Box 217, 7500 AE Enschede, The Netherlands
e-mail: jankuper@cs.utwente.nl

Abstract. The Genericity Lemma is one of the most important motivations to take in the untyped lambda calculus the notion of solvability as a formal representation of the informal notion of undefinedness. We generalise solvability towards typed lambda calculi, and we call this generalisation: *usability*. We then prove the Genericity Lemma for *un*-usable terms. The technique of the proof is based on *leftmost* reduction, which strongly simplifies the standard proof.

1 Introduction

In this paper we present an elementary and general proof of the Genericity Lemma. In the untyped lambda calculus this lemma says:

> Suppose M is an unsolvable term, and let N be a normal form.
> If $\boldsymbol{\lambda} \vdash FM = N$, then for all X we have $\boldsymbol{\lambda} \vdash FX = N$.

The informal meaning of this lemma is that if a term M is meaningless (undefined), and if a context containing M is convertible to a well-defined answer, then M did not have any influence on the computation of this answer and so M may be replaced by any term. In fact, the Genericity Lemma is one of the most important motivations to take in the untyped lambda calculus the notion of solvability as a formal representation of the informal notion of undefinedness.

Several proofs of the Genericity Lemma are known. For example, the standard proof (Barendregt 1984, chapter 14.3) uses a topological method, based on the tree topology (using Böhm trees, and showing that the compactification points in this topology are precisely the unsolvable terms). Takahashi gives a simpler proof in the untyped lambda calculus by exploiting the fact that the solvable terms are precisely the terms with a head normal form (Takahashi 1994). In (Kuper 1994, 1995) the Genericity Lemma is proved for a PCF-like calculus by generalising a technique introduced in (Barendregt 1971). This technique requires a tedious analysis of a reduction $FM \twoheadrightarrow N$.

Here, we prove the lemma for a more general situation than just untyped lambda calculus. Hence, we have to generalise the notion of solvability towards other lambda calculi, for example towards calculi with types, δ-reduction, μ-reduction. This generalisation can not be done directly from the definition of solvability in the untyped lambda calculus.

We will describe a generalisation of solvability (called *usability*) for a PCF-like calculus with product types and list types. We show that the Genericity Lemma holds for *un*usable terms. We prove the lemma by concentrating on the *leftmost* reduction $FM \twoheadrightarrow_{\ell} N$, which simplifies the proof strongly.

We remark that it is also possible to generalise Takahashi's proof to other calculi. However, this generalisation is not as simple as it might seem at first sight. There are two reasons for this. First, the connection between solvability and head normal forms which holds in the untyped lambda calculus, is in general not true for usability (see section 3).

Second, Takahashi's proof in fact proceeds by induction on the depth of the Böhm tree of a certain term, so a generalisation of this proof requires a generalisation of Böhm trees towards other calculi. This falls outside the scope of this paper.

2 The Lambda Calculus

We prove the Genericity Lemma for a PCF-like calculus, i.e., for a typed lambda calculus with full recursion and some constants, though the proof also works for untyped lambda calculus. We will refer to this calculus by $\boldsymbol{\lambda}$.

We will assume that $\boldsymbol{\lambda}$ has ground types **Nat** and **Bool**. Also, if σ and τ are types, then $\sigma \rightarrow \tau$ (functions), $\sigma \times \tau$ (products), and σ^* (lists) are types. The corresponding term formation rules are $\lambda x.M$, $\langle M, N \rangle$, and $[M, N]$ respectively. The final one stands for the cons-operation, where M is of type σ and N is of type σ^*. One might also choose for a constant **cons**, but because of the role of constants in forthcoming definitions (especially definition 2), we do not choose for this option (see section 3). The same holds for a constant **pair** in the case of product types.

Of course, the basic reduction rule of the calculus is the rule of β-reduction:

$$(\lambda x.M)N \rightarrow M[x{:=}N]$$

(where $M[x{:=}N]$ denotes substitution of N for all free occurrences of x in M, implicitly avoiding unintended bindings of variables).

The rule of η-reduction

$$\lambda x.Mx \rightarrow M$$

(where x is not free in M) may or may not hold.

Since non-termination is essential for the Genericity Lemma, we will assume that full recursion is possible in the calculus. We prefer to represent recursion by means of the μ-abstractor, accompanied with the rule

$$\mu x.M \rightarrow M[x{:=}\mu x.M],$$

but clearly, recursion might also be represented by means of fixed point combinators **Y**.

We assume that the following constants and the accompanying reduction rules (δ-rules) are present in the calculus:

- there are constants for natural numbers, booleans, successor function, test for zero. The accompanying δ-rules are standard,
- there is a conditional `if`, with accompanying rules

$$\texttt{if true} \rightarrow \lambda xy.x,$$

$$\texttt{if false} \rightarrow \lambda xy.y,$$

- there are constants for the projection functions $\boldsymbol{\pi}_i$ $(i = 1, 2)$:

$$\boldsymbol{\pi}_i\langle M_1, M_2\rangle \rightarrow M_i,$$

  and for lists:

$$\texttt{head}\,[M, N] \rightarrow M,$$

$$\texttt{tail}\,[M, N] \rightarrow N.$$

Notice that all δ-redexes are of the form fM, with f a constant.

Two essential properties of the calculus that are necessary for the proof of the Genericity Lemma given here, are the *Church-Rosser Property* and the *Standardisation Property.*

Church-Rosser property. For all terms M, N_1 and N_2 such that

$$N_1 \twoheadleftarrow M \twoheadrightarrow N_2,$$

there is a term L such that

$$N_1 \twoheadrightarrow L \twoheadleftarrow N_2.$$

A well-known corollary of the Church-Rosser property is that if a term M is convertible to a normal form N, then $M \twoheadrightarrow N$. Of course, $\boldsymbol{\lambda}$ has the Church-Rosser property.

Standardisation. Consider the following reduction:

$$M_0 \xrightarrow{\Delta_0} M_1 \xrightarrow{\Delta_1} \cdots$$

This reduction is a *standard reduction* if there is no $j < i$ such that Δ_i is a residual of a redex in M_j to the left of Δ_j.

Now the standardisation property says: for every $M \twoheadrightarrow N$, there is a standard reduction from M to N. A sufficient and elegant way to prove that a calculus has the standardisation property, is to show that the calculus is a left-normal combinatory reduction system (introduced in (Klop 1980), reformulated with some simplifications in (Kuper 1994)). It is easy to see that indeed $\boldsymbol{\lambda}$ is a left-normal combinatory reduction system.

An important corollary of the standardisation property is the normalisation theorem: if a term M has a normal form N, then $M \twoheadrightarrow N$ by the leftmost reduction, i.e., by contracting the leftmost redex in each reduction step (notation: $M \underset{\ell}{\twoheadrightarrow} N$).

3 Solvability and Usability

We start with the definition of solvability from the untyped lambda calculus, and generalise this notion towards some other calculi (for a more detailed treatment of this generalisation, see (Kuper 1994, 1995)). Recall that solvability is intended to formalise the concept of meaningfulness (or (un)definedness, cf. (Barendregt 1984)).

Definition 1 (Solvability). Let $\lambda\mathbf{x}.M$ be a closure of M. M is *solvable* if there is a sequence of terms $\mathbf{N}$ such that $(\lambda\mathbf{x}.M)\mathbf{N} \twoheadrightarrow \mathbf{I}$, where $\mathbf{I} \equiv \lambda x.x$.

This definition can not be directly generalised towards other calculi for two reasons: because of the restriction to contexts of the form $(\lambda\mathbf{x}.[\_])\mathbf{N}$, and because of the restriction of the result to $\mathbf{I}$. To overcome the first of these restrictions we introduce in $\boldsymbol{\lambda}$ the notion of *strict context*:

Definition 2 (Strict Context). A *strict context* $C[\_]$ is inductively defined as follows:

- the empty context, $[\_]$, is strict
- If $C[\_]$ is a strict context, and M a term, then

  (i) $(C[\_])M$,
  (ii) $\lambda x.C[\_]$

  are strict contexts.
- If f is a constant of function type, and $C[\_]$ is a strict context, then also

  (iii) $f(C[\_])$

  is a strict context.

In this definition we silently assume type correctness. Notice that each strict context has precisely one hole.

Intuitively, the strictness of a context $C[\_]$ means that in a (leftmost) reduction of $C[M]$ information from M is really used. This intuition can be considered as a starting point for a further generalisation of strict contexts towards other calculi.

Motivated by this intuition, we will use strict contexts in order to generalise the notion of solvability towards $\boldsymbol{\lambda}$. To overcome the second restriction in the definition of solvability (definition 1), we allow for any normal form instead of just $\mathbf{I}$. We call our generalisation of solvability: *usability*.

Definition 3 (Usability). A term M is *usable* if there is a strict context $C[\_]$ such that $C[M]$ has a normal form.

The intuition behind this definition is that a term M is meaningful (i.e., usable) if there is a terminating computation in which M is effectively used.

As examples of usable terms we mention that constants are usable. More general, all normal forms are usable. The standard example of a meaningless term, Ω, is not usable (in $\boldsymbol{\lambda}$ one can define Ω as $\mu x.x$). On the other hand, $\langle x\Omega, \Omega\rangle$ is usable: define the strict context

$$C[\_] \equiv (\lambda x.\boldsymbol{\pi}_1[\_])(\lambda xy.y)\underline{0},$$

and notice that $C[\langle x\Omega, \Omega\rangle] \twoheadrightarrow \underline{0}$.

Now we can motivate our choice to consider pairing as a syntactical term forming construction, and not by means of a constant **pair**: having a constant **pair** would make Ω usable ($\boldsymbol{\pi}_2$(**pair** $[\_]$ $\underline{0}$) is a strict context), which is not intended. To avoid this effect, one might add exceptions to definition 2, but this is rather clumsy. Having pairs by means of a syntactical term forming construction is much nicer.

In order to show that usability really is a generalisation of solvability, we restrict definition 2 to the untyped lambda calculus by removing clause (*iii*) from it.

Lemma 4. *In the untyped lambda calculus: M is usable iff M is solvable.*

Proof. Left to the reader. □

As a remarkable difference between the notions of solvability and usability, we mention that the usable terms are *not* precisely the terms with a head normal form. In $\boldsymbol{\lambda}$ one only has that usable terms have a weak head normal form, but not vice versa. In contrast to the situation in untyped lambda calculus, it is highly unlikely that there is a syntactical characterisation of usable terms at all. Consider the following two terms:

$$M_1 \equiv \mathbf{if}\ \mathtt{Zero?}(\mathtt{Pred}\ x)\ \mathbf{then}\ \underline{0}\ \mathbf{else}\ \Omega,$$
$$M_2 \equiv \mathbf{if}\ \mathtt{Zero?}(\mathtt{Succ}\ x)\ \mathbf{then}\ \underline{0}\ \mathbf{else}\ \Omega,$$

and notice that M_1 is usable, but M_2 is not.

As further evidence for the appropriateness of usability to formalise meaningfulness, we mention that all unusable terms can be consistently identified (respecting their types, of course). If in addition to this identification, a usable term is also identified to the unusable terms, then the calculus becomes inconsistent (Kuper 1994, 1995). Furthermore, usability has the genericity property (see theorem 10).

We will need the following lemma.

Lemma 5. *A is usable iff $\lambda x.A$ is usable.*

Proof. "⇒": If A is usable, then there is a strict context $C[\_]$ such that $C[A]$ has a normal form. Hence, $C[(\lambda x.A)x]$ ($\equiv C[A]$) has a normal form. Since $C[[\_]x]$ is a strict context, it follows that $\lambda x.A$ is usable.

"⇐": If $\lambda x.A$ is usable, then there is a strict context $C[\_]$ such that $C[\lambda x.A]$ has a normal form. Now $C[\lambda x.[\_]]$ is a strict context, so A is usable. □

4 The Genericity Lemma

The Genericity Lemma says: if M is not usable, and FM has a normal form, then for all X, FX has the same normal form.

In the proof of this lemma we will concentrate on the *leftmost* reduction $FM \underset{\ell}{\twoheadrightarrow} N$, and we show that M is not "in an essential way" involved in a leftmost reduction step. Then it is intuitively clear that M may be replaced by any X without changing the reduction steps.

In order to keep track of M during the reduction, we extend $\boldsymbol{\lambda}$ with a possibility to "pack" M. If M is in an essential way involved in a reduction step, then M must be "unpacked" first.

Definition 6 (The calculus $\boldsymbol{\lambda}\Box$). The *terms* of $\boldsymbol{\lambda}\Box$ are formed by the same term formation rules as the rules of the original calculus $\boldsymbol{\lambda}$. In addition, if M is a $\boldsymbol{\lambda}\Box$-term, then $\boxed{M}$ is a $\boldsymbol{\lambda}\Box$-term.

The *reduction rules* of $\boldsymbol{\lambda}\Box$ are the same as the rules of the original calculus $\boldsymbol{\lambda}$. In addition, there is the $\Box$-rule (the "unpack rule"):

$$\boxed{M} \to M.$$

The term $\boxed{M}$ is *to the left* of all subterms of M.

Substitution is defined in the usual way, extended with the clause

$$\boxed{M}\,[x:=N] \;\equiv\; \boxed{M[x:=N]}\,.$$

Strict contexts are defined by precisely the same clauses as in definition 2, and *usability* is defined as in definition 3.

We will use the following notations with the obvious interpretations: $\xrightarrow{\Box}$, $\overset{\Box}{\twoheadrightarrow}$, $\underset{\ell}{\overset{\Box}{\twoheadrightarrow}}$.

Remarks. If $\boxed{M}$ is involved in a $\boldsymbol{\lambda}$-reduction step in an essential way, then M must first be unpacked. For example, $(\lambda x.\ \boxed{M}\ )N$ and $(\lambda x.M)\ \boxed{N}$ are β-redexes, but $\boxed{\lambda x.M}\,N$ is *not*. To obtain a β-redex from this term, we first have to apply the $\Box$-rule on the subterm $\boxed{\lambda x.M}$. The same holds for terms of the form $\boxed{f}\,M$ and $f\,\boxed{M}$, whenever fM is a δ-redex.

Lemma 7. *$\boldsymbol{\lambda}\Box$ has the Church-Rosser property and the standardization property.*

Proof. Straightforward. One may consider $\Box$ as a new constant c for each type, with the rule

$$\mathrm{c}M \to M,$$

i.e., c behaves like the identity. It is clear that the resulting calculus is a left-normal CRS. Hence it has the indicated properties (Klop 1980). $\Box$

Lemma 8. $\boxed{M}$ *is usable iff* M *is usable.*

Proof. Immediate from the definition of usability, since $\boxed{M} = M$. □

The following lemma holds in both $\boldsymbol{\lambda}$ and $\boldsymbol{\lambda}\Box$.

Lemma 9. *If* $C[M]$ *has a normal form, and the displayed* M *is the leftmost redex in* $C[M]$*, then* M *is usable.*

Proof. Since the displayed M is the leftmost redex in $C[M]$, this M is not inside a box. For the same reason, M is not in the scope of a μ.

Let $C_s[\_]$ be the largest strict context inside $C[\_]$, where the hole of $C_s[\_]$ is the same as the hole of $C[\_]$. It is easily seen that $C[\_]$ is of one of the following four forms:

- $C[\_] \equiv C_s[\_]$. Then $C_s[M]$ has a normal form, i.e., M is usable.
- $C[\_] \equiv C'[F(C_s[\_])]$. Let N be the normal form of $C[M]$, then

  $$C'[F(C_s[M])] \underset{\ell}{\twoheadrightarrow} N \tag{1}$$

  by lemma 7. Since M is the leftmost redex and $C_s[\_]$ is the largest possible strict context, it follows that F is either a variable, or an application term of the form $GL_1 \cdots L_n$ $(n \geq 1)$, with G a constant or a variable. Hence, during reduction (1) reducts of $C_s[M]$ will not be a proper subterm of some redex. Hence, there is an $N' \subseteq N$ such that $C_s[M] \twoheadrightarrow N'$. Clearly, N' is a normal form, so M is usable.
- $C[\_] \equiv C'[\langle C_s[\_], X\rangle]$ or $C[\_] \equiv C'[\langle X, C_s[\_]\rangle]$. We consider the first case, the second case being analogous. Suppose $C'[\langle C_s[M], X\rangle] \twoheadrightarrow N$. Since M is the leftmost redex, it follows that $C_s[M] \twoheadrightarrow N' \subseteq N$. As above, M is usable.
- $C[\_] \equiv C'[\,[C_s[\_], X]\,]$ or $C[\_] \equiv C'[\,[X, C_s[\_]]\,]$. As above. □

Theorem 10 (The Genericity Lemma). *Suppose* M *is not usable, and* N *is a normal form. If* $\vdash FM = N$*, then for all* X

$$\vdash FX = N.$$

Proof. The proof consists of two steps:

(a) for closed M only,
(b) for arbitrary M.

Proof of (a): Consider in $\boldsymbol{\lambda}$ the leftmost reduction $\mathcal{R} : FM \underset{\ell}{\twoheadrightarrow} N$. It is straightforward to construct from $\mathcal{R}$ the leftmost reduction

$$\mathcal{R}' : F\boxed{M} \overset{\Box}{\underset{\ell}{\twoheadrightarrow}} N.$$

Since M is closed, and since there are no reductions inside boxed terms (the reduction $\mathcal{R}'$ is leftmost) it follows that for all terms of the form $\boxed{L}$ which arise during the reduction $\mathcal{R}'$, we have that $L \equiv M$. Now suppose there is an application of the $\Box$-rule in $\mathcal{R}'$, say

$$C[\boxed{M}] \xrightarrow{\Box} C[M] \underset{\ell}{\overset{\Box}{\twoheadrightarrow}} N.$$

Then $\boxed{M}$ is the leftmost redex in $C[\boxed{M}]$, and so (by lemma 9) $\boxed{M}$ is usable, i.e., M is usable. This is a contradiction, so there are no applications of the $\Box$-rule in $\mathcal{R}'$. Since N is $\Box$-free, it can now easily be shown (by induction on the length of $\mathcal{R}'$) that $\boxed{M}$ can be replaced by any X.

Proof of (b): Let $\mathbf{x}$ be the sequence of variables free in M. Then $\lambda\mathbf{x}.M$ is closed and not usable (by lemma 5). Hence,

$$\begin{aligned} FM &= F((\lambda\mathbf{x}.M)\mathbf{x}) \\ &= (\lambda y.F(y\mathbf{x}))(\lambda\mathbf{x}.M) \\ &= N, \end{aligned}$$

and so

$$\begin{aligned} FX &= F((\lambda\mathbf{x}.X)\mathbf{x}) \\ &= (\lambda y.F(y\mathbf{x}))(\lambda\mathbf{x}.X) \\ &= N. \quad \Box \end{aligned}$$

References

Barendregt, H.P., *Some extensional term models for combinatory logics and lambda calculi*, Ph.D. Thesis, Utrecht.

Barendregt, H.P., *The Lambda Calculus – Its Syntax and Semantics* (revised edition), North-Holland, Amsterdam.

Klop, J.W., *Combinatory Reduction Systems*, Ph.D. Thesis, Mathematical Centre, Amsterdam.

Kuper, J., *Partiality in Logic and Computation – Aspects of Undefinedness*, Ph.D. Thesis, Enschede.

Kuper, J., Usability: formalising (un)definedness in typed lambda calculus, CSL '94, Kazimierz, Poland, 1994. To appear in the proceedings, Springer-Verlag, Berlin, Heidelberg.

Takahashi, M., A simple proof of the Genericity Lemma, *in:* Neil D. Jones, Masami Hagiya, Masahiko Satu (Eds.): *Logic, Language and Computation – Festschrift in Honour of Satoru Takasu*, Springer-Verlag, Berlin, Heidelberg, pp. 117 – 118.

(Head-)Normalization of Typeable Rewrite Systems

Steffen van Bakel[1] and Maribel Fernández[2]

[1] Dipartimento di Informatica, Università degli Studi di Torino, Corso Svizzera 185, 10149 Torino, Italy. bakel@di.unito.it
[2] DMI - LIENS (CNRS URA 1327), Ecole Normale Supérieure, 45, rue d'Ulm, 75005 Paris, France. maribel@ens.ens.fr

Abstract. In this paper we study normalization properties of rewrite systems that are typeable using intersection types with ω and with sorts. We prove two normalization properties of typeable systems. On one hand, for all systems that satisfy a variant of the Jouannaud-Okada Recursion Scheme, every term typeable with a type that is not ω is head normalizable. On the other hand, non-Curryfied terms that are typeable with a type that does not contain ω, are normalizable.

Introduction

In the study of termination of reduction systems, the notion of types has played an important role. A well explored area in this aspect is that of the Lambda Calculus (LC). For LC, there exists a well understood notion of type assignment, known as the Curry Type Assignment System [8], which expresses abstraction and application, and introduces $\rightarrow$-types. A well-known result for this system is that all typeable terms are strongly normalizable. Another notion of type assignment for LC for which the relation between typeability and normalization has been studied profoundly, is the Intersection Type Discipline (the BCD-system), as presented in [7], that is an extension of Curry's system. The extension consists of allowing more than one type for term-variables and adding a type constant 'ω', and, next to the type constructor '$\rightarrow$', the type constructor '$\cap$'. The set of lambda terms having a head normal form, the set of lambda terms having a normal form, and the set of strongly normalizable lambda terms can all be characterized by the set of their assignable types.

In this paper, instead of studying the problem of termination using types in the setting of LC, the approach taken will be to study the desired property directly on the level of a programming language with patterns, i.e. in the world of term rewriting systems (TRS). For this purpose, in this paper we define a notion of type assignment on Curryfied TRS (CTRS) that uses intersection types (the system without the type constructors $\cap$ and ω is a particular case; more rules and terms are typeable in the intersection system). CTRS are defined as a slight extension of the TRS defined in [9, 12]. The language of the CTRS is first order (i.e. every function symbol has a fixed arity) but CTRS are assumed to contain a notion of application Ap, that allows partial application (Curryfication) of function symbols in the setting of a first order language.

Unlike typeable terms in LC, typeable terms in CTRS need not be normalizable (consider a typeable term t and a rule $t \to t$). In order to ensure head normalization of typeable terms in CTRS we will impose some syntactical restrictions on recursive rules, inspired by the recursive scheme defined by Jouannaud and Okada in [11]. This kind of recursive definitions was presented for the incremental definition of higher order functionals based on first order definitions, such that the whole system is terminating. The general scheme of Jouannaud and Okada was also used in [5] and [6] for defining higher order functions compatible with different lambda calculi, as well as in [4] to obtain a strong normalization result for CTRS that are typeable using ω-free intersection types. The main difference between the recursive scheme for CTRS defined in [4] and the one defined in this paper, is that here the patterns of recursive rules are constructor terms, which have sorts as types. It is worth noticing that without this condition, the recursive scheme of [4] does not ensure head normalization of typeable terms in intersection systems with ω (see examples in Section 3.1).

We will prove (using the well-known method of Computability Predicates [13, 10]) that for all typeable CTRS satisfying the variant of the scheme that is defined in this paper, every typeable term has a head normal form. We will also show that if Curryfication is not allowed, then terms whose type does not contain ω are normalizable. These results, together with the strong normalization result of [4], complete the study of the normalization properties of typeable CTRS in intersection type systems.

In [3] and [2] two partial intersection type assignment systems for TRS are presented. Apart from differences in syntax, the system we present here is the first one extended with type constants (called sorts). The system of [2] is a *decidable* restriction of the one presented here (the restriction lies in the structure of types). In [5] a partial type assignment system for higher order rewrite systems that uses intersection types and sorts (but not ω) is defined. It differs from ours in that function symbols are strongly-typed with sorts only, whereas we allow for types to contain type-variables as well, and in this way we can model polymorphism.

We define CTRS in Section 1, and the intersection type assignment system in Section 2. In Section 3 we study head normalization and normalization of typeable CTRS.

1 Curryfied Term Rewriting Systems

We will define Curryfied Term Rewriting Systems (CTRS) (an extension of the TRS defined in [9, 12]), as first order TRS that allow partial application of function symbols. They were first defined in [4].

Definition 1. An *alphabet* or *signature* Σ consists of:

1. A countable infinite set $\mathcal{X}$ of variables $x_1, x_2, x_3, \ldots$ (or $x, y, z, x', y', \ldots$).
2. A non-empty set $\mathcal{F}$ of *function symbols* $F, G, \ldots$, each equipped with an 'arity'.

3. A special binary operator, called *application* (Ap).

Definition 2. The set $T(\mathcal{F}, \mathcal{X})$ of *terms* (or *expressions*) is defined inductively:

1. $\mathcal{X} \subseteq T(\mathcal{F}, \mathcal{X})$.
2. If $F \in \mathcal{F} \cup \{Ap\}$ is an n-ary symbol ($n \geq 0$), and $t_1, \ldots, t_n \in T(\mathcal{F}, \mathcal{X})$, then $F(t_1, \ldots, t_n) \in T(\mathcal{F}, \mathcal{X})$. The t_i ($i = 1, \ldots, n$) are the arguments of the last term.

Definition 3. A *replacement* R is a map from $T(\mathcal{F}, \mathcal{X})$ to $T(\mathcal{F}, \mathcal{X})$ satisfying $\mathrm{R}(F(t_1, \ldots, t_n)) = F(\mathrm{R}(t_1), \ldots, \mathrm{R}(t_n))$ for every n-ary function symbol F (here $n \geq 0$). So, R is determined by its restriction to the set of variables, and sometimes we will use the notation $\{x_1 \mapsto t_1, \ldots, x_n \mapsto t_n\}$ to denote a replacement. We also write t^{R} instead of $\mathrm{R}(t)$.

Definition 4. 1. A *rewrite rule* is a pair (l, r) of terms in $T(\mathcal{F}, \mathcal{X})$. Often, a rewrite rule will get a name, e.g. $\mathbf{r}$, and we write $\mathbf{r} : l \to r$. Three conditions will be imposed: l is not a variable, the variables occurring in r are contained in l, and if Ap occurs in l, then $\mathbf{r}$ is of the shape:

$$Ap\,(F_{n-1}\,(x_1, \ldots, x_{n-1}), x_n) \;\to\; F\,(x_1, \ldots, x_n).$$

2. The systems we consider are Curry-closed, i.e. for every rewrite rule with left hand-side $F(t_1, \ldots, t_n)$ there are n additional rewrite rules:

$$\begin{array}{ll} Ap\,(F_{n-1}\,(x_1, \ldots, x_{n-1}), x_n) & \to\; F\,(x_1, \ldots, x_n) \\ \vdots & \\ Ap\,(F_0, x_1) & \to\; F_1\,(x_1) \end{array}$$

$F_{n-1}, \ldots, F_0$ are the *Curryfied versions of* F.

3. A rewrite rule $\mathbf{r} : l \to r$ determines a set of *reductions* $l^{\mathrm{R}} \to r^{\mathrm{R}}$ for all replacements R. The left hand side l^{R} is called a *redex*; it may be replaced by its '*contractum*' r^{R} inside a context C[ ]; this gives rise to *rewrite steps*: $\mathrm{C}[\,l^{\mathrm{R}}\,] \to_{\mathbf{r}} \mathrm{C}[\,r^{\mathrm{R}}\,]$. Concatenating rewrite steps we have *rewrite sequences* $t_0 \to t_1 \to t_2 \to \cdots$. If $t_0 \to \cdots \to t_n$ we also write $t_0 \to^* t_n$, and t_n is a *rewrite* or *reduct* of t_0.

4. A *Curryfied Term Rewriting System* (CTRS) is a pair $(\Sigma, \mathbf{R})$ of an alphabet Σ and a set $\mathbf{R}$ of rewrite rules. We write $t \to_{\mathbf{R}} t'$, if there is a $\mathbf{r} \in \mathbf{R}$ such that $t \to_{\mathbf{r}} t'$. The symbol $\to_{\mathbf{R}}^*$ denotes the reflexive and transitive closure of $\to_{\mathbf{R}}$. Terms that contain neither Curryfied versions of symbols, nor Ap, will be called *non-Curryfied terms*.

In this paper we will restrict ourselves to CTRS that satisfy the Church-Rosser property (also known as *confluence*), that is formulated by:

If $t \to^* u$, and $t \to^* v$, then there exists w such that $u \to^* w$, and $v \to^* w$.

There are several syntactic restrictions that can be posed in rules and reduction strategies so that this property is obtained (see e.g. [12]).

We take the view that in a rewrite rule a certain symbol is defined; it is this symbol to which the structure of the rewrite rule gives a type.

Definition 5. 1. In a rewrite rule, the leftmost, outermost symbol in the left hand side that is not an *Ap*, is called *the defined symbol* of that rule.
2. If the symbol F is the defined symbol of the rewrite rule $\mathbf{r}$, then $\mathbf{r}$ *defines* F.
3. F is *a defined symbol*, if there is a rewrite rule that defines F.
4. $Q \in \mathcal{F}$ is called a *constructor* or *constant symbol* if Q is not a defined symbol.

Definition 6. We assume that rewrite rules are *not* mutually recursive. A TRS whose dependency-graph is an ordered cycle-free graph, is called a *hierarchical* TRS. The rewrite rules of a such a TRS can be regrouped such that they are *incremental* definitions of the defined symbols $F^1, \ldots, F^k$, so that the rules defining F^i only depend on $F^1, \ldots, F^{i-1}$.

Example 1. Our definition of recursive symbols, using the notion of defined symbols, is different from the one normally considered. Since *Ap* is never a defined symbol, the following rewrite system

$$\begin{array}{lcl} D(x) & \rightarrow & Ap(x,x) \\ Ap(D_0,x) & \rightarrow & D(x) \end{array}$$

is *not* considered a recursive system. Notice that, for example, the term $D(D_0)$ has no normal form (these terms play the role of $(\lambda x.xx)(\lambda x.xx)$ in the LC). This means that, in the formalism of this paper, there exist non-recursive first-order rewrite systems that are not normalizing.

Definition 7. 1. A term is *neutral* if it is not of the form $F_i(t_1, \ldots, t_i)$, where F_i is a Curryfied version of a function symbol F.
2. A term is in *normal form* if it is irreducible.
3. A term t is in *head normal form* if for all t' such that $t \rightarrow^* t'$:
   (a) t' is not itself a redex, and
   (b) if $t' = Ap(v, u)$ then v is in head normal form.
4. A term is in *constructor-hat normal form* if either
   (a) it has the form $Ap(t_1, t_2)$ and t_1 is in constructor-hat normal form, or
   (b) it has the form $C[u_1, \ldots, u_n]$ where C is a context (possibly empty) that contains only constructor symbols, and for $1 \leq i \leq n$, u_i cannot be reduced to a term of the form $C'(s_1, \ldots, s_i)$ where C' is a constructor.
5. A term is *(head, respectively constructor-hat) normalizable* if it can be reduced to a term in (head, respectively constructor-hat) normal form. A rewrite system is *strongly normalizing* (or terminating) if all the rewrite sequences are finite; it is *(head, respectively constructor-hat) normalizing* if every term is (head, respectively constructor-hat) normalizable.

Example 2. Take the rules $F(G,H) \rightarrow A$, and $B(C) \rightarrow G$, then $F(B(C),H)$ is not a redex. But it is neither a head-normal form nor a constructor-hat normal form, since it reduces to $F(G,H)$, which is a redex and reduces to A which is a constructor.

The notations $Chnf(t)$, $Hnf(t)$ and $Nf(t)$ will indicate that t is in constructor-hat normal form, in head normal form, and in normal form, respectively. The notations $CHN(t)$, $HN(t)$ and $N(t)$ will indicate that t is constructor-hat normalizable, head normalizable, and normalizable, respectively.

Lemma 8. *1.* $HN(Ap(t,x)) \Rightarrow HN(t)$, *and* $CHN(Ap(t,x)) \Rightarrow CHN(t)$.
2. t *is neutral &* $Hnf(t) \Rightarrow \forall u\ [Hnf(Ap(t,u))]$.
t *is neutral &* $Chnf(t) \Rightarrow \forall u\ [Chnf(Ap(t,u))]$
3. t *is neutral* $\Rightarrow \forall u\ [Ap(t,u)$ *is neutral*].

2 Intersection types and type assignment

Strict types are the types that are strictly needed to assign a type to a term in the system presented in [7] (see also [1]). In the set of strict types, intersection type schemes and the type constant ω play a limited role. We will assume that ω is the same as an intersection over zero elements: if $n = 0$, then $\sigma_1 \cap \cdots \cap \sigma_n \equiv \omega$, so ω does not occur in an intersection subtype. Moreover, intersection type schemes (so also ω) occur in strict types only as subtypes of the left hand side of an arrow type scheme. In this paper we will consider strict types over a set S of sorts (constant types).

Definition 9. 1. $\mathcal{T}_s$, the set of *strict types*, is inductively defined by:
(a) All type-variables $\varphi_0, \varphi_1, \ldots \in \mathcal{T}_s$.
(b) All sorts $s_1, \ldots, s_n \in \mathcal{T}_s$.
(c) If $\tau, \sigma_1, \ldots, \sigma_n \in \mathcal{T}_s$ $(n \geq 0)$, then $\sigma_1 \cap \cdots \cap \sigma_n \rightarrow \tau \in \mathcal{T}_s$.
2. $\mathcal{T}_S$ is defined by: If $\sigma_1, \ldots, \sigma_n \in \mathcal{T}_s$ $(n \geq 0)$, then $\sigma_1 \cap \cdots \cap \sigma_n \in \mathcal{T}_S$.
3. On $\mathcal{T}_S$, the relation $\leq_S$ is defined by:
(a) $\forall 1 \leq i \leq n\ (n \geq 1)\ [\ \sigma_1 \cap \cdots \cap \sigma_n \leq_S \sigma_i\ ]$.
(b) $\forall 1 \leq i \leq n\ (n \geq 0)\ [\ \sigma \leq_S \sigma_i\ ] \Rightarrow \sigma \leq_S \sigma_1 \cap \cdots \cap \sigma_n$.
(c) $\sigma \leq_S \tau \leq_S \rho \Rightarrow \sigma \leq_S \rho$.
4. We define $\leq$ on $\mathcal{T}_S$ like $\leq_S$, by adding an extra alternative.
(d) $\rho \leq \sigma\ \&\ \tau \leq \mu \Rightarrow \sigma \rightarrow \tau \leq \rho \rightarrow \mu$.
5. We define the relation $\sim$ by: $\sigma \sim \tau \Leftrightarrow \sigma \leq \tau \leq \sigma$.
$\mathcal{T}_S$ may be considered modulo $\sim$. Then $\leq$ becomes a partial order, and in this paper we consider types modulo $\sim$.

Throughout this paper, the symbol φ (often indexed) will be a type-variable. Greek symbols like α, β, γ, μ, ν, η, ρ, σ and τ will range over types. Unless stated otherwise, if $\sigma_1 \cap \cdots \cap \sigma_n$ is used to denote a type, then all $\sigma_1, \ldots, \sigma_n$ are assumed to be strict. Notice that $\mathcal{T}_s$ is a proper subset of $\mathcal{T}_S$.

Definition 10. A *basis* is a set of statements of the shape '$x{:}\sigma$', where x is a term-variable, and $\sigma \in \mathcal{T}_S \backslash \omega$. If $B_1, \ldots, B_n$ are bases, then $\Pi\{B_1, \ldots, B_n\}$ is the basis defined as follows: $x{:}\sigma_1 \cap \cdots \cap \sigma_m \in \Pi\{B_1, \ldots, B_n\}$ if and only if $\{x{:}\sigma_1, \ldots, x{:}\sigma_m\}$ is the set of all statements about x that occur in $B_1 \cup \cdots \cup B_n$.
We will often write $B \cup \{x{:}\sigma\}$ for the basis $\Pi\{B, \{x{:}\sigma\}\}$, when x does not occur in B.

Partial intersection type assignment on a CTRS $(\Sigma, \mathbf{R})$ will be defined as the labelling of nodes and edges in the tree-representation of terms and rewrite rules with types in $\mathcal{T}_S$. All function symbols are assumed to have a type, that is produced by a mapping called ***environment***.

Our notion of type assignment is based on the definition of (chains of) three operations on types (pairs of basis and type), called substitution, expansion and lifting. Substitution is the operation that instantiates a type (i.e. that replaces type variables by types). The operation of expansion replaces types by the intersection of a number of copies of that type. The operation of lifting replaces basis and type by a smaller basis and a larger type, in the sense of $\leq$. See [3] for formal definitions.

Definition 11. Let $(\Sigma, \mathbf{R})$ be a CTRS, and $\mathcal{E}$ an environment.

1. We say that $t \in T(\mathcal{F}, \mathcal{X})$ *is typeable by* $\sigma \in \mathcal{T}_S$ *with respect to* $\mathcal{E}$, if there exists an assignment of types to edges and nodes that satisfies the following constraints:
   (a) The root edge of t is typed with σ; if $\sigma = \omega$, then the root edge is the only thing in the term-tree that is typed.
   (b) The type assigned to a function node containing $F \in \mathcal{F} \cup \{Ap\}$ (where F has arity $n \geq 0$) is $\tau_1 \cap \cdots \cap \tau_m$, if and only if for every $1 \leq i \leq m$ there are $\sigma^i_1, \ldots, \sigma^i_n \in \mathcal{T}_S$, and $\sigma_i \in \mathcal{T}_s$, such that $\tau_i = \sigma^i_1 \rightarrow \cdots \rightarrow \sigma^i_n \rightarrow \sigma_i$, the type assigned to the j-th $(1 \leq j \leq n)$ out-going edge is $\sigma^1_j \cap \cdots \cap \sigma^m_j$, and the type assigned to the incoming edge is $\sigma_1 \cap \cdots \cap \sigma_m$.

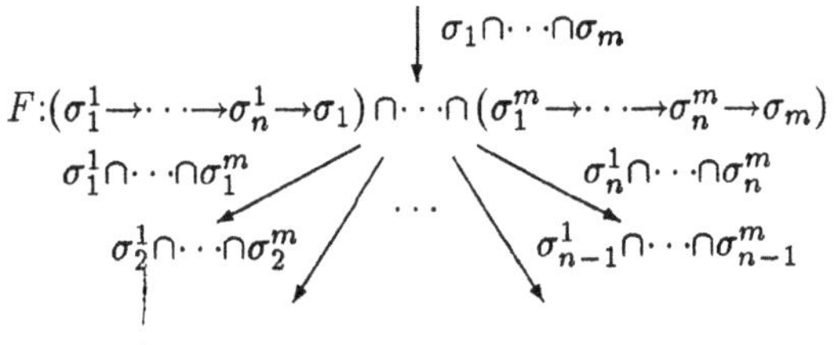

   (c) If the type assigned to a function node containing $F \in \mathcal{F} \cup \{Ap\}$ is τ, then there is a chain C, such that $C(\mathcal{E}(F)) = \tau$.
2. Let $t \in T(\mathcal{F}, \mathcal{X})$ be typeable by σ with respect to $\mathcal{E}$. If B is a basis such that for every statement $x{:}\tau$ occurring in the typed term-tree there is a $x{:}\tau' \in B$ such that $\tau' \leq \tau$, we write $B \vdash_{\mathcal{E}} t{:}\sigma$.

Example 3. For every occurrence of Ap in a term-tree, there are $\sigma_1, \ldots, \sigma_n$ and $\tau_1, \ldots, \tau_n$ such that the following is part of the term-tree.

$$\begin{array}{c} \downarrow \tau_1 \cap \cdots \cap \tau_n \\ Ap \colon ((\sigma_1 \to \tau_1) \to \sigma_1 \to \tau_1) \cap \cdots \cap ((\sigma_n \to \tau_n) \to \sigma_n \to \tau_n) \\ (\sigma_1 \to \tau_1) \cap \cdots \cap (\sigma_n \to \tau_n) \swarrow \quad \searrow \sigma_1 \cap \cdots \cap \sigma_n \end{array}$$

As in [3] we can prove:

Lemma 12. *1. If $B \vdash_{\mathcal{E}} t{:}\sigma$, and $B' \leq B$, then $B' \vdash_{\mathcal{E}} t{:}\sigma$.*
2. If $B \vdash_{\mathcal{E}} t{:}\sigma$, and $\sigma \leq \tau$, then $B \vdash_{\mathcal{E}} t{:}\tau$.
3. If $B \vdash_{\mathcal{E}} Ap(t,u){:}\sigma$, $\sigma \in \mathcal{T}_s$, then there exist τ, such that $B \vdash_{\mathcal{E}} t{:}\tau \to \sigma$, and $B \vdash_{\mathcal{E}} u{:}\tau$.
4. $B \vdash_{\mathcal{E}} t{:}\sigma_1 \cap \cdots \cap \sigma_n \Leftrightarrow \forall 1 \leq i \leq n \; [B \vdash_{\mathcal{E}} t{:}\sigma_i]$.
5. $B \vdash_{\mathcal{E}} F_n(t_1, \ldots, t_n){:}\sigma \;\&\; \sigma \in \mathcal{T}_s \Rightarrow \exists \alpha \in \mathcal{T}_s, \beta \in \mathcal{T}_s \; [\sigma = \alpha \to \beta]$.

In [3] a sufficient condition was formulated in order to ensure the subject reduction property: type assignment on rewrite rules was there defined using the notion of principal pair for a typeable term.

Definition 13. Let $t \in T(\mathcal{F}, \mathcal{X})$. A pair $\langle P, \pi \rangle$ is called *a principal pair for t with respect to $\mathcal{E}$*, if $P \vdash_{\mathcal{E}} t{:}\pi$ and for every B, σ such that $B \vdash_{\mathcal{E}} t{:}\sigma$ there is a chain C such that $C(\langle P, \pi \rangle) = \langle B, \sigma \rangle$.

Definition 14. Let $(\Sigma, \mathbf{R})$ be a CTRS, and $\mathcal{E}$ an environment.

1. We say that $l \to r \in \mathbf{R}$ with defined symbol F *is typeable with respect to $\mathcal{E}$*, if there are basis P, type $\pi \in \mathcal{T}_s$, and an assignment of types to nodes and edges such that:
   (a) $\langle P, \pi \rangle$ is a principal pair for l with respect to $\mathcal{E}$, and $P \vdash_{\mathcal{E}} r{:}\pi$.
   (b) In $P \vdash_{\mathcal{E}} l{:}\pi$ and $P \vdash_{\mathcal{E}} r{:}\pi$, all nodes containing F are typed with $\mathcal{E}(F)$.
2. We say that $(\Sigma, \mathbf{R})$ *is typeable with respect to $\mathcal{E}$*, if every $\mathbf{r} \in \mathrm{R}$ is typeable with respect to $\mathcal{E}$.

To guarantee the subject reduction property, in this paper we accept only those rewrite rules $l \to r$, that are typeable according to the above definition. As in [3], it is then possible to prove:

Theorem 15 Subject Reduction. *For all replacements R, bases B and types σ: if $B \vdash_{\mathcal{E}} l^{\mathrm{R}}{:}\sigma$, then $B \vdash_{\mathcal{E}} r^{\mathrm{R}}{:}\sigma$, so: if $B \vdash_{\mathcal{E}} t{:}\sigma$, and $t \to_{\mathbf{R}}^{*} t'$, then $B \vdash_{\mathcal{E}} t'{:}\sigma$.*

3 Normal Forms and Head Normal Forms

In [4] we introduced a class of CTRS that are strongly normalizing on terms that are typeable without using the constant ω. In this section we will study normalization properties of CTRS in a type assignment system with ω and sorts.

Since typeability alone is not sufficient to ensure any normalization property, we will impose syntactic restrictions on the rules. As a consequence of the results of this section, the class of typeable *non-recursive* CTRS is head-normalizing on terms whose type is not ω, and normalizing on non-Curryfied terms whose type does not contain ω. To appreciate the non-triviality of this statement, remember Example 1: a non-recursive CTRS may be not head-normalizing. In fact, the main result of this section (every term whose type is not ω is head-normalizable) shows that the term $D\,(D_0)$ is only typeable with ω.

But we will actually prove a stronger result: we will characterize a class of *recursive* definitions for which the same normalization properties hold. These results, together with the previous strong normalization result presented in [4], complete the study of the normalization properties of typeable rewrite systems in the intersection type assignment system.

The converse of our result does not hold: not every (head-)normalizable term is typeable. Take for example the strongly normalizing rewrite system

$I\,(x) \to x,$
$K\,(x,y) \to x,$
$F\,(I_0) \to I_0,$
$F\,(K_0) \to K_0.$

It is not possible to give an environment such that these rules can be typed, since there is no type σ that is a type for both I and K.

3.1 Head-normalization

In [4] we studied systems that define recursive functions satisfying the *general scheme* below. This scheme was inspired by [11] where generalized primitive recursive definitions were shown strongly normalizing when combined with typed LC. The same results were shown in [5] in the context of type assignment systems for LC and in [6] for the Calculus of Constructions.

Definition 16 General scheme. Let $\mathcal{F}_n = \mathcal{Q} \cup \{F^1, \ldots, F^n\}$, where $F^1, \ldots, F^n$ are the defined symbols of the signature that are not Curryfied-versions, and assume that $F^1, \ldots, F^n$ are defined in an incremental way. Suppose, moreover, that the rules defining $F^1, \ldots, F^n$ satisfy the *general scheme*:

$$F^i\,(\overrightarrow{\mathrm{C}}[\overline{x}],\overline{y}) \to \mathrm{C}'[F^i\,(\overrightarrow{\mathrm{C}_1}[\overline{x}],\overline{y}),\ldots,F^i\,(\overrightarrow{\mathrm{C}_m}[\overline{x}],\overline{y}),\overline{x},\overline{y}],$$

where $\overline{x}$, $\overline{y}$ are sequences of variables, and $\overline{x} \subseteq \overline{y}$. Also, $\overrightarrow{\mathrm{C}}[\,]$, $\mathrm{C}'[\,]$, $\overrightarrow{\mathrm{C}_1}[\,]$, and $\overrightarrow{\mathrm{C}_m}[\,]$ are sequences of contexts in $T(\mathcal{F}_{i-1},\mathcal{X})$, and $\overrightarrow{\mathrm{C}}[\overline{x}] \rhd_{mul} \overrightarrow{\mathrm{C}_j}[\overline{x}]$ $(1 \leq j \leq m)$, where $\lhd$ is the strict subterm ordering and *mul* denotes multiset extension.

The rules defining $F^1, \ldots, F^n$ and their Curry-closure together form a *safe recursive system*.

This general scheme imposes some restrictions on the definition of functions: the terms in every $\overrightarrow{\mathrm{C}_j}[\overline{x}]$ are subterms of terms in $\overrightarrow{\mathrm{C}}[\overline{x}]$ (this is the 'primitive recursive' aspect of the scheme), and the variables $\overline{x}$ must also appear as arguments in the left-hand side of the rule.

It is worthwhile noting that the rewrite rules of Combinatory Logic are *not* recursive, so, in particular, satisfy the scheme. Therefore, although the severe restriction imposed on rewrite rules, the systems satisfying the scheme still have full Turing-machine computational power, a property that first-order systems without Ap would not possess.

In a type assignment system without ω, the conditions imposed by the general scheme are sufficient to ensure strong normalization of typeable terms [4]. Unfortunately, the general scheme is not enough to ensure head normalization of typeable terms in a type system with ω: take the rewrite system

$$\begin{array}{ll} F(C(x)) & \rightarrow F(x), \\ A(x,y) & \rightarrow Ap(y, Ap(Ap(x,x),y)) \end{array}$$

that is typeable with respect to the environment

$$\begin{array}{l} \mathcal{E}(F) = \omega \rightarrow \sigma, \\ \mathcal{E}(C) = \omega \rightarrow \sigma, \\ \mathcal{E}(A) = ((\alpha \rightarrow \mu \rightarrow \beta) \cap \alpha) \rightarrow ((\beta \rightarrow \rho) \cap \mu) \rightarrow \rho, \end{array}$$

Then $B \vdash_{\mathcal{E}} F(A(A_0, C_0)){:}\sigma$, but $F(A(A_0, C_0)) \rightarrow^*_{\mathbf{R}} F(C(A(A_0, C_0))) \rightarrow_{\mathbf{R}} F(A(A_0, C_0))$. The underlying problem is that, using full intersection types, there are two kinds of typeable recursion in CTRS: the one explicitly present in the syntax, as well as the one obtained by the so-called *fixed-point combinators*; for every G that has type $\omega \rightarrow \sigma$, the term $A(A_0, G_0)$ has type σ, and $A(A_0, G_0) \rightarrow^*_{\mathbf{R}} G(A(A_0, G_0))$.

So, we need to impose stronger conditions on the scheme. We will consider those CTRS having an alphabet with a set $\mathcal{C}$ of constructors, such that constructors are given arrow-ground types in all the environments, i.e. for all environment $\mathcal{E}$, if $C \in \mathcal{C}$ then $\mathcal{E}(C) = s_1 \rightarrow \ldots \rightarrow s_n \rightarrow s$ where $s_1, \ldots, s_n, s$ are sorts.

Definition 17 Safety-scheme. A rewrite rule

$$F^i(\vec{C}[\vec{x}], \vec{y}) \rightarrow C'[F^i(\vec{C_1}[\vec{x}], \vec{y}), \ldots, F^i(\vec{C_m}[\vec{x}], \vec{y}), \vec{x}, \vec{y}]$$

satisfies the *Safety-scheme* in the environment $\mathcal{E}$, if it satisfies the conditions of the general scheme, where we replace the condition:
'for $1 \leq j \leq m$, $\vec{C}[\vec{x}] \rhd_{mul} \vec{C_j}[\vec{x}]$'
by the condition:
'for $1 \leq j \leq m$, $\vec{C}[\vec{x}] \rhd_{mul} \vec{C_j}[\vec{x}]$, $\vec{C}[\vec{x}]$, $\vec{C_j}[\vec{x}] \in T(\mathcal{C}, \mathcal{X})$, and the "patterns" $\vec{C}[\vec{x}]$ appear at positions where $\mathcal{E}(F^i)$ requires arguments of sort type'
From now on, we will call the systems that satisfy the Safety-scheme *safe*.

The rest of this section will be devoted to the proof of the head normalization theorem. We will use the well-known method of Computability Predicates [13, 10]. We will prove simultaneously that every typeable term is head normalizable and constructor-hat normalizable. The proof has two parts; in the first one we give the definition of a predicate *Comp* on bases, terms, and types, and prove some properties of *Comp*. The most important one states that if for a term t there are a basis B and type $\sigma \neq \omega$ such that $Comp(B, t, \sigma)$ holds, then $HN(t)$ and $CHN(t)$. In the second part *Comp* is shown to hold for each typeable term.

In the following we assume that $(\Sigma, \mathbf{R})$ is a typeable and safe CTRS in the environment $\mathcal{E}$.

Definition 18. 1. Let B be a basis, t a term, and σ a type such that $B \vdash_{\mathcal{E}} t{:}\sigma$. We define the Computability Predicate $Comp(B,t,\sigma)$ recursively on σ by:

(a) $\sigma = \varphi$, or $\sigma = s$ (sort).
$Comp(B,t,\sigma) \Leftrightarrow HN(t)\ \&\ CHN(t)$.

(b) $\sigma = \alpha \rightarrow \beta$.
$Comp(B,t,\alpha \rightarrow \beta) \Leftrightarrow \forall u \in T(\mathcal{F},\mathcal{X})\ [\ Comp(B',u,\alpha) \Rightarrow Comp(\Pi\{B,B'\}, Ap(t,u), \beta)\ ]$

(c) $\sigma = \sigma_1 \cap \cdots \cap \sigma_n$ $(n \geq 0)$.
$Comp(B,t,\sigma_1 \cap \cdots \cap \sigma_n) \Leftrightarrow \forall 1 \leq i \leq n\ [\ Comp(B,t,\sigma_i)\ ]$.

2. We say that a replacement R is *computable in a basis B* if there is a basis B' such that for every $x{:}\sigma \in B$, $Comp(B',x^{\mathrm{R}},\sigma)$ holds.

Notice that $Comp(B,t,\omega)$ holds as special case of part 1c. Also, since we use intersection types, and because of Definition 10, in part 2 we need not consider the existence of different bases for each $x{:}\sigma \in B$.

Property 19. C1. $Comp(B,t,\sigma)\ \&\ \sigma \neq \omega \Rightarrow HN(t)\ \&\ CHN(t)$.
C2. If $Comp(B,t,\sigma)$, and $t \rightarrow^*_{\mathrm{R}} t'$, then $Comp(B,t',\sigma)$.
C3. Let t be neutral. If $B \vdash_{\mathcal{E}} t{:}\sigma$, and there is a v such that $Comp(B,v,\sigma)$ and $t \rightarrow^*_{\mathrm{R}} v$, then $Comp(B,t,\sigma)$.
C4. Let t be neutral. If $B \vdash_{\mathcal{E}} t{:}\sigma$, $Hnf(t)$, and $Chnf(t)$, then $Comp(B,t,\sigma)$.

Proof. By simultaneous induction on the structure of types.

1. $\sigma = \varphi$, or $\sigma = s \in S$. By Definition 18.1a, using Theorem 15 and the Church-Rosser property for *C2*, and Theorem 15 for *C3* and *C4*.
2. $\sigma = \alpha \rightarrow \beta$.

C1. $Comp(B,t,\alpha\rightarrow\beta)\ \&\ x$ does not occur in $B \Rightarrow$ (*IH.C4*)
$Comp(B,t,\alpha\rightarrow\beta)\ \&\ Comp(\{x{:}\alpha\},x,\alpha) \Rightarrow$ (18.1b)
$Comp(B \cup \{x{:}\alpha\}, Ap(t,x), \beta) \Rightarrow$ (*IH.C1*)
$HN(Ap(t,x))\ \&\ CHN(Ap(t,x)) \Rightarrow$ (8.1)
$HN(t)\ \&\ CHN(t)$.

C2. $Comp(B,t,\alpha\rightarrow\beta) \Rightarrow$ (18.1b)
$(\ Comp(B',u,\alpha) \Rightarrow Comp(\Pi\{B,B'\},Ap(t,u),\beta)\ ) \Rightarrow$ (*IH.C2*)
$(\ Comp(B',u,\alpha) \Rightarrow Comp(\Pi\{B,B'\},Ap(t',u),\beta)\ ) \Rightarrow$ (18.1b)
$Comp(B,t',\alpha\rightarrow\beta)$.

C3. t is neutral $\&\ B \vdash_{\mathcal{E}} t{:}\alpha\rightarrow\beta\ \&\ \exists v\ [t \rightarrow^*_{\mathrm{R}} v\ \&\ Comp(B,v,\alpha\rightarrow\beta)] \Rightarrow$ (18.1b)
$(Comp(B',u,\alpha) \Rightarrow \exists v[t \rightarrow^*_{\mathrm{R}} v\ \&\ Comp(\Pi\{B,B'\},Ap(v,u),\beta)]) \Rightarrow$ (8.3)
$(Comp(B',u,\alpha) \Rightarrow \exists v[Ap(t,u) \rightarrow^*_{\mathrm{R}} v\ \&\ Comp(\Pi\{B,B'\},v,\beta)]) \Rightarrow$ (*IH.C3*)
$(\ Comp(B',u,\alpha) \Rightarrow Comp(\Pi\{B,B'\},Ap(t,u),\beta)\ ) \Rightarrow$ (18.1b)
$Comp(B,t,\alpha\rightarrow\beta)$.

C4. t is neutral $\&\ B \vdash_{\mathcal{E}} t{:}\alpha\rightarrow\beta\ \&\ Hnf(t)\ \&\ Chnf(t) \Rightarrow$ (8.2)
$(\ Comp(B',u,\alpha) \Rightarrow Ap(t,u)$ neutral $\&\ \Pi\{B,B'\} \vdash_{\mathcal{E}} Ap(t,u){:}\beta\ \&$
$Hnf(Ap(t,u))\ \&\ Chnf(Ap(t,u))\ ) \Rightarrow$ (*IH.C4*)
$(\ Comp(B',u,\alpha) \Rightarrow Comp(\Pi\{B,B'\},Ap(t,u),\beta)\ ) \Rightarrow$ (18.1b)
$Comp(B,t,\alpha\rightarrow\beta)$.

3. $\sigma = \sigma_1 \cap \cdots \cap \sigma_n$.
   C1. $Comp(B,t,\sigma_1 \cap \cdots \cap \sigma_n) \Rightarrow$ (18.1c)
   $\forall 1 \leq i \leq n\ [Comp(B,t,\sigma_i)] \Rightarrow$ (*IH.C1* & $n \neq 0$)
   $HN(t)$ & $CHN(t)$.
   C2. $Comp(B,t,\sigma_1 \cap \cdots \cap \sigma_n)$ & $t \rightarrow^* t' \Rightarrow$ (18.1c)
   $\forall 1 \leq i \leq n\ [Comp(B,t,\sigma_i)]$ & $t \rightarrow^* t' \Rightarrow$ (*IH.C2*)
   $\forall 1 \leq i \leq n\ [Comp(B,t',\sigma_i)] \Rightarrow$ (18.1c)
   $Comp(B,t',\sigma_1 \cap \cdots \cap \sigma_n)$.
   C3. t is neutral & $B \vdash_{\mathcal{E}} t{:}\sigma_1 \cap \cdots \cap \sigma_n$ & $\exists v\ [t \rightarrow_{\mathbf{R}} v$ &
   $Comp(B,v,\sigma_1 \cap \cdots \cap \sigma_n)] \Rightarrow$ (12.4 & 18.1c)
   $\exists v\ [t \rightarrow_{\mathbf{R}} v$ & $\forall 1 \leq i \leq n\ [Comp(B,v,\sigma_i)$ & $B \vdash_{\mathcal{E}} t{:}\sigma_i]] \Rightarrow$ (*IH.C3*)
   $\forall 1 \leq i \leq n\ [Comp(B,t,\sigma_i)] \Rightarrow$ (18.1c)
   $Comp(B,t,\sigma_1 \cap \cdots \cap \sigma_n)$.
   C4. t is neutral & $B \vdash_{\mathcal{E}} t{:}\sigma_1 \cap \cdots \cap \sigma_n$ & $Hnf(t)$ & $Chnf(t) \Rightarrow$ (12.4)
   t is neutral & $\forall 1 \leq i \leq n\ [B \vdash_{\mathcal{E}} t{:}\sigma_i$ & $Hnf(t)$ & $Chnf(t)] \Rightarrow$ (*IH.C4*)
   $\forall 1 \leq i \leq n\ [Comp(B,t,\sigma_i)] \Rightarrow$ (18.1c)
   $Comp(B,t,\sigma_1 \cap \cdots \cap \sigma_n)$.

In order to prove the head normalization theorem we shall prove a stronger property, for which we will need the following ordering and lemma.

Definition 20. Let $(\Sigma, \mathbf{R})$ be a CTRS.

1. Let $>_{\mathbb{N}}$ denote the standard ordering on natural numbers, $\trianglerighteq$ stand for the well-founded encompassment ordering, (i.e. $u \trianglerighteq v$ if $u \neq v$ and $u|_p = v^{\mathrm{R}}$ for some position $p \in u$ and replacement R), and *lex*, *mul* denote respectively the *lexicographic* and *multiset* extension of an ordering. Note that encompassment contains strict subterm (denoted by $\triangleright$).
2. We define the ordering $\gg$ on triples – consisting of a pair of natural numbers, a term, and a multiset of terms – as the object $((>_{\mathbb{N}}, >_{\mathbb{N}})_{lex}, \trianglerighteq, (\rightarrow_{Chnf} \cup \triangleright_{Chnf})_{mul})_{lex}$, where
   (a) $t \rightarrow_{Chnf} t'$ if $t \rightarrow^*_{\mathbf{R}} t'$, $\neg Chnf(t)$ and $Chnf(t')$,
   (b) $t \triangleright_{Chnf} t'$ if $t \triangleright t'$, $Chnf(t)$, $Chnf(t')$.
3. For computable R, we interpret a term t^{R} by the triple $\mathcal{I}(t^{\mathrm{R}}) = <(i,j),t,\{\mathrm{R}\}>$, where
   (a) i is the maximal super-index of the function symbols belonging to t,
   (b) j is the minimum of the differences $arity(F^i) - arity(F^i_k)$ such that F^i_k occurs in t,
   (c) $\{\mathrm{R}\}$ is the multiset $\{x^{\mathrm{R}} \mid x \in Var(\mathrm{t})$ & *the type of x in t is not ω*$\}$.

   These triples are compared in the ordering $\gg$, which is well-founded because R is computable, and we are taking the elements of its image that have a type different from ω.

Lemma 21. $Comp(B,t,\sigma)$, $\sigma \leq \rho \Rightarrow Comp(B,t,\rho)$.

We now come to the main theorem of this section, in which we show that for any typeable term and computable replacement R such that the term t^{R} is typeable, t^{R} is computable.

Property 22. Let t be a term such that $B \vdash_{\mathcal{E}} t{:}\sigma$, and R a computable replacement in B. Then there exists B' such that $Comp\,(B', t^{\mathrm{R}}, \sigma)$.

Proof. We will prove that t^{R} is computable or reduces to a computable term, by noetherian induction on $\gg$:

If t is a variable then t^{R} is computable by Lemma 21, since R is. Also, if σ is ω then t^{R} is trivially computable. Then, without loss of generality we assume that t is not a variable and that $\sigma \neq \omega$. Also, because of part 1c of Definition 18, we can restrict the proof to the case $\sigma \in \mathcal{T}_s$. We consider separatedly the cases:

1. t^{R} is neutral.
   (a) If $Hnf(t^{\mathrm{R}})$ and $Chnf(t^{\mathrm{R}})$ then t^{R} is computable by *C4*.
   (b) If $Hnf(t^{\mathrm{R}})$ but not $Chnf(t^{\mathrm{R}})$, then $t^{\mathrm{R}} = C(t_1, \ldots, t_n)$ where $C \in \mathcal{C}$ (constructor) or $t^{\mathrm{R}} = Ap(t_1, t_2)$ (in other case we would have $Chnf(t^{\mathrm{R}})$). For $1 \leq i \leq n$, t_i is computable, either because it is in R or by induction. In case $t^{\mathrm{R}} = C(t_1, \ldots, t_n)$, $t_1, \ldots, t_n$ have a type different from ω, then they have a constructor-hat normal form by *C1*, which implies $CHN(t^{\mathrm{R}})$. In case $t^{\mathrm{R}} = Ap(t_1, t_2)$, t_1 has a type different from ω, then by *C1*, $CHN(t_1)$, which implies $CHN(t^{\mathrm{R}})$. In both cases, t^{R} reduces to a neutral term t' such that $Chnf(t')$ and $Hnf(t')$. By *C3*, t' is computable, and again by *C3*, t^{R} is computable.
   (c) If not $Hnf(t^{\mathrm{R}})$, then there exists v such that $t^{\mathrm{R}} \rightarrow^*_{\mathrm{R}} v$ and v is itself a redex or $v = Ap(v_1, v_2)$ and v_1 is a redex. Without loss of generality, we can assume that $t^{\mathrm{R}} \rightarrow^*_{\mathrm{R}} v$ is the shortest derivation that satisfies this condition, then v can be decomposed into a non-variable term $u = F(z_1, \ldots, z_n)$ $(F \in \mathcal{F} \cup \{Ap\})$ and a computable replacement R', i.e. $v = u^{\mathrm{R}'}$. Note that R' is computable by induction and *C2*, since none of the reduction steps in the derivation $t^{\mathrm{R}} \rightarrow^*_{\mathrm{R}} v$ takes place at the root position (because it has minimal length).
      i. If $t \neq u$ modulo renaming of variables, then $\mathcal{I}(t^{\mathrm{R}}) \gg_2 \mathrm{I}(u^{\mathrm{R}'})$, then by induction, $u^{\mathrm{R}'}$ is computable and so is t^{R} by *C3*.
      ii. If $t = u$ modulo renaming of variables (without loss of generality, we can assume $t = u$); since t^{R} is neutral, v is neutral too, then only two cases are possible for u:
         A. $t = u = Ap(z_1, z_2)$.
         In this case, $t^{\mathrm{R}} \rightarrow^*_{\mathrm{R}} v = Ap(z_1, z_2)^{\mathrm{R}'}$. By the Subject Reduction Theorem, v has a type different from ω, then $z_1^{\mathrm{R}'}$ must have an arrow type, and since R' is computable, v is computable by 18.1b. Then t^{R} is computable by *C3*.
         B. $t = u = F^k(z_1, \ldots, z_n)$, and $v = t^{\mathrm{R}'}$ is reducible at the root position. Then, there is a rewrite rule

$$F^k(\overrightarrow{\mathrm{C}}[\overline{x}], \overline{y}) \rightarrow \mathrm{C'}[F^k(\overrightarrow{\mathrm{C}_1}[\overline{x}], \overline{y}), \ldots, F^k(\overrightarrow{\mathrm{C}_m}[\overline{x}], \overline{y}), \overline{x}, \overline{y}]$$

         such that $t^{\mathrm{R}} \rightarrow^* v = t^{\mathrm{R}'} = F^k(\overrightarrow{\mathrm{C}}[\overrightarrow{M}], \overrightarrow{N})$. By definition of safe CTRS, the patterns $\overrightarrow{\mathrm{C}}[\overline{x}]$ are constructor terms, and the terms $\overrightarrow{M}$ have sorts as types. Also, since $\overline{x} \subseteq \overline{y}$, we know that $\overrightarrow{M} \subseteq \overrightarrow{N}$, and then $\overrightarrow{M}$ are computable because $\overrightarrow{N}$ are (they are in R'). Then $CHN(\overrightarrow{M})$. Let R" be the computable replacement obtained from R' by putting $\overrightarrow{M}$ in CHNF (note that R" is computable by *C2*). There are two possible cases: either $\mathcal{I}(t^{\mathrm{R}}) \gg_3 \mathcal{I}(t^{R''})$, and then $t^{R''}$ is computable by induction, and so is t^{R} by *C3*, or

$$t^{\mathrm{R}} = t^{R''} = F^k(\overrightarrow{C}[\overrightarrow{M}], \overrightarrow{N}) \rightarrow_{\mathrm{R}}$$
$$\mathrm{C'}[F^k(\overrightarrow{C_1}[\overrightarrow{M}],\overrightarrow{N}),\ldots,F^k(\overrightarrow{C_m}[\overrightarrow{M}],\overrightarrow{N}), \overrightarrow{M}, \overrightarrow{N}].$$

Since $\overrightarrow{N}$ is computable (because it is in the image of R) and $\overrightarrow{M}$ is computable (because it is a subset of $\overrightarrow{N}$), the terms $\overrightarrow{C_i}[\overrightarrow{M}]$ are computable by induction (since by definition of the scheme, all function symbols in C_i have a superindex smaller than k). Also, by definition of the scheme, and because $Chnf(\overrightarrow{M})$, $\overrightarrow{C}[\overrightarrow{M}]$ $(\triangleright_{Chnf})_{mul}$ $\overrightarrow{C_i}[\overrightarrow{M}]$, then $F^k(\overrightarrow{C_i}[\overrightarrow{M},\overrightarrow{N}], \overrightarrow{N})$ is computable by induction. Again, by definition of the scheme and induction, $\mathrm{C'}[F^k(\overrightarrow{C_1}[\overrightarrow{M}], \overrightarrow{N}), \ldots, F^k(\overrightarrow{C_m}[\overrightarrow{M}], \overrightarrow{N}), \overrightarrow{M}, \overrightarrow{N}]$ is computable, since C' does not contain F^k. Then, by *C3*, t^{R} is computable since it is neutral.

2. t^{R} is not neutral. Then $t = F_i(t_1, \ldots, t_i)$, where F_i is the Curryfied version of some function symbol. In this case, since the type of t is not ω, t must have an arrow type, $\alpha \rightarrow \beta$. We have to prove that $Ap(F_i(t_1, \ldots, t_i), z)^{\mathrm{R}'}$ is computable for any replacement $R' = R \cup \{z \mapsto u\}$ such that u is computable of type α. But since $Ap(F_i(t_1, \ldots, t_i), z)^{\mathrm{R}'}$ is a neutral term, it is sufficient to prove that it reduces to a computable term (*C3*). Now, by definition of CTRS, $Ap(F_i(t_1, \ldots, t_i), z)^{\mathrm{R}'} \rightarrow_{\mathrm{R}} F_{i+1}(t_1, \ldots, t_i, z)^{\mathrm{R}'}$, and $\mathcal{I}(Ap(F_i(t_1, \ldots, t_i), z)^{\mathrm{R}'})$ $\gg_1 \mathcal{I}(F_{i+1}(t_1, \ldots, t_i, z)^{\mathrm{R}'})$, then $F_{i+1}(t_1, \ldots, t_i, z)^{\mathrm{R}'}$ is computable by induction, and we are finished.

Theorem 23 Head Normalization Theorem. *If $(\Sigma, \mathbf{R})$ is typeable in $\vdash_{\mathcal{E}}$, and safe, then every term t such that $B \vdash_{\mathcal{E}} t{:}\sigma$ and $\sigma \neq \omega$ has a head normal form.*

Proof. The theorem follows from Prop.22 and *C1*, taking R such that $x^{\mathrm{R}} = x$, which is computable by Prop.*C4*.

3.2 Normalization

In the intersection system for LC, it is well-known that terms that are typeable without ω in base and type are normalizable. This is not true in the rewriting framework, even if one considers safe recursive systems only. Take for instance the safe system:

$$Z(x, y) \rightarrow y,$$
$$D(x) \rightarrow Ap(x, x).$$

The term $Z_1(D(D_0))$ has type $\beta \rightarrow \beta$ in an environment where Z is typed with $\alpha \rightarrow \beta \rightarrow \beta$ and D with $\alpha \cap (\alpha \rightarrow \alpha) \rightarrow \alpha$, but is not normalizable.

We will then only study normalization of non-Curryfied terms in CTRS. Actually, to get a normalization result similar to that of LC we will also need to impose the following condition on CTRS:

Definition 24. A CTRS is *complete* if whenever a typeable non-Curryfied term t is reducible at a position p such that $t|_p$ has a type containing ω, t is reducible also at some $q < p$ such that $t|_q$ has a type without ω.

In order to be complete, a rewrite system must have rules that enable a reduction of $F(t_1, \ldots, t_n)$ at the root whenever there is a redex t_i the type of which contains ω. This means that the rules defining F cannot have non-variable patterns that have types with ω, and also that a constructor cannot accept arguments having a type which contains ω.

Constructors and recursive function symbols of safe systems satisfy these conditions. So, a safe recursive system is complete whenever the non-recursive defined symbols do not distinguish patterns that have a type containing ω.

From now on, we will consider only non-Curryfied CTRS that are safe and complete. This section will be devoted to the proof of the normalization theorem. We could use the method of Computability Predicates, as in the previous section, but since only non-Curryfied terms are considered, a direct proof is simpler. We will prove the theorem by noetherian induction, for which we will use the following ordering:

Definition 25. Let $(\Sigma, \mathbf{R})$ be a CTRS. Let $\succ$ denote the following well-founded ordering between terms: $t \succ t'$ if $t \rhd t'$ or t' is obtained from t by replacing the subterm $t|_p = F(t_1, \ldots, t_n)$ by the term $F(s_1, \ldots, s_n)$ where $\{t_1, \ldots, t_n\} \rhd_{mul} \{s_1, \ldots, s_n\}$. We define the ordering $\gg$ on triples composed of a natural number and two terms, as the object $(>_{\mathbb{N}}, \unrhd, \succ)_{lex}$.

Theorem 26 Normalization Theorem. *Let t be a non-Curryfied term in a typeable, complete, and safe CTRS. If $B \vdash_{\mathcal{E}} t{:}\sigma$ and ω does not appear in σ, then t is normalizable.*

Proof. By noetherian induction on $\gg$. We will interpret the term u by the triple $\mathcal{I}(u) = \langle i, u', u\rangle$ where i is the maximum of the super-indexes of the function symbols belonging to u that do not appear only in subterms in normal form or having a type with ω, and u' is the term obtained from u by replacing subterms in normal form with fresh variables. These triples are compared using $\gg$.

Assume t is not in normal form. All its strict subterms that have a type without ω are either already in normal form, or smaller than t with respect to $\gg$ and then normalizable by induction. Let v be the term obtained from t by reducing these subterms to normal form.

If $v \neq t$ then $\mathcal{I}(t) \gg_2 \mathcal{I}(v)$. Then v is normalizable by induction, and so is t.

If $v = t$ and it is a normal form, we are done. Otherwise, since the system is complete, t must be reducible at the root, and since it is a non-Curryfied term, the only possible reduction is:

$$t = F^i(\overrightarrow{C}[\overrightarrow{M}], \overrightarrow{N}) \rightarrow_{\mathbf{R}} C'[F^i(\overrightarrow{C_1}[\overrightarrow{M}], \overrightarrow{N}), \ldots, F^i(\overrightarrow{C_m}[\overrightarrow{M}], \overrightarrow{N}), \overrightarrow{M}, \overrightarrow{N}]$$

Now, the subterms of the right hand side of the form:

$$F^i(\overrightarrow{C_1}[\overrightarrow{M}], \overrightarrow{N}), \ldots, F^i(\overrightarrow{C_m}[\overrightarrow{M}], \overrightarrow{N}), \overrightarrow{M}, \overrightarrow{N}$$

that have a type without ω are normalizable by induction. Let $C''[\overline{u}]$ be the term obtained after normalizing those subterms, and including in the context the subterms that have a type with ω. By the Subject Reduction Theorem, this term has a type without ω, and by definition of the general scheme, it is smaller than t. Then, by the induction, it is normalizable.

References

1. S. van Bakel. Complete restrictions of the Intersection Type Discipline. *Theoretical Computer Science*, 102:135–163, 1992.
2. S. van Bakel. Partial Intersection Type Assignment of Rank 2 in Applicative Term Rewriting Systems. Technical Report 92-03, Department of Computer Science, University of Nijmegen, 1992.
3. S. van Bakel. Partial Intersection Type Assignment in Applicative Term Rewriting Systems. In *Proceedings of TLCA '93. International Conference on Typed Lambda Calculi and Applications*, Utrecht, the Netherlands, volume 664 of *LNCS*, pages 29–44. Springer-Verlag, 1993.
4. S. van Bakel and M. Fernández. Strong Normalization of Typeable Rewrite Systems. In *Proceedings of HOA '93. First International Workshop on Higher Order Algebra, Logic and Term Rewriting*, Amsterdam, the Netherlands. *Selected Papers*, volume 816 of *LNCS*, pages 20–39. Springer-Verlag, 1994.
5. F. Barbanera and M. Fernández. Combining first and higher order rewrite systems with type assignment systems. In *Proceedings of TLCA '93. International Conference on Typed Lambda Calculi and Applications*, Utrecht, the Netherlands, volume 664 of *LNCS*, pages 60–74. Springer-Verlag, 1993.
6. F. Barbanera, M. Fernández, and H. Geuvers. Modularity of Strong Normalization and Confluence in the λ-algebraic-cube. In *Proceedings of the ninth Annual IEEE Symposium on Logic in Computer Science*, Paris, France, 1994.
7. H. Barendregt, M. Coppo, and M. Dezani-Ciancaglini. A filter lambda model and the completeness of type assignment. *Journal of Symbolic Logic*, 48(4), 1983.
8. H.B. Curry and R. Feys. *Combinatory Logic*, volume 1. North-Holland, Amsterdam, 1958.
9. N. Dershowitz and J.P. Jouannaud. Rewrite systems. In *Handbook of Theoretical Computer Science*, volume B, chapter 6, pages 245–320. North-Holland, 1990.
10. J.-Y. Girard, Y. Lafont, and P. Taylor. *Proofs and Types*. Cambridge Tracts in Theoretical Computer Science. Cambridge University Press, 1989.
11. J.P. Jouannaud and M. Okada. Executable higher-order algebraic specification languages. In *Proceedings of the Sixth Annual IEEE Symposium on Logic in Computer Science*, pages 350–361, 1991.
12. J.W. Klop. Term Rewriting Systems. In *Handbook of Logic in Computer Science*, volume 2, chapter 1, pages 1–116. Clarendon Press, 1992.
13. W.W. Tait. Intensional interpretation of functionals of finite type I. *Journal of Symbolic Logic*, 32(2):198–223, 1967.

Explicit Substitutions with de Bruijn's Levels

Pierre LESCANNE and Jocelyne ROUYER-DEGLI

Centre de Recherche en Informatique de Nancy (CNRS) and INRIA-Lorraine
Campus Scientifique, BP 239,
F54506 Vandœuvre-lès-Nancy, France

email: {Pierre.Lescanne, Jocelyne.Rouyer}@loria.fr

1 Introduction

In the introduction of [7], Curry writes that substitution is the main issue in logic and that λ-calculus does not properly answer the problem because substitutions are outside the calculus. He pleads in favor of combinatory logic which offers a full treatment to substitutions through its use of a first order rewrite system in which substitution is cleanly handled. However, one may object that this formal system is not as natural as λ-calculus for describing the concept of function. In 1972, de Bruijn [8] proposed two notations he calls *indices* and *levels* that avoid α-conversion, i.e., renaming in terms, and later in 1978 [9, 10], he described a calculus based on his indices which nowadays we would call *explicit substitutions* and which proposes a full and correct treatment of substitution. Since that time, most of the formalisms for describing λ-calculus and explicit substitutions are based on de Bruijn's indices. Unlike our predecessors, in this paper we want to use de Bruijn's levels.

Calculus of explicit substitutions is a λ-calculus in which substitution is not external but is fully integrated at the same level as β reduction. This internalisation of the substitution calculus is achieved by rewrite rules which allow a full and easy mechanism for describing β reduction. The original goal of explicit substitutions is to provide the implementor of AUTOMATH (and later of functional programming languages) with a finer granularity in the description of the process of substitutions, as substitutions play the central role in the implementation of those systems and languages. This way, controls that postpone costly operations may be adopted and operations that will turn out to be unnecessary will never be performed *(lazy evaluation)*.

Except for a sketched attempt in [1] (*$\lambda\sigma$-calculus with names*), all the proposed calculi use *De Bruijn indices*. That notation has two drawbacks. First, terms are hard to read due to the use of numbers instead of names and due to the fact that the "same" variable is designated by different numbers according to the context. Second the association between variables and their values, the so-called environment, keeps changing whenever one leaves or enters an abstraction. To our knowledge "levels" were only mentioned twice in the literature, first by de Bruijn [8] and later by Crégut [4], but never in the framework of explicit substitutions. Both authors suggest a canonical indexing of the variables; Crégut calls it "reversed De Bruijn indexing". To make formulas as readable as in the classical λ-calculus, we give each variable a name made from its index and that name is the same everywhere in a pure term, i.e., a term without *closure* (see below) unlike classical De Bruijn index. Two important features of $\lambda\chi$ are the absence of variables of type *substitution* and the absence of composition of substitutions. Of course, the λ-calculus describes variables, but they are seen as constants by the rewrite system $\lambda\chi$ and they will be called *names* in what follows. The only variables of $\lambda\chi$ (those which play an actual role in the rewrite system) are of type *Term* and of type *Nat*.

To illustrate our approach and to allow the reader to make comparisons, Figure 1 gives a few examples in different notations. The first line is in the usual notation, the second is in $\lambda\chi$ notation, the third is in notation with de Bruijn's indices and the fourth is in level indices.

Yg	
Classical:	$(\lambda f\cdot(\lambda x\cdot f(xx))(\lambda x\cdot f(xx)))\ g$
Lambda Chi:	$(\lambda x_0\cdot(\lambda x_1\cdot x_0(x_1x_1))\,(\lambda x_1\cdot x_0(x_1x_1)))\ x_{-1}$
De Bruijn indices:	$(\lambda(\lambda(\underline{1}(\underline{0}\,\underline{0}))\ \lambda(\underline{1}(\underline{0}\,\underline{0}))))\ g$
level indices:	$(\lambda(\lambda\underline{0}(\underline{1}\,\underline{1}))\,(\lambda\underline{0}(\underline{1}\,\underline{1})))\ g$

Succ	
Classical:	$\lambda x\cdot x(\lambda y\cdot x\,y)$
Lambda Chi:	$\lambda x_0\cdot x_0(\lambda x_1\cdot x_0\,x_1)$
De Bruijn indices:	$\lambda\underline{0}(\lambda\underline{1}\,\underline{0})$
level indices:	$\lambda\underline{0}(\lambda\underline{0}\,\underline{1})$

Fig. 1. A few examples making notations explicit

Restrictions are imposed on names. In each term, each λ receives a level. λ's at the highest level (level 0) are associated with x_0 and λ's at level i are associated with x_i. Free variables, those that are not bound to any λ receive negative subscripts. In this framework, we can describe substitution and β reduction.

2 χ-terms, rules and examples

Let us call χ-terms, terms with explicit canonical variables and DB-terms, terms with De Bruijn's indices. The key to the correctness of $\lambda\chi$ is a translation from the first to the second. This translation relies on two operators σ_j and τ_i^j introduced for describing β reduction in DB-terms [4]. First let us introduce the important concept of level in χ-terms. In what follows $Term_i$ ($i \geq 0$) is the set of subterms of χ-terms at level i. A term at level i lies under i symbols λ and can contain only variables with indices up to $i-1$. The set of all χ-terms is $Term_0$.

The grammar of χ-terms is described by a set of inference rules, which we call the *level system*.

$$\frac{}{x_i : Term_{i+j+1}} \qquad \frac{a : Term_i \quad b : Term_i}{a\ b : Term_i}$$

$$\frac{a : Term_{i+1}}{\lambda x_i \cdot a : Term_i} \qquad \frac{a : Term_{i+j+1} \quad b : Term_i}{a[b/x_i]_j : Term_{i+j}}$$

Fig. 2. The level system: a description of χ-terms

Before giving the rules, let us look at the reduction of terms in this calculus.

$$\begin{aligned}
\mathsf{Y} \equiv\ & \lambda x_0 \cdot (\lambda x_1 \cdot x_0(x_1 x_1))\,(\lambda x_1 \cdot x_0(x_1 x_1)) && (1)\\
\underset{B}{\longrightarrow}\ & \lambda x_0 \cdot (x_0(x_1 x_1))[\lambda x_1 \cdot x_0(x_1 x_1)/x_1]_0 && (2)\\
\overset{+}{\underset{\chi}{\longrightarrow}}\ & \lambda x_0 \cdot x_0[\lambda x_1 \cdot x_0(x_1x_1)/x_1]_0\,(x_1[\lambda x_1 \cdot x_0(x_1x_1)/x_1]_0\, x_1[\lambda x_1 \cdot x_0(x_1x_1)/x_1]_0) && (3)\\
\underset{\chi}{\longrightarrow}\ & \lambda x_0 \cdot x_0\ (x_1[\lambda x_1 \cdot x_0(x_1x_1)/x_1]_0\ x_1[\lambda x_1 \cdot x_0(x_1x_1)/x_1]_0) && (4)\\
\overset{+}{\underset{\chi}{\longrightarrow}}\ & \lambda x_0 \cdot x_0\ (\lambda x_1 \cdot x_0(x_1 x_1))\,(\lambda x_1 \cdot x_0(x_1 x_1)) && (5)
\end{aligned}$$

Rewrite (2) creates an explicit substitution which when distributed through the whole term $x_0(x_1x_1)$ implements the β reduction. Rewrites (3) distribute the substitution through the term. Rewrite (4) performs the substitution on subterm x_0, i.e., leaves x_0 unchanged. Rewrites (5) perform the substitutions on subterm x_1.

$$\begin{aligned}
\mathsf{Y}\ \mathsf{Y} \equiv\ & (\lambda x_0 \cdot (\lambda x_1 \cdot x_0(x_1x_1))\,(\lambda x_1 \cdot x_0(x_1x_1)))\ \mathsf{Y} && (6)\\
\underset{B}{\longrightarrow}\ & ((\lambda x_1 \cdot x_0(x_1x_1))\ (\lambda x_1 \cdot x_0(x_1x_1)))[\mathsf{Y}/x_0]_0 && (7)\\
\underset{\chi}{\longrightarrow}\ & (\lambda x_1 \cdot x_0(x_1x_1))[\mathsf{Y}/x_0]_0\ (\lambda x_1 \cdot x_0(x_1x_1))[\mathsf{Y}/x_0]_0 && (8)\\
\underset{\chi}{\longrightarrow}\ & (\lambda x_0 \cdot (x_0(x_1x_1))[\mathsf{Y}/x_0]_1)\ (\lambda x_1 \cdot x_0(x_1x_1))[\mathsf{Y}/x_0]_0 && (9)\\
\overset{+}{\underset{\chi}{\longrightarrow}}\ & (\lambda x_0 \cdot (x_0[\mathsf{Y}/x_0]_1\ (x_1[\mathsf{Y}/x_0]_1\ x_1[\mathsf{Y}/x_0]_1)))\ (\lambda x_1 \cdot x_0(x_1x_1))[\mathsf{Y}/x_0]_0 && (10)\\
\overset{+}{\underset{\chi}{\longrightarrow}}\ & (\lambda x_0 \cdot (x_0[\mathsf{Y}/x_0]_1\ (x_0\ x_0)))\ (\lambda x_1 \cdot x_0(x_1x_1))[\mathsf{Y}/x_0]_0 && (11)
\end{aligned}$$

Rewrite (9) pushes the substitution under λ and so increments by 1 the index of the substitution and changes the names associated with this λ. Rewrite (10) distributes that substitution and rewrite (11) performs the substitutions on the x_1's which renames them to x_0.

In what a follows terms of the form $a[b/x_i]_j$ are called *closures*, terms of the form $a\ b$ are called *applications*, terms of the form $\lambda x_i \cdot a$ are called *abstractions*. From the above examples, we notice a few facts. Each substitution carries an index which we call the *depth* of the substitution. When a substitution $[a/x_i]_j$ goes through a λ, its index grows and the name associated with that λ decreases. When one actually performs a substitution, i.e., when one reduces a term of the form $x_i[a/x_k]_j$, if $i > k$ the index i of x_i is decremented, if $i < k$ the name is left unchanged. An actual replacement takes place only when a reduction of a term of the form $x_i[a/x_i]_j$ is performed. For instance, let us consider a reduction of the term $\lambda x_0 \cdot (x_0[\mathsf{Y}/x_0]_1\ (x_0\ x_0)))$. Performing the substitution $[\mathsf{Y}/x_0]_1$ requires replacement of the names x_0 by Y, in which names are modified. Its λx_0 will become a λx_1 and its λx_1 will become a λx_2 and its bound variables have to be modified accordingly. Therefore

$$\begin{aligned}&\lambda x_0 \cdot (x_0[\mathsf{Y}/x_0]_1\ (x_0\ x_0))\\ &\quad\equiv \lambda x_0 \cdot (x_0[\lambda x_0 \cdot (\lambda x_1 \cdot x_0(x_1 x_1))\,(\lambda x_1 \cdot x_0(x_1 x_1))/x_0]_1)\ (x_0\ x_0)\\ &\quad\underset{\chi}{\longrightarrow} \lambda x_0 \cdot (\mathit{rename}(\lambda x_0 \cdot (\lambda x_1 \cdot x_0(x_1 x_1))\,(\lambda x_1 \cdot x_0(x_1 x_1)), 0, 1)\ (x_0\ x_0))\\ &\quad\overset{+}{\underset{\chi}{\longrightarrow}} \lambda x_0 \cdot (\lambda x_1 \cdot (\lambda x_2 \cdot x_1(x_2 x_2))\,(\lambda x_2 \cdot x_1(x_2 x_2)))(x_0\, x_0).\end{aligned}$$

A function *rename* is introduced and in $\mathit{rename}(a, i, j)$, a is the term to be renamed, i is the level of the variable that created the renaming and j is the distance between a and the level where the substitution was created. In the previous case, we have

$$\begin{gathered}\mathit{rename}(\lambda x_0 \cdot (\lambda x_1 \cdot x_0(x_1 x_1))\,(\lambda x_1 \cdot x_0(x_1 x_1)), 0, 1)\\ \overset{+}{\underset{\chi}{\longrightarrow}}\\ \lambda x_1 \cdot (\lambda x_2 \cdot x_1(x_2 x_2))\,(\lambda x_2 \cdot x_1(x_2 x_2)).\end{gathered}$$

Rename can be seen as an "explicit" α conversion. Its behaviour is easily described by rewrite rules as well as the rest of the calculus (Fig. 3). The following rule needs to be added to the level system

$$\boxed{\frac{a : \mathit{Term}_{i+k}}{\mathit{rename}(a, i, j) : \mathit{Term}_{i+k+j}}}$$

to deal with *rename*. The system without the rule B, i.e., the system $\lambda\chi \setminus \{B\}$ that deals only with substitution removal, is called χ. The *normal form* of a is written $\chi(a)$, but we also use the notation $a \overset{!}{\underset{\chi}{\longrightarrow}}$ to say that b is the nomal form of a. $\lambda\chi$ preserves levels (a kind of subject reduction). In rule $\mathsf{Var}_>$, one has $0 \leq k < j$; that condition is naturally enforced by the level system. Moreover, $\lambda\chi$ uses the operator $+$ which is associative and commutative, such rewrite systems with associativity and commutativity are well known [2, 11] and allow us to avoid conditional rules. More precisely, we prefer to write

$$x_{i+k+1}[a/x_i]_j \;\rightarrow\; x_{i+k}$$

rather than

$$\frac{k > i}{x_k[a/x_i]_j \;\rightarrow\; x_{k-1}}$$

although in some proofs we prefer the equivalent second view. χ is not left-linear, but variables of sort Term_i (denoted by a, b and c) occur always once in left-hand sides (we will take advantage of this property) and the system has no critical pairs. In the next section we show that it correctly implements β reduction. χ does not contain composition of substitutions. There is a debate on its introduction. For some people it is a feature to have and for other it is a feature not to have. For us, the most natural way to handle this operation is by "concatenation" of substitutions getting expressions like $a[c_1/x_{i_1}]_{j_1}\cdots[c_p/x_{i_p}]_{j_p}$. Anyway calculi of explicit substitutions which contain composition are not strongly normalising on typed terms and we feel this feature is important.

(B)	$(\lambda x_i \cdot a)b \to a[b/x_i]_0$
(App)	$(a\ b)[c/x_i]_j \to a[c/x_i]_j\ b[c/x_i]_j$
(Lambda)	$(\lambda x_{i+j+1} \cdot a)[b/x_i]_j \to \lambda x_{i+j} \cdot (a[b/x_i]_{j+1})$
(Var$_>$)	$x_{i+k+1}[a/x_i]_j \to x_{i+k}$
(Var$_<$)	$x_i[a/x_{i+k+1}]_j \to x_i$
(Var$_=$)	$x_i[a/x_i]_j \to rename(a,i,j)$
(RenApp)	$rename(a\ b,i,j) \to rename(a,i,j)\ rename(b,i,j)$
(RenLambda)	$rename(\lambda x_{i+k} \cdot a,i,j) \to \lambda x_{i+k+j} \cdot rename(a,i,j)$
(RenVar$_\geq$)	$rename(x_{i+k},i,j) \to x_{i+k+j}$
(RenVar$_<$)	$rename(x_i,i+k+1,j) \to x_i$

Fig. 3. The calculus of explicit substitutions with levels $\lambda\chi$

3 Termination of χ

Consider the morphism:

$$\kappa: \quad \begin{aligned} \lambda x_i \cdot a &\mapsto \Lambda(\kappa(a)) \\ a\ b &\mapsto App(\kappa(a),\kappa(b)) \\ a[b/x_i]_j &\mapsto C(\kappa(a),\kappa(b)) \\ rename(a,i,j) &\mapsto ren(\kappa(a)) \\ x_i &\mapsto x \end{aligned}$$

and the system χ_1 of Fig.4. If $a \underset{\chi}{\longrightarrow} b$ then $\kappa(a) \underset{\chi_1}{\longrightarrow} \kappa(b)$.

Termination (or strong normalisation) of χ_1 implies this of χ and is proved by the recursive path ordering $>_{rpo}$ derived from the precedence $C > ren > \Lambda > App > x$. Remember that x is considered as a "constant".

Let us define rewsub an ordering based on $\underset{\chi}{\longrightarrow}$ and used in forthcoming proofs.

Definition 1 The rewsub ordering. The rewsub ordering is the ordering

$$(\underset{\chi}{\longrightarrow} \cup \sqsupset)^+,$$

where $\sqsupset$ is the subterm relation.

(App')	$C(App(a,b),c) \to App(C(a,c),C(b,c))$
(Lambda')	$C(\Lambda a,b) \to \Lambda(C(a,b))$
(Var$_>$')	$C(x,a) \to x$
(Var$_<$')	$C(x,a) \to x$
(Var$_=$')	$C(x,a) \to ren(a)$
(RenApp')	$ren(a\ b) \to ren(a)\ ren(b)$
(RenLambda')	$ren(\Lambda a) \to \Lambda(ren(a))$
(RenVar$_\geq$')	$ren(x) \to x$
(RenVar$_<$')	$ren(x) \to x$

Fig. 4. The rewrite system χ_1 for the proof of termination of χ

Rewsub is a well-founded ordering since $\underset{\chi}{\longrightarrow}$ is simply terminating [15], i.e., its termination is proved by a simplification ordering. Rewsub is actually the smallest simplification ordering that contains $\underset{\chi}{\longrightarrow}$.

4 Translation and correctness

As we have seen in the previous examples, β reduction is described by $a \underset{\beta}{\longrightarrow} b$ *if and only if* $a \underset{B}{\longrightarrow} b'$ *and* $b = \chi(b')$ *where is* $\chi(b')$ *is the normal form of* b' *by the rewrite system* χ. In this section, we prove the correctness of this definition. For this, we connect it with a definition known to be correct, namely a definition given in terms of De Bruijn indices [4, 5] and based on the two basic functions σ_n and τ_i^n.

$$\begin{array}{c} \sigma_n(ac,b) = \sigma_n(a,b)\sigma_n(c,b) \\ \sigma_n(\lambda a,b) = \lambda(\sigma_{n+1}(a,b)) \end{array} \quad \sigma_n(\underline{m},b) = \begin{cases} \underline{m-1} \text{ if } m > n+1 \\ \tau_0^n(b) \text{ if } m = n+1 \\ \underline{m} \text{ if } m \leq n \end{cases}$$

where:

$$\begin{array}{c} \tau_i^n(ab) = \tau_i^n(a)\tau_i^n(b) \\ \tau_i^n(\lambda a) = \lambda(\tau_{i+1}^n(a)) \end{array} \quad \tau_i^n(\underline{m}) = \begin{cases} \underline{m+n} \text{ if } m > i \\ \underline{m} \text{ if } m \leq i \end{cases}$$

$T(\_,i)$ is a translation of χ-terms of level i to DB-terms.

$$\begin{aligned} T(a\ b,i) &= T(a,i)\ T(b,i) \\ T(\lambda x_i \cdot a,i) &= \lambda T(a,i+1) \\ T(x_i,i+k+1) &= \underline{k+1} \\ T(a[b/x_i]_j,i+j) &= \sigma_j(T(a,i+j+1),T(b,i)) \\ T(rename(a,i,j),i+j+k) &= \tau_k^j(T(a,i+k)) \end{aligned}$$

The main proposition says that χ preserves T.

Proposition 2. *1. If $a \underset{\chi}{\longrightarrow} b$ then if a exists at level i, then b exists at level i and $T(a,i) = T(b,i)$.*
2. $T((\lambda x_i \cdot a)b, i) \underset{\beta}{\longrightarrow} T(a[b/x_i]_0, i) = T(\chi(a[b/x_i]_0), i)$.

Proof. We have to check the first assertion for each of the nine rules of χ at the correct level. The following equalities can be verified without difficulty using the definitions and the properties of T, σ_n and τ_i^n.

App

$$\begin{aligned} T((a\ b)[c/x_i]_j, i+j) &= \sigma_j(T(a, i+j+1), T(c,i))\ \sigma_j(T(b, i+j+1), T(c,i)) \\ &= T(a[c/x_i]_j b[c/x_i]_j, i+j) \end{aligned}$$

Lambda

$$\begin{aligned} T((\lambda x_{i+j+1} \cdot a)[c/x_i]_j, i+j) &= \lambda(\sigma_{j+1}(T(a, i+j+2), T(c,i))) \\ &= T(\lambda x_{i+j} \cdot (a[c/x_i]_{j+1}), i+j) \end{aligned}$$

$\mathsf{Var}_>$

$$\begin{aligned} T(x_{i+k+1}[a/x_i]_j, i+j) &= \underline{j-k} \\ &= T(x_{i+k}, i+j) \quad \text{for } k < j \end{aligned}$$

$\mathsf{Var}_<$

$$\begin{aligned} T(x_i[a/x_{i+k+1}]_j, i+k+1+j) &= \underline{k+j+1} \\ &= T(x_i, i+k+1+j) \end{aligned}$$

$\mathsf{Var}_=$

$$\begin{aligned} T(x_i[a/x_i]_j, i+j) &= \tau_0^j(T(a,i)) \\ &= T(rename(a,i,j), i+j) \end{aligned}$$

RenApp

$$\begin{aligned} T(rename(a\ b, i, j), i+j+k) &= \tau_k^j(T(a, i+k))\ \tau_k^j(T(b, i+k)) \\ &= T(rename(a,i,j)\ rename(b,i,j), i+j+k) \end{aligned}$$

RenLambda

$$\begin{aligned} T(rename(\lambda x_{i+k} \cdot, i, j), i+k+j) &= \lambda(\tau_{k+1}^j(T(a, i+k+1))) \\ &= T(\lambda x_{i+k+j} \cdot rename(a,i,j), i+k+j) \end{aligned}$$

$\mathsf{RenVar}_\leq$

$$\begin{aligned} T(rename(x_{i+k}, i, j), i+k+l+1+j) &= \underline{l+1} \\ &= T(x_{i+k+j}, i+k+j+l+1) \end{aligned}$$

RenVar$_<$

$$T(rename(x_i, i+k+1, j), i+k+l+1+j) = \underline{k+1+l+j}$$
$$= T(x_i, i+k+1+l+j)$$

For proving assertion 2, we use assertion 1.

$$\begin{array}{rcl} T((\lambda x_i \cdot a)\ b, i) & = & T(\lambda x_i \cdot a, i)\ T(b, i) \\ & = & \lambda(T(a, i+1)\ T(b, i)) \\ & \xrightarrow[\beta]{} & \sigma_0(T(a, i+1), T(b, i)) \\ & = & T(a[b/x_i]_0, i) \\ & & \quad \text{by definition of } T \\ & = & T(\chi(a[b/x_i]_0, i)) \\ & & \quad \text{by assertion 1.} \end{array}$$

5 Confluence of $\lambda\chi$

The goal of this section is to prove the confluence of $\lambda\chi$ over closed terms (terms without variables). The *confluence theorem* is based on a lemma classical when using Hardin's interpretation method which we call the *projection lemma*. It itself requires the $\lambda\chi$ version of an important classical lemma the *substitution lemma*. Before proving the substitution lemma, we state routine lemmas on *rename* expressions, namely Lemma 3 to Lemma 7, whose proofs by induction are straightforward providing Lemma 7 is invoked after Lemma 6.

The relation $\overset{\bullet}{\underset{\chi}{\longleftrightarrow}}$ is χ-convertibility. Since χ is confluent, it is defined by $a \overset{\bullet}{\underset{\chi}{\longleftrightarrow}} b$ if there exists c such that $a \overset{\bullet}{\underset{\chi}{\longrightarrow}} c$ and $b \overset{\bullet}{\underset{\chi}{\longrightarrow}} c$. In every proof of convertibility below, we will only examine terms where variables stand for terms in χ-normal form, i.e., terms without substitutions or *rename*. Indeed this will not change the generality of the result.

Lemma 3.

$$rename(rename(b, i+k, p), i, j) \overset{\bullet}{\underset{\chi}{\longleftrightarrow}} rename(rename(b, i, j), i+k+j, p)$$

Lemma 4. $rename(a[b/x_{i+k}]_p, i, j) \overset{\bullet}{\underset{\chi}{\longleftrightarrow}} rename(a, i, j)[rename(b, i, j)/x_{i+k+j}]_p$.

Lemma 5. $rename(c, p, i+j) \overset{\bullet}{\underset{\chi}{\longleftrightarrow}} rename(c, p, i+j+1)[d/x_{i+p}]_{j+q}$.

Lemma 6. $rename(c, p, i+k+j) \overset{\bullet}{\underset{\chi}{\longleftrightarrow}} rename(rename(c, p, i+k), p+i, j)$.

Lemma 7. $rename(b, i+p+1, j)[c/x_p]_{i+k+j} \overset{\bullet}{\underset{\chi}{\longleftrightarrow}} rename(b[c/x_p]_{i+k}, i+p, j)$.

Lemma Substitution lemma

$$a[b/x_{i+p+1}]_j[c/x_p]_{i+j} \overset{\bullet}{\underset{\chi}{\longleftrightarrow}} a[c/x_p]_{i+j+1}[b[c/x_p]_i/x_{i+p}]_j.$$

Proof. This lemma is proved by structural induction on a. The level of a is $i+p+j+2$.

- $a \equiv \lambda x_{i+p+j+2} \cdot a'$.
  $$\begin{aligned}(\lambda x_{i+p+j+2} \cdot a')[b/x_{i+p+1}]_j[c/x_p]_{i+j} &\overset{\bullet}{\underset{\chi}{\longrightarrow}} \lambda x_{i+p+j} \cdot (a'[b/x_{i+p+1}]_{j+1}[c/x_p]_{i+j+1})\\ &\overset{\bullet}{\underset{\chi}{\longleftrightarrow}} \lambda x_{i+p+j} \cdot (a'[c/x_p]_{i+j+2}[b[c/x_p]_i/x_{i+p}]_{j+1})\\ &\overset{\bullet}{\underset{\lambda\chi}{\longrightarrow}} (\lambda x_{i+p+j+2} \cdot a')[c/x_p]_{i+j+1}[b[c/x_p]_i/x_{i+p}]_j\end{aligned}$$
- $a \equiv a'\ b'$
  $$\begin{aligned}(a'\ b')[b/x_{i+p+1}]_j[c/x_p]_{i+j} &\underset{\chi}{\longrightarrow} a'[b/x_{i+p+1}]_j[c/x_p]_{i+j}\ b'[b/x_{i+p+1}]_j[c/x_p]_{i+j}\\ &\overset{\bullet}{\underset{\chi}{\longleftrightarrow}} a'[c/x_p]_{i+j+1}[b[c/x_p]_i/x_{i+p}]_j\ b'[c/x_p]_{i+j+1}[b[c/x_p]_i/x_{i+p}]_j\\ &\underset{\lambda\chi}{\longrightarrow} (a'\ b')[c/x_p]_{i+j+1}[b[c/x_p]_i/x_{i+p}]_j\end{aligned}$$
- $a \equiv x_k$ with $k < i+p+2$.
  We have to rewrite $x_k[b/x_{i+p+1}]_j[c/x_p]_{j+1}$ and $x_k[c/x_p]_{i+j+1}[b[c/x_p]_i/x_{i+p}]_j$. Only rules Var are applied and three cases have to be considered. Both terms rewrite in two steps to the same terms. Those terms are:
  - x_k for $k < p$,
  - x_{k-1} for $p < k < i+p+1$,
  - x_{k-2} for $k > i+p+1$.

  The other cases need lemmas. For $k = p$, we have
  $$x_k[b/x_{i+p+1}]_j[c/x_p]_{j+1} \overset{\bullet}{\underset{\chi}{\longrightarrow}} rename(c, p, i+j)$$
  and
  $$x_k[c/x_p]_{i+j+1}[b[c/x_p]_i/x_{i+p}]_j \overset{\bullet}{\underset{\chi}{\longrightarrow}} rename(c, p, i+j+1).$$
  and the result comes from Lemma 5.
  The last case $k = i+p+1$ gives us
  $$x_k[b/x_{i+p+1}]_j[c/x_p]_{i+j} \underset{\chi}{\longrightarrow} rename(b, i+p+1, j)[c/x_p]_{i+j}$$
  and
  $$x_k[c/x_p]_{i+j+1}[b[c/x_p]_i/x_{i+p}]_j \overset{\bullet}{\underset{\chi}{\longrightarrow}} rename(b, [c/x_p]_i, i+p, j)$$
  the result is a consequence of Lemma 7.

Lemma 8 Projection Lemma. *If* $a \underset{B}{\longrightarrow} b$ *then* $\chi(a) \overset{\bullet}{\underset{\beta}{\longrightarrow}} \chi(b)$.

Proof. The proof is very similar to that of [1] and [17]. The terms can be supposed of the form $a \equiv a'[c_1/x_{i_1}]_{j_1} \cdots [c_p/x_{i_p}]_{j_p}$ and $b \equiv b'[d_1/x_{i_1}]_{j_1} \cdots [d_p/x_{i_p}]_{j_p}$ where a' is not a closure. We proceed by **rewsub** induction on a and we distinguish cases according to the structure of a'.

- $a \equiv (a_1a_2)[c_1/x_{i_1}]_{j_1}\dots[c_p/x_{i_p}]_{j_p}$ and $b \equiv (b_1a_2)[c_1/x_{i_1}]_{j_1}\dots[c_p/x_{i_p}]_{j_p}$ and the B redex occurs inside a_1 with $a_1 \underset{B}{\rightarrow} b_1$, then

$$a_1[c_1/x_{i_1}]_{j_1}\dots[c_p/x_{i_p}]_{j_p} \underset{B}{\rightarrow} b_1[c_1/x_{i_1}]_{j_1}\dots[c_p/x_{i_p}]_{j_p}.$$

By induction,

$$\chi(a_1[c_1/x_{i_1}]_{j_1}\dots[c_p/x_{i_p}]_{j_p}) \overset{\bullet}{\underset{\beta}{\rightarrow}} \chi(b_1[c_1/x_{i_1}]_{j_1}\dots[c_p/x_{i_p}]_{j_p})$$

and

$$\begin{aligned}\chi(a) &= \chi((a_1a_2)[c_1/x_{i_1}]_{j_1}\dots[c_p/x_{i_p}]_{j_p})\\ &= \chi(a_1[c_1/x_{i_1}]_{j_1}\dots[c_p/x_{i_p}]_{j_p})\chi(a_2[c_1/x_{i_1}]_{j_1}\dots[c_p/x_{i_p}]_{j_p})\\ &\overset{\bullet}{\underset{\beta}{\rightarrow}} \chi(b_1[c_1/x_{i_1}]_{j_1}\dots[c_p/x_{i_p}]_{j_p})\chi(a_2[c_1/x_{i_1}]_{j_1}\dots[c_p/x_{i_p}]_{j_p})\\ &= \chi((b_1a_2)[c_1/x_{i_1}]_{j_1}\dots[c_p/x_{i_p}]_{j_p})\\ &= \chi(b).\end{aligned}$$

The case $a_2 \underset{B}{\rightarrow} b_2$ works similarly.
- If a' is the B redex, that is $a \equiv ((\lambda x_{i_0+p}\cdot a'_1)a_2)[c_1/x_{i_1}]_{j_1}\dots[c_p/x_{i_p}]_{j_p}$ then b' is a closure, that is $b \equiv a'_1[a_2/x_{i_0+p}]_0[c_1/x_{i_1}]_{j_1}\dots[c_p/x_{i_p}]_{j_p}$.

$$\begin{aligned}\chi(a) &= \chi(((\lambda x_{i_0+p}\cdot a'_1)a_2)[c_1/x_{i_1}]_{j_1}\dots[c_p/x_{i_p}]_{j_p})\\ &= \lambda x_{i_0}\cdot(\chi(a'_1[c_1/x_{i_1}]_{j_1+1}\dots[c_p/x_{i_p}]_{j_p+1}))\ \chi(a_2[c_1/x_{i_1}]_{j_1}\dots[c_p/x_{i_p}]_{j_p})\\ &\overset{\bullet}{\underset{\beta}{\rightarrow}} \chi(\chi(a'_1[c_1/x_{i_1}]_{j_1+1}\dots[c_p/x_{i_p}]_{j_p+1})[\chi(a_2[c_1/x_{i_1}]_{j_1}\dots[c_p/x_{i_p}]_{j_p})/x_{i_0}]_0)\\ &= \chi(a'_1[c_1/x_{i_1}]_{j_1+1}\dots[c_p/x_{i_p}]_{j_p+1}[a_2[c_1/x_{i_1}]_{j_1}\dots[c_p/x_{i_p}]_{j_p}/x_{i_0}]_0)\end{aligned}$$

and by repeating p times (zero times if $p = 0$) the Substitution Lemma

$$= \chi(a'_1[a_2/x_{i_0+p}]_0[c_1/x_{i_1}]_{j_1}\dots[c_p/x_{i_p}]_{j_p}) = \chi(b)$$

- If $a \equiv (\lambda x_{i_0}\cdot a_1)[c_1/x_{i_1}]_{j_1}\dots[c_p/x_{i_p}]_{j_p}$, then $a_1 \underset{B}{\rightarrow} b_1$ or $c_i \underset{B}{\rightarrow} d_i$,

$$a_1[c_1/x_{i_1}]_{j_1+1}\dots[c_p/x_{i_p}]_{j_p+1} \underset{B}{\rightarrow} b_1[d_1/x_{i_1}]_{j_1+1}\dots[d_p/x_{i_p}]_{j_p+1}$$

and

$$\lambda x_{i_0}\cdot(a_1[c_1/x_{i_1}]_{j_1+1}\dots[c_p/x_{i_p}]_{j_p+1}) \underset{B}{\rightarrow} \lambda x_{i_0}\cdot(b_1[d_1/x_{i_1}]_{j_1+1}\dots[d_p/x_{i_p}]_{j_p+1})$$

and we can apply the induction hypothesis, since the left-hand side is a χ rewrite of a.
- If $a \equiv x_{i_0}[c_1/x_{i_1}]_{j_1}\dots[c_p/x_{i_p}]_{j_p}$, then the B redex is in the substitution part, say in c_k.

$$x_{i_0}[c_1/x_{i_1}]_{j_1}\dots[c_k/x_{i_k}]\dots[c_p/x_{i_p}]_{j_p} \underset{B}{\rightarrow} x_{i_0}[c_1/x_{i_1}]_{j_1}\dots[d_k/x_{i_k}]\dots[c_p/x_{i_p}]_{j_p}.$$

If $k > 1$ and $x_{i_0}[c_1/x_{i_1}]_{j_1}\underset{\chi}{\rightarrow}f$ then

$$f[c_2/x_{i_2}]\dots[c_k/x_{i_k}]\dots[c_p/x_{i_p}]_{j_p} \underset{B}{\rightarrow} f[c_2/x_{i_2}]\dots[d_k/x_{i_k}]\dots[c_p/x_{i_p}]_{j_p}$$

and the result follows by induction. If $k = 1$ and $i_0 = i_1$ then

$$x_{i_0}[c_1/x_{i_1}]_{j_1} \dots [c_p/x_{i_p}]_{j_p} \underset{\chi}{\longrightarrow} rename(c_1, i_0, j_1)[c_2/x_{i_2}]_{j_2} \dots [c_p/x_{i_p}]_{j_p+1}$$

$$x_{i_0}[d_1/x_{i_1}]_{j_1} \dots [c_p/x_{i_p}]_{j_p} \underset{\chi}{\longrightarrow} rename(d_1, i_0, j_1)[c_2/x_{i_2}]_{j_2} \dots [c_p/x_{i_p}]_{j_p+1}.$$

Clearly, if $c_1 \underset{B}{\longrightarrow} d_1$, then $rename(c_1, i_0, j_1) \underset{B}{\longrightarrow} rename(d_1, i_0, j_1)$ and the result follows by induction. If $k = 1$ and $i_0 < i_1$ then

$$x_{i_0}[c_1/x_{i_1}]_{j_1} \dots [c_p/x_{i_p}]_{j_p} \underset{\chi}{\longrightarrow} x_{i_0}[c_2/x_{i_2}]_{j_2} \dots [c_p/x_{i_p}]_{j_p+1}$$

$$x_{i_0}[d_1/x_{i_1}]_{j_1} \dots [c_p/x_{i_p}]_{j_p} \underset{\chi}{\longrightarrow} x_{i_0}[c_2/x_{i_2}]_{j_2} \dots [c_p/x_{i_p}]_{j_p+1}$$

and both normal forms by χ are equal and the results holds trivially, the same if $k = 1$ and $i_0 > i_1$ except that x_{i_0} becomes $x_{i_0 - 1}$.

- If $a \equiv rename(a'', i, j)[c_1/x_{i_1}]_{j_1} \dots [c_p/x_{i_p}]_{j_p}$, then the B redex is either in a'' or in one of the c_i's. Two cases may occur, one rewrite of $rename(a'', i, j)$ preserves the B redex: $a'' = a_1\ a_2$ and the B redex occurs in a_1 or in a_2 or $a'' = \lambda x_i a_1$, then one proceeds by induction. The second case is when $a'' = (\lambda x_{i+k} a_1)\ a_2$, then $b' \equiv rename(a_1[a_2/x_{i+k}]_0, i, j)$,

$$\begin{aligned}
\chi(a) &= \chi(rename((\lambda x_{i+k} \cdot a_1)\ a_2, i, j)[c_1/x_{i_1}]_{j_1} \dots [c_p/x_{i_p}]_{j_p}) \\
&= \chi((\lambda x_{i+k+j} \cdot rename(a_1, i, j))rename(a_2, i, j)[c_1/x_{i_1}]_{j_1} \dots [c_p/x_{i_p}]_{j_p}) \\
&\underset{B}{\longrightarrow} \chi(rename(a_1, i, j)[rename(a_2, i, j)/x_{i+k+j}]_0[c_1/x_{i_1}]_{j_1} \dots [c_p/x_{i_p}]_{j_p}) \\
&= \chi(rename(a_1[a_2/x_{i+k}]_0, i, j)[c_1/x_{i_1}]_{j_1} \dots [c_p/x_{i_p}]_{j_p}) \\
&\equiv \chi(b).
\end{aligned}$$

The last equality follows from Lemma 4.

Theorem 9 Confluence. *$\lambda\chi$ is confluent.*

Proof. One uses the projection lemma and Hardin's interpretation method [14] illustrated by the following picture.

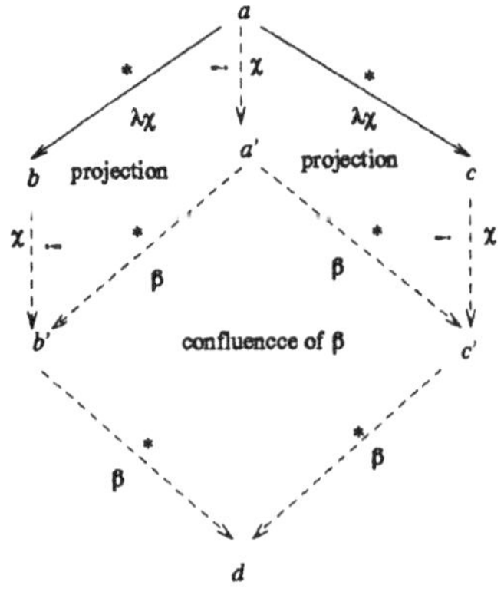

6 $\lambda\chi$ preserves strong normalisation

Not all calculi of explicit substitutions preserve strong β normalisation. For instance, Paul-André Mellies [18] has exhibited strongly β normalisable terms that are not strongly $\lambda\sigma$ normalisable or strongly $\lambda\sigma_{\Uparrow}$ normalisable. On the other hand, in [17, 3], we have proved that the calculus of substitutions $\lambda\upsilon$ preserves strong normalisation. A similar proof can be used for proving that $\lambda\chi$ preserves strong normalisation. This requires a few definitions an lemmas.

Definition 10 External position. A position p is *internal* in a term a, if the subterm $a_{|p}$ is a subterm of c where $d[c/x_i]_j$ is a subterm of a. A position is *external* if it is not an internal position.

Lemma 11. *If p is an external position and if $a \underset{B,p}{\longrightarrow} b$, then $\chi(a) \overset{+}{\underset{\beta}{\longrightarrow}} \chi(b)$. In particular, if $\chi(a)$ is strongly normalisable, then $\chi(a) \neq \chi(b)$.*

We are going only to sketch the proof which is similar to this described in [3]. It is by contradiction based on a minimal counter-example. More precisely if we suppose that $\lambda\chi$ does not preserve strong normalisation, there exists a pure term a which is strongly β normalisable and which is not strongly $\lambda\chi$ normalisable, in other words there exists infinite $\lambda\chi$ derivations starting from a. Among those infinite derivations there exists at least one which is minimal in the following sense (see Fig. 5). At each rewrite position in the derivation one rewrites at the lowest possible position that keeps the derivation infinite, that is one may rewrite at a lower position, but the derivations which continue that rewrite are finite. The proof relies on two important lemmas.

Lemma 12 Commutation Lemma. *If $\chi(a)$ is strongly β normalisable, $\chi(a) = \chi(b)$ and $a \underset{\lambda\chi,p}{\longrightarrow}^{int} \cdot \underset{\chi,q}{\longrightarrow}^{ext} b$ then $a \overset{+}{\underset{\chi}{\longrightarrow}}^{ext} \cdot \overset{*}{\underset{\lambda\chi}{\longrightarrow}}^{int} b$.*

Lemma 13. *Let a_1 be a strongly β normalisable term. In each infinite $\lambda\chi$ derivation of terms*

$$a_1 \underset{\lambda\chi}{\longrightarrow} a_2 \ldots a_n \underset{\lambda\chi}{\longrightarrow} a_{n+1} \ldots$$

there exists an N such that for $i \geq N$ all the $\lambda\chi$ rewrites are internal.

Theorem 14. *Each strongly β normalisable term is strongly $\lambda\chi$ normalisable.*

Proof. The proof works as follows. If there exists an infinite $\lambda\chi$ derivation starting with a β normalisable term, one considers a minimal such derivation $\mathcal{D}$ and one uses Lemmas 11-13 to show that a B rewrite which takes place after N (see Lemma 13) can be "lifted" at $J \leq N$. The lifted rewrite on a_J is at position q lower than p_J where p_J is the position of the Jth $\lambda\chi$ rewrite in $\mathcal{D}$. That rewrite is continued into an infinite derivation, which contradicts the minimality of $\mathcal{D}$.

Corollary 15. *$\lambda\chi$ is strongly normalising on typed terms.*

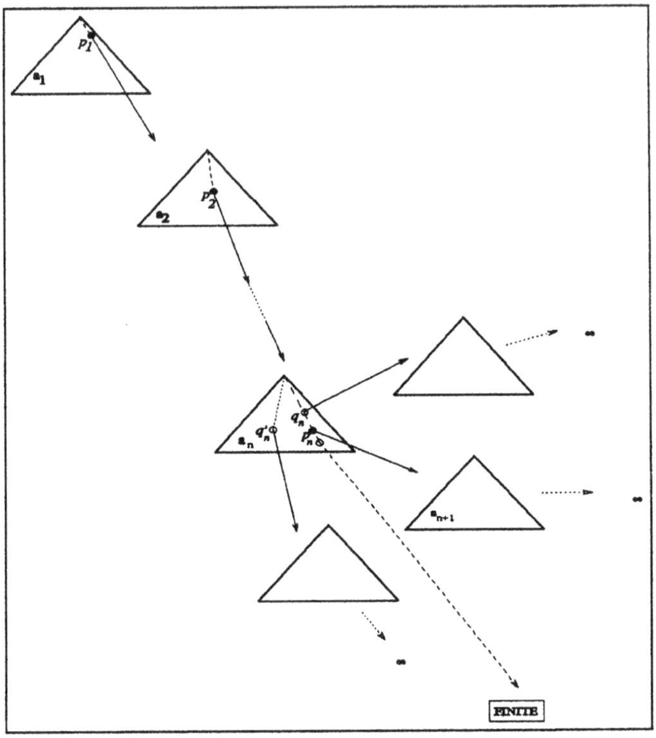

Fig. 5. A minimal $\lambda\chi$ derivation, $a_1, \ldots, a_n, a_{n+1}$

7 Conclusion and related works

In this paper we have presented a calculus of explicit substitutions with levels. We have proved three important properties.

- $\lambda\chi$ correctly implements β conversion in the standard λ calculus.
- $\lambda\chi$ is confluent on closed (ground) χ terms, i.e., terms without variables.
- $\lambda\chi$ is preserves strong β normalisation.

Since the first presentation by de Bruijn [10], all the other calculi proposed in the literature [1, 6, 12, 13, 16, 17, 19, 20] use De Bruijn indices (except one attempt in [1] already mentioned). Our approach is therefore original by its use of levels which improves readability. We attach much importance to confluence on ground terms and to strong normalisation on typed terms. The non confluence on open terms seems to us less fundamental because of the apparent impossibility of getting both properties, e.g., confluence on open terms and preservation of strong normalisation in the same system and the inability shown by Field [13, 12] to get optimality.

Our calculus raises interesting open issues on implementation and on higher order unification. We feel indeed that due to the non superposition of left-hand sides, our $\lambda\chi$ should entail a nice description of implementations, for instance, of graph reduction based implementations. In parallel, tools for higher order theorem provers like strong normalisation and higher unification should be easily described through $\lambda\chi$.

References

1. M. Abadi, L. Cardelli, P.-L. Curien, and J.-J. Lévy. Explicit substitutions. *Journal of Functional Programming*, 1(4):375–416, 1991.
2. L. Bachmair and N. Dershowitz. Completion for rewriting modulo a congruence. *Theoretical Computer Science*, 67(2-3):173–202, October 1989.
3. Z. Benaissa, D. Briaud, P. Lescanne, and J. Rouyer-Degli. λv, a calculus of explicit substitutions which preserves strong normalisation. Submitted, December 1994.
4. P. Crégut. An abstract machine for the normalization of λ-calculus. In *Proc. Conf. on Lisp and Functional Programming*, pages 333–340. ACM, 1990.
5. P. Crégut. *Machines à environnement pour la réduction symbolique et l'évaluation partielle*. PhD thesis, Université de PARIS 07, 1991.
6. P.-L. Curien, Th. Hardin, and J.-J. Lévy. Confluence properties of weak and strong calculi of explicit substitutions. RR 1617, INRIA, Rocquencourt, February 1992.
7. H. B. Curry, R. Feys, and W. Craig. *Combinatory Logic*, volume 1. Elsevier Science Publishers B. V. (North-Holland), Amsterdam, 1958.
8. N. G. de Bruijn. Lambda calculus with nameless dummies, a tool for automatic formula manipulation, with application to the Church-Rosser theorem. *Proc. Koninkl. Nederl. Akademie van Wetenschappen*, 75(5):381–392, 1972.
9. N. G. de Bruijn. Lambda calculus with namefree formulas involving symbols tha represent reference transforming mappings. *Proc. of the Koninklijke Nederlan Akademie*, 81(3):1–9, September 1978. Dedicated to A. Heyting at the occasion his 80th Birrthday on May 9, 1978.
10. N. G. de Bruijn. A namefree lambda calculus with facilities for internal definitic of expressions and segments. TH-Report 78-WSK-03, Technological Universi Eindhoven, Netherlands, Department of Mathematics, 1978.
11. N. Dershowitz and J.-P. Jouannaud. Rewrite Systems. In J. van Leeuwen, edi tor, *Handbook of Theoretical Computer Science*, chapter 6, pages 244–320. Elsevie Science Publishers B. V. (North-Holland), 1990.
12. J. Field. On laziness and optimality in lambda interpreters: Tools for specificatior and analysis. In *Proceedings of the 17th Annual ACM Symposium on Principles O Programming Languages, Orlando (Fla., USA)*, pages 1–15, San Fransisco, 1990. ACM.
13. J. Field. *Incremental Reduction in the Lambda Calculus and Related Reduction Systems*. PhD thesis, Cornell U., November 1991.

14. Th. Hardin. Confluence results for the pure strong categorical combinatory logic CCL: λ-calculi as subsystems of CCL. *Theoretical Computer Science*, 65:291–342, 1989.
15. M. Kurihara and A. Ohuchi. Modularity of simple termination of term rewriting systems. *Journal of IPS Japan*, 31(5):633–642, 1990.
16. P. Lescanne. From $\lambda\sigma$ to λv, a journey through calculi of explicit substitutions. In Hans Boehm, editor, *Proceedings of the 21st Annual ACM Symposium on Principles Of Programming Languages, Portland (Or., USA)*, pages 60–69. ACM, 1994.
17. P. Lescanne and J. Rouyer-Degli. The calculus of explicit substitutions λv. Technical Report RR-2222, INRIA-Lorraine, January 1994.
18. P.-A. Melliès. Typed λ-calculi with explicit substitutions may not terminate. In M. Dezani, editor, *Int. Conf. on Typed Lambda Calculus and Applications*, 1995.
19. A. Ríos. *Contributions à l'étude des λ-calculs avec des substitutions explicites.* Thèse de Doctorat d'Université, U. Paris VII, 1993.
20. T. Strahm. Partial applicative theories and explicit substitutions. Technical Report IAM 93-008, Univerität Bern, Institut für Informatik und angewandte Mathematik, June 1993.

A Restricted Form of Higher-Order Rewriting Applied to an HDL Semantics*

Richard J. Boulton

University of Cambridge Computer Laboratory, New Museums Site, Pembroke Street, Cambridge CB2 3QG, United Kingdom

Abstract. An algorithm for a restricted form of higher-order matching is described. The intended usage is for rewrite rules that use function-valued variables in place of some unknown term structure. The matching algorithm instantiates these variables with suitable λ-abstractions when given the term to be rewritten. Each argument of one of the variables is expected to match some unique substructure. Multiple solutions are avoided by making fixed choices when alternative ways to match arise. The algorithm was motivated by correctness proofs of designs written in a hardware description language. The feature of the language's semantics that necessitates the higher-order rewriting is described.

1 Background

Amongst other things, theorem provers are used for formal reasoning about computer systems, and rewriting plays a significant role in this activity. When a first-order logic is used, the rewriting is necessarily first-order. Even when a higher-order logic is used most of the equational reasoning required remains first-order. At least, it is often straightforward to avoid higher-order rewriting. However, circumstances do arise in which higher-order rewriting is the only practical solution. This paper describes one such circumstance and presents the rewriting algorithm developed to deal with it.

The HOL system [3] is a proof assistant for a higher-order logic, specifically a version of Church's Simple Theory of Types [2] with type variables in the object language. The HOL system is equipped with a programming language, ML, in which users can write their own proof procedures. The strong typing of ML is used to ensure that such procedures are sound. Recent versions of the system use Standard ML [7], a quite widely used functional programming language with strict evaluation.

Formal verification of hardware was the original application domain of the HOL system, but the higher-order logic has been found to be more generally applicable to formal reasoning. A common approach is to embed specialised formalisms (e.g., temporal logics, process calculi) and computer languages in higher-order logic by defining their semantics in the logic. The programmability

* Research supported by the Engineering and Physical Sciences Research Council (formerly the Science and Engineering Research Council) of Great Britain.

of the HOL system allows formalism-specific proof tools to be implemented. It is in using one such embedding, of the hardware description language ELLA[2], that the need for the higher-order rewriting procedure described in this paper arose.

ELLA [8] is a hardware design language with features that support both structural and algorithmic specification. Many of the language features are in a functional style and it is possible to isolate a useful subset of the language that has no imperative features. This was the approach taken for the embedding in HOL [1].

2 Motivation

The embedding of ELLA in HOL represents signals as functions from time (natural numbers) to the value of the signal at that time. Thus, functions operating over signals are higher-order. However, the need for higher-order rewriting arises from the mapping of λ-abstractions over the lists used to represent collections of signals. Consider the ELLA construct with syntax *unit*`[`int_1`..`int_2`]`. The *unit* is a collection of signals $(s_1, \ldots, s_n)$, and int_1 and int_2 are integers. The construct denotes $(s_{int_1}, \ldots, s_{int_2})$ provided $1 \leq int_1 \leq int_2 \leq n$.

Treating the integers as natural numbers for simplicity, the meaning of the above construct can be expressed in HOL as:

```
MAP (λn. EL (n-1) unit) (int₁ UPTO int₂)
```

where `EL` is a function such that `(EL 0 l)` returns the head (first element) of the list `l`, `(EL 1 l)` returns the second element, and so on. `UPTO` is an infix function such that `(m UPTO n)` returns the list `[m;m+1;...;n-1;n]`. If `n` is less than `m` then `(m UPTO n)` returns an empty list. The function `MAP` should be familiar to a functional programmer: it applies a function to each element of a list. The function being applied by `MAP` in this case is the λ-abstraction. `EL` is undefined when its first argument is greater than or equal to the size of the list. This ensures that the construct only has a meaning when int_1 and int_2 have values in the range of the *unit*.

Once *unit*, int_1 and int_2 have been instantiated, simplification is attempted. The application of `MAP` can only be expanded if int_1 and int_2 are explicit numeric constants. However, simplification inside the λ-abstraction may still be possible. This can be achieved by first-order rewriting if a general rewrite rule is first instantiated by hand, but that is very time consuming. To do the job mechanically a form of higher-order rewriting is required. To see this, consider the following term, which differs somewhat from the one above to make the point more clearly:

```
MAP (λn. (EL ((n-1)+1) l) + 2) (1 UPTO ((LENGTH l) - 1))
```

The function `LENGTH` computes the length of a list. The term 'evaluates' to the tail of the list `l` with 2 added to every element. If `l` is an empty list, the result is an empty list. Now suppose the term is to be simplified to:

[2] ELLA is a registered trademark of the Secretary of State for Defence (United Kingdom) acting through the Defence Research Agency.

```
MAP (λn. (EL n l) + 2) (1 UPTO ((LENGTH l) - 1))
```

For natural numbers it is not in general true that $((n-1)+1) = n$, for if n is 0, then $(n-1)$ is undefined or given some non-negative value. So, for the transformation to be valid it must be possible to show that **n** cannot be zero. This is the case in the example because the application of **UPTO** yields a list that cannot contain zero as an element. The following rewrite rule (a HOL theorem) can be proved:

```
⊢ MAP (λx. f ((x-1)+1)) (1 UPTO m) = MAP (λx. f x) (1 UPTO m)
```

It would suffice to prove the specific theorem required for the example, but a large number of similar theorems may have to be proved. It is better to have the general theorem above. However, `f((x-1)+1)` does not first-order match with `(EL ((n-1)+1) l) + 2`. It is necessary to use higher-order matching, so that `f` can become instantiated to `(λv. (EL v l) + 2)`.

More complicated examples require multiple variables to be higher-order matched. Such variables may appear in the arguments of others. The example below is a simplified version of one that arose in a real verification proof. Consider a term of the form:

```
MAP
(λn. ... (λt. ... (EL n (CONS (...t...) (...t...))) ...) ... n ...)
l
```

If **n** is greater than zero, the element selected by **EL** cannot be the first element of the list constructed by **CONS** and so only the tail of the list need be retained. However, **n** is bound, so in order to make this simplification an assertion is required that every value it can take on (in this case, every element of the list **l**) is greater than zero. The required theorem is:

```
⊢ ∀f g a d l.
    EVERY (λx. 0 < x) l ⟹
    (MAP (λx. f x (λt. g (EL x (CONS (a t) (d t))))) l =
     MAP (λx. f x (λt. g (EL (x-1) (d t)))) l)
```

There are two things to note about this theorem. First, **a** and **d** appear in the argument of **g** which itself appears in an argument of **f**. All four of these variables have to be higher-order matched. The variables **a** and **d** have to have **t** as an argument since it is a bound variable. The second point to note relates to this: **f** has **x** as an additional argument. This is present to match the occurrence of **n** that appears outside the abstraction over **t** in the term to be simplified. Without the extra argument the rewriting cannot succeed because **x** will then appear free in the abstraction with which **f** is instantiated. This **x** will not then correspond to the bound **x** in the rewrite rule. So, in general, a variable to be higher-order matched may require an extra argument for each bound variable.

The above analysis motivates the definition of a skeleton in the following section and the nature of the matching algorithm given in Sect. 5. The algorithm is a hybrid of first-order matching and higher-order matching.

3 Syntax of Terms and Skeletons

The matching algorithm in Sect. 5 operates over terms of higher-order logic. The terms are essentially those of the λ-calculus. The syntax is given by

$$t \ ::= \ x \mid c \mid \lambda x.t \mid t_1 \ t_2$$

where x denotes a variable (bound or free) and c denotes a constant. (The subscripts denote possibly distinct instances of the syntactic category.) The function Frees computes the free variables of a set of terms. The notation $t : \sigma$ is used to denote that term t has type σ. Types in the logic of the HOL system may be polymorphic, i.e., contain type variables (α). The syntax is:

$$\sigma \ ::= \ \alpha \mid (\sigma_1, \ldots, \sigma_n)tyc \ .$$

The number of arguments n to the type constructor tyc may be zero, in which case tyc is a base type.

Definition 1. A *skeleton* is a term in which certain free variables have been designated as candidates for higher-order matching. These variables are denoted by capital letters (e.g., G) and must satisfy the following conditions (when they are referred to as *gaps*):

1. Each gap may occur at most once.
2. Each gap must be applied to at least $m + n$ arguments where the first m arguments are precisely the variables bound at the point of occurrence (order irrelevant), and $n \geq 0$. (The size of n may be specified by the user.)
3. The n further arguments (called *the subskeletons*) may not be a variable or an application of a gap.

Intuitively, then, a skeleton is like a first-order pattern but with gaps that can be 'filled' by any structure. The 'filling' is achieved by higher-order matching. This can be seen in the examples in Sect. 2 as a correspondence between gaps and uses of '...'.

4 Additional Notation

In the description of the matching algorithm the following additional notation is used: σ and τ denote types, s skeletons, θ term substitutions, ϕ type substitutions, B bindings of bound variables, and u and v new term variables guaranteed to have a different name to all existing variables. All symbols may be indexed by subscripts, e.g., x_i, or appear primed, e.g., t'.

Tuples are denoted by $\langle \ldots \rangle$ with the components separated by commas, sets as $\{\ldots\}$, and lists as $[\ldots]$. Lists are ordered collections of elements which may not be distinct. A list with head x and tail X is denoted by $x :: X$. The notation $\overline{x_n}$ denotes the sequence of objects $x_1, \ldots, x_n$. So, $\{\overline{x_n}\}$ denotes the set containing these objects, $[\overline{x_n}]$ denotes the list of them in sequence, $\lambda \overline{x_n}.t$ denotes $\lambda x_1 \ldots \lambda x_n.t$, and $f\ (\overline{x_n})$ denotes the iterated application $(\ldots (f\ x_1) \ldots x_n)$.

Definition 2. A *term substitution* is a set of pairs $\left\{\overline{\langle x_n, t_n\rangle}\right\}$ where each x is a term variable, each t is a term, and the x's are distinct. Similarly, a *type substitution* is a set of pairs $\left\{\overline{\langle \alpha_n, \sigma_n\rangle}\right\}$.

An element pair of a substitution is written $x \mapsto t$ signifying that a substitution is a finite mapping from variables to terms (or types). The notation $\theta \setminus \{x \mapsto \ldots\}$ means the result of removing any pairs whose first component is x from the substitution θ. The notation θ/t means the result of replacing any elements of θ of the form $x \mapsto t$ by $x \mapsto u_i$ where the u's are new variables. Also, $t\phi$ denotes the result of instantiating the (polymorphic) types of t according to the type substitution ϕ, and $t\theta$ denotes the result of specialising the free variables of t according to the term substitution θ. The set of terms (second components) of a substitution θ is denoted by $\mathsf{Terms}(\theta)$.

The bindings denoted by B are sets of pairs in which the first component is a bound variable from the skeleton and the second component is the corresponding bound variable in the term to be matched (or a new unique variable). Much of the notation used for substitutions is also used for these bindings.

5 The Algorithm

This section describes the algorithm for restricted higher-order matching. The algorithm is defined as a collection of rules (Fig. 1) over two mutually-dependent relations. The first, $\stackrel{?}{=}_{\mathrm{F}}$, performs first-order matching until it encounters a gap, when the second, $\stackrel{?}{=}_{\mathrm{H}}$, is used to match the gap. While matching a gap, the first relation is used to match the subskeletons. Each rule should be interpreted as follows: if all the relations above the horizontal line (the *hypotheses*) hold then the relation below the line (the *conclusion*) also holds.

Three auxiliary relations also occur in the rules: $\stackrel{?}{=}_{\sigma}$, $\stackrel{?}{\cup}_{\mathrm{t}}$, and $\stackrel{?}{\cup}_{\sigma}$. These match types, merge term substitutions, and merge type substitutions, respectively. The matching of types is first-order. The merging relations are defined as follows:

$$\forall \theta_1, \theta_2, \theta.\ \theta_1 \stackrel{?}{\cup}_{\mathrm{t}} \theta_2 \longrightarrow \theta \text{ iff}$$
$$(\theta = \theta_1 \cup \theta_2\ \wedge\ (\forall x, t_1, t_2.\ \langle x, t_1\rangle \in \theta_1\ \wedge\ \langle x, t_2\rangle \in \theta_2\ \Rightarrow\ t_1 = t_2))$$
$$\forall \phi_1, \phi_2, \phi.\ \phi_1 \stackrel{?}{\cup}_{\sigma} \phi_2 \longrightarrow \phi \text{ iff}$$
$$(\phi = \phi_1 \cup \phi_2\ \wedge\ (\forall \alpha, \sigma_1, \sigma_2.\ \langle \alpha, \sigma_1\rangle \in \phi_1\ \wedge\ \langle \alpha, \sigma_2\rangle \in \phi_2\ \Rightarrow\ \sigma_1 = \sigma_2))$$

The notation $s \stackrel{?}{=}_{\mathrm{F}} t \stackrel{B}{\longrightarrow} \langle\theta, \phi\rangle$ denotes the fact that the 4-tuple $\langle s, t, B, \langle\theta,\phi\rangle\rangle$ is an element of the relation $\stackrel{?}{=}_{\mathrm{F}}$, and $\langle X, S\rangle \stackrel{?}{=}_{\mathrm{H}} t \stackrel{B}{\longrightarrow} \langle S', t', \theta, \phi\rangle$ that the 4-tuple $\langle\langle X, S\rangle, t, B, \langle S', t', \theta, \phi\rangle\rangle$ is an element of $\stackrel{?}{=}_{\mathrm{H}}$. X is a set of pairs where the first component of each pair is a new variable and the second component is a bound variable. S is a list of pairs where the first component is a new variable and the second is a skeleton. S' is a modified S, and t' is a term that will form part of the body of the λ-abstraction that will be bound to the gap whose subskeletons are in S.

$$\frac{\begin{array}{c}\left\langle\left\{\overline{\langle u_m : \sigma_m, x_m : \sigma_m\rangle}\right\}, \left[\overline{\langle v_n : \tau_n, s_n : \tau_n\rangle}\right]\right\rangle =^{?}_{\mathrm{H}} t \xrightarrow{B} \left\langle [\,], t' : \sigma', \theta, \phi\right\rangle \\ \sigma =^{?}_{\sigma} \sigma' \longrightarrow \phi' \quad \phi \cup^{?}_{\sigma} \phi' \longrightarrow \phi'' \quad \left\{G \mapsto \lambda\overline{u_m \phi''}.\lambda\overline{v_n \phi''}.t'\right\} \cup^{?}_{\mathrm{t}} \theta \longrightarrow \theta'\end{array}}{G\,(\overline{x_m : \sigma_m})\,(\overline{s_n : \tau_n}) : \sigma =^{?}_{\mathrm{F}} t \xrightarrow{B} \langle\theta', \phi''\rangle} \tag{1}$$

$$\frac{\sigma_s =^{?}_{\sigma} \sigma_t \longrightarrow \phi \quad (x_s : \sigma_s \mapsto x_t : \sigma_t) \in B}{x_s : \sigma_s =^{?}_{\mathrm{F}} x_t : \sigma_t \xrightarrow{B} \langle\{x_s \mapsto x_t\}, \phi\rangle} \tag{2}$$

$$\frac{\sigma_s =^{?}_{\sigma} \sigma_t \longrightarrow \phi \quad (x \mapsto \ldots) \notin B}{x : \sigma_s =^{?}_{\mathrm{F}} t : \sigma_t \xrightarrow{B} \langle\{x \mapsto t\}, \phi\rangle} \tag{3}$$

$$\frac{\sigma_s =^{?}_{\sigma} \sigma_t \longrightarrow \phi}{c : \sigma_s =^{?}_{\mathrm{F}} c : \sigma_t \xrightarrow{B} \langle\{\}, \phi\rangle} \tag{4}$$

$$\frac{\begin{array}{c}x_s =^{?}_{\mathrm{F}} x_t \xrightarrow{B'} \langle\theta_1, \phi_1\rangle \quad s =^{?}_{\mathrm{F}} t \xrightarrow{B'} \langle\theta_2, \phi_2\rangle \\ B' = \{x_s \mapsto x_t\} \cup (B \setminus \{x_s \mapsto \ldots\} / x_t) \quad \theta_1 \cup^{?}_{\mathrm{t}} \theta_2 \longrightarrow \theta \quad \phi_1 \cup^{?}_{\sigma} \phi_2 \longrightarrow \phi\end{array}}{\lambda x_s.s =^{?}_{\mathrm{F}} \lambda x_t.t \xrightarrow{B} \langle\theta \setminus \{x_s \mapsto \ldots\}, \phi\rangle} \tag{5}$$

$$\frac{\begin{array}{c}s_1 =^{?}_{\mathrm{F}} t_1 \xrightarrow{B} \langle\theta_1, \phi_1\rangle \quad s_2 =^{?}_{\mathrm{F}} t_2 \xrightarrow{B} \langle\theta_2, \phi_2\rangle \\ \theta_1 \cup^{?}_{\mathrm{t}} \theta_2 \longrightarrow \theta \quad \phi_1 \cup^{?}_{\sigma} \phi_2 \longrightarrow \phi\end{array}}{s_1\ s_2 =^{?}_{\mathrm{F}} t_1\ t_2 \xrightarrow{B} \langle\theta, \phi\rangle} \tag{6}$$

$$\frac{s =^{?}_{\mathrm{F}} t \xrightarrow{B} \langle\theta, \phi\rangle}{\langle X, \langle v, s\rangle :: S\rangle =^{?}_{\mathrm{H}} t \xrightarrow{B} \langle S, v\phi, \theta, \phi\rangle} \tag{7}$$

$$\frac{\sigma_i =^{?}_{\sigma} \sigma \longrightarrow \phi \quad (x_i : \sigma_i \mapsto x : \sigma) \in B}{\langle\{\ldots, \langle u_i, x_i : \sigma_i\rangle, \ldots\}, S\rangle =^{?}_{\mathrm{H}} x : \sigma \xrightarrow{B} \langle S, u_i\phi, \{x_i \mapsto x\}, \phi\rangle} \tag{8}$$

$$\frac{(\ldots \mapsto x) \notin B}{\langle X, S\rangle =^{?}_{\mathrm{H}} x \xrightarrow{B} \langle S, x, \{\}, \{\}\rangle} \tag{9}$$

$$\frac{}{\langle X, S\rangle =^{?}_{\mathrm{H}} c \xrightarrow{B} \langle S, c, \{\}, \{\}\rangle} \tag{10}$$

$$\frac{B' = B/x \quad \langle X, S\rangle =^{?}_{\mathrm{H}} t \xrightarrow{B'} \langle S', t', \theta, \phi\rangle \quad x \notin \mathsf{Frees}\,(\mathsf{Terms}\,(\theta))}{\langle X, S\rangle =^{?}_{\mathrm{H}} \lambda x.t \xrightarrow{B} \langle S', \lambda x.t', \theta, \phi\rangle} \tag{11}$$

$$\frac{\begin{array}{c}\langle X, S\rangle =^{?}_{\mathrm{H}} t_1 \xrightarrow{B} \left\langle S', t_1', \theta_1, \phi_1\right\rangle \quad \left\langle X, S'\right\rangle =^{?}_{\mathrm{H}} t_2 \xrightarrow{B} \left\langle S'', t_2', \theta_2, \phi_2\right\rangle \\ \theta_1 \cup^{?}_{\mathrm{t}} \theta_2 \longrightarrow \theta \quad \phi_1 \cup^{?}_{\sigma} \phi_2 \longrightarrow \phi\end{array}}{\langle X, S\rangle =^{?}_{\mathrm{H}} t_1\ t_2 \xrightarrow{B} \langle S'', t_1'\ t_2', \theta, \phi\rangle} \tag{12}$$

Fig. 1. The matching algorithm as a collection of rules

Definition 3. The relations $=^?_{\mathrm{F}}$ and $=^?_{\mathrm{H}}$ are defined inductively by the rules in Fig. 1. Rules (2) to (6) are only applicable if (1) is not.

Remark. It is necessary to do a type match between the gap and the value it is to be instantiated to (Rule (1)). The matching of the subskeletons to subterms takes care of much of this type matching, but not all of it.

Remark. When an argument of a gap matches successfully, the corresponding new unique variable is instantiated for types using the type substitution produced by the match (Rules (7) and (8)). This instantiated variable is then included in a new term, and the original term is rebuilt around it (and around any other variables inserted because of a successful match) (Rules (9) to (12)). When rebuilding an abstraction (11), a check is made to ensure that the bound variable does not appear free in any of the subterms bound by the matching of the body. If this is not the case, the variable is being passed out of its scope.

Definition 4 (Matching). A skeleton s matches a term t if and only if there exist term and type substitutions θ and ϕ such that $s =^?_{\mathrm{F}} t \xrightarrow{\{\}} \langle \theta, \phi \rangle$.

Remark. There may be more than one pair $\langle \theta, \phi \rangle$ satisfying the $=^?_{\mathrm{F}}$ relation. However, the aim is to obtain a unique solution when there is any, so that the rewriting will be deterministic. This can be achieved in a concrete implementation by making choices as to when to apply certain rules. The definition of skeletons and the form of the rules already impose restrictions. In particular, a list rather than a set is used for the subskeletons of a gap to force them to match in a unique order. The issue of restrictions is considered in more depth below and in later sections.

Lemma 5. *At any point, the set of first components of the binding B is equal to the set of enclosing bound variables in the skeleton.*

Proof sketch. Only rules (5) and (11) modify B but (11) has no effect on the set of first components. Rule (5) is the only rule that enters an abstraction in the skeleton. (Rule (11) enters an abstraction in the term but not the skeleton.) The binding B' used in matching the abstraction bodies in (5) has an element added for the bound variable of the skeleton and previous entries for a variable with the same name and type are removed. (Such a variable is invisible inside the body.) The hypothesis of (5) for matching the bound variables can only be satisfied by rule (2) or (3). The use of B' causes (2) to be chosen.

Lemma 6. *At any point, the set of second components of X is equal to the set of first components of B.*

Proof sketch. The set X is introduced by (1) and propagated by (11) and (12). No rule modifies it. Rule (1) makes the second components of X the bound variable arguments of the gap G. By condition 2 of Def. 1 these must be the variables bound in the surrounding skeleton. It follows from Lemma 5 that the second components of X equal the first components of B.

Lemma 7. *At any point, the second components of the binding B are distinct.*

Proof sketch. B is initially empty. Only rule (5) adds an element to B. Before adding $\{x_s \mapsto x_t\}$, any existing elements with x_t as the second component have that component replaced by a new variable distinct from all others. Rule (11) is the only other rule to modify B but any new second component introduced by it is also a new variable.

Definition 8 (Parallel substitution). For a term substitution θ and a type substitution ϕ, the parallel substitution $\theta \mid \phi$ is defined so that $t(\theta \mid \phi)$ is the result of simultaneously substituting free variables in t using θ and instantiating types elsewhere in t (bound variables and constants) using ϕ.

Remark. This form of substitution avoids the difficulties of applying the substitutions in sequence. If types are instantiated first then the term substitution must be instantiated for types before it can be used. Since type instantiation may force variables to be renamed the renaming has to be co-ordinated between the term and the term substitution if the latter is still to be applicable to the former following instantiation. As an example, consider:

$$f\ (x:\alpha)\ (x:\sigma) =^?_{\mathrm{F}} c\ (y:\sigma)\ z \xrightarrow{\{\}} \langle\{f \mapsto c, x:\alpha \mapsto y, x:\sigma \mapsto z\}, \{\alpha \mapsto \sigma\}\rangle\ .$$

The result of applying the type substitution to the skeleton is $f\ (x':\sigma)\ (x:\sigma)$, the polymorphic x being renamed to avoid conflict with the other x. The term substitution is no longer valid unless the types of its terms are instantiated and the same renaming occurs.

The alternative of applying the term substitution before the type substitution is problematic because the intermediate term may be ill-typed.

The above example demonstrates that substitution may cause variables to be renamed. Also, rule (5) removes the substitution entry for the bound variables, so that the bound variable name in the skeleton may be different from the name of the corresponding variable in the term even after the substitutions are applied. Hence, the instantiated skeleton may be only α-equivalent to the term.

Definition 9 (α-equivalence). A term t_1 is α-equivalent to a term t_2, denoted $t_1 =_\alpha t_2$, if and only if there exists some renaming of bound variables in t_1 that makes t_1 equal to t_2.

Lemma 10. *For all s, t, B, θ, ϕ satisfying $s =^?_{\mathrm{F}} t \xrightarrow{B} \langle\theta, \phi\rangle$, there exists an entry in θ for every free variable of s.*

Proof sketch. Rules (1) and (3) are the only ones dealing with free variables in the skeleton. They both introduce an entry for the free variable into θ. Only (5) removes entries from θ and then the entry is for a bound variable.

Corollary 11. *When $\theta \mid \phi$ is applied to s all renaming of free variables is explicit in θ. Renaming due to type instantiation only occurs for bound variables.*

Definition 12 (β-reduction). The notation $s_\downarrow$ denotes the result of β-reducing at occurrences of gaps in an instantiated skeleton s.

Conjecture 13 (Correctness). *For all s, t, B, θ, ϕ, if $s =^?_{\mathrm{F}} t \xrightarrow{B} \langle\theta, \phi\rangle$ then $s(\theta \mid \phi)_\downarrow =_\alpha t$.*

Remark. The fact that the binding B does not appear in the consequent of the above implication is indicative of its role only as a way of explicitly identifying which variables are bound. Its value affects whether rules are applicable but all the vital substitution information is also carried by θ.

Remark. Due to the mutual dependency of $=^?_{\mathrm{F}}$ and $=^?_{\mathrm{H}}$, a correctness result is required for $=^?_{\mathrm{H}}$ in order to prove the one for $=^?_{\mathrm{F}}$:

$$\left\langle\left\{\overline{\langle u_m, x_m\rangle}\right\}, \langle v_1, s_1\rangle :: \ldots :: \langle v_k, s_k\rangle :: S'\right\rangle =^?_{\mathrm{H}} t \xrightarrow{B} \langle S', t', \theta, \phi\rangle \quad\Rightarrow$$
$$t'\left\{\overline{u_m\phi \mapsto x_m(\theta \mid \phi)}\right\}\left\{\overline{v_k\phi \mapsto s_k(\theta \mid \phi)_\downarrow}\right\} =_\alpha t$$

where $k \geq 0$. Since the v's are new variables, the substitution involving them need not be restricted to those already matched (i.e., $v_1, \ldots, v_k$); those not matched do not appear in t', so substituting for them will have no effect.

Proof sketch. The correctness of type matching is assumed, i.e., for all σ_s, σ_t, ϕ, $\sigma_s =^?_\sigma \sigma_t \longrightarrow \phi$ implies $\sigma_s\phi = \sigma_t$. The proof proceeds by induction on the structure of the skeleton (for $=^?_{\mathrm{F}}$) and on the structure of the term (for $=^?_{\mathrm{H}}$).

Conjecture 14 (Finiteness). *The set $\left\{\langle\theta, \phi\rangle \;\middle|\; s =^?_{\mathrm{F}} t \xrightarrow{B} \langle\theta, \phi\rangle\right\}$ is finite.*

Proof sketch. Assuming that solution sets to $=^?_\sigma$, $\cup^?_{\mathrm{t}}$, and $\cup^?_\sigma$ are finite (Actually, they can be expected to have a unique solution, if any.), then there is only a finite number of ways of successfully applying each of the rules. Also, for every rule except (7) either the skeletons or the terms involved in the hypotheses are strictly smaller than the skeleton or term in the conclusion, and for (7) the hypothesis can only be satisfied by a rule that strictly decreases the skeleton. Hence, there can be no infinite sequence of rule applications.

Lemma 15. *For any skeleton, at most one of the rules (2) and (3) can have their hypotheses satisfied, and for any term to be matched, at most one of the rules (8) and (9) can have their hypotheses satisfied.*

Conjecture 16 (Uniqueness). *If rules (8) to (12) are rejected whenever (7) can be applied successfully then the successful match (if any) is unique.*

Remark. The condition causes subskeletons to be matched as early as possible in a top-down depth-first left-to-right order.

Remark. Under certain circumstances (7) may be applied successfully only to have the derivation fail due to the hypothesis $x \notin \mathsf{Frees}(\mathsf{Terms}(\theta))$ of (11). The condition of the conjecture may then prevent a successful match even when there is an alternative way to do it. Ideally, this hypothesis of (11) should be replaced by one for (3) which has the same effect at source.

Proof sketch. It is sufficient to show that for any s, t, B, there is at most one way of satisfying the $=^?_{\mathrm{F}}$ relation, and for any X, S, t, B, there is at most one way of satisfying the $=^?_{\mathrm{H}}$ relation. By the restriction on rules (2) to (6) in Def. 3 and by Lemma 15, only one of rules (1) to (6) can be applied successfully for any given s. By the condition of the conjecture and by Lemma 15, only one of rules (7) to (12) can be applied successfully for any given t. So, assuming that solution sets to $=^?_{\sigma}$, $\mathsf{U}^?_{\mathrm{t}}$, and $\mathsf{U}^?_{\sigma}$ are empty or singletons, an inductive argument proves the uniqueness for $=^?_{\mathrm{F}}$ and $=^?_{\mathrm{H}}$. (Rule (8) may appear to have more than one solution but Lemma 7 shows that there can be at most one element of B with x as its second component.)

6 An Example

Figure 2 is a successful application of the rules in Fig. 1 to the problem of matching $\lambda x.f\ x\ ((x-1)+1)$ (where f is a gap) to $\lambda n.(\mathtt{EL}\ ((n-1)+1)\ l)+n$. (See Sect. 2.) The rules are written with the hypotheses stacked and indented, e.g., (3) is written:

$$
\begin{array}{l}
x:\sigma_s =^?_{\mathrm{F}} t:\sigma_t \xrightarrow{B} \langle\{x \mapsto t\},\phi\rangle \\
\quad (3) \\
\quad \sigma_s =^?_{\sigma} \sigma_t \longrightarrow \phi \\
\quad (x \mapsto \ldots) \notin B\ .
\end{array}
$$

Due to lack of space, only major branches are shown. It is assumed that where there might be type variables these have been pre-instantiated to ()num, the type of numbers, so most of the type matching is omitted. The notation \$+ indicates the infix operator + applied as a prefix.

The term substitution $\{f \mapsto \lambda u_1.\lambda v_1.(\mathtt{EL}\ v_1\ l)+u_1\}$ is obtained. Instantiating the skeleton using this substitution gives:

$$\lambda x.(\lambda u_1.\lambda v_1.(\mathtt{EL}\ v_1\ l)+u_1)\ x\ ((x-1)+1)\ .$$

Applying β-reduction at the gap produces the term

$$\lambda x.(\mathtt{EL}\ ((x-1)+1)\ l)+x$$

which is α-equivalent to the term being matched.

$$\lambda x.f\ x\ ((x-1)+1) =^?_{\mathrm{F}} \lambda n.(\mathrm{EL}\ ((n-1)+1)\ l) + n \stackrel{\{\}}{\longrightarrow}$$
$$\langle\{f \mapsto \lambda u_1.\lambda v_1.(\mathrm{EL}\ v_1\ l) + u_1\}, \{\}\rangle$$

(5)

$$x : ()\mathbf{num} =^?_{\mathrm{F}} n : ()\mathbf{num} \stackrel{\{x\mapsto n\}}{\longrightarrow} \langle\{x \mapsto n\}, \{\}\rangle$$

(2)

$$()\mathbf{num} =^?_{\sigma} ()\mathbf{num} \longrightarrow \{\}$$
$$(x \mapsto n) \in \{x \mapsto n\}$$
$$f\ x\ ((x-1)+1) =^?_{\mathrm{F}} (\mathrm{EL}\ ((n-1)+1)\ l) + n \stackrel{\{x\mapsto n\}}{\longrightarrow}$$
$$\langle\{f \mapsto \lambda u_1.\lambda v_1.(\mathrm{EL}\ v_1\ l) + u_1, x \mapsto n\}, \{\}\rangle$$

(1)

$$\langle\{\langle u_1, x\rangle\}, [\langle v_1, (x-1)+1\rangle]\rangle =^?_{\mathrm{H}} (\mathrm{EL}\ ((n-1)+1)\ l) + n \stackrel{\{x\mapsto n\}}{\longrightarrow}$$
$$\langle[\,], (\mathrm{EL}\ v_1\ l) + u_1, \{x \mapsto n\}, \{\}\rangle$$

(12)

$$\langle\{\langle u_1, x\rangle\}, [\langle v_1, (x-1)+1\rangle]\rangle =^?_{\mathrm{H}} \$+\ (\mathrm{EL}\ ((n-1)+1)\ l) \stackrel{\{x\mapsto n\}}{\longrightarrow}$$
$$\langle[\,], \$+\ (\mathrm{EL}\ v_1\ l), \{x \mapsto n\}, \{\}\rangle$$

(12)

$$\langle\{\langle u_1, x\rangle\}, [\langle v_1, (x-1)+1\rangle]\rangle =^?_{\mathrm{H}} \$+ \stackrel{\{x\mapsto n\}}{\longrightarrow}$$
$$\langle[\langle v_1, (x-1)+1\rangle], \$+, \{\}, \{\}\rangle$$

(10)

$$\langle\{\langle u_1, x\rangle\}, [\langle v_1, (x-1)+1\rangle]\rangle =^?_{\mathrm{H}} \mathrm{EL}\ ((n-1)+1)\ l \stackrel{\{x\mapsto n\}}{\longrightarrow}$$
$$\langle[\,], \mathrm{EL}\ v_1\ l, \{x \mapsto n\}, \{\}\rangle$$

(12)

$$\langle\{\langle u_1, x\rangle\}, [\langle v_1, (x-1)+1\rangle]\rangle =^?_{\mathrm{H}} \mathrm{EL}\ ((n-1)+1) \stackrel{\{x\mapsto n\}}{\longrightarrow}$$
$$\langle[\,], \mathrm{EL}\ v_1, \{x \mapsto n\}, \{\}\rangle$$

(12)

$$\langle\{\langle u_1, x\rangle\}, [\langle v_1, (x-1)+1\rangle]\rangle =^?_{\mathrm{H}} \mathrm{EL} \stackrel{\{x\mapsto n\}}{\longrightarrow}$$
$$\langle[\langle v_1, (x-1)+1\rangle], \mathrm{EL}, \{\}, \{\}\rangle$$

(10)

$$\langle\{\langle u_1, x\rangle\}, [\langle v_1, (x-1)+1\rangle]\rangle =^?_{\mathrm{H}} (n-1)+1 \stackrel{\{x\mapsto n\}}{\longrightarrow}$$
$$\langle[\,], v_1\ \{\}, \{x \mapsto n\}, \{\}\rangle$$

(7)

$$(x-1)+1 =^?_{\mathrm{F}} (n-1)+1 \stackrel{\{x\mapsto n\}}{\longrightarrow} \langle\{x \mapsto n\}, \{\}\rangle$$

$$\vdots$$

$$\langle\{\langle u_1, x\rangle\}, [\,]\rangle =^?_{\mathrm{H}} l \stackrel{\{x\mapsto n\}}{\longrightarrow} \langle[\,], l, \{\}, \{\}\rangle$$

(9)

$$(\ldots \mapsto l) \notin \{x \mapsto n\}$$
$$\langle\{\langle u_1, x : ()\mathbf{num}\rangle\}, [\,]\rangle =^?_{\mathrm{H}} n : ()\mathbf{num} \stackrel{\{x\mapsto n\}}{\longrightarrow} \langle[\,], u_1\ \{\}, \{x \mapsto n\}, \{\}\rangle$$

(8)

$$()\mathbf{num} =^?_{\sigma} ()\mathbf{num} \longrightarrow \{\}$$
$$(x \mapsto n) \in \{x \mapsto n\}$$

Fig. 2. Example application of the matching rules

7 Implementation

A variant of the algorithm (with the restriction in Conjecture 16) has been implemented in the HOL system using Standard ML. Each relation becomes an ML function. The functions for $\stackrel{?}{=}_F$ and $\stackrel{?}{=}_H$ are mutually recursive. The functional program fails (raises an exception) when a relation cannot be satisfied. In most cases the hypotheses have to be processed in a specific order. As an example, here is ML code for rule (6):

```
let val {Rator = s1,Rand = s2} = dest_comb s
    and {Rator = t1,Rand = t2} = dest_comb t
    val (theta1,phi1) = match_F s1 t1 B
    and (theta2,phi2) = match_F s2 t2 B
in  (merge_term_subst theta1 theta2,merge_type_subst phi1 phi2)
end
```

where `dest_comb` breaks a term that is an application into its operator and operand. If the term is not an application, `dest_comb` fails.

Implementation of the rules only provides matching. To apply a rewrite rule its left-hand side (a skeleton) must be matched to the term to be rewritten. The right-hand side of the rule is then instantiated using the substitutions obtained. This produces a term in which there are λ-abstractions in place of the gaps. To complete the rewriting the β-redexes corresponding to the gaps are reduced. Of course, there may be other β-redexes, so the reduction must be done selectively.

As mentioned in Sect. 5, there is a tendency for bound variables to be renamed to names appearing in the rewrite rule. This can be largely dealt with by α-conversion. Correspondences are sought between bound variables in the original term and in the rewritten term.

The strategy for applying the basic rewriting procedure in the HOL system is essentially under the control of the user. A number of ML functions are provided that traverse the term looking for places to successfully apply rewrite rules. Some try repeated applications at the same subterm until the rules no longer apply. In practice it seems that when two simplifications are possible inside an abstraction, repeated rewriting of the term often simplifies both, even though the restricted matching only handles one at a time.

8 Variants

Let s be the skeleton $G\ (\mathtt{SUC}\ x)\ (\mathtt{SUC}\ y)$ where G is a gap with result type ()`num` (the type of natural numbers) and `SUC` is the successor function on natural numbers. Consider matching s to the term $\mathtt{SUC}\ (\mathtt{SUC}\ 0) + \mathtt{SUC}\ (\mathtt{SUC}\ 1)$. For the basic algorithm, the possible term substitutions are:

$$
\begin{array}{lll}
\{G \mapsto \lambda v_1.\lambda v_2.v_1 + v_2 & , x \mapsto \mathtt{SUC}\ 0, & y \mapsto \mathtt{SUC}\ 1\} \\
\{G \mapsto \lambda v_1.\lambda v_2.v_1 + \mathtt{SUC}\ v_2 & , x \mapsto \mathtt{SUC}\ 0, & y \mapsto 1 \quad\} \\
\{G \mapsto \lambda v_1.\lambda v_2.\mathtt{SUC}\ v_1 + v_2 & , x \mapsto 0 & , y \mapsto \mathtt{SUC}\ 1\} \\
\{G \mapsto \lambda v_1.\lambda v_2.\mathtt{SUC}\ v_1 + \mathtt{SUC}\ v_2, & x \mapsto 0 & , y \mapsto 1 \quad\}
\end{array}
$$

The type substitution is always empty. The restriction imposed by Conjecture 16 makes the first of these substitutions the unique solution. A problem with this restriction is that it prevents s from matching the term SUC (SUC $0 +$ SUC 1) because SUC x is forced to match the whole term leaving nothing for SUC y to match. An alternative is to restrict rule (12) instead so that when there are multiple ways to successfully apply it, only the one with the smallest intermediate list S' is accepted. However, this only reduces the number of solutions; it does not ensure uniqueness.

If uniqueness may be sacrificed then some other variants can be considered:

1. Use a set instead of a list for S.
2. Allow multiple occurrences of gaps.

Variant 1 allows the subskeletons to match in any order. They can then be made optionally matching by modifying rule (1) so that it does not require the modified S to be empty. They can be allowed to match multiple subterms by not removing one from S when it matches. (The substitutions generated will have to be consistent.)

Variant 2 is a more significant divergence, changing the nature of a skeleton. The λ-abstractions to which the separate occurrences of a gap become bound must be α-equivalent for a successful match. (Term merging has to be modified to test for α-equivalence rather than absolute equivalence.)

From an implementation point-of-view the problem with these variants is that backtracking is required to ensure that a match is found if it exists.

9 Related Work

Matching of terms is related to unification, so some research into higher-order unification is of interest when considering higher-order matching. In his well-known paper on higher-order unification [4], Huet describes an algorithm to determine whether two terms of (non-polymorphic) higher-order logic are unifiable. The algorithm consists of two alternating procedures SIMPL and MATCH. SIMPL deals with pairs of terms that are both *rigid*, that is when in β-normal form the top-level operator is either a constant or one of the enclosing bound variables. A term in β-normal form that is not rigid is *flexible*. MATCH searches for substitutions that will unify a flexible term with a rigid term using two rules, *imitation* and *projection*.

The relation $\stackrel{?}{=}_{\mathrm{F}}$ defined in this paper corresponds roughly to Huet's SIMPL in that it also deals with what are essentially rigid parts of the pattern (skeleton). This includes free variables that are not gaps. Correspondingly, relation $\stackrel{?}{=}_{\mathrm{H}}$ bears some resemblance to Huet's MATCH. MATCH tries to imitate the rigid term by incrementally instantiating the free variable (cf. gap) in the flexible term. Relation $\stackrel{?}{=}_{\mathrm{H}}$ also tries to imitate the rigid term but by building a term with which to instantiate the free variable in a single step. Beyond this the two algorithms differ substantially. The algorithm presented here is significantly restricted and is for matching, not unification.

Huet and Lang [5] adapt Huet's original algorithm to the process of matching second-order terms, which is known to be decidable. Their application area is program transformation which is close to the intended application of the algorithm presented here. Simon [12] also presents an algorithm for second-order matching but only for a subcase.

Miller's notion of a higher-order pattern [6] is of interest here because, like a skeleton, it is an example of a restricted form of term. A term in β-normal form is a *higher-order pattern* if every occurrence of a free variable f is in a subterm $(\ldots(f\ x_1)\ldots\ x_n)$ of the term such that, when reduced to η-normal form, the x_i's are distinct bound variables. The terms are of unrestricted order. Miller shows that it is decidable whether two higher-order patterns are unifiable and that, if so, a most general unifier can be computed. There are higher-order patterns that are not valid skeletons (e.g., ones with a multiply-occurring free variable) and there are skeletons that are not higher-order patterns (e.g., ones with something other than a bound variable as an argument of a gap).

Very recently Prehofer [11] has shown that unification is decidable for a larger class of terms. He defines a *relaxed higher-order pattern* (abbreviated to 'pattern') as a higher-order pattern in which the arguments of the free variables need not be distinct (but must still be bound variables). Also, a term is *linear* if no free variable occurs more than once. Prehofer shows that the unification of a linear pattern with an arbitrary second-order term that shares no variables with it is decidable. He also proves the decidability of a few extensions of this problem. There are still skeletons that do not satisfy the definition of a relaxed higher-order pattern.

Wolfram [13] presents a terminating procedure for higher-order matching. Whether or not this procedure is complete is not yet known. If it is complete then general higher-order matching is decidable. The results in this paper say little, if anything, about the decidability of higher-order matching when the pattern is restricted to being a skeleton. This is because the algorithm used here for matching is itself restricted.

10 Concluding Remarks

This paper has described an algorithm for a restricted form of higher-order rewriting motivated by real verification examples. The intended usage is for rewrite rules that use function-valued variables in place of some unknown term structure. The algorithm is a long way from being a complete procedure for higher-order rewriting. Indeed, in the HOL system, where types may be polymorphic, the options for full higher-order rewriting go beyond those possible in simple type theory. Nipkow [9] has considered this issue in the context of unification. However, full higher-order rewriting seems to be an unusual requirement in practical theorem proving and is difficult to handle because of the possibility of multiple solutions. Perhaps, then, it is better to have a number of more restricted procedures that are both predictable by the user and more efficient. This paper has described one such procedure.

An interesting topic for future work is consideration of rewrite systems based around the restricted higher-order matching, as for example Nipkow [10] has done for Miller's higher-order patterns.

Acknowledgements

The author thanks David Wolfram and the anonymous referees for their constructive comments on an earlier version of this paper.

References

1. R. J. Boulton. A HOL semantics for a subset of ELLA. Technical Report 254, University of Cambridge Computer Laboratory, April 1992.
2. A. Church. A formulation of the simple theory of types. *Journal of Symbolic Logic*, 5(1):56–68, 1940.
3. M. J. C. Gordon and T. F. Melham, editors. *Introduction to HOL: A theorem proving environment for higher order logic.* Cambridge University Press, 1993.
4. G. P. Huet. A unification algorithm for typed λ-calculus. *Theoretical Computer Science*, 1(1):27–57, 1975.
5. G. Huet and B. Lang. Proving and applying program transformations expressed with 2nd order patterns. *Acta Informatica*, 11:31–55, 1978.
6. D. Miller. A logic programming language with lambda-abstraction, function variables, and simple unification. In P. Schroeder-Heister, editor, *Extensions of Logic Programming: International Workshop, Tübingen, FRG.*, volume 475 of *Lecture Notes in Artificial Intelligence*, pages 253–281. Springer-Verlag, December 1989.
7. R. Milner, M. Tofte, and R. Harper. *The Definition of Standard ML.* MIT Press, 1990.
8. J. D. Morison and A. S. Clarke. *ELLA 2000: A Language for Electronic System Design.* McGraw-Hill Book Company, 1994.
9. T. Nipkow. Higher-order unification, polymorphism, and subsorts. Technical Report 210, University of Cambridge Computer Laboratory, November 1990.
10. T. Nipkow. Higher-order critical pairs. In *Proceedings of the 6th Annual IEEE Symposium on Logic in Computer Science*, pages 342–349, Amsterdam, The Netherlands, July 1991. IEEE Computer Society Press.
11. C. Prehofer. Decidable higher-order unification problems. In A. Bundy, editor, *Proceedings of the 12th International Conference on Automated Deduction*, volume 814 of *Lecture Notes in Artificial Intelligence*, pages 635–649, Nancy, France, June/July 1994. Springer-Verlag.
12. D. Simon. A linear time algorithm for a subcase of second order instantiation. In *Proceedings of the 7th International Conference on Automated Deduction*, volume 170 of *Lecture Notes in Computer Science*, pages 209–223. Springer-Verlag, 1984.
13. D. A. Wolfram. *The Clausal Theory of Types*, volume 21 of *Cambridge Tracts in Theoretical Computer Science.* Cambridge University Press, 1993.

Rewrite systems for integer arithmetic

H.R.Walters[1] and H.Zantema[2]

[1] H.R.Walters@cwi.nl
Centre for Mathematics and Computer Science,
P.O. Box 94079,
1090 GB Amsterdam

[2] hansz@cs.ruu.nl
Utrecht University, Computer Science Department,
P.O. Box 80.089,
3508 TB Utrecht

Abstract. We present three term rewrite systems for integer arithmetic with addition, multiplication, and, in two cases, subtraction. All systems are ground confluent and terminating; termination is proved by semantic labelling and recursive path order.
The first system represents numbers by successor and predecessor. In the second, which defines non-negative integers only, digits are represented as unary operators. In the third, digits are represented as constants. The first and the second system are complete; the second and the third system have logarithmic space and time complexity, and are parameterized for an arbitrary radix (binary, decimal, or other radices). Choosing the largest machine representable single precision integer as radix, results in unbounded arithmetic with machine efficiency.

Note: Partial support received from the Foundation for Computer Science Reasearch in the Netherlands (SION) under project 612-17-418, "Generic Tools for Program Analysis and Optimization".

1 Introduction

In [CW91] a term rewrite system is presented of base four integers with addition and multiplication. This rewrite system is proved to be locally confluent modulo associative-commutative multiplication and addition. Termination is not established and seems to be very hard to prove; it is listed as an open problem in [DJK93]. The system does not easily generalize to arbitrary number base. In [BW89] a rewrite system is presented of non-negative binary integers with addition, which is shown to be ground confluent and terminating. Both systems have logarithmic space and time complexity. Here, by space complexity we mean the space required to store a number as a function of its absolute value, and by time complexity we mean the number of steps required to reduce the addition, subtraction or multiplication of two such numbers, to normal form.

In this article we discuss term rewrite systems for integer arithmetic, where we desire (ground) confluence, termination, logarithmic complexity, a one-one correspondence between integers and ground normal forms, good readability of expressions, a minimal equational setting (non AC) and an arbitrary number base.

Unfortunately we shall not attain all these properties in a single rewrite system; we will present three rewrite systems which have most of the properties. All systems are pure term rewriting systems, unconditional and not taking terms modulo equations. All systems are ground confluent (which follows from having unique ground normal forms) and terminating (which is proved by recursive path order ([Der87]) and, where appropriate, semantic labelling ([Zan93]))

The first system is the common *successor-zero* system for natural numbers, extended with a *predecessor* function. This *successor-predecessor* system is referred to as SP. This system hardly exhibits the desired properties (no number base, bad readability, linear complexity); it is included only because it is well known, and commonly used to define integer arithmetic.

Confluence and termination are easily established. Drawbacks of this system are, firstly, that it has linear space and time complexity, or worse, in the case of multiplication, and secondly, that the successor-zero notation of numbers is hardly palatable to the human eye and is therefore mainly of theoretical importance.

The second system (DA, digit application) concerns non-negative integers only, but it does have all desired properties. It is confluent and terminating without assuming commutative-associativity of addition or multiplication; it has logarithmic complexity; and features human readable syntax of expressions (that is, the TRS itself is rather involved, but arithmetic expressions are simpele and easy to read).

In this system the digits occur as unary operators, written in postfix notation. An (invisible) constant of value zero is needed to represent numbers. This corresponds to the usual human syntax except for the number 0 which is represented by the invisible constant. The system needs some auxiliary functions.

The third system (JP, juxtaposition) defines integers consisting of sequences of digits, where digits are constants. It is ground confluent, but not confluent. It has logarithmic complexity, does not rely on auxiliary functions and features human-readable syntax of expressions. The system is terminating; the termination proof is quite involved. It is given in two levels: first the system obtained from ignoring all unary minus-signs is proved terminating, second the remaining rules for which this ignorance yields equality are proved terminating. Both of these termination proofs are given by transforming the system to an infinite labelled system by the technique of semantic labelling and proving termination of the labelled system by recursive path order (RPO) with status.

The systems DA and JP are presented for an arbitrary radix (number base); only the digits 0, and in one case 1, have special significance. All rules containing other digits are represented using *rule schemata*, which we will introduce in Section 3.

Apart from the formal logarithmic complexity, our rewrite systems have very good practical efficiency. For example, considering JP, discussed in Section 5:

- The binary version has 30 equations; the measured number of reduction steps for performing multiplication of two positive integers of b digits seems to be of the order of $b^2 \log b$.

- The decimal version of our rewrite system has essentially the same 30 equations, only some of them are parametrized. For instance, instead of the 81 rules for non-zero digit multiplication, a single rule schema is given. One of the instances is, for example, $7 * 8 \rightarrow 56$. The efficiency of multiplication is of the same order as in the binary version.
- Using the largest machine-representable single precision integer as radix, results in an implementation which provides unbounded integers at near-machine efficiency. This is discussed in Section 6.

Hence this system can be used as the basis for an implementation of integer arithmetic. Its efficiency is comparable with that of the more usual implementations. A major advantage of our approach is its extensibility: extension of the implementation by new functions like exponentiation can be done simply by only adding a few rewrite rules to the system.

In Section 2 we present the system SP. In an intermezzo in Section 3 we present rule schemata. Then, in Section 4 we present the system DA and in Section 5 the system JP. In Section 7 we present an overview of the various rewrite systems, and we discuss some conclusions.

Our notation and terminology are consistent with [Klo92]. We consider ordinary (non-AC) operators, and we are interested exclusively in rewriting finite terms.

2 Successor-predecessor integers

In the following rewrite system 0 is the constant *zero*, s and p are the functions representing successor (plus one) and predecessor (minus one), $+$, $*$ and $-$ have their usual meaning in arithmetic, and x, y and z are variables.

$\langle 1 \rangle\ x + 0 \rightarrow x$	$\langle 10 \rangle\ s(p(x)) \rightarrow x$
$\langle 2 \rangle\ x + s(y) \rightarrow s(x + y)$	$\langle 11 \rangle\ p(s(x)) \rightarrow x$
$\langle 3 \rangle\ x + p(y) \rightarrow p(x + y)$	
$\langle 4 \rangle\ x - 0 \rightarrow x$	$\langle 12 \rangle\ (x - y) + y \rightarrow x$
$\langle 5 \rangle\ x - s(y) \rightarrow p(x - y)$	$\langle 13 \rangle\ (x + y) - y \rightarrow x$
$\langle 6 \rangle\ x - p(y) \rightarrow s(x - y)$	$\langle 14 \rangle\ s(x) + y \rightarrow s(x + y)$
$\langle 7 \rangle\ x * 0 \rightarrow 0$	$\langle 15 \rangle\ s(x) - y \rightarrow s(x - y)$
$\langle 8 \rangle\ x * s(y) \rightarrow (x * y) + x$	$\langle 16 \rangle\ p(x) + y \rightarrow p(x + y)$
$\langle 9 \rangle\ x * p(y) \rightarrow (x * y) - x$	$\langle 17 \rangle\ p(x) - y \rightarrow p(x - y)$

The system SP

Termination of the TRS SP is established with recursive path ordering (RPO, [Der87]), taking the following precedence of operators: $* > \{+, -\} > \{s, p\}$.

Local confluence is verified using, for example, the Larch prover ([GG91]). Note that rules 12—17 are required only to establish local confluence; they are not required for ground confluence. Confluence follows from local confluence and termination.

One easily checks that 0, $s^n(0)$, $p^n(0)$ for $n = 1, 2, 3, \ldots$ are the ground normal forms. Hence ground normal forms correspond bijectively to integers.

The space complexity of this rewrite system is linear. That is, to represent a number n, a term of size $O[n]$ is required at least. Addition of two numbers, in absolute value not exceeding n, requires $O[n]$ reductions (worst case); multiplication requires $O[n^2]$ reductions. For example, the multiplication of 5 and 6 requires 46 reductions using the left-most inner-most reduction strategy.

3 Rule schemata

In the sequel we will consider rewrite systems in which non-zero digits occur, but we will do so for an arbitrary radix. The rules concerning these digits are very regular, and are similar for each radix.

In our system we express this fact by presenting such rules with *rule schemata*. A rule schema is a notational device defining a family of rules at once. The left-hand side is a term containing *schema variables* (e.g., δ); the right-hand side contains schema variables and *schema diagrams* representing common arithmetic operations (e.g., $\oplus$).

The meaning of a rule schema is the set of rules obtained by ranging all schema variables over, as the case may be, all digits or the non-zero digits, and by using a predetermined interpretation for the schema diagrams. We reserve the notation $\mathring{\delta}$, $\mathring{\delta}_1$, etc., for schema variables that range from 0 to the largest digit, and δ, δ_1, etc., for digits that range from 1 to the largest digit. That is, using $\mathcal{R}$ to represent the radix, we have $0 \leq \mathring{\delta} \leq \mathcal{R} - 1$ and $1 \leq \delta \leq \mathcal{R} - 1$.

Note that schema diagrams are meta notation and do not occur in the signature. We will discuss the individual schema diagrams where we use them. There, we will also discuss the specific properties we use to establish (ground) completeness.

4 Natural numbers with digits as unary operators

In our second rewrite system DA all digits are interpreted as unary postfix operators (hence, digit application). That is, a natural number followed by a digit is again a natural number. The base of this induction is a constant of value zero represented by the empty string. In the sequel we will make this string explicit – by surrounding it with parentheses, – whenever this is necessary for clarity. For example, we will write $x + () \rightarrow x$. A minor inconvienience is the fact that the number 0 is now represented as the empty string. Note that the *digit* 0 still appears as a postfix operator.

We will use the following three schema diagrams: $\oplus$, for addition modulo $\mathcal{R}$, © for addition carry, and $\odot$ for digit multiplication. Since digits are functions themselves, the meaning of schema diagrams is taken appropriately. We enclose schema diagrams that signify function applications rather than proper sub-terms in angle brackets.

For addition we define $\langle \mathring{\delta}_1 \oplus \mathring{\delta}_2 \rangle$ to be the postfix operator corresponding to last digit of $\mathring{\delta}_1 + \mathring{\delta}_2$. For example, $(x)\langle 5 \oplus 2 \rangle$ denotes $(x)7$ and $(x)\langle 7 \oplus 8 \rangle$ denotes $(x)5$. We denote this schema as a postfix operator.

In complement, the carry schema defines the addition carry. If $\mathring{\delta}_1 + \mathring{\delta}_2 < \mathcal{R}$ then $(x)\langle \mathring{\delta}_1 \copyright \mathring{\delta}_2 \rangle$ signifies x; if $\mathring{\delta}_1 + \mathring{\delta}_2 \geq \mathcal{R}$ then $(x)\langle \mathring{\delta}_1 \copyright \mathring{\delta}_2 \rangle$ signifies $s(x)$. Here s is an auxiliary function symbol defined in the system. For clarity, this schema is also denoted as a postfix operator, even though it signifies a function application, or no function application at all.

For multiplication, the schema diagram $\odot$ is introduced giving the result of multiplication of two digits. For example, $2 \odot 4$ denotes 8, while $5 \odot 6$ denotes 30.

In addition to the function s, the system defines one auxiliary unary operator for every digit. We will write these functions as $\star_{\mathring{\delta}}$. The meaning of $\star_{\mathring{\delta}}(x)$ is $\mathring{\delta} * x$. Note that each $\star_{\mathring{\delta}}$ is an auxiliary function; not a schema diagram.

$\langle 1 \rangle$	$()0 \rightarrow ()$
$\langle 2 \rangle$	$() + x \rightarrow x$
$\langle 3 \rangle$	$x + () \rightarrow x$
$\langle 4 \rangle$	$x\mathring{\delta}_1 + y\mathring{\delta}_2 \rightarrow (x+y)\langle \mathring{\delta}_1 \copyright \mathring{\delta}_2 \rangle \langle \mathring{\delta}_1 \oplus \mathring{\delta}_2 \rangle$
$\langle 5 \rangle$	$() * x \rightarrow ()$
$\langle 6 \rangle$	$x\mathring{\delta} * y \rightarrow (x * y)0 + \star_{\mathring{\delta}}(y)$
$\langle 7 \rangle$	$s() \rightarrow 1$
$\langle 8 \rangle$	$s(x\mathring{\delta}) \rightarrow (x)\langle 1 \copyright \mathring{\delta} \rangle \langle 1 \oplus \mathring{\delta} \rangle$
$\langle 9 \rangle$	$\star_{\mathring{\delta}}() \rightarrow ()$
$\langle 10 \rangle$	$\star_{\mathring{\delta}_1}(x\mathring{\delta}_2) \rightarrow (\star_{\mathring{\delta}_1}(x))0 + \mathring{\delta}_1 \odot \mathring{\delta}_2$

The system DA

Ground normal forms: The set of ground normal forms of this rewrite system is: $\mathcal{N} = \{()\} \cup \Phi$, where Φ is the smallest set satisfying $\Phi = \{()\delta\} \cup \{\alpha\mathring{\delta} \mid \alpha \in \Phi\}$. This fact is easily verified. Clearly ground normal forms correspond bijectively to non-negative integers.

Correctness: The natural numbers are a model for this rewrite system under the given interpretation; it is easily verified that all equations hold.

Termination: Termination is proved simply by recursive path order choosing the precedence

$$* > \{\star_{\mathring{\delta}}\} > + > s > \{\mathring{\delta}\} > ().$$

Confluence: The only overlap between the left hand sides of the rules is between the rules 2 and 3, only giving the trivial critical pair $((),())$. Hence the system is locally confluent, and hence, by termination, confluent.

5 Integers with digits as constants

A natural next step is to extend the system of the last section to integers by adding rules for the unary and binary minus operator. Both for termination and for confluence this gives some complications. For instance, one expects a rule $(-x)5 \to -(p(x))5$ for a predecessor operator p for which there is a rule $p() \to -()1$. This yields the self-embedding reduction $(-())5 \to -(p())5 \to -(-()1)5$. Similar problems arise for choosing other representations for predecessor. Even if termination can be achieved, then confluence is a big problem. Adding rules like $--x \to x$ yields many critical pairs that turn out to be not convergent.

From now on we leave the requirement of confluence, and only require ground confluence. For many applications, for example the implementation of ordinary arithmetic, ground confluence is sufficient.

We also leave the representation of digits being unary symbols; we choose digits to be constants and introduce a juxtaposition operator which combines digits into numbers. For example, the number 12 is defined by applying the (invisible) juxtaposition operator to the digits 1 and 2. Because of this juxtaposition operator this final system is called JP[3].

There are several reasons to change the representation. One reason is that it becomes even closer to human representation of integers. A second reason is that no auxiliary functions are needed any more. A third reason is that the tables for addition, subtraction and multiplication of two digits can be directly considered as rewite rules. For example, instead of the rule

$$\star_7(x8) \quad \to \quad (\star_7(x))0 + ((()5)6$$

we simply have the rule $7 * 8 \to 56$.

The interpretation of the juxtaposition operator is defined as follows: $[\![xy]\!] = \mathcal{R} * [\![x]\!] + [\![y]\!]$, where $\mathcal{R}$ is the radix. The operator is left-associative. For example, in base ten, $[\![123]\!] = [\![(12)3]\!] = 10 * [\![12]\!] + [\![3]\!] = 10 * (10 * [\![1]\!] + [\![2]\!]) + [\![3]\!] = 100 * [\![1]\!] + 10 * [\![2]\!] + [\![3]\!] = 123$. For our rewrite system we will keep this interpretation in mind, but the multiplication operator $*$ is not strictly needed as an auxiliary function. In normal forms, the right-hand argument of the juxtaposition operator is always a digit, but non-digits may occur in intermediate terms.

The schema diagrams we will use are all similar: they represent the normal form of a simple function applied to one or two digits, resulting in a digit, or two juxtaposed digits, possibly with a minus sign. The schemas are: $\oplus$ for addition, $\ominus$ for subtraction, $\odot$ for multiplication, $\tilde{\ }$ for $\mathcal{R}$ complement (i.e., $\tilde{\delta} = \mathcal{R} - \delta$) and $\breve{\ }$ for predecessor (i.e., $\breve{\delta} = \delta - 1$). For example, the decimal meaning of $5 \odot 6$ is 30, and the meaning of $\tilde{3}$ is 7. If δ_1 is 1, it is understood that $\breve{\delta}_1 \tilde{\delta}_2$ signifies $\tilde{\delta}_2$ rather than $0\tilde{\delta}_2$.

To avoid parentheses we introduce priorities: juxtaposition has highest priority; then unary minus; then multiplication, and then then addition and subtraction. Where necessary parentheses have been added to improve readability. Rule schemata have bold indices.

[3] The decimal version of this system was already presented in [Wal94]. There, an *ad-hoc* termination proof was given.

⟨1⟩	$0x \rightarrow x$	⟨17⟩	$0 - x \rightarrow -x$
⟨2⟩	$x(yz) \rightarrow (x+y)z$	⟨18⟩	$x - 0 \rightarrow x$
⟨3⟩	$x(-(yz)) \rightarrow -((y-x)z)$	⟨**19**⟩	$\delta_1 - \delta_2 \rightarrow \delta_1 \ominus \delta_2$
⟨**4**⟩	$\delta_1(-\delta_2) \rightarrow \check{\delta_1}\tilde{\delta_2}$	⟨20⟩	$xy - z \rightarrow x(y-z)$
⟨**5**⟩	$x0(-\delta) \rightarrow x(-1)\tilde{\delta}$	⟨21⟩	$x - yz \rightarrow -(y(z-x))$
⟨**6**⟩	$x\delta_1(-\delta_2) \rightarrow x\check{\delta_1}\tilde{\delta_2}$	⟨22⟩	$x - -y \rightarrow x + y$
⟨7⟩	$(-x)y \rightarrow -(x(-y))$	⟨23⟩	$-x - y \rightarrow -(x+y)$
⟨8⟩	$- - x \rightarrow x$		
⟨9⟩	$-0 \rightarrow 0$		
⟨10⟩	$0 + x \rightarrow x$	⟨24⟩	$0 * x \rightarrow 0$
⟨11⟩	$x + 0 \rightarrow x$	⟨25⟩	$x * 0 \rightarrow 0$
⟨**12**⟩	$\delta_1 + \delta_2 \rightarrow \delta_1 \oplus \delta_2$	⟨**26**⟩	$\delta_1 * \delta_2 \rightarrow \delta_1 \odot \delta_2$
⟨13⟩	$x + yz \rightarrow y(x+z)$	⟨27⟩	$x * yz \rightarrow (x*y)(x*z)$
⟨14⟩	$xy + z \rightarrow x(y+z)$	⟨28⟩	$xy * z \rightarrow (x*z)(y*z)$
⟨15⟩	$x + -y \rightarrow x - y$	⟨29⟩	$x * -y \rightarrow -(x*y)$
⟨16⟩	$-x + y \rightarrow y - x$	⟨30⟩	$-x * y \rightarrow -(x*y)$

The system JP

Ground normal forms: The set of ground normal forms of the system JP is: $\mathcal{N} = \{0\} \cup \Phi \cup \Phi^-$, where Φ is defined as the smallest set satisfying $\Phi = \Delta \cup \{\alpha\beta | \alpha \in \Phi \wedge \beta \in \{0\} \cup \Delta\}$, with Δ being the set of non-zero digits, and $\Phi^- = \{-\phi | \phi \in \Phi\}$. Hence ground normal forms correspond bijectively to integers.

This fact is easily verified. We will discuss the only non-trivial case. Consider a term of the form $\alpha\beta$.

- if α is 0 or of the form $-\eta$, the term can be reduced by rules 1 and 7, respectively.
- suppose α is a non-zero digit. Consider β:
  - if $\beta \in \{0\} \cup \Delta$, we have $\alpha\beta \in \Phi$.
  - if β is of the form $\sigma\tau$. Rule 2 applies.
  - suppose β is of the form $-\rho$. If $\rho \in \Delta$ rule 4 applies. If ρ is 0 or of the form $\sigma\tau$ or $-\sigma$, the term can be reduced with rule 9, 3 or 8, respectively.
- suppose α is of the form $\varepsilon\eta$. If β is a digit, or has the form $\sigma\tau$, the argument immediately above applies. Otherwise β is of the form $-\rho$. Again, if ρ is 0 or of the form $\sigma\tau$ or $-\sigma$ the term can be reduced as mentioned above. Hence ρ is a non-zero digit.
  To recapitulate: the remaining case is where $\alpha\beta$ is of the form $(\varepsilon\eta)(-\rho)$, where $\rho \in \Delta$.
  The left-hand side of this term has the same form as $\alpha\beta$, and our entire argumentation can be used recursively.

We see that the only irreducible terms not in $\mathcal{N}$ are of the form α_0, where $\alpha_n = \alpha_{n+1}(-d_n)$ and where each d_i is a non-zero digit. This is impossible for finite terms.

Correctness: The integer numbers are a model for this rewrite system under the given interpretation; it is easily verified that all equations hold.

Ground confluence: Observe that no two distinct normal forms have the same value, in this model. From this and the previous observation we can conclude that every term has a unique normal form. But then ground confluence is established if termination can be established; this will be shown in the next section. The system is not confluent, for example we have

$$x(\delta_1(-\delta_2)) \rightarrow x(\check{\delta_1}\tilde{\delta_2}) \rightarrow (x+\check{\delta_1})\tilde{\delta_2})$$

and

$$x(\delta_1(-\delta_2)) \rightarrow (x+\delta_1)(-\delta_2)$$

without having a common reduct. Also associativity and commutativity of + appear as critical pairs, hence the only chance to achieve confluence is by taking the + as an AC-operator. However, Knuth-Bendix completion of this system yields hundreds of critical pairs and seems not to terminate.

Restricting to natural numbers, i.e., taking only the rules in which no −-sign occurs, the system is locally confluent if + is taken as an AC-operator. However, modulo AC the termination proof is much more complicated, while for only the naturals we already had a confluent and terminating system DA without taking + modulo AC.

Termination: Termination of JP is not trivial. Standard techniques like recursive path order do not provide a termination proof. Termination of only the rules 2, 12 and 13 (these rules also occur in the system of [CW91]) can be proved by the Knuth-Bendix order, but this order can not handle duplicating rules like 27 and 28.

In the next section we give a termination proof of the full system using the technique of *semantic labelling* ([Zan93]).

5.1 Termination of JP

Outline of the proof In the proof no distinction is made between the binary plus and the binary minus, and no distinction is made between distinct digits. The proof is given in two levels: first we ignore the unary minus sign and prove termination of the rules for which the left hand side is different from the right hand side (R_1), next we prove termination of the rules for which equality is obtained (R_2). More precisely, termination of the full system JP follows from termination of both R_1 and R_2 and the observation

$$t \rightarrow_{\mathsf{JP}} u \implies \phi(t) \rightarrow_{R_1} \phi(u) \vee (\phi(t) = \phi(u) \wedge \psi(t) \rightarrow_{R_2} \psi(u))$$

Here, ϕ and ψ are defined inductively by

$$\begin{aligned} \phi(\delta^\circ) &= 0 \\ \phi(x) &= x \\ \phi(-t) &= \phi(t) \\ \phi(tu) &= \phi(t)\phi(u) \\ \phi(t+u) = \phi(t-u) &= \phi(t)+\phi(u) \\ \phi(t*u) &= \phi(t)*\phi(u) \end{aligned} \qquad \begin{aligned} \psi(\delta^\circ) &= 0 \\ \psi(x) &= x \\ \psi(-t) &= -\psi(t) \\ \psi(tu) &= \psi(t)\psi(u) \\ \psi(t+u) = \psi(t-u) &= \psi(t)+\psi(u) \\ \psi(t*u) &= \psi(t)*\psi(u) \end{aligned}$$

for all variables x, digits δ° and terms t, u. The systems R_1 and R_2 are constructed in such a way that the above property holds. For concluding termination of the full system JP it suffices to prove termination of both R_1 and R_2.

$$\begin{array}{ll} \langle 1 \rangle & 0x \to x \\ \langle 2 \rangle & x(yz) \to (x+y)z \\ \langle 3 \rangle & x(yz) \to (y+x)z \\ \langle 10,17 \rangle & 0+x \to x \\ \langle 11,18 \rangle & x+0 \to x \\ \langle 12,19 \rangle & 0+0 \to 0 \\ \langle 12 \rangle & 0+0 \to 00 \\ \langle 13 \rangle & x+yz \to y(x+z) \\ \langle 14,20 \rangle & xy+z \to x(y+z) \\ \langle 21 \rangle & x+yz \to y(z+x) \\ \langle 24 \rangle & 0*x \to 0 \\ \langle 25 \rangle & x*0 \to 0 \\ \langle 26 \rangle & 0*0 \to 00 \\ \langle 26 \rangle & 0*0 \to 0 \\ \langle 27 \rangle & x*yz \to (x*y)(x*z) \\ \langle 28 \rangle & xy*z \to (x*z)(y*z) \end{array}$$

The system R_1

$$\begin{array}{ll} \langle 4 \rangle & 0(-0) \to 00 \\ \langle 5 \rangle & x0(-0) \to x(-0)0 \\ \langle 6 \rangle & x0(-0) \to x00 \\ \langle 7 \rangle & (-x)y \to -(x(-y)) \\ \langle 8 \rangle & --x \to x \\ \langle 9 \rangle & -0 \to 0 \\ \langle 15,22 \rangle & x+-y \to x+y \\ \langle 16 \rangle & -x+y \to y+x \\ \langle 23 \rangle & -x+y \to -(x+y) \\ \langle 29 \rangle & x*(-y) \to -(x*y) \\ \langle 30 \rangle & (-x)*y \to -(x*y) \end{array}$$

The system R_2

For the termination proofs for both R_1 and R_2 we shall use the technique of semantic labelling. The version we need is described in the next section.

Semantic labelling This technique makes use of the fact that a TRS with some semantics can be transformed into another (labelled) TRS such that the original TRS terminates if and only if the labelled TRS terminates. The termination proof is then given by proving termination of the labelled TRS, which is often done by recursive path order.

Let R be any term rewrite system over a signature $\mathcal{F}$ and a set $\mathcal{X}$ of variable symbols. Let $\mathcal{M} = (M, \{f_\mathcal{M}\}_{f\in\mathcal{F}})$ be an $\mathcal{F}$-algebra. Let $\geq$ be any partial order on M for which all operations $f_\mathcal{M}$ are weakly monotone in all coordinates.

For $\sigma : \mathcal{X} \to M$ the term evaluation $[\sigma] : \mathcal{T}(\mathcal{F}, \mathcal{X}) \to M$ is defined inductively by

$$[\sigma](x) = x^\sigma,$$
$$[\sigma](f(t_1, \ldots, t_n)) = f_\mathcal{M}([\sigma](t_1), \ldots, [\sigma](t_n))$$

for $x \in \mathcal{X}, f \in \mathcal{F}, t_1, \ldots, t_n \in \mathcal{T}(\mathcal{F}, \mathcal{X})$.

We require that $\mathcal{M}$ is a *quasi-model* for R, i.e., $[\sigma](l) \geq [\sigma](r)$ for all $\sigma : \mathcal{X} \rightarrow M$ and all rules $l \rightarrow r$ of R (if we have $[\sigma](l) = [\sigma](r)$ then it is called a *model*).

Next we introduce labelling of operation symbols: choose for every $f \in \mathcal{F}$ a corresponding non-empty set S_f of labels. Now the new signature $\overline{\mathcal{F}}$ is defined by

$$\overline{\mathcal{F}} = \{f_s | f \in \mathcal{F}, s \in S_f\},$$

where the arity of f_s is defined to be the arity of f. An operation symbol f is called *labelled* if S_f contains more than one element. For unlabelled f the set S_f containing only one element can be left implicit; in that case we write f instead of f_s. Note that $\overline{\mathcal{F}}$ can be infinite, even if $\mathcal{F}$ is finite.

We assume that every set S_f is provided with a well-founded partial order $\geq$. Choose for every $f \in \mathcal{F}$ a map $\pi_f : M^n \rightarrow S_f$, where n is the arity of f. We require π_f to be weakly monotone in all coordinates. It describes how a function symbol is labelled depending on the values of its arguments as interpreted in $\mathcal{M}$. For unlabelled f this function π_f can be left implicit. We extend the labelling of operation symbols to a labelling of terms by defining $\mathsf{lab} : \mathcal{T}(\mathcal{F}, \mathcal{X}) \times M^{\mathcal{X}} \rightarrow \mathcal{T}(\overline{\mathcal{F}}, \mathcal{X})$ inductively by

$$\begin{aligned} \mathsf{lab}(x, \sigma) &= x, \\ \mathsf{lab}(f(t_1, \ldots, t_n), \sigma) &= f_{\pi_f([\sigma](t_1), \ldots, [\sigma](t_n))}(\mathsf{lab}(t_1, \sigma), \ldots, \mathsf{lab}(t_n, \sigma)) \end{aligned}$$

for $x \in \mathcal{X}, \sigma : \mathcal{X} \rightarrow M, f \in \mathcal{F}, t_1, \ldots, t_n \in \mathcal{T}(\mathcal{F}, \mathcal{X})$.

Now $\overline{R}$ is defined to be the TRS over $\overline{\mathcal{F}}$ consisting of the rules

$$\mathsf{lab}(l, \sigma) \rightarrow \mathsf{lab}(r, \sigma)$$

for all $\sigma : \mathcal{X} \rightarrow M$ and all rules $l \rightarrow r$ of R. Note that this system can be infinite, even if R is finite.

Finally the TRS Decr over $\overline{\mathcal{F}}$ is defined to consist of the rules

$$f_s(x_1, \ldots, x_n) \rightarrow f_{s'}(x_1, \ldots, x_n)$$

for all $f \in \mathcal{F}$ and all $s, s' \in S_f$ satisfying $s > s'$. Here $>$ denotes the strict part of $\geq$. Now we are ready to state the main result of semantic labelling:

Theorem

Let $\mathcal{M}$ be a quasi-model for a TRS R over $\mathcal{F}$. Let $\overline{R}$ and Decr be as above for any choice of S_f and π_f. Then R is terminating if and only if $\overline{R} \cup \mathsf{Decr}$ is terminating.

For the proof we refer to [Zan93].

For both R_1 and R_2 we shall give a quasi-model $\mathcal{M}$, and S_f and π_f such that the corresponding infinite system $\overline{R} \cup \mathsf{Decr}$ can be proved terminating by using RPO with status over a well-founded precedence.

Termination of R_1 We denote the juxtaposition operator by '$\cdot$', for clarity.

As the quasi-model $\mathcal{M}$ and the sets $S_.$ and S_+ we choose the strictly positive integers with the usual order. We choose

$$0_{\mathcal{M}} = 1, \quad x \cdot_{\mathcal{M}} y = x +_{\mathcal{M}} y = x + y, \quad x *_{\mathcal{M}} y = x * y$$

for all $x, y \in M$. One easily checks that indeed $[\sigma](l) \geq [\sigma](r)$ for all $\sigma : \mathcal{X} \to M$ and all rules $l \to r$ of R_1, hence indeed $\mathcal{M}$ is a *quasi-model* for R_1. Next we choose

$$\pi_.(x, y) = \pi_+(x, y) = x + y$$

for all $x, y \in M$. All functions involved are weakly monotone in all coordinates. Now the infinite system $\overline{R_1} \cup \mathsf{Decr}$ consists of the rules

$\langle 1 \rangle$	$0 \cdot_i x \to x$	for all $i > 1$
$\langle 2 \rangle$	$x \cdot_i (y \cdot_j z) \to (x +_k y) \cdot_i z$	for $j < i$ and $k < i$
$\langle 3 \rangle$	$x \cdot_i (y \cdot_j z) \to (y +_k x) \cdot_i z$	for $j < i$ and $k < i$
$\langle 10, 17 \rangle$	$0 +_i x \to x$	for all $i > 1$
$\langle 11, 18 \rangle$	$x +_i 0 \to x$	for all $i > 1$
$\langle 12, 19 \rangle$	$0 +_2 0 \to 0$	
$\langle 12 \rangle$	$0 +_2 0 \to 0 \cdot_2 0$	
$\langle 13 \rangle$	$x +_i (y \cdot_j z) \to y \cdot_i (x +_k z)$	for $j < i$ and $k < i$
$\langle 14, 20 \rangle$	$(x \cdot_j y) +_i z \to x \cdot_i (y +_k z)$	for $j < i$ and $k < i$
$\langle 21 \rangle$	$x +_i (y \cdot_j z) \to y \cdot_i (z +_k x)$	for $j < i$ and $k < i$
$\langle 24 \rangle$	$0 * x \to 0$	
$\langle 25 \rangle$	$x * 0 \to 0$	
$\langle 26 \rangle$	$0 * 0 \to 0 \cdot_2 0$	
$\langle 26 \rangle$	$0 * 0 \to 0$	
$\langle 27 \rangle$	$x * (y \cdot_i z) \to (x * y) \cdot_j (x * z)$	where j is a multiple of i
$\langle 28 \rangle$	$(x \cdot_i y) * z \to (x * z) \cdot_j (y * z)$	where j is a multiple of i
	$x \cdot_i y \to x \cdot_j y$	for $j < i$
	$x +_i y \to x +_j y$	for $j < i$.

For instance, for any $\sigma : \mathcal{X} \to M$ we obtain from rule 2 a labelled rule $x \cdot_i (y \cdot_j z) \to (x +_k y) \cdot_i z$ where $i = [\sigma](x) + [\sigma](y) + [\sigma](z)$, $j = [\sigma](y) + [\sigma](z)$, $k = [\sigma](x) + [\sigma](y)$, indeed satisfying $j < i$ and $k < i$.

This infinite system $\overline{R_1} \cup \mathsf{Decr}$ is proved terminating by RPO with status by choosing the well-founded precedence

$$* > \cdots > +_i > \cdot_i > +_{i-1} > \cdot_{i-1} > \cdots.$$

Here $\cdot_i$ has the lexicographic status from right to left.

Termination of R_2 As the quasi-model $\mathcal{M}$ and the set $S_.$ we again choose the positive integers with the usual order. Only juxtaposition will be labelled, the other operation symbols remain unlabelled. We choose

$$0_{\mathcal{M}} = 1, \quad -_{\mathcal{M}} x = x + 1, \quad x \cdot_{\mathcal{M}} y = x, \quad x +_{\mathcal{M}} y = x *_{\mathcal{M}} y = x + y$$

for all $x, y \in M$. One easily checks that indeed $[\sigma](l) \geq [\sigma](r)$ for all $\sigma : \mathcal{X} \to M$ and all rules $l \to r$ of R_2, hence indeed $\mathcal{M}$ is a *quasi-model* for R_2. Next we choose

$$\pi_{\cdot}(x, y) = x$$

for all $x, y \in M$. All functions involved are weakly monotone in all coordinates. Now the infinite system $\overline{R_2} \cup \mathsf{Decr}$ consists of the rules

$\langle 4 \rangle$	$0 \cdot_1 (-0) \to 0 \cdot_1 0$	
$\langle 5 \rangle$	$x \cdot_i 0 \cdot_i (-0) \to x \cdot_i (-0) \cdot_i 0$	for all i
$\langle 6 \rangle$	$x \cdot_i 0 \cdot_i (-0) \to x \cdot_i 0 \cdot_i 0$	for all i
$\langle 7 \rangle$	$(-x) \cdot_{i+1} y \to -(x \cdot_i (-y))$	for all i
$\langle 8 \rangle$	$- - x \to x$	
$\langle 9 \rangle$	$-0 \to 0$	
$\langle 15, 22 \rangle$	$x + -y \to x + y$	
$\langle 16 \rangle$	$-x + y \to y + x$	
$\langle 23 \rangle$	$-x + y \to -(x + y)$	
$\langle 29 \rangle$	$x * (-y) \to -(x * y)$	
$\langle 30 \rangle$	$(-x) * y \to -(x * y)$	
	$x \cdot_i y \to x \cdot_j y$	for $j < i$.

This system is proved terminating by RPO with status by choosing any precedence satisfying

$$\cdots > \cdot_{i+1} > \cdot_i > \cdot_{i-1} > \cdots > \cdot_1 > - \quad \text{and} \quad + > - \quad \text{and} \quad * > -.$$

For $\cdot_i$ we choose the lexicographic status from right to left and for $+$ we choose the multiset status.

5.2 Complexity

The common complexity measure for binary multiplication is the number of bit operations as a function of the number of bits in both arguments. If we focus on the number of digit multiplications we have to count the number of applications of rule 26. One can show that multiplication of two numbers of b_1 and b_2 digits respectively takes $b_1 * b_2$ of these digit multiplications. More precisely, any reduction of $\langle n_1 \rangle * \langle n_2 \rangle$ to normal form, where $\langle n_i \rangle$ represents the normal form of a positive number of b_i digits, takes $b_1 * b_2$ applications of rule 26. This coincides with common algorithms ([AHU84] p. 62).

In [AHU84] two algorithms are given with better asymptotical behavior. Firstly, the *divide-and-conquer* algorithm ([AHU84] pp. 62—65) is $O[b^{\log 3}]$. Detailed inspection of the algorithm, however, reveals that the implicit constants render our specification and the divide and conquer algorithm comparable for $b \lesssim 32$ bits. Secondly, the Shönhage-Strassen algorithm, which is based on fast Fourier transform, has complexity $O[b \cdot \log b \cdot \log \log b]$. However, the constant in this complexity result is so unwieldy that this result is of no practical use. Note that these complexity results are in the number b of bits, which is logarithmic in the value n of the argument.

In the context of term rewriting, the number of reductions is a more appropriate measure. A worst-case bound for this number could perhaps be established, but this bound would also regard exotic, irregular reduction strategies. Most implementations however, have a fixed, regular reduction strategy, which could have a better asymptotical behavior.

Considering the left-most innermost strategy only, our multiplication algorithm can be shown to be $O[\log^3 n] = O[b^3]$. Experimental measurements suggest however that the actual number of rewrite steps used for binary multiplication is close to $b^2 \log b$, which is for practical use comparable with the common algorithms. For example, the multiplication of two 10-bit numbers requires in the order of 500 reduction steps; that of two ten digit numbers in the order of 1000 reduction steps.

5.3 Discussion

The rewrite system JP is an efficient, practical system for integers arithmetic. Its human-readable syntax makes it easy to use, and the absence of auxiliary functions make it esthetically interesting.

There are two flaws to this system: firstly, it is not confluent and secondly, the fact that compound terms occur as the right-hand argument of the juxtaposition operator requires the intuitively unexpected rules 1—3.

6 Implementation of rule schemata

Rule schemata can be implemented in three ways:

1. Literal inclusion. This is in fact the method by which humans implement arithmetic: tables for multiplication and addition are committed to memory, together with various distributive laws.
   This method is suitable for small radices, but becomes impractical as the number of rules rises (in JP, base ten already requires 414 rules for all tables).
2. Auxiliary functions. The introduction (in JP) of successor and predecessor functions *on digits* allows basic operations to be specified using $O[\mathcal{R}]$ rules and only affects the complexity within a constant. This method decreases the practical efficiency, and renders the system less clear. Nevertheless, if 1 is impractical and 3 is impossible, it is the only alternative.
3. External functions. Every computer supports fixed precision integer arithmetic, which could be used to compute all needed instances of rule schemata on the fly. Clearly, this requires additional capabilities in the implementation. One framework for the correct use of external functions in term rewrite systems is discussed in [Wal90].

We will discuss this third alternative in some more detail.

Using built-in integers is profitable (if it is at all possible) under one condition: the time to apply a single rewrite rule should be in the same order as that to do the actual calculation (this is usually the case) and that to translate digits from their TRS representation to the machine representation, and to translate the result back (this depends on the implementation).

An interesting idea[4] is the following. If $\mathcal{R}$ is chosen to be the largest single precision integer plus one, then all schema diagrams and rule schemata can be implemented using double precision arithmetic. The TRS uses the machine integers directly for all practical values, but as soon as values larger than $\mathcal{R}$-1 occur, they are represented using the juxtaposition operator, or nested postfix application. This results in a highly efficient implementation of unbounded integers: *every* calculation on integers in the single-precision range (typically $[-2^{15}, 2^{15}-1]$, or $[-2^{31}, 2^{31}-1]$) requires a single reduction step and an amount of time comparable to only a few reduction steps in order to allow for conversions, and calculations on larger values require a number of steps which is cubic in the logarithm of the arguments, but now this logarithm is base 2^{15}, or even base 2^{31}. For example, if $\mathcal{R} = 2^{15}$ the number 9876543210 is represented as (9)(6496)(5866); the multiplication $9876543210 * 1234567890$ is computed using in the order of 100 reduction steps.

7 Conclusions

In the table below, we have listed the three presented rewrite systems with all discussed properties. We have included the system from [CW91] since the four systems are to some degree complementary. The indication of readability refers to readability of *expressions*, and reflects the authors' opinions. The indication of complexity refers to the order of number of rewrite steps required for the multiplication of two numbers not exceeding n.

The entry marked ‡, relating to the confluence of the system [CW91], must be interpreted as follows: if the system is indeed terminating (which is not yet proved) then it is confluent modulo associativity and commutativity of $*$ and $+$.

system	range	conf.	term.	compl.	readability	radix
[CW91]	Int	‡	unproved	$\log n$	reasonable	4
SP	Int	yes	yes	n	poor	–
DA	Nat	yes	yes	$\log n$	good	any
JP	Int	ground	yes	$\log n$	good	any

We have presented three rewrite systems for integer arithmetic with addition, multiplication, and in two cases subtraction. We have shown ground confluence and termination.

In the last system, the common, human readable notation for arithmetic can be used. The latter two systems have logarithmic complexity and can be used for any radix.

Termination of the last system turned out to be non-trivial. We gave a proof in which the system was split up into two levels. For both remaining systems we

[4] Thanks to J.F.Th.Kamperman.

gave a termination proof first by transforming it to an infinite labelled system by the technique of semantic labelling and finally proving termination of the labelled system by recursive path order (RPO) with status.

In our opinion, our second rewrite system is perfect for natural number arithmetic, and our final rewrite system is perfect for ground integer arithmetic. Richard Kennaway (personal communication) proposed a confluent and terminating rewrite system for integer arithmetic. Although it can be used for implementing addition, subtraction and multiplication it does not meet the basic requirement that ground normal forms correspond bijectively to integers. The construction of a complete rewrite system with logarithmic complexity satisfying this basic requirement remains an open problem.

References

[AHU84] A.V. Aho, J.E. Hopcroft, and J.D. Ullman. *The Design and Analysis of Computer Algorithms.* Addison Wesley, 1984.

[BW89] L.G. Bouma and H.R. Walters. Implementing algebraic specifications. In J.A. Bergstra, J. Heering, and P. Klint, editors, *Algebraic Specification*, ACM Press Frontier Series, pages 199–282. The ACM Press in co-operation with Addison-Wesley, 1989. Chapter 5.

[CW91] D. Cohen and P. Watson. An efficient representation of arithmetic for term rewriting. In R. Book, editor, *Proceeding of the Fourth International Conference on Rewriting Techniques and Application (Como, Italy)*, LNCS 488, pages 240–251. Springer Verlag, Berlin, 1991.

[Der87] N. Dershowitz. Termination of rewriting. *J. Symbolic Computation*, 3(1&2):69–115, Feb./April 1987. Corrigendum: 4, 3, Dec. 1987, 409-410.

[DJK93] N. Dershowitz, J.-P. Jouannaud, and J.W. Klop. More Problems in Rewriting. In C.Kirchner, editor, *Proceeding of the Fifth International Conference on Rewriting Techniques and Application (Montreal, Canada)*, LNCS 690. Springer Verlag, Berlin, 1993.

[GG91] S.J. Garland and J.V. Guttag. *A Guide to LP, The Larch Prover.* MIT, November 1991.

[Klo92] J.W. Klop. Term rewriting systems. In S. Abramsky, D. Gabbay, and T. Maibaum, editors, *Handbook of Logic in Computer Science, Volume 2.*, pages 1–116. Oxford University Press, 1992.

[Wal90] H.R. Walters. Hybrid implementations of algebraic specifications. In H. Kirchner and W. Wechler, editors, *Proceedings of the Second International Conference on Algebraic and Logic Programming*, volume 463 of *Lecture Notes in Computer Science*, pages 40–54. Springer-Verlag, 1990.

[Wal94] H.R. Walters. A complete term rewriting system for decimal integer arithmetic. Technical Report CS-9435, Centrum voor Wiskunde en Informatica, 1994. Available by *ftp* from ftp.cwi.nl:/pub/gipe as Wal94.ps.Z.

[Zan93] H. Zantema. Termination of term rewriting by semantic labelling. Technical Report RUU-CS-93-24, Utrecht University, July 1993. *Accepted for publication in Fundamenta Informaticae.*

General Solution of Systems of Linear Diophantine Equations and Inequations

Habib Abdulrab **Marianne Maksimenko**

LIR, INSA de Rouen, BP 08, 76131 Mont Saint Aignan Cedex
E.mail: {habib.abdulrab, marianne.maksimenko}@insa-rouen.fr

Abstract:

Given a system Σ of linear diophantine equations and inequations of the form $L_i \#i M_i$, *$i=1, ..., n$, where* $\#i \in \{=, <, >, \neq, \geq, \leq\}$, *we compute a finite set S of numerical and parametric solutions describing the set of all the solutions of Σ (i.e. its* general solution*). Our representation of the general solution gives direct and simple functions generating the set of all the solutions: this is obtained by giving all the nonnegative natural values to the integer variables of the right hand-side of the parametric solutions, without any linear combination.*

In particular, unlike the usual representation based on minimal *solutions, our representation of the general solution is nonambiguous: given any solution s of Σ, it can be deduced from a unique numerical or parametric solution of S.*

1. Introduction, Definitions and Notations

Finding the set of all the solutions of systems of linear diophantine equations Σ: Ax = b, has several applications in automatic demonstration, unification theory, constraint logic programming... For example, it is the main part in AC-unification [3], it corresponds to A-unification with one variable [4]...

The set S of all the solutions is usually represented via a set S_0 of all the *minimal* solutions of Σ, and a set S_1 of all the nonnull *minimal* solutions of Σ': Ax = 0. Each solution of S is expressed as a solution of S_0 and a linear combination of some solutions of Σ'. A variety of works show how to compute these minimal sets. A non exhaustive list of important works can be found in [5,6,7,8,9,10,11,12, 13, 14].

In this paper we show how to use another representation (parametric representation) of S in the case of equations and inequations. Unlike the first usual representation, the second gives direct functions representing the general solution, without any linear combination. For example, the general solution of the equation $2x_1 = x_2 + x_3$ is given by the two parametric solutions:

$$\begin{cases} x_1 \to y_1 + y_2 \\ x_2 \to 2y_1 \\ x_3 \to 2y_2 \end{cases}$$ (corresponding to the case where x_2, x_3 are even), and

$$\begin{cases} x_1 \to y_1 + y_2 + 1 \\ x_2 \to 2y_1 + 1 \\ x_3 \to 2y_2 + 1 \end{cases}$$ (corresponding to the case where x_2, x_3 are odd).

The general solution is obtained here by giving to y_1 and y_2 all the nonnegative integer values.

Unlike the usual representation, the second is nonambiguous, in the sense that each solution can be deduced from only one of the two parametric solutions. On the other hand, the solution (3,3,3) can be deduced from $S_1 = \{(1,0,2), (1,2,0), (1,1,1)\}$ in an ambiguous manner.

Observe that such a nonambiguous representation may be nonminimal. It is called *minimal* when it has the smallest number of numerical and parametric solutions. For example, the system of two equations: $3z_1 + 2z_3 = 2z_2 + 8$, and $2z_1 + z_2 = 4z_3 + 7$, has a nonambiguous nonminimal representation consisting of the following numerical solution:

$$\begin{cases} z_1 \to 4 \\ z_2 \to 3 \\ z_3 \to 1 \end{cases}$$ and the three parametric (non linearly independent) solutions:

$$\begin{cases} z_1 \to 10 + 18a \\ z_2 \to 19 + 48a \\ z_3 \to 8 + 21a \end{cases} \qquad \begin{cases} z_1 \to 16 + 18a \\ z_2 \to 35 + 48a \\ z_3 \to 15 + 21a \end{cases} \qquad \begin{cases} z_1 \to 22 + 18a \\ z_2 \to 51 + 48a \\ z_3 \to 22 + 21a \end{cases}$$

On the other hand, the representation consisting of the following numerical and parametric solutions:

$$\begin{cases} z_1 \to 4 \\ z_2 \to 3 \\ z_3 \to 1 \end{cases} \qquad \begin{cases} z_1 \to 10 + 6a \\ z_2 \to 19 + 16a \\ z_3 \to 8 + 7a \end{cases}$$

is minimal.

Here is an example of a system of three inequations: $3z_1 + 2z_2 > 8$, $2z_1 < 4z_2 + 7$, and $z_1 > 2$. It has the tow following parametric solutions:

$$\begin{cases} z_1 \to 3 + 2a \\ z_2 \to a+b \end{cases} \qquad \begin{cases} z_1 \to 4 + 2a \\ z_2 \to a+b+1 \end{cases}$$

Here we use the same notations as [1]. Let $z_1, z_2, ..., z_n$ be natural variables and $\mathfrak{L}$ the set of all the linear polynomials of the form:

$$\zeta_1 z_1 + \zeta_2 z_2 + \ldots + \zeta_n z_n + \zeta_{n+1}, \qquad (1)$$

where ζ_i are natural numbers.

The coefficients of a polynomial will always be designed by $\zeta_1, \zeta_2, \ldots, \zeta_n$. Here we don't include ζ_{n+1} when we talk about the coefficients.

Let Σ be a system of linear diophantine equations and inequations

$$L_i \ \#i \ M_i \qquad (i=1, \ldots, p),$$

where $L_i, M_i \in \mathfrak{L}$, $\#i \in \{=, <, >, \neq, \geq, \leq\}$, and let ω be a transformation of the variables $z_1, z_2, \ldots, z_n$, defined by the following formulas:

$$z_i \to N_i \qquad (i=1, \ldots, n), \qquad (2)$$

where $N_i \in \mathfrak{L}$. Remark that we use in N_i the same variables as those of Σ. But, if z_i occurs in N_i, it should be interpreted as a new variable z_i'.

A transformation ω will be called a *parametric solution* of the system Σ, if for each $i=1, \ldots, p$, we have $\omega(L_i) \ \#i \ \omega(M_i)$, where $\omega(L_i)$ and $\omega(M_i)$ are linear polynomials. Thus, for each $z_1, z_2, \ldots, z_n$ the condition $\omega(\zeta_{i,1}z_1+\zeta_{i,2}z_2+\ldots+\zeta_{i,n}z_n+\zeta_{i,n+1}) \ \#i \ \omega(\eta_{i,1}z_1+\eta_{i,2}z_2+\ldots+\eta_{i,n}z_n+\eta_{i,n+1})$ is satisfied, where for each *i* we have $L_i=\zeta_{i,1}z_1+\zeta_{i,2}z_2+\ldots+\zeta_{i,n}z_n+\zeta_{i,n+1}$, $M_i=\eta_{i,1}z_1+\eta_{i,2}z_2+\ldots+\eta_{i,n}z_n+\eta_{i,n+1}$.

If in the solution (2) each polynomial N_i is a natural number, we call (2) a *numerical solution* of the system Σ, or simply a *solution*.

If we assign any natural values to the parameters $z_1, z_2, \ldots, z_n$ of a parametric solution (2), this becomes a numerical solution. Thus, a parametric solution describes a certain class of numerical solutions.

We call a *general solution* of Σ the set of all the numerical solutions.

The *index* of the system Σ is a pair of natural numbers (p, r), where p is the number of the equations and inequations in the system Σ, r is the number of the positive coefficients in L_1 and M_1.

We say that $(p_1, r_1)<(p_2, r_2)$, iff $(p_1<p_2)$ or $(p_1=p_2 \ \& \ r_1<r_2)$.

2. General Solution of Systems of Equations and Inequations

The following theorem 1 due to Makanin [1] is the base of our work. Theorem 2 is an extension of theorem 1 for systems of linear diophantine equations and inequations. The proof of theorem 2 uses the ideas of the proof of theorem 1, but includes new cases related to inequations.

Moreover, in the case of equations, our proof uses some better bounds, and modifies slightly the transformations of the proof of theorem 1 [1], in order to associate with each solution of Σ a unique numerical or parametric solution, as proved in theorem 3. This non ambiguity is ensured for any system of equations and inequations.

Theorem 1

There exists an algorithm that computes the general solution of any system of linear diophantine equations, by means of a finite number of numerical and parametric solutions.

Theorem 2

There exists an algorithm that computes the general solution of any system of linear diophantine equations and inequations, by means of a finite number of numerical and parametric solutions.

Proof

It is sufficient to prove this theorem for the case where $\#i \in \{=, <\}$, because: a) If an inequation of the system has the form $L_i > M_i$, we can exchange L_i and M_i and obtain the inequation $L_i < M_i$.

b) If an inequation has the form $L_i \neq M_i$, we can replace the system by two systems, by substituting the i-th inequation by $L_i < M_i$ in the first one and by $L_i > M_i$ in the second one. The solution of the system Σ is obtained by the union of the solutions of these systems.

c) If an inequation has the form $L_i \leq M_i$, we can replace the system by two systems, by substituting the i-th inequation by $L_i < M_i$ in the first one and by $L_i = M_i$ in the second one. The solution of the system Σ is the union of the solutions of these systems.

d) The case $L_i \geq M_i$ is analogous to c).

Now, prove the theorem when $\#i \in \{=, <\}$.

Prove this fact by induction on the index of the system Σ.

If the index equals (0, 0), then the system Σ is empty. Its general solution is $z_i \rightarrow z_i$ ($i=1, \ldots, n$).

Suppose that the theorem is true for each system whose index is smaller than (p, r), and consider any system whose index is (p, r).

The following cases are possible.

Case I

The number of the positive coefficients in both hand-sides of $L_1\#1M_1$ is equal to zero.

In this case, the index equals (p, 0), and $L_1\#_1M_1$ has the form

$$\zeta_{n+1} < \eta_{n+1}$$

or

$$\zeta_{n+1} = \eta_{n+1}.$$

If such an expression is false, then the system has no solution. Otherwise, the system Σ is equivalent to Σ' defined by the equations and the inequations

$$L_i \,\#i\, M_i \qquad\qquad (i=2, \ldots, p).$$

The index of the system Σ' equals (p-1, s), where s is the number of the positive coefficients in $L_2 \#_2 M_2$. According to the induction hypothesis, the general solution of Σ' can be described by means of a finite number of numerical and parametric solutions. But, because Σ and Σ' are equivalent, this description is the same as that of Σ.

Case II

The number of the positive coefficients in one hand-side of $L_1 \#1 M_1$ is equal to zero and it is greater than zero in the other hand-side.

In this case we have two subcases, depending on if the number of the positive coefficients is equal to zero in L_1 or in M_1. For each subcase we prove separately that the general solution of $L_1 \#1 M_1$ can be described by means of a finite number of numerical and parametric solutions. And after, we will prove the theorem's fact for both subcases in the same time, without this division.

Subcase II.1

The number of the positive coefficients in L_1 is greater than zero and that of M_1 is equal to zero.

In this case, $L_1 \#1 M_1$ has the form

$$\zeta_{i1}z_{i1}+\zeta_{i2}z_{i2}+ \ldots +\zeta_{ir}z_{ir}+\zeta_{n+1}<\eta_{n+1},$$

or

$$\zeta_{i1}z_{i1}+\zeta_{i2}z_{i2}+ \ldots +\zeta_{ir}z_{ir}+\zeta_{n+1}=\eta_{n+1},$$

where $r>0$, $\zeta_{i1}>0, \ldots, \zeta_{ir}>0$, $\zeta_{n+1}\geq 0$, $\eta_{n+1}\geq 0$.

The inequation $L_1<M_1$ or the equation $L_1=M_1$ in the natural variables $z_{i1}, z_{i2}, \ldots, z_{ir}$ can have only a finite number of solutions, all numerical. If $L_1 \#1 M_1$ has no solution, the system Σ has no solution too. If $L_1 \#1 M_1$ has a solution, we can find all the solutions of $L_1 \#1 M_1$ by enumerating all the integers z_{ij} of the interval $[0, (\eta_{n+1}-\zeta_{n+1})/\zeta_{ij}]$. All these solutions are numerical.

Subcase II.2

The number of the positive coefficients in M_1 is greater than zero and that of L_1 is equal to zero. Exchange L_1 and M_1. If we have an equation, we obtain the subcase II.1. So, we consider the case of an inequation.

In this case the inequation $L_1>M_1$ has the form

$$\zeta_{i1}z_{i1}+\zeta_{i2}z_{i2}+ \ldots +\zeta_{ir}z_{ir}+\zeta_{n+1}>\eta_{n+1}, \tag{3}$$

where $r>0$, $\zeta_{i1}>0, \ldots, \zeta_{ir}>0$, $\zeta_{n+1}\geq 0$, $\eta_{n+1}\geq 0$.

If $\zeta_{n+1}>\eta_{n+1}$, (3) is always true. The system Σ is equivalent to Σ', defined by the equations and the inequations

$$L_i \#i M_i \qquad (i=2, \ldots, p).$$

The index of the system Σ' equals $(p-1, s)$, where s is the number of the positive coefficients in $L_2 \#_2 M_2$. According to the induction hypothesis the general solution of Σ' can be described by means of a finite number of numerical and parametric solutions. But, as Σ and Σ' are equivalent, this description is the same as that of Σ. Thus, the theorem is proved when $\zeta_{n+1}>\eta_{n+1}$.

Consider now that $\zeta_{n+1}\leq \eta_{n+1}$.

Prove by induction on the number k of positive coefficients that the general solution of the inequation $L_1>M_1$ can be described by means of a finite number of numerical and parametric solutions.

If k=1, the inequation has the form: $\zeta_{i1}z_{i1}+\zeta_{n+1}>\eta_{n+1}$, where $\zeta_{i1}>0$, $\zeta_{n+1}\geq 0$, $\eta_{n+1}\geq 0$.

Thus, $\zeta_{i1}z_{i1}>\eta_{n+1}-\zeta_{n+1}$, where $\zeta_{i1}>0$, $\eta_{n+1}-\zeta_{n+1}\geq 0$.

The solution of this inequation is given by the transformation:

$z_{i1}\to z_{i1}+c_1$, where $c_1=[(\eta_{n+1}-\zeta_{n+1})/\zeta_{i1}+1]$ ([k] denotes the greatest integer less than or equal to k)

$z_j\to z_j$, where $j\neq i_1$.

Suppose that the proposition is true when $i_r<k$, and consider the case when $i_r=k$.

Rewrite the inequation (3) as follows:

$$\zeta_{i1}z_{i1}+\zeta_{i2}z_{i2}+ \ldots +\zeta_{ir}z_{ir}>\eta_{n+1}-\zeta_{n+1}, \qquad (4)$$

where $r>0$, $\zeta_{i1}>0$, ..., $\zeta_{ir}>0$, $\zeta_{n+1}\geq 0$, $\eta_{n+1}\geq 0$.

If the values of z_{ir} are $> c_r$, where $c_r=[(\eta_{n+1}-\zeta_{n+1})/\zeta_{ir}+1]$, the inequality is true. The solution of (4) in this case is given by the transformation:

$z_{ir}\to z_{ir}+c_r$,

$z_j\to z_j$, where $j\neq i_r$.

If the values of z_{ir} are $\leq c_r$, the number of such values is finite. Call them β_1, β_2, ..., β_γ. After the substitution of all these possible values in the inequation (4), we obtain a finite number of inequations; the union of their solutions is the solution of the inequation (4) with the condition $z_{ir}\leq c_r$. Each of these obtained inequations satisfies the induction hypothesis and can be described by means of a finite number of numerical and parametric solutions: S_1, S_2, ..., S_γ. The solution of the inequation (4) in this case is obtained by the union of the solutions

$$\begin{cases}\beta_1\\ S_1\end{cases}, \begin{cases}\beta_2\\ S_2\end{cases}, \ldots, \begin{cases}\beta_\gamma\\ S_\gamma.\end{cases}$$

We proved that $L_1\#1M_1$ can be described by means of a finite number of numerical and parametric solutions.

Now, suppose that $L_1\#1M_1$ has the solutions ω_1, ω_2, ..., ω_α. After the substitution of each solution ω_i in the system

$$\begin{cases}L_2\,\#2\,M_2\\ \ldots\ldots\\ L_p\,\#p\,M_p,\end{cases}$$

we obtain the systems Σ_1, Σ_2, ..., Σ_α.

For each i=1, ..., α, the index of the system Σ_i is smaller than (p, r). According to the induction hypothesis, the general solution of each system Σ_i, i=1, ..., α can be described by the parametric solutions: $R_{i,1}$, $R_{i,2}$, ..., R_{i,y_α}.

Thus, the general solution of the system is described by means of the numerical and parametric solutions:

$$\begin{cases}\omega_1\\ R_{1,1}\end{cases}, \begin{cases}\omega_1\\ R_{1,2}\end{cases}, \ldots, \begin{cases}\omega_1\\ R_{1,y_1}\end{cases};$$

$$\begin{cases}\omega_2\\ R_{2,1}\end{cases}, \begin{cases}\omega_2\\ R_{2,2}\end{cases}, \ldots, \begin{cases}\omega_2\\ R_{2,y_2}\end{cases};$$

....

$$\begin{cases}\omega_\alpha\\ R_{\alpha,1}\end{cases}, \begin{cases}\omega_\alpha\\ R_{\alpha,2}\end{cases}, \ldots, \begin{cases}\omega_\alpha\\ R_{\alpha,y_\alpha}\end{cases}.$$

Case III

The number of the positive coefficients in both hand-sides of $L_1\#1M_1$ is greater than zero. Two subcases can appear.

Subcase III.1

There exists a variable z_u, having a positive coefficients in both L_1 and M_1.

In this case, $L_1\#1M_1$ has the form:

$$L_1'+\zeta_u z_u+L_1''<M_1'+\eta_u z_u+M_1'' \tag{5}$$

or

$$L_1'+\zeta_u z_u+L_1''=M_1'+\eta_u z_u+M_1'', \tag{6}$$

where $\zeta_u>0$, $\eta_u>0$.

If we have $\zeta_u<\eta_u$, we construct the system Σ' by replacing (5) or (6) by

$$L_1'+L_1''<M_1'+(\eta_u-\zeta_u)z_u+M_1'',$$

or

$$L_1'+L_1''=M_1'+(\eta_u-\zeta_u)z_u+M_1''.$$

The systems Σ and Σ' are obviously equivalent, but the index of Σ' is smaller than that of Σ. The general solution of Σ is the same as that of Σ', which can be described, by induction, by a finite number of numerical and parametric solutions.

When $\zeta_u\geq\eta_u$, we construct the system Σ' by replacing (5) and (6) by

$$L_1'+(\zeta_u-\eta_u)z_u+L_1''<M_1'+M_1'',$$

or

$$L_1'+(\zeta_u-\eta_u)z_u+L_1''=M_1'+M_1''.$$

and apply the induction hypothesis.

Subcase III.2

The variables of the positive coefficients of L_1 are different from those of M_1.

In this case L_1 #1 M_1 has the form

$$\zeta_i z_i+L_1' \#1\ \eta_j z_j+M_1',$$

where $\zeta_i>0$, $\eta_j>0$ and $i\neq j$.

Apply to the system Σ the transformation

$$z_i\to(\eta_j/k)z_i+\beta \tag{7}$$

$$z_j\to(\zeta_i/k)z_j+\gamma,$$

where $k = \gcd(\eta_j, \zeta_i)$, β and γ are any integers that satisfy $0 \leq \beta < \eta_j/k$, $0 \leq \gamma < \zeta_i/k$. We obtain a system Σ_1 whose index equals that of Σ.

We can construct $\eta_j\zeta_i/k^2$ systems of type Σ_1, by giving different values to β and γ. Construct all the systems of this type, and denote them by the list:

$\Sigma_1, \Sigma_2, ..., \Sigma_q.$ (8)

The general solution of the system Σ can be constructed by substituting the general solutions of each system Σ_i in the formula (7), where the numbers β and γ correspond to those of Σ_i, and by the union of all these solutions.

To construct the general solution of each system Σ_m of (8), we apply two transformations for each one of these systems:

1) We apply the transformation $z_i \to z_j+z_i$ and reduce $\zeta_i\eta_j z_j/k$ from both hand-sides of the first equation, and then we obtain the system Σ'_m.

2) Similarly, we apply the transformation $z_j \to z_j+z_i+1$ and reduce $\zeta_i\eta_j z_i/k$ from both hand-sides, and we obtain the system Σ''_m.

The general solution of Σ_m can be obtained from the substitution of the general solution in the formulas $z_i \to z_j+z_i$ and $z_j \to z_j+z_i+1$, and by the union of these solutions.

Thus, the general solution of Σ can be described by the help of the formulas associated with the general solutions of the systems

$\Sigma'_1, \Sigma''_1, \Sigma'_2, \Sigma''_2, ..., \Sigma'_q, \Sigma''_q.$ (9)

Because in each Σ_i of the list (8) the coefficient of z_i in L_1 is equal to that of z_j in M_1, the index of each system of the list (9) is smaller than that of Σ. Hence, according to the induction hypothesis, the general solution of each system of (9) can be described by a finite number of numerical and parametric solutions.

♦

3. Example

The following example aims to show how to describe the computation of the solutions of Σ by a finite tree whose nodes are systems of equations, whose arcs are labelled by the transformations, and whose leafs are numerical and parametric solutions. A technique of intervals is introduced to denote a sequence of equations of the form $L_i=n$, for all $n = n_1,..., n_2$, by $L_i=[n_1, n_2]$. Here we show how the solutions are obtained from such a tree by the successive applications of the transformations.

Consider the system of the unique equation $2x_1 = y_1 + y_2$ (see Figure 1).

The case (e) which corresponds to the case where the last coefficient in (a) is 0, gives the following parametric solution.

$$\begin{array}{llllll} & \text{(a)} & \text{(b)} & \text{(c)} & \text{(d)} & \text{(e)} \\ \left\{\begin{array}{l} x_1 \to \\ y_1 \to \\ y_2 \to \end{array}\right. & \begin{array}{l} x_1 \\ 2y_1 \\ y_2 \end{array} & \begin{array}{l} \to x_1+y_1 \\ \to 2y_1 \\ \to y_2 \end{array} & \begin{array}{l} \to x_1+y_1 \\ \to 2y_1 \\ \to 2y_2 \end{array} & \begin{array}{l} \to x_1+y_1+y_2 \\ \to 2y_1 \\ \to 2y_2 \end{array} & \begin{array}{l} \to y_1+y_2 \\ \to 2y_1 \\ \to 2y_2 \end{array} \end{array}$$

The case (f) which corresponds to the case where the last coefficient in (a) is 1, gives the following parametric solution.

$$
\begin{array}{llllll}
 & \text{(a)} & \text{(b)} & \text{(c)} & \text{(d)} & \text{(f)} \\
\left\{\begin{array}{l} x_1 \to x_1 \\ y_1 \to 2y_1+1 \\ y_2 \to y_2 \end{array}\right. &
\begin{array}{l} \to \\ \to \\ \to \end{array} &
\begin{array}{l} x_1+y_1 \\ 2y_1+1 \\ y_2 \end{array} \begin{array}{l} \to \\ \to \\ \to \end{array} &
\begin{array}{l} x_1+y_1 \\ 2y_1+1 \\ 2y_2+1 \end{array} \begin{array}{l} \to \\ \to \\ \to \end{array} &
\begin{array}{l} x_1+y_1+y_2 \\ 2y_1+1 \\ 2y_2+1 \end{array} \begin{array}{l} \to \\ \to \\ \to \end{array} &
\begin{array}{l} \mathbf{y_1+y_2+1} \\ \mathbf{2y_1+1} \\ \mathbf{2y_2+1} \end{array}
\end{array}
$$

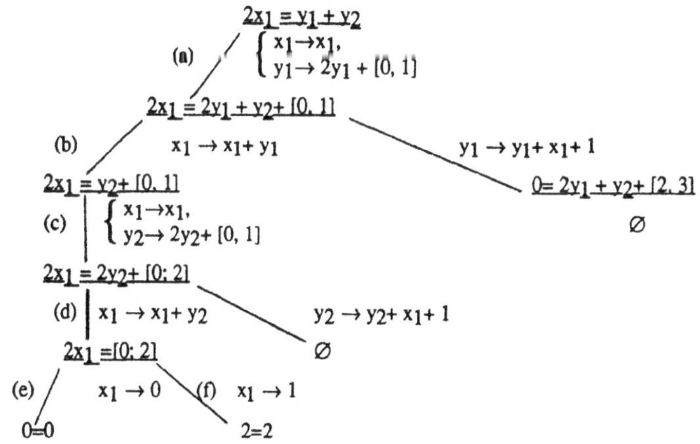

Figure 1

◆

The following theorem shows that the set of numerical and parametric solutions given by theorem 2 is nonambiguous.

Theorem 3

Each solution of the system of equations and inequations Σ can be deduced from a unique numerical or parametric solution generated by theorem 2.

Proof

Prove this fact by induction on the index of Σ. Let $\alpha=(s_1, ..., s_n)$ be a solution of Σ.

If the index of the system equals (0, 0), then Σ has no equation. Its general solution is $z_i \to z_i$ ($i=1, ..., n$). It is a unique parametric solution, and α can be deduced from it by: $z_i \to s_i$ ($i=1, ..., n$).

Suppose that the theorem is true for each system whose index is smaller than (p, r), and consider then any system whose index is (p, r). The following cases are possible.

a) $L_1 \#1\ M_1$ is always true.

Σ is equivalent to Σ', defined by the equations and the inequations

$L_i \#i\ M_i$ $\qquad\qquad$ ($i=2, ..., p$).

The index of the system Σ' equals (p-1, s), where s is the number of the positive coefficients in $L_2 \#2\ M_2$. According to the induction hypothesis α can be deduced from a unique numerical or parametric solution of Σ'. So, it can be deduced for Σ.

b) $L_1 \#1\ M_1$ has only numerical solutions.

Thus, if $j_1, ..., j_k$ ($k \leq n$) are the positive coefficients in L_1 (for M_1, all the coefficients are equal to zero in the considered cases), then $\alpha_1=(s_{j1}, ..., s_{jk})$ belongs to the set of the solutions of $L_1\#1\ M_1$ in the natural variables $z_{j1}, ..., z_{jk}$. The solution α_1 is obviously unique. The substitution of α_1 in Σ' gives a system in the natural variables $\{z_1, ..., z_n\}\setminus\{z_{j1}, ..., z_{jk}\}$, where $\alpha\setminus\alpha_1$ is a solution of Σ' whose index is smaller than (p, r). This solution can be deduced from a unique numerical or parametric solution of Σ' by the induction hypothesis. Consequently, α is deduced from a unique numerical or parametric solution of Σ.

c) $L_1\#1\ M_1$ has at least one parametric solution.

This case corresponds to the cases III, and II.2 with $\zeta_{n+1}\leq\eta_{n+1}$ of theorem 2. Call these two cases 1) and 2), respectively.

Case 1)

In the subcase III.1, we reason in the same manner as in the proof of theorem 2, and we apply the induction hypothesis. In the subcase III.2, $L_1\ \#1\ M_1$ has the form

$$\zeta_i z_i+L'_1\ \#1\ \eta_j z_j+M'_1,$$

where $\zeta_i>0$, $\eta_j>0$ and $i\neq j$.

Apply to the system Σ the transformation (7): $z_i\rightarrow(\eta_j/k)z_i+\beta$, $z_j\rightarrow(\zeta_i/k)z_j+\gamma$. We obtain $\eta_j\zeta_i/k^2$ systems, by giving all the different values to β and γ. Only one of these systems corresponds to

$$s_i=(\eta_j/k)s_i'+\beta$$
$$s_j=(\zeta_i/k)s_j'+\gamma.$$

If $s_i\geq s_j$, α is deduced from the unique solution associated with the transformation $z_i\rightarrow z_j+z_i$. If $s_i<s_j$, α is deduced from the unique solution associated with the transformation $z_j\rightarrow z_j+z_i+1$. In both cases we can obtain our result by the induction hypothesis.

Case 2)

Prove by induction on the number k of the positive coefficients of the inequation $L_1>M_1$ that α can be deduced from a unique numerical or parametric solution.

If k=1, the inequation has the form $\zeta_{i1}z_{i1}+\zeta_{n+1}>\eta_{n+1}$, and α is deduced from the unique parametric transformation of this case, and from a unique numerical or parametric solution of the system (with smaller index) of the remaining equations.

Suppose that the proposition is true when $i_r<k$, and consider the case when $i_r=k$. If the value s_t of z_{i_r} is greater than c_r associated with (4), α is deduced in the same manner as k = 1. Otherwise, s_t is equal to only one of the different integers values $\beta_1, \beta_2, ..., \beta_\gamma$ of z_{i_r}. So, $\alpha\setminus(s_t)$ can be deduced from a unique numerical or parametric solution of the system (with smaller index) of the remaining equations. Thus, α can be deduced from a unique numerical or parametric solution.

4. Implementation

A first direct implementation of this algorithm has been realised in Common Lisp by our students S. Gattepaille and S. Robidoux. A new and more efficient implementation is in preparation, where some improvements, such as the followings, are conceived:

1) All the equations of the form of case II are solved as a system, and not by solving the first equation and substituting its solutions to the rest of the system, and solving new systems... The resolution of the whole system of case II is to be realised by using the best known bounds and the usual heuristics used in the numeration of minimal solutions.

2) Optimising the resolution of such systems, where the second numbers are intervals $[n_1, ..., n_2]$ is also studied. In this case, the solutions of the systems associated with n_1+i, $i=1, ..., n_2$ are based on those of n_1+j, $j<i$.

3) The solutions where $\#i \in \{=, <, >, \neq, \geq, \leq\}$ are computed directly, in the same manner as $\#i \in \{=, <\}$, without duplicating the equations and inequations of the initial system, which are given here to simplify the proofs.

5. Conclusion

Solving systems of linear diophantine equations and inequations, with nonambiguous representation of the general solution, is presented in this paper. There are some advantages of such a representation. At first, it is functional and very convenient: the generation of the solutions can be done here only by giving all the nonnegative natural values to the integer variables of the right hand-side of the parametric solution. It is not based on solving one equation and reporting its solutions in the others, where colossal expressions are usually generated in the intermediary phases, even if there is a small number of solutions. It is completely incremental: adding a new equation does not imply at all to resolve the initial system.

Minimal parametric representation may allow a concise representation of the usual *minimal* solutions. For example, the general solution of the equation $x_1=y_1+ ... +y_n$ is $x_1 \rightarrow y_1+ ... +y_n$ which "replaces" the n minimal solutions (1, 1, 0, ..., 0), (1, 0, 1, 0, ..., 0), ..., (1, 0, ..., 0, 1).

Some directions devoted to a new and efficient implementation of the algorithm described in this paper are also given.

6. Comparison with Other Work

It is possible to deduce the existence of a nonambiguous representation of the solutions of systems of linear diophantine equations (in the framework of the usual representation by the set S_0 of all the minimal solutions of Σ, and the set S_1 of all the nonnull minimal solutions of Σ': $Ax = 0$) from a general result due to Eilenberg and Schutzenberger [2], which states that every semi-linear set (i.e. a finite union of linear sets $S_0 + S_1{}^*$) is semi-simple (i.e. each solution can be deduced in a nonambiguous way). For example, the solutions of the equation $2x_1 = x_2 + x_3$ given in 1. can be deduced from the nonambiguous parametrisation: $n_1(1,0,2) + n_2(1,2,0) + n_3(1,1,1)$; $n_1, n_2 = 0, 1, \ldots$; $n_3 = 0, 1$.

But the proof of that result is not constructive, because it proceeds by proving that the complementary of a semi-linear set is semi-simple, where it is sufficient to complement twicc.

Acknowledgement

We would like to express our great thanks to J.-L. Lambert for his very useful remarks and suggestions, and for all RTA's referees for their corrections, and for their very careful comments.

References

[1] *The Systems of Standard Word Equations in the n -layer Alphabet of Variables,* (*in Russian*) G.S. Makanin, Sibirskiï matematictheskiï journal, AN SSSR, Sibirskoje otdelenije, vol. XIX, N° 3, 1978.

[2] *Rational Sets in Commutative monoids*, S. Eilenberger, and M.P. Schutzenberger. J. Algebra, 1969, 173-191.

[3] *AC-unification Race: the System Solving Approach*, M. Adi and C. Kirchner, 1989.

[4] *On General Solution of Word Equations,* G.S. Makanin, and H. Abdulrab, Proceedings of "Important results and trends in Theoretical Computer Science" Graz, Austria 10-12/6/94, LNCS.

[5] *Complexity of Unification in Free Groups and Free Semi-groups*, A. Koscielski and L. Pacholski. Technical report, Institute of Computer Science, University of Wroclaw and Institute of Mathematics, Polish Academy of Sciences, 1989.

[6] *Efficient Solution of Linear Diophantine Equations*. M. Clausen and A. Fortenbacher. JSC, 8:201-216, 1989. Special issue on Unification.

[7] *On effecient Algorithm for Solving Systems of Linear Diophantine Equations*. E. Contejean. Information and Computation, vol. 113, n. 1, pp. 143-172, 1994.

[8] *Une Borne pour les Générateurs des Solutions Entières Positives d'une Equation Diophantienne Linéaire*. J-L. Lambert, Compte rendu de L'académie des Sciences de Paris, 305:39-40, 1978.

[9] *Minimal Solutions of Linear Diophantien Systems: Bounds and Algorithms*. L. Pottier. Proceedings of the fourth International Conference on Rewritting Technics and Applications, Como, Italy, pp 162-173, 1991.

[10] *Solutions of Linear Diophantine Systems*. J.F. Romeuf. Rapport LITP 88-76.

[11] *A polynomial Algorithm for Solving Systems of Two Linear Diophantine Equations*. J.F. Romeuf. Technical report, LIR, Rouen, 1990.

[12] *Solving Systems of Linear Diophantine Equations: An Algebraic Approach*. E. Domenjoud, Proceedings of 16th Mathematical Foundation of Computer Sciene. Warsaw, LNCS 520, 1991, Springer Verlag.

[13] *On linear homogeneous diophantine equations*. Elliot. Quartely J. of Pure and Applied Maths, 136, 1903.

[14] *A Syzygetic Theory*. P. MacMahon. Combinatory Analysis, volume 2, chapter II, pp. 111-114. Reprinted by Chelsea in 1960, 1919.

Combination of Constraint Solving Techniques: An Algebraic Point of View

Franz Baader[1] and Klaus U. Schulz*[2]

[1] Lehr- und Forschungsgebiet Theoretische Informatik, RWTH Aachen, Ahornstraße 55, 52074 Aachen, Germany, baader@informatik.rwth-aachen.de
[2] CIS, Universität München, Wagmüllerstraße 23, 80538 München, Germany, schulz@cis.uni-muenchen.de

Abstract. In a previous paper we have introduced a method that allows one to combine decision procedures for unifiability in disjoint equational theories. Lately, it has turned out that the prerequisite for this method to apply—namely that unification with so-called linear constant restrictions is decidable in the single theories—is equivalent to requiring decidability of the positive fragment of the first order theory of the equational theories. Thus, the combination method can also be seen as a tool for combining decision procedures for positive theories of free algebras defined by equational theories. Complementing this logical point of view, the present paper isolates an abstract algebraic property of free algebras—called combinability—that clarifies why our combination method applies to such algebras. We use this algebraic point of view to introduce a new proof method that depends on abstract notions and results from universal algebra, as opposed to technical manipulations of terms (such as ordered rewriting, abstraction functions, etc.) With this proof method, the previous combination results for unification can easily be extended to the case of constraint solvers that also take relational constraints (such as ordering constraints) into account. Background information from universal algebra about free structures is given to clarify the algebraic meaning of our results.

1 Introduction

In most of the applications of unification modulo an equational theory E, a unification algorithm for elementary E-unification (which treats just the terms built over the signature of E) is not sufficient. Usually, there are at least additional free function symbols present, or even symbols defined by another equational theory. For this reason, the *combination problem* for unification algorithms is an important research topic in unification theory. Informally, this problem can be described as follows: Let E and F be equational theories over disjoint signatures, and assume that unification algorithms for E and for F are given. How can we combine these algorithms to obtain a unification algorithm for $E \cup F$. Originally,

* This author was supported by the EC Working Group CCL, EP6028.

the term "unification algorithm" referred to an algorithm that computes a complete set of unifiers (see [SS89, Bou93] for the most recent results on combining such algorithms). With the development of constraint approaches to theorem proving [Bür91, NiR94] and term rewriting [KK89], the role of algorithms that compute complete sets of unifiers is more and more taken on by algorithms that decide solvability of the unification problems. In this setting, more general constraints than the equational constraints $s = t$ of unification problems become important as well. For example, one might be interested in ordering constraints of the form $s \leq t$ on terms [CT94], where the predicate $\leq$ could be interpreted as the subterm ordering or as a reduction ordering.

For unification, the problem of combining decision procedures has been solved in [BS92] in a rather general way. The main tool of this combination method is a ***decomposition algorithm***, which separates a given unification problem Γ of the joined theory (i.e., an $(E \cup F)$-unification problem) into pure unification subproblems Γ_E and Γ_F of the single theories. Solutions of these pure problems must satisfy additional conditions, called ***linear constant restrictions*** in [BS92], to yield a solution of Γ. The main result of [BS92] is that solvability of unification problems in the combined theory $E \cup F$ is decidable, provided that solvability of unification problems with linear constant restrictions is decidable in E and F. It should be noted that this result can easily be lifted to solvability of $(E \cup F)$-unification problems with linear constant restrictions. This combination result has been generalized to disunification [BS93a] and to unification in the union of theories with shared constant symbols [Rin92]. In both cases, the decomposition algorithm of [BS92] could be adapted to the new problem without serious modifications. An important goal of the present paper is to give an abstract characterization of the situations in which this seemingly ubiquitous decomposition method can be applied. In addition to the better understanding of the underlying principles, this could also yield a better basis for further generalizations. The proof method used in [BS92, BS93a, Rin92]—which depends on an infinite ordered rewrite system obtained by unfailing completion, term abstraction functions, etc.—seems not to facilitate such an abstract view (see, e.g., the rather technical "shared constructor" condition in [DKR94]).

At first sight, the notion of "unification with linear constant restrictions" is just a technical notion that makes our combination machinery work, but seems to have little further significance. In [BS93] it is shown, however, that E-unification with linear constant restrictions is decidable iff the positive fragment of the first-order theory of E is decidable. Since the positive theory of E coincides with the positive theory of the E-free Σ-algebra $\mathcal{T}(\Sigma, X)/\!=_E$ over infinitely many generators X, the combination result of [BS92] can be reformulated as follows: Let E and F be equational theories over disjoint signatures Σ and Δ, and let X be a countably infinite set of generators. The positive theory of $\mathcal{T}(\Sigma \cup \Delta, X)/\!=_{E \cup F}$ is decidable, provided that the positive theories of $\mathcal{T}(\Sigma, X)/\!=_E$ and $\mathcal{T}(\Delta, X)/\!=_F$ are decidable.

In the present paper, this observation is used as the starting point of a more abstract, algebraic approach to formulating and solving the combination problem. Starting with two algebras over disjoint signatures, the goal is to construct a "combined" algebra such that validity of positive formulae in this algebra can be decided by using a decomposition algorithm and decision procedures for the positive theories of the original algebras. Obviously, this can only be achieved if the algebras satisfy some additional properties. *We will call an algebra $\mathcal{A}$* combinable *iff it is generated by a countably infinite set X such that any mapping from a finite subset of X to $\mathcal{A}$ can be extended to a surjective endomorphism of $\mathcal{A}$.* For combinable algebras $\mathcal{A}$ and $\mathcal{B}$ over disjoint signatures Σ and Δ, we can construct the so-called *free amalgamated product* $\mathcal{A} \odot \mathcal{B}$, which is a $(\Sigma \cup \Delta)$-algebra.[3] Now a simple modification of the decomposition algorithm of [BS92] can be used to show that the positive theory of $\mathcal{A} \odot \mathcal{B}$ is decidable iff the positive theories of $\mathcal{A}$ and of $\mathcal{B}$ are decidable.

Obviously, the free algebras $\mathcal{T}(\Sigma, X)/{=_E}$ and $\mathcal{T}(\Delta, X)/{=_F}$ over a countably infinite set of generators X are combinable. In this case, the free amalgamated product yields an algebra that is isomorphic to the combined free algebra $\mathcal{T}(\Sigma \cup \Delta, X)/{=_{E \cup F}}$. Thus, the combination result of [BS92] is obtained as a corollary. As described until now, the amalgamation of combinable algebras does not yield a real generalization of this result. Indeed, one can use well-known results from universal algebra to show that an algebra is combinable (as defined above) iff it is a free algebra over countably many generators for an equational theory. What is new, though, is the proof method, which—in contrast to the original proof—only depends on elementary notions from universal algebra, and thus clarifies the role played by the combinability condition. This new proof can be seen as an adaptation of the proof ideas in [SS89] to the combination of decision procedures. Unlike in [SS89], however, everything is done on the abstract algebraic level instead of on the term level. Interestingly, on this level it is also very easy to prove completeness of an optimized version of the decomposition algorithm of [BS92], which significantly reduces the number of nondeterministic choices.

In addition, the abstract algebraic approach allows for an easier generalization of the results. In fact, instead of algebras we will consider algebraic structures in the following. This means that the signatures may contain both function symbols and predicate symbols, and these additional predicate symbols may occur in the constraint problems to be solved. With the usual notion of homomorphism for structures, most of the results from universal algebra carry over to structures. The combination result for combinable algebras sketched above thus holds for free structures as well. This yields a combination method for constraint solvers of more general constraints than just equational constraints.

The next section recalls some results from universal algebra for free structures. In Section 3 we construct the free amalgamated product of free structures, and show that it again yields a free structure. Section 4 describes the decomposition algorithm and proves that it is sound and complete for existential positive input formulae. In Subsection 4.3, this result is extended to positive formulae with arbitrary quantifier prefix.

[3] This construction is similar to the one made in [SS89] for free algebras.

2 Free Structures

Let Σ be a signature consisting of a finite set Σ_F of function symbols and a finite set Σ_P of predicate symbols, where each symbol has a fixed arity. We assume that equality $=$ is an additional predicate symbol that does not occur in Σ_P. An *atomic Σ-formula* is an *equation* $s = t$ between Σ_F-terms s, t, or a *relational atomic formula* of the form $p[s_1, \ldots, s_m]$ where p is a predicate symbol in Σ_P of arity m and $s_1, \ldots, s_m$ are Σ_F-terms. A positive Σ-matrix is any Σ-formula obtained from atomic Σ-formulae using conjunction and disjunction only. A positive Σ-formula is obtained from a positive Σ-matrix by adding an arbitrary quantifier prefix, and an existential positive Σ-formula is a positive formula where the prefix consists of existential quantifier only. As usual, we shall sometimes write $t(v_1, \ldots, v_n)$ (resp. $\varphi(v_1, \ldots, v_n)$) to express that t (resp. φ) is a term (resp. formula) whose (free) variables are a subset of $\{v_1, \ldots, v_n\}$. Sentences are formulae without free variables.

A *Σ-structure* $\mathcal{A}$ has a non-empty carrier set A, and it interprets each $f \in \Sigma_F$ of arity n as an n-ary function $f_\mathcal{A}$ on A, and each $p \in \Sigma_P$ of arity m as an m-ary relation $p_\mathcal{A}$. For a formula $\varphi = \varphi(v_1, \ldots, v_n)$, we write $\mathcal{A} \models \varphi(a_1, \ldots, a_n)$ to express that φ is true in $\mathcal{A}$ under the evaluation $\{v_1 \mapsto a_1, \ldots, v_n \mapsto a_n\}$.

Usually, *Σ-constraints* are formulae of the form $\varphi(v_1, \ldots, v_n)$ with free variables. A solution of such a constraint (in a fixed Σ-structure $\mathcal{A}$) is an evaluation $\{v_1 \mapsto a_1, \ldots, v_n \mapsto a_n\}$ such that $\mathcal{A} \models \varphi(a_1, \ldots, a_n)$. Obviously, the constraint $\varphi(v_1, \ldots, v_n)$ has a solution in $\mathcal{A}$ iff the formula $\exists v_1 \ldots \exists v_n\, \varphi(v_1, \ldots, v_n)$ is valid in $\mathcal{A}$. In the present paper, we are only interested in solvability of constraints, and will thus usually take this logical point of view. In the following, tuples of variables will often be abbreviated by $\vec{v}, \vec{u}, \vec{w}$, and tuples of elements of a structure by $\vec{a}, \vec{b}$, etc. Substructures and direct products of structures are defined in the usual way. If the Σ-substructure $\mathcal{B}$ of $\mathcal{A}$ is generated by $X \subseteq A$, we write $\mathcal{B} = \langle X \rangle_\Sigma$. Later on, we will consider several signatures simultaneously. If Δ is a subset of the signature Σ, then any Σ-structure $\mathcal{A}$ can be considered as a Δ-structure (called the Δ-reduct of $\mathcal{A}$) by just forgetting about the interpretation of the additional symbols. To make clear with respect to which signature a given Σ-structure $\mathcal{A}$ is currently considered we will sometimes write $\mathcal{A}^\Sigma$ for the full Σ-structure and $\mathcal{A}^\Delta$ for its Δ-reduct.

A Σ-homomorphism is a mapping h between Σ-structures $\mathcal{A}$ and $\mathcal{B}$ such that

$$\begin{aligned} h(f_\mathcal{A}(a_1, \ldots, a_n)) &= f_\mathcal{B}(h(a_1), \ldots, h(a_n)) \\ p_\mathcal{A}[a_1, \ldots, a_n] &\Rightarrow p_\mathcal{B}[h(a_1), \ldots, h(a_n)] \end{aligned}$$

for all $f \in \Sigma_F$, $p \in \Sigma_P$, $a_1, \ldots, a_n \in A$. A Σ-isomorphism is a bijective Σ-homomorphism whose inverse is also a Σ-homomorphism.

There is an interesting (well-known) connection between surjective homomorphisms and positive formulae. The following lemma (see [Mal73], pp. 143, 144, for a proof), and its relationship to the concept of combinability, turns out to be crucial for the new proof method introduced in Section 4.

Lemma 1. *Let $h : \mathcal{A} \rightarrow \mathcal{B}$ be a surjective homomorphism between the Σ-structures $\mathcal{A}$ and $\mathcal{B}$, $\varphi(v_1, \ldots, v_m)$ be a positive Σ-formula, and $a_1, \ldots, a_m$ be elements of A. Then $\mathcal{A} \models \varphi(a_1, \ldots, a_m)$ implies $\mathcal{B} \models \varphi(h(a_1), \ldots, h(a_m))$.*

As for the case of algebras, *Σ-varieties* are defined as classes of Σ-structures that are closed under direct products, substructures, and homomorphic images. The well-known Birkhoff Theorem says that a class of Σ_F-algebras is a variety iff it is an equational class, i.e., the class of models of a set of equations. For structures, a similar characterization is possible [Mal71]: A class $\mathcal{V}$ of Σ-structures is a Σ-variety if, and only if, there exists a set E of atomic Σ-formulae[4] such that $\mathcal{V}$ is the class of models of E. In this situation, we say that $\mathcal{V}$ is the Σ-variety defined by E, and we write $\mathcal{V} = \mathcal{V}(E)$.

As in the case of varieties of algebras, varieties of structures always have free objects. Recall that a Σ-structure $\mathcal{A}$ is *free for the class of Σ-structures $\mathcal{K}$ over the set X* iff *(1)* $\mathcal{A} \in \mathcal{K}$, *(2)* $\mathcal{A}$ is generated by X, and *(3)* every mapping from X into the carrier of a Σ-structure $\mathcal{B} \in \mathcal{K}$ can be extended to a Σ-homomorphism of $\mathcal{A}$ into $\mathcal{B}$.

If $\mathcal{A}$ and $\mathcal{B}$ are free Σ-structures for the same class $\mathcal{K}$, and if their sets of generators have the same cardinality then these structures are isomorphic. Every non-trivial variety contains free structures with sets of generators of arbitrary cardinality [Mal71]. Conversely, free structures are always free for some variety [Mal71, Coh65].

Theorem 2. *Let $\mathcal{A}$ be a Σ-structure that is generated by X. Then $\mathcal{A}$ is free over X for $\{\mathcal{A}\}$ iff $\mathcal{A}$ is free over X for some Σ-variety.*

In the following, a Σ-structure $\mathcal{A}$ will be called free (over X) iff it is free (over X) for $\{\mathcal{A}\}$. Let us now analyze how free Σ-structures look like (see [Mal71, Wea93] for more information). Obviously, the Σ_F-reduct of such a structure is a free Σ_F-algebra, and thus it is (isomorphic to) an E-free Σ_F-algebra $\mathcal{T}(\Sigma_F, X)/{=_E}$ for an equational theory E. In particular, the $=_E$-equivalence classes $[s]$ of Σ_F-terms constitute the carrier of $\mathcal{A}$. It remains to be shown how the predicate symbols are interpreted on this carrier. Since $\mathcal{A}$ is free over X, any mapping from X into $\mathcal{T}(\Sigma_F, X)/{=_E}$ can be extended to a Σ-endomorphism of $\mathcal{A}$. This, together with the definition of homomorphisms of structures, shows that the interpretation of the predicates must be closed under substitution, i.e., for all $p \in \Sigma_P$, all substitutions σ, and all terms $s_1, \ldots, s_m$, if $p[[s_1], \ldots, [s_m]]$ holds in $\mathcal{A}$ then $p[[s_1\sigma], \ldots, [s_m\sigma]]$ must also hold in $\mathcal{A}$. Conversely, it is easy to see that any extension of the Σ_F-algebra $\mathcal{T}(\Sigma_F, X)/{=_E}$ to a Σ-structure that satisfies this property is a free Σ-structure over X.

Example 1. Let Σ_F be an arbitrary set of function symbols, and assume that Σ_P consists of a single binary predicate symbol $\leq$. Consider the (absolutely free) term algebra $\mathcal{T}(\Sigma_F, X)$. We can extend this algebra to a Σ-structure by

[4] As usual, open formulae are here considered as implicitly universally quantified.

interpreting $\leq$ as subterm ordering. Another possibility would be to take a reduction ordering [Der87] such as the lexicographic path ordering. In both cases, we have closure under substitution, which means that we obtain a free Σ-structure.

Free structures over countably infinite sets of generators are canonical for the positive theory of their variety in the following sense:

Theorem 3. *Let $\mathcal{A}$ be free over the countably infinite set X for a Σ-variety $\mathcal{V}(E)$, and let ϕ be a positive Σ-formula. Then ϕ is valid in all elements of $\mathcal{V}(E)$ (i.e., ϕ is a logical consequence of E) iff ϕ is valid in $\mathcal{A}$.*

For the purpose of this paper, the following characterization of free structures is useful (see [BS94] for a proof).

Lemma 4. *Let $\mathcal{A}$ be a Σ-structure that is generated by the countably infinite set X. Then the following conditions are equivalent:*

1. *$\mathcal{A}$ is free over X.*
2. *For every finite subset X_0 of X, every mapping $h_0 : X_0 \to A$ can be extended to a* surjective *endomorphism of $\mathcal{A}$.*

Condition 2 is the *combinability condition* mentioned in the introduction. Thus Theorem 2 shows that *a structure is free for some variety iff it is combinable.* This observation, together with Lemma 1, can be used to obtain the following lemma, which is the key tool in the new proof of correctness of the combination method.

Lemma 5. *Let $\mathcal{A}$ be a free Σ-structure over the countably infinite set of generators X, and let $\gamma = \forall\vec{u}_1\exists\vec{v}_1 \ldots \forall\vec{u}_k\exists\vec{v}_k\ \varphi(\vec{u}_1,\vec{v}_1,\ldots,\vec{u}_k,\vec{v}_k)$ be a positive Σ-sentence. Then the following conditions are equivalent:*

1. *$\mathcal{A} \models \forall\vec{u}_1\exists\vec{v}_1 \ldots \forall\vec{u}_k\exists\vec{v}_k\ \varphi(\vec{u}_1,\vec{v}_1,\ldots,\vec{u}_k,\vec{v}_k)$.*
2. *There exist tuples $\vec{x}_1 \in \vec{X}, \vec{e}_1 \in \vec{A}, \ldots, \vec{x}_k \in \vec{X}, \vec{e}_k \in \vec{A}$ and finite subsets $Z_1, \ldots, Z_k$ of X such that*
   (a) *$\mathcal{A} \models \varphi(\vec{x}_1,\vec{e}_1,\ldots,\vec{x}_k,\vec{e}_k)$,*
   (b) *all generators occurring in the tuples $\vec{x}_1,\ldots,\vec{x}_k$ are distinct,*
   (c) *for all $j, 1 \leq j \leq k$, the components of $\vec{e}_j$ are generated by Z_j, i.e., are elements of $\langle Z_j\rangle_\Sigma$, and*
   (d) *for all $j, 1 < j \leq k$, no component of $\vec{x}_j$ occurs in $Z_1 \cup \ldots \cup Z_{j-1}$.*

As an example, assume that $\mathcal{A}$ is the (absolutely free) term algebra $\mathcal{T}(\{g\},X)$ over a signature consisting of a unary symbol g. The formula $\exists v\forall u\ g(u) = g(v)$ is not valid. In fact, Condition 2 is not satisfied since for all $x \in X$ and $e \in \mathcal{T}(\{g\},X)$, $g(x) = g(e)$ is only satisfied in $\mathcal{T}(\{g\},X)$ if $e = x$, and thus x is contained in any generating set of e.

Readers who are familiar with the notion of "unification with linear constant restrictions" should note the close connection between (c) and (d) of Condition 2 (which say that a generator $x \in X$ must not be contained in the generating set Z for $e \in A$ if e comes before x in the sequence $\vec{x}_1,\vec{e}_1,\ldots,\vec{x}_k,\vec{e}_k$) and a linear constant restriction (which says that the constant a must not occur in the image e of a variable u if u comes before a in the linear constant restriction).

3 Amalgamation of Free Structures

Let Σ and Δ be disjoint signatures, and let X be a countably infinite set (of generators). Let $\mathcal{A}$ be a free Σ-structure over X and and let $\mathcal{B}$ be a free Δ-structure over X. Equivalently, $\mathcal{A}$ is free over X for some Σ-variety $\mathcal{V}(E)$ and $\mathcal{B}$ is free over X for some Δ-variety $\mathcal{V}(F)$ (by Theorem 2). The following construction yields a $(\Sigma \cup \Delta)$-structure $\mathcal{A} \odot \mathcal{B}$ that is free over X for the $(\Sigma \cup \Delta)$-variety $\mathcal{V}(E \cup F)$.

We consider two countably infinite supersets X_∞ and Y_∞ of $X_0 := Y_0 := X$ such that $X_\infty \cap Y_\infty = X$ and $X_\infty \setminus X_0$ and $Y_\infty \setminus Y_0$ are infinite. Let $\mathcal{A}_\infty$ be free for $\mathcal{V}(E)$ over X_∞, and let $\mathcal{B}_\infty$ be free for $\mathcal{V}(F)$ over Y_∞. Obviously, $\mathcal{A}$ is the substructure of $\mathcal{A}_\infty$ that is generated by $X_0 \subseteq X_\infty$. Since both structures are free for the same variety, and since their generating sets X_0 and X_∞ have the same cardinality, $\mathcal{A}$ and $\mathcal{A}_\infty$ are isomorphic. The same holds for $\mathcal{B}$ and $\mathcal{B}_\infty$.

We shall make a zig-zag construction that defines ascending towers of Σ-structures $\mathcal{A}_n$ and Δ-structures $\mathcal{B}_n$. These structures are connected by bijective mappings h_n and g_n. The free amalgamated product $\mathcal{A} \odot \mathcal{B}$ will be obtained as the limit structure, which obtains its functional and relational structure from both towers by means of the limits of the mappings h_n and g_n.

$n = 0$: Let $\mathcal{A}_0 := \mathcal{A} = \langle X_0 \rangle_\Sigma$. We interpret the "new" elements in $A_0 \setminus X_0$ as generators in $\mathcal{B}_\infty$. For this purpose, select a subset $Y_1 \subseteq Y_\infty$ such that $Y_1 \cap Y_0 = \emptyset$, $|Y_1| = |A_0 \setminus X_0|$, and the remaining complement $Y_\infty \setminus (Y_0 \cup Y_1)$ is countably infinite. Choose any bijection $h_0 : Y_0 \cup Y_1 \to A_0$ where $h_0|_{Y_0} = id_{Y_0}$.

Let $\mathcal{B}_0 := \langle Y_0 \rangle_\Delta$. As for $\mathcal{A}_0$, we interpret the "new" elements in $B_0 \setminus Y_0$ as generators in $\mathcal{A}_\infty$. Select a subset $X_1 \subseteq X_\infty$ such that $X_1 \cap X_0 = \emptyset$, $|X_1| = |B_0 \setminus Y_0|$ and the remaining complement $X_\infty \setminus (X_0 \cup X_1)$ is countably infinite. Choose any bijection $g_0 : X_0 \cup X_1 \to B_0$ where $g_0|_{X_0} = id_{X_0}$.

$n \to n+1$: Suppose that $\mathcal{A}_n = \langle \bigcup_{i=0}^n X_i \rangle_\Sigma$ and $\mathcal{B}_n = \langle \bigcup_{i=0}^n Y_i \rangle_\Delta$ are already defined, and that subsets X_{n+1} of X_∞ and Y_{n+1} of Y_∞ are already given. We assume that the complements $X_\infty \setminus \bigcup_{i=0}^{n+1} X_i$ and $Y_\infty \setminus \bigcup_{i=0}^{n+1} Y_i$ are infinite, and that the sets X_i (resp. Y_i) are pairwise disjoint. In addition, we assume that bijections $h_n : B_{n-1} \cup Y_n \cup Y_{n+1} \to A_n$ and $g_n : A_{n-1} \cup X_n \cup X_{n+1} \to B_n$ are defined such that

$$(*)\quad g_n(h_n(b)) = b \text{ for } b \in B_{n-1} \cup Y_n \text{ and } h_n(g_n(a)) = a \text{ for } a \in A_{n-1} \cup X_n$$
$$(**)\quad h_n(Y_{n+1}) = A_n \setminus (A_{n-1} \cup X_n) \text{ and } g_n(X_{n+1}) = B_n \setminus (B_{n-1} \cup Y_n).$$

Note that $(**)$ implies that $h_n(B_{n-1} \cup Y_n) = A_{n-1} \cup X_n$ and $g_n(A_{n-1} \cup X_n) = B_{n-1} \cup Y_n$.

We define $\mathcal{A}_{n+1} = \langle \bigcup_{i=0}^{n+1} X_i \rangle_\Sigma$ and $\mathcal{B}_{n+1} = \langle \bigcup_{i=0}^{n+1} Y_i \rangle_\Delta$, and select subsets $Y_{n+2} \subseteq Y_\infty$ and $X_{n+2} \subseteq X_\infty$ such that $Y_{n+2} \cap \bigcup_{i=0}^{n+1} Y_i = \emptyset = X_{n+2} \cap \bigcup_{i=0}^{n+1} X_i$. In addition, the cardinalities must satisfy $|Y_{n+2}| = |A_{n+1} \setminus (A_n \cup X_{n+1})|$ and $|X_{n+2}| = |B_{n+1} \setminus (B_n \cup Y_{n+1})|$, and the remaining complements $Y_\infty \setminus \bigcup_{i=0}^{n+2} Y_i$ and $X_\infty \setminus \bigcup_{i=0}^{n+2} X_i$ must be countably infinite. Let

$$v_{n+1} : Y_{n+2} \to A_{n+1} \setminus (A_n \cup X_{n+1}) \quad\text{and}\quad \xi_{n+1} : X_{n+2} \to B_{n+1} \setminus (B_n \cup Y_{n+1})$$

be arbitrary bijections. We define $h_{n+1} := v_{n+1} \cup g_n^{-1} \cup h_n$ and $g_{n+1} := \xi_{n+1} \cup$

$h_n^{-1} \cup g_n$. In more detail:

$$h_{n+1}(b) = \begin{cases} v_{n+1}(b) & \text{for } b \in Y_{n+2} \\ h_n(b) & \text{for } b \in B_{n-1} \cup Y_n \cup Y_{n+1} \\ g_n^{-1}(b) & \text{for } b \in B_n \setminus (B_{n-1} \cup Y_n) \end{cases}$$

and

$$g_{n+1}(a) = \begin{cases} \xi_{n+1}(a) & \text{for } a \in X_{n+2} \\ g_n(a) & \text{for } a \in A_{n-1} \cup X_n \cup X_{n+1} \\ h_n^{-1}(a) & \text{for } a \in A_n \setminus (A_{n-1} \cup X_n). \end{cases}$$

Without loss of generality we may assume (for notational convenience) that the construction eventually covers all generators in X_∞ and Y_∞; in other words, we assume that $\bigcup_{i=0}^{\infty} X_i = X_\infty$ and $\bigcup_{i=0}^{\infty} Y_i = Y_\infty$, and thus $\bigcup_{i=0}^{\infty} A_i = A_\infty$ and $\bigcup_{i=0}^{\infty} B_i = B_\infty$. We define the limit mappings

$$h_\infty := \bigcup_{i=0}^{\infty} h_i : B_\infty \to A_\infty, \quad \text{and} \quad g_\infty := \bigcup_{i=0}^{\infty} g_i : A_\infty \to B_\infty.$$

It is easy to see that h_∞ and g_∞ are bijections that are inverse to each other. They may be used to carry the Δ-structure of $\mathcal{B}_\infty$ to $\mathcal{A}_\infty$ and to carry the Σ-structure of $\mathcal{A}_\infty$ to $\mathcal{B}_\infty$: let f (f') be an n-ary function symbol of Δ (Σ), let p (p') be an n-ary predicate symbol of Δ (Σ), and $a_1, \ldots, a_n \in A_\infty$ ($b_1, \ldots, b_n \in B_\infty$). We define

$$\begin{aligned} f_{\mathcal{A}_\infty}(a_1, \ldots, a_n) &:= h_\infty(f_{\mathcal{B}_\infty}(g_\infty(a_1), \ldots, g_\infty(a_n))), \\ f'_{\mathcal{B}_\infty}(b_1, \ldots, b_n) &:= g_\infty(f'_{\mathcal{A}_\infty}(h_\infty(b_1), \ldots, h_\infty(b_n))), \\ p_{\mathcal{A}_\infty}[a_1, \ldots, a_n] &:\Longleftrightarrow p_{\mathcal{B}_\infty}[g_\infty(a_1), \ldots, g_\infty(a_n)], \\ p'_{\mathcal{B}_\infty}[b_1, \ldots, b_n] &:\Longleftrightarrow p'_{\mathcal{A}_\infty}[h_\infty(b_1), \ldots, h_\infty(b_n)]. \end{aligned}$$

With this definition, the mappings h_∞ and g_∞ are inverse isomorphisms between the $(\Sigma \cup \Delta)$-structures $\mathcal{A}_\infty$ and $\mathcal{B}_\infty$. Identifying isomorphic structures, we call $\mathcal{A}_\infty^{\Sigma\cup\Delta} \simeq \mathcal{B}_\infty^{\Sigma\cup\Delta}$ the *free amalgamated product* $\mathcal{A} \odot \mathcal{B}$ *of* $\mathcal{A}$ *and* $\mathcal{B}$. The layering of the domain A_∞ (resp. B_∞) of this structure into the sets A_n (resp. B_n) will become important in the proof of our combination result. As a Σ-structure, $\mathcal{A} \odot \mathcal{B}$ is isomorphic to $\mathcal{A}$, which is free over X for $\mathcal{V}(E)$, and as a Δ-structure it is isomorphic to $\mathcal{B}$, which is free over X for $\mathcal{V}(F)$. In [BS94] it is shown that as a $(\Sigma \cup \Delta)$-structure it is free over X for $\mathcal{V}(E \cup F)$.

Theorem 6. *Let Σ and Δ be disjoint signatures, and let $\mathcal{A}$ be free over X for the Σ-variety $\mathcal{V}(E)$ and $\mathcal{B}$ be free over X for the Δ-variety $\mathcal{V}(F)$, where X is countably infinite. Then $\mathcal{A} \odot \mathcal{B}$ is free over X for the $(\Sigma \cup \Delta)$-variety $\mathcal{V}(E \cup F)$.*

4 Combination results

As in the previous section, let $\mathcal{V}(E)$ be a Σ-variety and $\mathcal{V}(F)$ be a Δ-variety, where Σ and Δ are disjoint signatures. For a countably infinite set of generators X, let $\mathcal{A}$ be free for $\mathcal{V}(E)$ over X, and let $\mathcal{B}$ be free for $\mathcal{V}(F)$ over X. We know that the positive theories of $\mathcal{V}(E)$ and $\mathcal{A}$ (resp. $\mathcal{V}(F)$ and $\mathcal{B}$) coincide (by Theorem 3), and that the free amalgamated product $\mathcal{A} \odot \mathcal{B}$ is free for $\mathcal{V}(E \cup F)$ over X (by Theorem 6).

In the first part of this section, we consider only existential positive $(\Sigma \cup \Delta)$-sentences. The decomposition algorithm described below can be used to reduce validity of such sentences in $\mathcal{A} \odot \mathcal{B}$ (or, equivalently, in $\mathcal{V}(E \cup F)$) to validity of positive sentences in $\mathcal{A}$ and in $\mathcal{B}$. At the end of the section we shall sketch how this result can be extended to positive sentences with arbitrary quantifier prefix.

Before we can describe the algorithm, we must introduce some notation. In the following, V denotes an infinite set of variables used by the first order languages under consideration. Let t be a $(\Sigma \cup \Delta)$-term. This term is called ***pure*** iff it is either a Σ-term or a Δ-term. An equation is pure iff it is an equation between pure terms of the same signature. A relational formula $p[s_1, \ldots, s_m]$ is pure iff $s_1, \ldots, s_m$ are pure terms of the signature of p. Now assume that t is a non-pure term whose topmost function symbol is in Σ. A subterm s of t is called *alien subterm* of t iff its topmost function symbol belongs to Δ and every proper superterm of s in t has its top symbol in Σ. Alien subterms of terms with top symbol in Δ are defined analogously. For a relational formula $p[s_1, \ldots, s_m]$, alien subterms are defined as follows: if s_i has a top symbol whose signature is different from the signature of p then s_i itself is an alien subterm; otherwise, any alien subterm of s_i is an alien subterm of $p[s_1, \ldots, s_m]$.

4.1 The Decomposition Algorithm

Let φ_0 be a positive existential $(\Sigma \cup \Delta)$-sentence. Without loss of generality, we may assume that φ_0 has the form $\exists \vec{u}_0 \; \gamma_0$, where γ_0 is a conjunction of atomic formulae. Indeed, since existential quantifiers distribute over disjunction, a sentence $\exists \vec{u}_0 \; (\gamma_1 \vee \gamma_2)$ is valid iff $\exists \vec{u}_0 \; \gamma_1$ or $\exists \vec{u}_0 \; \gamma_2$ is valid.

Step 1: Transform non-pure atomic formulae.

(1) Equations $s = t$ of γ_0 where s and t have topmost function symbols belonging to different signatures are replaced by (the conjunction of) two new equations $u = s, u = t$, where u is a new variable. The quantifier prefix is extended by adding an existential quantification for u.

(2) As a result, we may assign a unique label Σ or Δ to each atomic formula that is not an equation between variables. The label of an equation $s = t$ is the signature of the topmost function symbols of s and/or t. The label of a relational formula $p[s_1, \ldots, s_m]$ is the signature of p.

(3) Now alien subterms occurring in atomic formulae are successively replaced by new variables. For example, assume that $s = t$ is an equation in the current formula, and that s contains the alien subterm s_1. Let u be a variable not occurring in the current formula, and let s' be the term obtained from s by replacing s_1 by u. Then the original equation is replaced by (the conjunction of) the two equations $s' = t$ and $u = s_1$. The quantifier prefix is extended by adding an existential quantification for u. The equation $s' = t$ keeps the label of $s = t$, and the label of $u = s_1$ is the signature of the top

symbol of s_1. Relational atomic formulae with alien subterms are treated analogously. This process is iterated until all atomic formulae occurring in the conjunctive matrix are pure. It is easy to see that this is achieved after finitely many iterations.

Step 2: Remove atomic formulae without label.
Equations between variables occurring in the conjunctive matrix are removed as follows: If $u = v$ is such an equation then one removes $\exists u$ from the quantifier prefix and $u = v$ from the matrix. In addition, every occurrence of u in the remaining matrix is replaced by v. This step is iterated until the matrix contains no equations between variables.

Let φ_1 be the new sentence obtained this way. The matrix of φ_1 can be written as a conjunction $\gamma_{1,\Sigma} \wedge \gamma_{1,\Delta}$, where $\gamma_{1,\Sigma}$ is a conjunction of all atomic formulae from φ_1 with label Σ, and $\gamma_{1,\Delta}$ is a conjunction of all atomic formulae from φ_1 with label Δ. There are three different types of variables occurring in φ_1: shared variables occur both in $\gamma_{1,\Sigma}$ and in $\gamma_{1,\Delta}$; Σ-variables occur only in $\gamma_{1,\Sigma}$; and Δ-variables occur only in $\gamma_{1,\Delta}$. Let $\vec{u}_{1,\Sigma}$ be the tuple of all Σ-variables, $\vec{u}_{1,\Delta}$ be the tuple of all Δ-variables, and $\vec{u}_1$ be the tuple of all shared variables.[5] Obviously, φ_1 is equivalent to the sentence

$$\exists \vec{u}_1 \, (\exists \vec{u}_{1,\Sigma} \; \gamma_{1,\Sigma} \wedge \exists \vec{u}_{1,\Delta} \; \gamma_{1,\Delta}) \, .$$

The next two steps of the algorithm are nondeterministic, i.e., a given sentence is transformed into finitely many new sentences. Here the idea is that the original sentence is valid iff at least one of the new sentences is valid.

Step 3: Variable identification.
Choose (nondeterministically) a partition of the set of all shared variables. The variables in each class of the partition are "identified" with each other by choosing an element of the class as representative, and replacing in the sentence all occurrences of variables of the class by this representative. Quantifiers for replaced variables are removed.

Let $\exists \vec{u}_2 \, (\exists \vec{u}_{1,\Sigma} \; \gamma_{2,\Sigma} \wedge \exists \vec{u}_{1,\Delta} \; \gamma_{2,\Delta})$ denote one of the sentences obtained by Step 3.

Step 4: Choose signature labels and ordering.
We choose a label Σ or Δ for every (shared) variable in $\vec{u}_2$, and a linear ordering $<$ on these variables.

For each of the choices made in Step 3 and 4, the algorithm yields a pair (α, β) of sentences as output.

Step 5: Generate output sentences.
The sentence $\exists \vec{u}_2 (\exists \vec{u}_{1,\Sigma} \; \gamma_{2,\Sigma} \wedge \exists \vec{u}_{1,\Delta} \; \gamma_{2,\Delta})$ is split into two sentences

$$\alpha = \forall \vec{v}_1 \exists \vec{w}_1 \ldots \forall \vec{v}_k \exists \vec{w}_k \exists \vec{u}_{1,\Sigma} \; \gamma_{2,\Sigma} \text{ and } \beta = \exists \vec{v}_1 \forall \vec{w}_1 \ldots \exists \vec{v}_k \forall \vec{w}_k \exists \vec{u}_{1,\Delta} \; \gamma_{2,\Delta}.$$

[5] The order in these tuples can be chosen arbitrarily.

Here $\vec{v}_1\vec{w}_1 \ldots \vec{v}_k\vec{w}_k$ is the unique re-ordering of $\vec{u}_2$ along $<$. The variables $\vec{v}_i$ ($\vec{w}_i$) are the variables with label Δ (label Σ).

Thus, the overall output of the algorithm is a finite set of pairs of sentences. Note that the sentences α and β are positive formulae, but they need no longer be existential positive formulae.

This algorithm is a straightforward adaptation of the decomposition algorithm described in [BS92] to existential positive formulae with equations and relational constraints. Note, however, that it optimizes the previous algorithm in one significant way: the nondeterministic steps—which are responsible for the NP-complexity of the algorithm—are applied only to shared variables and not to all variables occurring in the system. For the case of algorithms computing complete sets of unifiers, this optimization is already implicitly present in [Bou93]. Steps similar to Step 1, 3, and the labelling in Step 4 are present in most methods for combining unification algorithms. Nelson & Oppen's combination method for universal theories [NO79] explicitly uses Step 1, and implicitly, Step 3 is also present.

4.2 Correctness of the decomposition algorithm

First, we show soundness of the algorithm, i.e., if one of the output pairs is valid then the original sentence was valid.

Lemma 7. *$\mathcal{A} \odot \mathcal{B} \models \varphi_0$ if $\mathcal{A} \models \alpha$ and $\mathcal{B} \models \beta$ for some output pair (α, β).*

Proof. Since $\mathcal{A}^\Sigma$ and $\mathcal{A}^\Sigma_\infty$ are isomorphic Σ-structures, we know that $\mathcal{A}^\Sigma_\infty \models \alpha$. Accordingly, we also have $\mathcal{B}^\Delta_\infty \models \beta$. More precisely, this means

$$(*)\quad \mathcal{A}^\Sigma_\infty \models \forall\vec{v}_1\exists\vec{w}_1 \ldots \forall\vec{v}_k\exists\vec{w}_k\exists\vec{u}_{1,\Sigma}\ \gamma_{2,\Sigma}(\vec{v}_1, \vec{w}_1, \ldots, \vec{v}_k, \vec{w}_k, \vec{u}_{1,\Sigma})$$
$$(**)\quad \mathcal{B}^\Delta_\infty \models \exists\vec{v}_1\forall\vec{w}_1 \ldots \exists\vec{v}_k\forall\vec{w}_k\exists\vec{u}_{1,\Delta}\ \gamma_{2,\Delta}(\vec{v}_1, \vec{w}_1, \ldots, \vec{v}_k, \vec{w}_k, \vec{u}_{1,\Delta}).$$

Because of the existential quantification over $\vec{v}_1$ in $(**)$, there exist elements $\vec{b}_1 \in \vec{B}_\infty$ such that

$$(***)\quad \mathcal{B}^\Delta_\infty \models \forall\vec{w}_1 \ldots \exists\vec{v}_k\forall\vec{w}_k\exists\vec{u}_{1,\Delta}\ \gamma_{2,\Delta}(\vec{b}_1, \vec{w}_1, \ldots, \vec{v}_k, \vec{w}_k, \vec{u}_{1,\Delta}).$$

We consider $\vec{a}_1 := h_\infty(\vec{b}_1)$. Because of the universal quantification over $\vec{v}_1$ in $(*)$ we have

$$\mathcal{A}^\Sigma_\infty \models \exists\vec{w}_1 \ldots \forall\vec{v}_k\exists\vec{w}_k\exists\vec{u}_{1,\Sigma}\ \gamma_{2,\Sigma}(\vec{a}_1, \vec{w}_1, \ldots, \vec{v}_k, \vec{w}_k, \vec{u}_{1,\Sigma}).$$

Because of the existential quantification over $\vec{w}_1$ in this formula there exist elements $\vec{c}_1 \in \vec{A}_\infty$ such that

$$\mathcal{A}^\Sigma_\infty \models \forall\vec{v}_2\exists\vec{w}_2 \ldots \forall\vec{v}_k\exists\vec{w}_k\exists\vec{u}_{1,\Sigma}\ \gamma_{2,\Sigma}(\vec{a}_1, \vec{c}_1, \vec{v}_2, \vec{w}_2, \ldots, \vec{v}_k, \vec{w}_k, \vec{u}_{1,\Sigma}).$$

We consider $\vec{d}_1 := g_\infty(\vec{c}_1)$. Because of the universal quantification over $\vec{w}_1$ in $(***)$ we have

$$\mathcal{B}^\Delta_\infty \models \exists\vec{v}_2\forall\vec{w}_2 \ldots \exists\vec{v}_k\forall\vec{w}_k\exists\vec{u}_{1,\Delta}\ \gamma_{2,\Delta}(\vec{b}_1, \vec{d}_1, \vec{v}_2, \vec{w}_2, \ldots, \vec{v}_k, \vec{w}_k, \vec{u}_{1,\Delta}).$$

Iterating this argument, we thus obtain

$$\mathcal{A}_\infty^\Sigma \models \exists \vec{u}_{1,\Sigma}\ \gamma_{2,\Sigma}(\vec{a}_1, \vec{c}_1, \ldots, \vec{a}_k, \vec{c}_k, \vec{u}_{1,\Sigma}),$$
$$\mathcal{B}_\infty^\Delta \models \exists \vec{u}_{1,\Delta}\ \gamma_{2,\Delta}(\vec{b}_1, \vec{d}_1, \ldots, \vec{b}_k, \vec{d}_k, \vec{u}_{1,\Delta}),$$

where $\vec{a}_i = h_\infty(\vec{b}_i)$ and $\vec{d}_i = g_\infty(\vec{c}_i)$ (for $1 \le i \le k$). Since h_∞ is a $(\Sigma \cup \Delta)$-isomorphism that is the inverse of g_∞, we also know that

$$\mathcal{A}_\infty^\Delta \models \exists \vec{u}_{1,\Delta}\ \gamma_{2,\Delta}(\vec{a}_1, \vec{c}_1, \ldots, \vec{a}_k, \vec{c}_k, \vec{u}_{1,\Delta})$$

It follows that

$$\mathcal{A}_\infty^{\Sigma\cup\Delta} \models \exists \vec{u}_{1,\Sigma}\ \gamma_{2,\Sigma}(\vec{a}_1, \vec{c}_1, \ldots, \vec{a}_k, \vec{c}_k, \vec{u}_{1,\Sigma}) \wedge \exists \vec{u}_{1,\Delta}\ \gamma_{2,\Delta}(\vec{a}_1, \vec{c}_1, \ldots, \vec{a}_k, \vec{c}_k, \vec{u}_{1,\Delta}).$$

Obviously, this implies that $\mathcal{A} \odot \mathcal{B} \simeq \mathcal{A}_\infty^{\Sigma\cup\Delta} \models \exists \vec{u}_2\,(\exists \vec{u}_{1,\Sigma}\ \gamma_{2,\Sigma} \wedge \exists \vec{u}_{1,\Delta}\ \gamma_{2,\Delta})$, i.e., one of the sentences obtained after Step 3 of the algorithm holds in $\mathcal{A} \odot \mathcal{B}$. It is easy to see that this implies that $\mathcal{A} \odot \mathcal{B} \models \varphi_0$. □

Next, we show completeness of the decomposition algorithm, i.e., if the input sentence was valid then there exists a valid output pair.

Lemma 8. *If $\mathcal{A} \odot \mathcal{B} \models \varphi_0$ then $\mathcal{A} \models \alpha$ and $\mathcal{B} \models \beta$ for some output pair (α, β).*

Proof. Assume that $\mathcal{A} \odot \mathcal{B} \simeq \mathcal{B}_\infty^{\Sigma\cup\Delta} \models \exists \vec{u}_0 \gamma_0$. Obviously, this implies that $\mathcal{B}_\infty^{\Sigma\cup\Delta} \models \exists \vec{u}_1\,(\exists \vec{u}_{1,\Sigma}\ \gamma_{1,\Sigma}(\vec{u}_1, \vec{u}_{1,\Sigma}) \wedge \exists \vec{u}_{1,\Delta}\ \gamma_{1,\Delta}(\vec{u}_1, \vec{u}_{1,\Delta}))$, i.e., $\mathcal{B}_\infty^{\Sigma\cup\Delta}$ satisfies the sentence that is obtained after Step 2 of the decomposition algorithm. Thus there exists an assignment $\nu : V \to B_\infty$ such that $\mathcal{B}_\infty^{\Sigma\cup\Delta} \models \exists \vec{u}_{1,\Sigma}\ \gamma_{1,\Sigma}(\nu(\vec{u}_1), \vec{u}_{1,\Sigma}) \wedge \exists \vec{u}_{1,\Delta}\ \gamma_{1,\Delta}(\nu(\vec{u}_1), \vec{u}_{1,\Delta})$.

In Step 3 of the decomposition algorithm we identify two shared variables u and u' of $\vec{u}_1$ if, and only if, $\nu(u) = \nu(u')$. With this choice,

$$\mathcal{B}_\infty^{\Sigma\cup\Delta} \models \exists \vec{u}_{1,\Sigma}\ \gamma_{2,\Sigma}(\nu(\vec{u}_2), \vec{u}_{1,\Sigma}) \wedge \exists \vec{u}_{1,\Delta}\ \gamma_{2,\Delta}(\nu(\vec{u}_2), \vec{u}_{1,\Delta}),$$

and all components of $\nu(\vec{u}_2)$ are distinct.

In Step 4, a shared variable u in $\vec{u}_2$ is labeled with Δ if $\nu(u) \in B_\infty \setminus (\bigcup_{i=1}^\infty Y_i)$, and with Σ otherwise. In order to choose the linear ordering on the shared variables, we partition the range B_∞ of ν as follows:

$$B_0,\ \ Y_1,\ \ B_1 \setminus (B_0 \cup Y_1),\ \ Y_2,\ \ B_2 \setminus (B_1 \cup Y_2),\ \ Y_3,\ \ B_3 \setminus (B_2 \cup Y_3),\ \ldots$$

Now, let $\vec{v}_1, \vec{w}_1, \ldots, \vec{v}_k, \vec{w}_k$ be a re-ordering of the tuple $\vec{u}_2$ such that the following holds:

1. The tuple $\vec{v}_1$ contains exactly the shared variables whose ν-images are in B_0.
2. For all $i, 1 \le i \le k$, the tuple $\vec{w}_i$ contains exactly the shared variables whose ν-images are in Y_i.
3. For all $i, 1 < i \le k$, the tuple $\vec{v}_i$ contains exactly the shared variables whose ν-images are in $B_{i-1} \setminus (B_{i-2} \cup Y_{i-1})$.

Obviously, this implies that the variables in the tuples $\vec{w}_i$ have label Σ, whereas the variables in the tuples $\vec{v}_i$ have label Δ. Note that some of these tuples may be of dimension 0. The re-ordering determines the linear ordering we choose in Step 4. Let

$$\alpha = \forall\vec{v}_1\exists\vec{w}_1\ldots\forall\vec{v}_k\exists\vec{w}_k\exists\vec{u}_{1,\Sigma}\ \gamma_{2,\Sigma} \text{ and } \beta = \exists\vec{v}_1\forall\vec{w}_1\ldots\exists\vec{v}_k\forall\vec{w}_k\exists\vec{u}_{1,\Delta}\ \gamma_{2,\Delta}$$

be the output pair that is obtained by these choices. Let $\vec{y}_i := \nu(\vec{w}_i) \in \vec{Y}$ and $\vec{b}_i := \nu(\vec{v}_i) \in \vec{B}_\infty$. The sequence $\vec{b}_1, \vec{y}_1, \ldots, \vec{b}_k, \vec{y}_k$ satisfies Condition 2 of Lemma 5 for $\varphi = \exists\vec{u}_{1,\Delta}\ \gamma_{2,\Delta}$, the structure $\mathcal{B}_\infty^\Delta$, and appropriate sets $Z_1, \ldots, Z_k$ (see [BS94]). Thus, we obtain $\mathcal{B} \simeq \mathcal{B}_\infty^\Delta \models \beta$. In order to show $\mathcal{A} \models \alpha$, we use the fact that $h_\infty : \mathcal{B}_\infty \to \mathcal{A}_\infty$ is a $(\Sigma \cup \Delta)$-isomorphism. Thus, $\mathcal{B}_\infty^{\Sigma\cup\Delta} \models \exists\vec{u}_{1,\Sigma}\ \gamma_{2,\Sigma}(\nu(\vec{u}_2), \vec{u}_{1,\Sigma})$ implies that $\mathcal{A}_\infty^\Sigma \models \exists\vec{u}_{1,\Sigma}\ \gamma_{2,\Sigma}(h_\infty(\nu(\vec{u}_2)), \vec{u}_{1,\Sigma})$.

Let $\vec{x}_i := h_\infty(\vec{b}_i) = h_\infty(\nu(\vec{v}_i))$ and $\vec{a}_i := h_\infty(\vec{y}_i) = h_\infty(\nu(\vec{w}_i))$ (for $i = 1, \ldots, k$). In [BS94] it is shown that the sequence $\vec{x}_1, \vec{a}_1, \ldots, \vec{x}_k, \vec{a}_k$ satisfies Condition 2 of Lemma 5 for $\varphi = \exists\vec{u}_{1,\Sigma}\ \gamma_{2,\Sigma}$, the structure $\mathcal{A}_\infty^\Sigma$, and appropriate sets $Z_1', \ldots, Z_k'$. Thus, $\mathcal{A} \simeq \mathcal{A}_\infty^\Sigma \models \alpha$. □

The two lemmas obviously imply the next theorem.

Theorem 9. *Let $\mathcal{V}(E)$ be a Σ-variety and $\mathcal{V}(F)$ be a Δ-variety for disjoint signatures Σ and Δ. The positive existential theory of the $(\Sigma\cup\Delta)$-variety $\mathcal{V}(E\cup F)$ is decidable, provided that the positive theories of $\mathcal{V}(E)$ and of $\mathcal{V}(F)$ are decidable.*

If the signatures contain no predicate symbols, this theorem is a reformulation of Theorem 2.1 of [BS92]. What is new here is the algebraic proof method and the fact that relational constraints can be treated as well.

4.3 Decision Procedures for Positive Theories

A disadvantage of Theorem 9 is that it does not show modularity of decidability of the positive theory of varieties of structures. Indeed, the prerequisites of the theorem (decidability of the *full* positive theories of $\mathcal{V}(E)$ and $\mathcal{V}(F)$) are stronger than its consequence (decidability of the *existential* positive theory of $\mathcal{V}(E\cup F)$).

In [BS94] we describe an algorithm that can be used to reduce decidability of the *full* positive theory of $\mathcal{V}(E \cup F)$ to decision procedures for the positive theories of $\mathcal{V}(E)$ and $\mathcal{V}(F)$. The main idea is to transform positive sentences (with arbitrary quantifier prefix) into existential positive sentences by Skolemizing the universally quantified variables.[6] In addition to the theories E and F one thus obtains a free theory (for the new Skolem functions). In principle, the decomposition algorithm for existential positive sentences is now applied twice to decompose the input sentence into three positive sentences α, β, ρ, whose validity must respectively be decided in E, F, and the free theory. Note that it is well-known that the whole first-order theory of absolutely free term algebras is decidable [Mal71, Mah88, CL89].

[6] We are Skolemizing *universally* quantified variables since we are interested in validity of the sentence and not in satisfiability.

Correctness of this way of proceeding can be shown with the help of the following lemma, which exhibits an interesting connection between Skolemization and amalgamation with an absolutely free algebra (see [BS94] for the proof).

Lemma 10. *Let $\mathcal{A}$ be a Σ-structure that is free in $\mathcal{V}(E)$ over the countably infinite set of generators X, and let γ be a positive Σ-sentence. Suppose that the (positive) existential sentence γ' is obtained from γ via Skolemization of the universally quantified variables in γ, introducing the set of Skolem function symbols Γ. Then $\mathcal{A} \models \gamma$ if, and only if, $\mathcal{A} \odot \mathcal{T}(\Gamma, X) \models \gamma'$.*

Thus, we obtain the desired modularity result:

Theorem 11. *Let $\mathcal{V}(E)$ be a Σ-variety and $\mathcal{V}(F)$ be a Δ-variety for disjoint signatures Σ and Δ. The positive theory of the $(\Sigma \cup \Delta)$-variety $\mathcal{V}(E \cup F)$ is decidable, provided that the positive theories of $\mathcal{V}(E)$ and of $\mathcal{V}(F)$ are decidable.*

5 Conclusion and Outlook

We have presented an abstract algebraic approach to the problem of combining constraint solvers for constraint languages over disjoint signatures. The constraints that can be handled this way are built from atomic equational *and relational* constraints with the help of conjunction, disjunction, and both universal and existential quantifiers. Solvability means validity of such (closed) constraint formulae in a free structure, or equivalently in a variety of structures.

Simple examples of free structures with a non-trivial relational part are (absolutely free) term algebras that are equipped with an ordering that is invariant under substitution, such as the lexicographic path ordering or the subterm ordering. For our combination result to apply, however, the positive theory of these structures must be decidable. For a total lexicographic path ordering, this is not the case. For the subterm ordering, the existential theory is decidable, but the full first-order theory is undecidable [CT94]. Decidability of the positive theory is still an open problem. For partial lexicographic path orderings, even decidability of the existential theory is unknown.

Combination of constraint solving techniques in the presence of predicate symbols other than equality have independently been considered by H. Kirchner and Ch. Ringeissen [KR94]. However, their approach is based on the rewriting and abstraction techniques mentioned in the introduction (see, e.g., [BS92, Bou93]). Consequently, the interpretation of the predicate symbols in the combined structure is defined in a rather technical way, and it is not a priori clear what this definition means in an intuitive algebraic sense. We conjecture that, for free structures, the combined structure of [KR94] coincides with our free amalgamated product.

We are currently working on a generalization of the notion of "combinable structure" that considerably extends the notion of a "free structure." An example of a structure that is not a free structure, but nevertheless satisfies the generalized combinability condition, is the algebra of rational trees.

References

[Bou93] A. Boudet. Combining unification algorithms. *J. Symbolic Computation*, 16:597–626, 1993.

[BS92] F. Baader and K.U. Schulz. Unification in the union of disjoint equational theories: Combining decision procedures. In *Proceedings of CADE-11*, LNCS 607, 1992.

[BS93] F. Baader and K.U. Schulz. Unification in the union of disjoint equational theories: Combining decision procedures, 1993. Extended version, submitted for publication.

[BS93a] F. Baader and K.U. Schulz. Combination techniques and decision problems for disunification. In *Proceedings of RTA-93*, LNCS 690, 1993.

[BS94] F. Baader and K.U. Schulz. *Combination of Constraint Solving Techniques: An Algebraic Point of View.* Research Report CIS-Rep-94-75, CIS, University Munich, 1994. This report is available via anonymous ftp from "cantor.informatik.rwth-aachen.de" in the directory "pub/papers."

[Bür91] H.-J. Bürckert. *A Resolution Principle for a Logic with Restricted Quantifiers*, LNCS 568, 1991.

[Coh65] P.M. Cohn. *Universal Algebra*. Harper & Row, New York, 1965.

[CL89] H. Comon and P. Lescanne. Equational problems and disunification. *J. Symbolic Computation*, 7:371–425, 1989.

[CT94] H. Comon and R. Treinen. Ordering constraints on trees. In *Colloquium on Trees in Algebra and Programming (CAAP)*, LNCS, 1994.

[Der87] N. Dershowitz. Termination of rewriting. *J. Symbolic Computation*, 3:69–116, 1987.

[DKR94] E. Domenjoud, F. Klay, and Ch. Ringeissen. Combination techniques for non-disjoint theories. In *Proceedings of CADE-12*, LNCS 814, 1994.

[KK89] C. Kirchner and H. Kirchner. Constrained equational reasoning. In *Proceedings of SIGSAM 1989 International Symposium on Symbolic and Algebraic Computation*. ACM Press, 1989.

[KR94] H. Kirchner and Ch. Ringeissen. Combining symbolic constraint solvers on algebraic domains. *J. Symbolic Computation*, 18(2):113–155, 1994.

[Mah88] M.J. Maher. Complete axiomatizations of the algebras of finite, rational and infinite trees. In *Proceedings of LICS'88*, IEEE Computer Society, 1988.

[Mal71] A.I. Mal'cev. *The Metamathematics of Algebraic Systems*, volume 66 of *Studies in Logic and the Foundation of Mathematics*. North Holland, Amsterdam, London, 1971.

[Mal73] A.I. Mal'cev. *Algebraic Systems*, volume 192 of *Die Grundlehren der mathematischen Wissenschaften in Einzeldarstellungen*. Springer, Berlin, 1973.

[NO79] G. Nelson and D.C. Oppen. Simplification by cooperating decision procedures. *ACM TOPLAS*, 1(2):245 257, 1979.

[Rin92] Ch. Ringeissen. Unification in a combination of equational theories with shared constants and its application to primal algebras. In *Proceedings of LPAR'92*, LNCS 624, 1992.

[NiR94] R. Nieuwenhuis and A. Rubio, "AC-superposition with constraints: No AC-unifiers needed," in: *Proceedings CADE-12*, Springer LNAI 814, 1994.

[SS89] M. Schmidt-Schauß. Unification in a combination of arbitrary disjoint equational theories. *J. Symbolic Computation*, 8(1,2):51–99, 1989.

[Wea93] N. Weaver. Generalized varieties. *Algebra Universalis*, 30:27–52, 1993.

Some Independence Results for Equational Unification

Friedrich Otto[1], Paliath Narendran[2], Daniel J. Dougherty[3]

[1] Fachbereich Mathematik/Informatik, Universität-GH Kassel, 34109 Kassel, Germany Internet:otto@theory.informatik.uni-kassel.de
[2] Institute of Programming and Logics, Department of Computer Science, State University of New York, Albany, NY 12222, U.S.A. Internet:dran@cs.albany.edu
[3] Mathematics Department, Wesleyan University, Middletown, CT 06459, U.S.A. Internet:ddougherty@wesleyan.edu

Abstract. For finite convergent term-rewriting systems the *equational unification problem* is shown to be recursively independent of the *equational matching problem*, the *word matching problem*, and the (simultaneous) 2nd*-order equational matching problem*. We also present some new decidability results for *simultaneous equational unification* and 2nd-order equational matching.

1 Introduction

Syntactic unification is the problem of solving an equation in the free algebra $T(F, X)$ of terms generated from a set F of function symbols and constants and a set X of variables. Such unification is an essential tool in logic programming and machine-oriented logic, where it is the basic mechanism underlying the resolution principle, and it is fundamental in the rewrite-rule based approach to equational reasoning as embodied in the Knuth-Bendix completion procedure [1].

Often we are interested in reasoning about terms modulo a 1st-order equational theory, and it has proved fruitful in many cases to "build-in" such a theory in the unification process. This leads to the notion of *equational unification*, E-unification for short, the problem of solving an equation in (the initial model for) the given equational theory. Because of its importance for logic programming and equational reasoning, this problem has received a great deal of attention in the literature; see [2] for an extensive survey of the area.

We can generalize the syntactic unification problem in another direction, by considering terms which may contain function variables, eligible for instantiation. For example, if a is an individual constant, f a function symbol, and v a function variable, then the unification problem $f(a) = v(a)$ has (at least) the solution in which v is instantiated by f, *and* the solution in which v is instantiated by the constant function $f(a)$. This is an example of a 2nd*-order-unification* problem. Note that there was not an equational theory in the background here, but the relevant equality was not simply syntactic equality, either: some information about how functions behave is built into the theory. In particular, in 2nd-order unification it is customary to take function variables as ranging over the functions definable in the lambda-calculus, and to take equality between functions to be axiomatized by the (β) and (η) equations of the lambda-calculus.

We are naturally led to combine these generalizations, allowing function variables and postulating certain equations as axioms. Naturally enough, this is called 2nd*-order E-unification.*

Certain variations are significant in practice. One must often consider *simultaneous* unification problems, in which we seek a solution to a set of equations. Finally, it is sometimes important to consider "one-sided" unification, in which only one, designated, term in a pair is eligible for instantiation. This is the *matching* problem. Needless to say, each of these variations can be considered in an equational and/or a 2nd-order framework.

Clearly each of E-unification and 2nd-order-unification are harder than syntactic unification, in an intuitive sense. One way to make this precise is to observe that syntactic unification is easily decidable, while each of E-unification and 2nd-order-unification are undecidable in general (see [2] for a discussion). But beyond this the situation is not so straightforward.

In this paper we present some results which help to clarify the relationships among these paradigms, typically comparing them with respect to decidability. To derive these results it has turned out to suffice to look at equational theories that are generated by finite convergent term-rewriting systems in which only unary function symbols appear. A term-rewriting system of this form can easily be interpreted as a *string-rewriting system*, and we will make heavy use of string-rewriting terminology.

Summary of results

Let F denote a signature in which all function symbols occurring have arity 1, and let $\mathcal{E}$ be a set of equations which do *not* involve any individual constants, and in each of which the same individual variable occurs. Thus, $\mathcal{E}$ expresses some universal relationships among the functions in F (such as "f and g commute," or "f is idempotent"). Such equations may conveniently be presented as equations between strings.

A model for $\mathcal{E}$, then, essentially consists of an appropriate monoid $\mathcal{M}$ of functions over a set M. In such a model we may ask whether an equation $s = t$ has a solution, when the variables range over the individuals and (perhaps) the functions of $\mathcal{M}$. But there are two natural choices for what we mean by "the functions of $\mathcal{M}$." One is to consider the functions which are lambda-definable from the elements of F — this leads to the full 2nd-order E-unification problem. Another point of view would be to take function variables as denoting only functions that are generated (by composition) from the interpretations of the elements of F. This latter situation is precisely the *Word Unification* problem for $\mathcal{E}$, when $\mathcal{E}$ is treated simply as a set of string-equations.

So the same set of equations $\mathcal{E}$ defines three different notions of unification: the 1st-order E-unification problem, the 2nd-order E-unification problem, and the Word E-unification problem. Of course we may also consider the analogous *matching* problems, and for each of these the *simultaneous* versions.

Say that two problems are *(recursively) independent* if there exist theories $\mathcal{E}$ for which the one problem is decidable while the other problem is undecidable, and conversely. We show that the equational unification problem is independent of:

- the equational matching problem (Section 3),
- the word matching problem (Section 4), and
- the (simultaneous) 2nd-order equational matching problem (Section 5).

In addition, we prove that the simultaneous E-unification problem is decidable for theories $\mathcal{E}$ which are defined by finite, monadic, and confluent string-rewriting systems (Section 3), and that for arbitrary $\mathcal{E}$ as above the (single-pair) 2nd-order E-matching problem is recursively reducible to the (1st-order) E-matching problem (Theorem 13).

The paper is organized as follows. In Section 2 some basic definitions are given, and the terminology used is introduced. In Sections 3 to 5 the results mentioned above are derived. The paper closes with a short summary and some related problems that remain open at this time.

2 Preliminaries

Here we present the basic definitions concerning term-rewriting systems, equational unification, and string-rewriting systems that we will need throughout the paper. We keep the definitions given to a minimum - more information and discussion of the notions introduced can be found in the literature. For term-rewriting systems our main reference is Dershowitz and Jouannaud [5], for equational unification it is Baader and Siekmann [2], and for string-rewriting systems it is Book and Otto [3].

Let F be a finite set of function symbols, each $f \in F$ having a fixed arity $\alpha(f) \in \mathbb{N}$, and let X be a countable set of variables. As usual, the function symbols of arity 0 will be called constants. Then $T = T(F, X)$ denotes the set of *terms* generated by F and X.

Let t be a term from $T(F, X)$. Following standard notation we denote the set of *occurrences* of the term t by $O(t)$. The length of the longest sequence in $O(t)$ is called the *depth* of the term t (denoted as $depth(t)$), and the number of sequences in $O(t)$ is the *size* of t (denoted as $size(t)$). For $p \in O(t)$, $t|_p$ denotes the subterm of t at occurrence p. If s is another term, then $t[s]_p$ denotes the term that is obtained by replacing the subterm of t at occurrence p by the term s. For a term $t \in T(F, X)$, $Var(t)$ denotes the set of variables that have occurrences in t.

A *substitution* is a mapping $\sigma : X \to T(F, X)$ such that $\sigma(x) = x$ holds for almost all variables x. It can uniquely be extended to a morphism $\sigma : T(F, X) \to T(F, X)$.

A *term-rewriting system* R is a (finite) set of *rules* $\{\ell_i \to r_i \mid i \in I\}$, where ℓ_i and r_i are terms from $T(F, X)$ such that $Var(r_i) \subseteq Var(\ell_i)$ $(i \in I)$.

A term t is *reducible* modulo R if there is a rule $\ell \to r$ in R, an occurrence $p \in O(t)$, and a substitution σ such that $\sigma(\ell) = t|_p$. The term $t[\sigma(r)]_p$ is the result of *reducing* t by $\ell \to r$ at p. By $\to_R$ we denote the *single-step reduction* *relation* defined by the term-rewriting system R. Its reflexive and transitive closure $\rightarrow_R^*$ is the *reduction relation* induced by R. A term t is said to be in *normal form* or *irreducible* modulo R if no reduction can be applied to t. By $IRR(R)$ we denote the set of all irreducible terms.

The equational theory that is associated with a term-rewriting system R is the congruence $=_R$ that is generated by the reduction relation $\rightarrow_R$, that is, it is the congruence $\leftrightarrow_R^* := (\rightarrow_R \cup \leftarrow_R)^*$.

A term-rewriting system R is said to be *noetherian* if there are no infinite sequences of reductions, that is, for each term t, each sequence $t \rightarrow_R t_1 \rightarrow_R \cdots$ is finite. It is *confluent* if, for all terms s, t, and u, $s \rightarrow_R^* t$ and $s \rightarrow_R^* u$ imply that, for some term w, $t \rightarrow_R^* w$ and $u \rightarrow_R^* w$. Finally, R is called *convergent* (or *complete*) if it is both noetherian and confluent. In this case each term has a unique normal form with respect to R, that is, for each term t, there exists one and only one irreducible term t_0 such that $t =_R t_0$. In addition, from t the term t_0 can be determined effectively by reduction. The system R is *depth-reducing* if $depth(\ell) > depth(r)$ holds for each rule $\ell \rightarrow r$ of R.

Let R be a term-rewriting system on $T(F,X)$. Two terms $s,t \in T(F,X)$ are said to be *unifiable modulo* R if there exists a substitution σ such that $\sigma(s) =_R \sigma(t)$ holds. The substitution σ is then called an *R-unifier* of s and t. We say that there exists an *R-match* from s onto t if there exists a substitution σ such that $\sigma(s) =_R t$. As indicated in the introduction our main results will be concerned with term-rewriting systems that only involve function symbols of arity one. In fact, this class of term-rewriting systems is essentially just the class of *string-rewriting systems*.

Let Σ be an alphabet and Σ^* be the set of all strings over Σ including the empty string λ. For $w \in \Sigma^*$, $|w|$ denotes the length of w. Obviously, Σ^* is in one-to-one correspondence to the set of terms $T(\Sigma, \{x\})$, where each letter from Σ is simply interpreted as a unary function symbol.

A string-rewriting system S on an alphabet Σ is a finite set of pairs of strings from Σ^*. A pair (ℓ, r) is often referred to as a *rule*. Usually, the rule (ℓ, r) will be denoted as $(\ell \rightarrow r)$. The set of all right-hand sides of rules of the string-rewriting system S is denoted by $range(S)$. Under the isomorphism between Σ^* and $T(\Sigma, \{x\})$ the string-rewriting system S corresponds to the term-rewriting system $R_S := \{\ell(x) \rightarrow r(x) \mid (\ell \rightarrow r) \in S\}$. Thus, the notions of being noetherian, confluent, and convergent immediately carry over to string-rewriting systems.

We close this section by introducing some additional notation that we will only need for the case of string-rewriting systems.

Let S be a string-rewriting system on Σ. For $u \in \Sigma^*$, $[u]_S$ denotes the congruence class $[u]_S = \{w \in \Sigma^* \mid u \longleftrightarrow_S^* w\}$, and for a language $L \subset \Sigma^*$, $[L]_S = \bigcup_{u \in L} [u]_S$.

A string-rewriting system S is called *length-reducing* if each rule $(\ell \rightarrow r)$ of S satisfies $|\ell| > |r|$. Observe that S is length-reducing if and only if the term-rewriting system R_S is depth-reducing. It is called *monadic* if it is length-reducing, and if $range(S) \subseteq \Sigma \cup \{\lambda\}$, that is, if the right-hand side of each rule of S is a single letter or the empty string. Finally, it is called ***special*** if it is length-reducing, and if $range(S) = \{\lambda\}$.

A string-rewriting system S is called ***interreduced*** if $range(S) \subset IRR(S)$, and if $\ell \in IRR(S \setminus \{\ell \to r\})$ holds for each rule $(\ell \to r)$ of S. For each finite convergent system an equivalent finite convergent system can be computed effectively such that the latter system is also interreduced. Here two systems are called ***equivalent*** if they generate the same congruence relation. Therefore, we can always assume in the following that the finite convergent systems considered are interreduced.

3 Equational Matching and Unification Problems

Let Σ be a finite alphabet, let S be a string-rewriting system on Σ, and let $V := \{v_i \mid i \in \mathbb{N}\}$ be a set of string variables such that $\Sigma \cap V = \emptyset$. We consider existential sentences of the following form:

$$\exists v_1, \ldots, v_n : g_1 \sim h_1 \textit{ and } \ldots \textit{ and } g_m \sim h_m,$$

where $g_i, h_i \in (\Sigma \cup \{v_1, \ldots, v_n\})^*$, $i = 1, \ldots, m$. We say that this sentence has a ***solution for*** S if there exists a mapping $\phi : \{v_1, \ldots, v_n\} \to \Sigma^*$ such that $\phi(g_i) \leftrightarrow_S^* \phi(h_i)$ holds for all $i = 1, \ldots, m$. Here ϕ is extended to $(\Sigma \cup \{v_1, \ldots, v_n\})^*$ in the obvious way. By varying the syntactic form of the existential sentences considered we can define various equational matching and unification problems for S.

First of all, if each symbol $a \in \Sigma$ is interpreted as a unary function symbol $a(.)$, and if S is considered as the term-rewriting system on the signature $F_\Sigma := \{a(.) \mid a \in \Sigma\}$, then the resulting equational matching and unification problems for S can be defined as usual. Using existential sentences they can be expressed as follows.

(1.) The **Equational (1st-Order) Matching Problem** for S:
INSTANCE : Two strings $g, h \in \Sigma^*$.
QUESTION : Does the sentence $\exists v : gv \sim h$ have a solution for S?

(2.) The **Equational (1st-Order) Unification Problem** for S:
INSTANCE : Two strings $g, h \in \Sigma^*$.
QUESTION : Does $\exists v_1, v_2 : gv_1 \sim hv_2$ have a solution for S?

Here it is also possible that the two variables v_1 and v_2 coincide, that is, the question could be whether $\exists v : gv \sim hv$ has a solution.

By considering more than one pair of strings at a time, we obtain the simultaneous versions of the above problems. Recall from [9] that the simultaneous E-matching and E-unification problems are in general more difficult than their non-simultaneous counterparts.

On the other hand, we have the "classical" word matching and unification problems, where each symbol $a \in \Sigma$ is considered as a constant, and where an additional binary function symbol ("concatenation") is used that is associative. These problems can be stated as follows.

(3.) The **Word Matching Problem** for S:

INSTANCE : Two strings $g \in (\Sigma \cup V)^*$ and $h \in \Sigma^*$.

QUESTION : Does $\exists v_1, \ldots, v_k : g \sim h$ have a solution for S, where $\{v_1, \ldots, v_k\} = \{v \in V \mid |g|_v > 0\}$?

(4.) The **Word Unification Problem** for S:

INSTANCE : Two strings $g, h \in (\Sigma \cup V)^*$.

QUESTION : Does $\exists v_1, \ldots, v_k : g \sim h$ have a solution for S, where $\{v_1, \ldots, v_k\} = \{v \in V \mid |g|_v + |h|_v > 0\}$?

Here $|w|_v$ denotes the v-length of the string w, that is, the number of occurrences of the symbol v in w.

Again, by considering more than one pair of strings at a time, we obtain the simultaneous versions of these problems. While for the empty system the simultaneous versions of these problems are reducible to the non-simultaneous versions [13], it is not known whether this also holds for non-empty systems.

Recall that word unification generalizes 1st-order unification and specializes 2nd-order unification in that function variables are allowed to be instantiated, but only by functions definable explicitly from F. Concerning these problems the following results are known.

Proposition 1.

(a) The Equational Matching Problem is decidable in polynomial time for finite, monadic, and confluent string-rewriting systems [3].

(b) The Word Unification Problem is decidable for the empty system $S = \emptyset$ [8].

(c) There is a finite, monadic, and confluent system for which the Word Matching Problem is undecidable [3].

(d) There is a finite, special, and confluent system for which the Word Unification Problem is undecidable [11].

We now turn to the relationship between the E-Matching Problem and the E-Unification Problem for finite and convergent string-rewriting systems. In [9] a finite, length-reducing, confluent string-rewriting system $T_2(S)$ is constructed such that the E-Unification Problem for $T_2(S)$ is undecidable. Exploiting the fact that the rules of the system $T_2(S)$ are of a very restricted form only, it is not hard to show that the E-Matching Problem for $T_2(S)$ is decidable. Thus, we have the following result.

Theorem 2. *There is a finite, length-reducing, and confluent string-rewriting system S_1 such that the E-Matching Problem for S_1 is decidable, while the E-Unification Problem for S_1 is undecidable.*

On the other hand, we have the following result.

Theorem 3. *There is a finite, length-reducing, and confluent string-rewriting system S_2 such that the E-Matching Problem for S_2 is undecidable, while the E-Unification Problem for S_2 is decidable.*

Proof. The E-Matching Problem for a string-rewriting system S is called the *Right-Divisibility Problem* for S in [3]. There an example of a finite, length-reducing, and confluent string-rewriting system S_0 on a finite alphabet Σ_0 is given such that this problem is undecidable for S_0 (Corollary 5.2.5). We construct the system S_2 as an extension of S_0. Let $\Sigma = \Sigma_0 \cup \{Z\}$, where Z is a new symbol, and let S_2 denote the following string-rewriting system on Σ:

$$S_2 = S_0 \cup \{aZ \to Z \mid a \in \Sigma\}.$$

Then S_2 is a finite length-reducing system, and since S_0 is confluent, and Z is a new symbol, it is easily seen that S_2 is confluent, too. Obviously, the E-Unification Problem for S_2 is decidable.

On the other hand, let $g, h \in \Sigma_0^*$. Then there exists a string $w \in \Sigma^*$ such that $gw \leftrightarrow_{S_2}^* h$ if and only if there exists a string $w \in \Sigma_0^*$ such that $gw \leftrightarrow_{S_0}^* h$. However, this problem is undecidable by the choice of S_0. Thus, the E-Matching Problem for S_2 is undecidable. □

Theorems 2 and 3 show that for the class of finite, length-reducing, and confluent string-rewriting systems, the E-Matching Problem and the E-Unification Problem are independent. This is not really a new result for term-rewriting systems (see [4]), but we add it here for the sake of completeness. However, if there is an uninterpreted function symbol, that is, a letter that does not occur in any rule of the system under consideration, the situation changes.

Theorem 4. *Let S be a string-rewriting system on some alphabet Σ. If there exists a letter in Σ that does not occur in any rule of S, then the E-Matching Problem for S is reducible to the E-Unification Problem for S.*

The corresponding result also holds for the simultaneous E-matching and E-unification problems.

We close this section with some positive results on the (simultaneous) E-matching and E-unification problems for the class of finite, monadic, and confluent string-rewriting systems. These results are based on a careful analysis of the reduction process with respect to these systems.

Let S be a finite, monadic, and confluent string-rewriting system on Σ. For $g, h \in IRR(S)$ we want to characterize those strings $u, w \in IRR(S)$ that satisfy the congruence $gu \leftrightarrow_S^* hw$.

For $x \in IRR(S)$ and $a \in \Sigma \cup \{\lambda\}$, let $RF(x, a)$ denote the set $RF(x, a) = \{y \in IRR(S) \mid xy \to_S^* a\}$. Then $RF(x, a)$ is a regular language, and from x and a, a nondeterministic finite-state acceptor (nfa) can be constructed in polynomial time for $RF(x, a)$ ([10], Theorem 5.1). Using sets of this form we get the following characterization.

Lemma 5. *Let S be a finite, monadic, and confluent string-rewriting system on Σ, and let $g, h \in IRR(S)$. Then the following two statements are equivalent:*

(a) $\exists u, w \in \Sigma^* : gu \leftrightarrow_S^* hw$.

(b) $\exists g_1, g_2, h_1, h_2, y \in \Sigma^*, \exists a, b \in \Sigma \cup \{\lambda\} : g = g_1g_2$ and $h = h_1h_2$ and $RF(g_2, a) \neq \emptyset \neq RF(h_2, b)$ and (i) $g_1 = h_1$ and ($a = b$ or $\lambda \in \{a, b\}$) or (ii) $g_1 = h_1by$ or (iii) $h_1 = g_1ay$.

For $g, h \in IRR(S)$ we are also interested in those strings $w \in IRR(S)$ that satisfy the congruence $gw \leftrightarrow_S^* hw$. Here we get the following characterization.

Lemma 6. *Let S be a finite, monadic, and confluent string-rewriting system on Σ, and let $g, h \in IRR(S)$. Then the following two statements are equivalent:*

(a) $\exists w \in \Sigma^* : gw \leftrightarrow_S^* hw$.
(b) $\exists g_1, g_2, h_1, h_2, y \in \Sigma^*, \exists a, b \in \Sigma \cup \{\lambda\} : g = g_1 g_2$ *and* $h = h_1 h_2$ *and*
  (i) $g_1 a = h_1 b$ *and* $RF(g_2, a) \cap RF(h_2, b) \neq \emptyset$ *or*
  (ii) $h_1 b = g_1 a y$ *and* $RF(g_2, a) \cdot y \cap RF(h_2, b) \neq \emptyset$ *or*
  (iii) $g_1 a = h_1 b y$ *and* $RF(g_2, a) \cap RF(h_2, b) \cdot y \neq \emptyset$.

For $g, h \in IRR(S)$, there are only $|g| \cdot |h|$ factorizations of the form $g = g_1 g_2$ and $h = h_1 h_2$. Thus, Lemma 5 and Lemma 6 give the following result.

Theorem 7. *The E-Unification Problem is decidable in polynomial time for each finite, monadic, and confluent string-rewriting system.*

The same result holds for the E-Matching Problem by [3], Corollary 4.3.2. Actually, the above lemmata suffice to prove the following.

Theorem 8. *The simultaneous E-Matching Problem is decidable for each finite, monadic, and confluent string-rewriting system.*

We do not currently know the complexity of this problem. The uniform version of it, where the string-rewriting system is considered as a part of the problem instance, is PSPACE-complete.

Using the above lemmata a finite collection of word equations with variables and regular constraints for these variables can be determined from an instance of the simultaneous E-unification problem such that this instance has a solution mod S if and only if the word equations determined from it have a solution that satisfies the given constraints. Since it is decidable whether a word equation has a solution satisfying regular constraints [14], this yields the main result of this section.

Theorem 9. *The simultaneous E-Unification Problem is decidable for each finite, monadic, and confluent string-rewriting system.*

The complexity of this problem is open. Note that even though *unifiability* is decidable, complete sets of unifiers may be infinite. In fact, there are convergent monadic systems whose unification is of type *nullary*, that is, minimal, complete sets may not exist in some cases. A simple example is the string-rewriting system $\{ab \rightarrow a, ac \rightarrow a, ad \rightarrow a, bd \rightarrow \lambda, cd \rightarrow \lambda\}$ and the terms ax and ay. ($a(x)$ and $a(y)$ when viewed as terms over unary function symbols.)

4 Equational Unification and Word Matching

As described in [9] there exists a finite set $S := \{(x_i, y_i) \mid i = 2, \ldots, k\}$ of pairs of non-empty strings $x_i, y_i \in \{a, b\}^+$ $(i = 2, \ldots, k)$ such that the following version of the modified Post Correspondence Problem (MPCP) is undecidable:

INSTANCE : Two strings $x_1, y_1 \in \{a, b\}^+$.
QUESTION : Is there a sequence of integers $i_1, \ldots, i_m \in \{2, \ldots, k\}$ such that $x_1 x_{i_1} \ldots x_{i_m} = y_1 y_{i_1} \ldots y_{i_m}$?

If such a sequence of integers exists, then it is called a *solution* of the instance $\{(x_1, y_1)\} \cup S$ of the MPCP. We say that MPCP(x_1, y_1) has a solution to express the fact that such a solution exists.

From the above finite set S we now construct a string-rewriting system $T_m(S)$ on the alphabet $\Sigma = \{a, b, c_2, \ldots, c_k, ¢, \$, \#\}$ as follows:

$$T_m(S) = \{x_i\$c_i \to \$, y_i¢c_i \to ¢ \mid i = 2, \ldots, k\}.$$

This system, which will also be used in the next section, has the following properties.

Lemma 10.

(a) The string-rewriting system $T_m(S)$ is finite, monadic, confluent, and interreduced.
(b) For all $x_1, y_1 \in \{a,b\}^+$, the following two statements are equivalent:
  (i) MPCP(x_1, y_1) has a solution.
  (ii) There are strings $g, h \in \Sigma^*$ such that $g\$h\#g¢h \leftrightarrow^*_{T_m(S)} x_1\$\#y_1¢$.

From the choice of the set S and from Theorem 9 we thus obtain the following separation result.

Corollary 11. *The string-rewriting system $T_m(S)$ has a decidable simultaneous E-Unification Problem, but an undecidable Word Matching Problem.*

Obviously, the word matching problem is a special case of the word unification problem. The following result shows that in general the latter is more difficult than the former.

Theorem 12. *There exists a finite, length-reducing, and confluent string-rewriting system for which the Word Matching Problem is decidable, while the E-Unification Problem is undecidable.*

Proof. Let $S = \{(x_i, y_i) \mid i = 2, \ldots, k\} \subseteq \{a,b\}^+ \times \{a,b\}^+$ be chosen as above, let $\Sigma = \{a, b, e_2, \ldots, e_k, c, d, \$, §\}$, let $n = max\{|x_i|, |y_i| \mid i = 2, \ldots, k\} + 1$, and let $T_c(S)$ denote the following string-rewriting system on Σ:

$$T_c(S) = \{ce_i^n \to x_ic, de_i^n \to y_id \mid i = 2, \ldots, k\} \cup \{c\$ \to §, d\$ \to §\}.$$

Then $T_c(S)$ is length-reducing and confluent.

It is easily verified that, for all $x_1, y_1 \in \{a,b\}^+$, MPCP(x_1, y_1) has a solution if and only if the sentence "$\exists v : x_1cv \sim y_1dv$" has a solution mod $T_c(S)$. Hence, the E-Unification Problem is undecidable for the system $T_c(S)$.

On the other hand, consider the existential sentence

$$\exists v_1, \ldots, v_k : g_0v_{i_1}g_1 \ldots v_{i_m}g_m \sim h,$$

where $g_0, g_1, \ldots, g_m, h \in \Sigma^*$ are irreducible, and $v_{i_1}, \ldots, v_{i_m} \in \{v_1, \ldots, v_k\}$. If $w_1, \ldots, w_k \in \Sigma^*$ is a solution, then $g_0w_{i_1}g_1 \ldots w_{i_m}g_m \to^*_{T_c(S)} h$. However, the set of ancestors $\{w \in \Sigma^* \mid w \to^*_{T_c(S)} h\}$ of h mod $T_c(S)$ is finite. Thus, we can decide whether or not the existential sentence above has a solution mod $T_c(S)$, and so, the Word Matching Problem for $T_c(S)$ is decidable. □

5 The 2nd-Order E-Matching Problem

We finally turn to the 2nd-order equational matching problem. Let S be a string-rewriting system on Σ; as before we will interpret S as embodying an equational theory of unary functions, and will consider unification and matching problems modulo S. The sense in which the problems in this section are "higher-order" is this: substitutions may replace function variables by any term definable in the lambda-calculus, using only unary second-order variables; and there is an additional, "built-in" notion of equality between terms, that is generated by the familiar (β) and (η) axioms. However, we will adopt a notation in which, as explained below, ($\beta\eta$) equality can be handled *implicitly*.

An observation on notation: in the logic literature systems with one-place functions and predicates are called "monadic." Since this has nothing to do with the notion of "monadic string-rewriting system," we will avoid confusion and consistently use the term "unary" to delimit function-arity.

In order to incorporate the intuitions and technical results on string-rewriting systems, it will be convenient to use a concrete syntax for terms and substitutions which does *not* use explicit abstraction. Indeed, the notation we use is essentially that of Goldfarb [7] and Farmer [6]. We wish to emphasize that the difference between our presentation and currently standard presentations of second-order logic are purely superficial, and so we detail the correspondence below.

So assume as before that we have a set Σ of symbols, whose elements will be treated as (unary) function constants, and a set V of (unary) function variables. From the point of view of lambda-calculus, these are the atoms of functional type $\iota \rightarrow \iota$, where ι is the base type. In order to build terms of base type ι, we need to have some atoms of base type, so we assume our language includes a set Δ of individual constants and a set X of individual variables. We may now build lambda-terms by closing the atoms under application $f(t)$ and abstraction over a variable $\Lambda x.t$ (here t is any term, and we use the capital Λ here to avoid confusion with the earlier use of λ to denote the empty string). It is easy to check that (under the constraints imposed by the "unary" character of the atoms) the terms of base type which are in normal form under $\beta\eta$-reduction are precisely those of the form $(a_1(a_2\cdots(a_n(c)\cdots)))$, where the a_i are atoms from $\Sigma \cup V$ and c is an atom from $\Delta \cup X$. (It is well-known that any term in our calculus is reducible to one in normal form).

Hence, the 2nd-order terms in $T_2(F_\Sigma, V, X)$ are in 1-to-1 correspondence with the strings in the language $(\Sigma \cup V)^* \cdot (\Delta \cup X)$. For the remainder of this paper we will interpret such strings as lambda-terms, without additional comment.

Now return to the string-rewriting system S on Σ. We have seen that S can be considered as a term-rewriting system on the 1st-order terms $T(F_\Sigma, X)$, and we extend S to be a rewriting system on the second-order terms simply by treating the function variables V as free function symbols. Thus, for $s, t \in (\Sigma \cup V)^* \cdot (\Delta \cup X)$ we have $s \rightarrow_S t$ if and only if $\exists g, h \in (\Sigma \cup V)^*, \exists d \in X \cup \Delta, \exists(\ell \rightarrow r) \in S : s = g\ell h(d)$ and $t = grh(d)$.

The problems of higher-order matching and unification differ from their 1st-order counterparts in the complexity of the substitutions allowed. A substitution may replace a function variable by any term of function-type; for example, if $u, v \in V$, $a, b \in \Sigma$, and $x \in X$, one might replace u by the term $\Lambda x.a(v(a(b(x))))$, which simply designates the function-composition more readily denoted by the string $avab$. But consider the function-term $\Lambda x.a(v(a(b(y))))$, where y is an individual variable. This term denotes a constant function, and one that cannot be described as a composition of functions from $\Sigma \cup V$. Note that after *applying* a substitution to a base-type term, no matter what form the substitution takes, the resulting term will be Λ-free. But in order to manipulate the substitutions themselves we must have an extended notation.

To this end, let $\square$ be an additional symbol that will be used as a "place holder." By interpreting $\square$ as an additional individual constant we can form the set $T_2(F_\Sigma \cup \{\square\}, V, X)$ of *extended* 2nd*-order terms*, which corresponds to the language $(\Sigma \cup V)^* \cdot (\Delta \cup \{\square\} \cup X)$. A 2nd*-order substitution* is a mapping $\phi : V \cup X \to (\Sigma \cup V)^* \cdot (\Delta \cup \{\square\} \cup X) \cup V$ that satisfies the following three conditions:

(1) $\text{dom}(\phi) = \{x \in V \cup X \mid \phi(x) \neq x\}$ is finite,
(2) $\phi(w) \in (\Sigma \cup V)^* \cdot (\Delta \cup X)$ for all $w \in X$, and
(3) $\phi(v) \in (\Sigma \cup V)^* \cdot (\Delta \cup \{\square\} \cup X)$ for all $v \in \text{dom}(\phi) \cap V$.

The intended interpretation is that for $v \in V$, $\phi(v) \in (\Sigma \cup V)^* \cdot (\Delta \cup X)$ means that the 1-place function variable v is to be replaced by a constant function (and $\phi(v)$ denotes the value obtained for any argument), while $\phi(v) = w(\square)$ means that v will be replaced by the function of arity 1 that is defined by w.

With this in mind we may extend ϕ to a mapping $\phi_e : (\Sigma \cup V)^* \cdot (\Delta \cup X) \to (\Sigma \cup V)^* \cdot (\Delta \cup X)$ as follows:

- if $g \in \Delta$, then $\phi_e(g) = g$,
- if $g \in X$, then $\phi_e(g) = \phi(g)$,
- if $g = ag_1$ for some $a \in \Sigma$, then $\phi_e(g) := a(\phi_e(g_1))$,
- if $g = vg_1$ for some $v \in V$ such that $v \notin \text{dom}(\phi)$, then $\phi_e(g) = v(\phi_e(g_1))$,
- if $g = vg_1$ for some $v \in V \cap \text{dom}(\phi)$ such that $\phi(v) \in (\Sigma \cup V)^* \cdot (\Delta \cup X)$, then $\phi_e(g) = \phi(v)$, and
- if $g = vg_1$ for some $v \in V \cap \text{dom}(\phi)$ such that $\phi(v) = w(\square)$ for some $w \in (\Sigma \cup V)^*$, then $\phi_e(g) := w(\phi_e(g_1))$.

To simplify the notation we will denote the extension ϕ_e of ϕ simply by ϕ.

We may now consider second-order matching and unification modulo a string-rewriting system S. More precisely (in this paper) we are reasoning modulo an equational theory which has equations only between second-order terms with no individual constants. For example we may express that f is idempotent by: $ff = f$. To say that f is a left-inverse for g we would write: $fg = \lambda$.

The **2nd-Order E-Matching Problem** for S is now defined as follows:

INSTANCE : Two strings $g, h \in (\Sigma \cup V)^* \cdot (\Delta \cup X)$.
QUESTION : Is there a 2nd-order substitution ϕ such that $\phi(g) \leftrightarrow_S^* h$?

Accordingly, the **2nd-Order E-Unification Problem** for S is defined by:

INSTANCE : Two strings $g, h \in (\Sigma \cup V)^* \cdot (\Delta \cup X)$.
QUESTION : Is there a 2nd-order substitution ϕ such that $\phi(g) \leftrightarrow_S^* \phi(h)$?

Although 2nd-order unification is undecidable in general, even for the empty theory [7], it is decidable [6] whether two terms s, t in our "unary" language are 2nd-order unifiable (in the empty equational theory). As for the 2nd-Order Unification Problem for string-rewriting systems, we observe that it is reducible to the Word Unification Problem for the same string-rewriting system [12], using a technique of Farmer [6]. We will not address 2nd-Order unification further here, but will concentrate on matching problems.

Our first result is that 2nd-order matching is no harder than 1st-order matching.

Theorem 13. *The 2nd-Order E-Matching Problem for a string-rewriting system S on Σ is effectively reducible to the E-Matching Problem for S, where S is considered as a string-rewriting system on $\Sigma \cup V$.*

Proof. Let $g, h \in (\Sigma \cup V)^* \cdot (\Delta \cup X)$. If $|g|_V = 0$, then $g = g_1 b$ for some $g_1 \in \Sigma^*$ and $b \in \Delta \cup X$. If $b \in X$, then we actually have an instance of the E-Matching Problem for the string-rewriting system S on $\Sigma \cup V$. If $b \in \Delta$, then there exists a substitution ϕ satisfying $\phi(g) \leftrightarrow_S^* h$ if and only if $h = h_1 b$ for some $h_1 \in \Sigma^*$ and $g_1 \leftrightarrow_S^* h_1$ holds.

If $|g|_V > 0$, then $g = f u g_1$ for some string $f \in \Sigma^*$ and some function variable $u \in V$. Let $h_1 \in (\Sigma \cup V)^*$ and $b \in (\Delta \cup X)$ such that $h = h_1 b$.

Claim. There exists a 2nd-order substitution ϕ satisfying $\phi(g) \leftrightarrow_S^* h$ if and only if there is a string $w \in (\Sigma \cup V)^*$ such that $fw \leftrightarrow_S^* h_1$, that is, if and only if the existential sentence "$\exists v : fv \sim h_1$" has a solution mod S.

Proof of claim. First, assume that ϕ is a 2nd-order substitution that satisfies $\phi(g) \leftrightarrow_S^* h$. Since $g = fug_1$, we see that $\phi(g) = f\phi(ug_1)$. Let $w_1 := \phi(ug_1) \in (\Sigma \cup V)^* \cdot (\Delta \cup X)$. Then w_1 can be factored as $w_1 = wd$ with $w \in (\Sigma \cup V)^*$ and $d \in \Delta \cup X$. Since $fwd = fw_1 = \phi(g) \leftrightarrow_S^* h = h_1 b$, we conclude that $b = d$ and $fw \leftrightarrow_S^* h_1$.

Conversely, assume that $w \in (\Sigma \cup V)^*$ satisfies $fw \leftrightarrow_S^* h_1$. Define a 2nd-order substitution ϕ through $\phi(u) := wb$ and $\phi(u') := u'$ for all $u' \in (V \setminus \{u\}) \cup X$. Then $\phi(g) = \phi(fug_1) = f\phi(ug_1) = f\phi(u) = fwb \leftrightarrow_S^* h_1 b = h$. □ □

Corollary 14. *For finite, monadic, and confluent string-rewriting systems the 2nd-Order E-Matching Problem is decidable in polynomial time.*

Thus, for the finite, monadic, and confluent string-rewriting system $T_m(S)$ of Section 4 we have the following situation:

(i) the Simultaneous E-Matching and E-Unification Problems are decidable,
(ii) the Word Matching and Unification Problems are undecidable, and
(iii) the 2nd-Order E-Matching Problem is decidable.

Also Theorem 2 yields the following.

Corollary 15. *There exists a finite, length-reducing, and confluent string-rewriting system S_1 such that the 2nd-Order E-Matching Problem for S_1 is decidable, while the E-Unification Problem for S_1 is undecidable.*

One consequence of Theorem 13 is the observation that in order to obtain equational theories (rewriting systems) for which the 2nd-order E-matching problem is strictly more difficult than the (1st-order) E-matching and E-unification problems, we must at least consider the *simultaneous* versions of these problems. We may achieve the same effect by adding a free binary function constant to the signature. Indeed, we return to one of the theories investigated in Section 4.

Let $T_m(S)$ be the finite, monadic, and confluent string-rewriting system on the alphabet $\Sigma = \{a, b, c_0, \ldots, c_k, ¢, \$, \#\}$ from Section 1. From Σ we obtain a set of function symbols $F = \Sigma \cup \{c, f\}$, where each letter from Σ is interpreted as a unary function symbol, c is a new constant, and f is a new binary function symbol. By T_1 we denote the set of 1st-order terms $T(F, X)$, and by T_2 we denote the set of 2nd-order terms $T_2(F, V, X)$. (The modification to the definition needed to incorporate binary function constants should be obvious.) In the following the string-rewriting system $T_m(S)$ is interpreted as a term-rewriting system on T_1 as well as on T_2. In the latter case the function variables $v \in V$ are treated as free unary function symbols with respect to the reduction relation induced by $T_m(S)$.

Theorem 16. *The 2nd-Order E-Matching Problem is undecidable for $T_m(S)$, when the system $T_m(S)$ is considered as a term-rewriting system on T_2.*

Proof. In Lemma 10 we have seen that a certain restricted version of the Word Matching Problem is undecidable for the string-rewriting system $T_m(S)$. Here we will reduce this problem to the 2nd-Order E-Matching Problem for the term-rewriting system $T_m(S)$ on T_2. Let $x_1, y_1 \in \{a, b\}^+$. We form the terms $s := f(v\$(x), v¢(x))$ and $t := f(x_1\$(c), y_1¢(c))$, where $x \in X$ and $v \in V$. Here $x_1\$(c)$ denotes the term that is built from the string $x_1\$$ by interpreting each letter as a unary function symbol, and accordingly for $y_1¢(c)$. We consider the instance (s, t) of the 2nd-Order E-Matching Problem for $T_m(S)$, that is, we ask whether there exists a 2nd-order substitution $\phi : \{v, x\} \to T(F \cup \{\Box\}, V, X)$ satisfying $\phi(s) \Leftrightarrow^* t$. Here $\Leftrightarrow^*$ denotes the congruence relation on T_2 that is induced by the term-rewriting system $T_m(S)$. Observe that $t \in T_1$, and hence, $t' \in T_1$ holds for all $t' \in T_2$ satisfying $t \Leftrightarrow^* t'$. Hence, if a 2nd-order substitution ϕ is to satisfy $\phi(s) \Leftrightarrow^* t$, then $\phi(x) \in T_1$ and $\phi(v) \in T(F \cup \{\Box\}, X)$ must hold necessarily.

Claim. The following two statements are equivalent:
(i) There exists a 2nd-order substitution ϕ such that $\phi(s) \Leftrightarrow^* t$.
(ii) There exist strings $g, h \in \Sigma^*$ such that $g\$h\#g¢h \leftrightarrow^*_{T_m(S)} x_1\$\#y_1¢$.

Proof of claim. (ii) $\Rightarrow$ (i): Let $g, h \in \Sigma^*$ such that $g\$h\#g¢h \leftrightarrow^*_{T_m(S)} x_1\$\#y_1¢$. We define a 2nd-order substitution ϕ through $\phi(x) := h(c)$ and $\phi(v) := g(\Box)$. Then $\phi(s) = \phi(f(v\$(x), v¢(x))) = f(g\$h(c), g¢h(c)) \Leftrightarrow^* f(x_1\$(c), y_1¢(c)) = t$.

(i) $\Rightarrow$ (ii): Let $\phi : \{v, x\} \to T(F \cup \{\Box\}, V, X)$ be a 2nd-order substitution such that $\phi(s) \Leftrightarrow^* t$. As observed above this means that $\phi(x) \in T_1$ and $\phi(v) \in T(F \cup \{\Box\}, X)$.

First assume that $\phi(v) = t_1 \in T_1$. Then $\phi(s) = \phi(f(v\$(x), v¢(x))) = f(t_1, t_1) \Leftrightarrow^* t = f(x_1\$(c), y_1¢(c))$, which implies that $x_1\$(c) \Leftrightarrow^* t_1 \Leftrightarrow^* y_1¢(c)$, since no rule of $T_m(S)$ contains an occurrence of the function symbol f. This in turn means that $x_1\$ \leftrightarrow^*_{T_m(S)} y_1 ¢$, which contradicts the fact that $T_m(S)$ is confluent, since $x_1\$$ and $y_1 ¢$ are both irreducible mod $T_m(S)$. Thus, $\phi(v) \notin T_1$, that is, $\phi(v)$ contains some occurrences of the special "place holder" $\Box$. Since $|s|_f = 1 = |t|_f$, and since f is a free function symbol for $T_m(S)$, $|\phi(v)|_f$ must be 0, that is, $\phi(v) = g(\Box)$ for some $g \in \Sigma^*$. Let $\phi(x) := h_1 \in T_1$. Again it follows that $|h_1|_f = 0$, that is, $h_1 = h(z)$ for some $h \in \Sigma^*$ and $z \in X \cup \{c\}$. Hence, $f(x_1\$(c), y_1 ¢(c)) = t \Leftrightarrow^* \phi(s) = f(g\$h(z), g¢h(z))$ and so $g\$h(z) \Leftrightarrow^* x_1\$(c)$ and $g¢h(z) \Leftrightarrow^* y_1 ¢(c)$. Thus $z = c$, $g\$h \leftrightarrow^*_{T_m(S)} x_1\$$, and $g¢h \leftrightarrow^*_{T_m(S)} y_1 ¢$. □

From the choice of the system $T_m(S)$ we conclude that the 2nd-Order E-Matching Problem for $T_m(S)$ is undecidable, when $T_m(S)$ is considered as a term-rewriting system on T_2. □

In contrast to this undecidability result it can be shown that the simultaneous E-Unification problem is decidable for the term-rewriting system $T_m(S)$ on T_2. For doing so we need some technical results on the relation $\Leftrightarrow^*$ on T_2.

Since for the reduction process mod $T_m(S)$ the function variables from V are treated as free function symbols, we can restrict our attention to the 1st-order terms in $T_1 = T(F, X)$. Hence, in what follows we consider $T_m(S)$ as a term-rewriting system on T_1. The following technical results are easily verified.

Lemma 17. *Let $g = w_0(f(w_1, w_2))$ and $h = z_0(f(z_1, z_2))$ be terms from T_1 such that $w_0, z_0 \in \Sigma^*$. Then $g \Leftrightarrow^* h$ if and only if $w_0 \leftrightarrow^*_{T_m(S)} z_0$ and $w_i \Leftrightarrow^* z_i$ for $i = 1, 2$.*

Lemma 18. *If $g = w_0(f(w_1, w_2))$ and $h = z_0(x)$ are unifiable* mod $T_m(S)$, *where $w_0, z_0 \in \Sigma^*$ and $x \in X$, then the variable x does not occur in g, and $w_0 \leftrightarrow^*_{T_m(S)} z_0 y_0$ for some string $y_0 \in \Sigma^*$, that is, the mapping $x \mapsto y_0(f(w_1, w_2))$ is a match from h onto g for $T_m(S)$.*

Based on these two technical lemmata the following decidability result can be established.

Theorem 19. *For the term-rewriting system $T_m(S)$ the simultaneous E-Unification Problem is decidable .*

Combining the above results on the term-rewriting system $T_m(S)$ we obtain the following.

Corollary 20. *The finite, depth-reducing, and confluent term-rewriting system $T_m(S)$ has a decidable (simultaneous) E-Unification Problem, while the 2nd-Order E-Matching Problem for $T_m(S)$ is undecidable.*

We can also interpret the system $T_m(S)$ as a term-rewriting system on the signature $\Sigma \cup \{c\}$, that is, we can delete the binary function symbol f. Theorem 9 and the proof of Theorem 16 show the following.

Corollary 21. *The finite, depth-reducing, and confluent term-rewriting system $T_m(S)$ on the signature $\Sigma \cup \{c\}$ has a decidable (simultaneous) E-Unification Problem, while the simultaneous 2nd-Order E-Matching Problem for $T_m(S)$ is undecidable.*

6 Conclusion

We have established three independence results for the problem of equational unification. However, many related problems remain open at this time. In [11] a finite, special, and confluent string-rewriting system is presented for which the Word Unification Problem is undecidable. What can be said about the Word Matching Problem for this type of systems?

Finally, we might further consider the 2nd-Order Unification Problem for string-rewriting systems. Certainly it is reducible to the Word Unification Problem for the same string-rewriting system [12], but is it recursively equivalent to that problem?

References

1. Avenhaus, J., and K. Madlener, "Term Rewriting and Equational Reasoning," in: R.B. Banerji (ed.), *Formal Techniques in Artificial Intelligence*, North-Holland, Amsterdam, 1990, pp. 1–43.
2. Baader, F., and J. Siekmann, "Unification Theory," in: D.M. Gabbay, C.J. Hogger and J.A. Robinson (eds.), *Handbook of Logic in Artificial Intelligence and Logic Programming*, Oxford University Press, Oxford, UK, 1993.
3. Book, R.V., and F. Otto, *String-Rewriting Systems*, Springer : New York, 1993.
4. Bürckert, H.-J., "Matching–a special case of unification?" *Journal of Symbolic Computation* 8 (1989) 523-536.
5. Dershowitz, N., and J.P. Jouannaud, "Rewrite Systems," in: J. van Leeuwen (ed.), *Handbook of Theoretical Computer Science, Vol. B: Formal Models and Semantics*, Elsevier, Amsterdam, 1990, pp. 243–320.
6. Farmer, W.M., "A unification algorithm for second-order monadic terms," *Annals of Pure and Applied Logic* 39 (1988) 131–174.
7. Goldfarb, W.D., "The undecidability of the second-order unification problem," *Theoretical Computer Science* 13 (1981) 225–230.
8. Makanin, G.S., "The problem of solvability of equations in a free semigroup," *Math. USSR Sbornik* 32 (1977) 129–198.
9. Narendran, P., and F. Otto, "Some results on equational unification," in: M. Stickel (ed.), *10th Int. Conf. on Automated Deduction*, Proceedings, Lecture Notes in Artificial Intelligence 449, Springer, Berlin, 1990, pp. 276–291.
10. Otto, F., "On two problems related to cancellativity," *Semigroup Forum* 33 (1986) 331–356.
11. Otto, F., "Solvability of word equations modulo finite special and confluent string-rewriting systems is undecidable in general," *Information Processing Letters*, to appear.
12. Otto, F., *The 2nd-Order E-Unification Problem for Finite Monadic String-Rewriting Systems*, in preparation.
13. Pécuchet, J.-P., *Equations avec Constantes et Algorithme de Makanin*, Thèse 3e Cycle, Université de Rouen, France, 1981.
14. Schulz, K.U., "Makanin's Algorithm for Word Equations – Two Improvements and a Generalization," in: K.U. Schulz (ed.), *Word Equations and Related Topics*, Proceedings, Lecture Notes in Computer Science 572, Springer, Berlin, 1990, pp. 85–150.

Regular Substitution Sets: A Means of Controlling E-Unification **

Jochen Burghardt

GMD Forschungsinstitut für Rechnerarchitektur und Softwaretechnik
(German National Research Center for Computer Science)
Rudower Chaussee 5, D-12489 Berlin
Tel: *49-30-6392-1867, Fax: -1805
E-Mail: jochen@first.gmd.de

Abstract. A method for selecting solution constructors in narrowing is presented. The method is based on a sort discipline that describes regular sets of ground constructor terms as sorts. It is extended to cope with regular sets of ground substitutions, thus allowing different sorts to be computed for terms with different variable bindings. An algorithm for *computing* signatures of equationally defined functions is given that allows potentially infinite overloading. Applications to formal program development are sketched.

1 Motivation

Solving equations by narrowing has important applications, e.g. in the area of formal software development. However, the usual narrowing strategies are only able to restrict the set of application positions[1]. Ordered paramodulation [1] is able to provide a succession in which the defining equations have to be selected, but it cannot guarantee that an appropriate one is selected first. Bockmayr [2] has shown, under some general conditions, that narrowing strategies essentially enumerate the whole term universe rather than specifically selecting the appropriate equations of a defined function to narrow with or the appropriate constructor to insert into the solution. In this paper, we present an approach to restrict the set of applicable defining equations in a narrowing step that is based on dynamic computation of function signatures, rather than their declaration by a user.

The main idea is as follows[2]: As e.g. in [7], we distinguish between constructors and equationally defined functions; each well-defined ground term can be reduced to a ground constructor term, viz. its unique normal form. For a term v, let V be the set of all possible values of v, i.e., the set of all normal forms of admitted ground constructor instances of v. Then, a goal equation $v_1 = v_2$ cannot be solved if $V_1 \cap V_2 = \{\}$; in this case, it can be pruned from the search

** This work was partially funded by the BMFT project KORSO, contract ITS9001A7.

[1] Cf. e.g. the mathematical definition of the notion of strategy in [6].

[2] Notations and naming conventions are consistent with Def. 1 below.

space of narrowing. Unfortunately, V_1 and V_2 are undecidable in general; to overcome this problem, we will define computable upper approximations $\overline{V_1} \supset V_1$ and $\overline{V_2} \supset V_2$, respectively, and base the pruning decision on the consideration of $\overline{V_1} \cap \overline{V_2}$.

To this end, we provide a framework of "extended sorts" to describe infinite sets of ground constructor terms like $\overline{V}$ in a closed form, which is based on regular tree grammars. It is essential that extended sorts are closed wrt. intersection and that their inhabitance can be decided in order to conduct the above disjointness test. Moreover, set equality and subsort property can be decided, and we always have $\overline{V} = V$ if v is a constructor term.

An algorithm for computing the extended sort $\overline{V}$ from a term v is presented. In terms of conventional order-sorted rewriting we thereby achieve potentially infinite overloading, since for an arbitrary input sort S we can *compute* a signature $f : S \rightarrow \overline{f[S]}$ rather than being limited to a few user-defined signatures which are generally too coarse to successfully apply the disjointness test. It is clear that the impact of this test on search space reduction depends on the expressiveness of the sort framework and on the quality of signature approximation.

Consider, for example, the theory consisting of equations *a.* to *i.* in Fig. 4 (Sect. 5). When trying to solve a goal equation like $val(x) = s^5(0)$ wrt. this theory, conventional strategies are unable to decide which of the equations *g.*, *h.*, *i.* is to be used for a first narrowing step. Narrowing (at root position) with equation *g.*, *h.*, and *i.* results in the new goal equations $0 = s^5(0)$, $dup(val(x')) = s^5(0)$, and $s(dup(val(x'))) = s^5(0)$, respectively. While the first one is obviously false, the unsatisfiability of the second one can be detected as our algorithm computes the sort of its left-hand side as *Even* and recognizes that this is disjoint from its right-hand side's sort, $\{s^5(0)\}$; in a similar way, the third one is considered as "possibly satisfiable" by the disjointness test. Hence, narrowing only makes sense with equation *i.*, and any solution to the above goal equation must take the form $x = x' :: i$. In Sect. 5, examples of the pruning of infinite search-tree branches are given. Note that if a user were to declare the signatures $+ : Nat \times Nat \rightarrow Nat$, $dup : Nat \rightarrow Nat$, and $val : Bin \rightarrow Nat$, the disjointness test would allow narrowing with equations *h.* and *i.* In more complicated applications, a user cannot know in advance which signatures might become essential to disjointness tests in the course of the narrowing proof.

This example also shows that it is important to consider variable bindings during the computation of a term's sort in order to get good approximations. For example, when computing a signature for *dup*, the term $x+x$ should be assigned the sort *Even*, whereas $x + y$ can only be assigned *Nat*, assuming that x and y range over *Nat*. In conventional order-sorted approaches, the mapping from a term to its sort is usually a homomorphic extension of the sort assignment of variables, thus necessarily ignoring variable bindings, e.g.:

$$\begin{aligned}
& sortof(x + x) \\
= \; & get_range_from_signatures(+, sortof(x), sortof(x)) \\
= \; & get_range_from_signatures(+, sortof(x), sortof(y)) \\
= \; & sortof(x + y).
\end{aligned}$$

Instead, we will use infinite sets of ground substitutions to denote sorts of variables, e.g. $\{[x := s^i(0), y := s^j(0)] \mid i, j \in \mathbb{N}\}$ to indicate that x and y range over Nat. The mapping from a term to its set of possible values can then be achieved by applying each element of the substitution set, e.g.:

$$\begin{aligned} &\{[x := s^i(0)] \mid i \in \mathbb{N}\} \quad (x + x) \\ =\ &\{[x := s^i(0)]\,(x + x) \mid i \in \mathbb{N}\} \\ =\ &\{s^i(0) + s^i(0) \mid i \in \mathbb{N}\}. \end{aligned}$$

Similarly, $\{[x := s^i(0), y := s^j(0)] \mid i, j \in \mathbb{N}\}\ (x + y) = \{s^i(0) + s^j(0) \mid i, j \in \mathbb{N}\}$. Both sets are different, hence the chance of finding different approximations for them within our extended sort framework is not forfeited[3]. In Fig. 3, we show that in fact $Even$ can be obtained as the sort of $x + x$; obtaining Nat for $x + y$ is similar.

In order to have finite descriptions of such ground substitution sets, we express ground substitutions as ground constructor terms ("t-substitutions") in a lifted algebra, allowing sets of them to be treated as regular tree languages ("t-sets"). T-sets can also express simple relations between distinct variables, allowing e.g. the representation of certain conditional equations by unconditional ones.

Regular string languages have been used e.g. by Mishra [11] as a basis for sort inference on Horn clauses. Owing to the restriction to *string* languages describing admissible paths in term trees, he is only able to express infinite sets that are closed wrt. all constructors; e.g. the set of all lists of naturals containing at least one 0 cannot be modeled. Comon [5] uses regular tree languages to describe sets of ground constructor terms as sorts, and the corresponding automaton constructions to implement sort operations. He provides a transformation system to decide first-order formulas with equality and sort membership as the only predicates. He shows the decision of inductive reducibility as an application. However, he does not consider equationally defined functions, e.g. $(\forall x, y\ x{+}y = y{+}x) \rightarrow 0{+}1 = 1{+}0$ reduces to $(\forall x, y\ x{+}y = y{+}x) \rightarrow \mathit{false}$ in his calculus.

This paper is organized as follows. After a short introduction on regular sorts in Sect. 2, regular substitution sets and extended sorts are presented in Sect. 3. In Sect. 4, the algorithm for computing signatures of equationally defined functions is given. It is shown that an unsorted root narrowing calculus from [9] remains complete if extended by appropriate sort restrictions. Section 5 sketches an application to formal program construction and draws some conclusions. For a full version including all proofs, see [3].

2 Regular Sorts

Definition 1 (Notations). Let $\mathcal{V}$ be a countable set of variables, $\mathcal{CR}$ a finite set of term constructor symbols, each with fixed arity, $\mathcal{F}$ a finite set of symbols

[3] Schmidt-Schauß [12] admits "term declarations", allowing the user to declare different sorts for terms with different bindings. In our approach, however, the sorts are to be computed automatically.

for non-constructor functions, and $\mathcal{S}$ a countable set of sort names. Let $ar(g)$ denote the arity of a function symbol g. For a set[4] of symbols $X \subset \mathcal{V} \cup \mathcal{CR} \cup \mathcal{F} \cup \mathcal{S}$, let $\mathcal{T}_X$ be the set of terms formed of symbols from X; we abbreviate $\mathcal{T}_{X \cup Y}$ to $\mathcal{T}_{X,Y}$. For example, the elements of $\mathcal{T}_{\mathcal{CR}}$, $\mathcal{T}_{\mathcal{CR},\mathcal{V}}$, and $\mathcal{T}_{\mathcal{CR},\mathcal{F},\mathcal{V}}$ are called ground constructor terms, constructor terms, and terms, respectively. Let identifiers like $u, u', u_i, \ldots$ always denote members of $\mathcal{T}_{\mathcal{CR},\mathcal{V}}$, similar for $v \in \mathcal{T}_{\mathcal{CR},\mathcal{F},\mathcal{V}}$, $w \in \mathcal{T}_{\mathcal{CR},\mathcal{F},\mathcal{V},\mathcal{S}}$, $x, y, z \in \mathcal{V}$, $f \in \mathcal{F}$, $g \in \mathcal{F} \cup \mathcal{CR}$, $cr \in \mathcal{CR}$, and $S \in \mathcal{S}$.

$\langle v_1, \ldots, v_n \rangle$ denotes an n tuple, $\langle v_i \mid p(v_i), \; i = 1, \ldots, n \rangle$ denotes a tuple containing each v_i such that $p(v_i)$ holds. We assume the existence of at least one nullary (e.g. *nil*) and one binary constructor (e.g. *cons*), so we can model arbitrary tuples as constructor terms. We define the *elementwise extension* of a function $f : A \to B$ to a set $A' \subset A$ by $f[A'] := \{f(a) \mid a \in A'\}$. $A \times B$ denotes the cartesian product of sets A and B. We tacitly extend notations like $\bigcup_{i=1}^{n} A_i$ to several binary operators defined in this paper, e.g. $\big|_{i=1}^{n} S_i := S_1 \mid \ldots \mid S_n$.

Let $vars(v_1, \ldots, v_n)$ denote the set of variables occurring in any of the terms v_i. A term is called *linear* if it contains no multiple occurrences of the same variable. We distinguish between "ordinary" substitutions defined as usual (denoted by $\beta, \gamma, \ldots$) and "t-substitutions" defined as constructor terms in Sect. 3 below. For an ordinary substitution β, let $dom(\beta) := \{x \in \mathcal{V} \mid \beta x \neq x\}$, and $ran(\beta) := \bigcup_{x \in dom(\beta)} vars(\beta x)$. $[x_1 := v_1, \ldots, x_n := v_n]$ denotes the substitution that maps each x_i to v_i. $\beta|_V$ denotes the domain restriction of β to a set V of variables. We assume all substitutions to be idempotent. We use the common notions of renaming substitution, and of most general unifier $mgu(u_1, u_2)$, however, we will additionally assume that u_1 and u_2 have disjoint variables. mgu is tacitly extended to finite sets of terms.

We follow the approach of [3] in describing regular sets of ground constructor terms as fixed points of sort equations, which is equivalent to the approach using finite tree automata [5], but provides a unique methodology for algorithms and proofs.

Definition 2. We allow sort definitions of the following syntax:

$$\text{SortName} \doteq \text{SortName} \mid \ldots \mid \text{SortName},$$
$$\text{SortName} \doteq \text{Constructor}(\text{SortName}, \ldots, \text{SortName})$$

Let $\overset{\bullet}{<}$ be the transitive closure of the relation $S_i \overset{\bullet\bullet}{<} S \; :\Leftrightarrow \; S \doteq S_1 \mid \ldots \mid S_n$. We admit finite systems of sort definitions such that $\overset{\bullet}{<}$ is an irreflexive partial, hence well-founded, order. For example, the sort system consisting of $A \doteq B$ and $B \doteq A$ is forbidden. Each occurring sort name has to be defined. In examples, we use arbitrary *sort expressions* built from sort names, constructors, and "|" on the right-hand side of a sort definition, cf. *Nat* and *Bin* in Fig. 4 and *Even* in Fig. 3. Each such sort system can be transformed to obey the above requirements while maintaining the least-fixed-point semantics below.

As usual, we distinguish between the syntactic and the semantic level, the former

[4] "$\subset$" denotes subset or equality, "$\subsetneq$" denotes proper subset.

being concerned with sort names, sort expressions, sort definitions, algorithms, etc., while the latter deals with properties of sets of ground constructor terms. The semantics S^M of a sort expression S is the smallest subset of the initial algebra of ground constructor terms such that

$$\begin{array}{ll} S^M = S_1^M \cup \ldots \cup S_n^M & \text{if } S \doteq S_1 \mid \ldots \mid S_n \\ S^M = cr[S_1^M \times \ldots \times S_n^M] & \text{if } S \doteq cr(S_1, \ldots, S_n) \\ S^M = \{cr\} & \text{if } S \doteq cr \end{array}$$

A subset $T \subset \mathcal{T}_{\mathcal{CR}}$ is called *expressible* or *regular* if a system of sort definitions exists such that $T = S^M$ for some sort expression S. Note that $u^M = \{u\}$ for all $u \in \mathcal{T}_{\mathcal{CR}}$, e.g., $s(0)^M = \{s(0)\}$. It can be shown that each admitted sort system has exactly one fixed point [3], thus validating the following induction principle which is used in almost all correctness proofs of sort algorithms, cf. Alg. 7.

Theorem 3. *Let p be a family of unary predicates, indexed over the set of all defined sort names. Show for each defined sort name S:*

$$\begin{array}{ll} \forall u \in \mathcal{T}_{\mathcal{CR}} \; p_S(u) \leftrightarrow p_{S_1}(u) \vee \ldots \vee p_{S_n}(u) & \text{if } S \doteq S_1 \mid \ldots \mid S_n \\ \forall u \in \mathcal{T}_{\mathcal{CR}} \; p_S(u) \leftrightarrow \exists u_1, \ldots, u_n \in \mathcal{T}_{\mathcal{CR}} \; u = cr(u_1, \ldots, u_n) & \\ \qquad \wedge p_{S_1}(u_1) \wedge \ldots \wedge p_{S_n}(u_n) & \text{if } S \doteq cr(S_1, \ldots, S_n) \\ \forall u \in \mathcal{T}_{\mathcal{CR}} \; p_S(u) \leftrightarrow u = cr & \text{if } S \doteq cr \end{array}$$

Then, $\forall u \in \mathcal{T}_{\mathcal{CR}} \; u \in S^M \leftrightarrow p_S(u)$ *holds for each defined sort name S.*

Algorithms for computing the intersection and the relative complement of two regular sorts as well as for deciding the inhabitance of a sort, and thus of the subsort and sort equivalence property, are given in [3]. They essentially consist of distributivity rules, constructor-matching rules, and loop-checking rules. The latter stop the algorithm when it calls itself recursively with the same arguments, and generate a corresponding new recursive sort definition.

3 Regular Substitution Sets

In this section, we apply the formalism from Sect. 2 to define possibly infinite regular sets of ground substitutions. We define suitable free constructors from which ground substitutions can be built as terms. Note that the classical approach, constructing substitutions by functional composition from simple substitutions, cannot be used, since functional composition is not free but obeys e.g. the associativity law.

Informally, to build a substitution term corresponding to $[x_1 := u_1, \ldots, x_n := u_n]$ with u_i ground, we "overlay" the u_i to obtain the substitution term; to the right, an example is shown for $[x := cons(0, nil), y := s(s(0))]$.

$$\begin{array}{lllll} x := & cons & (\; 0 & , nil \;) \\ y := & s & (\; s \; (0) &) \\ \hline & \multicolumn{4}{l}{cons_x s_y (0_x s_y (0_y), nil_x)} \end{array}$$

Formally, given a set $V \subset \mathcal{V}$ of variables, define the constructors for t-substitutions with domain V as the set $(V \to \mathcal{CR})$ of all total mappings from V to

$\mathcal{CR}$. T-substitution constructors are denoted by $\mathbf{cr}, \mathbf{cr}', \ldots$, the empty mapping by ε. Function application is written as $\mathbf{cr}_x$, the arity is defined as $ar(\mathbf{cr}) := \max_{x \in dom(\mathbf{cr})} ar(\mathbf{cr}_x)$. In examples, we write e.g. $0_x s_y$ to denote the mapping $(x \mapsto 0, y \mapsto s)$, having $0_x s_y \in (\{x, y\} \to \mathcal{CR})$ and $ar(0_x s_y) = max(0, 1) = 1$.

Once we have defined t-substitution constructors, we inherit the initial term algebra $\mathcal{T}_{(V \to \mathcal{CR})}$ over them. However, we have to exclude some nonsense terms: Define the subset $\mathcal{T}^*_{(V \to \mathcal{CR})} \subset \mathcal{T}_{(V \to \mathcal{CR})}$ of *admissible t-substitutions* with domain V as the least set such that

$$\mathbf{cr}(\sigma'_1, \ldots, \sigma'_{ar(\mathbf{cr})}) \in \mathcal{T}^*_{(V \to \mathcal{CR})}$$
$$\text{if } \mathbf{cr} \in (V \to \mathcal{CR}) \text{ and } \sigma'_i \in \mathcal{T}^*_{(\{x \in V \mid ar(\mathbf{cr}_x) \geqslant i\} \to \mathcal{CR})} \text{ for } i = 1, \ldots, ar(\mathbf{cr}).$$

The definition implies that $\mathbf{cr} \in \mathcal{T}^*_{(V \to \mathcal{CR})}$ if $\mathbf{cr} \in (V \to \mathcal{CR})$ is a nullary t-substitution constructor. For example, we have $0_y \in \mathcal{T}^*_{(\{y\} \to \mathcal{CR})}$, and hence $0_x s_y(0_y) \in \mathcal{T}^*_{(\{x,y\} \to \mathcal{CR})}$, but neither $0_y \in \mathcal{T}^*_{(\{x,y\} \to \mathcal{CR})}$, nor $0_x s_y(0_x 0_y) \in \mathcal{T}^*_{(V \to \mathcal{CR})}$ for any V. Figure 1 shows some more t-substitutions together with their intended semantics. We denote t-substitutions by $\sigma', \tau', \mu', \ldots$. Since $\mathcal{T}^*_{(V_1 \to \mathcal{CR})} \cap \mathcal{T}^*_{(V_2 \to \mathcal{CR})} = \{\}$ for $V_1 \neq V_2$, we may define $dom(\sigma') := V$ iff $\sigma' \in \mathcal{T}^*_{(V \to \mathcal{CR})}$. Let $(V \hookrightarrow \mathcal{CR})$ be the set of all partial mappings from V to $\mathcal{CR}$, define the set of admissible t-substitutions with a subset of V as domain by $\mathcal{T}^*_{(V \hookrightarrow \mathcal{CR})} := \bigcup_{V' \subset V, V' \text{ finite}} \mathcal{T}^*_{(V' \to \mathcal{CR})}$.

We can now describe regular sets of ground substitutions as subsets of the initial term algebra over $(\mathcal{V} \hookrightarrow \mathcal{CR})$ using the mechanisms and algorithms of Sect. 2, i.e. for intersection, relative complement, and inhabitance. We call such sets *t-sets*. The remarks on the syntactic and the semantic level from Sect. 2 apply similarly to t-sets; we will denote – syntactic – names of t-sets as well as – semantic – subsets of $\mathcal{T}_{(\mathcal{V} \hookrightarrow \mathcal{CR})}$ by $\sigma, \tau, \mu, \ldots$. We will only consider t-sets with unique domain $dom(\sigma') = V_\sigma$ for all $\sigma' \in \sigma$, define $dom(\sigma) := V_\sigma$. The empty t-set is denoted by $\perp$. For each finite V, $\top_V := \mathcal{T}^*_{(V \hookrightarrow \mathcal{CR})}$ is expressible as a regular set. We write $\top$ for $\top_V$ when V is clear from the context; note that $\mathcal{T}^*_{(\mathcal{V} \hookrightarrow \mathcal{CR})}$ is not expressible since infinitely many t-substitution constructors exist. Figure 1 shows some example t-sets.

T-substitutions, built as constructor terms:

$s_x(s_x(0_x))$	$\hat{=}[x := s(s(0))]$
$s_x s_y(0_x 0_y)$	$\hat{=}[x := s(0), y := s(0)]$
$s_x 0_y(0_x)$	$\hat{=}[x := s(0), y := 0]$

T-sets, described as regular sets:

$Nat_x \doteq 0_x \mid s_x(Nat_x)$	$\hat{=}\{[x := s^i(0)] \mid i \in \mathbb{N}\}$,	similar Nat_y
$Nat_{x=y} \doteq 0_x 0_y \mid s_x s_y(Nat_{x=y})$	$\hat{=}\{[x := s^i(0), y := s^i(0)] \mid i \in \mathbb{N}\}$	
$Nat_{x,y} \doteq 0_x 0_y \mid s_x s_y(Nat_{x,y}) \mid 0_x s_y(Nat_y) \mid s_x 0_y(Nat_x)$	$\hat{=}\{[x := s^i(0), y := s^j(0)] \mid i, j \in \mathbb{N}\}$	
$Nat_{x<y} \doteq 0_x s_y(Nat_y) \mid s_x s_y(Nat_{x<y})$	$\hat{=}\{[x := s^i(0), y := s^j(0)] \mid i, j \in \mathbb{N}, i < j\}$	

Fig. 1. Examples of t-substitutions and t-sets

Definition 4. Define the t-substitution application $\sigma' u$ by

$$\begin{array}{lll} \sigma'(cr(u_1,\ldots,u_k)) & := cr[\sigma' u_1 \times \ldots \times \sigma' u_k] & \\ \sigma'(cr) & := \{cr\} & \\ (\mathbf{cr}(\sigma'_1,\ldots,\sigma'_n))(x) & := \mathbf{cr}_x[\sigma'_1 x \times \ldots \times \sigma'_{ar(\mathbf{cr}_x)} x] & \text{if } x \in dom(\mathbf{cr}),\ n > 0 \\ (\mathbf{cr})(x) & := \{\mathbf{cr}_x\} & \text{if } x \in dom(\mathbf{cr}),\ n = 0 \\ (\mathbf{cr}(\sigma'_1,\ldots,\sigma'_n))(x) & := \{\} & \text{if } x \notin dom(\mathbf{cr}) \end{array}$$

$\sigma' u$ yields a set with at most one ground constructor term. Application is extended elementwise to t-sets by $\sigma u := \bigcup_{\sigma' \in \sigma} \sigma' u$.

Note that, in contrast to an ordinary substitution β, a t-substitution σ' is undefined outside its domain, i.e. it returns the empty set. For example, $(0_x s_y(0_y))\ (x) = \{0\}$, $(0_x s_y(0_y))\ (y) = s[0_y(y)] = s[\{0\}] = \{s(0)\}$, and $(0_x s_y(0_y))\ (z) = \{\}$. We have $\varepsilon u = \{\}$ if u contains variables, $\varepsilon u = u$ if u is ground, and always $\bot u = \{\}$.

Although t-substitutions are homomorphic wrt. all constructors in $\mathcal{CR}$, t-sets are generally not, e.g. $Nat^M_{x<y}(\langle x, y\rangle) \subsetneqq \langle Nat^M_{x<y}(x), Nat^M_{x<y}(y)\rangle$. We call a t-set σ *independent* if it is homomorphic on linear terms, i.e. if $\sigma\langle x_1,\ldots,x_n\rangle = \langle \sigma x_1,\ldots,\sigma x_n\rangle$. An independent t-set assigns the value of one variable independently of the value of the others, e.g. $Nat_{x,y}$ in Fig. 1, while a "dependent" t-set allows dependencies between variables to be expressed, e.g. $Nat_{x<y}$ and $Nat_{x=y}$.

Lemma 5. *$\sigma' = \tau'$ iff $\sigma' x = \tau' x$ for all $x \in \mathcal{V}$, where "$=$" on the left-hand side denotes the syntactic equality in $\mathcal{T}^*_{(\mathcal{V}\hookrightarrow\mathcal{CR})}$.*

This desired equivalence of term equality and function equality is the reason for restricting t-substitutions to a subset $\mathcal{T}^*_{(\mathcal{V}\hookrightarrow\mathcal{CR})}$ of the initial algebra $\mathcal{T}_{(\mathcal{V}\hookrightarrow\mathcal{CR})}$, excluding nonsense terms like e.g. $0_x s_y(0_x 0_y)$ and $0_x s_y(0_x 1_y)$ which would contradict the freeness requirement $\mathbf{cr}(\sigma') = \mathbf{cr}(\tau') \Rightarrow \sigma' = \tau'$.

Lemma 6. *$\mathcal{T}^*_{(\mathcal{V}\hookrightarrow\mathcal{CR})}$ corresponds to the set of all ordinary ground substitutions in the following sense: For each σ' there exists a β, such that $\sigma' u = \{\beta u\}$ whenever $vars(u) \subset dom(\sigma')$. Vice versa, for each β there exists a σ' with the respective property. Cf. Fig. 1 which shows some example correspondences.*

Algorithm 7. *The following algorithm computes the elementwise application of a t-set to a variable. Let σ be the name of a t-set, and let S be a new sort name. Define $apply(\sigma, x) = S$, where the algorithm introduces a new sort definition for S:*

1. *If $apply(\sigma, x)$ has already been called earlier, S is already defined (loop check).*
2. *If $\sigma \doteq \sigma_1 \mid \ldots \mid \sigma_n$, define $S \doteq apply(\sigma_1, x) \mid \ldots \mid apply(\sigma_n, x)$*
3. *If $\sigma \doteq \mathbf{cr}(\sigma_1,\ldots,\sigma_n)$ with $x \in dom(\mathbf{cr})$, define $S \doteq \mathbf{cr}_x(apply(\sigma_1, x),\ldots, apply(\sigma_{ar(\mathbf{cr}_x)}, x))$,*
4. *Else, define $S \doteq \bot$, where $\bot$ denotes the empty sort.*

Using Thm. 3 it can be shown that $apply(\sigma, x)^M = \sigma^M x$.

Definition 8. Using Lemma 6, we can define counterparts to the usual operations on ordinary substitutions and extend them elementwise to t-sets: Restriction $\sigma'|_V$ for $V \subset \mathcal{V}$, parallel composition $\sigma' \diamond \tau'$ for $\sigma'|_{dom(\tau')} = \tau'|_{dom(\sigma')}$, factorization[5] $\sigma'/_\beta$ for $dom(\beta) \subset dom(\sigma')$, and lifting $[x := u]$ can be inductively defined such that the properties below hold. They are extended elementwise from t-substitutions to t-sets. Restriction and parallel composition of t-sets, and lifting of regular sorts can be computed by appropriate algorithms.

$$\begin{array}{lll}
dom(\sigma'|_V) = dom(\sigma') \cap V, & \sigma'|_V\, x = \sigma' x & \text{if } x \in V, \quad = \{\}, \text{ else} \\
dom(\sigma' \diamond \tau') = dom(\sigma') \cup dom(\tau'), & (\sigma' \diamond \tau')\, x = \sigma' x & \text{if } x \in dom(\sigma'), \text{ similar } \tau' \\
dom(\sigma'/_\beta) = ran(\beta), & \sigma'/_\beta \beta u = \sigma' u & \text{if } \sigma'/_\beta \neq \{\} \\
dom([x := u]) = \{x\}, & [x := u]\, x = \{u\} &
\end{array}$$

If σ' and τ' don't agree on their domain intersection, $\sigma' \diamond \tau' = \{\}$, thus, $\sigma \diamond \tau$ yields a kind of relation join. We have $\sigma \diamond \{\} = \{\}$ and $\sigma \diamond \{\varepsilon\} = \sigma$. For example, $Nat^M_{x<y}|_{\{x\}} = Nat^M_x$, $Nat^M_x \diamond Nat^M_y = Nat^M_{x,y}$, and $[x := Nat^M] = Nat^M_x$.

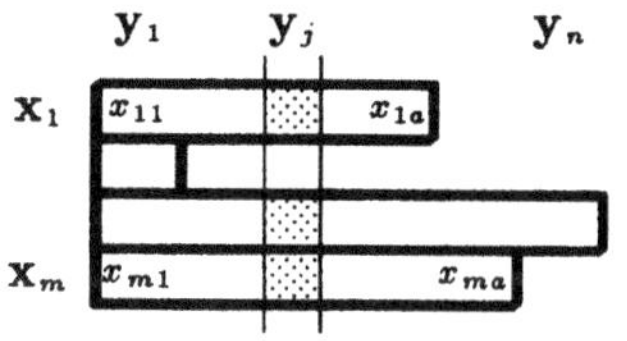

Each t-set σ with $dom(\sigma) = \{x_1, \ldots, x_n\}$ corresponds to a predicate p_σ in the sense that for all ground terms $u_1, \ldots, u_n$, $p_\sigma(u_1, \ldots, u_n)$ holds iff $[x_1 := u_1] \diamond \ldots \diamond [x_n := u_n] \in \sigma$. A predicate p can be represented as p_σ for some σ iff it is definable by Horn clauses of the following form: $p(cr_1(\mathbf{x}_1), \ldots, cr_m(\mathbf{x}_m)) \leftarrow p_1(\mathbf{y}_1) \wedge \ldots \wedge p_1(\mathbf{y}_n)$ where $\mathbf{x}_i := \langle x_{i1}, \ldots, x_{ia_i} \rangle$ for $i = 1, \ldots, m$, and $\mathbf{y}_j := \langle x_{ij} \mid i = 1, \ldots, m, j \leqslant ar(cr_i) \rangle$ for $j = 1, \ldots, n$ with $n := max_{i=1,\ldots,m} ar(cr_i)$. The relation between $\mathbf{x}_i$ and $\mathbf{y}_j$ is shown in the above diagram. If all term constructors cr_i have the same arity n, the $\mathbf{y}_j$ are the column vectors of an $m \times n$ matrix built from the $\mathbf{x}_i$ as line vectors. For example, the definition[6] $Lgth_{x,y} \doteq 0_x nil_y \mid s_x snoc_y(Lgth_{x,y}, Nat_y)$ corresponds to the Horn clauses $lgth(0, nil)$ and $lgth(s(x), snoc(y_1, y_2)) \leftarrow lgth(x, y_1) \wedge nat(y_2)$. We have the following corollary which is an extension of the "SOGEN" algorithm from [12] to a class of n-ary predicates; in particular, it is not covered by [10].

Corollary 9. *The satisfiability of any predicate defined by Horn clauses of the above form can be decided. The set of such predicates is closed wrt. conjunction, disjunction, and negation.*

In classical order-sorted approaches, each variable in a term is assigned a sort, e.g. $x^{:\mathrm{Nat}} + x^{:\mathrm{Nat}}$. We will, instead, use a t-set to specify the set of possible ground instances, written e.g. ${}^{(Nat_x)}(x+x)$, with the informal meaning that each ("admissible") substitution instantiating $x + x$ must be extendible to a ground substitution contained in Nat^M_x. This approach allows variable bindings in a term to be reflected by its sort, as sketched in Sect. 1. Moreover, dependent t-sets can express certain relations between distinct variables, e.g. the conditional equation $x^{:\mathrm{Nat}} < y^{:\mathrm{Nat}} \rightarrow f(x, y) = g(x, y)$ can be expressed unconditionally by

[5] The factorization $\sigma'/_\beta$ of σ' wrt. β corresponds to a kind of root pattern matching of β in σ'; factorization plays a central role in Thms. 10 and 11.

[6] $snoc : \mathbf{list}_\alpha \times \alpha \rightarrow \mathbf{list}_\alpha$ denotes a reversed $cons$.

Expressible relations e.g.:		Operations on relations e.g.:	
σ_{pr}	prefix x of length y of a *snoc*-list z	relation join	$\sigma \diamond \tau$
σ_{lx}	lexicographical order on *cons*-lists (wrt. expressible element ordering)	function evaluation	(join)
σ_{lf}	lifting of element relation to list	restriction	$\sigma\vert_V$
σ_{mt}	matching of tree x at the root of tree y	bounded renaming	$\sigma/_\beta$
σ_{zp}	zip z of *cons*-lists x and y	conjunction	$\sigma \cap \tau$
		disjunction	$\sigma \mid \tau$
		negation	$\top \setminus \sigma$

E.g. $\sigma_{pr}|_{\{x,y\}}$ denotes the length function on *snoc*-lists,
$(\sigma_{pr} \diamond \{[y := s^3(0)]\})\ (x)$ denotes the regular sort of all *snoc*-lists of length 3.

Fig. 2. Some relations and operations on relations expressible by t-sets

${}^{(Nat_{x<y})} f(x,y) = {}^{(Nat_{x<y})} g(x,y)$. Figure 2 shows some more nontrivial relations that are expressible by t-sets.

More formally, we define a *sorted term* to be a pair of an expressible t-set σ and an (unsorted) term v; it is written as ${}^{\sigma}v$. The t-set σ denotes the admitted instances of v, cf. the use of ${}^{\sigma}v$ in Def. 12 and 17 below. For a constructor term u, the set of ground constructor terms σu denotes the set of all admitted values of ${}^{\sigma}u$, i.e. σu relates to ${}^{\sigma}u$ as V to v in Sect. 1. We call an expression of the general form $\sigma_1 u_1 \cup \ldots \cup \sigma_n u_n$ with expressible σ_i an *extended sort*[7]. Any regular sort S^M from Sect. 2 can be expressed as an extended sort, viz. $[x := S^M]\ x$, but the converse is false, e.g. $Nat_x^M\ \langle x, x\rangle$ is not a regular sort, as can be shown using a pumping lemma [3]. Thus, regular sorts are a proper subclass of extended sorts, and we need new algorithms to compute with the latter which are provided in the following; Theorems 10 and 11 reduce the computation of intersection and the decision of subset property of extended sorts to the respective problems of t-sets, which can be solved by "lifted" versions of the algorithms mentioned in Sect. 2.

Theorem 10. *Extended sorts are closed wrt. intersection, i.e. let* $\beta = mgu(u_1, u_2)$, $dom(\beta) = vars(u_1, u_2)$, $vars(u_i) \subset dom(\sigma_i)$, $dom(\sigma_1) \cap dom(\sigma_2) = \{\}$, *and* $u = \beta u_1$, *then* $\sigma_1 u_1 \cap \sigma_2 u_2 = (\sigma_1 \diamond \sigma_2)/_\beta\ u$.

Theorem 11. *Let* $u, u_1, \ldots, u_n$ *have pairwise disjoint variables, let* u *be unifiable with each* u_i. *For* $I \subset \{1, \ldots, n\}$ *let* $\beta_I = mgu(\{u\} \cup \{u_i \mid i \in I\})$ *if it exists, let* J *be the set of all* I *with existing mgu. Let* $\gamma_I := \beta_I|_{vars(u)}$ *and* $\gamma_{I,i} := \beta_I|_{vars(u_i)}$. *Let* $\sigma_I := \{\sigma' \in \sigma \mid \sigma'/_{\gamma_{\{i\}}} \neq \{\} \leftrightarrow i \in I\}$.
Then $\sigma u \subset \tau_1 u_1 \cup \ldots \cup \tau_n u_n$ *iff* $\sigma_I/_{\gamma_I} \subset \bigcup_{i \in I} \tau_i/_{\gamma_{I,i}}$ *for all* $I \in J$.

Domain conditions as in Theorem 10 can always be satisfied by bounded renaming, factorizing by a renaming substitution. If u_1 and u_2 cannot be unified, $\sigma_1 u_1$ and $\sigma_2 u_2$ are always disjoint. Theorems 10 and 11 provide the basis for the

[7] Finite unions are necessary to ensure certain closedness properties. Sets of the form σv, where v contains non-constructor functions, will be approximated by extended sorts, cf. Sect. 4.

necessary algorithms to conduct the disjointness test from Sect. 1 on extended sorts. However, we have the problem that $\sigma/_{\beta}$ is not always expressible as a t-set. It *is* expressible if $\beta x \in \mathcal{V}$ for each $x \in dom(\beta)$ or if $dom(\beta)$ is a singleton set. If $\beta\langle x_1, \dots, x_n\rangle$ is a linear term where $\{x_1, \dots, x_n\} = dom(\beta)$, we can at least achieve the representation of an extended sort $\sigma/_{\beta}u$ as a finite union $\tau_1(\gamma_1 u) \cup \dots \cup \tau_m(\gamma_m u)$ with expressible t-sets τ_i and ordinary substitutions γ_i. For arbitrary β, however, $\sigma/_{\beta}u$ can in general not be expressed as an extended sort. It is an open problem whether a superclass of extended sorts exists which is closed wrt. the needed operations, especially intersection.

Hence, we restrict ourselves to the subclass of independent t-sets where simple algorithms for computing with extended sorts exist. Finite unions of independent t-sets are closed wrt. all important operations, such as intersection, factorization, union, restriction, and parallel composition. Extended sorts $\sigma_1 u_1 \cup \dots \cup \sigma_n u_n$ with independent σ_i are closed under intersection; inclusion, and hence equivalence, can be decided. Note that the ability to reflect different variable bindings by different sorts is still preserved; what we lose is the possibility to express nontrivial relations between variables.

4 Equational Theories

In this section, we extend the previous formalism to allow equationally defined functions f. We allow defining equations of the following form which ensures the "executability" of f: $^{\mu_1}f(u_{11}, \dots, u_{1n}) = {}^{\mu_1}v_1, \ \dots \ , {}^{\mu_m}f(u_{m1}, \dots, u_{mn}) = {}^{\mu_m}v_m$, where $vars(v_i) \subset vars(u_{i1}, \dots, u_{in})$. We assume that the variables of different defining equations are disjoint. Signatures of such a function are computed from its defining equations by the *rg* algorithm presented below in Alg. 16, which will play a central role in pruning the search space of narrowing. The algorithm takes a t-set and a term with non-constructor functions and computes an upper approximation by an extended sort, e.g. $rg([x, y := Nat], x + y) = [z := Nat]\ z$. In terms of Sect. 1, we have $rg(\sigma, v) = \overline{V}$ where σ denotes the values over which the variables in v may range. The *rg* algorithm consists of local transformations like rewriting and some simplification rules (cf. Def. 14), global transformations looking at a sequence of local transformation steps and recognizing certain kinds of self-references (cf. Lemma 15), and an approximation rule. Only the main rules can be discussed here; the complete algorithm is given in [3].

In Theorem 18, a narrowing calculus from [9] is equipped with sorts. In [3], the calculus is shown to remain complete if the applicability of its main rule is restricted by the disjointness test from Sect. 1.

Definition 12. Define the rewrite relation induced by the defining equations by: $^{\sigma}v_1 \to {}^{\sigma}v_2$ iff

1. a defining equation $^{\mu}f(u_1, \dots, u_n) = {}^{\mu}v$, a substitution β, and a term $v'(x)$ linear in x exist such that $v_1 = v'(\beta f(u_1, \dots, u_n))$, $v_2 = v'(\beta v)$,
2. and for all $\sigma' \in \sigma$ exists $\mu' \in \mu$ such that for all $x \in vars(u_1, \dots, u_n)$ $^{\varepsilon}\sigma'\beta x \to^* {}^{\varepsilon}\mu' x$ if $^{\varepsilon}\sigma'\beta x$ is well-defined.

While the former condition is just rewriting by pattern matching, the latter one is an analogue to the classical well-sortedness requirement for β, requiring any well-defined variable instance to be admitted by the defining equation's sort. A ground term is called *well-defined* if it is reducible to a ground constructor term. $\rightarrow^*$ and $\leftrightarrow^*$ are defined as usual; the definition of $\rightarrow$ is recursive, but well-founded. We require confluence and termination of $\rightarrow$, ensuring $\mathcal{T}_{\mathcal{CR}} \subset \mathcal{T}_{\mathcal{CR},\mathcal{F}}/\leftrightarrow^*$, that is, $\leftrightarrow^*$ does not identify terms in $\mathcal{T}_{\mathcal{CR}}$, but new irreducible terms like $nil + nil$ may arise which we will consider as "junk terms" and exclude from equation solutions. For a well-defined ground term v, let $nf(v) \in \mathcal{T}_{\mathcal{CR}}$ denote its unique normal form; for $A \subset \mathcal{T}_{\mathcal{CR},\mathcal{F}}$, let $nf[A] := \{nf(v) \mid v \in A, v \text{ well-defined}\}$.

The applicability of $\rightarrow$ is not decidable in general owing to the well-sortedness condition 12.2. It is possible to compute sufficiently large t-sets μ_i for the defining equations such that 12.2 becomes trivial; however, if the μ_i are too large, well-sorted terms arise that are not well-defined. As in any order-sorted term rewriting approach, we cannot overcome both problems simultaneously.

Range sorts are computed using expressions of the form $(w_1 : u_1) \ldots (w_n : u_n)$ which can intuitively be thought of as generalized equation systems; the semantic is the set of all t-substitutions making each u_i equal ($\leftrightarrow^*$) to an element of w_i. For example, $(Nat : x)$ denotes $\{[x := s^i(0)] \mid i \in \mathbb{N}\}$, and $(x + x : z)$ can be evaluated to $\{[x := s^i(0), z := s^{2 \cdot i}(0)] \mid i \in \mathbb{N}\}$.

Definition 13. Define $w^M \subset \mathcal{T}_{\mathcal{CR},\mathcal{F},\mathcal{V},\mathcal{S}}$ by:

$$
\begin{array}{rll}
S^M & := S^M \text{ as in Def. 2} & \text{for } S \in \mathcal{S} \\
g(v_1, \ldots, v_n)^M & := \{g(v_1', \ldots, v_n') \mid v_1' \in v_1^M, \ldots, v_n' \in v_n^M\} & \text{for } g \in \mathcal{CR} \cup \mathcal{F} \\
x^M & := \{x\} & \text{for } x \in \mathcal{V}
\end{array}
$$

For $w \in \mathcal{T}_{\mathcal{CR},\mathcal{S}}$, w^M agrees with Def. 2. For $w \in \mathcal{T}_{\mathcal{CR},\mathcal{F},\mathcal{V}}$ we always have $w^M = \{w\}$. We tacitly extend the operations of Sect. 3 to $\mathcal{T}_{\mathcal{CR},\mathcal{F},\mathcal{V}}$ by treating function symbols from $\mathcal{F}$ like constructors from $\mathcal{CR}$, e.g. $(0_x)\ (x + x) = \{0 + 0\}$.
Let $(w : u)^M := \{\sigma' \mid dom(\sigma') = vars(w, u), \exists w' \in w^M, u' \in u^M\ \sigma' w' \leftrightarrow^* \sigma' u'\}$ and $((w_1 : u_1)\ (w_2 : u_2))^M := (w_1 : u_1)^M \diamond (w_2 : u_2)^M$. We write (σ) to denote an expression $(w_1 : u_1) \ldots (w_n : u_n)$ such that $((w_1 : u_1) \ldots (w_n : u_n))^M = \sigma$, e.g. $(Nat_{x,y})$ denotes $(Nat : x)(Nat : y)$, but note that σ need neither be independent nor even expressible. The terms are unsorted in order to deal with t-sets explicitly; $({}^\sigma w : {}^\tau u)$ can be written as $(\sigma)\ (\tau)\ (w : u)$. Note that the t-substitutions in $(w : u)^M$ always yield ground constructor terms.

Definition 14. The following local transformation rules for rg are defined (excerpt):

1. $(f(v_1', \ldots, v_n') : u) = \big|_{i=1}^{m} (v_i : u)\ (v_1' : u_{i1})\ \ldots\ (v_n' : u_{in})\ (\mu_i)$
   if $f \in \mathcal{F}$ has the defining equations
   ${}^{\mu_1}f(u_{11}, \ldots, u_{1n}) = {}^{\mu_1}v_1\ , \ldots ,\ {}^{\mu_m}f(u_{m1}, \ldots, u_{mn}) = {}^{\mu_m}v_m$.
2. $(u' : x)\ (v : u) = (u' : x)\ ([x := u']\ v : [x := u']\ u)$ if $x \notin vars(u')$,
3. $(S : x)\ (S' : x) = (S \cap S' : x)$,

Rules 2. and 3. also show "$(\cdot : \cdot)$" as a generalization of term equality and sort membership, respectively. All local rules satisfy $lhs^M|_{vars(lhs)} = rhs^M|_{vars(rhs)}$.

Only one proper approximation rule is needed, viz. $(w_1 : u_1) \ \ldots \ (w_n : u_n)^M \subset (w_2 : u_2) \ \ldots \ (w_n : u_n)^M$; all other rules can be made exact by including the left-hand side in the right-hand side.

Applying local transformations creates a *derivation tree* with *alternatives* (separated by "|") as nodes, each alternative having a unique *derivation path* from the root, cf. Fig. 3. Global transformations operate on such derivation trees. A proof methodology is provided in [3] for their verification that also allows the introduction of new global rules if necessary for some class of applications. As an example, a simplified version of the loop-checking rule is shown in Lemma 15.

Since not every function signature can be expressed by extended sorts, termination of the *rg* algorithm has to be enforced artificially, using the above approximation rule. In [3], a termination heuristic is discussed which seems to yield a good trade-off between computation time and the precision of the result.

Lemma 15. *(Global Transformation: Loop-Checking Rule)*
Assume $z \notin dom(\sigma) \supset vars(v) \not\ni x$ *and a derivation tree of the form*

$(\sigma)\ (v : z) \quad = \ldots$
$= (\sigma)\ (v : x)\ (u_1(x) : z) \mid \ldots \mid (\sigma)\ (v : x)\ (u_n(x) : z) \mid (u_{n+1} : z) \mid \ldots \mid (u_m : z)$

where in each alternative's path at least one application of rule 14.1 occurred. Then, $((\sigma)\ (v : z))^M \subset (S : z)^M$, *where* S *is a new sort name defined by* $S \doteq u_1(S) \mid \ldots \mid u_n(S) \mid u_{n+1} \mid \ldots \mid u_m$. *If all* $u_i(x)$ *are linear in* x, *we have equality.*

Algorithm 16. *To compute* $rg(\sigma, v)$, *start with the expression* $(\sigma)\ (v : z)$ *where* z *is new, and repeatedly apply rules with the following priorities: global rules, approximation rule, simplifying local rules (like Def. 14.2 and 3), and rewriting (Def. 14.1). On termination, an expression* $(\sigma_1)\ (u_1 : z) \mid \ldots \mid (\sigma_n)\ (u_n : z)$ *with expressible t-sets* σ_i *is obtained. The final result is then* $rg(\sigma, v) := \sigma_1 u_1 \mid \ldots \mid \sigma_n u_n$, *satisfying* $nf[\sigma^M v] \subset rg(\sigma, v)^M$.

The precision of the computed range sort can be enhanced by supplying redundant equations, thus getting a similar effect to that obtained by term declarations in [12]. To this end, Def. 14.1 can be generalized such that it is sufficient to consider an arbitrary subset I of defining equations that covers the whole domain of f, i.e., using the notions from Def. 14.1,

$$nf[\mathsf{T}\langle v_1', \ldots, v_n' \rangle] \cap \bigcup_{i \in I} \mu_i \langle u_{i1}, \ldots, u_{in} \rangle = nf[\mathsf{T}\langle v_1', \ldots, v_n' \rangle] \cap \bigcup_{i=1}^{m} \mu_i \langle u_{i1}, \ldots, u_{in} \rangle$$

The choice of these subsets can be postponed to enable optimal application of global transformations, cf. the example in Fig. 3.

Definition 17. A substitution β is called a solution of an equation ${}^{\sigma}v_1 = {}^{\sigma}v_2$ iff a t-set τ exists which denotes the sorts of variables in the $ran(\beta)$ and such that

1. ${}^{\tau}\beta v_1 \leftrightarrow^* {}^{\tau}\beta v_2$,
2. $\forall \tau' \in \tau\ \exists \sigma' \in \sigma\ \forall x \in vars(v_1, v_2)\ \ \tau'\beta x$ well-defined $\Rightarrow \tau'\beta x \leftrightarrow^* \sigma' x$ similar to the classical well-sortedness requirement for β, and
3. $nf[\tau\beta v_1] \neq \{\}$, i.e. the solution has at least one well-defined ground instance.

Assuming the definitions in Fig. 4, we can compute $rg(Nat_x, x+x)$:

$$
\begin{aligned}
& (Nat : x)\,(x + x : z) \\
= & (Nat : x)\,(x_1 : z)\,(x : x_1)\,(x : 0) \mid (Nat : x)\,(x_1 : z)\,(x : 0)\,(x : x_1) \mid \\
& (Nat{:}x)\,(s(x_1 + y_1){:}z)\,(x{:}x_1)\,(x{:}sy_1) \mid (Nat{:}x)\,(s(x_1{+}y_1){:}z)\,(x{:}sx_1)\,(x{:}y_1) \\
= & (0 : z) \mid (Nat : y_1)\,(sy_1 + y_1 : z_1)\,(sz_1 : z) \mid (Nat : x_1)\,(x_1 + sx_1 : z_1)\,(sz_1{:}z) \\
= & (0 : z) \mid (Nat : y_2)\,(ssy_2{+}y_2 : z_2)\,(ssz_2{:}z) \mid (Nat : y_2)\,(y_2 + y_2 : z_2)\,(ssz_2 : z) \\
& \qquad \mid (Nat : y_2)\,(y_2 + y_2 : z_2)\,(ssz_2 : z) \mid (Nat{:}x_2)\,(x_2{+}ssx_2{:}z_2)\,(ssz_2{:}z) \\
= & (0 : z) \mid (Nat : y_2)\,(y_2 + y_2 : z_2)\,(ssz_2 : z) \\
= & (Even : z)
\end{aligned}
$$

where the new sort definition $Even \doteq 0 \mid s(s(Even))$ is generated. The performed steps are: Rule 14.1 with equations a.-d.; simplification; Rule 14.1 with c.-d. twice in parallel, including simplification; deletion of the 2nd, 4th, and 5th alternative, since they are covered by the 3rd one; thus making Lemma 15 applicable as the final step.

Fig. 3. Example rg computation

Theorem 18. *An arbitrary narrowing calculus preserving solution sets remains complete if restricted appropriately by sorts. For example, for lazy narrowing [9], abbreviating* $\tau := \sigma \diamond \top_{vars(u_1,\ldots,u_n)}$*, we get for the main rule:*

$$
\frac{{}^{\tau}v_1 = {}^{\tau}u_1 \wedge \ldots \wedge {}^{\tau}v_n = {}^{\tau}u_n \wedge {}^{\tau}v = {}^{\tau}v'}{{}^{\sigma}f(v_1,\ldots,v_n) = {}^{\sigma}v}
\qquad
\begin{array}{l}
{}^{\top}f(u_1,\ldots,u_n) = {}^{\top}v' \text{ defng. eqn.}, \\
rg(\tau, v)^M \cap rg(\tau, v')^M \neq \{\}, \\
rg(\tau, v_1)^M \cap rg(\tau, u_1)^M \neq \{\}, \ldots, \\
rg(\tau, v_n)^M \cap rg(\tau, u_n)^M \neq \{\}
\end{array}
$$

Note that the variables in defining equations have to be assigned the sort $\top$. *Starting from a conditional narrowing calculus, nontrivially sorted defining equations become possible.*

Lemma 19. *Let* σ *be independent, let* x *be new, Then, the equation* ${}^{\sigma}v_1 = {}^{\sigma}v_2$ *has a solution iff* $rg(\sigma, \langle v_1, v_2\rangle)^M \cap \top\langle x, x\rangle \neq \{\}$*, provided the approximation rule was not used in* rg *computation.*

Lemma 19 shows that the amount of search space reduction by the sorts depends only on the quality of approximations by rg and the expressiveness of our sort language. Without reflecting variable bindings, such a result is impossible even if no non-constructor functions and no "occur check" are involved, e.g. $\langle x, y\rangle = \langle 0, s(0)\rangle$ is solvable, but $\langle x, x\rangle = \langle 0, s(0)\rangle$ is not.

5 Application to Formal Program Development

As an example, consider the formal development of an addition algorithm *add* for binary numbers, using a binary constructor "::" as infix *snoc*, and two nullary constructors "*o*" and "*i*" as digits. Consider the sort and function definitions in Fig. 4. All terms are sorted by the t-set $[x := Nat] \diamond [y := Nat] \diamond [z := Bin]$ which is omitted in the equations for sake of brevity. Equations marked by "*" are redundant ones and can be proven by structural induction. The rightmost column contains the sort computed by rg for each equation; all sorts happen to be regular.

$Nat \doteq 0 \mid s(Nat)$
$Bin \doteq nil \mid Bin::o \mid Bin::i$

E.g.
$val(nil::i::o::i) = s^5(0)$

Prove
$\forall x, y\ \exists z\ \ val(x) + val(y) = val(z)$

Implicit t-set
$[x := Nat] \otimes [y := Nat] \otimes [z := Bin]$

a.	$x + 0 = x$	Nat
*b.**	$0 + x = x$	Nat
c.	$x + s(y) = s(x + y)$	$s(Nat)$
*d.**	$s(x) + y = s(x + y)$	$s(Nat)$
e.	$dup(x) = x + x$	$Even$
*f.**	$dup(x + y) = dup(x) + dup(y)$	$Even$
g.	$val(nil) = 0$	0
h.	$val(z::o) = dup(val(z))$	$Even$
i.	$val(z::i) = s(dup(val(z)))$	$s(Even)$

Fig. 4. Sort and function definitions for synthesis of binary arithmetic algorithm

The main contribution of the sorts is the computation of $rg(Nat_x, dup(x)) = [z := Even]\ z = Even$, where the sort definition $Even \doteq 0 \mid s(s(Even))$ is automatically introduced. Although only independent t-sets are involved in the example, the variable bindings in $x + x$ are reflected by its sort, viz. $Even$. For this range-sort computation, the redundant equation *d.* is necessary, cf. Fig. 3.

In order to synthesize an algorithm for *add*, the goal $\forall x, y\ \exists z\ \ val(x) + val(y) = val(z)$ is proved by structural induction on x and y, the appropriate induction scheme being provided by a specialization of Thm. 3. For example, in case $x = x'::i$, $y = y'::o$, we have to solve the equation $val(x'::i) + val(y'::o) = val(z)$ wrt. z, and the sorted narrowing rule from Thm. 18 is only applicable to equation *i.* since the left-hand side's sort is computed as $s(Even)$. Note that, in order to get the full benefit of the sort calculus, narrowing should be applied only at the root of a term, since then additional sort information is supplied from the other side of the equation; this is the reason for using lazy narrowing. The employed calculus' drawback of admitting only trivially sorted defining equations is overcome by checking the found solutions subsequently for well-sortedness.

Narrowing with equation *h.* instead of *i.* would lead into an infinite branch, trying to solve an equation $s(dup(\ldots)) = dup(\ldots)$. Such infinite branches are cut off by the sorts, especially by the global transformation rules which detect certain kinds of recursion loops. This seems to justify the computational overhead of sort computation. Thanks to the provided proof methodology based on t-sets, new global rules for detecting new recursion patterns can easily be added if required.

The control information provided by the sort calculus acquires particular importance in "proper" narrowing steps, that is, the ones really contributing to the solution term. While conventional narrowing procedures essentially enumerate each element of the constructor term algebra and test whether it is a solution, the presented sort calculus directly approaches the solutions, dependent on the precision of computed range sorts.

The sort algorithms, especially rg, perform in fact simple induction proofs. For example, it is easy to prove by induction that $x + x$ always has sort $Even$, and that sorts $Even$ and $s(Even)$ are disjoint, once these claims have been guessed or intuitively recognized. However, while a conventional induction prover would not propose these claims as auxiliary lemmas during the proof of $\forall x, y\ \exists z\ \ val(x) + val(y) = val(z)$, they are implicitly generated by the sort algorithms. The sort calculus allows the "recognition of new concepts" so to speak, although only within the rather limited framework given by the sort language. In [8], an approach to the automatic generation of more complex auxiliary lemmas is presented based on E-generalization using regular sorts, too.

A prototype support system written in Quintus-Prolog takes a total of 41 seconds user time on a Sparc 1 to automatically conduct the 9 induction proofs, with 135 narrowing subgoals necessary for the development of incrementation, addition, and multiplication algorithms on binary numbers. As a case study from the area of compiler construction, an implementation of sets of lists of natural numbers by ordered son-brother trees has been proved. The algorithm for inserting a new list into a tree is used to construct comb vectors for parse table compression; it is specified as an implementation of $(\{\cdot\} \cup \cdot)$. The use of sorts reduces the search space of the synthesis proof to that of a verification proof, i.e. it uniquely determines all proper narrowing steps or solution constructors. The computed signatures are too complex to be likely to be declared by a user who doesn't know the proof in advance. The application of the sort calculus in the area of formal software development is given more detailed treatment in [4].

6 References

[1] Bachmair, L., Ganzinger, H., On restrictions of ordered paramodulation with simplification, Proc. 10th CADE, LNAI 449, Jul 1990

[2] Bockmayr, A., Beiträge zur Theorie des logisch-funktionalen Programmierens, Dissertation, University Karlsruhe, 1991

[3] Burghardt, J., Eine feinkörnige Sortendisziplin und ihre Anwendung in der Programmkonstruktion, Dissertation, GMD Report 212, Oldenbourg, 1993

[4] Burghardt, J., A fine-grain sort discipline and its application to formal program construction, in: M. Broy, S. Jähnichen (eds.), KORSO – Correct software by formal methods, LNCS, to appear

[5] Comon, H., Equational formulas in order-sorted algebras, Proc. ICALP, Warwick, Springer, July 1990

[6] Echahed, R., On completeness of narrowing strategies, LNCS 298, 1988

[7] Fribourg, L., A narrowing procedure for theories with constructors, Proc. 7. CADE, LNCS 170, p. 259-279, 1984

[8] Heinz, B., Lemma discovery by anti-unification of regular sorts, TU Berlin, Technical Report 94-21, 1994

[9] Hölldobler, S., Foundations of equational programming, LNAI 353, Aug 1989

[10] Leitsch, A., Deciding Horn clauses by hyperresolution, Proc. 2nd Workshop in Computer Science Logic, LNCS 440, p. 225-241, 1989

[11] Mishra, P., Towards a theory of types in PROLOG, Proc. 1984 Int. Symp. on Logic Programming

[12] Schmidt-Schauß, M., Computational aspects of an order-sorted logic with term declarations, Univ. Kaiserslautern, Dissertation, April 1988

DISCOUNT: A system for distributed equational deduction

Jürgen Avenhaus, Jörg Denzinger, Matthias Fuchs

Fachbereich Informatik, Universität Kaiserslautern
Postfach 3049, 67653 Kaiserslautern
E-mail: { avenhaus, denzinge, fuchs }@informatik.uni-kl.de

1 Distributed deduction based on the teamwork method

The teamwork method ([De93], [AD93]) is a concept for distributing deduction systems on several processors. Its behavior is inspired by the behavior of a project team in a company. The general idea is, for a given problem, to explore the search space simultaneously in different directions and to allow for cross-fertilizations. This is accomplished by running a team of *experts* independently on several processors during a *working phase*. Each working phase is followed by a *team meeting*. Communication among the experts is restricted to these team meetings.

During a team meeting each expert is evaluated by a *referee*. The task of a referee is twofold. Firstly, it evaluates the success of its associated expert regarding the given problem in general. Secondly, it extracts the best results produced by its expert. This information is passed to the *supervisor*. The supervisor then determines the best expert of the preceding working phase (the winner), incorporates the selected results of the losers into the winner's data base, creates a new team, transmits the updated data base of the winner to the members of the new team and this way starts a new working phase. For the first working phase the initial problem is transmitted.

The combination of cooperation (good results are shared) and competition (only the winner can keep all its results) are the distinctive properties of the teamwork method. On the one hand, the "explosion" of the search space is avoided by suppressing most of the results of the inferior experts. On the other hand, in many cases it becomes possible to gain access to important points in the search space that cannot be reached by one expert alone in a reasonable amount of time.

In section 2 of the paper we describe the DISCOUNT system ([DP92]), an implementation of the teamwork method for purely equational deduction. We demonstrate the power of the system by reporting experimental results in section 3.

2 The DISCOUNT system

The DISCOUNT system (**DIS**tributed **CO**mpletion **U**si**N**g **T**eamwork, [DP92]) is a distributed prover for equational theorems written in C. It is based on the *unfailing Knuth-Bendix completion procedure* (UKB-procedure, [KB70], [HR87],

[BDP89]). DISCOUNT can be used for proving in the "usual" sequential way, i.e., employing a single expert, and for applying the teamwork method, i.e., employing a team of experts. We shall now discuss the implementation issues concerning the realization of the teamwork concept.

DISCOUNT has originally been designed for a network of work stations (SPARCstation ELC), but recent works have shown that it can also be implemented on multiprocessor architectures with the expected gains in efficiency because of faster communication. Communication is an important component of all parallelized and distributed procedures and has as such played a major role in the development of the teamwork concept. Communication must be kept below a reasonable level in order to prevent the loss of efficiency gains obtained by distribution. We shall now sketch the realization of the principal procedure underlying DISCOUNT in the light of efficient distribution so that we can see where communication effort occurs.

When DISCOUNT is invoked with n experts it starts as a single process P_0. P_0 assumes the role of the (current) supervisor. After reading the *problem file*, which contains all information concerning the proof task (axioms, goal, reduction ordering etc.), and the *configuration file*, which holds all information concerning the configuration of the team (experts that may work in the team, the associated referees etc.), P_0 starts $n-1$ processes (in general each on a different processor). In order to be flexible and to minimize transmission of data, each process can work in the following modes: "supervisor", "expert" and "referee".

During a working phase, each process works in mode "expert" realizing an expert. An expert is a basic prover, here the UKB-procedure equipped with a specific heuristic for selecting the next equation (critical pair) to be worked on. At the beginning of a team meeting each process switches to mode "referee" realizing the referee of the respective expert. This architecture renders superfluous the transmission of data between an expert and its referee. The evaluation of the overall success of an expert is expressed by an integer number (computed by its referee). These numbers are transmitted to the process hosting the current supervisor, i.e., this process switches to mode "supervisor". (The communication effort of this transmission is negligible.) The supervisor then determines the winner using the numbers passed to it, whereupon the process that hosted the winning expert takes over the role of the supervisor. This way there is no need for transmitting the data base of the winning expert at this point. All other processes stay in or switch back to mode "referee" in order to select good results. These results are sent to the current supervisor which incorporates them into its (i.e., the winner's) data base and composes a new team. At this moment the main communication occurs: The current supervisor has to send the updated data base (the actual problem) to all members of the team. This data base may grow to be as large as tens or even hundreds of thousands of equations. The transmission is accomplished via broadcast ([Li93]) in order to be independent of the size n of the team. Before the transmission starts, all experts switch to mode "expert" so that they can preprocess incoming data according to their heuristics. After having completed the transmission the supervisor also switches to mode "expert" and a new working phase begins with all n processes working as experts.

The description above shows that communication is restricted to team meetings. This centralization offers a high degree of transparency. Furthermore, by refraining from exchanging information during working phases experts can do their work in a highly efficient manner. Nevertheless the DISCOUNT system (as well as the teamwork approach in general) captures the advantages of parallel computation.

Figure 1 illustrates the data flow occurring between two team meetings. In this example, the team consists of three processes hosted by three (distinct) processors. Dotted lines represent virtual data flow which is conceptually present, but does not take place in reality because it is effected by changing the role of the respective process. Straight lines represent real data flow, i.e., the transmission of data from one process to the other.

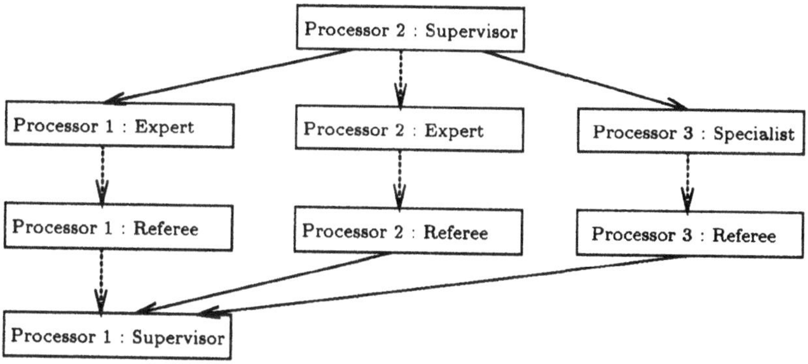

Figure 1: real and virtual data flow between two team meetings

We have now laid down the fundamental working method of DISCOUNT. Before concluding this section, we shall take a closer look at the main components of DISCOUNT.

First of all, experts are the only components that actually work on finding a proof. As we already mentioned above, all experts are based on the UKB-procedure. They only differ in the heuristics employed for selecting the next critical pair (and possibly in the reduction ordering used). At the very beginning, the initial supervisor (P_0) retrieves a set of m experts ($m \geq n$) from the configuration file. P_0 sends the complete set of m experts to each of the other processes $P_1, \ldots, P_{n-1}$, which are also informed by P_0 which one of these experts they are to host, respectively. The current supervisor is allowed to select n of these m experts when composing a (new) team. At first, P_i hosts expert i. Later on, experts may be substituted with experts currently not in use (if $m > n$). But two different processes never host identical experts, i.e., if P_{i_1} and P_{i_2} host experts j_1 and j_2, respectively, then $j_1 \neq j_2$ if $i_1 \neq i_2$. Replacing experts is controlled by the supervisor during team meetings and is based on the assessments of the referees (see [De93] for details). Exchanging an expert is simply effected by the supervisor sending a new index k to the process concerned.

In our current implementation we have 14 generic experts with many parameters that have to be instantiated to get "real" experts. Besides statistical criteria (like sizes of terms) some of these generic experts also use (semantic) interpretations of terms or goal-oriented selection methods (see [DF94]) for choosing the next critical pair.

There is a referee associated with each of the m experts. This information about expert–referee pairs is also stored in the configuration file and is made available to all processes as well. However, the referee of a given expert can be changed by the supervisor. As previously sketched, referees perform two tasks. First, they assess the system of their associated expert, i.e., they evaluate the quality of the set of rules, equations and critical pairs. This evaluation is based on statistical criteria such as the sizes of these sets and data gathered during the most recent working phase such as, for instance, the number of reductions performed. On account of these assessments the supervisor names the winning expert, whereupon the second task of the referees associated with the "inferior" experts begins. Good results, i.e. good rules and equations, must be selected. This selection is also based mainly on statistical data. For instance, rules and equations that entailed many simplifications (e.g., reductions or subsumptions) are generally to be considered as profitable.

Finally, we would like to point out that DISCOUNT can fully protocol all its actions during a proof run using either only a single expert (*sequential mode*) or a team of experts (*team mode*). In the latter case, we prefer to write this protocol during a so-called reproduction run that uses a small log file written during the team meetings. In case a proof could be found the resulting files can be sequentialized, analyzed and transformed. We have implemented a component (see [DS94a] and [DS94b]) that structures the proof by determining appropriate lemmas, thus generating a proof easy to read by humans (ASCII or LaTeX output). The component responsible for generating lemmas is supported by the referees since the results they select very often are promising candidates. Structuring the proof with lemmas does not only enhance readability significantly, but also is indispensable if the proof is complex. In that case merely writing down the chain linking the left side of the goal with the right side may fill several pages and hence is very difficult to comprehend.

3 Experimental results

In this section we want to present some results obtained with DISCOUNT in team mode. They show that the teamwork method is not only able to lead to significant speed-ups for many examples, but also to improve the power of a prover by finding proofs that could not be found in sequential mode.

Each row of the following table lists the name of an example, the run time of the best known team (using two processors) and the run time of the best known single expert of our system (which is not necessarily a member of the respective team). An entry '—' indicates that no proof could be found within three hours. Run times are given in seconds. Timing was done on SPARCstations ELC.

Example	**team**	**best single expert**
bool5b	72.9	—
sa2	10.7	—
herky3	6.8	16.1
luka3	81.7	—
ring	308.0	3019.0
p2.a	5.4	79.5
p2.b	5.4	—
p8.b	56.8	—
p9.a	8.7	19.68
p9.b	8.4	51.0
p10	23.2	—

Table 1: run times for teams and single experts

The examples listed in table 1 cover a wide range of problems. Example *bool5b* states that associativity follows from the remaining axioms in boolean rings. *sa2* (single axiom for a group) can be found in [BH95], *herky3* (ternary boolean algebra) in [Zh93]. The example *luka3* uses the equational axiomatization of the propositional calculus presented in [Ta56] to prove $(\neg x \rightarrow \neg y) \rightarrow (y \rightarrow x)$ (see also [AD93]). Example *ring* states that a ring satisfying $x^3 = x$ is commutative (see [St84]). The examples *p2.a*, *p2.b*, *p8.b*, *p9.a*, *p9.b* and *p10* are taken from the domain of lattice ordered groups ([DF94]). We must emphasize that DISCOUNT does not use AC-completion or any other form of specialized treatment of the AC property. This is relevant for both the example *ring* and the examples related to lattice ordered groups.

The table shows that there are many examples which we could not solve sequentially in a reasonable time, but which could be solved very fast using the teamwork method. This indicates that the cooperation of experts exploring different parts of a search space leads to cross-fertilizations that are very valuable. Furthermore, we found that the combination of competition and cooperation among experts really leads to manageable search spaces for the team. For an extended discussion of the benefits of teamwork for theorem provers see [DS94a]. A special benefit for equational deduction, namely goal-oriented deduction, is discussed in [DF94].

The executables of the DISCOUNT system (the prover and various tools for analyzing and processing proofs) for Sun SPARCstations (operating system SunOS 4.1) are available via ftp. For further information please contact Matthias Fuchs (fuchs@informatik.uni-kl.de).

References

[AD93] **Avenhaus, J.; Denzinger, J.:** *Distributing equational theorem proving,* Proc. 5th RTA, Montreal, LNCS 690, 1993, pp. 62-76.

[BDP89] **Bachmair, L.; Dershowitz, N.; Plaisted, D.A.:** *Completion without Failure,* Coll. on the Resolution of Equations in Algebraic Structures, Austin (1987), Academic Press, 1989.

[BH95] **Bonacina, M.P.; Hsiang, J.:** *The clause-diffusion methodology for distributed deduction,* Fundamenta Informaticae, Special issue on term rewriting systems, D.A. Plaisted (ed.), in press

[De93] **Denzinger, J.:** *Teamwork : A method to design distributed knowledge based theorem provers (in German),* Ph.D. thesis, University of Kaiserslautern, 1993.

[DF94] **Denzinger, J.; Fuchs, M.:** *Goal oriented equational theorem proving using teamwork,* Proc. 18th KI-94, Saarbrücken, LNAI 861, 1994, pp. 343-354; also available as SEKI-Report SR-94-04, University of Kaiserslautern, 1994.

[DP92] **Denzinger, J.; Pitz, W.:** *Das DISCOUNT-System: Benutzerhandbuch,* SEKI working paper SWP-92-16, Universität Kaiserslautern, 1992.

[DS94a] **Denzinger, J.; Schulz, S.:** *Analysis and Representation of Equational Proofs Generated by a Distributed Completion Based Proof System,* SEKI-Report SR-94-05, University of Kaiserslautern, 1994.

[DS94b] **Denzinger, J.; Schulz, S.:** *Recording, Analyzing and Presenting Distributed Deduction Processes,* Proc. PASCO '94, Linz, 1994, pp. 114-123.

[HR87] **Hsiang, J.; Rusinowitch, M.:** *On word problems in equational theories,* Proc. 14th ICALP, Karlsruhe, LNCS 267, 1987, pp. 54-71.

[KB70] **Knuth, D.E.; Bendix, P.B.:** *Simple Word Problems in Universal Algebra,* Computational Algebra, J. Leech, Pergamon Press, 1970, pp. 263-297.

[Li93] **Lind, J.:** *Sicheres Broadcasting,* Projektarbeit, Fachbereich Informatik, Universität Kaiserslautern, 1993.

[St84] **Stickel, M.E.:** *A case study of theorem proving by the Knuth-Bendix method: Discovering that $x^3 = x$ implies ring commutativity,* Proc. CADE 7, Napa, CA, USA, 1984, LNCS 170, pp. 248-258

[Ta56] **Tarski, A.:** *Logic, Semantics, Metamathematics,* Oxford University Press, 1956

[Zh93] **Zhang, H.:** *Automated proofs of equality problems in Overbeek's competition,* JAR 11, 1993, pp. 333-351.

ASTRE: Towards a Fully Automated Program Transformation System

F. Bellegarde*

Pacific Software Research Center
Oregon Graduate Institute of Science & Technology
bellegar@cse.ogi.edu

1 Introduction

It has often been said that functional programs are constructed using functions as pieces. Data structures such as lists and trees are the glue to hold them together. This compositional style of programming produces many intermediate data structures. One way to circumvent this problem is to perform *fusion or deforestation* on programs. Deforestation algorithms (elimination of useless intermediate data-structures) [6, 10] do not recognize that an expression contains two or more functions that consume the same data structure. These functions can be put together in a tuple as a single function that traverses the data structure only once. This tactical is usually called two-loops fusion or two-loops tupling since it is implemented by using a tupling technique. It has been pointed out by Dershowitz [8] that an fold-unfold methodology [5] can be controlled by a completion procedure. Following this idea, the transformation system Astre [2] is based on completion procedures.

2 The Astre System

The transformational approach to the development of programs is attractive for writing small components of large software systems. This approach, to be effective, must be fully automated so that it is not necessary to be an expert in transformation strategies to use the transformational approach for software design. A prototype of a fully automated mode of Astre is a component of the tool suite that support a Method for Software Design for Reliability and Reuse developed in the Pacific Software Research Center [1]. The tool suite provides a translation of ML programs into a rewrite system input of Astre. It includes an implementation of the Chin and Darlington's specialization algorithm [7]. for conversion to first-order. The result is always an orthogonal (left-linear and non-overlapping) constructor-based (a constructor-based system of equalities is similar to set of definition equalities with pattern-matching arguments in functional programming) rewrite system R_0. In Astre, synthesis by completion is used as

* The author was supported by a contract with Air Force Materiel Command (F19628-93-C-0069).

a mechanism to transform R_0 into a sequence of orthogonal, terminating and constructor based rewrite systems $R_1, R_2, \ldots, R_n$ to get a new, semantically e-quivalent ML program P_n which is more efficient. In P_n functions are presented by a set of mutually recursive functions with pattern-matching arguments. A fully automatic version of Astre automatizes deforestation and two-loops fusion strategies. A semi-automatic mode authorizes the user to input laws to facilitate a deforestation. Number of issues occur to automate the synthesis process.

- Generation of useless critical pairs is the major drawback for using completion in its application to synthesis. Astre carefully controls the production of critical pairs hence ensuring termination of the completion [4].
- Astre controls the orientation of the critical pairs into rules as required by the transformation strategy. It guarantees that termination is preserved for a constructor-based orthogonal rewrite system [4].
- Furthermore, given a tactical for transformation, Astre ensures the termination of the sequence of syntheses from the source rewrite system R_0 into the succession of synthesized rewrite systems $R_1, R_2, \ldots$ [4].
- Synthesis rules introduce a new function to synthesize. In the fold-unfold methodology synthesis rules are called *definition rules* or *eurekas rules* because they are introduced through the insight of a clever user. Mechanisms to generate automatically a synthesis rule for deforestation and two-loops fusion strategies are presently implemented in Astre.
- Moreover a set rewrite rules (inductive theorems of R) can be input to the synthesis process in the semi-automatic mode. These *laws* usually facilitate the process in the fold-unfold method.

However, we have noticed that most of the situations that requires laws in a deforestation can be simply handled by introducing additional synthesis rules like $length(append(x, y)) \rightarrow h(x, length(y))$. Such a synthesis rule helps the symbol *length* to go down to consume a term that substitutes the variable y as well as the inductive law $length(append(x, y)) = append(length(x), length(y))$ can do. This technique does not work for pushing *length* to go down towards the inductive variable x. If this is needed the deforestation fails. Let us compare the results given by the fully automatic version of Astre and the semi-automatic mode on the following example.

3 Example

The pencil and paper transformation of the functional program is presented by S. Thompson [9]. The problem solved by this program is stated as follows by S. Thompson:

> Given a finite list of numbers, find the maximum value for the sum of a (contiguous) sublist of the list.

Numbers can be positive as well as negative integers. Let us begin with the quadratic in the length of the list first-order ML program in Figure 1. where

```
fun My_append My_nil x          = x
    | My_append (C(x, xs)) y    = (C(x, My_append xs y));
fun map_cons x My_nil           = My_nil
    | map_cons x (C(y, ys))     = (C(C(x, y), map_cons x ys));
fun frontlists My_nil           = C(My_nil, My_nil)
    | frontlists (C(x, xs))     = My_append (map_cons x (frontlists xs))
                                  (C(My_nil, My_nil));
fun sublists My_nil             = C(My_nil, My_nil)
    | sublists (C(x, xs))       = My_append (map_cons x(frontlists xs))
                                  (sublists xs);
fun sum My_nil                  = 0
    | sum (C(x, xs))            = I_plus x (sumxs);
fun map_sum My_nil              = My_nil
    | map_sum (C(x, xs))        = (C(sum x, map_sum xs));
fun fold_max (C(x, My_nil))     = x
    | fold_max (C(x, C(y, z)))  = I_max x (fold_max (C(y, z)));
fun maxsub(x)                   = fold_max(map_sum(sublists(x)));
```

Fig. 1. Source Program

I_plus, and *I_max* are library functions. The automatic mode of Astre yields the result presented in Figure 2 in 11.29*s*. Astre performs nine syntheses from which seven successful deforestations, one additional synthesis to help deforestation, and one two-loops fusion. One deforestations fails. It corresponds to an attempt to eliminate the intermediary list produced by *A_sym3* and consumed by *A_sym2_1* in the definition of *A_sym5*. Astre does not consider for deforestation an intermediary list produced by a recursive call like *A_sym3* and consumed by *A_sym1*. Such situations are more relevant to a derecursion tactical and are the sources of failure in the absence of laws. Except these two compositions, the program in Figure 2 is completely deforested. In other words, there is no other intermediary lists produced by a function and consumed by another one. Run-time difference between the source and the transformed program are not significative. Recall that the goal of a deforestation in not to improve the run-time but only elimination of useless intermediate data structures without loss of run-time efficiency. However, if, using the semi-automatic mode, the user cleverly provides the following inductive laws at the level of the second synthesis:

$$\begin{aligned}
map_sum\ (A_sym1\ x\ y\ z) &= My_append\ (map_plus\ x\ (map_sum\ y)) \\
&\quad (map_sum\ z) \\
fold_max\ (My_append\ x\ y) &= I_max\ (fold_max\ x)\ (fold_max\ y) \\
fold_max\ (map_plus\ x\ y) &= I_plus\ x\ (fold_max\ y)
\end{aligned}$$

where *map_plus* is defined by:

$$map\_plus\ x\ My\_nil = My\_nil$$

$$map\_plus\ x\ (C(y, ys)) = My\_append\ (I\_plus\ x\ y)\ (mapplus\ ys)$$

$$
\begin{array}{ll}
\textbf{fun } A\_sym1\ x1\ (C(x2,x3))\ x4 & = C((C(x1,x2)),(A\_sym1\ x1\ x3\ x4)) \\
\quad |\ A\_sym1\ x2\ My\_nil\ x1 & = x1; \\
\textbf{fun } sum\ My\_nil & = 0 \\
\quad |\ sum\ (C(x1,x2)) & = I\_plus\ x1\ (sumx2); \\
\textbf{fun } A\_sym3\ x1\ My\_nil\ x2 & = C((C(x1,My\_nil)),x2) \\
\quad |\ A\_sym3\ x1\ (C(x2,x3))\ x4 & = A\_sym1\ x1\ (A\_sym3\ x2\ x3 \\
 & \quad (C(My\_nil,My\_nil)))\ x4; \\
\textbf{fun } A\_sym6\ x1\ (C(x2,x3)) & = I\_max\ x1\ (A\_sym6\ (sum\ x2)\ x3) \\
\quad |\ A\_sym6\ x1\ My\_nil & = x1; \\
\textbf{fun } A\_sym4\ x1\ x2\ (C(x3,x4))\ x5 & = I\_max\ x1 \\
 & \quad (A\_sym4\ (I\_plus\ x2\ (sum\ x3))\ x2\ x4\ x5) \\
\quad |\ A\_sym4\ x1\ x2\ My\_nil\ x3 & = A\_sym6\ x1\ x3; \\
\textbf{fun } A\_sym2\_1\ x1\ (C(x2,x3))\ x4\ x5 & = A\_sym4\ (I\_plus\ x1\ (sum\ x2))\ x1\ x3\ x4 \\
\quad |\ A\_sym2\_1\ x1\ My\_nil\ x2\ x3 & = x3; \\
\textbf{fun } A\_sym5\ x1\ My\_nil\ x2\ x3 & = A\_sym6\ (I\_plus\ x1\ 0)\ x2 \\
\quad |\ A\_sym5\ x1\ (C(x2,x3))\ x4\ x5 & = A\_sym2\_1\ x1 \\
 & \quad (A\_sym3\ x2\ x3\ (C(My\_nil,My\_nil)))\ x4\ x5; \\
\textbf{fun } A\_sym8\ My\_nil & = ((C(My\_nil,My\_nil)),0) \\
\quad |\ A\_sym8\ (C(x1,x2)) & = \textbf{let } (u,v) = A\_sym8\ x2 \textbf{ in} \\
 & \quad ((A\_sym3\ x1\ x2\ u),(A\_sym5\ x1\ x2\ u\ v))\textbf{end}; \\
\textbf{fun } maxsub\ x1 & = second\ (A\_sym8\ x1);
\end{array}
$$

Fig. 2. Fully Automatic Output

Astre outputs:

$$
\begin{array}{ll}
\textbf{fun } A\_sym4\ My\_nil & = (0,0) \\
\quad |\ A\_sym4\ (C(x1,x2)) & = \textbf{let } (u,v)\ =\ A\_sym4\ x2 \textbf{ in} \\
 & \quad ((I\_max\ (I\_plus\ x1\ u)\ 0), \\
 & \quad (I\_max\ (I\_plus\ x1\ u)\ v))\ \textbf{end}; \\
\textbf{fun } maxsub\ x1 & = second\ (A\_sym4\ x1);
\end{array}
$$

after three deforestations and one two-loops fusion. No deforestation fails. Moreover, the laws allow to improve the complexity. The program is linear in the length of the list. Thompson's pencil and paper's transformation gives the same result modulo the two-loops fusion.

$$
\begin{array}{ll}
\textbf{fun } maxfrontMy\_nil & = 0 \\
\quad |\ maxfront\ (C(x,y)) & = bimax\ 0\ (I\_plus\ x\ (maxfront\ y)); \\
\textbf{fun } maxsubMy\_nil & = 0 \\
\quad |\ maxsub\ (C(x,y)) & = \max\ (maxsub\ y)\ (I\_plus\ x\ (maxfront\ y));
\end{array}
$$

4 Conclusion

Astre has achieved its initial goal: it is fully automatic for deforestation and two-loops fusion of functional programs. Other fully automatic algorithms for deforestation does not include two-loops fusion. They reject all deforestations that necessitate laws but Astre can perform most of them using additional synthesis. Moreover they do not extend easily to include laws or other strategies. The limitation of Astre is the termination obligation of the input rewrite system. Also Astre does not process easily a large amount of rules. The present prototype has been used so far up-to 500 rules input. At this size, it becomes intractable to perform all the syntheses. We plan to automatize derecursion tactical and automatic insertion of laws [3] in a near future.

References

1. J. Bell et al. Software Design for Reliability and Reuse: A proof-of-concept demonstration. In *TRI-Ada '94 Proceedings*, pages 396–404. ACM, November 1994.
2. F. Bellegarde. Program Transformation and Rewriting. In *Proceedings of the fourth conference on Rewriting Techniques and Applications*, volume 488 of *LNCS*, pages 226–239. Springer-Verlag, 1991.
3. F. Bellegarde. A transformation system combining partial evaluation with term rewriting. In *Higher Order Algebra, Logic and Term Rewriting (HOA '93)*, volume 816 of *LNCS*, pages 40–58. Springer-Verlag, September 1993.
4. F. Bellegarde. Termination issues in automated syntheses. Technical report, Department of Computer Science and Engineering, Oregon Graduate Institute, September 1994.
5. R. M. Burstall and J. Darlington. A Transformation System for Developing Recursive Programs. *Journal of the ACM*, 24:44–67, 1977.
6. W. Chin. Safe Fusion of Functional Expressions II: Further Improvements. *Journal of Functional Programming*, 11:1–40, 1994.
7. W. Chin and J. Darlington. Higher-Order Removal: A modular approach. Unpublished work, 1993.
8. N. Dershowitz. Completion and its Applications. In *Resolution of Equations in Algebraic Structures*. Academic Press, New York, 1988.
9. S. Thompson. *Type Theory and Functional Programming*. Addison Wesley, 1991.
10. P. Wadler. Deforestation: Transforming Programs to eliminate trees. In *Proceedings of the second European Symposium on Programming ESOP'88*, volume 300 of *LNCS*. Springer-Verlag, 1988.

Parallel ReDuX → PaReDuX*

Reinhard Bündgen Manfred Göbel Wolfgang Küchlin

Wilhelm-Schickard-Institut, Universität Tübingen,
D-72076 Tübingen, Germany
({buendgen,goebel,kuechlin}@informatik.uni-tuebingen.de)

The PaReDuX system is a collection of parallelized programs to experiment with term completion procedures. It is designed to run on a shared memory multiprocessor and exploits the fine-grained parallelism provided in a multi-threaded environment. PaReDuX features programs for a plain Knuth-Bendix procedure (ptc), a Peterson-Stickel procedure for AC-theories (pac) and an unfailing completion procedure (puc). These programs are parallelized versions of the corresponding ReDuX programs [2].
The implementation of PaReDuX is based on the virtual S-thread system of PARSAC-2 [10] and demonstrates a discipline for programming software for symbolic computation. This discipline releases the programmer from the burden of scheduling parallel tasks and for caring about the subtleties of parallel memory management. Thus the implementer may concentrate on exploiting the logical parallelism of the algorithms.
Our experiences with the parallelization of the Knuth-Bendix procedure and the Peterson-Stickel procedure are described in greater detail in [3, 4] and in [5], respectively.

Parallel Term Completion

Completion and parallel computation each have a great many facets, and the combination of both opens a complex decision space. We think it unlikely that a simple optimal solution exists. Instead, we believe that completion needs to be parallelized on all levels of granularity in such a way that a variety of parallel (and sequential) methods can be combined from a library.
Coarse-grained methods (cf. [1]) run several completion processes in parallel or select several pairs in parallel as new rules. They sometimes exhibit large super-linear speed-ups due to strategy effects. They are typically suitable for distributed memory machines, and these may scale to a large number of processors. However, massively parallel completion systems appear to be hard to build, maintain, and utilize. Moreover, recent analyses [8] of the related Buchberger algorithm indicate that the coarse-grained parallelism in completion may be limited, i.e. the method may not scale. Usually, a long sequence of lemmas must be derived before they lead to a persistent rule. In this case, fine grained methods are an important source of additional speed-ups.
PaReDuX allows for strategy compliant fine-grained parallel completion. For example, several pairs are created, or several terms are reduced, in parallel without changing the overall work. These methods need shared memory machines, which do not scale beyond a few dozen processors. However, these methods exhibit many pleasant features such as reproducibility of results and predictability of performance as processors are added or

* This work was supported by a grant from Deutsche Forschungsgemeinschaft.

taken away. Therefore they can unconditionally replace sequential methods on modern parallel workstations.
Below, we give an indication of performance gains possible under this approach on a four processor workstation. In future work, we plan to use some of this work as a building block within a more coarse-grained, network oriented, completion method. The ultimate target architecture of our system is a network of parallel workstations.

Implementation Paradigms

A major concern of our work is to develop a programming discipline for implementing parallel symbolic software. This is particularly challenging because software for symbolic computation suffers from extreme irregularities in its demand of resources — both time and space. Our system environment relieves the programmer from scheduling parallel tasks and from memory management on a parallel machine. Thus he or she can concentrate on the *logical parallelization* of our problems at hand: concurrent tasks are just *forked* instead of called as a procedure, and memory allocation and garbage collection are hidden as in any sequential LISP-system.
Since our software is intended to run on shared memory machines, the particular advantages of these architectures should be exploited. For symbolic computation this means that we want to have parallel access (if possible without synchronization overhead) to objects stored in the global shared heap. This is important because copying arguments costs extra overhead in time and space. Note that extra space consumed during copying may trigger additional garbage collections that very likely act as sequential bottlenecks. Our programming paradigm also allows to manipulate data structures destructively if this is essential for efficiency.
Another issue is porting existing code to a parallel environment. That is, we want to enable sequential code to run in parallel by changing only a few data structures and adding instructions describing which parts of the code can run independently. We also require that our parallel code can run on a sequential machine. Typically, only a fraction of a program is worth parallelizing. Therefore it is natural to demand that parallel and sequential code can be combined.

The Parallel System Environment

PaReDuX uses the virtual S-thread system developed for PARSAC-2 as an interface to an operating system providing light-weight processes like C-threads [7]. A C thread is an execution context for a C procedure, providing a private register file and a private C stack. For efficient parallel symbolic computation additionally a private portion of the heap is needed together with an appropriate (parallel) garbage collection facility. In PARSAC-2 this is provided by the S-threads system [11]. In addition, the *virtual S-thread system (VS-threads)* [12] allows for lazy S-thread creation. That is, a VS-thread may either be run as a parallel S-thread or as a procedure call depending on the availability of processors. VS-threads are in general very light-weight. They are used to provide

efficient context switches and to overcome system dependent limitations to the maximal number of parallel tasks. For PaReDuX, the memory management of VS-threads had to be modified to allow for global garbage collections. This includes extensions that allow the user to stop all threads. The last feature is also essential for implementing or-parallelism.
Due to automatic processor allocation to virtual threads and the support for list processing in the S-thread system we may abstract from the technical details of parallelization as it is required by our implementation paradigms.
VS-threads have been installed under Unix System V using PCR [15], under Mach using Mach's C-threads and under Solaris 2.3 with a native port to Solaris threads. PaReDuX was tested on a Solbourne 704 with 4 processors and on a 2 processor Sun 10-512.

Parallelizing ReDuX

PaReDuX has been developed by parallelizing the sequential code of ReDuX. To use the parallel system environment described above, ReDuX was automatically translated to SACLIB compatible C-code. That is, we use the SACLIB/PACLIB memory model [9] instead of the ALDES memory model. The changes to the sequential code are twofold: (1) the data structure for terms had to be modified and (2) directives for forking procedures in parallel had to be added.

1. The only modifications of the data structures comprised the representation of variable bindings after an applied substitution. In ReDuX each variable has a binding field that may contain a pointer to a term indicating that the variable is bound by a substitution. In order to allow several parallel threads to bind substitutions (matchers) simultaneously to variables occurring in a rule, an additional level of indirection had to be introduced. Now each (parallel) thread owns a private binding array and each variable of a term associates with an index into the array. These changes allow a shared instance of a rule to be used simultaneously in different normalizations and critical pair computations.
2. To keep different parallel processes balanced, we parallelize using a divide and conquer scheme such that one of the subprocedures can be forked as a parallel thread and the other subprocedure is processed on the local thread. This means that iterative procedures had into be transformed to recursive divide and conquer style procedures. The minimal size of a procedure that shall be parallelized can be specified by two kinds of grain size parameters (recursion depth and input size).

We parallelized reducibility tests and normalizations of a set of equations. Further the computation of the critical pairs of a set of rules has been parallelized. For AC theories we also recursively parallelized the normalization of single terms. These parallelized pieces of code have been incorporated into different completion procedures: plain Knuth-Bendix completion, Peterson-Stickel completion modulo AC, and unfailing completion.
So far our parallel completion schemes are *strategy compliant*. That is, for each task in the sequential program a corresponding task in the parallel program must be performed. This allows us to evaluate our design decisions concerning our programming paradigms

and low level parallelization techniques. It also ensures that important program properties, like correctness and fairness of the strategy, are inherited from the sequential program.

Using the PaReDuX System

The PaReDuX programs expect ReDuX style input for equational specifications and term orderings. In addition, several parallelization parameters may be set to influence the performance of the completion. One such parameter determines the minimum number of critical pairs to be normalized by one thread. Other grain size parameters determine the minimum number of critical pairs to which a subconnectedness criterion is to be applied, and the minimum number of rules for the deduction of new critical pairs, which one thread has to process. The last two parameters act as filters for the normalization procedure. An additional grain size parameter is used to fix an upper bound for the recursion levels in our divide and conquer approach.
Besides the completion result the programs output statistics like completion profiles and the number of threads generated that characterize the parallelizability of the problems. All I/O is done on stdin and stdout. Thus the PaReDuX programs may be used interactively and (using I/O redirection) in batch mode. For experiments with recursively parallelized AC-normalizations a random term generator is provided.

Experiences with PaReDuX

For problems of an appropriate size[2] we obtain speed-ups of about three on a four processor machine compared with the sequential reference implementation. Often tens of thousands of virtual threads were generated during a completion. This means that there is a lot of potential for parallelism on the one hand but on the other hand the size of different parallel tasks may vary extremely. These variations were mostly absorbed by the fine grained parallelism (with lazy task generation).
It also turned out that due to the VS-thread system our parallelization is very robust w. r. t. modifications of grain size parameters (cf. [3]). In particular the restriction of the recursion depth for parallel subtasks is superfluous in practice.
Table 1 shows some runtime experiments on a Solbourne 5/704 computer with four 33 MHz Sparc 2 processors and 48 MByte main memory. The first column specifies the input specifications, the second one the PaReDuX-programs used and the third column shows the times needed by the respective sequential programs. All times given are wall clock times in seconds. Then follow the speed-ups of the parallel programs with 1–4 processors measured against the sequential programs. The last column contains the total number of virtual threads generated during a completion. The specifications used are the following: Z_{22} is a finitely presented group specified as a simulated string rewriting

[2] Too small problems like completing a free group that do not bear enough potential for parallelism resulted in poor speed-ups.

system (generators become unary functions) [1], P_7 [6] and M_{15} [3] are instances of parameterized specifications, both specifications contain binary operators and are non-left-linear. BRG [5] is a finitely presented Boolean ring with atoms describing the relations $a > b$, $a \leq b$, $a = b$, $a \geq b$, $a < b$ and $a \neq b$. AZ_{22} [5] is a finitely presented Abelian group with the same presentation equations as Z_{22}. R_{35} [5] is a specification of those rational numbers generated by the integers and the constants $\frac{1}{3}$ and $\frac{1}{5}$. $Luka_2$ is taken from [14]. The input equations are an equational axiomatisation for propositional calculus by Frege. Lukasiewicz gave another set of axioms of which $Luka_2$ is the second one. $Lusk_4$ proves that in a group with $x^3 = e$ the commutator h(h(x,y),y) is equal to the neutral element e [13]. Z_{22W} [3] specifies the same group as Z_{22} but with a binary group operator and a unary inversion function. For each program all experiments were run with the same set of parallelization parameters. Note that due to strategy compliance there is no super-linear speed-up component.

input specification	program	sequential time	speed-ups on n processors $n=1$	$n=2$	$n=3$	$n=4$	# VS threads
Z_{22}	ptc	83.8	0.96	1.66	2.30	2.80	3112
P_7	ptc	423.7	0.97	1.62	2.20	2.67	34565
M_{15}	ptc	753.8	0.95	1.61	2.14	2.62	54806
BRG	pac	1598.3	0.97	1.91	2.79	3.46	101222
AZ_{22}	pac	232.5	0.91	1.71	2.47	2.97	47412
R_{35}	pac	214.7	0.90	1.76	2.46	2.97	9480
$Luka_2$	puc	742.9	0.93	1.85	2.74	3.55	28678
$Lusk_4$	puc	620.5	0.97	1.91	2.83	3.66	10660
Z_{22W}	puc	2248.0	0.97	1.85	2.62	3.28	26914

Table1. Some benchmarks

Availability

The Solbourne/PCR version (for multiprocessor Sparc systems running under UNIX System V) of PaReDuX is available via anonymous ftp from

ftp.informatik.uni-tuebingen.de in */wsi/SR/PaReDuX*.

Conclusion

Within PaReDuX we have parallelized different completion procedures using a fine-grained multi-threaded parallelization scheme. Experiments with strategy compliant parallelizations have verified our choice of our system environment and our parallelization techniques. Future extensions of PaReDuX will concentrate both on non-strategy-compliant parallelizations and very fine grained parallelizations.

Acknowledgments

We thank Patrick Maier, Michael Sperber and Jochen Walter for their contributions to PaReDuX and the anonymous referees for their comments.

References

1. J. Avenhaus and J. Denzinger. Distributing equational theorem proving. In C. Kirchner, editor, *Rewriting Techniques and Applications (LNCS 690)*, pages 62–76. Springer-Verlag, 1993. (Proc. RTA'93, Montreal, Canada, June 1993).
2. R. Bündgen. Reduce the redex → ReDuX. In C. Kirchner, editor, *Rewriting Techniques and Applications (LNCS 690)*, pages 446–450. Springer-Verlag, 1993. (Proc. RTA'93, Montreal, Canada, June 1993).
3. R. Bündgen, M. Göbel, and W. Küchlin. Experiments with multi-threaded Knuth-Bendix completion. Technical Report 94–05, Wilhelm-Schickard-Institut, Universität Tübingen, D-72076 Tübingen, 1994.
4. R. Bündgen, M. Göbel, and W. Küchlin. A fine-grained parallel completion procedure. In *International Symposium on Symbolic and Algebraic Computation*, pages 269–277. ACM Press, 1994. (Proc. ISSAC'94, Oxford, England, July 1994).
5. R. Bündgen, M. Göbel, and W. Küchlin. Multi-threaded AC term rewriting. In H. Hong, editor, *First International Symposium on Parallel Symbolic Computation PASCO'94*, pages 84–93. World Scientific, 1994. (Proc. PASCO'94, Linz, Austria, September 1994).
6. J. Christian. Fast Knuth-Bendix completion: Summary. In N. Dershowitz, editor, *Rewriting Techniques and Applications (LNCS 355)*, pages 551–555. Springer-Verlag, 1989. (Proc. RTA'89, Chapel Hill, NC, USA, April 1989).
7. E. C. Cooper and R. P. Draves. C threads. Technical Report CMU-CS-88-154, Computer Science Department, Carnegie Mellon University, Pittsburgh, PA 15213, June 1988.
8. J. Faugère. Parallelization of Gröbner basis. In H. Hong, editor, *First International Symposium on Parallel Symbolic Computation PASCO'94*, pages 124–132. World Scientific, 1994. (Proc. PASCO'94, Linz, Austria, September 1994).
9. H. Hong, A. Neubacher, and W. Schreiner. The design of the SACLIB/PACLIB kernels. In A. Miola, editor, *Design and Implementation of Symbolic Computation Systems (LNCS 722)*, pages 288–302. Springer-Verlag, 1993. (Proc. DISCO'93, Gmunden, Austria, September 1993).
10. W. W. Küchlin. PARSAC-2: A parallel SAC-2 based on threads. In S. Sakata, editor, *Applied Algebra, Algebraic Algorithms, and Error-Correcting Codes: 8th International Conference, AAECC-8*, volume 508 of *LNCS*, pages 341–353, Tokyo, Japan, Aug. 1990. Springer-Verlag.
11. W. W. Küchlin. The S-threads environment for parallel symbolic computation. In R. Zippel, editor, *Computer Algebra and Parallelism*, volume 584 of *LNCS*, pages 1–18, Ithaca, NY, Mar. 1992. Springer-Verlag. (Proc. CAP'90, Ithaca, NY, May 1990).
12. W. W. Küchlin and J. A. Ward. Experiments with virtual C Threads. In *Proc. Fourth IEEE Symp. on Parallel and Distributed Processing*, pages 50–55, Dallas, TX, Dec. 1992. IEEE Press.
13. E. L. Lusk and R. A. Overbeek. Reasoning about equality. *Journal of Automated Reasoning*, 1:209–228, 1985.
14. A. Tarski. *Logic, Semantics, Metamathematics*. Oxford University Press, 1956.
15. M. Weiser, A. Demers, and C. Hauser. The portable common runtime approach to interoperability. In *12th ACM SOSP*, pages 114–122, 1989.

STORM: A Many-to-One Associative-Commutative Matcher

Ta Chen*[1] and Siva Anantharaman**[2]

[1] Department of Computer Science
SUNY at Stony Brook, Stony Brook, NY 11794 (U.S.A.)
e-mail: tchen@cs.sunysb.edu
[2] Département d'Informatique
LIFO, Université d'Orléans, 45067-Orléans Cédex 02 (France)
e-mail: siva@lifo.univ-orleans.fr

1 Introduction

Term matching is a fundamental operation in equational and functional programming; and the power of a theorem prover depends heavily on how efficiently this operation is performed, especially when a match has to be found for a given 'subject' from among several 'patterns'. This is *a fortiori* true for provers based on the many variants of the completion procedures (cf. e.g. [3, 4]), where the number of patterns changes constantly. Those resorting to associative-commutative rewrite systems require in fact a suitably modified operation, called (many-to-one) *associative-commutative matching* (or *AC-matching*). In this paper, we describe STORM (StonyBrook Orleans Matcher), which is an implementation of a many-to-one *AC*-matching algorithm proposed in [1]. The essential feature here is a mechanism of *discrimination nets*, often used in non-*AC* theorem proving.

On the one hand, STORM functions as an independent system that accepts input terms and outputs their matches. On the other, it can also be incorporated into theorem provers or term-rewriting systems, and function as a matching subsystem. We will refer to persons who incorporate STORM into their systems as *programmers* and others as *users*.

2 The STORM system

Figure 1 shows the organization of the STORM system. We give a brief description of each component below.

AC-discrimination net. The main data structure in the system is an *AC*-discrimination net. A standard discrimination net for a set of patterns factors out common parts of the patterns and is used as a matching automaton. An *AC*-discrimination is a collection of hierarchically structured standard discrimination

* Research supported in part by NSF grants CCR-9102159, CCR-9404921, CDA-9303181 and INT-9314412 and a grant from the GDR-Programmation (CNRS).
** Research supported in part by a grant from the GDR-Programmation (CNRS).

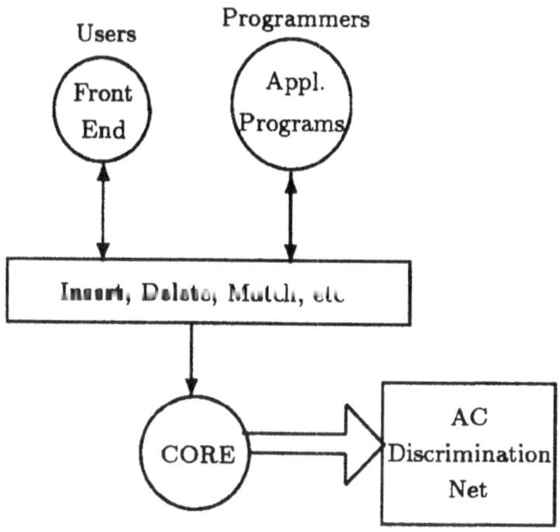

Fig. 1. Organization of STORM

nets. The matching automaton corresponding to an AC-discrimination net uses bipartite matching to combine results from the lower level of the net so as to make them available to the higher level. Such a net is constructed from the patterns that are to match against the subject terms. (cf. [1] for details.)

Application Programs. From the programmer's point of view, STORM is a library of matching functions written in C. Our goal of building this library is two-fold: (i) new theorem proving or term rewriting systems can be developed rapidly without having to be concerned about the basic operations such as term-matching (AC or not). Instead, they can concentrate on proving strategies; (ii) existing systems can also take advantage of the new efficient AC-matching algorithm to improve their performance on AC problems. Since the internal representation of terms in existing (or new) systems may be different from that of STORM, an interface between the two representations must be provided. There are at least two methods to do this. In the first method, programmers are required to provide translation procedures to translate back and forth between the two representations. Though this appears to be simple, it may hamper the performance, since for every call to the subsystem there are two translations. For the version of STORM released earlier (June 1994), such translations were necessary.

The new release is implemented differently so as to cut down this overhead, as follows. Programmers first have to supply a header file `interface.h` which includes macro definitions for symbol and term handling procedures, such as testing whether a symbol is AC, getting the arguments of a term etc. Procedures in the matching library are written using macro names. Each macro in `interface.h` gives the real name of a programmer's procedure or a C-expression. For example, the macro name for testing whether a symbol is AC is `_SYM_IsAC(s)`. Suppose the corresponding procedure supplied by the programmer is `is_varyadic(s)`, then the following definition must be present in *interface.h*.

```
#define    _SYM_IsAC(s)        is_varyadic(s)
```

The CORE is then recompiled and linked with the main system.

CORE. AC-discrimination nets are used in many-to-one AC-matching. Note that we can view the process of using discrimination nets to do matching, as gradually eliminating patterns from a matchset, which initially is equal to the entire pattern set. A pattern may be eliminated because of symbol mismatch or variable substitution inconsistencies. In standard discrimination nets, it is easy to do consistency checking during the matching process. This is a much more complex task for AC-nets, since each occurrence of an AC-variable[3] can acquire several substitutions. To declare that no consistent substitution exists, all these substitutions must be constructed and examined. On the other hand, even if some substitutions are consistent, the pattern can still be eliminated because of a symbol mismatch later in the matching process.

For instance, let f be an AC symbol, $f(x, g(f(x, y, a), h))$ the pattern, and $f(a, g(f(b, c, d, a), k))$ the subject. Suppose terms are traversed left-to-right, the first x can only be instantiated to a while the second is instantiated to b, c, d, $f(b, c)$, $f(b, d)$ then $f(c, d)$. None of them is consistent with the first x, so the matching fails. However, if we ignore the substitutions for x for the moment, the mismatch between h and k also cause a matching failure later. So, the work done in constructing all those substitutions can be saved if we delay consistency checking until all symbols in the patterns and the subject have been compared. Besides although AC-matching is an NP-complete problem, it can be solved in polynomial time if patterns are restricted to linear terms [2] . To exploit these observations, we have designed AC-matching as a two-phase process. In the first phase, the net is used as a filter to obtain patterns which AC-match modulo non-linearity with a subject term. In AC-matching modulo nonlinearity, multiple occurrences of AC-variables are ignored. It does *not explicitly compute* a consistent substitution, but only checks for its existence. However, an AC-discrimination net checks for consistency of substitutions computed for non-AC-variables. In the second phase we verify that substitutions for AC-variables in the filtered set of patterns are consistent. This is where the NP-completeness of the problem comes in. If we can eliminate most of the patterns during the first phase, then the expensive second phase will be invoked sparingly. In such a case, the two phase approach can give substantial performance gains.

Front end. The front end of STORM accepts command lines from the users and accesses the following services provided by the CORE: insertion and deletion of patterns from the AC-discrimination net, finding matching patterns of a subject term and estimates of the number of matching patterns. For example, the

[3] Variables which are arguments of an AC-symbol are called *AC-variables*, others are non-AC-variables

following is a sample session. By default the system gives only one matching substitution for each matching pattern. It can also be directed to give all matches : either one at a time or all at once.

```
% storm                     /* start up */
STORM> ac f.                /* declare f as an AC symbol */

STORM> insert               /* insert the following patterns */
g(a,b), g(a,c), g(a,f(b,x,y)).

STORM> list                 /* list all patterns in the system */
[3] g(a,f(b,x,y))
[2] g(a,c)
[1] g(a,b)

STORM> match                /* match the following term */
g(a,f(a,b,c)).
-----> g(a,f(b,x,y))        /* only one matching pattern found */

  #1                        /* the first matching substitution */
        x -> a
        y -> c
There may be more.          /* a reminder */

STORM> quit
%                           /* back to UNIX shell */
```

Users can also prepare a file of commands and run the system in batch mode. An editor can be called up within the front end to edit such command files. Several unix shell commands such as `ls` are also available.

3 Performance

In this section, we give performance statistics for two series of experiments. (The experiments were all done on a Sparc-10 with 16MB of memory ; and the timing figures are in seconds).

The first experiment is on a collection of term sets taken from typical OTTER applications[5]. In this experiment, the same term set is used both as an indexed set and a query set, i.e. each rule in the set is used to find matching patterns in the same set with and without discrimination nets. In addition to the entire set, we also tested a smaller subset (100 patterns vs. 1000). None of the terms contain AC-symbols in the original sets ; in our experiments however, each term set is first run with all symbols being non-AC then with the topmost symbol of each pattern declared AC. In each run, we give one timing figure for finding the first matching pattern and another for finding all matching patterns. The table n° 1 below lists the results, where column 2 gives the time needed to construct the (AC-)discrimination nets.

Problem	Net	Without Net	With Net	Speedup
CL_pos_100				
First	0.01	6.53	0.24	27.21
All		6.88	0.24	28.67
CL_pos_100, eq in AC				
First	0.02	8.38	1.05	7.98
All		8.83	1.08	8.18
CL_pos_1000				
First	0.14	34.32	1.33	25.80
All		67.98	1.31	51.89
CL_pos_1000, eq in AC				
First	0.15	43.74	11.48	3.81
All		87.62	11.44	7.65
CL_neg_100				
First	0.05	0.16	0.05	3.20
All		0.27	0.06	4.50
CL_neg_100, eq in AC				
First	0.03	0.48	0.15	3.20
All		0.93	0.12	7.75
CL_neg_1000				
First	0.45	13.27	2.01	6.60
All		24.70	1.56	15.80
CL_neg_1000, eq in AC				
First	0.47	45.45	12.74	3.57
All		89.63	12.20	7.35

Table 1. Timings for some benchmark matching problems

The second experiment concerns problems with a dynamic set of patterns, as encountered in completion or theorem proving. It was carried out with the prover 'REVEAL' (into which STORM has been incorporated); the objective is to compare the number of matching attempts needed in obtaining the proof, or the final canonical system, for several AC-benchmark problems, with or without the AC-discrimination net.

Three different strategies were employed for finding redices, that we will label here 'Net', 'Naive-1' and 'Naive-2'. Naive-1 and Naive-2 are approaches for finding redices without the help of discrimination nets. Their difference is that given a term to be simplified, Naive-1 first selects a position in the term, tries all the rules available at that position, and if no redex is found, it goes on to the next position. On the contrary, Naive-2 picks a rule and tries it at all positions of the term to be simplified before trying the next rule. Clearly, Naive-1 uses the same strategy as the Net approach except that it does it without the help of discrimination nets. Table 2 below lists the number of matching attempts for a few benchmark equational (AC-)problems, w.r.t. each one of these strategies.

Here is a brief description of the problems concerned : i) 'grobner' computes by KB-completion the grobner base of a certain ideal in the polynomial ring in 2 variables over integers; ii) 'grpfini30' gives a 'complete set of reductions' for an abelian group of order 30, specified with 3 generators; iii) 'jacobson' is

Problem	Naive-1	Naive-2	Net
grobner	99549	107436	10639
grpfini30	8203	8366	2146
jacobson	357824	225015	63971
moufang	184273	-	23572
robbinh	15762	15395	2807
uqsl2	70179	66562	3651

Table 2. Comparison of number of matching attempts

the usual jacobson's problem, where commutativity is proved for rings with $X^3 = X$; iv) 'moufang' is the problem on (non associative) 'alternative rings' where the sesquilinearity of the associator is proved; v) 'robbinh' proves that any Robbins algebra is a Huntington algebra (so a Boolean algebra) if there exists a c, such that $c + c = c$; and vi) 'uqsl2' proves some complex algebraic relations on certain unitary quantum groups. The *total speedup* with the Net strategy on these problems, is about 1.8 on the average.

4 Remarks and Further Development

The results and speedups that we have obtained, in particular with the integration of the AC-net approach into REVEAL, provide strong evidence that : (i) the two-phase approach described above indeed improves significantly the performance of provers in (many-to-one) AC-matching problems, and (ii) the performance of our AC-discrimination nets compares well with that of standard discrimination nets, when no AC symbols are present.

STORM (like REVEAL) is available from *ftp-lifo.univ-orleans.fr*, via anonymous ftp. Its front end should be soon graphical, and we also expect to add some more basic functionalities, such as AC-subterm matching.

References

1. L. Bachmair, T. Chen, and I.V. Ramakrishnan. Associative-commutative discrimination nets. In *Proceedings of the 4th International Joint Conference on Theory and Practice of Software Development, CAPP/FASE*, pages 61–74, Orsay, France, April 1993. Springer-Verlag LNCS 668.
2. D. Benanav, D. Kapur, and P. Narendran. Complexity of matching problems. *Journal of Symbolic Computation*, 3:203–216, 1987.
3. N. Dershowitz and J.-P. Jouannaud. Rewrite systems. In *Handbook of Theoretical Computer Science*, volume B, chapter 6, pages 243–309. Elsevier, 1990.
4. D. E. Knuth and P. B. Bendix. Simple world problems in universal algebras. In J. Leech, editor, *Computational Problems in Abstract Algebra*, pages 263–297. Pergamon Press, Oxford, 1970.
5. W. McCune. Experiments with discrimination-tree indexing and path indexing for term retrieval. *Journal of Automated Reasoning*, 9:147–167, 1992.

LEMMA: a System for Automated Synthesis of Recursive Programs in Equational Theories

Jacques CHAZARAIN and Serge MULLER

Laboratoire d'Informatique I3S
Université de Nice Sophia Antipolis
Bat 4- 250 Av. Albert Einstein
06560 Valbonne, FRANCE
e-mail:{jmch, smuller}@unice.fr
Phone: 33 93 52 98 23 Fax: 33 93 52 98 21

1 Introduction

The goal of program synthesis is the derivation of a program from a given specification. Most work on program synthesis has focused on program transformation. Here we present LEMMA, a system for automated synthesis of Prolog-like programs from an *implicit* specification. It's based on test set and term rewriting.

A program specification is a description of the input-output relation that defines the legal output for each legal input. A specification may or may not indicates a method for computing the output. We distinguish between *explicit* versus *implicit* specification.

- In the explicit case, the specification already provides a computational method to get the output, the purpose is to transform this method into another one with better properties (efficiency,...).
- In the implicit case, the specification is written down using first order logic of the following form:

  Given: an input predicate $\Psi(x)$ and an output predicate $\Phi(x, y)$ which constitute the specification of the program to be written.
  Determine: a program computing a function $y = f(x)$ such that if x is a vector satisfying $\Psi(x)$, then $f(x)$ is defined and $\Phi(x, f(x))$ is true in the background theory that expresses the known properties of the function symbols in $\Phi(x, y)$.
  We restrict our study to the case of total function ($\Psi(x)$ is the identity) and we assume that the background theory is an equational theory.

To our knowledge, LEMMA is one of the first systems which is able to provide an *automated* synthesis of a variety of recursive programs in equational theories.

2 Basic notions

We briefly recall some basic notions in order to explain our method.

2.1 Equational reasoning

A set E of equations defines a *variety* $Vart(E)$, that is, the class of algebras which are the models of the equations considered as axioms. An equation $l = r$ is a logical consequence of E iff it is valid in all models of E.

2.2 Inductive theorems

The initial model $INIT(E)$ is defined to be the quotient of the algebra $T(F)$ by restriction of the congruence $\vdash_{equ}$ to closed terms. So each congruence class can be represented by a closed term. Initiallity of $INIT(E)$ means that there is a unique F- homomorphism from $INIT(E)$ to every algebra in $Vart(E)$.

The validity of an equation $l = r$ in the initial model requires more than just equational reasoning, some kind of induction is necessary. Unfortunately, there is no general proof theory which captures the semantic notion of the initial model. To overcome this problem, one uses formulas schemata to formulate induction axioms which are used to prove the validity of formulas in the initial model. The validity of a formula Φ in the initial model of E is denoted by $E \vdash_{ind} \Phi$.

2.3 Test set of a rewriting system

A test set is a set of terms which gives a suitable finite description of the closed substitutions to be considered for proofs in the initial model of a set E of equations, provided that E is transformed into a rewriting system $\mathcal{E}$.

The construction of test sets $TS(\mathcal{E})$ for a set of left-linear rules is decidable and can be performed in a relatively efficient way [9].

3 Related work on program synthesis

3.1 Transformational approach for explicit specifications

The goal is to find a sequence of transformations which, when applied successively to the output relation, yields a sequence of equivalent descriptions leading to the desired program ([3]). The mathematical basis for transformation rules is the substitution of equals for equals.

Newer approaches employing both deductive and inductive methods to transform, in an *interactive* way, programs into more efficient ones are proposed by [5] and recently improved by [13], [6] and [10].

3.2 Deductive approach for implicit specifications

The transformation rules are completed by logical rules to derive interactively programs from specifications. Since mid-eighties various approaches to the synthesis of functional programs have been proposed: [11, 12], [7]). Much of Manna and Waldinger's work is concentrated on the combination of resolution-based methods and transformation or rewriting rules. Inductive based methods have also been used to accomplish this task see [14]. More recently there is the work of Biundo [1, 2] who derives recursive programs from specifications using tools and heuristics à la Boyer and Moore. In all these works, except the last one, the proposed methods require interaction with the user. In the Prolog community results about synthesis of Prolog programs exist, but the proof must be given or directed by the user (see for instance [8] and its bibliography).

4 Overview of our method

Due to the six page limit, we'll briefly describe our synthesis process and illustrate it on an example. We refer to [4] for a theorem which is at the basis of our synthesis process.

Our method consists in first computing a test set for the rewriting system which defines the domain. Then we replace each universal variable by each test set element and search for a variable or a Test Set element at the place of existential variables in order to simplify the formula to an equational tautology or to a "smaller" instance of the initial formula.

Suppose we want to construct a program for computing the euclidian division by 2 in the natural integers. It is usually specified by the formula:

$$\forall x \, \exists y, r \quad (x = y + (y + r) \quad \wedge \quad r \leq S(0) = True) \tag{0}$$

We consider the following rewriting system:

$$0 + x \rightarrow x \qquad S(x) + y \rightarrow S(x + y) \qquad x + S(y) \rightarrow S(x + y)$$

$$0 \leq x \rightarrow True \qquad S(x) \leq 0 \rightarrow False \qquad S(x) \leq S(y) \rightarrow x \leq y$$

We also need to introduce the *logical rule* (see [4]): $S(x) = S(y) \Rightarrow x = y$.

We first compute a test set (which, in essence, is a finite description of the initial model of our background theory).

In this case the set $\{0, S(x), True, False\}$ is a test set for R.

Remark. In this example, we need to consider two sorts: integer and boolean. Of course, variables of integer sort can only be replaced by test set terms of the same sort: $0, S(x')$.

We now explain, by hand, the steps which are in fact generated automatically by our system.

The substitution of test set terms in the universal variable x gives the following two formulas to reduce:

$$\exists y, r \quad (0 = y + (y + r) \quad \wedge \quad r \leq S(0) = True) \tag{1}$$

$$\forall x' \, \exists y, r \quad (S(x') = y + (y + r) \quad \wedge \quad r \leq S(0) = True) \tag{2}$$

Case of formula (1). By substituting the existential variable y by 0 we get:

$$\exists r \quad (0 = 0 + (0 + r) \quad \wedge \quad r \leq S(0) = True)$$

which reduces to $\exists r \quad (0 = r \quad \wedge \quad r \leq S(0) = True)$. If now we try $r = 0$ for the the existential variable r then, we get $(0 = 0 \quad \wedge \quad True = True)$, a tautology. Thus we generate the clause: $div2(0,0,0) \longleftarrow$;.

Case of formula (2).

$$\forall x' \exists y, r \quad (S(x') = y + (y + r) \quad \wedge \quad r \leq S(0) = True)$$

There is no successful simplification and it is also a $\forall\exists$-formula. Hence, we regard it as a new lemma and we apply our method to it. We generate the clause:

$$div2(S(x'), y, r) \longleftarrow div2'(x', y, r).$$

In this case two new formulas, corresponding to the test set values for x': 0 and $S(x'')$, are obtained:

$$\exists y, r \quad (S(0) = y + (y + r) \quad \wedge \quad r \leq S(0) = True) \qquad (2.1)$$

$$\forall x'' \exists y, r \quad (S(S(x'')) = y + (y + r) \quad \wedge \quad r \leq S(0) = True) \qquad (2.2)$$

- Since (2.1) cannot be reduced, we try 0 for the existential variable y. We get after simplification:

  $$\exists r \quad (S(0) = r \quad \wedge \quad r \leq S(0) = True).$$

  Now, the first subformula gives $r = S(0)$, which succeeds also in the second subformula. Thus we generate the clause:

  $$div2'(0, 0, S(0)) \longleftarrow ;$$

- It remains to treat formula (2.2)

  $$\forall x'' \exists y, r \quad (S(S(x'')) = y + (y + r) \quad \wedge \quad r \leq S(0) = True).$$

  If we skip some dead end when $y = 0$ and substitute for y the test set value $S(y')$, we get after simplifications:

  $$\forall x'' \exists y', r \quad (S(S(x'')) = S(S(y' + (y' + r))) \quad \wedge \quad r \leq S(0) = True).$$

  By using the logical reduction for S we get:

  $$\forall x'' \exists y', r \quad (x'' = y' + (y' + r) \quad \wedge \quad r \leq S(0) = True).$$

  Now this formula is the initial formula (with alpha renaming).
  So, the formula (0) and the auxiliary lemma (2) are inductive theorems.
  The synthesis process generates the clause:

  $$div2'(S(x''), S(y'), r) \longleftarrow div2(x'', y', r)) ;$$

To sum up, the following set of rules is the desired target Prolog program for computing the euclidian division by 2 in the integers $\mathbb{N}$.

$$\begin{array}{ll} div2(0,0,0) & \longleftarrow ; \\ div2(S(x),y,r) & \longleftarrow div2'(x,y,r) ; \\ div2'(0,0,S(0)) & \longleftarrow ; \\ div2'(S(x),S(y),r) & \longleftarrow div2(x,y,r) ; \end{array}$$

The important point, is that the auxiliary predicate $div2'$ has been generated by our proof method, we didn't have to guess it.

5 Implementation

LEMMA is implemented in Common Lisp. In order to make the synthesis process work, the user has to give the specification, the rules defining the domain knowledge and the signatures of the functions symbols. Then, LEMMA provides the automated process: the proof tree is displayed, so are the resulting clauses if the proof terminates. We've chosen to generate Horn clauses as programs instead of classical rewriting rules. This choice has been dictates by the fact that we can't claim the existence of a *function* f such that $E \vdash_{ind} \Phi(x, f(x))$. Indeed, the specification more generaly represents a *relation* between x and y. Here are some examples

Comparison of two natural integers	$\forall\, x,y\ \exists z$ $(x+z=y) \vee (y+z=x)$	$\Phi(0,0,0) \leftarrow$ $\Phi(0,S(y),S(y)) \leftarrow ;$ $\Phi(S(x),0,S(x)) \leftarrow ;$ $\Phi(S(x),S(y),S(z)) \leftarrow \Phi(x,y,z);$
Opposite in integers	$\forall x\ \exists y$ $x+y=0$	$opp(0,0) \leftarrow ;$ $opp(S(x),P(y)) \leftarrow opp(x,y);$ $opp(P(x),S(y)) \leftarrow opp(x,y);$
Maximum of two natural integers	$\forall x,y\ \exists z$ $(x=z \ \vee\ y=z)$ $\wedge\,(z \geq x = True)$ $\wedge\,(z \geq y = True)$	$max(0,0,0) \leftarrow ;$ $max(0,S(y),S(y)) \leftarrow ;$ $max(S(x),0,S(x)) \leftarrow ;$ $max(S(x),S(y),S(z)) \leftarrow max(x,y,z) ;$
Last element of a list	$\forall L\ \exists M$ $butlast(L)@M = L$	$last(nil,nil) \leftarrow ;$ $last(x.nil,x.nil) \leftarrow ;$ $last(x.y.nil,l) \leftarrow last(y.nil,l) ;$

6 Conclusion

We have presented a system, called LEMMA, for synthesis of program from an implicit specification of the form $\forall x\ \exists y\ \Phi(x,y)$ when the Domain Knowledge (background theory) is given as a set of Equations. Our fully automated method uses purely algebraic simplifications as the only engine to generate target programs. This approach regards program synthesis as a proof-based task and combines features of deduction and induction within a single framework. Our implementation prototype has been applied to the synthesis of a variety of programs computing on domains such as integers or lists. It can be retrieved by anonymous ftp from **piano.unice.fr**, directory **/pub**.

References

1. S. Biundo. A synthesis system mechanizing proofs by induction. In B. Du Boulay et al, editor, *Advances in Artificial Intelligence*, volume 2, pages 287–296. Elsevier Science Publishers B.V. (North-Holland), 1987.
2. S. Biundo. Automated synthesis of recursive algorithms as a theorem prooving tool. In *8th European Conference on Artificial Intelligence*, pages 553–558, Munich (Germany), August 1988.
3. R. Burstall and J. Darlington. A transformational system for developing recursive programs. *JACM*, 24(1):44–67, January 1977.
4. J. Chazarain and E. Kounalis. On proving formulas $\forall x \exists y\ \phi(x,y)$. In *12th Conference on Automated Deduction, CADE-94*, volume 814 of *LNCS*, pages 118–132, Nancy (France), 1994.
5. N. Dershowitz. Synthesis by completion. In *9th International Joint Conference on Artificial Intelligence*, volume 1, pages 208–214, Los Angeles (USA), 1985.
6. N. Dershowitz and E. Pinchover. Inductive synthesis of equational programs. In *8th National Conference on Artificial Intelligence (AAAI-90)*, pages 234–239, Boston (USA), 1990. MIT Press.
7. N. Dershowitz and U. Reddy. Deductive and inductive synthesis of equational programs. *Journal of Symbolic Computation*, 15:467–494, 1993.
8. L. Fribourg. Extracting logic programs from proofs that use extended prolog execution and induction. In J.-M. Jacquet, editor, *Constructing Logic Programs*, chapter 2, pages 39–66. John Wiley & Sons Ltd, 1993.
9. E. Kounalis. Testing for inductive (co)-reducibility in rewrite systems. In A. Arnold, editor, *15th Colloquium on Trees in Algebra and Programming (CAAP 90)*, volume 431 of *LNCS*, pages 221–238, Copenhagen (Denmark), May 1990. Springer-Verlag. Full paper in Theoretical Computer Science, 1992.
10. E. Kounalis. A simplification-based approach to program synthesis. In *10th European Conference on Artificial Intelligence (ECAI 92)*, pages 82–86, Vienna (Austria), August 1992.
11. Z. Manna and R. Waldinger. A deductive approach to program synthesis. *ACM Transactions on Programming Languages and Systems*, 2(1):90–121, 1980.
12. Z. Manna and R. Waldinger. Fundamentals of deductive program synthesis. *IEEE Transactions on Software Engineering*, 18(8):674–704, 1992.
13. U. Reddy. Rewriting techniques for program synthesis. In N. Dershowitz, editor, *3rd International Conference on Rewriting Techniques and Applications*, volume 355 of *LNCS*, pages 388–403, Chapell Hill (USA), 1989. Springer-Verlag.
14. D. Smith. Derived preconditions and their use in program synthesis. In D. W. Loveland, editor, *6th Conference on Automated Deduction*, volume 138 of *LNCS*, New York (USA), 1982. Springer-Verlag.

Generating Polynomial Orderings for Termination Proofs

Jürgen Giesl

FB Informatik, Technische Hochschule Darmstadt,
Alexanderstr. 10, 64283 Darmstadt, Germany
Email: giesl@inferenzsysteme.informatik.th-darmstadt.de

Abstract. Most systems for the automation of termination proofs using polynomial orderings are only *semi-automatic*, i.e. the "right" polynomial ordering has to be given by the user. We show that a variation of Lankford's partial derivative technique leads to an easier and slightly more powerful method than most other semi-automatic approaches. Based on this technique we develop a method for the *automated synthesis* of a suited polynomial ordering.

1 Introduction

A term rewriting system (trs) $\mathcal{R}$ is *terminating* for a set of terms $\mathcal{T}$ if there exists no infinite derivation of terms in $\mathcal{T}$. While in general this problem is undecidable [HL78], several methods for proving termination have been developed, cf. [Der87]. We present a method for *automated* termination proofs using polynomial orderings, which is based on a variant of Lankford's partial derivative technique (section 2). Our method can be used both in a *semi-automatic* and a *fully automated* way (section 3).

2 A Termination Criterion with Variable Coefficients

The use of *polynomial orderings* for termination proofs has been suggested by D. S. Lankford [Lan79] and has been extended to *real* polynomials by N. Dershowitz [Der82]. A *polynomial interpretation* τ associates a real multivariate polynomial $f_\tau(x_1,\ldots,x_n)$ with each n-ary function symbol f. The ordering implicitly defined by a polynomial interpretation τ is called the *corresponding* polynomial ordering $\succ_\tau$ (i.e. $t \succ_\tau s$ iff $\tau(t) > \tau(s)$). To use $\succ_\tau$ for termination proofs, $\succ_\tau$ must be the strict part of a *quasi-simplification ordering* (i.e. $\succeq_\tau$ must be *monotonic* and must satisfy the *subterm property*).

In order to compare *non-ground* terms, τ is extended to interpret variables as variables over the reals. To prove the termination of a trs $\mathcal{R}$ (with *finitely* many rules), $\mathcal{R}$ has to be *compatible with* a polynomial ordering; i.e. for each rule $l \to r$ in $\mathcal{R}$, $\tau(l) > \tau(r)$ must hold for all instantiations of the variables with numbers n that are *greater or equal than the minimal value of a ground term* (i.e. numbers n with $n \geq \min\{\tau(t) \mid t \text{ ground term}\}$) [DJ90][1].

[1] We always assume that there exist ground terms in $\mathcal{T}$.

Consider the trs $\mathcal{R}$ for associativity and endomorphism from [Bel84] and [BL87]. Here $\mathcal{T}$ consists of all terms constructed from the constant a, the unary function symbol map and the binary function symbol $\circ$.

$$(x \circ y) \circ z \to x \circ (y \circ z), \tag{1}$$
$$\mathsf{map}(x) \circ \mathsf{map}(y) \to \mathsf{map}(x \circ y), \tag{2}$$
$$\mathsf{map}(x) \circ (\mathsf{map}(y) \circ z) \to \mathsf{map}(x \circ y) \circ z. \tag{3}$$

To generate a polynomial interpretation we first have to decide on the *maximum degree* of the polynomials. We follow a heuristic from [Ste91] and associate a *simple-mixed*[2] polynomial with each function symbol. So in our example the constant a is associated with a polynomial a_0, the unary function symbol map is associated with a polynomial $\mathsf{map}_\tau(x) = m_0 + m_1 x$ (or $m_0 + m_2 x^2$) and $\circ$ is associated with $\circ_\tau(x, y) = c_0 + c_1 x + c_2 y + c_3 xy$. Here we use a polynomial interpretation τ which maps function symbols to polynomials with *variable coefficients* $a_0, m_0, m_1, c_0, c_1, c_2, c_3$.

Now we have to find an instantiation of the variable coefficients $a_0, \ldots, c_3$ such that $\tau(l) - \tau(r) > 0$ holds for each rule $l \to r$ in $\mathcal{R}$. For the first rule (1) we obtain the following inequality.

$$c_0c_1 - c_0c_2 + (c_1^2 - c_1 - c_0c_3)\, x + (c_2 - c_2^2 + c_0c_3)\, z + (c_1c_3 - c_2c_3)\, xz > 0. \tag{4}$$

The problem is that we cannot directly check whether an instantiation of the variable coefficients $c_0, \ldots, c_3$ makes this inequality valid *for all* $x, z \geq \min\{\tau(t) \mid t$ ground term$\}$. Note that in general this question is undecidable [Lan79]. Therefore we will transform (4) into new inequalities which *do not contain the rule variables* x and z any more. Then for each instantiation of the variable coefficients it is trivial to check whether they satisfy these new inequalities. The invariant of this transformation is that every instantiation of $c_0, \ldots, c_3$ satisfying the new inequalities also satisfies the original inequality *for all* $x, z \geq \min\{\tau(t) \mid t$ ground term$\}$.

Let μ be a new variable and let us assume for the moment that μ is instantiated with a value less or equal than $\min\{\tau(t) \mid t$ ground term$\}$. Then instead of demanding that inequality (4) should hold for all $x, z \geq \mu$, it is sufficient if this inequality holds for $x = \mu$ and if the polynomial on the right hand side of inequality (4) is not decreasing when x is increasing. In other words, the partial derivative of this polynomial with respect to x should be non-negative. Therefore we can replace (4) by the inequalities

$$c_0c_1 - c_0c_2 + (\ldots)\mu + (\ldots)z + (\ldots)\mu z > 0 \text{ (resulting from } x = \mu\text{) and} \tag{5}$$
$$c_1^2 - c_1 - c_0c_3 + (c_1c_3 - c_2c_3)z \geq 0 \text{ (resulting from partial derivation).} \tag{6}$$

By further application of this technique (i.e. demanding that (5) and (6) hold for $z = \mu$ and that their partial derivatives with respect to z are non-negative) we

[2] A non-unary polynomial p is *simple-mixed* iff all its exponents are not greater than 1. A unary polynomial p is simple-mixed if it has the form $\alpha_0 + \alpha_1 x$ or $\alpha_0 + \alpha_2 x^2$.

obtain inequalities without the variables x and z. We proceed analogously for the other rules (2), (3) of $\mathcal{R}$ and obtain inequalities which only contain the variable coefficients $a_0, \ldots, c_3$ and μ, but not the rule variables x, y, z. If an instantiation satisfies these inequalities, then the trs $\mathcal{R}$ is compatible with the corresponding polynomial ordering.

For the elimination of the rule variables x, y, z we have repeatedly used the following two *differentiation rules*.

$$\frac{p(\ldots x \ldots) > 0}{p(\ldots \mu \ldots) > 0, \; \frac{\partial p(\ldots x \ldots)}{\partial x} \geq 0} \qquad \text{(Diff1)}$$

$$\frac{p(\ldots x \ldots) \geq 0}{p(\ldots \mu \ldots) \geq 0, \frac{\partial p(\ldots x \ldots)}{\partial x} \geq 0} \qquad \text{(Diff2)}$$

The differentiation rules (Diff1) and (Diff2) are based on the partial derivative method of Lankford [Lan76]. But Lankford's method can only prove that a polynomial is *eventually positive* (i.e. $p(x_1, \ldots, x_n) > 0$ holds for large enough x_i). Note that it is not sufficient for the termination of $\mathcal{R}$ if there exists a polynomial interpretation τ such that $\tau(l) - \tau(r)$ is eventually positive for each rule $l \to r$ in $\mathcal{R}$. For instance, the trs with the rule $x \to \mathsf{a}$ is not terminating although $\tau(x) - \tau(\mathsf{a})$ is eventually positive for every polynomial interpretation τ.

For $\mathcal{R}$'s termination proof we furthermore have to ensure the *subterm property* and *monotonicity* of the corresponding quasi-ordering. To guarantee the subterm property we demand that $\mathsf{map}_\tau(x) - x \geq 0$ (and the corresponding inequalities for $\circ_\tau$) hold and eliminate the variables x, y by application of the differentiation rule (Diff2).

For the monotonicity, we have to ensure that if x is increasing, $\circ_\tau(x, y)$ is not decreasing. So we demand that the partial derivative of $\circ_\tau(x, y)$ with respect to x is non-negative. In our example we have $\frac{\partial \circ_\tau(x,y)}{\partial x} = c_1 + c_3 y$ and therefore we demand $c_1 + c_3 y \geq 0$. Now we can use (Diff2) again to eliminate the remaining rule variable y. We proceed analogously for map_τ and for the the monotonicity of $\circ_\tau(x, y)$ in its second argument.

Still we have to ensure that the variable μ is really instantiated with a value less or equal than the minimal value of a ground term. Because of the subterm property, the requirement $\mu \leq \min\{\tau(t) \mid t \text{ ground term}\}$ is equivalent to the condition $\mu \leq c_\tau$ for all constants c of the signature. Therefore in our example the instantiation of the variables also has to satisfy the inequality $a_0 - \mu \geq 0$. The following theorem summarizes our termination criterion using polynomial interpretations with (possibly variable) coefficients.

Theorem 1 (Termination Criterion with Real Variable Coefficients). *Let $\mathcal{R}$ be a trs, let τ be a polynomial interpretation with (possibly variable) coefficients. Repeated application of (Diff1) and (Diff2) to*

$$\begin{aligned} \tau(l) - \tau(r) &> 0 && \textit{for all rules } l \to r \textit{ in } \mathcal{R}, \\ f_\tau(\ldots x \ldots) - x &\geq 0 && \textit{for all function symbols } f, \\ \tfrac{\partial f_\tau(\ldots x \ldots)}{\partial x} &\geq 0 && \textit{for all function symbols } f, \\ c_\tau - \mu &\geq 0 && \textit{for all constants } c \end{aligned}$$

yields a unique set of inequalities containing no rule variables any more. If there exists an instantiation of the variable coefficients and the variable μ with real numbers which satisfies the resulting inequalities, then $\mathcal{R}$ is terminating.

In our example the resulting inequalities are satisfied by the instantiation $\mu = 2$, $a_0 = 2$, $m_0 = 0$, $m_1 = 2$, $c_0 = 0$, $c_1 = 1$, $c_2 = 0$, $c_3 = 1$. (This corresponds to the polynomial interpretation given by $\mathsf{a}_{\tau} = 2$, $\mathsf{map}_{\tau}(x) = 2x$ and $\mathsf{o}_{\tau}(x, y) = xy + x$.) Therefore by theorem 1 the termination of $\mathcal{R}$ is proved.

Most systems for "automated" termination proofs using polynomial orderings are *semi-automatic*, i.e. the user has to provide a polynomial interpretation and the system checks whether the trs is compatible with the corresponding polynomial ordering. Of course the termination criterion of theorem 1 can also be applied in a semi-automatic way. Then we use associations to polynomials whose coefficients are *numbers* (instead of variables) and we replace the variable μ by a *number* μ. A comparison with the semi-automatic methods of Ben Cherifa and Lescanne [BL87] and Steinbach [Ste92] leads to the following results[3].

- *If [Ste92] and [BL87] can prove a polynomial p positive, then our method can do so as well.*
- *If our method can prove p positive for all $x_1, \ldots, x_n \geq \mu$, then there exists a $\mu' \geq \mu$ such that the methods of [Ste92] and [BL87] can prove p positive for all $x_1, \ldots, x_n \geq \mu'$. Choosing $\mu' = \mu$ is not always possible.*
- *While the worst case complexity of the systems in [Ste92] and [BL87] is exponential in the number of monomials in p, our method is exponential in the number of its variables.*

3 A Fully Automated Termination Proof Procedure

In theorem 1 we introduced a method to automatically generate a set of inequalities only containing variable coefficients and the variable μ. To prove the termination of a trs $\mathcal{R}$ mechanically we now have to synthesize an instantiation of these variables satisfying the inequalities.

When examining term rewriting systems occurring in the literature we noticed that most termination proofs with polynomial interpretations only use polynomials whose coefficients are 0, 1 or 2. Checking whether a certain instantiation of variables with numbers satisfies the inequalities resulting from theorem 1 can be done very efficiently. Therefore we suggest to apply a "generate and test" approach first which generates all instantiations of the variables with numbers from $\{0, 1, 2\}$ until one of these instantiations satisfies the inequalities. This results in a fully automated termination proof procedure which succeeds for most of those term rewriting systems which are compatible with a polynomial ordering.

[3] In this comparison we do not consider the additional use of the arithmetic-mean-geometric-mean inequality in [Ste92] and extend the method of [BL87] by backtracking and arbitrary minimal value μ.

Nevertheless there do exist term rewriting systems which require a polynomial ordering with coefficients other than 0, 1 or 2. It is decidable whether there exists an instantiation with real numbers satisfying a set of inequalities [Tar51]. But even the most efficient known decision method (the *cylindrical algebraic decomposition* algorithm by G. E. Collins [Col75]) is very time-consuming. For that reason these methods have been rarely used for automated termination proofs.

Therefore we suggest an *incomplete*, more efficient modification of Collins' algorithm. As we know of no trs whose termination proof requires a polynomial interpretation with non-rational real coefficients, we have restricted the algorithm to *rational* instead of *real* numbers which eases the implementation considerably. Moreover, we have introduced execution time limits for each step of Collins' algorithm. If the time limit for the actual step is exceeded, then the algorithm can only use the results of the actual step computed so far and has to carry on with the next step. Now Collins' algorithm is no longer used as a decision method, but only as a *heuristic*.

To sum up, we propose the following termination proof procedure:

1. Construct a set of inequalities as described in theorem 1 (using a polynomial interpretation with possibly variable coefficients).
2. Check whether these inequalities are satisfied by an instantiation with numbers from $\{0, 1, 2\}$.
3. If not, try to prove their satisfiability by a modified version of Collins' algorithm.

Instead of the differentiation rules we could also use a technique from [Ste92] for the elimination of the rule variables x, y, z. But while Steinbach's technique introduces *several* new variables, the advantage of (Diff1) and (Diff2) is that these rules introduce only *one* new variable μ. For the generation of a polynomial ordering compatible with $\mathcal{R}$ we therefore only have to find an instantiation of the variable coefficients and μ.

An alternative approach for the automated generation of the "right" polynomial interpretation has been presented in [Ste91] which can be useful if the number of variable coefficients is small. In these cases Steinbach's method may also be used to search for an instantiation that satisfies the inequalities resulting from theorem 1.

We have presented an efficient, powerful and easy to implement algorithm for termination proofs using polynomial orderings which can be used both in a semi-automatic and in a fully automated way. Our termination proof procedure has been implemented[4] in Common Lisp on a Sun SPARC-2. Table 1 illustrates its performance with some examples. The second row contains the execution time our algorithm needs to generate a polynomial interpretation which is compatible with the trs in the first row.

[4] The implementation and an extended version of this paper are available by anonymous ftp from `kirmes.inferenzsysteme.informatik.th-darmstadt.de` under `pub/termination`.

Example	Time
Nested Function Symbols ([Ste91, Example 8.1])	0.1 sec.
Endomorphism & Associativity ([Bel84], [BL87])	0.1 sec.
Running Example 6.1 in [Ste91] (by A. Middeldorp)	0.2 sec.
Binomial Coefficients ([Ste91, Example 8.8], [Ste92, Example 13])	1.6 sec.
Distributivity & Associativity ([Der87, p. 78])	1.9 sec.

Table 1. Performance of our method.

Acknowledgements

I would like to thank Jürgen Brauburger, Caroline Claus, Alexander Friedl, Stefan Gerberding, Thomas Kolbe, Jens Marschner, Martin Protzen and Christoph Walther for comments and support.

References

[Bel84] F. Bellegarde. Rewriting Systems on FP Expressions that reduce the Number of Sequences they yield. *Symp. LISP & Funct. Prog.*, ACM, Austin, TX, 1984.

[BL87] A. Ben Cherifa & P. Lescanne. Termination of Rewriting Systems by Polynomial Interpretations and its Implementation. *Science of Computer Programming*, 9(2):137-159, 1987.

[Col75] G. E. Collins. Quantifier Elimination for Real Closed Fields by Cylindrical Algebraic Decomposition. In *Proc. 2nd GI Conf. on Automata Theory and Formal Languages*, Kaiserslautern, Germany, 1975.

[Der82] N. Dershowitz. Orderings for Term-Rewriting Systems. *Theoretical Computer Science*, 17:279-301, 1982.

[Der87] N. Dershowitz. Termination of Rewriting. *Journal of Symbolic Computation*, 3(1, 2):69-115, 1987.

[DJ90] N. Dershowitz & J.-P. Joannaud. Rewrite Systems. *Handbook of Theoretical Comp. Science*, J. van Leuwen, Ed., vol. B, ch. 6, 243-320, Elsevier, 1990.

[HL78] G. Huet & D. S. Lankford. On the Uniform Halting Problem for Term Rewriting Systems. Rapport Laboria 283, Institut de Recherche d'Informatique et d'Automatique, Le Chesnay, France, 1978.

[Lan76] D. S. Lankford. A Finite Termination Algorithm. Internal Memo, Southwestern University, Georgetown, TX, 1976.

[Lan79] D. S. Lankford. On Proving Term Rewriting Systems are Noetherian. Technical Report Memo MTP-3, Louisiana Tech. Univ., Ruston, LA, 1979.

[Ste91] J. Steinbach. Termination Proofs of Rewriting Systems — Heuristics for Generating Polynomial Orderings. SEKI-Report SR-91-14, Univ. Kaiserslautern, Germany, 1991.

[Ste92] J. Steinbach. Proving Polynomials Positive. In *Proc. 12th Conf. Foundations Software Technology & Theoretical Comp. Sc.*, New Delhi, India, 1992.

[Tar51] A. Tarski. *A Decision Method for Elementary Algebra and Geometry*. University of California Press, Berkeley, 1951.

Disguising recursively chained rewrite rules as equational theorems, as implemented in the prover EFTTP Mark 2 *

M. Randall Holmes

Boise State University, Boise, Idaho, USA

1 Introduction

This paper describes an approach to writing a tactic language for an equational theorem prover which is implemented in the author's theorem prover EFTTP Mark 2 (research on this prover is supported by a grant from the US Army Research Office, to which the author is very grateful). Since the official name of the prover is long, we abbreviate it as Mark2 hereafter.

Mark2 is written in ML (SML/NJ and Caml Light) and we expected at the outset to follow the lead of other provers (HOL,Nuprl; see [3], [1], respectively) in using ML to write tactics. This turned out not to be needed. Tactics in Mark2 are expressed as equational theorems. They are proven in the same way that any equation is proven and they are stored in the same way as other theorems. An interesting side effect of this is that Mark2 is independent of ML; we are in the process of implementing it in C++ with an eye to improving efficiency, which would not be possible if we were dependent on ML for tactic writing.

The source code for the SML/NJ version of the prover and a limited manual are available at (http://math.idbsu.edu/faculty/holmes.html); the author may be contacted at (holmes@math.idbsu.edu).

2 A Brief Introduction to the Mark2 Prover as a Dumb Equational Prover

The philosophy of the Mark2 prover is to implement purely algebraic reasoning with as few distractions as possible. Purely equational reasoning is as powerful in principle as the apparently more complex forms of reasoning embodied in first order and higher order logics, in the presence of suitable axioms (see, for example, [8], [7], [2] or our [5]). The premise of our project is that, in the presence of automated assistance, purely equational reasoning should prove as effective in practice as other logical frameworks. We do not claim to have proven our point with the present implementation of Mark2 (which lacks various obvious optimizations)!

We designed the prover to keep the notion of substitution as simple as possible. We avoid the use of bound variables. There is no built-in system of absolute

* Supported by the U. S. Army Research Office, grant no. DAAH04-94-G-0247

types, although the definition facility enforces a system of relative typing analogous to that in Quine's "New Foundations" (see our [4] or unpublished [6]).

The syntax of the input language is straightforward; we note that infix precedence is not supported at the moment (all operators associate to the right as far as possible); we will supply parentheses for clarification that the prover itself might not give. Two notational conventions which are not usual are noted: function application is represented by the infix @, and the constant function with value a term T is represented by $[T]$. Variables begin with question marks.

A session with the prover (in its simplest form) begins with the user entering a term, intended to serve as the left side of a theorem to be proved, then proceeding to apply equational theorems as rewrite rules until a final term is reached, at which point the user issues a command recording a theorem, which will then be usable in the same way as the theorems already available.

One problem with this very basic kind of proving is the difficulty of controlling the application of theorems to subterms. We provide the ability to move around the tree representing the term in order to make this easier. The user has the option of applying a theorem in the current theory as a rewrite rule (in either direction) to the currently accessible subterm (only at the top level, not to any of its subterms) or of moving his attention to a different subterm: the basic commands provided are "move to the left subterm", "move to the right subterm", "move up (to the smallest term properly containing the current subterm)" or "move to the top"; more powerful "movement" commands are also available.

This is a very laborious way to prove theorems. Our original intention was to add tactics written in ML which manipulated subterms in more powerful but still safe ways; this turned out to be unnecessary for the most part.

3 The Transformation of a Dumb Equational Prover into a Programming Environment

The underlying idea which led to the development of the tactic-writing method used in Mark2 is simple. We decided to add a device for introducing names of theorems into terms, to signal our eventual intention of applying a given theorem to a given subterm. Suppose "COMM" is the name of the theorem $?x+?y=?y+?x$, the commutative law of addition. The term $COMM => (?a+?b)+?c$ would have the same denotation as the term $(?a+?b)+?c$, but would convey to the reader the additional information of our intention to apply the theorem COMM to it; the infix => was introduced to construct this kind of term, along with an infix <= which signals the intention to apply a theorem in reverse. A term like $COMM =>?x*?y$ would have the same denotation as $?x*?y$ but would express the odd intention to apply the commutative law of addition to this term. Our original intention went no farther than to allow the introduction of embedded theorem names at various points in a term prior to issuing a command "execute" which would apply all of the theorems thus embedded in the current subterm (where application was possible). The "execute" command simply removed embedded theorem names which did not apply or which it did not recognize. Note that embedded theorems are applied only to the top level term to which they are attached, not to subterms as is more usual (though this effect can be achieved using the tactic-writing method described below; it usually proves more efficient to exercise some control over what subterms are taken to be targets of the rule).

Unexpectedly, it becomes possible to prove "theorems" with interesting behavior. For example, consider the theorem ZERO: $0+?x = ?x$, which one might expect to find in a theory containing the axiom COMM cited above. One can certainly prove a theorem COMMZERO: $?x + 0 = ?x$, using COMM and ZERO together. But the theorem EITHERZERO: $?x = (ZERO => COMMZERO => ?x)$ is a different matter. Certainly it is true—embedded theorem names have no effect on the values of terms. If the command "execute" is defined so as to aggressively carry out all theorem applications it encounters, including those introduced in the course of applying previously applied theorems then the effect of the theorem EITHERZERO will be to apply the identity axiom for addition in either its left or right form, if appropriate (actually, it is possible that it will be applied twice). This is rather surprising behaviour for what appears to be a single equational theorem.

The following is certainly true: $(0+?x) = (ZEROES => ?x)$ So we can prove this theorem and give it the name ZEROES (a declaration of ZEROES as a prospective theorem is required before the proof). The effect of this "theorem" when applied to a term is to eliminate any number of zeroes added on the left and then (equally importantly) stop. The reason that it continues to be applied as long as it sees a zero added on the left is that it introduces an application of itself each time it is successfully applied. This recursion is well-founded because an application of ZEROES eventually encounters a subterm not of this form and fails to be applicable, whereupon it is simply removed by the "execute" process.

The exact behaviour of such theorems depends on the way in which the "execute" command is implemented. The current version proceeds in a "depth-first" manner, applying every embedded theorem that is present initially or is generated by earlier theorem applications, always applying innermost embedded theorems first. A referee pointed out correctly that there is an analogy between the choice of execution order here and PROLOG implementation issues. Confluence holds, so execution order will not affect results of computations which terminate, but termination and efficiency could be affected by the choice of a different order. We are investigating a different execution order more appropriate for a parallel machine.

Since all steps allowed in the execution of theorems interpreted as programs are applications of theorems in the current theory, programs are safe in the sense that they will not lead the prover to prove false theorems. They may fail to prove *any* theorem, by failing to terminate.

4 An example: the construction of a limited abstraction algorithm

The first major application of this technique was the implementation of abstraction and reduction algorithms. These were needed to support the avoidance of bound variables in Mark2, which requires the use of synthetic abstraction terms instead of λ-terms. We describe the way in which an abstraction algorithm for a limited class of terms is implemented.

Before we describe this algorithm, we need to describe two refinements of the tactic-writing language.

We mentioned a problem which arises with the theorem EITHERZERO defined above: if EITHERZERO is applied to a suitable term (e.g., $0+?x+0$), two successive applications of the identity for addition may occur (on different sides) which is a little untidy. The solution is to introduce variants =>> and <<= of the theorem embedding infixes with the property that the indicated theorem is to be applied only if the immediately preceding theorem application fails. This is used to build lists of theorems to be applied as alternatives, suppressing the danger that more than one of them might be applied in sequence. EITHERZERO would now have the form $?x = (COMMZERO =>> ZERO =>?x)$, where we could be certain that there would be only one application of the identity for addition.

The second refinement is the introduction of parameterized theorems. Consider the theorem CONST which defines the expected behavior of constant functions: CONST: $[?x]@?y =?x$. Now consider application of this theorem in reverse: the result of $CONST <=?a$ is $[?a]@???y$ (the system supples the two additional question marks when it is forced to create a variable, in order to avoid unintended collisions of variables). A more sophisticated version of the converse of CONST would be the following: $(REVCONST@?y) : ?x = [?x]@?y$. The parameterized theorem is treated as a "function" applied to its argument. An example: the term $(REVCONST@2) =>?x$ would become $[?x]@2$. The use of parameters is convenient in other contexts where it is desirable to provide information to a tactic which is not contained in the term to which it is applied. Tactics can take both objects of the theory and other theorems or tactics as arguments; they can also take multiple arguments (in different styles: $(TACTIC@arg1)@arg2$ and $TACTIC@(arg1, arg2)$ are both possible forms).

The tactic ABSTRACT is intended to have the following effect: a term $(ABSTRACT@?x)@T$ (where T is a complex term, usually containing occurrences of $?x$) should take the form $U@?x$ when executed, where U is expected to contain no occurrences of $?x$. One should think of U as $(\lambda ?x)(T)$. If T is not a term of a suitable form, the algorithm may fail, in the sense that U will not be free of occurrences of $?x$. It should also be noted that the argument passed to ABSTRACT does not need to be a variable; one term can be expressed as a function of another term of arbitrary form using the ABSTRACT tactic.

Recall that the infix "@" is used to represent function application. The theory in which ABSTRACT is implemented also includes an identity function

Id, an infix "," implementing pairing, a infix ";" implementing function product (with defining axiom PROD: $(?f;?g)@?x = (?f@?x), (?g@?x))$ and an infix "@@" implementing composition (with defining axiom COMP: $(?f@@?g)@?x = ?f@(?g@?x)$.

The theorem ABSTRACT has the form $(ABSTRACT@?x) : ?y = (ABSCONST@?x) \Rightarrow\!> (ABSCOMP@?x) \Rightarrow\!> (ABSPROD@?x) \Rightarrow\!> (ABSID@?x) \Rightarrow ?y$. This is not very informative; it tells us that one of a sequence of alternative tactics to which the same argument will be passed will be applied to the target term (which may have any form).

The theorem ABSID first applied has the form $(ABSID@?x) : ?x = Id@?x$, where Id is the identity function. This is an obvious base case of the abstraction algorithm.

The theorem ABSPROD next applied has the form $(ABSPROD@?x) : (?y, ?z) = PROD <= ((ABSTRACT@?x) => ?y), ((ABSTRACT@?x) => ?z))$. The intention is that the recursive applications of ABSTRACT will yield something of the form $PROD <= ((?Y@?x), (?Z@?x))$ which the reverse application of PROD will convert to $(?Y;?Z)@?x$.

The theorem ABSCOMP next applied has the form $(ABSCOMP@?x) : (?f@?y) = COMP <= (?f@((ABSTRACT@?x) => ?y))$; the intention is that a term of the form $COMP <= ?f@(?Y@?x)$ result, which the reverse application of COMP converts to $(?f@@?Y)@?x$. Note that it is assumed that the term matching $?f$ will not contain any occurrences of the term matching $?x$ (this is one of the restrictions on the class of terms for which this algorithm works).

Finally the theorem ABSCONST has the form $(ABSCONST@?x) :$ $?y = [?y]@?x$ (the same as REVCONST above). It will only be applied if the target term is neither the same as the argument nor composite in the sense of being a pair or a function application. Note that if we did not use the alternative forms of the theorem application infixes, this would *always* be applied, which would not give the desired results!

The full implementation of ABSTRACT uses a built-in "theorem" which automatically generates theorems parallel in form to PROD for other infixes (the theorem for addition would read $(?f :+ ?g)@?x = (?f@?x) + (?g@?x)$, for example). The tactic REDUCE which reverses the effect of ABSTRACT is somewhat simpler to implement.

5 Closing Remarks

The effect of the theorems ABSTRACT and REDUCE has been to make it possible to make complex substitution processes invisible to the user, supporting the avoidance of bound variables. The synthetic abstraction terms constructed by ABSTRACT are fairly readable, since they are exactly parallel in structure to the original term, but this is seldom an issue, because ABSTRACT and REDUCE are most often used together in a way which makes it unnecessary for the user of complex theorems built with their help ever to see a function abstraction term. (We are considering the introduction of bound variables to the prover, to make the notation more readable; we are considering how the presence of variable-binding constructions would interact with the tactic-writing strategy discussed in this paper).

Examples of tactics which we have implemented include a full automatic tautology-checker (not very efficient), as well as algebraic expansion and simplification algorithms for the usual algebra and Boolean algebra. A nice example of a tactic taking another tactic as a parameter is a tactic $DUAL$ for a theory of Boolean algebras which generates the dual of its argument (even if its argument is a tactic rather than a simple equation)!

It may be of interest to note that Mark2 allows one to bind a theorem to a function, so that that theorem is applied automatically wherever that function appears applied to an appropriate number of arguments (determined by the form of the theorem bound to it). This works for higher-order functions as well. Thus functional programming can be simulated under the prover.

References

1. R. Constable and others, *Implementing Mathematics with the Nuprl Proof Development System*, Prentice-Hall, Englewood Cliffs, 1986.
2. H. B. Curry and R. Feys, *Combinatory Logic*, Vol. I, North Holland, Amsterdam, 1958.
3. Mike Gordon, "*HOL*, a Proof Generating System for Higher Order Logic", in *VLSI Specification, Verification, and Synthesis*, edited by Birtwistle and Subrahmanyam, Kluwer, 1987.
4. Holmes, M. R. "Systems of combinatory logic related to Quine's 'New Foundations'". *Annals of Pure and Applied Logic*, vol. 53 (1991), pp. 103-133.
5. M. Randall Holmes, "A Functional Formulation of First-Order Logic 'With Infinity' Without Bound Variables", preprint.
6. M. Randall Holmes, "Untyped λ-calculus with relative typing", preprint.
7. W. V. O. Quine, "Algebraic Logic and Predicate Functors", Bobbs-Merrill, 1971 (booklet).
8. Alfred Tarski and Steven Givant, *A Formalization of Set Theory Without Variables*, American Mathematical Society, Providence, 1988.

Prototyping completion with constraints using computational systems

Hélène Kirchner and Pierre-Etienne Moreau

CRIN-CNRS & INRIA-Lorraine
BP 239
54506 Vandœuvre-lès-Nancy Cedex
E-mail: Helene.Kirchner@loria.fr

Abstract. We use computational systems to express a completion with constraints procedure that gives priority to simplifications. Computational systems are rewrite theories enriched by strategies. The implementation of completion in ELAN, an interpretor of computational systems, is especially convenient for experimenting with different simplification strategies, thanks to the powerful strategy language of ELAN.

1 Motivations

Completion procedures, as many computational processes, can be formulated as instances of a schema that consists of applying rewrite rules on formulas with some strategy, until getting specific normal forms. In this sense they can be understood as computational systems [1], i.e. rewrite theories in rewriting logic, enriched by a notion of strategy for selecting interesting derivations. Symbolic constraint solvers can also be described by computational systems that rewrite constraints to their solved forms. Putting together completion and constraint solving rules leads to a specification of completion with constraints using exclusively computational systems. It should be emphasized that each computation step, including for instance substitution or search for positions in a term, is specified by a computional system. This specification is actually executable in the system ELAN, an interpretor of computational systems. Checking this expressive power of computational systems was a first motivation for this work.

Using symbolic constraints in completion processes is an attractive idea but leads to non-trivial problems. For efficiency reasons, simplification is essential in completion but more difficult in the context of constrained equalities. As solutions are not computed, a constrained equality is simplifiable if all its instances are simplifiable. But this is not enough and redundancy of the simplified formula has to be checked. A second motivation for this work was to design a completion process with constraints that gives priority to simplifications, and to compare it with completion without constraints. In order to clarify the combination of weakening by propagation and simplification, we wanted to perform experimentations with different strategies for simplification and propagation.

This particular implementation of completion with constraints was not aimed at efficiency but rather at studying different strategies. Thanks to the powerful strategy language of ELAN, it appears very easy to experiment various possibilities in a flexible and high-level way.

2 Completion with constraints

Full definitions and further references can be found in [2]. A constraint is a first-order formula built on a signature Σ. Elementary constraints are two constants T and F, and equations $(t =^? t')$ with t, t' in the set of terms $\mathcal{T}(\Sigma, \mathcal{X})$. Non-elementary constraints are obtained by conjunction. The notation σ_c is used to denote the substitution which is the most general solution of the constraint c, unique up to renaming.

A *constrained equality*, denoted $(l = r \parallel c)$, is given by two terms l and r and a constraint c. When a constrained equality $(l = r \parallel c)$ satisfies, for a given simplification ordering $>$ total on ground terms, $\sigma_c(l) > \sigma_c(r)$, it is called a constrained rewrite rule and written $(l \to r \parallel c)$.

Completion with constraints is based on the superposition rule that deduces from two constrained rules $(l_1 \to r_1 \parallel c_1)$ and $(l_2 \to r_2 \parallel c_2)$ a new constrained equality $(l_2[r_1]_\omega = r_2 \parallel c_3)$ where ω is a non-variable position in l_2, provided $c_3 = c_1 \wedge c_2 \wedge (l_{2|\omega} =^? l_1)$ is satisfiable. Here the constraint $(l_{2|\omega} =^? l_1)$ is crucial to implement the basic strategy, since the part $l_2[r_1]_\omega$ in the deduced formula is only made of basic positions. Writing this equation as a constraint prevents further inference steps to be applied in the substitutions solutions of this constraint.

While the superposition rule generates new constrained formulas, simplification is crucial to eliminate redundant equalities or rewrite rules. Informally an equality is redundant if it can be proved from smaller ones. Three simplifications can be defined for a given set R of constrained rewrite rules. The constrained equality $(g = d \parallel c_1)$ is *simplified* by $(l \to r \parallel c) \in R$, into $(u = v \parallel c_2)$,

- *With constrained simplification:* if there exist a non-variable position ω in g, and a match μ from $\sigma_c(l)$ to $\sigma_{c_1}(g_{|\omega})$. Then $u = g[\mu(\sigma_c(r))]_\omega$, $v = d$ and $c_2 = c_1$.
- *With instantiated constrained simplification:* if there exist a non-variable position ω in $\sigma_{c_1}(g)$, and a match μ from $\sigma_c(l)$ to $\sigma_{c_1}(g)_{|\omega}$. Then $u = \sigma_{c_1}(g)[\mu(\sigma_c(r))]_\omega$, $v = \sigma_{c_1}(d)$ and $c_2 = T$.
- *With standard simplification:* if $c = T$ and there exist a non-variable position ω in g, and a match μ from l to $g_{|\omega}$. Then $u = g[\mu(r)]_\omega$, $v = d$ and $c_2 = c_1$.

Constrained simplification does not simplify the constraint part and constrained equalities with a reducible constraint have to be eliminated with an additional simplification process. Instantiated constrained simplification replaces a constrained equality $(g = d \parallel c_1)$ by an unconstrained one $(u = v \parallel T)$, which has the disadvantage to lose basic positions, but covers the case of reducible constraints. For example, consider a set of constrained rewrite rules that contains $(x \to 0 \parallel x =^? i(0))$ and the constrained equality $(y * 0 = 0 \parallel y =^? i(i(0)))$. With constrained simplification, the rule cannot be used to simplify the equality due to the restriction to non-variable positions. With instantiated constrained simplification, the rule simplifies the equality into $(i(0) * 0 = 0 \parallel T)$.

The application of a simplification rule consists in the replacement of the simplifiable constrained equality by the result of the simplification steps. However, to ensure redundancy of the simplifiable equality, another condition has to be satisfied by constrained simplification. If an instance $(l = r \parallel \sigma)$ is used to simplify an instance $(g = d \parallel \sigma_1)$ of $(g = d \parallel c_1)$, each term of the co-domain $\mathcal{R}an(\sigma)$ of σ has to be included in a term of the co-domain of σ_1. Formally, given two sets of terms S and S', let us define the relation $S \sqsubseteq S'$ if any term in S is a subterm of a term in S'. Let us call the condition $\mathcal{R}an(\mu\sigma_c) \sqsubseteq \mathcal{R}an(\sigma_{c_1})$ the *redundancy criterion*. It is proved in [2] that for the three defined simplifications $(g = d \parallel c_1)$ is redundant if either c is equivalent to T, or $\mathcal{R}an(\mu\sigma_c) \sqsubseteq \mathcal{R}an(\sigma_{c_1})$.

To check that a constrained equality is redundant, two special situations may occur. If the rewrite rule has a constraint equivalent to T, redundancy is always satisfied. This means that unconstrained rewrite rules can be used for simplification without checking the redundancy criterion. On the other hand, if the constrained equality $(g = d \parallel c_1)$ has a constraint equivalent to T the redundancy criterion can never be satisfied. Moreover unconstrained equalities are not the only ones for which the criterion does not apply. Let us consider for instance the constrained equality $(k(f(h(y))) = k(y) \parallel y =^? b)$. It is simplifiable using $(f(x) \rightarrow x \parallel x =^? h(b))$ into $(k(h(b)) = k(y) \parallel y =^? b)$. But the redundancy criterion is not satisfied. Propagation is then needed to prove that such equalities are redundant. Brute force propagation simply generates from $(l = r \parallel c)$, the unconstrained formula $(\sigma_c(l) = \sigma_c(r) \parallel T)$. Indeed basic positions are then lost.

The problem is to find a practical tradeoff between simplification and propagation and to experiment with three approaches.

- A first possibility is to use propagation and standard simplification. Once the constraint has been propagated, the instantiated (unconstrained) rule can be used to simplify in a standard way any term in another formula. But if we propagate too often, benefit from constraints is lost. By instantiation, basic positions are forgotten and the completion with constraints is roughly equivalent to standard completion.
- A second possibility is to use constrained simplification and check redundancy of the simplified formula. Some redundancies may remain but can be eliminated as late as possible using propagation. Note that the completion process can diverge if propagation is not applied.
- A third possibility is to use instead instantiated constrained simplification and standard simplification whenever possible.

3 Implementation in ELAN

In this section, we give an overview of how completion with constraints is implemented in ELAN using computational systems [1]. A computational system is given by a signature providing the syntax, a set of conditional rewrite rules describing the deduction mechanism, and a strategy to guide application of rewrite rules. Formally, this is the combination of a rewrite theory in rewriting logic, together with a notion of strategy to efficiently compute with given rewrite rules.

The underlying framework is first-order, since formulas, sets of formulas and proofs are considered as first-order terms. Computation is exactly application of rewrite rules on a term and a strategy describes the intended set of computations, or equivalently in rewriting logic, a subset of proof terms. A computational system may be generic in the sense that it is parameterized by a class of signatures and axioms. In this case to each program, i.e. each pair of a signature and a set of axioms, corresponds by instantiation, a computational system.

For the specification of completion with constraints, the syntax, the rewrite rules and the strategies are described in different modules that are just partially developed here. A module has a name and a list of parameters. It can import other modules and contains sort declarations, operator declarations, rules and strategies.

The main data structure is a **compute_state** described in a module parameterized by a set of variables **vars**, a set of function symbols **fss**, ordered by a precedence relation **prec**, a set of objects **X** (that will be constrained equalities). A **compute_state** is implemented using six lists.

```
module  compute_state[vars,fss,prec,X]
import  list[X]
sort    compute_state ;
op
    @*@*@*@*@*@ : (list[X] list[X] list[X] list[X] list[X] list[X])
                  compute_state;
endop ...
end of module
```

The top-level computational system imports this data structure and contains the rules for completion with constraints. It also imports other modules describing constrained equalities, constraints, unification and matching, rewriting and a module defining the used simplification strategy. Rules, adapted from those describing the ANS-completion in the ORME system, are applied to a **compute_state** until a specific form is obtained where only the first component A is non-empty. Variables used in these rules are declared with their sorts. These rules exemplify the syntax of ELAN. In the **where** part, local variables are instantiated by the result of evaluating sub-strategies, for instance **simplify_ANCT** in the first rule, calling itself simplification by the rule **s** on lists **A,N,C,T**.

In ELAN, a successful application of a rewrite rule on a given term is processed in the expected way. The left-hand side of the rule has to match a subterm of the given term; this causes instantiation of variables occurring in the left-hand side. Then the local assignments and conditions are evaluated, in reverse order, so that the evaluation starts with the last one and stops with the first one. The conditions have to be evaluated to a constant *true* and the local assignments must instantiate all remaining variables. At the end, the original subterm is replaced by the instantiated right-hand side of the rule. In the **where** assignments, the terms are reduced with respect to strategies and the results are then assigned to the variables on the left-hand sides of the := assignment signs.

```
module cans_completion[vars,fss,prec,syst]
import  simplify_basique[vars,fss,prec]
        eq_constraint[vars,fss,prec]
        unif_match[vars,fss]
        constraint[vars,fss]
        rewrite[vars,fss,prec]
        compute_state[vars,fss,prec,eq_constraint]
rules for compute_state
declare
        A,N,C,T,S,E          : list[eq_constraint];
        n,c,t,s,e            : eq_constraint;
        ln,lt,le             : list[eq_constraint];
        cs1,cs2,cs3          : compute_state;
bodies
  [Rule1] A * N * C * T * s.nil * E             => S_to_T(cs1)
        where cs1:=(simplify_ANCT) A * N * C * T * s.nil * E
  end
  [Rule2] A * N * C * T * nil * e.le => cs3
        where cs3:=(orient_E) cs2
        where cs2:=(delete_E) cs1
        where cs1:=(clean_E) A * N * C * T * nil * e.le
  end
  [Rule3] A * n.ln * c.nil * T * nil * nil      => cs1
        where cs1:=(deduce_NC) A * n.ln * c.nil * T * nil * nil
  end
  [Rule4] A * nil * c.nil * T * nil * nil       => AC_to_N(cs1)
        where cs1:=(internal_deduce) A * nil * c.nil * T * nil * nil
  end
  [Rule5] A * N * nil * t.lt * nil * nil        => cs1
        where cs1:=(choice_C) A * N * nil * t.lt * nil * nil
  end
```

The general strategy called **completion** consists in applying deterministically these five rules in the given order until no rule applies anymore.

```
strategy completion
  while
    dont care choose(Rule1 Rule2 Rule3 Rule4 Rule5)
  endwhile
end of strategy
```

Strategy definitions in ELAN are expressed in a strategy language, that includes the rule names, a concatenation operator (omitted in the syntax), iterator operators **while** and **repeat** and choice operators. Note that rule names are always encapsulated in a choice operator, because a rule name represents in general a set of rules. The **dont know choose** operator is a choice operator that, given n arguments, provides one of them and if its evaluation fails, provides another one. The **dont care choose** operator, contrary to **dont know choose**, selects only one of sub-strategies giving a non-empty set of results. To increase the expressive power of expressions built on this vocabulary, there is the possibility to express application of a strategy on subterms, thanks to the **where** mechanism in rewrite rules. Moreover strategies can be named and these names can be used in the definition of other strategies.

To complete a system of equalities, the user has to provide it in the following form.

```
specification p
Vars     x   y   z ;
Ops      f:2 e:0 i:1 ,          % operators with their arity
Prec     e:1 f:2 i:3 ;          % precedence e<f<i
System
         f(f(x,y),z) = f(x,f(y,z)) .
         f(e,x)=x .
         f(x,i(x))=e .
         nil
end of specification
```

In order to make experiments, several strategies for simplification and propagation have been tested. For instance, the following strategy **simplify** applies as long as possible, standard simplification, then propagation.

```
strategy simplify
  while
    dont care choose(st_simplify propagate)
  endwhile
end of strategy
```

Other strategies can be defined in a similar way to experiment with constrained simplification and instantiated constrained simplification. We also compared completion with constraints (`cans_completion`) with standard completion on several examples, in order to check the gain of using constraints. This gain is measured by the number of applications of the superposition rule. For instance, on the classical example of groups, `cans_completion` applied to the previous specification computes the result by applying 128 superposition steps instead of 219 in the case of a standard completion without constraints.

How to get the system:

ELAN is available by ftp at ftp.loria.fr:/pub/loria/protheo/softwares/Elan.

References

1. Claude Kirchner, Hélène Kirchner, and M. Vittek. Designing constraint logic programming languages using computational systems. In P. Van Hentenryck and V. Saraswat, editors, *Principles and Practice of Constraint Programming. The Newport Papers.*, pages 131–158. MIT press, 1995.
2. H. Kirchner and P-E. Moreau. Prototyping completion with constraints using computational systems. Technical Report 94-R-201, CRIN, 1994.

Guiding Term Reduction Through a Neural Network: Some Preliminary Results for the Group Theory

Alberto Paccanaro
Dipartimento di Scienze dell'Informazione
Università degli Studi di Milano
Milano, Italy
&
Laboratorio de Electrónica Digital
Universidad Católica "Nuestra Señora de la de Asunción"
Asunción, Paraguay
Tel: +595-21-334650
Fax: +595-21-310587
apaccana@ledip.py

Abstract. Some experiments have been carried out in order to build Neural Networks which, given a term belonging to an Equational Theory, could suggest which rewrite rules belonging to the Completed TRS for that theory represent the best choice at each reduction step in order to minimize the number of reductions needed to reach the normal form. For the Groups Theory a net was built which had an accuracy of 61%. Moreover the same net in the 71% of the cases could correctly suggest a rule applicable to the term.

1 Presentation of the Problem

The reduction of a term to its normal form with respect to a complete TRS can be done applying different sequences of rewrite rules. Obviously some sequences are shorter than other ones.

Purpose of this research is to build a Neural Network which should suggest at each reduction step which rewrite rule of a complete TRS represents the best choice for reducing a term in order to minimize the number of steps needed to reach the normal form. Although it may be argued that the solution of this problem is not so relevant in practice due to the fact that the TRS is ensured to be terminating, nevertheless it allows to throw some light on many problems which will have to be faced in order to solve a more general problem: that of building a neural network which can suggest the shortest path in the proof tree of a theorem. The most important question is: *can a neural network recognize "similarities" among terms?* This is a fundamental issue, since the feasibility of the approach of using neural networks for improving the performance of an Automated Theorem Prover will strictly depend on the ability of the networks to learn to interrelate and find similarities among terms. In the particular experiment proposed above, the network will have to detect the similarities existing between the term to be reduced and the term constituting the left hand side of a rewrite rule that reduces it.

2 Experimental Setup

In this paper some particular networks which have been built for guiding term reduction for the completed Group Theory are presented. The Neural Network architecture chosen was the Backpropagation Network for its utility to deal with problems that require complex pattern recognition and no trivial mapping functions (Freeman et al.[1991], Wasserman [1989]). Terms and rewrite rules were codified in order to be suitable as inputs to a neural network.

Terms Codification Terms had to be codified as the input vectors for the net. The size of such vectors was set to 18. The codification for terms that we tried was the following: an integer number was assigned to each symbol belonging to the theory and each element of the vector contained the integer number associated to the symbol in that position of the term . The vector was filled up starting from its last position, while the elements not used for codifying the term were assigned value 0.

In the following we shall present results of experiments which make use of 3 particular codifications:

$$f = 500, i = 700, e = 900, X_n = n*10 \quad \text{(vars are assigned integer multiples of 10)} \tag{1}$$

$$f = 500, i = 700, e = 900, X_n = 0 \quad \text{(vars are assigned to 0)} \tag{2}$$

$$f = 500, i = 700, e = 900, X_n = 20 \quad \text{(vars are assigned to 20)} \tag{3}$$

where X_n indicates the n-th variable of the theory. For example, using codification (2), the input vector corresponding to the term $f(i(X), e)$ would turn out to be:

$$[0,0,0,0,0,0,0,0,0,0,0,0,0,0,500,700,0,900].$$

Rules Codification The 10 rules of the completed Group Theory had to be codified as the output vectors for the net. To do this an integer number between 1 and 10 was assigned to each rule. The size of the output vectors was set to 10; their elements were binary digits which all but one were set to 0. The value of the integer number the vector represented was given by the position of the digit which was set to 1. For example the output vector corresponding to rule 4 would be:

$$[0,0,0,1,0,0,0,0,0,0].$$

In the following, the notation:

$$18/n_1/n_2/n_3/10$$

will indicate a fully connected network with 18 input nodes, 10 output nodes and 3 hidden layers with respectively n_1, n_2 and n_3 nodes.

3 Building a Net to Suggest Applicable Rules for Term Reduction

The first set of experiments was aimed to build a Neural Network which, given a term belonging to the Group Theory, could suggest a rule of the Complete TRS for that theory which could simplify it.

For this purpose all the possible sequences of rewrite rules which reduced 3000 terms to their normal form were built. These were used to extract 4540 pairs whose first element was a term and the second a rule that reduced it. When for a given term there existed more than one rule satisfying this condition, one of them was randomly chosen. Of these pairs 4000 were used for training the networks, and 540 were kept for testing purposes.

The hyperbolic tangent was used as output nodes transfer function. Nets were trained with a number of iterations varying between 200,000 and 300,000.

Table 1 shows the percentage of accuracy obtained during the testing for the 3 codifications(1),(2) and (3) and for 7 net configurations; these results were obtained assigning a value of 1 to the output vector component with highest value in a winner take all fashion. All other components were assigned a value of 0.

	Codification 1	Codification 2	Codification 3
18/6/6/6/10	63	61	63
18/6/9/6/10	65	64	64
18/6/6/9/10	63	65	64
18/9/6/6/10	65	64	63
18/7/7/10	68	71	69
18/7/10/10	65	67	66
18/10/7/10	67	68	67

Table 1.

During testing, the net output vectors were also used to build ordered lists of rules in the following way: the rule corresponding to the highest output vector component, supposedly the one that the net *"most strongly recommended"*, was placed first in the list; then the one corresponding to the second highest component was placed as second element; and so on. For example the output vector:

$$[0.086032, 0.016569, 0.776402, 0.000677, -0.020795,$$
$$0.031142, 0.163710, 0.033489, 0.015419, 0.017692]$$

would be interpreted as representing the list of rules:

$$\{2, 6, 0, 7, 5, 9, 1, 8, 3, 4\} \quad (4)$$

Rules were then tried in the order in which appeared in the lists, in a F.I.F.O. fashion. Table 2 presents the percentage of accuracy obtained during testing for each of the elements in the lists and for 7 net configurations.

	18/7/7/10	18/10/7/10	18/7/10/10	18/6/6/6/10	18/9/6/6/10	18/6/9/6/10	18/6/6/9/10
1	70,7407	67,5926	66,6666	61,4814	64,4444	63,8889	64,6295
2	17,5926	18,3333	20,5556	20,9259	20,0000	19,4444	19,2593
3	6,1111	7,5925	7,2222	9,2592	7,7777	8,8888	7,5925
4	3,1481	3,8888	2,4074	5,1851	4,2592	3,3333	5,5555
5	0,3703	1,1111	1,2963	0,9259	1,4814	1,4814	0,9259
6	0,7407	0,3703	1,2963	0,3703	0,3703	1,4814	0,9259
7	0,7407	0,1851	0,3703	0,5555	0,5555	0,3703	0,3703
8	0,3703	0,7407	0,0000	0,5555	0,5555	0,5555	0,3703
9	0,1851	0,1851	0,1851	0,5555	0,5555	0,3703	0,3703
10	0,0000	0,0000	0,0000	0,1851	0,0000	0,1851	0,0000

Table 2.

It is possible to see that in most cases an applicable rule can be found among the first very few suggestions of the net. As a matter of fact, just trying the first two suggestions of most nets in about the 85% of the cases we are able to encounter a rule that reduces a non-canonical term.

Moreover the percentage of correctness of the rules at the various positions in the lists, disminuishes as we try rules in positions further from the beginning of the list, as we would expect.

4 Building a Net To Suggest Applicable Rules for Optimized Term Simplification

Finally, we began to face the problem planned in the first section of this paper: the construction of a Neural Network which should suggest the shortest sequence of rewrite rules for reducing a term to its normal form. For this purpose again we used the complete reduction trees for the 3000 terms described above. This time the trees were used to extract pairs whose first element was a term and the second a rule that reduced it to a term belonging to one of the shortest simplification paths. When for a given term there existed more than one rule satisfying this condition, one of them was randomly chosen. Of these pairs 4000 were used for training the nets, and 540 were kept for testing purposes.

As before, the hyperbolic tangent was used as output nodes transfer function. Nets were trained with a number of iterations varying from 200.000 and 300.000.

Table 3 shows the percentage of accuracy obtained during the testing for the 3 codifications(1), (2) and (3) and for 7 net configurations; these results were obtained assigning a value of 1 to the output vector component with highest value in a winner take all fashion. All other components were assigned a value of 0.

	Codification 1	Codification 2	Codification 3
18/6/6/6/10	54	53	54
18/6/9/6/10	55	55	55
18/6/6/9/10	54	55	55
18/9/6/6/10	54	55	54
18/7/7/10	59	61	59
18/7/10/10	55	57	55
18/10/7/10	56	59	57

Table 3.

Again, during testing the net output vectors were also used to build ordered lists of rules as described for the experiment presented in the above section; rules were then tried in the order in which appeared in these lists, in a FIFO fashion.

Table 4 presents the percentage of accuracy obtained during testing for each of the elements in the list and for the 7 net configurations used above.

	18/7/7/10	18/10/7/10	18/7/10/10	18/6/6/6/10	18/9/6/6/10	18/6/9/6/10	18/6/6/9/10
1	61,1111	58,7036	56,6666	52,7777	54,8148	55,3703	55,3703
2	18,8888	18,8889	22,4074	21,1111	21,8519	20,5556	20,5556
3	9,0740	10,5556	7,9629	9,8148	9,2592	10,1852	8,8888
4	2,7777	4,8148	4,4444	7,9629	4,8148	3,8888	6,4814
5	2,5925	2,0370	2,4074	2,0370	2,7777	2,4074	1,1111
6	1,4814	1,1111	2,2222	1,6666	2,2222	2,7777	3,8888
7	1,8518	1,6666	1,1111	1,2963	1,1111	1,6666	0,7407
8	1,2963	0,7407	0,9259	1,6666	0,7407	0,7407	0,5555
9	0,3703	0,9259	0,5555	1,2963	1,4814	1,4814	1,2963
10	0,5555	0,5555	1,2963	0,3703	0,9259	0,9256	1,1111

Table 4.

Again, Table 4 shows that in most cases one of the rules we are looking for, can be found among the first very few suggestions of the net and the percentage of correctness of the rules at the various positions in the lists, disminuishes as we try rules in positions further from the beginning of the list, as we would expect. As a matter of fact, just trying the first two suggestions of most nets in about the 80% of the cases we are able to encounter a rule that reduces a non-canonical term and belongs to one of the shortest reduction sequences for that term.

5 Conclusions and Further Developments

The purpose of our research is to build a Neural Network which should suggest at each step which rewrite rule belonging to a completed theory represents the best choice in order to minimize the number of reduction steps needed to reach the normal form.

For the Groups Theory it was possible to identify a set of networks and a codification method for rules and terms which has given good results. This answers positively to the question posed in the beginning, that is if a neural network can recognize similarities among terms belonging to a theory. Therefore we shall try to extend these results to other theories, and then possibly build a general net which should be able to treat any Completed Theory.

Acknowledgements

Thanks to Clara Willigs for participating to some of the experimentations which have been presented here.

My special thanks to Giovanni Degli Antoni for always supporting me during this work, and constantly betting on the result.

References

1. N. Dershowitz [1989], "Completion and its applications", in Resolution of equations in algebraic structures, pp. 31-48.
2. J.A. Freeman, D.M. Skapura [1991], "Neural Networks - Algorithms, Applications, and Programming Techniques", Addison-Wesley.
3. J. Hsiang, M. Rusinovitch [1987], "On Word Problems in Equational Theories", 14th Int. Conf. on Automata Languages and Programming, Karlsrhue.
4. J.M. Hullot [1980], "A catalogue of canonical term rewriting systems", Rept. CSL-113, SRI International, Menlo Park, CA.
5. D.E. Knuth, P. Bendix [1970], "Simple Word Problems in Universal Algebras", Proceedings of the Conf. on Computational Problems in Abstract Algebras, Pergamon Press, pp. 263-298.
6. P.D. Wasserman [1989], "Neural Computing: Theory and Practice", van Nostrand Reinhold, New York

Studying Quasigroup Identities by Rewriting Techniques: Problems and First Results

Mark E. Stickel*[1] and Hantao Zhang**[2]

[1] Artificial Intelligence Center, SRI International
Menlo Park, California 94025 *stickel@ai.sri.com*
[2] Computer Science Department, The University of Iowa
Iowa City, IA 52242 *hzhang@cs.uiowa.edu*

Abstract. Finite quasigroups in the form of Latin squares have been extensively studied in design theory. Some quasigroups satisfy constraints in the form of equations, called *quasigroup identities*. In this note, we propose some questions concerning quasigroup identities that can sometimes be answered by the rewriting techniques.

1 Introduction

This note discusses problems in quasigroups, whose multiplication tables are Latin squares. The information on Latin squares provided here is mainly drawn from a survey paper by Bennett and Zhu [2]; the interested reader may refer to that work for more information on Latin squares, their importance in design theory, and some other related applications.

Recently, automated model-generation programs have been used to solve the existence problem of quasigroups with specified size and properties. Several dozen open cases were first solved by these programs [10, 4, 6, 9, 5, 7, 8]. The finite enumeration methods used did not employ equality reasoning and were limited to finding quasigroups of specific (small) size. This note presents a complementary line of research: using rewriting techniques to prove general properties of quasigroups. For example, we might wish to show that if a quasigroup satisfies an equation, one of its conjugates will satisfy a second equation, regardless of size. This may help mathematicians to gain insights in attacking open problems in quasigroups.

In this note, we list some problem areas for possible investigation of quasigroups using equational reasoning. We also present some results that were obtained after we submitted the problems to RTA.

2 Quasigroups and Conjugates

A quasigroup is simply a cancellative groupoid. That is, a quasigroup is an ordered pair $\langle S, * \rangle$ where S is a finite set and $*$ is a binary operation on S such

* Partially supported by the National Science Foundation under Grant CCR-8922330.
** Partially supported by the National Science Foundation under Grants CCR-9202838 and CCR-9357851.

that $a_1 * b = a_2 * b \Rightarrow a_1 = a_2$ and $a * b_1 = a * b_2 \Rightarrow b_1 = b_2$. The cardinality of S, $|S|$, is called the *order* of the quasigroup. The "multiplication table" for the operation $*$ forms a Latin square, of which each row and each column is a permutation of S. Many classes of quasigroups are of interest, partly because they are very natural objects in their own right, and partly because of their relationship to design theory.

People are interested in Latin squares that satisfy a set of constraints. These constraints are often expressed in terms of the quasigroup operator $*$ plus some universally quantified variables. For example, idempotent Latin squares are those that satisfy $x * x = x$.

Evidently, whenever $\langle S, * \rangle$ is a quasigroup, given values of any two variables in $x * y = z$, we can uniquely determine the value of the third variable. For example, we may therefore associate with $\langle S, * \rangle$ an operation $\star$ such that $x \star z = y$ iff $x * y = z$. It is easy to see that $\langle S, \star \rangle$ is also a quasigroup. $\langle S, \star \rangle$ is one of the six *conjugates* of $\langle S, * \rangle$. These are defined via the six operations $*_{ijk}$ where i, j, and k are distinct members of $\{1, 2, 3\}$:

$$(x_i *_{ijk} x_j = x_k) \Longleftrightarrow (x_1 * x_2 = x_3)$$

We shall refer to $\langle S, *_{ijk} \rangle$ as the (i, j, k)-conjugate of $\langle S, * \rangle$. Where S is understood to be common, we will simply refer to $*_{ijk}$ as the (i, j, k)-conjugate of $*$. Needless to say, the $(1, 2, 3)$-conjugate is the same as the original quasigroup.

Example 1. Here are the six conjugates of a small Latin square:

(a)	(b)	(c)	(d)	(e)	(f)
1 4 2 3	1 2 4 3	1 3 2 4	1 2 4 3	1 3 4 2	1 3 2 4
2 3 1 4	4 3 1 2	2 4 1 3	3 4 2 1	3 1 2 4	3 1 4 2
4 1 3 2	2 1 3 4	4 2 3 1	2 1 3 4	2 4 3 1	4 2 3 1
3 2 4 1	3 4 2 1	3 1 4 2	4 3 1 2	4 2 1 3	2 4 1 3

(a) a Latin square; (b) its $(2, 1, 3)$-conjugate; (c) its $(3, 2, 1)$-conjugate; (d) its $(2, 3, 1)$-conjugate; (e) its $(1, 3, 2)$-conjugate; (f) its $(3, 1, 2)$-conjugate.

We say that constraint C' is a *conjugate-implicant* of constraint C if whenever a quasigroup satisfies C, one of its conjugates satisfies C'. We say two constraints are *conjugate-equivalent* if they are conjugate-implicants of each other. For example, the identity $x*(x*y) = y*x$ is conjugate-equivalent to $(y*x)*x = x*y$, since the latter can be obtained by taking $*_{213}$ for $*$ in the former. A constraint is said to be *conjugate-invariant* if whenever a quasigroup satisfies the constraint, every conjugate satisfies the constraint. For example, the idempotency law, $x * x = x$, is conjugate-invariant.

3 Conjugate-Equivalence of Quasigroup Identities

From the viewpoint of rewriting techniques, we are mostly interested in constraints in the form of equations, called *quasigroup identities.* A quasigroup identity is said to be *nontrivial* if it is consistent with the specification of a Latin square. A quasigroup identity is called a *short conjugate-orthogonal identity* in [3] if it is of form $a(x, y) * b(x, y) = x$, where $a, b \in \{*_{123}, *_{213}, *_{132}, *_{312}, *_{231}, *_{321}\}$.

Theorem 1 [3, 1]. *Any nontrivial short conjugate-orthogonal identity is conjugate-equivalent to one of the following:*

Code Name[3]	Identity
$QG3$	$(x * y) * (y * x) = x$
$QG4$	$(y * x) * (x * y) = x$
$QG5$	$((y * x) * y) * y = x$
$QG6$	$(x * y) * y = x * (x * y)$
$QG7$	$(y * x) * y = x * (y * x)$
$QG8$	$x * (x * y) = y * x$
$QG9$	$((x * y) * y) * y = x$

In fact, the above theorem can be verified by a completion based theorem prover. The conjugate operations $*$, $*_{132}$, $*_{213}$, $*_{231}$, $*_{312}$, and $*_{321}$ can be specified by equations in the following way. For each equation $x *_{ijk} y = z$, solve for y in terms of x and z and solve for x in terms of y and z to produce the following twelve equations:

1: $x * (x *_{132} z) = z$	7: $x *_{213} (x *_{231} z) = z$
2: $(y *_{231} z) * y = z$	8: $(y *_{132} z) *_{213} y = z$
3: $x *_{132} (x * z) = z$	9: $x *_{312} (z *_{231} x) = z$
4: $(z *_{231} y) *_{132} y = z$	10: $(y * z) *_{312} y = z$
5: $x *_{231} (z * x) = z$	11: $x *_{321} (z *_{132} x) = z$
6: $(z *_{132} y) *_{231} y = z$	12: $(z * y) *_{321} y = z$

Applying the Knuth-Bendix completion procedure to this set of equations results in the replacement of 7–12 above by 7–9 below:

1: $x * (x *_{132} z) = z$	7: $x *_{213} y = y * x$
2: $(y *_{231} z) * y = z$	8: $x *_{312} y = y *_{132} x$
3: $x *_{132} (x * z) = z$	9: $x *_{321} y = y *_{231} x$
4: $(z *_{231} y) *_{132} y = z$	
5: $x *_{231} (z * x) = z$	
6: $(z *_{132} y) *_{231} y = z$	

This set of equations is terminating and confluent when read as a set of left-to-right reductions. That these equations suffice to characterize quasigroups can be shown by proving the cancellation laws, i.e., $a_1 = a_2$ and $b_1 = b_2$ can be proved from $a_1 * b = a_2 * b$ and $a * b_1 = a * b_2$, respectively.

We have verified Theorem 1 by the Knuth-Bendix procedure with the results in Table 1. Each location in the table corresponds to a short conjugate-orthogonal identity; its value is either the code name "QGi" of the identity that is conjugate-equivalent to it or "—" if the identity is trivial. For example, the

[3] This extends the nomenclature $QG1$–$QG7$ introduced in [4].

identity $(x *_{312} y) * (x *_{321} y) = x$ is conjugate-equivalent to $QG9$. Construction of this table required 57 proofs using the Knuth-Bendix procedure (none took more than 5 seconds of CPU time). Each trivial identity required the derivation of $x = y$ from the quasigroup axioms plus the short conjugate-orthogonal identity. Each nontrivial identity required two proofs: that QGi is a conjugate-implicant of the identity, and that the identity is a conjugate-implicant of QGi. For instance, to prove the identity $(x *_{312} y) * (x *_{321} y) = x$ is a conjugate-implicant of $QG9$, we assume $QG9$ and equations 1-9 as axioms and prove that the identity is a theorem (after $*$ is replaced by one of the six conjugate operations).

	$*$	$*_{132}$	$*_{213}$	$*_{231}$	$*_{312}$	$*_{321}$
$*$	—	$QG8$	$QG3$	$QG4$	—	$QG8$
$*_{132}$	$QG5$	—	$QG4$	$QG7$	—	$QG5$
$*_{213}$	$QG4$	$QG7$	—	$QG5$	—	$QG5$
$*_{231}$	$QG7$	$QG6$	$QG8$	—	—	$QG8$
$*_{312}$	$QG5$	$QG8$	$QG8$	$QG5$	—	$QG9$
$*_{321}$	—	—	—	—	—	—

Table 1. Short Conjugate-Orthogonal Identities

The forms of these identities raise a number of questions. Let $|t|$ denote the number of variable occurrences in term t. We call $(|u|, |v|)$ the ***type*** of the identity $u = v$. For example, four identities in Theorem 1 are of type $(4, 1)$, one of type $(3, 2)$ and two of type $(3, 3)$.

Below are some open questions whose answers would extend Theorem 1.

1. Is it true that every nontrivial quasigroup identity (containing two variables) of type (a, b), $a + b = 5$, is conjugate-equivalent to one of the seven identities in Theorem 1? A positive answer will increase the scope of Theorem 1; a negative answer by a counterexample will give us a new identity.
2. For every pair of integers a, b, $a + b \leq 6$, can we find seven identities of type (a, b) that are conjugate-equivalent respectively to the seven identities in Theorem 1? An answer to this question will give us a better picture of the classification of identities of the same type. For example, we are able to find identities of type $(4, 1)$ that are conjugate-equivalent to $QG3$–$QG9$, except possibly $QG6$.
3. What quasigroup identities are conjugate-invariant? For instance, $QG3$ and $QG6$ are conjugate-invariant.

4 Conjugate Orthogonality

Two quasigroups $\langle S, * \rangle$ and $\langle S, \star \rangle$ over the same set S are said to be *orthogonal* iff for any two elements u, v of S, the set $\{\langle x, y \rangle \mid x * y = u, x \star y = v\}$ is singleton, or equivalently, for all elements x, y, z, w of S

$$((x * y = z * w) \wedge (x \star y = z \star w)) \Rightarrow (x = z \wedge y = w).$$

It sometimes happens that one conjugate of a quasigroup is orthogonal to one of its other conjugates. Following convention [2], we refer to a Latin square (quasigroup) of order v that is orthogonal to its (i, j, k)-conjugate as an (i, j, k)-COLS(v) (one that is also idempotent is referred to as an (i, j, k)-COILS(v)). For example, the Latin square (a) in Example 1 is both a $(1,3,2)$-COLS(4) and a $(3,1,2)$-COLS(4) since it is orthogonal to (e) and (f), its $(1,3,2)$- and $(3,1,2)$-conjugates.

Thus, $(2,1,3)$-COLS, $(3,2,1)$-COLS, and $(3,1,2)$-COLS are those quasigroups that satisfy $QG0$, $QG1$, and $QG2$, respectively.

$QG0$: $(x * y = z * w \wedge x *_{213} y = z *_{213} w) \Rightarrow (x = z \wedge y = w)$
$QG1$: $(x * y = z * w \wedge x *_{321} y = z *_{321} w) \Rightarrow (x = z \wedge y = w)$
$QG2$: $(x * y = z * w \wedge x *_{312} y = z *_{312} w) \Rightarrow (x = z \wedge y = w)$

These constraints can be rephrased uniformly in $*$ as:

$QG0$: $(x * y = z * w \wedge y * x = w * z) \Rightarrow (x = z \wedge y = w)$
$QG1$: $(x * y = z * w \wedge u * y = x \wedge u * w = z) \Rightarrow (x = z \wedge y = w)$
$QG2$: $(x * y = z * w \wedge y * u = x \wedge w * u = z) \Rightarrow (x = z \wedge y = w)$

	$*$	$*_{132}$	$*_{213}$	$*_{231}$	$*_{312}$	$*_{321}$
$*$	—	$QG1'$	$QG0$	$QG2'$	$QG2$	$QG1$
$*_{132}$	$QG1'$	—	$QG2$	$QG1''$	$QG0''$	$QG2''$
$*_{213}$	$QG0$	$QG2$	—	$QG1$	$QG1'$	$QG2'$
$*_{231}$	$QG2'$	$QG1''$	$QG1$	—	$QG2''$	$QG0'$
$*_{312}$	$QG2$	$QG0''$	$QG1'$	$QG2''$	—	$QG1''$
$*_{321}$	$QG1$	$QG2''$	$QG2'$	$QG0'$	$QG1''$	—

Table 2. Conjugate-Orthogonality Constraints

The orthogonality of a pair of conjugates can be logically equivalent to, or conjugate-equivalent to, orthogonality of other pairs of conjugates. For example, $QG1$ and $QG2$ are conjugate-equivalent to the definitions of $(1,3,2)$-COLS and $(2,3,1)$-COLS, respectively. These relationships are summarized in Table 2. Each table entry is a code name for a constraint that is defined to be logically equivalent to the orthogonality of its row and column labels. For example, $QG1$ is defined by orthogonality of $*$ and $*_{321}$, but could have been defined equivalently by orthogonality of $*_{213}$ and $*_{231}$. The constraints $QG0$, $QG0'$, and $QG0''$ are conjugate-equivalent; so are $QG1$, $QG1'$, and $QG1''$; and so are $QG2$, $QG2'$, and $QG2''$.

Given a quasigroup identity, people are interested in whether some of the three constraints, $QG0$, $QG1$ and $QG2$, are its conjugate-implicants. It has been shown in [3] (see Theorem 2 below) that if a Latin square satisfies the short conjugate-orthogonal identity $(x *_{ijk} y) * (x *_{abc} y) = x$, then its (i, j, k)-conjugate is orthogonal to its (a, b, c)-conjugate. Superimposing Tables 1 and 2, we can determine which of $QG0$–$QG2$ are conjugate-implicants of $QG3$–$QG9$ (Table 3).

Identity	**Conjugate-Implicants**	**Orthogonal Conjugates**
$QG3$	$QG0$	$* \perp *_{213}$, $*_{231} \perp *_{321}$, $*_{132} \perp *_{312}$
$QG4$	$QG0, QG2$	$* \perp *_{213}$, $*_{231} \perp *_{312}$, $*_{132} \perp *_{321}$
$QG5$	$QG1, QG2$	$* \perp *_{231} \perp *_{213} \perp *_{321} \perp *$, $* \perp *_{312}$, $*_{132} \perp *_{213}$
$QG6$	$QG1$	$* \perp *_{132} \perp *_{231} \perp *_{213} \perp *_{312} \perp *_{321} \perp *$
$QG7$	$QG1, QG2$	$* \perp *_{231} \perp *_{132} \perp *_{213} \perp *_{321} \perp *_{312} \perp *$
$QG8$	$QG0, QG1$	$* \perp *_{132} \perp *_{231} \perp *_{321} \perp *_{312} \perp *_{213} \perp *$
$QG9$	$QG1$	$* \perp *_{321}$, $*_{213} \perp *_{231}$

Table 3. Conjugate-Orthogonality Results for $QG3$–$QG9$

These relationships had not been exhaustively studied before and two of these results are noteworthy. The fact that a conjugate of $QG4$ satisfies $QG2$ was observed by Stickel in 1994. This observation lead to the positive solution of the previously open problem of the existence of $(3, 1, 2)$-COILS(12). Bennett then proved the hitherto unnoticed theorem that $QG2$ is a conjugate-implicant of $QG4$, which we have now verified by the Knuth-Bendix procedure. The fact that a conjugate of $QG7$ satisfies $QG1$ is a new result, which was discovered by the Knuth-Bendix procedure in this study.

The third column of Table 3 shows the situation in more detail. It lists for each identity which pairs of conjugates we found to be orthogonal by the Knuth-Bendix procedure. The expression $a_1 \perp \cdots \perp a_n$ means a_i is conjugate orthogonal to a_{i+1} $(1 \leq i < n)$. We used our model-generation programs to show that no other pairs are certain to be orthogonal for each of $QG3$–$QG9$, except possibly $QG7$.

So far, we have only examined short conjugate-orthogonal identities. The following theorem describes an infinite set of identities that guarantee orthogonality of at least one pair of conjugates.

Theorem 2 [3]. *Let $*_{ijk}$ and $*_{abc}$ be conjugate operations on S. Then $*_{ijk}$ is orthogonal to $*_{abc}$ if and only if there is a quasigroup word $w(u, v)$ such that $w((x *_{ijk} y), (x *_{abc} y)) = x$ holds.*

Short conjugate-orthogonal identities are those for which $w(u, v) = uv$ (i.e., $u * v$). The next ones to try might be $w(u, v) = u(uv)$, $w(u, v) = u(vu)$, etc. Just as we were able to verify properties of short conjugate-orthogonal identities, and even discover minor new properties, automated term rewriting techniques should be valuable for discovering properties of large numbers of identities not previously examined.

5 Conclusion

In this note, we have identified two classes of problems for which rewrite techniques can help: (i) finding identities of certain type that are (or are not) conjugate-equivalent to some (or any) known identities; (ii) finding which pairs of conjugates are orthogonal for an identity. These two classes of problems concern relations among quasigroup constraints. We have also demonstrated the applicability of automated term rewriting techniques to reasoning about quasigroup identities. We believe that automated term rewriting techniques will allow us to usefully explore classes of identities not previously considered.

References

1. Bennett, F.: The spectra of a variety of quasigroups and related combinatorial designs. *Discrete Math.* **34** (1987): 43-64.
2. Bennett, F., Zhu, L.: Conjugate-orthogonal Latin squares and related structures, J. Dinitz & D. Stinson (eds), *Contemporary Design Theory: A Collection of Surveys.* John Wiley & Sons, 1992.
3. Evans, T.: Algebraic structures associated with Latin squares and orthogonal arrays. *Proc. of Conf. on Algebraic Aspects of Combinatorics. Congr. Numer.* **13** (1975): 31-52.
4. Fujita, M., Slaney, J., Bennett, F.: Automatic generation of some results in finite algebra, *Proc. International Joint Conference on Artificial Intelligence*, 1993.
5. McCune, W.: A Davis-Putnam program and its application to finite first-order model search: quasigroup existence problems. Preprint, Division of MCS, Argonne National Laboratory, 1994.
6. Slaney, J., Fujita, M., Stickel, M.: Automated reasoning and exhaustive search: Quasigroup existence problems. To appear in *Computers and Mathematics with Applications*, 1994.
7. Zhang, H., Bonacina, M. P.: Cumulating search in a distributed computing environment: a case study in parallel satisfiability. Proc. of the First International Symposium on Parallel Symbolic Computation. Sept. 26–28, 1994, Linz, Austria.
8. Zhang, H., Hsiang, J.: Solving open quasigroup problems by propositional reasoning. *Proc. of International Computer Symposium*, Taiwan, December 1994.
9. Zhang, H., Stickel, M.: Implementing the Davis-Putnam algorithm by tries. Technical Report, Dept. of Computer Science, The University of Iowa, 1994.
10. Zhang, J.: Search for idempotent models of quasigroup identities, Typescript, Institute of Software, Academia Sinica, Beijing, 1991.

Problems in Rewriting III*

Nachum Dershowitz[1], Jean-Pierre Jouannaud[2], and Jan Willem Klop[3]

[1] Department of Computer Science, University of Illinois, 1304 West Springfield Avenue, Urbana, IL 61801, U.S.A, nachum@cs.uiuc.edu
[2] Laboratoire de Recherche en Informatique, Bat. 490, Université de Paris Sud, 91405 Orsay, France, jouannau@lri.lri.fr
[3] CWI, Kruislaan 413, 1098 SJ Amsterdam, The Netherlands
Department of Mathematics and Computer Science, Free University, de Boelelaan 1081, 1081 HV Amsterdam, The Netherlands, jwk@cwi.nl

1 Introduction

We presented lists of open problems in the theory of rewriting in the proceedings of the previous two conferences [36; 37]. We continue with that tradition this year. We give references to solutions to eleven problems from the previous lists, report on progress on several others, provide a few reformulations of old problems, and include ten new problems.

2 Old Problems

Some progress has been made on previously listed problems. For convenience, we repeat the problems about which we are able to report progress.

Problem 4. One of the outstanding open problems in typed lambda calculi is the following: Given a term in ordinary untyped lambda calculus, is it decidable whether it can be typed in the second-order $\lambda 2$ calculus? See [11; 48].

This question has been solved in the negative. In [105] J.B. Wells proves that given a closed, type-free lambda term, the question whether it is typable in second-order $\lambda 2$ calculus, is undecidable. Moreover, given a closed type-free lambda term M and a type σ, then it is also undecidable in second-order $\lambda 2$ calculus whether M has type σ.

Problem 6 (A. Middeldorp [73]). If R and S are two term-rewriting systems with disjoint vocabularies, such that for each of R and S any two convertible normal forms must be identical, then their union $R \cup S$ also enjoys this property [73]. Accordingly, we say that unicity of normal forms (UN) is a "modular" property of term-rewriting systems. "Unicity of normal

* The first author was supported in part by the National Science Foundation under Grants CCR-90-07195 and CCR-90-24271; the second author was partially supported by the ESPRIT working groups COMPASS and CCL; the third author's work was partially supported by ESPRIT BRA project 6454: Confer.

forms with respect to reduction" ($UN^{\rightarrow}$) is the weaker property that any two normal forms of the same term must be identical. For non-left-linear systems, this property is not modular. The question remains: Is $UN^{\rightarrow}$ a modular property of left-linear term-rewriting systems?

A positive solution is given in [70].

Problem 23 (E. A. Cichon [23]). The following system [35], based on the "Battle of Hydra and Hercules" in [60], is terminating, but not provably so in Peano Arithmetic:

$$\begin{aligned} h(z, e(x)) &\rightarrow h(c(z), d(z, x)) \\ d(z, g(0,0)) &\rightarrow e(0) \\ d(z, g(x, y)) &\rightarrow g(e(x), d(z, y)) \\ d(c(z), g(g(x, y), 0)) &\rightarrow g(d(c(z), g(x, y)), d(z, g(x, y))) \\ g(e(x), e(y)) &\rightarrow e(g(x, y)) \end{aligned}$$

Transfinite (ϵ_0-) induction is required for a proof of termination. Must any termination *ordering* have the Howard ordinal as its order type, as conjectured in [23]?

If the notion of termination ordering is formalized by using ordinal notations with variables, then a termination proof using such orderings yields a slow growing bound on the lengths of derivations. If the order type is less than the Howard-Bachmann ordinal then, by Girard's Hierarchy Theorem, the derivation lengths are provably total in Peano Arithmetic. Hence a termination proof for this particular rewrite system for the Hydra game cannot be given by such an ordering [A. Weiermann, personal communication].

Problem 24. The existential fragment of the first-order theory of the "recursive path ordering" (with multiset and lexicographic "status") is decidable when the precedence on function symbols is total [25; 57], but is undecidable for arbitrary formulas. Is the existential fragment decidable for partial precedences? The Σ_4 ($\exists^*\forall^*\exists^*\forall^*$) fragment is undecidable, in general [101]. The positive existential fragment for the empty precedence (that is, for homeomorphic tree embedding) is decidable [13]. One might also ask whether the first-order theory of *total* recursive path orderings is decidable. Related results include the following: The existential fragment of the subterm ordering is decidable, but its Σ_3 ($\exists^*\forall^*\exists^*$) fragment is not [102]. The first-order theory of encompassment (the instance-of-subterm relation) is decidable [19]. Once we're at it, we might as well ask what the complexity of the satisfiability test for the existential fragment is—in the total case.

Though the first-order theory of encompassment is decidable [19], the first-order (Σ_2) theory of the recursive (lexicographic status) path ordering, assuming certain simple conditions on the precedence, is not [27].

Rephrased Problem 25 (R. Treinen [100]). Consider a finite set of function symbols containing at least one AC (associative-commutative) function symbol. Let T be the corresponding set of terms (modulo the AC properties). It is known from [101] that the first-order theory (Σ_3 fragment) of T is undecidable when F contains at least a non-constant symbol (besides the AC symbol). When F only contains an AC symbol and constants, the theory reduces to Presburger's arithmetic and is hence decidable. On the other hand the Σ_1 fragment of T is always decidable [26]. The decidability of the Σ_2 fragment of the theory of T remains open. Even more, the solvability of the following important particular case is open: given $t, t_1, \ldots, t_n \in T(F, X)$, is there an instance of t which is not an instance of $t_1, \ldots, t_n$ modulo the AC axioms? This is known as *complement problems* modulo AC.

Several special cases have been solved [40; 67], and in unpublished work in progress.

Problem 35. Huet's proof [47] of the "completeness" of completion is predicated on the assumption that the ordering supplied to completion does not change during the process. Assume that at step i of completion, the ordering used is able to order the current rewriting relation $\rightarrow_{R_i}$, but not necessarily $\rightarrow_{R_k}$ for $k < i$ (since old rules may have been deleted by completion). Is there an example showing that completion is then incomplete (the persisting rules are not confluent)?

The answer is yes, even when completion terminates with finitely many rules [93].

Problem 37 (U. Reddy, F. Bronsard). In [17] a rewriting-like mechanism for clausal reasoning called "contextual deduction" was proposed. It specializes "ordered resolution" by using pattern matching in place of unification, only instantiating clauses to match existing clauses. Does contextual deduction always terminate? (In [17] it was taken to be obvious, but that is not clear; see also [79].) It was shown in [17] that the mechanism is complete for refuting ground clauses using a theory that contains all its "strong-ordered" resolvents. Is there a notion of "complete theory" (like containing all strong-ordered resolvents not provable by contextual refutation) for which contextual deduction is complete for refutation of ground clauses?

Contextual deduction as defined in [17] does not terminate. Bronsard and Reddy have gone on to solve this [18] by using a more restricted, decidable mechanism. A completeness proof, incorporating equational inference with complete systems, is given in [16].

Problem 38 (J. Siekmann). Is satisfiability of equations in the theory of distributivity (unification modulo modulo one right- and one left-distributivity axiom) decidable? (With just one of these, the problem had already been solved in [97].) A partial positive solution is given in [29], based on a striking result on the structure of certain proofs modulo distributivity. Although many more cases are described in [28; 30], the general case remains open.

This theory is decidable [95; 94].

Problem 43. Design a framework for combining constraint solving algorithms. Some particular cases have been attacked: In [4] it was shown how decision procedures for solvability of unification problems can be combined. In [5] a similar technique is applied to (unquantified) systems of equations and disequations. In [90] the combination of unification algorithms is extended to the case where alphabets share constants. In related work [12], unification is performed in the combination of an equational theory and membership constraints.

Some progress is in [91].

Problem 44 (H. Comon). "Syntactic" theories enjoy the property that a (semi) unification algorithm can be derived from the axioms [53; 61]. This algorithm terminates for some particular cases (for instance, if all variable occurrences in the axioms are at depth at most one, and cycles have no solution) but does not in general. For the case of associativity and commutativity (AC), with a seven-axiom syntactic presentation, the derivation tree obtained by the non-deterministic application of the syntactic unification rules (*Decompose, Mutate, Merge, Coalesce, Check*, Delete*) in [53] can be pruned so as to become finite in most cases. The basic idea is that one unification problem (up to renaming) must appear infinitely times on every infinite branch of the tree (since there are finitely many axioms in the syntactic presentation). Hence, it should be possible to prune or freeze every infinite branch from some point on. The problem is to design such pruning rules so as to compute a finite derivation tree (hence, a finite complete set of unifiers) for every finitary unification problem of a syntactic equational theory.

The core of this problem has been solved [14].

Problem 46 (D. Kapur). Ground reducibility of extended rewrite systems, modulo congruences like associativity and commutativity (AC), is undecidable [59]. For left-linear AC systems, on the other hand, it is decidable [55]. What can be said more generally about restrictions on extended rewriting that give decidability? This problem is related to number 2.

Progress has been made in [63], where it is proven that ground reducibility remains undecidable when a single non-constant function symbol is associative.

Problem 50. Combinations of typed λ-calculi with term-rewriting systems have been studied extensively in the past few years [7; 15; 38; 39]. The strongest termination result allows first-order rules as well as higher-order rules defined by a generalization of primitive recursion. Suppose all rules for functional constant F follow the schema:

$$F(\bar{l}[\bar{X}],\bar{Y}) \rightarrow v[F(\bar{r}_1[\bar{X}],\bar{Y}),\ldots,F(\bar{r}_m[\bar{X}],\bar{Y}),\bar{Y})]$$

where the (not necessarily disjoint) variables in $\bar{X}$ and $\bar{Y}$ are of arbitrary order, each of $\bar{l},\bar{r}_1,\ldots,\bar{r}_m$ is in $\mathcal{T}(\mathcal{F},\{\bar{X}\})$, $v[\bar{z},\bar{Y}]$ is in $\mathcal{T}(\mathcal{F},\{\bar{Y},\bar{z}\})$, for new variables $\bar{z}$ of appropriate types, and $\bar{r}_1, \ldots, \bar{r}_m$ are each less than $\bar{l}$ in the multiset extension of the strict subterm ordering. If $\mathcal{T}(\mathcal{F}, \mathcal{X})$ is the term-algebra which includes only *previously* defined functional constants—forbidding the use of mutually recursive functional constants—termination is ensured [56]. Does termination also hold when there are mutually recursive definitions? Does this also hold when the subterm assumption is unfulfilled? (In [56] an alternative schema is proposed, with the subterm assumption weakened at the price of having only first-order variables in $\bar{X}$.) Questions of confluence of combinations of typed λ-calculi and higher-order systems also merit investigation. These results have been extended to combinations with more expressive type systems [9; 8].

An extension to the Calculus of Constructions has been reported in [10]. One can also allow the use of lexicographic and other "statuses" for the higher-order constants when comparing the subterms of F in left and right hand sides [Jouannaud and Okada, unpublished]. Finally, this can also be done when the rewrite rules follow from the induction schema in the initial algebra of the constructors [106].

Rephrased Problem 51 (H. Comon, M. Dauchet). Given an arbitrary finite term rewriting system R, is the first order theory of one-step rewriting ($\rightarrow_R$) decidable? Decidability would imply the decidability of the first-order theory of encompassment (that is, being an instance of a subterm) [19], as well as several known decidability results in rewriting. (It is well known that the theory of $\rightarrow_R^*$ is in general undecidable.)

Problem 56 (V. van Oostrom). An abstract reduction system is "decreasing Church-Rosser", if there exists a labelling of the reduction relation by a well-founded set of labels, such that all local divergences can be completed to form a "decreasing diagram" (see [84] for precise definitions). Does the Church-Rosser property imply decreasing Church-Rosser? That is, is it always possible to localize the Church-Rosser property? This is known to be the case for (weakly) normalizing and finite systems.

It is now known to hold for countable systems [68],[85, Cor. 2.3.30].

Rephrased Problem 57 (F. Baader [3]). Does there exist a semigroup theory (without constants in the equations) for which there is a reduced canonical term-rewriting system (with the right-hand side and subterms of the left in normal form) that is not length decreasing?

Problem 58 (M. Oyamaguchi). Is any "strongly" non-overlapping right-linear term-rewriting system confluent? ("Strong" in the sense that left-hand sides are non-overlapping even when the occurrences of variables have been renamed apart [21].) On the one hand, strongly non-overlapping systems need not be confluent [46]; on the other hand, strongly non-overlapping right-ground systems are [88].

A partial positive solution is given in [83; 99], namely, any strongly non-overlapping right-linear term-rewriting system is confluent if it satisfies the condition that for any rewrite rule, no variables occurring more than once in the left-hand-side occur in the right-hand-side.

Problem 60 (H. Zantema). Let R be a many-sorted term-rewriting system and R' the one-sorted system consisting of the same rules, but in which all operation symbols are considered to be of the same sort. Any rewrite in R is also a rewrite in R'. The converse does not hold, since terms and rewrite steps in R' are allowed that are not well-typed in R. In [108] it was shown that termination of R is in general not equivalent to termination of R', but it is if R does not contain both collapsing and duplicating rules. Are termination of R and of R' equivalent in the case where all variables occurring in R are of the same sort? If this statement holds, it would follow that simulating operation symbols of arity n greater than 2 by $n-1$ binary symbols in a straightforward way does not affect termination behavior.

A positive solution has recently been claimed [M. Marchiori, personal communication].

Problem 61 (T. Nipkow, M. Takahashi). For higher-order rewrite formats as given by combinatory reduction systems [62] and higher-order rewrite systems [80; 96], confluence has been proved in the restricted case of orthogonal systems. Can confluence be extended to such systems when they are weakly orthogonal (all critical pairs are trivial)? When critical pairs arise only at the root, confluence is known to hold.

Weakly orthogonal higher-order rewriting systems are confluent. This has been shown both via the Tait-Martin-Löf method and via finite developments [86, Sec. 3].

Problem 62 (V. van Oostrom). Let R and S be two left-linear, confluent combinatory reduction systems with the same alphabet. Suppose the rules of R do not overlap the rules of S. Is $R \cup S$ confluent? This is true for the restricted case when R is a term-rewriting system (an easy generalization of a result by F. Müller [77]), or if neither system has critical pairs. (The restriction to the same alphabet is essential, since confluence is in general not preserved under the addition of function symbols, not even for left-linear systems.)

The answer is yes [86, Thm. 3.13].

Problem 63 (M. Oyamaguchi).
Is confluence of right ground term rewriting systems decidable? Compare [87; 33; 34; 88].

Related is [76].

Problem 65 (D. Cohen, P. Watson [24]). An interesting system for doing arithmetic by rewriting was presented in [24]. Unfortunately, its termination has not been proved.

Termination of a related system is proved in [103].

Problem 68 (H. Comon). Consider the existential fragment of the theory defined by a binary predicate symbol $\subseteq$, a finite set of function symbols $f_1, \ldots, f_n$, the function symbols $\cap, \cup, \neg$, and the projection symbols $f_{i,j}^{-1}$ for $j \leq arity(f_i)$. Variables are interpreted as subsets of the Herbrand Universe. With the obvious interpretation of these symbols, is satisfiability of such formulæ decidable? Special cases have been solved in [44; 2; 6; 42].

This has been solved positively [43; 20; 1].

3 New Problems

Problem 78 (P. Lescanne). There are confluent calculi of explicit substitutions, but these do not preserve termination (strong normalization) [31; 72], and there are calculi that are not confluent on open terms, but which do preserve termination [65]. Is there a calculus of explicit substitution that is both confluent and preserves termination?

Problem 79 (M. Ogawa). Does a system that is nonoverlapping under unification with infinite terms (unification without "occur-check" [71]) have unique normal forms? This conjecture was originally proposed in [81] with an incomplete proof, as an extension of the result on strongly nonoverlapping systems [62; 21]. Related results appear in [88; 99; 69], but the original conjecture is still open. This is related to Problem 2. This problem is also related with modularity of confluence of systems sharing constructors, see [82].

Problem 80 (H. Comon). *Strong sequentiality* is a property of rewrite systems introduced in [49] (see [51]), which ensures the existence of optimal reduction strategies. Is strong sequentiality decidable for arbitrary rewrite systems? What is the complexity of strong sequentiality in the linear case? in the orthogonal case? Decidability results for particular rewrite systems are given in [52; 98; 58], among others.

Problem 81 (A. Weiermann). If the termination of a finite rewrite system over a finite signature can be proved using a simplification ordering, then the derivation lengths are bounded by a Hardy function of ordinal level less than the small Veblen number $\phi_{\Omega^\omega}0$. (See [104].) Is it possible to lower this bound by replacing the Hardy function by a slow growing function? That is, is it possible to bound the derivation lengths by a multiply recursive function?

Problem 82 (J. Zhang). Is there a convergent extended rewrite system for ternary boolean algebra, for which the following permutative equations hold:

$$f(x,y,z) = f(x,z,y) = f(y,x,z) = f(y,z,x) = f(z,x,y) = f(z,y,x)$$
$$f(f(x,y,z),u,x) = f(x,y,f(z,u,x))$$

See [107; 110; 22; 66].

Problem 83. A collection of rewrite orderings operating on disjoint signatures can be extended to an ordering operating on the union of the signatures, while still preserving part of the properties [92]. Such constructions can be used for proving modular termination properties of rewrite systems. Do they extend to the case where one of the starting orderings is given by $\beta\eta$ reductions on typed lambda terms?

Problem 84. Unification of patterns (à la [75]) modulo associativity and commutativity has been shown decidable [89]. Does it extend to equational theories whose axioms have the same set of variables on left and right hand side?

Problem 85 (M. Rusinowitch). Ordered paramodulation is known to be complete for simplification orderings that are total on ground terms [45]. Other theorem proving strategies are similarly restricted. How can these restrictions be relaxed?

Problem 86 (H. Zantema). When there exists a monotonic well-ordering ("monotonic" means that replacing a subterm with a smaller one decreases the whole term) of ground terms that shows termination of a rewrite system, the system is called "totally terminating." The union of two totally terminating rewrite systems which do not share any symbols is totally terminating if at least one of them does not contain a rule that has more occurrences of some variable on the right than on the left [41]. What if variables are duplicated?

Problem 87 (H. Zantema). Termination of string-rewriting systems is known to be undecidable [49]. Termination of a single term-rewriting rule was proved undecidable in [32; 64]. It is also undecidable whether there exists a simplification ordering that proves termination of a single term rewriting rule [74] (cf. [54]). Is it decidable whether a single term rewrite rule can be proved terminating by a monotonic ordering that is total on ground terms? (With more rules it is not [109].)

4 Coda

Please send any contributions by electronic or ordinary mail to any of us. We hope to continue periodically publicizing new problems and solutions to old ones.

Acknowledgements

We thank all the individuals who contributed questions, updates and solutions. Many thanks to Stefan Blom for substantial help producing this list.

References

1. A. Aiken, D. Kozen, and E. Wimmers. Decidability of systems of set constraints with negative constraints. Technical Report 93-1362, Computer Science Department, Cornell University, 1993.
2. A. Aiken and E. Wimmers. Solving systems of set constraints. In *Proceedings of the Seventh Symposium on Logic in Computer Science*, pages 329–340, Santa Cruz, CA, June 1992. IEEE.

3. F. Baader. Rewrite systems for varieties of semigroups. In M. Stickel, editor, *Proceedings of the Tenth International Conference on Automated Deduction (Kaiserslautern, West Germany)*, volume 449 of *Lecture Notes in Computer Science*, pages 381–395, Berlin, July 1990. Springer-Verlag.
4. F. Baader and K. Schulz. Unification in the union of disjoint equational theories: Combining decision procedures. In D. Kapur, editor, *Proceedings of the Eleventh International Conference on Automated Deduction (Saratoga Springs, NY)*, volume 607 of *Lecture Notes in Artificial Intelligence*, Berlin, June 1992. Springer-Verlag.
5. F. Baader and K. Schulz. Combination techniques and decision problems for disunification. In C. Kirchner, editor, *Proceedings of the Fifth International Conference on Rewriting Techniques and Applications (Montreal, Canada)*, volume 690 of *Lecture Notes in Computer Science*, Berlin, 1993. Springer-Verlag.
6. L. Bachmair, H. Ganzinger, and U. Waldmann. Set constraints are the monadic class. In *Proceedings of the Symposium on Logic in Computer Science (Montreal, Canada)*, pages 75–83. IEEE, 1993.
7. F. Barbanera. Combining term rewriting and type assignment systems. *IJFCS*, 1:165–184, 1990.
8. F. Barbanera and M. Fernández. Combining first and higher order rewrite systems with type assignment systems. In *Proceedings of the International Conference on Typed Lambda Calculi and Applications, Utrecht, Holland*, 1993.
9. F. Barbanera and M. Fernández. Modularity of termination and confluence in combinations of rewrite systems with λ_ω. In *Proceedings of the 20th International Colloquium on Automata, Languages, and Programming*, 1993.
10. F. Barbanera, M. Fernández, and H. Geuvers. Modularity of strong normalization and confluence in the λ-algebraic-cube. In *lics94*, 1994.
11. H. P. Barendregt. Lambda calculi with types. In S. Abramsky, D. M. Gabbay, and T. S. E. Maibaum, editors, *Handbook of Logic in Computer Science*. Oxford University Press, Oxford, 1991. To appear.
12. A. Boudet. Unification in order-sorted algebras with overloading. In D. Kapur, editor, *Proceedings of the Eleventh International Conference on Automated Deduction (Saratoga Springs, NY)*, volume 607 of *Lecture Notes in Artificial Intelligence*, Berlin, June 1992. Springer-Verlag.
13. A. Boudet and H. Comon. About the theory of tree embedding. In J.-P. Jouannaud, editor, *Proceedings of the Colloquium on Trees in Algebra and Programming (Orsay, France)*, Lecture Notes in Computer Science, Berlin, April 1993. Springer-Verlag.
14. A. Boudet and E. Contejean. "Syntactic" AC-unification. In J.-P. Jouannaud, editor, *Proc. CCL*, pages 136–151, Munich, September 1994. Springer-Verlag.
15. V. Breazu-Tannen and J. Gallier. Polymorphic rewriting conserves algebraic strong normalization. In *Proceedings of the Sixteenth International Colloquium on Automata, Languages and Programming (Stresa, Italy)*, volume 372 of *Lecture Notes in Computer Science*, pages 137–150, Berlin, July 1989. European Association of Theoretical Computer Science, Springer-Verlag.
16. F. Bronsard. *Using Term Orders to Control Deductions*. PhD thesis, University of Illinois, 1995. Forthcoming.
17. F. Bronsard and U. S. Reddy. Conditional rewriting in Focus. In M. Okada, editor, *Proceedings of the Second International Workshop on Conditional and Typed Rewriting Systems (Montreal, Canada)*, volume 516 of *Lecture Notes in Computer Science*, Berlin, 1991. Springer-Verlag.
18. F. Bronsard and U. S. Reddy. Reduction techniques for first-order reasoning. In M. Rusinowitch and J. L. Rémy, editors, *Proceedings of the Third International Workshop on Conditional Rewriting Systems (Pont-a-Mousson, France, July 1992)*, volume

656 of *Lecture Notes in Computer Science*, pages 242–256. Springer-Verlag, Berlin, January 1993.

19. A.-C. Caron, J.-L. Coquidé, and M. Dauchet. Encompassment properties and automata with constraints. In C. Kirchner, editor, *Proceedings of the Fifth International Conference on Rewriting Techniques and Applications (Montreal, Canada)*, volume 690 of *Lecture Notes in Computer Science*, Berlin, 1993. Springer-Verlag.
20. W. Charatonik and L. Pacholski. Negative set constraints with equality. In *Proceedings of 9th IEEE Symposium on Logic in Computer Science*, Paris, 1994. in press.
21. P. Chew. Unique normal forms in term rewriting systems with repeated variables. In *Proceedings of the Thirteenth Annual Symposium on Theory of Computing*, pages 7–18. ACM, 1981.
22. J. Christian. Problem corner: An experiment with Grau's ternary Boolean algebra. Submitted.
23. E. A. Cichon. Bounds on derivation lengths from termination proofs. Technical Report CSD-TR-622, Department of Computer Science, University of London, Surrey, England, June 1990.
24. D. Cohen and P. Watson. An efficient representation of arithmetic for term rewriting. In R. Book, editor, *Proceedings of the Fourth International Conference on Rewriting Techniques and Applications (Como, Italy)*, volume 488 of *Lecture Notes in Computer Science*, pages 240–251, Berlin, April 1991. Springer-Verlag.
25. H. Comon. Solving inequations in term algebras (Preliminary version). In *Proceedings of the Fifth Annual Symposium on Logic in Computer Science*, pages 62–69, Philadelphia, PA, June 1990. IEEE.
26. H. Comon. Complete axiomatizations of some quotient term algebras. *Theoretical Computer Science*, 118(2), September 1993.
27. H. Comon and R. Treinen. The first-order theory of lexicographic path orderings is undecidable. Rapport de Recherche 867, Laboratoire de Recherche en Informatique, Universite de Paris-Sud, Orsay, France, November 1993.
28. E. Contejean. *Eléments pour la Décidabilité de l'Unification modulo la Distributivité.* PhD thesis, Univ. Paris-Sud, Orsay, France, April 1992.
29. E. Contejean. A partial solution for D-unification based on a reduction to AC1-unification. In *Proceedings of the EATCS International Conference on Automata, Languages and Programming*, pages 621–632, Lund, Sweden, July 1993. Springer-Verlag.
30. E. Contejean. Solving linear Diophantine constraints incrementally. In D. S. Warren, editor, *Proc. of the Tenth Int. Conf. on Logic Programming*, Logic Programming, pages 532–549, Budapest, Hungary, June 1993. MIT Press.
31. P.-L. Curien, T. Hardin, and J.-J. Lévy. Confluence properties of weak and strong calculi of explicit substitutions. RR 1617, Institut National de Rechereche en Informatique et en Automatique, Rocquencourt, February 1992.
32. M. Dauchet. Simulation of Turing machines by a regular rewrite rule. *Theoretical Computer Science*, 103(2):409–420, 1992.
33. M. Dauchet, T. Heuillard, P. Lescanne, and S. Tison. Decidability of the confluence of finite ground term rewriting systems and of other related term rewriting systems. *Information and Computation*, 88(2):187–201, October 1990.
34. M. Dauchet and S. Tison. The theory of ground rewrite systems is decidable. In *Proceedings of the Fifth Symposium on Logic in Computer Science*, pages 242–248, Philadelphia, PA, June 1990.
35. N. Dershowitz and J.-P. Jouannaud. Rewrite systems. In J. van Leeuwen, editor, *Handbook of Theoretical Computer Science*, volume B: Formal Methods and Semantics, chapter 6, pages 243–320. North-Holland, Amsterdam, 1990.

36. N. Dershowitz, J.-P. Jouannaud, and J. W. Klop. Open problems in rewriting. In R. Book, editor, *Proceedings of the Fourth International Conference on Rewriting Techniques and Applications (Como, Italy)*, volume 488 of *Lecture Notes in Computer Science*, pages 445–456, Berlin, April 1991. Springer-Verlag.

37. N. Dershowitz, J.-P. Jouannaud, and J. W. Klop. More problems in rewriting. In C. Kirchner, editor, *Proceedings of the Fifth International Conference on Rewriting Techniques and Applications (Montreal, Canada)*, volume 690 of *Lecture Notes in Computer Science*, pages 468–487, Berlin, June 1993. Springer-Verlag.

38. N. Dershowitz and M. Okada. A rationale for conditional equational programming. *Theoretical Computer Science*, 75:111–138, 1990.

39. D. Dougherty. Adding algebraic rewriting to the untyped lambda calculus (extended abstract). In R. Book, editor, *Proceedings of the Fourth International Conference on Rewriting Techniques and Applications (Como, Italy)*, volume 488 of *Lecture Notes in Computer Science*, pages 37–48, Berlin, April 1991. Springer-Verlag.

40. M. Fernández. AC-complement problems: Validity and negation elimination. In C. Kirchner, editor, *Proceedings of the Fifth International Conference on Rewriting Techniques and Applications (Montreal, Canada)*, volume 690 of *Lecture Notes in Computer Science*, Berlin, 1993. Springer-Verlag.

41. M. C. F. Ferreira and H. Zantema. Total termination of term rewriting. In C. Kirchner, editor, *Proceedings of the Fifth Conference on Rewriting Techniques and Applications*, volume 690 of *Lecture Notes in Computer Science*, pages 213–227. Springer, 1993.

42. R. Gilleron, S. Tison, and M. Tommasi. Solving systems of set constraints using tree automata. In *Proceedings of the Symposium on Theoretical Aspects of Computer Science (Würzburg, Germany)*, Lecture Notes in Computer Science, Berlin, 1993. Springer-Verlag.

43. R. Gilleron, S. Tison, and M. Tommasi. Solving systems of set constraints with negated subset relationships. In *Proc. 34th Symposium on Foundations of Computer Science*, pages 372–380, Palo Alto, CA, November 1993. IEEE Computer Society Press.

44. N. Heintze and J. Jaffar. A decision procedure for a class of set constraints. In *Proceedings of the Fifth Symposium on Logic in Computer Science (Philadelphia, PA)*, pages 42–51. IEEE, June 1990.

45. J. Hsiang and M. Rusinowitch. A new method for establishing refutational completeness in theorem proving. In J. H. Siekmann, editor, *Proceedings of the Eighth International Conference on Automated Deduction (Oxford, England)*, volume 230 of *Lecture Notes in Computer Science*, pages 141–152, Berlin, July 1986. Springer-Verlag.

46. G. Huet. Confluent reductions: Abstract properties and applications to term rewriting systems. *J. of the Association for Computing Machinery*, 27(4):797–821, October 1980.

47. G. Huet. A complete proof of correctness of the Knuth-Bendix completion algorithm. *J. Computer and System Sciences*, 23(1):11–21, 1981.

48. G. Huet, editor. *Logical Foundations of Functional Programming*. University of Texas at Austin Year of Programming. Addison-Wesley, Reading, MA, 1990.

49. G. Huet and D. S. Lankford. On the uniform halting problem for term rewriting systems. Rapport laboria 283, Institut de Recherche en Informatique et en Automatique, Le Chesnay, France, March 1978.

50. G. Huet and J.-J. Lévy. Call by need computations in non-ambiguous linear term rewriting systems. Rapport Laboria 359, Institut National de Recherche en Informatique et en Automatique, Le Chesnay, France, August 1979.

51. G. Huet and J.-J. Lévy. Computations in orthogonal rewriting systems, I and II. In J.-L. Lassez and G. Plotkin, editors, *Computational Logic: Essays in Honor of Alan Robinson*, pages 395–443. MIT Press, Cambridge, MA, 1991. This is a revision of [50].
52. G. Huet and J.-J. Lévy. Computations in orthogonal rewriting systems, II. In J.-L. Lassez and G. Plotkin, editors, *Computational Logic: Essays in Honor of Alan Robinson*, chapter 12, pages 415–443. MIT Press, Cambridge, MA, 1991.
53. J.-P. Jouannaud and C. Kirchner. Solving equations in abstract algebras: A rule-based survey of unification. In J.-L. Lassez and G. Plotkin, editors, *Computational Logic: Essays in Honor of Alan Robinson*. MIT-Press, 1991.
54. J.-P. Jouannaud and H. Kirchner. Construction d'un plus petit ordre de simplification. *RAIRO Theoretical Informatics*, 18(3):191–207, 1984.
55. J.-P. Jouannaud and E. Kounalis. Automatic proofs by induction in equational theories without constructors. *Information and Computation*, 81(1):1–33, 1989.
56. J.-P. Jouannaud and M. Okada. Executable higher-order algebraic specification languages. In *Proceedings of the Sixth Symposium on Logic in Computer Science*, pages 350–361, Amsterdam, The Netherlands, 1991. IEEE.
57. J.-P. Jouannaud and M. Okada. Satisfiability of systems of ordinal notations with the subterm property is decidable. In J. L. Albert, B. Monien, and M. R. Artalejo, editors, *Proceedings of the Eighteenth EATCS Colloquium on Automata, Languages and Programming (Madrid, Spain)*, volume 510 of *Lecture Notes in Computer Science*, pages 455–468, Berlin, July 1991. Springer-Verlag.
58. J.-P. Jouannaud and W. Sadfi. Strong sequentiality of left-linear overlapping rewrite systems. In N. Dershowitz and N. Lindenstrauss, editors, *Proceedings of the Fourth International Workshop on Conditional Rewriting Systems (Jerusalem, Israel, July 1994)*, Berlin, 1995. Springer-Verlag. To appear.
59. D. Kapur, P. Narendran, and H. Zhang. On sufficient completeness and related properties of term rewriting systems. *Acta Informatica*, 24(4):395–415, August 1987.
60. L. Kirby and J. Paris. Accessible independence results for Peano arithmetic. *Bulletin London Mathematical Society*, 14:285–293, 1982.
61. C. Kirchner. Computing unification algorithms. In *Proceedings of the First Symposium on Logic in Computer Science*, pages 206–216, Cambridge, Massachussets, June 1986. IEEE.
62. J. W. Klop. *Combinatory Reduction Systems*, volume 127 of *Mathematical Centre Tracts*. Mathematisch Centrum, Amsterdam, 1980.
63. G. Kucherov and M. Rusinowitch. On the ground reducibility problem for word rewriting systems with variables. *Information Processing Letters*, 1994. To appear. Earlier version appeared in the Proceedings of 1994 ACM/SIGAPP Symposium on Applied Computing, Phoenix, AZ.
64. P. Lescanne. On termination of one rule rewrite systems. *Theoretical Computer Science*, 132:395–401, 1994.
65. P. Lescanne and J. Rouyer-Degli. The calculus of explicit substitutions λv. Technical Report RR-2222, INRIA-Lorraine, January 1994.
66. L.Fribourg. A superposition oriented theorem prover. *Theoretical Computer Science*, 35:161, 1985.
67. D. Lugiez and J.-L. Moysset. Complement problems and tree automata in AC-like theories. In *Proceedings of the Symposium on Theoretical Aspects of Computer Science (Würzburg, Germany)*, Lecture Notes in Computer Science, Berlin, 1993. Springer-Verlag.
68. K. Mano, September 1993. Personal communication.
69. K. Mano and M. Ogawa. A new proof of Chew's theorem. Technical report, IPSJ PRG94-19-7, 1994.

70. M. Marchiori. Modularity of $UN^{\rightarrow}$ for left-linear term rewriting systems. Technical report, CWI, Amsterdam, 1994.
71. A. Martelli and G. Rossi. Efficient unification with infinite terms in logic programming. In *International conference on fifth generation computer systems*, pages 202–209, 1984.
72. P.-A. Melliès. Typed λ-calculi with explicit substitutions may not terminate, 1995. To appear.
73. A. Middeldorp. Modular aspects of properties of term rewriting systems related to normal forms. In N. Dershowitz, editor, *Proceedings of the Third International Conference on Rewriting Techniques and Applications (Chapel Hill, NC)*, volume 355 of *Lecture Notes in Computer Science*, pages 263–277, Berlin, April 1989. Springer-Verlag.
74. A. Middeldorp and B. Gramlich. Simple termination is difficult. *Applicable Algebra in Engineering, Communication and Computing*, 6(2):115–128, 1995.
75. D. Miller. A logic programming language with lambda-abstraction, function variables, and simple unification. In P. Schroeder-Heister, editor, *Extensions of Logic Programming*, volume 690 of *Lecture Notes in Computer Science*. Springer-Verlag, 1991.
76. M.Oyamaguchi. On the word problem for right-ground term-rewriting systems. In *Trans. IEICE*, volume E73-5, pages 718–723, 1990.
77. F. Müller. Confluence of the lambda calculus with left-linear algebraic rewriting. *Information Processing Letters*, 41:293–299, April 1992.
78. A. Nerode and Y. V. Matiyasevich, editors. *Logical Foundations of Computer Science, Third International Symposium, LFCS'94, St. Petersburg, Russia, July 1994, Proceedings*, volume 813 of *Lecture Notes in Computer Science*. Springer-Verlag, 1994.
79. R. Nieuwenhuis and F. Orejas. Clausal rewriting. In S. Kaplan and M. Okada, editors, *Extended Abstracts of the Second International Workshop on Conditional and Typed Rewriting Systems*, pages 81–88, Montreal, Canada, June 1990. Concordia University. Revised version to appear in *Lecture Notes in Computer Science*, Springer-Verlag, Berlin.
80. T. Nipkow. Higher-order critical pairs. In *Proceedings of the Sixth Symposium on Logic in Computer Science*, pages 342–349, Amsterdam, The Netherlands, 1991. IEEE.
81. M. Ogawa and S. Ono. On the uniquely converging property of nonlinear term rewriting systems. Technical report, IEICE COMP89-7, 1989.
82. E. Ohlebusch. On the modularity of confluence of constructor-sharing term rewriting systems. In *Proceedings of the Colloquium on Trees in Algebra and Programming*, 1994.
83. Y. Ohta, M. Oyamaguchi, and Y. Toyama. On the Church-Rosser property of simple-right-linear term rewriting systems. *Trans. IEICE*, to appear.
84. V. v. Oostrom. Confluence by decreasing diagrams. IR 298, Vrije Universiteit, Amsterdam, The Netherlands, August 1992. To appear in *Theoretical Computer Science*.
85. V. v. Oostrom. *Confluence for Abstract and Higher-Order Rewriting*. PhD thesis, Vrije Universiteit, Amsterdam, March 1994.
86. V. v. Oostrom and F. v. Raamsdonk. Weak orthogonality implies confluence: the higher-order case. In [78, pp. 379–392], 1994.
87. M. Oyamaguchi. The Church-Rosser property for ground term rewriting systems is decidable. *Theoretical Computer Science*, 49(1):43–79, 1987.
88. M. Oyamaguchi and Y. Ohta. On the confluent property of right-ground term rewriting systems. *Trans. IEICE*, J76-D-I:39–45, 1993.
89. Z. Quian and K. Wang. Modular ac-unification of higher-order patterns. In J.-P. Jouannaud, editor, *CCL94*, pages 105–120, Munich, September 1994. Springer-Verlag.

90. C. Ringeissen. Unification in a combination of equational theories with shared constants and its application to primal algebras. In A. Voronkov, editor, *Proceedings of the Conference on Logic Programming and Automated Reasoning (St. Petersburg, Russia)*, volume 624 of *Lecture Notes in Artificial Intelligence*, Berlin, July 1992. Springer-Verlag.
91. C. Ringeissen. Combinaison de résolutions de contraintes. Master's thesis, Nancy, December 1993. Thèse.
92. A. Rubio. *Automated deduction with constrained clauses.* PhD thesis, Univ. de Catalunya, 1994.
93. A. Sattler-Klein. About changing the ordering during Knuth-Bendix completion. In *Proceedings of the Symposium on Theoretical Aspects of Computer Science*, pages 175–186, 1994.
94. M. Schmidt-Shauß. An algorithm for distributive unification. Research report 13/94, Fachbereich Informatik, Universität Franckfurt, Germany, December 1994.
95. M. Schmidt-Shauß. Unification of stratified second-order terms. Research report 12/94, Fachbereich Informatik, Universität Franckfurt, Germany, December 1994.
96. M. Takahashi. λ-calculi with conditional rules. In M. Bezem and J. F. Groote, editors, *Proceedings of the International Conference on Typed Lambda Calculi and Applications (Utrecht, The Netherlands)*, volume 664 of *Lecture Notes in Computer Science*, pages 406–417, Berlin, 1993. Springer-Verlag.
97. E. Tiden and S. Arnborg. Unification problems with one-sided distributivity. *J. Symbolic Computation*, 3:183–202, 1987.
98. Y. Toyama. Strong sequentiality of left linear overlapping term rewriting systems. In *Proc. 7th IEEE Symp. on Logic in Computer Science*, Santa Cruz, CA, 1992.
99. Y. Toyama and M. Oyamaguchi. Church-Rosser property and unique normal form property of non-duplicating term rewriting systems. In N. Dershowitz and N. Lindenstrauss, editors, *Workshop on Conditional Term Rewriting Systems (Jerusalem, July 1994)*, Lecture Notes in Computer Science. Springer-Verlag, to appear.
100. R. Treinen. A new method for undecidability proofs of first order theories. In K. V. Nori and C. E. V. Madhavan, editors, *Proceedings of the Tenth Conference on Foundations of Software Technology and Theoretical Computer Science*, volume 472 of *Lecture Notes in Computer Science*, pages 48–62. Springer-Verlag, 1990.
101. R. Treinen. A new method for undecidability proofs of first order theories. *J. Symbolic Computation*, 14(5):437–458, November 1992.
102. K. N. Venkataraman. Decidability of the purely existential fragment of the theory of term algebras. *J. of the Association for Computing Machinery*, 34(2):492–510, 1987.
103. H. R. Walters and H. Zantema. Rewrite systems for integer arithmetic. In J. Hsiang, editor, *Proceedings of the 6th Conference on Rewriting Techniques and Applications (this proceedings)*, Lecture Notes in Computer Science. Springer, 1995.
104. A. Weiermann. Bounding derivation lengths with functions from the slow growing hierarchy. Preprint Münster, 1993.
105. J. B. Wells. Typability and type checking in the second-order λ-calculus are equivalent and undecidable. In *Proceedings of 9th IEEE Symposium on Logic in Computer Science*, Paris, 1994.
106. B. Werner. *Méta-théorie du Calcul des Constructions Inductives.* Thèse Univ. Paris VII, France, 1994.
107. L. Wos. Automated reasoning: 33 basic research problems.
108. H. Zantema. Termination of term rewriting: interpretation and type elimination. *Journal of Symbolic Computation*, 17:23–50, 1994.

109. H. Zantema. Total termination of term rewriting is undecidable. Technical Report UU-CS-1994-55, Utrecht University, December 1994.
110. J. Zhang. A 3-place commutative operator from TBA. *AAR Newsletters*, to appear.

Index

Abdulrab, H., 339
Alouini, I., 132
Anantharaman, S., 163, 414
Asperti, A., 102
Avenhaus, J., 397

Baader, F., 352
van Bakel, S., 279
Bellegarde, F., 403
Boulton, R.J., 309
Bündgen, R., 408
Burghardt, J., 382

Chazarain, J., 420
Chen, T., 414
Corradini, A., 225

Denzinger, J., 397
Dershowitz, N., 457
Dougherty, D.J., 367

Fernández, M., 279
Fettig, R., 86
Fuchs, M., 397

Göbel, M., 408
Gadducci, F., 225
Gehrke, W., 210
Geser, A., 41
Giesl, J., 426
Graf, P., 117

Holmes, M.R., 432

Ida, T., 179

Jouannaud, J.-P., 457

Küchlin, W., 408
Kahrs, S., 241
Kamperman, J.F.Th., 147
Kennaway, R., 257
Kirchner, H., 438
Klop, J.W., 257, 457
Kondo, H., 71
Kuper, J., 271
Kurihara, M., 71

Lescanne, P., 294
Lysne, O., 26

Maksimenko, M., 339
Marchiori, M., 2
Matiyasevich, Y., 1
Middeldorp, A., 179
Montanari, U., 225
Moreau, P.-E., 438
Muller, S., 420

Narendran, P., 367
Nieuwenhuis, R., 56
Nipkow, T., 256

Ohuchi, A., 71
Otto, F., 367

Paccanaro, A., 444
Piris, J., 26

Richard, G., 163
Rouyer-Degli, J., 294

Sénizergues, G., 194
Schmid, K., 86
Schulz, K.U., 352
Sleep, R., 257
Steinbach, J., 11
Stickel, M.E., 101, 450
Suzuki, T., 179

de Vries, F.-J., 257

Walters, H.R., 147, 324

Zantema, H., 41, 324
Zhang, H., 450

Lecture Notes in Computer Science

For information about Vols. 1–835
please contact your bookseller or Springer-Verlag

Vol. 836: B. Jonsson, J. Parrow (Eds.), CONCUR '94: Concurrency Theory. Proceedings, 1994. IX, 529 pages. 1994.

Vol. 837: S. Wess, K.-D. Althoff, M. M. Richter (Eds.), Topics in Case-Based Reasoning. Proceedings, 1993. IX, 471 pages. 1994. (Subseries LNAI).

Vol. 838: C. MacNish, D. Pearce, L. Moniz Pereira (Eds.), Logics in Artificial Intelligence. Proceedings, 1994. IX, 413 pages. 1994. (Subseries LNAI).

Vol. 839: Y. G. Desmedt (Ed.), Advances in Cryptology - CRYPTO '94. Proceedings, 1994. XII, 439 pages. 1994.

Vol. 840: G. Reinelt, The Traveling Salesman. VIII, 223 pages. 1994.

Vol. 841: I. Prívara, B. Rovan, P. Ružička (Eds.), Mathematical Foundations of Computer Science 1994. Proceedings, 1994. X, 628 pages. 1994.

Vol. 842: T. Kloks, Treewidth. IX, 209 pages. 1994.

Vol. 843: A. Szepietowski, Turing Machines with Sublogarithmic Space. VIII, 115 pages. 1994.

Vol. 844: M. Hermenegildo, J. Penjam (Eds.), Programming Language Implementation and Logic Programming. Proceedings, 1994. XII, 469 pages. 1994.

Vol. 845: J.-P. Jouannaud (Ed.), Constraints in Computational Logics. Proceedings, 1994. VIII, 367 pages. 1994.

Vol. 846: D. Shepherd, G. Blair, G. Coulson, N. Davies, F. Garcia (Eds.), Network and Operating System Support for Digital Audio and Video. Proceedings, 1993. VIII, 269 pages. 1994.

Vol. 847: A. L. Ralescu (Ed.) Fuzzy Logic in Artificial Intelligence. Proceedings, 1993. VII, 128 pages. 1994. (Subseries LNAI).

Vol. 848: A. R. Krommer, C. W. Ueberhuber, Numerical Integration on Advanced Computer Systems. XIII, 341 pages. 1994.

Vol. 849: R. W. Hartenstein, M. Z. Servít (Eds.), Field-Programmable Logic. Proceedings, 1994. XI, 434 pages. 1994.

Vol. 850: G. Levi, M. Rodríguez-Artalejo (Eds.), Algebraic and Logic Programming. Proceedings, 1994. VIII, 304 pages. 1994.

Vol. 851: H.-J. Kugler, A. Mullery, N. Niebert (Eds.), Towards a Pan-European Telecommunication Service Infrastructure. Proceedings, 1994. XIII, 582 pages. 1994.

Vol. 852: K. Echtle, D. Hammer, D. Powell (Eds.), Dependable Computing – EDCC-1. Proceedings, 1994. XVII, 618 pages. 1994.

Vol. 853: K. Bolding, L. Snyder (Eds.), Parallel Computer Routing and Communication. Proceedings, 1994. IX, 317 pages. 1994.

Vol. 854: B. Buchberger, J. Volkert (Eds.), Parallel Processing: CONPAR 94 – VAPP VI. Proceedings, 1994. XVI, 893 pages. 1994.

Vol. 855: J. van Leeuwen (Ed.), Algorithms – ESA '94. Proceedings, 1994. X, 510 pages.1994.

Vol. 856: D. Karagiannis (Ed.), Database and Expert Systems Applications. Proceedings, 1994. XVII, 807 pages. 1994.

Vol. 857: G. Tel, P. Vitányi (Eds.), Distributed Algorithms. Proceedings, 1994. X, 370 pages. 1994.

Vol. 858: E. Bertino, S. Urban (Eds.), Object-Oriented Methodologies and Systems. Proceedings, 1994. X, 386 pages. 1994.

Vol. 859: T. F. Melham, J. Camilleri (Eds.), Higher Order Logic Theorem Proving and Its Applications. Proceedings, 1994. IX, 470 pages. 1994.

Vol. 860: W. L. Zagler, G. Busby, R. R. Wagner (Eds.), Computers for Handicapped Persons. Proceedings, 1994. XX, 625 pages. 1994.

Vol: 861: B. Nebel, L. Dreschler-Fischer (Eds.), KI-94: Advances in Artificial Intelligence. Proceedings, 1994. IX, 401 pages. 1994. (Subseries LNAI).

Vol. 862: R. C. Carrasco, J. Oncina (Eds.), Grammatical Inference and Applications. Proceedings, 1994. VIII, 290 pages. 1994. (Subseries LNAI).

Vol. 863: H. Langmaack, W.-P. de Roever, J. Vytopil (Eds.), Formal Techniques in Real-Time and Fault-Tolerant Systems. Proceedings, 1994. XIV, 787 pages. 1994.

Vol. 864: B. Le Charlier (Ed.), Static Analysis. Proceedings, 1994. XII, 465 pages. 1994.

Vol. 865: T. C. Fogarty (Ed.), Evolutionary Computing. Proceedings, 1994. XII, 332 pages. 1994.

Vol. 866: Y. Davidor, H.-P. Schwefel, R. Männer (Eds.), Parallel Problem Solving from Nature - PPSN III. Proceedings, 1994. XV, 642 pages. 1994.

Vol 867: L. Steels, G. Schreiber, W. Van de Velde (Eds.), A Future for Knowledge Acquisition. Proceedings, 1994. XII, 414 pages. 1994. (Subseries LNAI).

Vol. 868: R. Steinmetz (Ed.), Multimedia: Advanced Teleservices and High-Speed Communication Architectures. Proceedings, 1994. IX, 451 pages. 1994.

Vol. 869: Z. W. Raś, Zemankova (Eds.), Methodologies for Intelligent Systems. Proceedings, 1994. X, 613 pages. 1994. (Subseries LNAI).

Vol. 870: J. S. Greenfield, Distributed Programming Paradigms with Cryptography Applications. XI, 182 pages. 1994.

Vol. 871: J. P. Lee, G. G. Grinstein (Eds.), Database Issues for Data Visualization. Proceedings, 1993. XIV, 229 pages. 1994.

Vol. 872: S Arikawa, K. P. Jantke (Eds.), Algorithmic Learning Theory. Proceedings, 1994. XIV, 575 pages. 1994.

Vol. 873: M. Naftalin, T. Denvir, M. Bertran (Eds.), FME '94: Industrial Benefit of Formal Methods. Proceedings, 1994. XI, 723 pages. 1994.

Vol. 874: A. Borning (Ed.), Principles and Practice of Constraint Programming. Proceedings, 1994. IX, 361 pages. 1994.

Vol. 875: D. Gollmann (Ed.), Computer Security – ESORICS 94. Proceedings, 1994. XI, 469 pages. 1994.

Vol. 876: B. Blumenthal, J. Gornostaev, C. Unger (Eds.), Human-Computer Interaction. Proceedings, 1994. IX, 239 pages. 1994.

Vol. 877: L. M. Adleman, M.-D. Huang (Eds.), Algorithmic Number Theory. Proceedings, 1994. IX, 323 pages. 1994.

Vol. 878: T. Ishida; Parallel, Distributed and Multiagent Production Systems. XVII, 166 pages. 1994. (Subseries LNAI).

Vol. 879: J. Dongarra, J. Waśniewski (Eds.), Parallel Scientific Computing. Proceedings, 1994. XI, 566 pages. 1994.

Vol. 880: P. S. Thiagarajan (Ed.), Foundations of Software Technology and Theoretical Computer Science. Proceedings, 1994. XI, 451 pages. 1994.

Vol. 881: P. Loucopoulos (Ed.), Entity-Relationship Approach – ER'94. Proceedings, 1994. XIII, 579 pages. 1994.

Vol. 882: D. Hutchison, A. Danthine, H. Leopold, G. Coulson (Eds.), Multimedia Transport and Teleservices. Proceedings, 1994. XI, 380 pages. 1994.

Vol. 883: L. Fribourg, F. Turini (Eds.), Logic Program Synthesis and Transformation – Meta-Programming in Logic. Proceedings, 1994. IX, 451 pages. 1994.

Vol. 884: J. Nievergelt, T. Roos, H.-J. Schek, P. Widmayer (Eds.), IGIS '94: Geographic Information Systems. Proceedings, 1994. VIII, 292 pages. 19944.

Vol. 885: R. C. Veltkamp, Closed Objects Boundaries from Scattered Points. VIII, 144 pages. 1994.

Vol. 886: M. M. Veloso, Planning and Learning by Analogical Reasoning. XIII, 181 pages. 1994. (Subseries LNAI).

Vol. 887: M. Toussaint (Ed.), Ada in Europe. Proceedings, 1994. XII, 521 pages. 1994.

Vol. 888: S. A. Andersson (Ed.), Analysis of Dynamical and Cognitive Systems. Proceedings, 1993. VII, 260 pages. 1995.

Vol. 889: H. P. Lubich, Towards a CSCW Framework for Scientific Cooperation in Europe. X, 268 pages. 1995.

Vol. 890: M. J. Wooldridge, N. R. Jennings (Eds.), Intelligent Agents. Proceedings, 1994. VIII, 407 pages. 1995. (Subseries LNAI).

Vol. 891: C. Lewerentz, T. Lindner (Eds.), Formal Development of Reactive Systems. XI, 394 pages. 1995.

Vol. 892: K. Pingali, U. Banerjee, D. Gelernter, A. Nicolau, D. Padua (Eds.), Languages and Compilers for Parallel Computing. Proceedings, 1994. XI, 496 pages. 1995.

Vol. 893: G. Gottlob, M. Y. Vardi (Eds.), Database Theory – ICDT '95. Proceedings, 1995. XI, 454 pages. 1995.

Vol. 894: R. Tamassia, I. G. Tollis (Eds.), Graph Drawing. Proceedings, 1994. X, 471 pages. 1995.

Vol. 895: R. L. Ibrahim (Ed.), Software Engineering Education. Proceedings, 1995. XII, 449 pages. 1995.

Vol. 896: R. N. Taylor, J. Coutaz (Eds.), Software Engineering and Human-Computer Interaction. Proceedings, 1994. X, 281 pages. 1995.

Vol. 897: M. Fisher, R. Owens (Eds.), Executable Modal and Temporal Logics. Proceedings, 1993. VII, 180 pages. 1995. (Subseries LNAI).

Vol. 898: P. Steffens (Ed.), Machine Translation and the Lexicon. Proceedings, 1993. X, 251 pages. 1995. (Subseries LNAI).

Vol. 899: W. Banzhaf, F. H. Eeckman (Eds.), Evolution and Biocomputation. VII, 277 pages. 1995.

Vol. 900: E. W. Mayr, C. Puech (Eds.), STACS 95. Proceedings, 1995. XIII, 654 pages. 1995.

Vol. 901: R. Kumar, T. Kropf (Eds.), Theorem Provers in Circuit Design. Proceedings, 1994. VIII, 303 pages. 1995.

Vol. 902: M. Dezani-Ciancaglini, G. Plotkin (Eds.), Typed Lambda Calculi and Applications. Proceedings, 1995. VIII, 443 pages. 1995.

Vol. 903: E. W. Mayr, G. Schmidt, G. Tinhofer (Eds.), Graph-Theoretic Concepts in Computer Science. Proceedings, 1994. IX, 414 pages. 1995.

Vol. 904: P. Vitányi (Ed.), Computational Learning Theory. EuroCOLT'95. Proceedings, 1995. XVII, 415 pages. 1995. (Subseries LNAI).

Vol. 905: N. Ayache (Ed.), Computer Vision, Virtual Reality and Robotics in Medicine. Proceedings, 1995. XIV, 567 pages. 1995.

Vol. 906: E. Astesiano, G. Reggio, A. Tarlecki (Eds.), Recent Trends in Data Type Specification. Proceedings, 1995. VIII, 523 pages. 1995.

Vol. 907: T. Ito, A. Yonezawa (Eds.), Theory and Practice of Parallel Programming. Proceedings, 1995. VIII, 485 pages. 1995.

Vol. 908: J. R. Rao Extensions of the UNITY Methodology: Compositionality, Fairness and Probability in Parallelism. XI, 178 pages. 1995.

Vol. 910: A. Podelski (Ed.), Constraint Programming: Basics and Trends. Proceedings, 1995. XI, 315 pages. 1995.

Vol. 911: R. Baeza-Yates, E. Goles, P. V. Poblete (Eds.), LATIN '95: Theoretical Informatics. Proceedings, 1995. IX, 525 pages. 1995.

Vol. 913: W. Schäfer (Ed.), Software Process Technology. Proceedings, 1995. IX, 261 pages. 1995.

Vol. 914: J. Hsiang (Ed.), Rewriting Techniques and Applications. Proceedings, 1995. XII, 473 pages. 1995.